GENETICS ESSENTIALS

Concepts and Connections

Benjamin A. Pierce

Southwestern University

W. H. Freeman and Company / New York

Executive Editor: Susan Winslow
Development Editor: Beth McHenry
Media and Supplements Editor: Anna Bristow
Senior Project Editor: Georgia Lee Hadler
Manuscript Editor: Patricia Zimmerman
Associate Director of Marketing: Debbie Clare
Cover Designer: Blake Logan
Text Designer: Marsha Cohen/Parallelogram Graphics
Illustrations: Dragonfly Media Group
Senior Illustration Coordinator: Bill Page
Photo Editor: Ted Szczepanski
Production Coordinator: Paul Rohloff
Composition: Preparé
Printing and Binding: RR Donnelley

Library of Congress Control Number: 2009936816

ISBN-13: 978-1-4292-3040-7
ISBN-10: 1-4292-3040-1

Printed in the United States of America

Second printing

W. H. Freeman and Company
41 Madison Avenue
New York, NY 10010
Houndsmills, Basingstoke RG21 6XS. England

www.whfreeman.com

To the students who enroll in my genetics class each year and continually inspire me with their intelligence, curiosity, and enthusiasm

Brief Contents

Contents

Letter from the Author

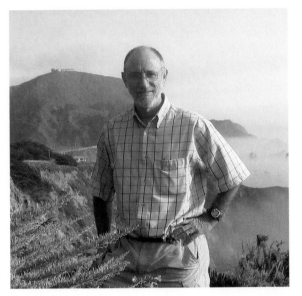

Genetics is among the most exciting and important biology courses that you will take. Almost daily, we are bombarded with examples of the relevance of genetics: the discovery of genes that influence human diseases, traits, and behaviors; the use of DNA testing to trace disease transmission and solve crimes; the use of genetic technology to develop new products. And, today, genetics is particularly important to the student of biology, serving as the foundation for many biological concepts and processes. It is truly a great time to be learning genetics!

Although genetics is important and relevant, mastering the subject is a significant challenge for many students. The field encompasses complex processes and is filled with detailed information. Genetics is often the first biology course in which students must develop problem-solving skills and apply what they have learned to novel situations.

My goal as author of your textbook is to help you overcome these challenges and to excel at genetics. As we make our journey together through introductory genetics, I'll share what I've learned in my 29 years of teaching genetics, give advice and encouragement, motivate you with stories of the people, places, and experiments of genetics, and help to keep our focus on the major concepts.

Genetics Essentials: Concepts and Connections has been written in response to requests from instructors and students for a more streamlined and focused genetics textbook that covers less content. It builds on the solid foundation of my full-length genetics textbook, *Genetics: A Conceptual Approach*, which is now in its third edition.

At Southwestern University, my office door is always open, and my own students frequently drop by to share their own approaches to learning, as well as their experiences, concerns, and triumphs. I would love to hear from you—by email (pierceb@southwestern.edu), by telephone (512-863-1974), or in person (Southwestern University, Georgetown, Texas).

Ben Pierce
Professor of Biology and holder of the Lillian Nelson Pratt Chair
Southwestern University

Preface

Welcome to *Genetics Essentials: Concepts and Connections*, a brief genetics textbook designed specifically for your one-semester course. Throughout the three editions of my more comprehensive text, *Genetics: A Conceptual Approach*, my goal was to help students concentrate on the big picture of genetics. In writing *Genetics Essentials*, I wanted to continue to use key concepts to guide students in mastering genetics, but with a more focused approach. Each chapter of *Genetics Essentials* has been streamlined, but the text still maintains the features that have made *Genetics: A Conceptual Approach* successful: seamlessly merged text and illustrations, a strong emphasis on problem solving, and, most importantly, a strong focus on the concepts and connections that make genetics meaningful for students.

HALLMARK FEATURES

■ **Key Concepts and Connections** Throughout the book, I've included pedagogical devices to help students focus on the major concepts of each topic.

Concepts boxes summarize the important take-home messages and key points of the chapter. All of the key concepts in the chapter are also listed at the end of the chapter in the ***Concepts Summary.***

Concept Check questions—some open ended, others multiple choice—allow students to assess their understanding of the take-home message of the preceding section. Answers to the Concept Checks are included in the end-of-chapter material.

Connecting Concepts sections help students see how key ideas within a chapter relate to one another. These sections integrate preceding discussions, showing how processes are similar, where they differ, and how one process informs another. After reading Connecting Concepts sections, students will better understand how newly learned concepts fit into the bigger picture of genetics.

■ **Accessibility** I have intentionally used a friendly and conversational writing style, so that students will find the book inviting and informative. The **stories** at the beginning of every chapter draw students into the material. These stories highlight the relevance of genetics to the student's daily life and feature new research in genetics, the genetic basis of human disease, hereditary oddities, and other interesting topics.

■ **Clear, Simple Illustration Program** The attractive and instructive illustration program continues to play a pivotal role in reinforcing the

---

Connecting Concepts

Recognizing Sex-Linked Inheritance

What features should we look for to identify a trait as sex linked? A common misconception is that any genetic characteristic in which the phenotypes of males and females differ must be sex linked. In fact, the expression of m[...] between males and female[...] teristics are the same in b[...] enced by sex hormones. Th[...] females cause the same ge[...] males and females.

Another misconceptio[...] more frequently in one se[...] traits are expressed more c[...] These traits are said to be s[...] expressed in only one sex; [...] Both sex-influenced and sex[...] in more detail later in the c[...]

does not guarantee that a t[...] mal characteristics are expre[...] unique, however, in that all t[...] will express the father's phenotype, and a Y-linked trait can be inherited only from the father's side of the family. Thus, a Y-linked trait can be inherited only from the paternal grandfather (the father's father), never from the maternal grandfather (the mother's father).

X-linked characteristics also exhibit a distinctive pattern of inheritance. X linkage is a possible explanation when the results of reciprocal crosses differ. If a characteristic is X linked, a cross between an affected male and an unaffected female will not give the same results as a cross between an affected female and an unaffected male. For almost all autosomal characteristics, the results of reciprocal crosses are the same. We should not conclude, however, that, when the reciprocal crosses give different results, the characteristic is X linked. Other sex-associated forms of inheritance, discussed later in the chapter, also produce different results in reciprocal crosses. The key to recognizing X-linked inheritance is to remember that a male always inherits his X chromosome from his mother, not from his father. Thus, an X-linked characteristic is not passed directly from father to son; if a male clearly inherits a characteristic from his father—and the mother is not heterozygous—it cannot be X linked.

Concepts

Cells reproduce by copying and separating their genetic information and then dividing. Because eukaryotes possess multiple chromosomes, mechanisms exist to ensure that each new cell receives one copy of each chromosome. Most eukaryotic cells are diploid, and their two chromosome sets can be arranged in homologous pairs. Haploid cells contain a single set of chromosomes.

✔ **Concept Check 2**

Diploid cells have

a. two chromosomes.

b. two sets of chromosomes.

c. one set of chromosomes.

d. two pairs of homologous chromosomes.

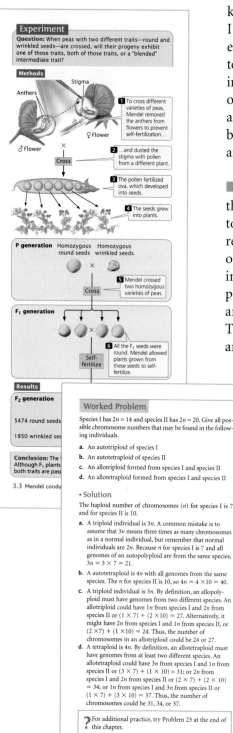

Experiment

Question: When peas with two different traits—round and wrinkled seeds—are crossed, will their progeny exhibit one of those traits, both of those traits, or a "blended" intermediate trait?

Methods

Anthers   Stigma

♀ Flower

♂ Flower

Cross

1 To cross different varieties of peas, Mendel removed the anthers from flowers to prevent self-fertilization...

2 ...and dusted the stigma with pollen from a different plant.

3 The pollen fertilized ova, which developed into seeds.

4 The seeds grew into plants.

P generation   Homozygous   Homozygous
                   round seeds   wrinkled seeds

Cross

5 Mendel crossed two homozygous varieties of peas.

F₁ generation

Self-
fertilize

6 All the F₁ seeds were round. Mendel allowed plants grown from these seeds to self-fertilize.

Results

F₂ generation

5474 round seeds

1850 wrinkled see

Conclusion: The
Although F₁ plants
both traits are pass

3.3 Mendel condu

Worked Problem

Species I has $2n = 14$ and species II has $2n = 20$. Give all possible chromosome numbers that may be found in the following individuals.

a. An autotriploid of species I
b. An autotetraploid of species II
c. An allotriploid formed from species I and species II
d. An allotetraploid formed from species I and species II

• Solution

The haploid number of chromosomes (n) for species I is 7 and for species II is 10.

a. A triploid individual is $3n$. A common mistake is to assume that $3n$ means three times as many chromosomes as in a normal individual, but remember that normal individuals are $2n$. Because n for species I is 7 and all genomes of an autopolyploid are from the same species, $3n = 3 \times 7 = 21$.

b. A autotetraploid is $4n$ with all genomes from the same species. The n for species II is 10, so $4n = 4 \times 10 = 40$.

c. A triploid individual is $3n$. By definition, an allopolyploid must have genomes from two different species. An allotriploid could have $1n$ from species I and $2n$ from species II or $(1 \times 7) + (2 \times 10) = 27$. Alternatively, it might have $2n$ from species I and $1n$ from species II, or $(2 \times 7) + (1 \times 10) = 24$. Thus, the number of chromosomes in an allotriploid could be 24 or 27.

d. A tetraploid is $4n$. By definition, an allotetraploid must have genomes from at least two different species. An allotetraploid could have $3n$ from species I and $1n$ from species II or $(3 \times 7) + (1 \times 10) = 31$; or $2n$ from species I and $2n$ from species II or $(2 \times 7) + (2 \times 10) = 34$; or $1n$ from species I and $3n$ from species II or $(1 \times 7) + (3 \times 10) = 37$. Thus, the number of chromosomes could be 31, 34, or 37.

For additional practice, try Problem 23 at the end of this chapter.

key concepts presented in each chapter. Because many students are visual learners, I have worked closely with the illustrators to make sure that the main point of each illustration is easily identified and understood. Many illustrations are in color to help students orient themselves as they study experiments and processes. Most include narratives that take students step-by-step through a process or that point out important features of a structure or experiment. Throughout the book, there are illustrations that facilitate student understanding of the experimental process by posing a question, describing experimental methodology, presenting results, and drawing a conclusion that reinforces the major concept being addressed.

■ **Emphasis on Problem Solving** I believe that problem solving is essential to the mastery of genetics. It is also one of the most difficult skills for a student to learn. In-text **Worked Problems** walk students through a key problem and review important strategies for students to consider when tackling a problem of a similar type. The book also includes extensive problem sets, broken down into three categories: comprehension questions; application questions and problems; and challenge questions. Many problems are designated as data-analysis problems that are based on real data from the scientific literature. These end-of-chapter problems reinforce the concepts covered in the chapter and enable students to apply their knowledge and to practice problem solving.

■ **Streamlined Content** To provide students taking a brief genetics course with the most important concepts, I've shortened the book considerably. *Genetics Essentials* is more than 250 pages shorter than *Genetics: A Conceptual Approach*, a reduction of more than 35%.

MEDIA AND SUPPLEMENTS

The complete package of media resources and supplements is designed to provide instructors and students with the most innovative tools to aid in a broad variety of teaching and learning approaches—including e-learning. All the available resources are fully integrated with the textbook's style and goals, enabling students to connect concepts in genetics and to think as geneticists, as well as develop their problem-solving skills.

Instructors are provided with a comprehensive set of teaching tools, carefully developed to support lecture and individual teaching styles. The following resources are made available to adopters using the printed textbook:

■ **Clicker Questions,** by Steven Gorsich, Central Michigan University, allow instructors to integrate active learning in the classroom and to assess student understanding of key concepts during lecture. Available in Microsoft Word and PowerPoint, numerous questions are based on the Concepts Check questions featured in the textbook.

■ The **Instructors' Resource DVD** contains all textbook images in PowerPoint slides and as high-resolution JPEG files, all animations, clicker questions, the solutions manual, and the test bank in Microsoft Word format.

■ All **Textbook Images and Tables** are offered as high-resolution **JPEG** files in **PowerPoint**. Each image has been fully optimized to increase type sizes and adjust color saturation.

■ The **Test Bank,** prepared by Brian W. Schwartz, Columbus State University; Alex Georgakilas, East Carolina University; Gregory Copenhaver, University of North Carolina at Chapel Hill; Rodney Mauricio, University of Georgia; and Ravinshankar Palanivelu, University of Arizona, contains multiple-choice, true-or-false, and short-answer questions. The test bank, available on the Instructors' Resource DVD and on the book companion Web site (www.whfreeman.com/pierceessentials1e), consists of chapter-by-chapter Microsoft Word files that are easy to download, edit, and print.

Students are provided with media designed to help them grasp genetic concepts and improve their problem-solving ability, including:

■ **Podcasts,** adapted from the **Tutorial** presentations listed below, are available for download from the book companion Web site (www.whfreeman.com/pierceessentials1e). Students can review important genetics processes and concepts at their convenience by downloading the animations to their MP3 players.

■ **Interactive Animated Tutorials** illuminate important concepts in genetics. These tutorials help students understand key processes in genetics by outlining these processes in a step-by-step manner. The tutorials are available on the book companion Web site. The animated concepts are:

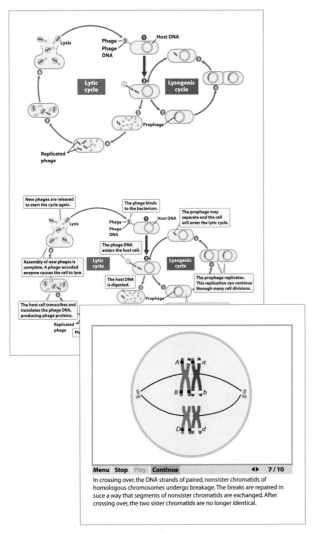

2.1	Cell Cycle and Mitosis	9.5	Mechanism of Homologous Recombination
2.2	Meiosis	10.1	Bacterial Transcription
2.3	Genetic Variation in Meiosis	10.2	Overview of mRNA Processing
3.1	Genetic Crosses Including Multiple Loci	10.3	Overview of Eukaryotic Gene Expression
4.1	X-Linked Inheritance	10.4	RNA Interference
5.1	Determining Gene Order by Three-Point Cross	11.1	Bacterial Translation
6.1	Bacterial Conjugation	12.1	The *lac* Operon
8.1	Levels of Chromatin Structure	13.1	DNA Mutations
9.1	Overview of Replication	14.1	Plasmid Cloning
9.2	Bidirectional Replication of DNA	14.2	Dideoxy Sequencing of DNA
9.3	Coordination of Leading- and Lagging-Strand Synthesis	14.3	Polymerase Chain Reaction
9.4	Nucleotide Polymerization by DNA Polymerase	17.1	The Hardy–Weinberg Law and the Effects of Inbreeding and Natural Selection

■ **Solutions and Problem-Solving Manual,** by Jung Choi, Georgia Institute of Technology, and Mark McCallum, Pfeiffer University, contains complete answers and worked-out solutions to all questions and problems that appear in the textbook.

ACKNOWLEDGMENTS

Most teachers are motivated by their students and I am no exception. My professional career as a university teacher and scholar has been vastly enriched by the thousands of students who have filled my classes in the past 29 years, first at Connecticut College, then at Baylor University, and now at Southwestern University. The intelligence, enthusiasm, curiosity, and humor of these students have been a source of inspiration and pleasure throughout my professional life. I thank my own teachers, Dr. Raymond Canham and Dr. Jeffrey Mitton, for introducing me to genetics and serving as mentors and role models.

I am indebted to Southwestern University for providing an environment in which quality teaching and research flourish. My colleagues in the Biology Department continually sustain me with friendship, collegiality, and advice. I am grateful to James Hunt, Provost of Southwestern University and Dean of the Brown College, who has been a valued friend, colleague, supporter, and role model.

Modern science textbooks are a team effort, and I have been blessed to work with an outstanding team at W. H. Freeman and Company. Acquisitions Editor Jerry Correa had the original vision for this book. Senior Acquisitions Editor Susan Winslow superbly managed the project, providing encouragement, creative ideas, support, and advice throughout. Development Editor Beth McHenry was my daily partner in crafting the book; she kept me focused and on schedule while providing great creative and editorial advice. Beth's good humor, hard work, and professional attitude made working on the book a pleasure. Lisa Samols, my editor on *Genetics: A Conceptual Approach,* served as development editor in the early stages of writing and remained engaged throughout the project. As always, Lisa was professional, upbeat, competent, and fun.

I am indebted to Georgia Lee Hadler at W. H. Freeman for expertly managing the book's production. Patricia Zimmerman was an outstanding manuscript editor, keeping a close watch on details and contributing many valuable editorial suggestions. I thank Dragonfly Media Group for creating and revising the book's outstanding illustration program and Bill Page for coordinating this process. Additional thanks to Paul Rohloff at W. H. Freeman and Pietro Paolo Adinolfi at Preparé for ably coordinating the composition and manufacturing phases of production. Blake Logan developed the book's design and worked with Ted Szczepanski to develop the outstanding cover for the book. Anna Bristow managed the supplements. I am grateful to Brian Schwartz and Alex Georgakilas for writing the Test Bank. Debbie Clare brought energy and many creative ideas to the marketing of the book.

I extend special thanks to the W. H. Freeman sales representatives, regional managers, and regional sales specialists. To know and work with them has been a pleasure and privilege. Ultimately, their hard work and good service account for the success of Freeman books.

A number of colleagues served as reviewers of the textbook, kindly lending me their technical expertise and teaching experience. Their assistance is gratefully acknowledged; any remaining errors are entirely my responsibility.

It is impossible to express my indebtedness to my family—Marlene, Sarah, and Michael—for their inspiration, love, and support.

My gratitude goes to the reviewers of *Genetics Essentials* and earlier editions of *Genetics: A Conceptual Approach:*

JEANNE M. ANDREOLI
Marygrove College

BRIAN P. ASHBURNER
University of Toledo

MELISSA ASHWELL
North Carolina State University

ANDREA BAILEY
Brookhaven College

GEORGE W. BATES
Florida State University

EDWARD BERGER
Dartmouth University

DANIEL BERGEY
Black Hills State University

ANDREW J. BOHONAK
San Diego State University

GREGORY C. BOOTON
Ohio State University

NICOLE BOURNIAS
California State University, Channel Islands

NANCY L. BROOKER
Pittsburgh State University

ROBB T. BRUMFIELD
Louisiana State University

JILL A. BUETTNER
Richland College

GERALD L. BULDAK
Loyola University Chicago

ZENAIDO TRES CAMACHO
Western New Mexico University

CATHERINE CARTER
South Dakota State University

J. AARON CASSILL
University of Texas, San Antonio

HENRY C. CHANG
Purdue University

CAROL J. CHIHARA
University of San Francisco

HUI-MIN CHUNG
University of West Florida

MARY C. COLAVITO
Santa Monica College

DEBORAH A. EASTMAN
Connecticut College

LEHMAN L. ELLIS
Our Lady of Holy Cross College

BERT ELY
University of South Carolina

F. LES ERICKSON
Salisbury State University

ROBERT FARRELL
Penn State University

WAYNE C. FORRESTER
Indiana University

ROBERT G. FOWLER
San Jose State University

GAIL FRAIZER
Kent State University

LAURA L. FROST
Point Park University

JACK R. GIRTON
Iowa State University

ELLIOT S. GOLDSTEIN
Arizona State University

JESSICA L. GOLDSTEIN
Barnard College

STEVEN W. GORSICH
Central Michigan University

PATRICK GUILFOILE
Bemidji State University

ASHLEY A. HAGLER
University of North Carolina, Charlotte

GARY M. HAY
Louisiana State University

STEPHEN C. HEDMAN
University of Minnesota, Duluth

KENNETH J. HILLERS
California Polytechnic State University

ROBERT D. HINRICHSEN
Indian University of Pennsylvania

STAN HOEGERMAN
College of William and Mary

MARGARET HOLLINGSWORTH
State University of New York, Buffalo

LI HUANG
Montana State University

CHERYL L. JORCYK
Boise State University

ELENA L. KEELING
California Polytechnic State University

ANTHONY KERN
Northland College

MARGARET J. KOVACH
University of Tennessee at Chattanooga

BRIAN KREISER
University of Southern Mississippi

CATHERINE B. KUNST
University of Denver

MARY ROSE LAMB
University of Puget Sound

MELANIE J. LEE-BROWN
Guilford College

PATRICK H. MASSON, *University of Wisconsin, Madison*

SHAWN MEAGHER
Western Illinois University

MARCIE H. MOEHNKE
Baylor University

JESSICA L. MOORE
University of South Florida

NANCY MORVILLO
Florida Southern College

HARRY NICKLA
Creighton University

ANN V. PATERSON
Williams Baptist College

TRISH PHELPS
Austin Community College, Eastview

GREG PODGORSKI
Utah State University

WILLIAM A. POWELL
State University of New York, College of Environmental Science and Forestry

SUSAN K. REIMER
Saint Francis University

CATHERINE A. REINKE
Carleton College

DEEMAH N. SCHIRF
University of Texas, San Antonio

KATHERINE T. SCHMEIDLER
Irvine Valley Community College

JON SCHNORR
Pacific University

STEPHANIE C. SCHROEDER
Webster University

BRIAN W. SCHWARTZ
Columbus State University

RODNEY J. SCOTT
Wheaton College

BARKUR S. SHASTRY
Oakland University

WENDY A. SHUTTLEWORTH
Lewis-Clark State College

THOMAS SMITH
Southern Arkansas University

WALTER SOTERO-ESTEVA
University of Central Florida

ERNEST C. STEELE JR.
Morgan State University

FUSHENG TANG
University of Arkansas, Little Rock

DOUGLAS THROWER
University of California, Santa Barbara

DANIEL P. TOMA
Minnesota State University, Mankato

DOROTHY E. TUTHILL
University of Wyoming

TZVI TZFIRA
University of Michigan

NANETTE VAN LOON
Borough of Manhattan Community College

ERIK VOLLBRECHT
Iowa State University

DANIEL WANG
University of Miami

YI-HONG WANG
Penn State University, Erie-Behrend College

WILLIAM R. WELLNITZ
Augusta State University

CINDY L. WHITE, PH.D.
University of Colorado

STEVEN D. WILT
Bellarmine University

KATHLEEN WOOD
University of Mary Hardin-Baylor

BRIAN C. YOWLER
Geneva College

JIANZHI ZHANG
University of Michigan, Ann Arbor

1 Introduction to Genetics

Albinism among the Hopis

Hopi bowl, early twentieth century. Albinism, a genetic condition, arises with high frequency among the Hopi people and occupies a special place in the Hopi culture. *[The Newark Museum/Art Resource, NY.]*

Rising a thousand feet above the desert floor, Black Mesa dominates the horizon of the Enchanted Desert and provides a familiar landmark for travelers passing through northeastern Arizona. Black Mesa is not only a prominent geological feature; more significantly, it is the ancestral home of the Hopi Native Americans. Fingers of the mesa reach out into the desert, and alongside or on top of each finger is a Hopi village. Most of the villages are quite small, filled with only a few dozen inhabitants, but they are incredibly old. One village, Oraibi, has existed on Black Mesa since 1150 A.D. and is the oldest continually occupied settlement in North America.

In 1900, Aleš Hrdliĕka, an anthropologist and physician working for the American Museum of Natural History, visited the Hopi villages of Black Mesa and reported a startling discovery. Among the Hopis were 11 white people—not Caucasians, but actually white Hopi Native Americans. These persons had a genetic condition known as albinism (**Figure 1.1**).

Albinism is caused by a defect in one of the enzymes required to produce melanin, the pigment that darkens our skin, hair, and eyes. People with albinism don't produce melanin or they produce only small amounts of it and, consequently, have white hair, light skin, and no pigment in the irises of their eyes. Melanin normally protects the DNA of skin cells from the damaging effects of ultraviolet radiation in sunlight, and melanin's presence in the developing eye is essential for proper eyesight.

The genetic basis of albinism was first described by Archibald Garrod, who recognized in 1908 that the condition was inherited as an autosomal recessive trait, meaning that a person must receive two copies of an albino mutation—one from each parent—to have albinism. In recent years, the molecular natures of the mutations that lead to albinism have been elucidated. Albinism in humans is caused by defects in any one of four different genes that control the synthesis and storage of melanin; many different types of mutations can occur at each gene, any one of which may lead to albinism. The form of albinism found among the Hopis is most likely oculocutaneous albinism type 2, due to a defect in the *OCA* gene on chromosome 15.

The Hopis are not unique in having albinos among the members of their tribe. Albinism is found in almost all human ethnic groups and is described in ancient writings; it has probably been present since humankind's beginnings. What is unique about the Hopis is the high frequency of albinism. In most human groups, albinism is rare, present

1.1 Albinism among the Hopi Native Americans. In this photograph, taken about 1900, the Hopi girl in the center has albinism. *[The Field Museum/Charles Carpenter.]*

in only about 1 in 20,000 persons. In the villages on Black Mesa, it reaches a frequency of 1 in 200, a hundred times as frequent as in most other populations.

Why is albinism so frequent among the Hopi Native Americans? The answer to this question is not completely known, but geneticists who have studied albinism among the Hopis speculate that the high frequency of the albino gene is related to the special place that albinism occupied in the Hopi culture. For much of their history, the Hopis considered members of their tribe with albinism to be important and special. People with albinism were considered pretty, clean, and intelligent. Having a number of people with albinism in one's village was considered a good sign, a symbol that the people of the village contained particularly pure Hopi blood. Albinos performed in Hopi ceremonies and assumed positions of leadership within the tribe, often becoming chiefs, healers, and religious leaders.

Hopi albinos were also given special treatment in everyday activities. The Hopis farmed small garden plots at the foot of Black Mesa for centuries. Every day throughout the growing season, the men of the tribe trek to the base of Black Mesa and spend much of the day in the bright southwestern sunlight tending their corn and vegetables. With little or no melanin pigment in their skin, people with albinism are extremely susceptible to sunburn and have increased incidences of skin cancer when exposed to the sun. Furthermore, many don't see well in bright sunlight. But the male Hopis with albinism were excused from this normal male labor and allowed to remain behind in the village with the women of the tribe, performing other duties.

Geneticists have suggested that these special considerations given to albino members of the tribe are partly responsible for the high frequency of albinism among the Hopis. Throughout the growing season, the albino men were the only male members of the tribe in the village during the day with all the women and, thus, they enjoyed a mating advantage, which helped to spread their albino genes. In addition, the special considerations given to albino Hopis allowed them to avoid the detrimental effects of albinism—increased skin cancer and poor eyesight. The small size of the Hopi tribe probably also played a role by allowing chance to increase the frequency of the albino gene. Regardless of the factors that led to the high frequency of albinism, the Hopis clearly had great respect and appreciation for the members of their tribe who possessed this particular trait. Unfortunately, people with genetic conditions in other societies are more often subject to discrimination and prejudice.

Genetics is one of the frontiers of modern science. Pick up almost any major newspaper or news magazine and chances are that you will see something related to genetics: the discovery of cancer-causing genes; the use of gene therapy to treat diseases; or reports of possible hereditary influences on intelligence, personality, and sexual orientation. These findings often have significant economic and ethical implications, making the study of genetics relevant, timely, and interesting.

This chapter introduces you to genetics and reviews some concepts that you may have encountered briefly in a preceding biology course. We begin by considering the importance of genetics to each of us, to society at large, and to students of biology. We then turn to the history of genetics, how the field as a whole developed. The final part of the chapter reviews some fundamental terms and principles of genetics that are used throughout the book.

1.1 Genetics Is Important to Individuals, to Society, and to the Study of Biology

Albinism among the Hopis illustrates the important role that genes play in our lives. This one genetic defect, among the 20,000 genes that humans possess, completely changes the life of a Hopi who possesses it. It alters his or her occupation, role in Hopi society, and relations with other members of the tribe. We all possess genes that influence our lives in significant ways. Genes affect our height, weight, hair color, and skin pigmentation. They influence our susceptibility to many diseases and disorders (**Figure 1.2**) and even contribute to our intelligence and personality. Genes are fundamental to who and what we are.

Although the science of genetics is relatively new compared with many other sciences, people have understood the

(a)   **(b)**

Laron dwarfism

Susceptibility to diphtheria

Low-tone deafness

Limb–girdle muscular dystrophy

Diastrophic dysplasia

Chromosome 5

1.2 Genes influence susceptibility to many diseases and disorders. (a) An X-ray of the hand of a person suffering from diastrophic dysplasia (bottom), a hereditary growth disorder that results in curved bones, short limbs, and hand deformities, compared with an X-ray of a normal hand (top). (b) This disorder is due to a defect in a gene on chromosome 5. Braces indicate regions on chromosome 5 where genes giving rise to other disorders are located. [Part a: (top) Biophoto Associates/Science Source/Photo Researchers; (bottom) courtesy of Eric Lander, Whitehead Institute, MIT.]

hereditary nature of traits and have practiced genetics for thousands of years. The rise of agriculture began when people started to apply genetic principles to the domestication of plants and animals. Today, the major crops and animals used in agriculture have undergone extensive genetic alterations to greatly increase their yields and provide many desirable traits, such as disease and pest resistance, special nutritional qualities, and characteristics that facilitate harvest. The Green Revolution, which expanded food production throughout the world in the 1950s and 1960s, relied heavily on the application of genetics (**Figure 1.3**). Today, genetically engineered corn, soybeans, and other crops constitute a significant proportion of all the food produced worldwide.

The pharmaceutical industry is another area in which genetics plays an important role. Numerous drugs and food additives are synthesized by fungi and bacteria that have been genetically manipulated to make them efficient producers of these substances. The biotechnology industry employs molecular genetic techniques to develop and mass-produce substances of commercial value. Growth hormone, insulin, and clotting factor are now produced commercially

by genetically engineered bacteria (**Figure 1.4**). Techniques of molecular genetics have also been used to produce bacteria that remove minerals from ore, break down toxic chemicals, and inhibit damaging frost formation on crop plants.

Genetics plays a critical role in medicine. Physicians recognize that many diseases and disorders have a hereditary component, including genetic disorders such as sickle-cell anemia and Huntington disease as well as many common diseases such as asthma, diabetes, and hypertension. Advances in molecular genetics have resulted not only in important insights into the nature of cancer but also in the development of many diagnostic tests. Gene therapy—the direct alteration of genes to treat human diseases—has now been carried out on thousands of patients.

The Role of Genetics in Biology

Although an understanding of genetics is important to all people, it is critical to the student of biology. Genetics provides one of biology's unifying principles: all organisms use genetic systems that have a number of features in common. Genetics also undergirds the study of many other biological disciplines. Evolution, for example, is genetic change taking place through time; so the study of evolution requires an understanding of genetics. Developmental biology relies heavily on genetics: tissues and organs form through the

(a)

(b)

1.3 In the Green Revolution, genetic techniques were used to develop new high-yielding strains of crops. (a) Norman Borlaug, a leader in the development of new strains of wheat that led to the Green Revolution. Borlaug was awarded the Nobel Peace Prize in 1970. (b) Modern, high-yielding rice plant (left) and traditional rice plant (right). [Part a: UPI/Corbis-Bettman. Part b: IRRI.]

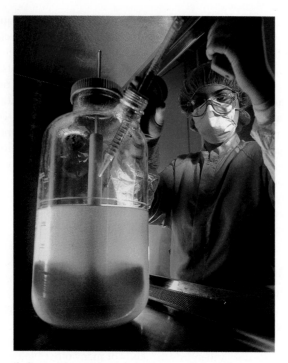

1.4 The biotechnology industry uses molecular genetic methods to produce substances of economic value. *[James Holmes/Celltech Ltd./Science Photo Library/Photo Researchers.]*

regulated expression of genes (**Figure 1.5**). Even such fields as taxonomy, ecology, and animal behavior are making increasing use of genetic methods. The study of almost any field of biology or medicine is incomplete without a thorough understanding of genes and genetic methods.

Genetic Diversity and Evolution

Life on Earth exists in a tremendous array of forms and features that occupy almost every conceivable environment. Life is also characterized by adaptation: many organisms are exquisitely suited to the environment in which they are found.

1.5 The key to development lies in the regulation of gene expression. This early fruit-fly embryo illustrates the localized production of proteins from two genes that determine the development of body segments in the adult fly. *[From Peter Lawrence, The Making of a Fly (Blackwell Scientific Publications, 1992).]*

The history of life is a chronicle of new forms of life emerging, old forms disappearing, and existing forms changing.

Despite their tremendous diversity, living organisms have an important feature in common: all use similar genetic systems. A complete set of genetic instructions for any organism is its **genome,** and all genomes are encoded in nucleic acids—either DNA or RNA. The coding system for genomic information also is common to all life: genetic instructions are in the same format and, with rare exceptions, the code words are identical. Likewise, the processes by which genetic information is copied and decoded are remarkably similar for all forms of life. These common features of heredity suggest that all life on Earth evolved from the same primordial ancestor that arose between 3.5 billion and 4 billion years ago. Biologist Richard Dawkins describes life as a river of DNA that runs through time, connecting all organisms past and present.

That all organisms have similar genetic systems means that the study of one organism's genes reveals principles that apply to other organisms. Investigations of how bacterial DNA is copied (replicated), for example, provide information that applies to the replication of human DNA. It also means that genes will function in foreign cells, which makes genetic engineering possible. Unfortunately, these similar genetic systems are also the basis for diseases such as AIDS (acquired immune deficiency syndrome), in which viral genes are able to function—sometimes with alarming efficiency—in human cells.

Life's diversity and adaptation are products of evolution, which is simply genetic change through time. Evolution is a two-step process: first, genetic variants arise randomly and, then, the proportion of particular variants increases or decreases. Genetic variation is therefore the foundation of all evolutionary change and is ultimately the basis of all life as we know it. Genetics, the study of genetic variation, is critical to understanding the past, present, and future of life.

Concepts

Heredity affects many of our physical features as well as our susceptibility to many diseases and disorders. Genetics contributes to advances in agriculture, pharmaceuticals, and medicine and is fundamental to modern biology. All organisms use similar genetic systems, and genetic variation is the foundation of the diversity of all life.

✔ Concept Check 1

What are some of the implications of all organisms having similar genetic systems?

a. That all life forms are genetically related

b. That research findings on one organism's gene function can often be applied to other organisms

c. That genes from one organism can often exist and thrive in another organism

d. All of the above

Divisions of Genetics

Traditionally, the study of genetics has been divided into three major subdisciplines: transmission genetics, molecular genetics, and population genetics (**Figure 1.6**). Also known as classical genetics, **transmission genetics** encompasses the basic principles of heredity and how traits are passed from one generation to the next. This area addresses the relation between chromosomes and heredity, the arrangement of genes on chromosomes, and gene mapping. Here, the focus is on the individual organism—how an individual organism inherits its genetic makeup and how it passes its genes to the next generation.

Molecular genetics concerns the chemical nature of the gene itself: how genetic information is encoded, replicated, and expressed. It includes the cellular processes of replication, transcription, and translation—by which genetic information is transferred from one molecule to another—and gene regulation—the processes that control the expression of genetic information. The focus in molecular genetics is the gene—its structure, organization, and function.

Population genetics explores the genetic composition of groups of individual members of the same species (populations) and how that composition changes over time and geographic space. Because evolution is genetic change, population genetics is fundamentally the study of evolution. The focus of population genetics is the group of genes found in a population.

1.6 Genetics can be subdivided into three interrelated fields. [Top left: Alan Carey/Photo Researchers. Top right: Mona file M0214602tif. Bottom: J. Alcock/Visuals Unlimited.]

Division of the study of genetics into these three groups is convenient and traditional, but we should recognize that the fields overlap and that each major subdivision can be further divided into a number of more specialized fields, such as chromosomal genetics, biochemical genetics, quantitative genetics, and so forth. Alternatively, genetics can be subdivided by organism (fruit fly, corn, or bacterial genetics), and each of these organisms can be studied at the level of transmission, molecular, and population genetics. Modern genetics is an extremely broad field, encompassing many interrelated subdisciplines and specializations.

Model Genetic Organisms

Through the years, genetic studies have been conducted on thousands of different species, including almost all major groups of bacteria, fungi, protists, plants, and animals. Nevertheless, a few species have emerged as **model genetic organisms**—organisms having characteristics that make them particularly useful for genetic analysis and about which a tremendous amount of genetic information has accumulated. Six model organisms that have been the subject of intensive genetic study are: *Drosophila melanogaster,* the fruit fly; *Escherichia coli,* a bacterium present in the gut of humans and other mammals; *Caenorhabditis elegans,* a nematode worm (also called a roundworm); *Arabidopsis thaliana,* the thale cress plant; *Mus musculus,* the house mouse; and *Saccharomyces cerevisiae,* baker's yeast (**Figure 1.7**). These species are the organisms of choice for many genetic researchers, and their genomes were sequenced as a part of the Human Genome Project.

At first glance, this group of lowly and sometimes despised creatures might seem unlikely candidates for model organisms. However, all possess life cycles and traits that make them particularly suitable for genetic study, including a short generation time, manageable numbers of progeny, adaptability to a laboratory environment, and the ability to be housed and propagated inexpensively. The life cycles, genomic characteristics, and features that make these model organisms useful for genetic studies are included in special model-organism illustrations in later chapters for five of the six species. Other species that are frequently the subject of genetic research and are also considered genetic models include bread mold (*Neurospora crassa*), corn (*Zea mays*), zebrafish (*Danio rerio*), and clawed frog (*Xenopus laevis*). Although not generally considered a genetic model, humans also have been subjected to intensive genetic scrutiny.

The value of model genetic organisms is illustrated by the use of zebrafish to identify genes that affect skin pigmentation in humans. For many years, geneticists have recognized that differences in pigmentation among human ethnic groups (**Figure 1.8a**) are genetic, but the genes causing these differences were largely unknown. Zebrafish have recently become an important model in genetic studies because they are small vertebrates that produce many offspring and are

(a)

Drosophila melanogaster
Fruit fly (pp. 76–78)

(b)

Escherichia coli
Bacterium (pp. 152–153)

(c)

Caenorhabditis elegans
Nematode worm (pp. 263–265)

1.7 Model genetic organisms are species having features that make them useful for genetic analysis.
[Part a: SPL/Photo Researchers. Part b: Gary Gaugler/Visuals Unlimited. Part c: Natalie Pujol/Visuals Unlimited. Part d: Peggy Greb/ARS. Part e: Joel Page/AP. Part f: T. E. Adams/Visuals Unlimited.]

easy to rear in the laboratory. The zebrafish *golden* mutant, caused by a recessive mutation, has light pigmentation due to the presence of fewer, smaller, and less-dense pigment-containing structures called melanosomes in its cells (**Figure 1.8b**). Light skin in humans is similarly due to fewer and less-dense melanosomes in pigment-containing cells.

Keith Cheng and his colleagues at Pennsylvania State University College of Medicine hypothesized that light skin in humans might result from a mutation that is similar to the *golden* mutation in zebrafish. Taking advantage of the ease with which zebrafish can be manipulated in the laboratory, they isolated and sequenced the gene responsible for the *golden* mutation and found that it encodes a protein that takes part in calcium uptake by melanosomes. They then

searched a database of all known human genes and found a similar gene called *SLC24A5*, which encodes the same function in human cells. When they examined human populations, they found that light-skinned Europeans typically possessed one form of this gene, whereas darker-skinned Africans, Eastern Asians, and Native Americans usually possessed a different form of the gene. Many other genes also affect pigmentation in humans, as illustrated by mutations in the *OCA* gene that produce albinism among the Hopi Native Americans (discussed in the introduction to this chapter). Nevertheless, *SLC24A5* appears to be responsible for 24% to 38% of the differences in pigmentation between Africans and Europeans. This example illustrates the power of model organisms in genetic research.

(a)

(b)

Normal zebrafish

Golden mutant

1.8 The zebrafish, a genetic model organism, has been instrumental in helping to identify genes encoding pigmentation differences among humans. (a) Human ethnic groups differ in degree of skin pigmentation. (b) The zebrafish *golden* mutation is caused by a gene that controls the amount of melanin pigment in melanosomes. [Part a: PhotoDisc. Part b: K. Cheng/J. Gittlen, Cancer Research Foundation, Pennsylvania State College of Medicine.]

(d)

Arabidopsis thaliana
Thale cress plant (pp. 312–314)

(e)

Mus musculus
House mouse (pp. 365–367)

(f)

Saccharomyces cerevisiae
Baker's yeast

Concepts

The three major divisions of genetics are transmission genetics, molecular genetics, and population genetics. Transmission genetics examines the principles of heredity; molecular genetics deals with the gene and the cellular processes by which genetic information is transferred and expressed; population genetics concerns the genetic composition of groups of organisms and how that composition changes over time and geographic space. Model genetic organisms are species that have received special emphasis in genetic research; they have characteristics that make them useful for genetic analysis.

✔ **Concept Check 2**

Would the horse make a good model genetic organism? Why or why not?

1.2 Humans Have Been Using Genetics for Thousands of Years

Although the science of genetics is young—almost entirely a product of the past 100 years or so—people have been using genetic principles for thousands of years.

The Early Use and Understanding of Heredity

The first evidence that people understood and applied the principles of heredity in earlier times is found in the domestication of plants and animals, which began between approximately 10,000 and 12,000 years ago. The world's first agriculture is thought to have developed in the Middle East, in what is now Turkey, Iraq, Iran, Syria, Jordan, and Israel, where domesticated plants and animals were major dietary components of many populations by 10,000 years ago. The first domesticated organisms included wheat, peas, lentils, barley, dogs, goats, and sheep (**Figure 1.9a**). By 4000 years ago, sophisticated genetic techniques were already in use in the Middle East. Assyrians and Babylonians developed several hundred varieties of date palms that differed in fruit size, color, taste, and time of ripening (**Figure 1.9b**). Other crops and domesticated animals were developed by cultures in Asia, Africa, and the Americas in the same period.

Concepts

Humans first applied genetics to the domestication of plants and animals between approximately 10,000 and 12,000 years ago. This domestication led to the development of agriculture and fixed human settlements.

The ancient Greeks gave careful consideration to human reproduction and heredity. The dissection of animals by the Greek physician Alcmaeon (circa 520 B.C.) sparked a long philosophical debate about where semen was produced that culminated in the concept of **pangenesis.** This concept suggested that specific pieces of information travel from various parts of the body to the reproductive organs, from which they are passed to the embryo (**Figure 1.10a**). Pangenesis led the ancient Greeks to propose the notion of the **inheritance of acquired characteristics,** in which traits acquired in one's lifetime become incorporated into one's hereditary information and are passed on to offspring; for example, people who developed musical ability through diligent study would produce children who are innately endowed with musical ability. Although incorrect, these ideas persisted through the twentieth century.

Dutch eyeglass makers began to put together simple microscopes in the late 1500s, enabling Robert Hooke (1635–1703) to discover cells in 1665. Microscopes provided

1.9 Ancient peoples practiced genetic techniques in agriculture. (a) Modern wheat, with larger and more numerous seeds that do not scatter before harvest, was produced by interbreeding at least three different wild species. (b) Assyrian bas-relief sculpture showing artificial pollination of date palms at the time of King Assurnasirpalli II, who reigned from 883 to 859 B.C. *[(Part a): Scott Bauer/ARS/USDA. Part b: The Metropolitan Museum of Art, gift of John D. Rockefeller, Jr., 1932. (32.143.3).]*

naturalists with new and exciting vistas on life, and perhaps it was excessive enthusiasm for this new world of the very small that gave rise to the idea of **preformationism.** According to preformationism, inside the egg or sperm there exists a tiny miniature adult, a *homunculus,* which simply enlarges during development (**Figure 1.11**). Preformationism meant that all traits would be inherited from only one parent—from the father if the homunculus was in the sperm or from the mother if it was in the egg. Although many observations suggested that offspring

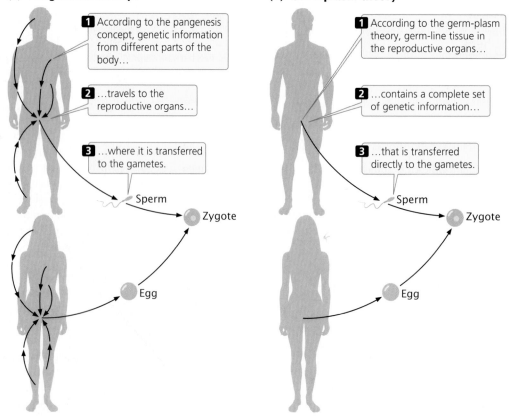

(a) Pangenesis concept

1 According to the pangenesis concept, genetic information from different parts of the body…

2 …travels to the reproductive organs…

3 …where it is transferred to the gametes.

Sperm

Zygote

Egg

(b) Germ-plasm theory

1 According to the germ-plasm theory, germ-line tissue in the reproductive organs…

2 …contains a complete set of genetic information…

3 …that is transferred directly to the gametes.

Sperm

Zygote

Egg

1.10 **Pangenesis, an early concept of inheritance, compared with the modern germ-plasm theory.**

1.11 Preformationism was a popular idea of inheritance in the seventeenth and eighteenth centuries. Shown here is a drawing of a homunculus inside a sperm. *[Science VU/Visuals Unlimited.]*

composed of cells, cells arise only from preexisting cells, and the cell is the fundamental unit of structure and function in living organisms. Biologists began to examine cells to see how traits were transmitted in the course of cell division.

Charles Darwin (1809–1882), one of the most influential biologists of the nineteenth century, put forth the theory of evolution through natural selection and published his ideas in *On the Origin of Species* in 1859. Darwin recognized that heredity was fundamental to evolution, and he conducted extensive genetic crosses with pigeons and other organisms. However, he never understood the nature of inheritance, and this lack of understanding was a major omission in his theory of evolution.

Walther Flemming (1843–1905) observed the division of chromosomes in 1879 and published a superb description of mitosis. By 1885, it was generally recognized that the nucleus contained the hereditary information.

Near the close of the nineteenth century, August Weismann (1834–1914) finally laid to rest the notion of the inheritance of acquired characteristics. He cut off the tails of mice for 22 consecutive generations and showed that the tail length in descendants remained stubbornly long. Weismann proposed the **germ-plasm theory,** which holds that the cells in the reproductive organs carry a complete set of genetic information that is passed to the egg and sperm (**Figure 1.10b**).

possess a mixture of traits from both parents, preformationism remained a popular concept throughout much of the seventeenth and eighteenth centuries.

Another early notion of heredity was **blending inheritance,** which proposed that offspring are a blend, or mixture, of parental traits. This idea suggested that the genetic material itself blends, much as blue and yellow pigments blend to make green paint. Once blended, genetic differences could not be separated out in future generations, just as green paint cannot be separated out into blue and yellow pigments. Some traits do *appear* to exhibit blending inheritance; however, thanks to Gregor Mendel's research with pea plants, we now understand that individual genes do not blend.

The Rise of the Science of Genetics

In 1676, Nehemiah Grew (1641–1712) reported that plants reproduce sexually by using pollen from the male sex cells. With this information, a number of botanists began to experiment with crossing plants and creating hybrids, including Gregor Mendel (1822–1884; **Figure 1.12**), who went on to discover the basic principles of heredity.

Developments in cytology (the study of cells) in the 1800s had a strong influence on genetics. Building on the work of others, Matthias Jacob Schleiden (1804–1881) and Theodor Schwann (1810–1882) proposed the concept of the **cell theory** in 1839. According to this theory, all life is

1.12 Gregor Mendel was the founder of modern genetics. Mendel first discovered the principles of heredity by crossing different varieties of pea plants and analyzing the pattern of transmission of traits in subsequent generations. *[Hulton Archive/Getty Images.]*

1.13 The human genome was completely sequenced in 2003. Each of the colored bars represents one nucleotide base in the DNA.

The year 1900 was a watershed in the history of genetics. Gregor Mendel's pivotal 1866 publication on experiments with pea plants, which revealed the principles of heredity, was rediscovered, as discussed in more detail in Chapter 3. The significance of his conclusions was recognized, and other biologists immediately began to conduct similar genetic studies on mice, chickens, and other organisms. The results of these investigations showed that many traits indeed follow Mendel's rules.

Walter Sutton (1877–1916) proposed in 1902 that genes are located on chromosomes. Thomas Hunt Morgan (1866–1945) discovered the first genetic mutant of fruit flies in 1910 and used fruit flies to unravel many details of transmission genetics. The foundation for population genetics was laid in the 1930s when geneticists begin to synthesize Mendelian genetics and evolutionary theory.

Geneticists began to use bacteria and viruses in the 1940s; the rapid reproduction and simple genetic systems of these organisms allowed detailed study of the organization and structure of genes. At about this same time, evidence accumulated that DNA was the repository of genetic information. James Watson (b. 1928) and Francis Crick (1916–2004), along with Maurice Wilkins (1916–2004) and Rosalind Franklin (1920–1958), described the three-dimensional structure of DNA in 1953, ushering in the era of molecular genetics.

By 1966, the chemical structure of DNA and the system by which it determines the amino acid sequence of proteins had been worked out. Advances in molecular genetics led to the first recombinant DNA experiments in 1973, which touched off another revolution in genetic research. Methods for rapidly sequencing DNA were first developed in 1977, which later allowed whole genomes of humans and other organisms to be determined. The polymerase chain reaction, a technique for quickly amplifying tiny amounts of DNA,

was developed by Kary Mullis (b. 1944) and others in 1983. This technique is now the basis of numerous types of molecular analysis. In 1990, the Human Genome Project was launched. By 1995, the first complete DNA sequence of a free-living organism—the bacterium *Haemophilus influenzae*—was determined, and the first complete sequence of a eukaryotic organism (yeast) was reported a year later. A rough draft of the human genome sequence was reported in 2000, with the sequence essentially completed in 2003, ushering in a new era in genetics (**Figure 1.13**). Today, the genomes of numerous organisms are being sequenced, analyzed, and compared.

The Future of Genetics

Numerous advances in genetics are being made today, and genetics is at the forefront of biological research. For example, the information content of genetics is increasing at a rapid pace, as the genome sequences of many organisms are added to DNA databases every year. New details about gene structure and function are continually expanding our knowledge of how genetic information is encoded and how it specifies phenotypic traits.

Information about sequence differences among individual organisms is a source of new insights about evolution and helps to locate genes that affect complex traits such as hypertension in humans and weight gain in cattle. In recent years, our understanding of the role of RNA in many cellular processes has expanded greatly; RNA has a role in many aspects of gene function. New genetic microchips that simultaneously analyze thousands of RNA molecules are providing information about the activity of thousands of genes in a given cell, allowing a detailed picture of how cells respond to external signals, environmental stresses, and disease states such as cancer. In the emerging field of proteomics, powerful computer programs are being used to model the structure and function of proteins from DNA sequence information. All of this information provides us with a better understanding of numerous biological processes and evolutionary relationships. The flood of new genetic information requires the continuous development of sophisticated computer programs to store, retrieve, compare, and analyze genetic data and has given rise to the field of bioinformatics, a merging of molecular biology and computer science.

In the future, the focus of DNA-sequencing efforts will shift from the genomes of different species to individual differences within species. In the not too distant future, each person may possess a copy of his or her entire genome sequence, which can be used to assess the risk of acquiring various diseases and to tailor their treatment should they arise. The use of genetics in the agricultural, chemical, and health-care fields will continue to expand. This ever-widening scope of genetics will raise significant ethical, social, and economic issues.

This brief overview of the history of genetics is not intended to be comprehensive; rather it is designed to provide a sense of the accelerating pace of advances in genetics. In the chapters to come, we will learn more about the experiments and the scientists who helped shape the discipline of genetics.

Concepts

Developments in plant hybridization and cytology in the eighteenth and nineteenth centuries laid the foundation for the field of genetics today. After Mendel's work was rediscovered in 1900, the science of genetics developed rapidly and today is one of the most active areas of science.

✔ **Concept Check 3**

How did developments in cytology in the nineteenth century contribute to our modern understanding of genetics?

1.3 A Few Fundamental Concepts Are Important for the Start of Our Journey into Genetics

Undoubtedly, you learned some genetic principles in other biology classes. Let's take a few moments to review some of the fundamental genetic concepts.

- **Cells are of two basic types: eukaryotic and prokaryotic.** Structurally, cells consist of two basic types, although, evolutionarily, the story is more complex (see Chapter 2). Prokaryotic cells lack a nuclear membrane and possess no membrane-bounded cell organelles, whereas eukaryotic cells are more complex, possessing a nucleus and membrane-bounded organelles such as chloroplasts and mitochondria.

- **The gene is the fundamental unit of heredity.** The precise way in which a gene is defined often varies, depending on the biological context. At the simplest level, we can think of a gene as a unit of information that encodes a genetic characteristic. We will enlarge this definition as we learn more about what genes are and how they function.

- **Genes come in multiple forms called alleles.** A gene that specifies a characteristic may exist in several forms, called alleles. For example, a gene for coat color in cats may exist in an allele that encodes black fur or an allele that encodes orange fur.

- **Genes confer phenotypes.** One of the most important concepts in genetics is the distinction between traits and genes. Traits are not inherited directly. Rather, genes are inherited and, along with environmental factors, determine the expression of traits. The genetic information that an individual organism possesses is its genotype; the trait is its phenotype. For example, the A blood type is a phenotype; the genetic information that encodes the blood-type-A antigen is the genotype.

- **Genetic information is carried in DNA and RNA.** Genetic information is encoded in the molecular structure of nucleic acids, which come in two types: deoxyribonucleic acid (DNA) and ribonucleic acid (RNA). Nucleic acids are polymers consisting of repeating units called nucleotides; each nucleotide consists of a sugar, a phosphate, and a nitrogenous base. The nitrogenous bases in DNA are of four types (abbreviated A, C, G, and T), and the sequence of these bases encodes genetic information. DNA consists of two complementary nucleotide strands. Most organisms carry their genetic information in DNA, but a few viruses carry it in RNA. The four nitrogenous bases of RNA are abbreviated A, C, G, and U.

- **Genes are located on chromosomes.** The vehicles of genetic information within a cell are chromosomes (**Figure 1.14**), which consist of DNA and associated proteins. The cells of each species have a characteristic number of chromosomes; for example, bacterial cells normally possess a single chromosome; human cells possess 46; pigeon cells possess 80. Each chromosome carries a large number of genes.

- **Chromosomes separate through the processes of mitosis and meiosis.** The processes of mitosis and meiosis ensure that each daughter cell receives a complete set of an organism's chromosomes. Mitosis is the separation of replicated chromosomes in the division of somatic (nonsex) cells. Meiosis is the pairing and separation of replicated chromosomes in the division of sex cells to produce gametes (reproductive cells).

1.14 Genes are carried on chromosomes. A chromosome, shown here, consists of a DNA complexed to protein and may carry genetic information for many traits. *[Biophoto Associates/Science Source/Photo Researchers.]*

- **Genetic information is transferred from DNA to RNA to protein.** Many genes encode traits by specifying the structure of proteins. Genetic information is first transcribed from DNA into RNA, and then RNA is translated into the amino acid sequence of a protein.
- **Mutations are permanent, heritable changes in genetic information.** Gene mutations affect the genetic information of only a single gene; chromosome mutations alter the number or the structure of chromosomes and therefore usually affect many genes.

- **Some traits are affected by multiple factors.** Some traits are influenced by multiple genes that interact in complex ways with environmental factors. Human height, for example, is affected by hundreds of genes as well as environmental factors such as nutrition.
- **Evolution is genetic change.** Evolution can be viewed as a two-step process: first, genetic variation arises and, second, some genetic variants increase in frequency, whereas other variants decrease in frequency.

Concepts Summary

- Genetics is central to the life of every person: it influences a person's physical features, susceptibility to numerous diseases, personality, and intelligence.
- Genetics plays important roles in agriculture, the pharmaceutical industry, and medicine. It is central to the study of biology.
- All organisms use similar genetic systems. Genetic variation is the foundation of evolution and is critical to understanding all life.
- The study of genetics can be divided into transmission genetics, molecular genetics, and population genetics.
- Model genetic organisms are species having characteristics that make them particularly amenable to genetic analysis and about which much genetic information exists.
- The use of genetics by humans began with the domestication of plants and animals.
- The ancient Greeks developed the concepts of pangenesis and the inheritance of acquired characteristics.
- Preformationism suggested that a person inherits all of his or her traits from one parent. Blending inheritance proposed that offspring possess a mixture of the parental traits.

- By studying the offspring of crosses between varieties of peas, Gregor Mendel discovered the principles of heredity. Developments in cytology in the nineteenth century led to the understanding that the cell nucleus is the site of heredity.
- In 1900, Mendel's principles of heredity were rediscovered. Population genetics was established in the early 1930s, followed closely by biochemical genetics and bacterial and viral genetics. The structure of DNA was discovered in 1953, stimulating the rise of molecular genetics.
- Cells are of two basic types: prokaryotic and eukaryotic.
- The genes that determine a trait are termed the genotype; the trait that they produce is the phenotype.
- Genes are located on chromosomes, which are made up of nucleic acids and proteins and are partitioned into daughter cells through the process of mitosis or meiosis.
- Genetic information is expressed through the transfer of information from DNA to RNA to proteins.
- Evolution requires genetic change in populations.

Important Terms

genome (p. 4)
transmission genetics (p. 5)
molecular genetics (p. 5)
population genetics (p. 5)
model genetic organism (p. 5)

pangenesis (p. 7)
inheritance of acquired characteristics (p. 7)
preformationism (p. 8)
blending inheritance (p. 9)

cell theory (p. 9)
germ-plasm theory (p. 9)

Answers to Concept Checks

1. d

2. No, because horses are expensive to house, feed, and propagate, they have too few progeny, and their generation time is too long.

3. Developments in cytology in the 1800s led to the identification of parts of the cell, including the cell nucleus and chromosomes. The cell theory focused the attention of biologists on the cell, which eventually led to the conclusion that the nucleus contains the hereditary information.

Comprehension Questions

Answers to questions and problems preceded by an asterisk can be found at the end of the book.

Section 1.1

*1. How does the Hopi culture contribute to the high incidence of albinism among members of the Hopi tribe?

*2. Give at least three examples of the role of genetics in society today.

3. Briefly explain why genetics is crucial to modern biology.

*4. List the three traditional subdisciplines of genetics and summarize what each covers.

5. What are some characteristics of model genetic organisms that make them useful for genetic studies?

Section 1.2

6. When and where did agriculture first arise? What role did genetics play in the development of the first domesticated plants and animals?

*7. Outline the notion of pangenesis and explain how it differs from the germ-plasm theory.

8. What does the concept of the inheritance of acquired characteristics propose and how is it related to the notion of pangenesis?

*9. What is preformationism? What did it have to say about how traits are inherited?

10. Define blending inheritance and contrast it with preformationism.

11. How did developments in botany in the seventeenth and eighteenth centuries contribute to the rise of modern genetics?

*12. Who first discovered the basic principles that laid the foundation for our modern understanding of heredity?

13. List some advances in genetics that have been made in the twentieth century.

Section 1.3

14. What are the two basic cell types (from a structural perspective) and how do they differ?

*15. Outline the relations between genes, DNA, and chromosomes.

Application Questions and Problems

Section 1.1

16. What is the relation between genetics and evolution?

*17. For each of the following genetic topics, indicate whether it focuses on transmission genetics, molecular genetics, or population genetics.

   a. Analysis of pedigrees to determine the probability of someone inheriting a trait

   b. Study of the genetic history of people on a small island to determine why a genetic form of asthma is so prevalent on the island

   c. The influence of nonrandom mating on the distribution of genotypes among a group of animals

   d. Examination of the nucleotide sequences found at the ends of chromosomes

   e. Mechanisms that ensure a high degree of accuracy during DNA replication

   f. Study of how the inheritance of traits encoded by genes on sex chromosomes (sex-linked traits) differs from the inheritance of traits encoded by genes on nonsex chromosomes (autosomal traits)

Section 1.2

*18. Genetics is said to be both a very old science and a very young science. Explain what is meant by this statement.

19. Match the description (*a* through *d*) with the correct theory or concept listed below.

   | Preformationism | Germ-plasm theory |
   | Pangenesis | Inheritance of acquired characteristics |

   a. Each reproductive cell contains a complete set of genetic information.

   b. All traits are inherited from one parent.

   c. Genetic information may be altered by the use of a characteristic.

   d. Cells of different tissues contain different genetic information.

*20. Compare and contrast the following ideas about inheritance.

   a. Pangenesis and germ-plasm theory

   b. Preformationism and blending inheritance

   c. The inheritance of acquired characteristics and our modern theory of heredity

Section 1.3

*21. Compare and contrast the following terms:

   a. Eukaryotic and prokaryotic cells

   b. Gene and allele

   c. Genotype and phenotype

   d. DNA and RNA

   e. DNA and chromosome

Challenge Questions

Section 1.1

22. We now know as much or more about the genetics of humans as we know about that of any other organism, and humans are the focus of many genetic studies. Do you think humans should be considered a model genetic organism? Why or why not?

23. Describe some of the ways in which your own genetic makeup affects you as a person. Be as specific as you can.

24. Describe at least one trait that appears to run in your family (appears in multiple members of the family). Do you think this trait runs in your family because it is an inherited trait or because is caused by environmental factors that are common to family members? How might you distinguish between these possibilities?

Section 1.3

*25. Suppose that life exists elsewhere in the universe. All life must contain some type of genetic information, but alien genomes might not consist of nucleic acids and have the same features as those found in the genomes of life on Earth. What do you think might be the common features of all genomes, no matter where they exist?

26. Pick one of the following ethical or social issues and give your opinion on the issue. For background information, you might read one of the articles on ethics listed and marked with an asterisk in the Suggested Readings section for Chapter 1 at www.whfreeman.com/pierce.

  a. Should a person's genetic makeup be used in determining his or her eligibility for life insurance?

  b. Should biotechnology companies be able to patent newly sequenced genes?

  c. Should gene therapy be used on people?

  d. Should genetic testing be made available for inherited conditions for which there is no treatment or cure?

  e. Should governments outlaw the cloning of people?

*27. Suppose that you could undergo genetic testing at age 18 for susceptibility to a genetic disease that would not appear until middle age and has no available treatment.

  a. What would be some of the possible reasons for having such a genetic test and some of the possible reasons for not having the test?

  b. Would you personally want to be tested? Explain your reasoning.

2 Chromosomes and Cellular Reproduction

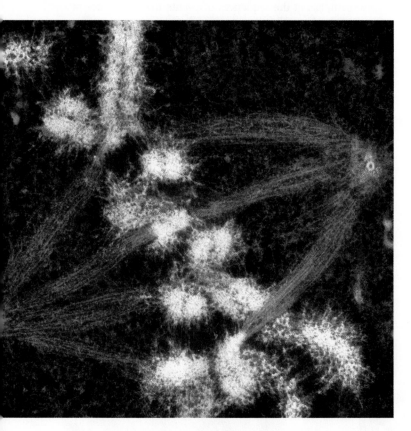

Chromosomes in mitosis, the process whereby each new cell receives a complete copy of the genetic material.
[Conly L. Reider/Biological Photo Service.]

The Blind Men's Riddle

In a well-known riddle, two blind men by chance enter a department store at the same time, go to the same counter, and both order five pairs of socks, each pair of different color. The sales clerk is so befuddled by this strange coincidence that he places all ten pairs (two black pairs, two blue pairs, two gray pairs, two brown pairs, and two green pairs) into a single shopping bag and gives the bag with all ten pairs to one blind man and an empty bag to the other. The two blind men happen to meet on the street outside, where they discover that one of their bags contains all ten pairs of socks. How do the blind men, without seeing and without any outside help, sort out the socks so that each man goes home with exactly five pairs of different colored socks? Can you come up with a solution to the riddle?

By an interesting coincidence, cells have the same dilemma as that of the blind men in the riddle. Most organisms possess two sets of genetic information, one set inherited from each parent. Before cell division, the DNA in each chromosome replicates; after replication, there are two copies—called sister chromatids—of each chromosome. At the end of cell division, it is critical that each new cell receives a complete copy of the genetic material, just as each blind man needed to go home with a complete set of socks.

The solution to the riddle is simple. Socks are sold as pairs; the two socks of a pair are typically connected by a thread. As a pair is removed from the bag, the men each grasp a different sock of the pair and pull in opposite directions. When the socks are pulled tight, it is easy for one of the men to take a pocket knife and cut the thread connecting the pair. Each man then deposits his single sock in his own bag. At the end of the process, each man's bag will contain exactly two black socks, two blue socks, two gray socks, two brown socks, and two green socks.[1]

Remarkably, cells employ a similar solution for separating their chromosomes into new daughter cells. As we will learn in this chapter, the replicated chromosomes line up at the center of a cell undergoing division and, like the socks in the riddle, the sister chromatids of each chromosome are pulled in opposite directions. Like the thread connecting two socks of a pair, a molecule called cohesin holds the sister chromatids together until severed by a molecular knife called separase. The two resulting chromosomes separate and the cell divides, ensuring that a complete set of chromosomes is deposited in each cell.

---

[1]This analogy is adapted from K. Nasmyth. Disseminating the genome: Joining, resolving, and separating sister chromatids during mitosis and meiosis. *Annual Review of Genetics* 34:673–745, 2001.

In this analogy, the blind men and cells differ in one critical regard: if the blind men make a mistake, one man ends up with an extra sock and the other is a sock short, but no great harm results. The same cannot be said for human cells. Errors in chromosome separation, producing cells with too many or too few chromosomes, are frequently catastrophic, leading to cancer, reproductive failure, or—sometimes—a child with severe handicaps.

This chapter explores the process of cell reproduction and how a complete set of genetic information is transmitted to new cells. In prokaryotic cells, reproduction is simple, because prokaryotic cells possess a single chromosome. In eukaryotic cells, multiple chromosomes must be copied and distributed to each of the new cells, and so cell reproduction is more complex. Cell division in eukaryotes takes place through mitosis and meiosis, processes that serve as the foundation for much of genetics.

Grasping mitosis and meiosis requires more than simply memorizing the sequences of events that take place in each stage, although these events are important. The key is to understand how genetic information is apportioned in the course of cell reproduction through a dynamic interplay of DNA synthesis, chromosome movement, and cell division. These processes bring about the transmission of genetic information and are the basis of similarities and differences between parents and progeny.

	Prokaryotic cells	Eukaryotic cells
Nucleus	Absent	Present
Cell diameter	Relatively small, from 1 to 10 μm	Relatively large, from 10 to 100 μm
Genome	Usually one circular DNA molecule	Multiple linear DNA molecules
DNA	Not complexed with histones in eubacteria; some histones in archaea	Complexed with histones
Amount of DNA	Relatively small	Relatively large
Membrane-bounded organelles	Absent	Present
Cytoskeleton	Absent	Present

2.1 **Prokaryotic and eukaryotic cells differ in structure.** *[Photographs (left to right) by T. J. Beveridge/ Visuals Unlimited; W. Baumeister/Science Photo Library/Photo Researchers; G. Murti/Phototake; Biophoto Associates/ Photo Researchers.]*

(a) **(b)**

2.2 Prokaryotic and eukaryotic DNA compared. (a) Prokaryotic DNA (shown in red) is neither surrounded by a nuclear membrane nor complexed with histone proteins. (b) Eukaryotic DNA is complexed to histone proteins to form chromosomes (one of which is shown) that are located in the nucleus. [Part a: A. B. Dowsett/Science Photo Library/Photo Researchers. Part b: Biophoto Associates/Photo Researchers.]

2.1 Prokaryotic and Eukaryotic Cells Differ in a Number of Genetic Characteristics

Biologists traditionally classify all living organisms into two major groups, the *prokaryotes* and the *eukaryotes* (**Figure 2.1**). A **prokaryote** is a unicellular organism with a relatively simple cell structure. A **eukaryote** has a compartmentalized cell structure having components bounded by intracellular membranes; eukaryotes may be unicellular or multicellular.

Research indicates that a division of life into two major groups, the prokaryotes and eukaryotes, is not so simple. Although similar in cell structure, prokaryotes include at least two fundamentally distinct types of bacteria: the **eubacteria** (true bacteria) and the **archaea** (ancient bacteria). An examination of equivalent DNA sequences reveals that eubacteria and archaea are as distantly related to one another as they are to the eukaryotes. Although eubacteria and archaea are similar in cell structure, some genetic processes in archaea (such as transcription) are more similar to those in eukaryotes, and the archaea are actually closer evolutionarily to eukaryotes than to eubacteria. Thus, from an evolutionary perspective, there are three major groups of organisms: eubacteria, archaea, and eukaryotes. In this book, the prokaryotic–eukaryotic distinction will be made frequently, but important eubacterial–archaeal differences also will be noted.

From the perspective of genetics, a major difference between prokaryotic and eukaryotic cells is that a eukaryote has a *nuclear envelope*, which surrounds the genetic material to form a **nucleus** and separates the DNA from the other cellular contents. In prokaryotic cells, the genetic material is in close contact with other components of the cell—a property that has important consequences for the way in which genes are controlled.

Another fundamental difference between prokaryotes and eukaryotes lies in the packaging of their DNA. In eukaryotes, DNA is closely associated with a special class of proteins, the **histones**, to form tightly packed chromosomes. This complex of DNA and histone proteins is termed **chromatin**, which is the stuff of eukaryotic chromosomes. Histone proteins limit the accessibility of enzymes and other proteins that copy and read the DNA, but they enable the DNA to fit into the nucleus. Eukaryotic DNA must separate from the histones before the genetic information in the DNA can be accessed. Archaea also have some histone proteins that complex with DNA, but the structure of their chromatin is different from that found in eukaryotes. However, eubacteria do not possess histones; so their DNA does not exist in the highly ordered, tightly packed arrangement found in eukaryotic cells (**Figure 2.2**). The copying and reading of DNA are therefore simpler processes in eubacteria.

Genes of prokaryotic cells are generally on a single, circular molecule of DNA—the chromosome of a prokaryotic cell. In eukaryotic cells, genes are located on multiple, usually linear DNA molecules (multiple chromosomes). Eukaryotic cells therefore require mechanisms that ensure that a copy of each chromosome is faithfully transmitted to each new cell. This generalization—a single, circular chromosome in prokaryotes and multiple, linear chromosomes in eukaryotes—is not always true. A few bacteria have more than one chromosome, and important bacterial genes are frequently found on other DNA molecules called *plasmids* (see Chapter 6). Furthermore, in some eukaryotes, a few genes are located on circular DNA molecules, as in mitochondria and chloroplasts.

Concepts

Organisms are classified as prokaryotes or eukaryotes, and prokaryotes consist of archaea and eubacteria. A prokaryote is a unicellular organism that lacks a nucleus, its DNA is not complexed to histone proteins, and its genome is usually a single chromosome. Eukaryotes are either unicellular or multicellular, their cells possess a nucleus, their DNA is complexed to histone proteins, and their genomes consist of multiple chromosomes.

✔ Concept Check 1

List several characteristics that eubacteria and archaea have in common and that distinguish them from eukaryotes.

Viruses are simple structures composed of an outer protein coat surrounding nucleic acid (either DNA or RNA; **Figure 2.3**). Viruses are neither cells nor primitive forms of life: they can reproduce only within host cells, which means that they must have evolved after, rather than before, cells evolved. In addition, viruses are not an evolutionarily distinct group but are most closely related to their hosts—the genes of a plant virus are more similar to those in a plant cell than to those in animal viruses, which suggests that viruses evolved from their hosts, rather than from other viruses. The close relationship between the genes of virus and host makes viruses useful for studying the genetics of host organisms.

2.2 Cell Reproduction Requires the Copying of the Genetic Material, Separation of the Copies, and Cell Division

For any cell to reproduce successfully, three fundamental events must take place: (1) its genetic information must be copied, (2) the copies of genetic information must be separated from each other, and (3) the cell must divide. All cellular reproduction includes these three events, but the processes that lead to these events differ in prokaryotic and eukaryotic cells because of their structural differences.

Prokaryotic Cell Reproduction

When prokaryotic cells reproduce, the circular chromosome of the bacterium is replicated. Replication usually begins at a specific place on the bacterial chromosome, called the origin of replication. In a process that is not fully understood, the origins of the two newly replicated chromosomes move away from each other and toward opposite ends of the cell. In at least some bacteria, proteins bind near the replication origins and anchor the new chromosomes to the plasma membrane at opposite ends of the cell. Finally, a new cell wall

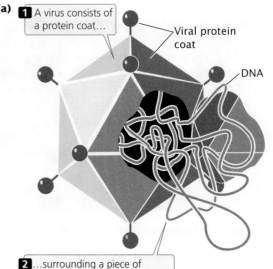

(a)

1 A virus consists of a protein coat…

Viral protein coat

DNA

2 …surrounding a piece of nucleic acid—in this case, DNA.

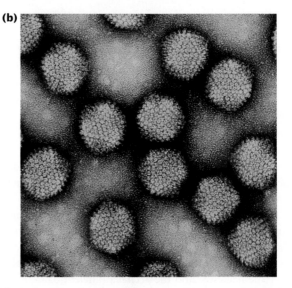

(b)

2.3 A virus is a simple replicative structure consisting of protein and nucleic acid. Part *b* is a micrograph of adenoviruses. *[Hans Gelderblom/Visuals Unlimited.]*

forms between the two chromosomes, producing two cells, each with an identical copy of the chromosome. Under optimal conditions, some bacterial cells divide every 20 minutes. At this rate, a single bacterial cell could produce a billion descendants in a mere 10 hours.

Eukaryotic Cell Reproduction

Like prokaryotic cell reproduction, eukaryotic cell reproduction requires the processes of DNA replication, copy separation, and division of the cytoplasm. However, the presence of multiple DNA molecules requires a more complex mechanism to ensure that exactly one copy of each molecule ends up in each of the new cells.

Eukaryotic chromosomes are separated from the cytoplasm by the nuclear envelope. The nucleus was once thought to be a fluid-filled bag in which the chromosomes floated, but we now know that the nucleus has a highly organized internal scaffolding called the *nuclear matrix*. This matrix consists of a network of protein fibers that maintains precise spatial relations among the nuclear components and takes part in DNA replication, the expression of genes, and the modification of gene products before they leave the nucleus. We will now take a closer look at the structure of eukaryotic chromosomes.

Eukaryotic chromosomes   Each eukaryotic species has a characteristic number of chromosomes per cell: potatoes have 48 chromosomes, fruit flies have 8, and humans have 46. There appears to be no special significance between the complexity of an organism and its number of chromosomes per cell.

In most eukaryotic cells, there are two sets of chromosomes. The presence of two sets is a consequence of sexual reproduction: one set is inherited from the male parent and the other from the female parent. Each chromosome in one set has a corresponding chromosome in the other set, together constituting a homologous pair (**Figure 2.4**). Human cells, for example, have 46 chromosomes, constituting 23 homologous pairs.

The two chromosomes of a **homologous pair** are usually alike in structure and size, and each carries genetic information for the same set of hereditary characteristics. (An exception is the sex chromosomes, which will be discussed in Chapter 4.) For example, if a gene on a particular chromosome encodes a characteristic such as hair color, another copy of the gene (each copy is called an *allele*) at the same position on that chromosome's homolog *also* encodes hair color. However, these two alleles need not be identical: one

of them might encode red hair and the other might encode blond hair. Thus, most cells carry two sets of genetic information; these cells are **diploid**. But not all eukaryotic cells are diploid: reproductive cells (such as eggs, sperm, and spores) and even nonreproductive cells in some organisms may contain a single set of chromosomes. Cells with a single set of chromosomes are **haploid**. A haploid cell has only one copy of each gene.

Concepts

Cells reproduce by copying and separating their genetic information and then dividing. Because eukaryotes possess multiple chromosomes, mechanisms exist to ensure that each new cell receives one copy of each chromosome. Most eukaryotic cells are diploid, and their two chromosome sets can be arranged in homologous pairs. Haploid cells contain a single set of chromosomes.

✔ Concept Check 2

Diploid cells have

a. two chromosomes.

b. two sets of chromosomes.

c. one set of chromosomes.

d. two pairs of homologous chromosomes.

Chromosome structure   The chromosomes of eukaryotic cells are larger and more complex than those found in prokaryotes, but each unreplicated chromosome nevertheless consists of a single molecule of DNA. Although linear, the DNA molecules in eukaryotic chromosomes are highly folded and condensed; if stretched out, some human chromosomes would be several centimeters long—thousands of times as long as the span of a typical nucleus. To package such a tremendous length of DNA into this small volume,

Humans have 23 pairs of chromosomes, including the sex chromosomes, X and Y. Males are XY, females are XX.

(a)

A *diploid* organism has two sets of chromosomes organized as *homologous* pairs.

(b)

Allele *A*   Allele *a*

These two versions of a gene encode a trait such as hair color.

2.4 Diploid eukaryotic cells have two sets of chromosomes. (a) A set of chromosomes from a female human cell. Each pair of chromosomes is hybridized to a uniquely colored probe, giving it a distinct color. (b) The chromosomes are present in homologous pairs, which consist of chromosomes that are alike in size and structure and carry information for the same characteristics. [*Part a: Courtesy of Dr. Thomas Ried and Dr. Evelin Schrock.*]

each DNA molecule is coiled again and again and tightly packed around histone proteins, forming a rod-shaped chromosome. Most of the time, the chromosomes are thin and difficult to observe but, before cell division, they condense further into thick, readily observed structures; it is at this stage that chromosomes are usually studied.

A functional chromosome has three essential elements: a centromere, a pair of telomeres, and origins of replication. The *centromere* is the attachment point for *spindle microtubules*, which are the filaments responsible for moving chromosomes during cell division (**Figure 2.5**). The centromere appears as a constricted region. Before cell division, a protein complex called the *kinetochore* assembles on the centromere; later spindle microtubules attach to the kinetochore. Chromosomes lacking a centromere cannot be drawn into the newly formed nuclei; these chromosomes are lost, often with catastrophic consequences to the cell. On the basis of the location of the centromere, chromosomes are classified into four types: metacentric, submetacentric, acrocentric, and telocentric (**Figure 2.6**). One of the two arms of a chromosome (the short arm of a submetacentric or acrocentric chromosome) is designated by the letter p and the other arm is designated by q.

Telomeres are the natural ends, the tips, of a linear chromosome (see Figure 2.5); they serve to stabilize the chromosome ends. If a chromosome breaks, producing new ends, these ends have a tendency to stick together, and the chromosome is degraded at the newly broken ends. Telomeres provide chromosome stability. The results of research (discussed in Chapter 8) suggest that telomeres also participate in limiting cell division and may play important roles in aging and cancer.

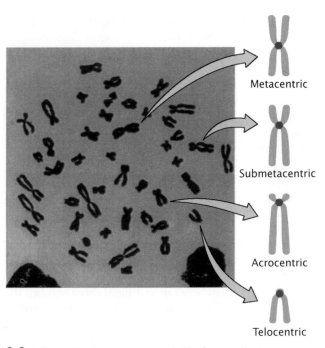

2.6 Eukaryotic chromosomes exist in four major types based on the position of the centromere. [*Micrograph by L. Lisco, D. W. Fawcett/Visuals Unlimited.*]

Origins of replication are the sites where DNA synthesis begins; they are not easily observed by microscopy. In preparation for cell division, each chromosome replicates, making a copy of itself, as already mentioned. These two initially identical copies, called **sister chromatids**, are held together at the centromere (see Figure 2.5). Each sister chromatid consists of a single molecule of DNA.

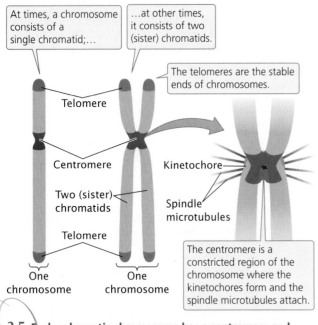

2.5 Each eukaryotic chromosome has a centromere and telomeres.

Concepts

Sister chromatids are copies of a chromosome held together at the centromere. Functional chromosomes contain centromeres, telomeres, and origins of replication. The kinetochore is the point of attachment for the spindle microtubules; telomeres are the stabilizing ends of a chromosome; origins of replication are sites where DNA synthesis begins.

✔ **Concept Check 3**

What are three essential elements required for a chromosome to function?

The Cell Cycle and Mitosis

The **cell cycle** is the life story of a cell, the stages through which it passes from one division to the next (**Figure 2.7**). This process is critical to genetics because, through the cell cycle, the genetic instructions for all characteristics are passed from parent to daughter cells. A new cycle begins after

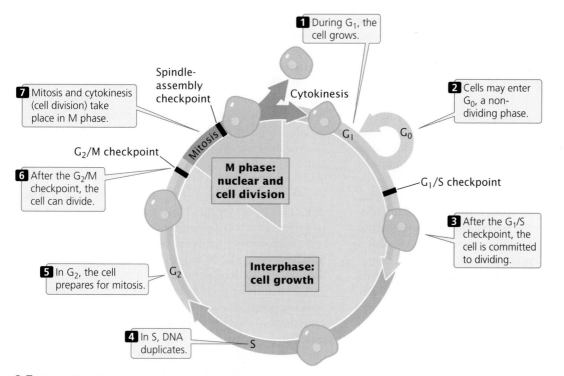

1 During G₁, the cell grows.

Spindle-assembly checkpoint

Cytokinesis

7 Mitosis and cytokinesis (cell division) take place in M phase.

2 Cells may enter G₀, a non-dividing phase.

G₁   G₀

G₂/M checkpoint

Mitosis

M phase: nuclear and cell division

G₁/S checkpoint

6 After the G₂/M checkpoint, the cell can divide.

3 After the G₁/S checkpoint, the cell is committed to dividing.

5 In G₂, the cell prepares for mitosis.

G₂

Interphase: cell growth

4 In S, DNA duplicates.

S

2.7 **The cell cycle consists of interphase and M phase.**

a cell has divided and produced two new cells. Each new cell metabolizes, grows, and develops. At the end of its cycle, the cell divides to produce two cells, which can then undergo additional cell cycles. Progression through the cell cycle is regulated at key transition points called **checkpoints**.

The cell cycle consists of two major phases. The first is **interphase**, the period between cell divisions, in which the cell grows, develops, and prepares for cell division. The second is the **M phase** (mitotic phase), the period of active cell division. The M phase includes **mitosis**, the process of nuclear division, and **cytokinesis**, or cytoplasmic division. Let's take a closer look at the details of interphase and the M phase.

Interphase   Interphase is the extended period of growth and development between cell divisions. Interphase includes several checkpoints, which regulate the cell cycle by allowing or prohibiting the cell's division. These checkpoints, like the checkpoints in the M phase, ensure that all cellular components are present and in good working order before the cell proceeds to the next stage. Checkpoints are necessary to prevent cells with damaged or missing chromosomes from proliferating. Defects in checkpoints can lead to unregulated cell growth, as is seen in some cancers.

By convention, interphase is divided into three phases: G₁, S, and G₂ (see Figure 2.7). Interphase begins with *G₁* (for gap 1). In G₁, the cell grows, and proteins necessary for cell division are synthesized; this phase typically lasts several hours. There is a critical point termed the *G₁/S checkpoint* near the end of G₁. The G₁/S checkpoint holds the cell in G₁

until the cell has all of the enzymes necessary for the replication of DNA. After this checkpoint has been passed, the cell is committed to divide.

Before reaching the G₁/S checkpoint, cells may exit from the active cell cycle in response to regulatory signals and pass into a nondividing phase called *G₀*, which is a stable state during which cells usually maintain a constant size. They can remain in G₀ for an extended period of time, even indefinitely, or they can reenter G₁ and the active cell cycle. Many cells never enter G₀; rather, they cycle continuously.

After G₁, the cell enters the *S* phase (for DNA synthesis), in which each chromosome duplicates. Although the cell is committed to divide after the G₁/S checkpoint has been passed, DNA synthesis must take place before the cell can proceed to mitosis. If DNA synthesis is blocked (by drugs or by a mutation), the cell will not be able to undergo mitosis. Before the S phase, each chromosome is composed of one chromatid; after the S phase, each chromosome is composed of two chromatids (see Figure 2.5).

After the S phase, the cell enters *G₂* (gap 2). In this phase, several additional biochemical events necessary for cell division take place. The important *G₂/M checkpoint* is reached near the end of G₂. This checkpoint is passed only if the cell's DNA is undamaged. Damaged DNA can inhibit the activation of some proteins that are necessary for mitosis to take place. After the G₂/M checkpoint has been passed, the cell is ready to divide and enters the M phase. Although the length of interphase varies from cell type to cell type, a typical dividing mammalian cell spends about 10 hours in G₁, 9 hours in S, and 4 hours in G₂ (see Figure 2.7).

Throughout interphase, the chromosomes are in a relaxed, but by no means uncoiled, state, and individual chromosomes cannot be seen with the use of a microscope. This condition changes dramatically when interphase draws to a close and the cell enters the M phase.

M phase   The M phase is the part of the cell cycle in which the copies of the cell's chromosomes (sister chromatids) separate and the cell undergoes division. The separation of sister chromatids in the M phase is a critical process that results in a complete set of genetic information for each of the resulting cells. Biologists usually divide the M phase into six stages: the five stages of mitosis (prophase, prometaphase, metaphase, anaphase, and telophase), illustrated in **Figure 2.8**, and cytokinesis. It's important to keep in mind that the M phase is a continuous process, and its separation into these six stages is somewhat arbitrary.

During interphase, the chromosomes are relaxed and are visible only as diffuse chromatin, but they condense during **prophase**, becoming visible under a light microscope. Each chromosome possesses two chromatids because the chromosome was duplicated in the preceding S phase. The *mitotic spindle,* an organized array of microtubules that move the chromosomes in mitosis, forms. In animal cells, the spindle grows out from a pair of *centrosomes* that migrate to opposite sides of the cell. Within each centrosome is a special organelle, the *centriole,* which also is composed of microtubules. Some plant cells do not have centrosomes or centrioles, but they do have mitotic spindles.

Disintegration of the nuclear membrane marks the start of **prometaphase**. Spindle microtubules, which until now have been outside the nucleus, enter the nuclear region. The ends of certain microtubules make contact with the chromosomes. For each chromosome, a microtubule from one of the centrosomes anchors to the kinetochore of *one* of the sister chromatids; a microtubule from the opposite centrosome then attaches to the other sister chromatid, and so the chromosome is anchored to both of the centrosomes. The microtubules lengthen and shorten, pushing and pulling the chromosomes about. Some microtubules extend from each centrosome toward the center of the spindle but do not attach to a chromosome.

During **metaphase**, the chromosomes become arranged in a single plane, the *metaphase plate,* between the two centrosomes. The centrosomes, now at opposite ends of the cell with microtubules radiating outward and meeting in the middle of the cell, center at the spindle poles. A *spindle-assembly checkpoint* ensures that each chromosome is aligned on the metaphase plate and attached to spindle fibers from opposite poles.

Anaphase begins when the sister chromatids separate and move toward opposite spindle poles. After the chromatids have separated, each is considered a separate chromosome. **Telophase** is marked by the arrival of the chromosomes at the spindle poles. The nuclear membrane reforms around each set of chromosomes, producing two separate nuclei within the cell. The chromosomes relax and lengthen, once again disappearing from view. In many cells, division of the cytoplasm (cytokinesis) is simultaneous with telophase. The major features of the cell cycle are summarized in **Table 2.1**.

Table 2.1	Features of the cell cycle
Stage	**Major Features**
G_0 phase	Stable, nondividing period of variable length.
Interphase	
G_1 phase	Growth and development of the cell; G_1/S checkpoint.
S phase	Synthesis of DNA.
G_2 phase	Preparation for division; G_2/M checkpoint.
M phase	
Prophase	Chromosomes condense and mitotic spindle forms.
Prometaphase	Nuclear envelope disintegrates, and spindle microtubules anchor to kinetochores.
Metaphase	Chromosomes align on the spindle-assembly checkpoint.
Anaphase	Sister chromatids separate, becoming individual chromosomes that migrate toward spindle poles.
Telophase	Chromosomes arrive at spindle poles, the nuclear envelope re-forms, and the condensed chromosomes relax.
Cytokinesis	Cytoplasm divides; cell wall forms in plant cells.

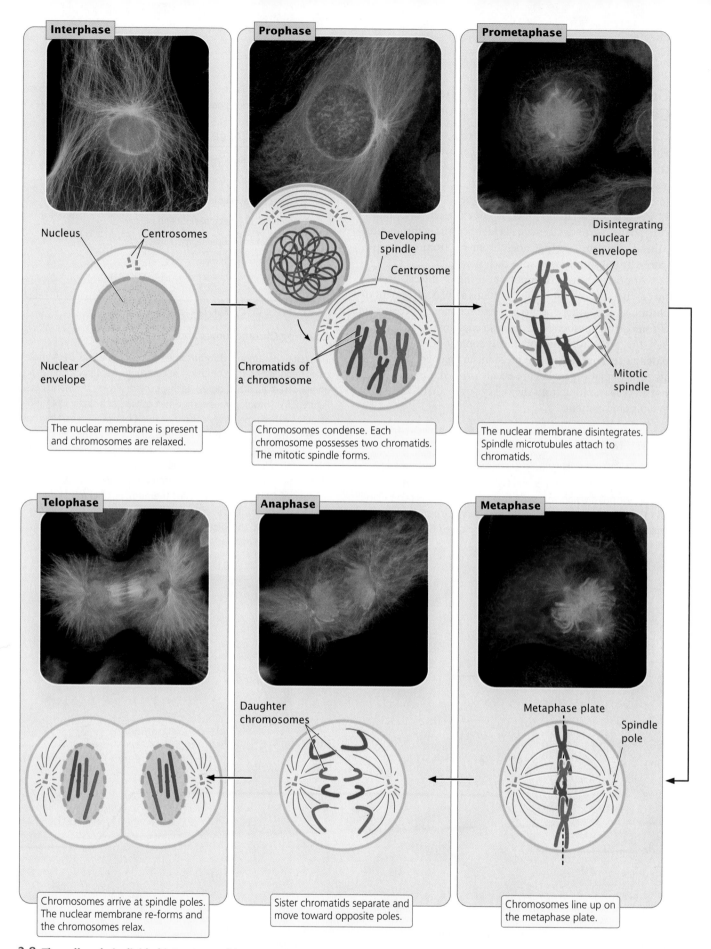

Interphase

Nucleus

Centrosomes

Nuclear envelope

The nuclear membrane is present and chromosomes are relaxed.

Prophase

Developing spindle

Centrosome

Chromatids of a chromosome

Chromosomes condense. Each chromosome possesses two chromatids. The mitotic spindle forms.

Prometaphase

Disintegrating nuclear envelope

Mitotic spindle

The nuclear membrane disintegrates. Spindle microtubules attach to chromatids.

Telophase

Chromosomes arrive at spindle poles. The nuclear membrane re-forms and the chromosomes relax.

Anaphase

Daughter chromosomes

Sister chromatids separate and move toward opposite poles.

Metaphase

Metaphase plate

Spindle pole

Chromosomes line up on the metaphase plate.

2.8 The cell cycle is divided into stages. *[Photographs by Conly L. Rieder/Biological Photo Service.]*

Genetic Consequences of the Cell Cycle

What are the genetically important results of the cell cycle? From a single cell, the cell cycle produces two cells that contain the same genetic instructions. These two cells are genetically identical with each other and with the cell that gave rise to them. They are genetically identical because DNA synthesis in the S phase creates an exact copy of each DNA molecule, giving rise to two genetically identical sister chromatids. Mitosis then ensures that one chromatid from each replicated chromosome passes into each new cell.

Another genetically important result of the cell cycle is that each of the cells produced contains a full complement of chromosomes—there is no net reduction or increase in chromosome number. Each cell also contains approximately half the cytoplasm and organelle content of the original parental cell, but no precise mechanism analogous to mitosis ensures that organelles are evenly divided. Consequently, not all cells resulting from the cell cycle are identical in their cytoplasmic content.

Concepts

The active cell-cycle phases are interphase and the M phase. Interphase consists of G_1, S, and G_2. In G_1, the cell grows and prepares for cell division; in the S phase, DNA synthesis takes place; in G_2, other biochemical events necessary for cell division take place. Some cells enter a quiescent phase called G_0. The M phase includes mitosis and cytokinesis and is divided into prophase, prometaphase, metaphase, anaphase, and telophase.

✔ Concept Check 4

Which is the correct order of stages in the cell cycle?

a. G_1, S, prophase, metaphase, anaphase

b. S, G_1, prophase, metaphase, anaphase

c. Prophase, S, G_1, metaphase, anaphase

d. S, G_1, anaphase, prophase, metaphase

Connecting Concepts

Counting Chromosomes and DNA Molecules

The relations among chromosomes, chromatids, and DNA molecules frequently cause confusion. At certain times, chromosomes are unreplicated; at other times, each possesses two chromatids (see Figure 2.5). Chromosomes sometimes consist of a single DNA molecule; at other times, they consist of two DNA molecules. How can we keep track of the number of these structures in the cell cycle?

There are two simple rules for counting chromosomes and DNA molecules: (1) to determine the number of chromosomes, count the number of functional centromeres; (2) to determine the number of DNA molecules, count the number of chromatids. Let's examine a hypothetical cell as it passes through the cell cycle (**Figure 2.9**). At the beginning of G_1, this diploid cell has two

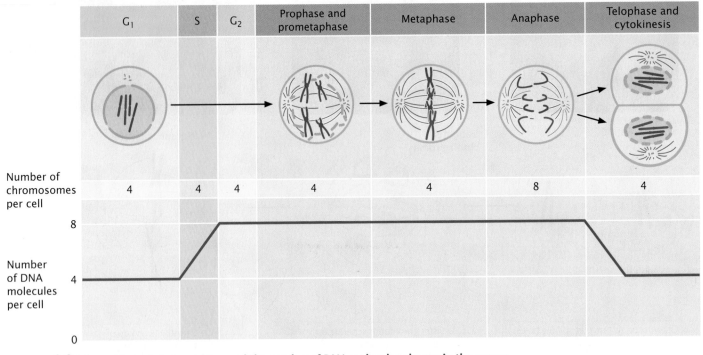

2.9 The number of chromosomes and the number of DNA molecules change in the course of the cell cycle. The number of chromosomes per cell equals the number of functional centromeres, and the number of DNA molecules per cell equals the number of chromatids.

complete sets of chromosomes, inherited from its parent cell. Each chromosome consists of a single chromatid—a single DNA molecule—and so there are four DNA molecules in the cell during G_1. In the S phase, each DNA molecule is copied. The two resulting DNA molecules combine with histones and other proteins to form sister chromatids. Although the amount of DNA doubles in the S phase, the number of chromosomes remains the same, because the two sister chromatids are tethered together and share a single functional centromere. At the end of the S phase, this cell still contains four chromosomes, each with two chromatids; so there are eight DNA molecules present.

Through prophase, prometaphase, and metaphase, the cell has four chromosomes and eight DNA molecules. At anaphase, however, the sister chromatids separate. Each now has its own functional centromere, and so each is considered a separate chromosome. Until cytokinesis, the cell contains eight chromosomes, each consisting of a single chromatid; thus, there are still eight DNA molecules present. After cytokinesis, the eight chromosomes (eight DNA molecules) are distributed equally between two cells; so each new cell contains four chromosomes and four DNA molecules, the number present at the beginning of the cell cycle.

In summary, the number of chromosomes increases briefly only in anaphase, when the two chromatids of a chromosome separate, and decreases only through cytokinesis. The number of DNA molecules increases only in the S phase and decreases only through cytokinesis.

2.3 Sexual Reproduction Produces Genetic Variation Through the Process of Meiosis

If all reproduction were accomplished through mitosis, life would be quite dull, because mitosis produces only genetically identical progeny. With only mitosis, you, your children, your parents, your brothers and sisters, your cousins, and many people you don't even know would be clones—copies of one another. Only the occasional mutation would introduce any genetic variability. All organisms reproduced in this way for the first 2 billion years of Earth's existence (and it is the way in which some organisms still reproduce today). Then, some 1.5 billion to 2 billion years ago, something remarkable evolved: cells that produce genetically variable offspring through sexual reproduction.

The evolution of sexual reproduction is one of the most significant events in the history of life. By shuffling the genetic information from two parents, sexual reproduction greatly increases the amount of genetic variation and allows for accelerated evolution. Most of the tremendous diversity of life on Earth is a direct result of sexual reproduction.

Sexual reproduction consists of two processes. The first is **meiosis**, which leads to *gametes* in which chromosome number is reduced by half. The second process is **fertiliza-**

tion, in which two haploid gametes fuse and restore chromosome number to its original diploid value.

Meiosis

The words *mitosis* and *meiosis* are sometimes confused. They sound a bit alike, and both refer to chromosome division and cytokinesis. But don't be deceived. The outcomes of mitosis and meiosis are radically different, and several unique events that have important genetic consequences take place only in meiosis.

How does meiosis differ from mitosis? Mitosis consists of a single nuclear division and is usually accompanied by a single cell division. Meiosis, on the other hand, consists of two divisions. After mitosis, chromosome number in newly formed cells is the same as that in the original cell, whereas meiosis causes chromosome number in the newly formed cells to be reduced by half. Finally, mitosis produces genetically identical cells, whereas meiosis produces genetically variable cells. Let's see how these differences arise.

Like mitosis, meiosis is preceded by an interphase stage that includes G_1, S, and G_2 phases. Meiosis consists of two distinct processes: *meiosis I* and *meiosis II*, each of which includes a cell division. The first division, which comes at the end of meiosis I, is termed the reduction division because the number of chromosomes per cell is reduced by half (**Figure 2.10**). The second division, which comes at the end of meiosis II, is sometimes termed the equational division. The events of meiosis II are similar to those of mitosis. However, meiosis II differs from mitosis in that chromosome number has already been halved in meiosis I, and the cell does not begin with the same number of chromosomes as it does in mitosis (see Figure 2.10).

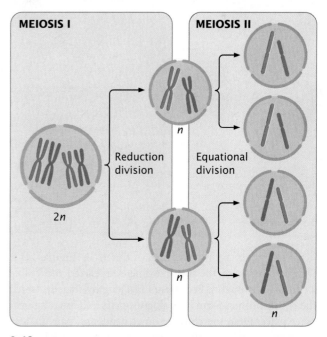

2.10 Meiosis includes two cell divisions. In this illustration, the original cell is $2n = 4$. After two meiotic divisions, each resulting cell is $1n = 2$.

Meiosis I

Middle Prophase I

Centrosomes

Chromosomes begin to condense, and the spindle forms.

Late Prophase I

Pairs of homologs

Homologous chromosomes pair.

Late Prophase I

Chiasmata

Crossing over takes place, and the nuclear membrane breaks down.

Meiosis II

Prophase II

The chromosomes recondense.

Metaphase II

Equatorial plate

Individual chromosomes line up on the equatorial plate.

Anaphase II

Sister chromatids separate and move toward opposite poles.

The stages of meiosis are outlined in **Figure 2.11**. During interphase, the chromosomes are relaxed and visible as diffuse chromatin. **Prophase I** is a lengthy stage in which the chromosomes form homologous pairs and crossing over takes place. First, the chromosomes condense, pair up, and begin **synapsis**, a very close pairing association. Each homologous pair of synapsed chromosomes consists of four chromatids called a **bivalent** or **tetrad**. The chromosomes become shorter and thicker, and a three-part *synaptonemal complex* develops between homologous chromosomes. **Crossing over** takes place, in which homologous chromosomes exchange genetic information. The centromeres of the paired chromosomes move apart; the two homologs remain attached at each *chiasma* (plural, *chiasmata*), which is the result of crossing over. Finally, the chiasmata move toward the ends of the chromosomes as the strands slip apart; so the

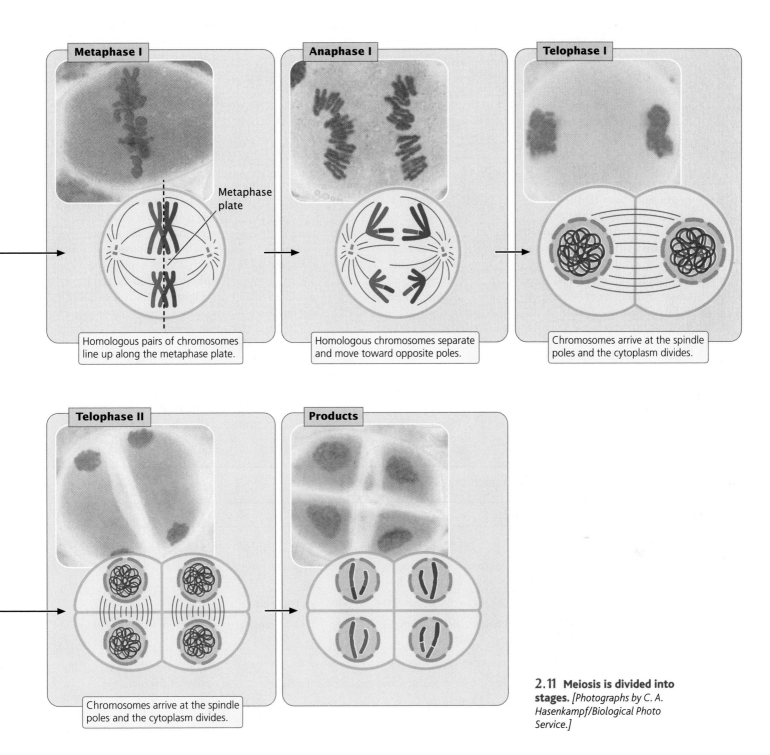

Metaphase I

Homologous pairs of chromosomes line up along the metaphase plate.

Metaphase plate

Anaphase I

Homologous chromosomes separate and move toward opposite poles.

Telophase I

Chromosomes arrive at the spindle poles and the cytoplasm divides.

Telophase II

Chromosomes arrive at the spindle poles and the cytoplasm divides.

Products

2.11 Meiosis is divided into stages. [Photographs by C. A. Hasenkampf/Biological Photo Service.]

homologs remain paired only at the tips. Near the end of prophase I, the nuclear membrane breaks down and the spindle forms.

Metaphase I is initiated when homologous pairs of chromosomes align along the metaphase plate (see Figure 2.11). A microtubule from one pole attaches to one chromosome of a homologous pair, and a microtubule from the other pole attaches to the other member of the pair.

Anaphase I is marked by the separation of homologous chromosomes. The two chromosomes of a homologous pair are pulled toward opposite poles. Although the homologous chromosomes separate, the sister chromatids remain attached and travel together. In **telophase I**, the chromosomes arrive at the spindle poles and the cytoplasm divides.

The period between meiosis I and meiosis II is **interkinesis**, in which the nuclear membrane re-forms around the chromosomes clustered at each pole, the spindle breaks down, and the chromosomes relax. These cells then pass through **prophase II**, in which the events of interkinesis are reversed: the chromosomes recondense, the spindle re-forms, and the nuclear envelope once again breaks down. In interkinesis in some types of cells, the chromosomes remain condensed, and the spindle does not break down. These cells move directly from cytokinesis into **metaphase II**, which is similar to metaphase of mitosis: the individual chromosomes line up on the metaphase plate, with the sister chromatids facing opposite poles.

In **anaphase II**, the kinetochores of the sister chromatids separate and the chromatids are pulled to opposite poles. Each chromatid is now a distinct chromosome. In **telophase II**, the chromosomes arrive at the spindle poles, a nuclear envelope re-forms around the chromosomes, and the cytoplasm divides. The chromosomes relax and are no longer visible. The major events of meiosis are summarized in **Table 2.2**.

Consequences of Meiosis

What are the overall consequences of meiosis? First, meiosis comprises two divisions; so each original cell produces four cells (there are exceptions to this generalization, as, for example, in many female animals; see Figure 2.15b). Second, chromosome number is reduced by half; so cells produced by meiosis are haploid. Third, cells produced by meiosis are genetically different from one another and from the parental cell.

Genetic differences among cells result from two processes that are unique to meiosis. The first is crossing over, which takes place in prophase I. Crossing over refers to the exchange of genes between nonsister chromatids (chromatids from different homologous chromosomes). After crossing over has taken place, the sister chromatids may no longer be identical. Crossing over is the basis for intrachromosomal **recombination**, creating new combinations of alleles on a chromatid. To see how crossing over produces genetic variation, consider two pairs of alleles, which we will abbreviate *Aa* and *Bb*. Assume that one chromosome possesses the *A* and *B* alleles and its homolog possesses the *a* and *b* alleles (**Figure 2.12a**). When DNA is replicated in the S phase, each chromosome duplicates, and so the resulting sister chromatids are identical (**Figure 2.12b**).

In the process of crossing over, there are breaks in the DNA strands and the breaks are repaired in such a way that segments of nonsister chromatids are exchanged (**Figure 2.12c**). The molecular basis of this process will be described in more detail in Chapter 9; the important thing here is that, after crossing over has taken place, the two sister chromatids are no longer identical—one chromatid has alleles *A* and *B*, whereas its sister chromatid (the chromatid that underwent

Table 2.2	Major events in each stage of meiosis
Stage	**Major Events**
Meiosis I	
Prophase I	Chromosomes condense, homologous chromosomes synapse, crossing over takes place, nuclear envelope breaks down, and mitotic spindle forms.
Metaphase I	Homologous pairs of chromosomes line up on the metaphase plate.
Anaphase I	The two chromosomes (each with two chromatids) of each homologous pair separate and move toward opposite poles.
Telophase I	Chromosomes arrive at the spindle poles.
Cytokinesis	The cytoplasm divides to produce two cells, each having half the original number of chromosomes.
Interkinesis	In some types of cells, the spindle breaks down, chromosomes relax, and a nuclear envelope re-forms, but no DNA synthesis takes place.
Meiosis II	
Prophase II*	Chromosomes condense, the spindle forms, and the nuclear envelope disintegrates.
Metaphase II	Individual chromosomes line up on the metaphase plate.
Anaphase II	Sister chromatids separate and move as individual chromosomes toward the spindle poles.
Telophase II	Chromosomes arrive at the spindle poles; the spindle breaks down and a nuclear envelope re-forms.
Cytokinesis	The cytoplasm divides.

*Only in cells in which the spindle has broken down, chromosomes have relaxed, and the nuclear envelope has re-formed in telophase I. Other types of cells proceed directly to metaphase II after cytokinesis.

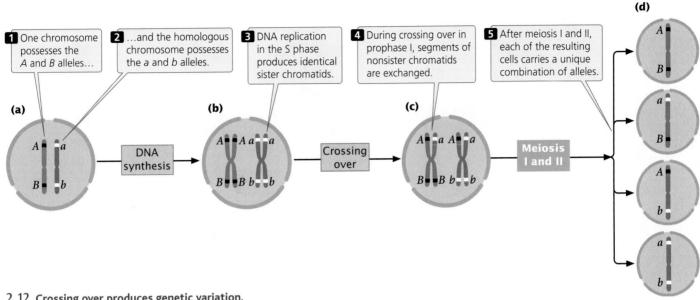

1 One chromosome possesses the A and B alleles...

2 ...and the homologous chromosome possesses the a and b alleles.

3 DNA replication in the S phase produces identical sister chromatids.

4 During crossing over in prophase I, segments of nonsister chromatids are exchanged.

5 After meiosis I and II, each of the resulting cells carries a unique combination of alleles.

2.12 **Crossing over produces genetic variation.**

crossing over) has alleles *a* and *B*. Likewise, one chromatid of the other chromosome has alleles *a* and *b*, and the other has alleles *A* and *b*. Each of the four chromatids now carries a unique combination of alleles: $\underline{A\ B}$, $\underline{a\ B}$, $\underline{A\ b}$, and $\underline{a\ b}$. Eventually, the two homologous chromosomes separate, each going into a different cell. In meiosis II, the two chromatids of each chromosome separate, and thus each of the four cells resulting from meiosis carries a different combination of alleles (**Figure 2.12d**).

The second process of meiosis that contributes to genetic variation is the random distribution of chromosomes in anaphase I of meiosis after their random alignment in metaphase I. To illustrate this process, consider a cell with three pairs of chromosomes, I, II, and III (**Figure 2.13a**). One chromosome of each pair is maternal in origin (I_m, II_m, and III_m); the other is paternal in origin (I_p, II_p, and III_p). The chromosome pairs line up in the center of the cell in metaphase I; and, in anaphase I, the chromosomes of each homologous pair separate.

How each pair of homologs aligns and separates is random and independent of how other pairs of chromosomes align and separate (**Figure 2.13b**). By chance, all the maternal chromosomes might migrate to one side, with all the paternal chromosomes migrating to the other. After division, one cell would contain chromosomes I_m, II_m, and III_m, and the other, I_p, II_p, and III_p. Alternatively, the I_m, II_m, and III_p chromosomes might move to one side, and the I_p, II_p, and III_m chromosomes to the other. The different migrations would produce different combinations of chromosomes in the resulting cells (**Figure 2.13c**). There are four ways in which a diploid cell with three pairs of chromosomes can divide, producing a total of eight different combinations of chromosomes in the gametes. In general, the number of possible combinations is 2^n, where *n* equals the number of

homologous pairs. As the number of chromosome pairs increases, the number of combinations quickly becomes very large. In humans, who have 23 pairs of chromosomes, there are 8,388,608 different combinations of chromosomes possible from the random separation of homologous chromosomes. Through the random distribution of chromosomes in anaphase I, alleles located on different chromosomes are sorted into different combinations. The genetic consequences of this process, termed independent assortment, will be explored in more detail in Chapter 3.

In summary, crossing over shuffles alleles on the *same* chromosome into new combinations, whereas the random distribution of maternal and paternal chromosomes shuffles alleles on *different* chromosomes into new combinations. Together, these two processes are capable of producing tremendous amounts of genetic variation among the cells resulting from meiosis.

Concepts

Meiosis consists of two distinct processes: meiosis I and meiosis II. Meiosis (usually) produces four haploid cells that are genetically variable. The two mechanisms responsible for genetic variation are crossing over and the random distribution of maternal and paternal chromosomes.

✔ Concept Check 5

Which of the following events takes place in meiosis II but not meiosis I?

a. Crossing over

b. Contraction of chromosomes

c. Separation of homologous chromosomes

d. Separation of chromatids

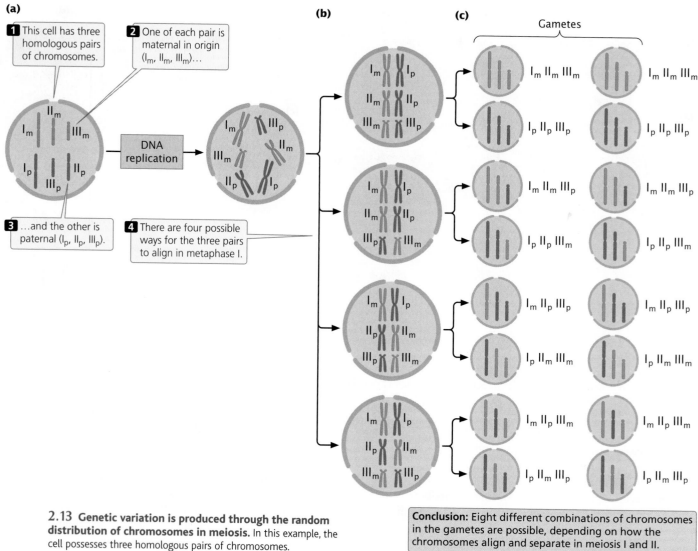

(a)

1 This cell has three homologous pairs of chromosomes.

2 One of each pair is maternal in origin (I_m, II_m, III_m)…

3 …and the other is paternal (I_p, II_p, III_p).

4 There are four possible ways for the three pairs to align in metaphase I.

DNA replication

(b)

(c)   Gametes

2.13 Genetic variation is produced through the random distribution of chromosomes in meiosis. In this example, the cell possesses three homologous pairs of chromosomes.

Conclusion: Eight different combinations of chromosomes in the gametes are possible, depending on how the chromosomes align and separate in meiosis I and II.

Connecting Concepts

Mitosis and Meiosis Compared

Now that we have examined the details of mitosis and meiosis, let's compare the two processes (**Figure 2.14**). In both mitosis and meiosis, the chromosomes contract and become visible; both processes include the movement of chromosomes toward the spindle poles, and both are accompanied by cell division. Beyond these similarities, the processes are quite different.

Mitosis results in a single cell division and usually produces two daughter cells. Meiosis, in contrast, comprises two cell divisions and usually produces four cells. In diploid cells, homologous chromosomes are present before both meiosis and mitosis, but the pairing of homologs takes place only in meiosis.

Another difference is that, in meiosis, chromosome number is reduced by half as a consequence of the separation of homologous pairs of chromosomes in anaphase I, but no chromosome reduction takes place in mitosis. Furthermore, meiosis is characterized by two processes that produce genetic variation: crossing over (in prophase I) and the random distribution of maternal and paternal chromosomes (in anaphase I). There are normally no equivalent processes in mitosis.

Mitosis and meiosis also differ in the behavior of chromosomes in metaphase and anaphase. In metaphase I of meiosis, *homologous pairs* of chromosomes line up on the metaphase plate, whereas *individual chromosomes* line up on the metaphase plate in metaphase of mitosis (and in metaphase II of meiosis). In anaphase I of meiosis, *paired chromosomes* separate and migrate toward opposite spindle poles, each chromosome possessing two chromatids attached at the centromere. In contrast, in anaphase of mitosis (and in anaphase II of meiosis), *sister chromatids* separate, and each chromosome that moves toward a spindle pole consists of a single chromatid.

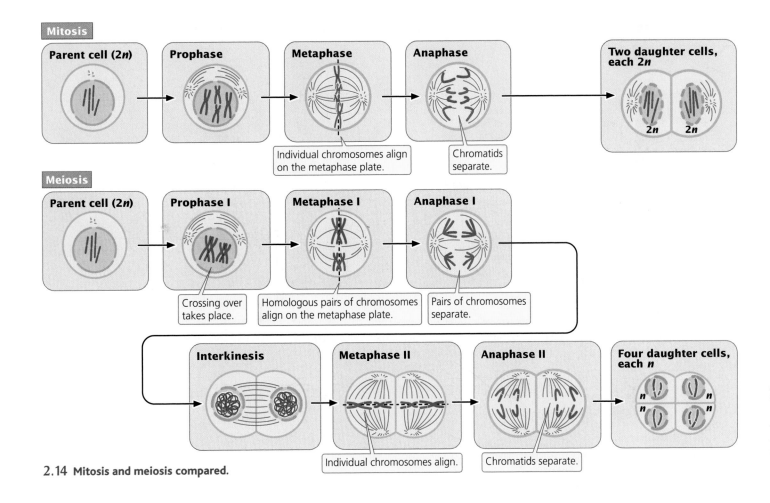

2.14 **Mitosis and meiosis compared.**

Meiosis in the Life Cycles of Animals and Plants

The overall result of meiosis is four haploid cells that are genetically variable. Let's now see where meiosis fits into the life cycles of a multicellular animal and a multicellular plant.

Meiosis in animals   The production of gametes in a male animal, called **spermatogenesis**, takes place in the testes. There, diploid primordial germ cells divide mitotically to produce diploid cells called **spermatogonia** (**Figure 2.15a**). Each spermatogonium can undergo repeated rounds of mitosis, giving rise to numerous additional spermatogonia. Alternatively, a spermatogonium can initiate meiosis and enter into prophase I. Now called a **primary spermatocyte**, the cell is still diploid because the homologous chromosomes have not yet separated. Each primary spermatocyte completes meiosis I, giving rise to two haploid **secondary spermatocytes** that then undergo meiosis II, with each producing two haploid **spermatids**. Thus, each primary spermatocyte produces a total of four haploid spermatids, which mature and develop into sperm.

The production of gametes in a female animal, called **oogenesis**, begins much like spermatogenesis. Within the ovaries, diploid primordial germ cells divide mitotically to produce **oogonia** (**Figure 2.15b**). Like spermatogonia, oogonia can undergo repeated rounds of mitosis or they can enter into meiosis. When they enter prophase I, these still-diploid cells are called **primary oocytes**. Each primary oocyte completes meiosis I and divides.

Here, the process of oogenesis begins to differ from that of spermatogenesis. In oogenesis, cytokinesis is unequal: most of the cytoplasm is allocated to one of the two haploid cells, the **secondary oocyte**. The smaller cell, which contains half of the chromosomes but only a small part of the cytoplasm, is called the **first polar body**; it may or may not divide further. The secondary oocyte completes meiosis II, and, again, cytokinesis is unequal—most of the cytoplasm passes into one of the cells. The larger cell, which acquires most of the cytoplasm, is the **ovum**, the mature female gamete. The smaller cell is the **second polar body**. Only the ovum is capable of being fertilized, and the polar bodies usually disintegrate. Oogenesis, then, produces a single mature gamete from each primary oocyte.

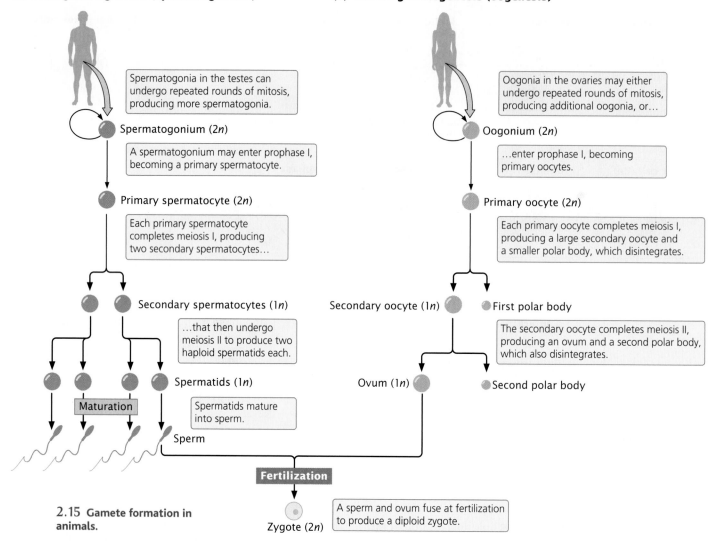

(a) Male gametogenesis (spermatogenesis)

Spermatogonia in the testes can undergo repeated rounds of mitosis, producing more spermatogonia.

Spermatogonium (2*n*)

A spermatogonium may enter prophase I, becoming a primary spermatocyte.

Primary spermatocyte (2*n*)

Each primary spermatocyte completes meiosis I, producing two secondary spermatocytes…

Secondary spermatocytes (1*n*)

…that then undergo meiosis II to produce two haploid spermatids each.

Spermatids (1*n*)

Maturation

Spermatids mature into sperm.

Sperm

2.15 Gamete formation in animals.

(b) Female gametogenesis (oogenesis)

Oogonia in the ovaries may either undergo repeated rounds of mitosis, producing additional oogonia, or…

Oogonium (2*n*)

…enter prophase I, becoming primary oocytes.

Primary oocyte (2*n*)

Each primary oocyte completes meiosis I, producing a large secondary oocyte and a smaller polar body, which disintegrates.

Secondary oocyte (1*n*)   First polar body

The secondary oocyte completes meiosis II, producing an ovum and a second polar body, which also disintegrates.

Ovum (1*n*)   Second polar body

Fertilization

Zygote (2*n*)

A sperm and ovum fuse at fertilization to produce a diploid zygote.

Concepts

In the testes, a diploid spermatogonium undergoes meiosis, producing a total of four haploid sperm cells. In the ovary, a diploid oogonium undergoes meiosis to produce a single large ovum and smaller polar bodies that normally disintegrate.

✔ Concept Check 6

A secondary spermatocyte has 12 chromosomes. How many chromosomes will be found in the primary spermatocyte that gave rise to it?

a. 6

b. 12

c. 18

d. 24

Meiosis in plants Most plants have a complex life cycle that includes two distinct generations (stages): the diploid *sporophyte* and the haploid *gametophyte*. These two stages alternate; the sporophyte produces haploid spores through meiosis, and the gametophyte produces haploid gametes through mitosis (**Figure 2.16**). This type of life cycle is sometimes called *alternation of generations*. In this cycle, the immediate products of meiosis are called *spores*, not gametes; the spores undergo one or more mitotic divisions to produce gametes. Although the terms used for this process are somewhat different from those commonly used in regard to animals (and from some of those employed so far in this chapter), the processes in plants and animals are basically the same: in both, meiosis leads to a reduction in chromosome number, producing haploid cells.

In flowering plants, the sporophyte is the obvious, vegetative part of the plant; the gametophyte consists of only a

1 Through meiosis, the diploid (2*n*) sporophyte produces haploid (1*n*) spores, which become the gametophyte.

Mitosis

Spores

♀ gamete

♂ gamete

2 Through mitosis, the gametophytes produce haploid gametes…

Gametophyte (haploid, *n*)

Meiosis

Fertilization

Sporophyte (diploid, 2*n*)

Zygote

3 …that fuse during fertilization to form a diploid zygote.

Mitosis

4 Through mitosis, the zygote becomes the diploid sporophyte.

2.16 Plants alternate between diploid and haploid life stages (female, ♀ ; male, ♂).

few haploid cells within the sporophyte. The flower, which is part of the sporophyte, contains the reproductive structures. The male part of the flower, the stamen, contains diploid reproductive cells called **microsporocytes**, each of which undergoes meiosis to produce four haploid **microspores** (**Figure 2.17a**). Each microspore divides mitotically, producing an immature pollen grain consisting of two haploid nuclei. One of these nuclei, called the tube nucleus, directs the growth of a pollen tube. The other, termed the generative nucleus, divides mitotically to produce two sperm cells. The pollen grain, with its two haploid nuclei, is the male gametophyte.

The female part of the flower, the ovary, contains diploid cells called **megasporocytes**, each of which undergoes meiosis to produce four haploid **megaspores** (**Figure 2.17b**), only one of which survives. The nucleus of the surviving megaspore divides mitotically three times, producing a total of eight haploid nuclei that make up the female gametophyte, the embryo sac. Division of the cytoplasm then produces separate cells, one of which becomes the *egg*.

When the plant flowers, the stamens open and release pollen grains. Pollen lands on a flower's stigma—a sticky platform that sits on top of a long stalk called the style. At the base of the style is the ovary. If a pollen grain germinates, it grows a tube down the style into the ovary. The two sperm cells pass down this tube and enter the embryo sac (**Figure 2.17c**). One of the sperm cells fertilizes the egg cell, producing a diploid zygote, which develops into an embryo. The other sperm cell fuses with two nuclei enclosed in a single cell, giving rise to a 3*n* (triploid)

endosperm, which stores food that will be used later by the embryonic plant. These two fertilization events are termed *double fertilization.*

Concepts

In the stamen of a flowering plant, meiosis produces haploid microspores that divide mitotically to produce haploid sperm in a pollen grain. Within the ovary, meiosis produces four haploid megaspores, only one of which divides mitotically three times to produce eight haploid nuclei. After pollination, one sperm fertilizes the egg cell, producing a diploid zygote; the other fuses with two nuclei to form the endosperm.

✔ Concept Check 7

Which structure is diploid?

a. Microspore   c. Megaspore

b. Egg   d. Microsporocyte

We have now examined the place of meiosis in the sexual cycle of two organisms, a typical multicellular animal and a flowering plant. These cycles are just two of the many variations found among eukaryotic organisms. Although the cellular events that produce reproductive cells in plants and animals differ in the number of cell divisions, the number of haploid gametes produced, and the relative size of the final products, the overall result is the same: meiosis gives rise to haploid, genetically variable cells that then fuse during fertilization to produce diploid progeny.

(a) Stamen

Flower

(b) Pistil

Ovary

Microsporocyte (diploid)

Megasporocyte (diploid)

Diploid, 2n

1 In the stamen, diploid microsporocytes undergo meiosis…

Meiosis

Meiosis

6 In the ovary, diploid megasporocytes undergo meiosis…

Haploid, 1n

2 …to produce four haploid microspores.

Four microspores (haploid)

Four megaspores (haploid)

Only one survives

7 …to produce four haploid megaspores, but only one survives.

3 Each undergoes mitosis to produce a pollen grain with two haploid nuclei.

Mitosis

Mitosis

8 The surviving megaspore divides mitotically three times…

Haploid generative nucleus

Pollen grain

Haploid tube nucleus

2 nuclei

4 The tube nucleus directs the growth of a pollen tube.

4 nuclei

Mitosis

8 nuclei

9 …to produce eight haploid nuclei.

Pollen tube

Two haploid sperm cells

Tube nucleus

10 The cytoplasm divides, producing separate cells,…

Division of cytoplasm

11 …one of which becomes the egg.

5 The generative nucleus divides mitotically to produce two sperm cells.

Polar nuclei

Embryo sac

12 Two of the nuclei become polar nuclei…

Sperm

Polar nuclei

Egg

Egg

13 …and the other nuclei are partitioned into separate cells.

(c)

Double fertilization

14 Double fertilization takes place when the two sperm cells of a pollen grain enter the embryo sac.

Endosperm, (triploid, 3n)

16 The other sperm cell fuses with the binucleate cell to form triploid endosperm.

15 One sperm cell fertilizes the egg cell, producing a diploid zygote.

Embryo (diploid, 2n)

2.17 Sexual reproduction in flowering plants.

Concepts Summary

- A prokaryotic cell possesses a simple structure, with no nuclear envelope and usually a single, circular chromosome. A eukaryotic cell possesses a more complex structure, with a nucleus and multiple linear chromosomes consisting of DNA complexed to histone proteins.

- Cell reproduction requires the copying of the genetic material, separation of the copies, and cell division.

- In a prokaryotic cell, the single chromosome replicates, each copy moves toward opposite sides of the cell, and the cell divides. In eukaryotic cells, reproduction is more complex than in prokaryotic cells, requiring mitosis and meiosis to ensure that a complete set of genetic information is transferred to each new cell.

- In eukaryotic cells, chromosomes are typically found in homologous pairs. Each functional chromosome consists of a centromere, telomeres, and multiple origins of replication. After a chromosome has been copied, the two copies remain attached at the centromere, forming sister chromatids.

- The cell cycle consists of the stages through which a eukaryotic cell passes between cell divisions. It consists of (1) interphase, in which the cell grows and prepares for division and (2) the M phase, in which nuclear and cell division take place. The M phase consists of (1) mitosis, the process of nuclear division, and (2) cytokinesis, the division of the cytoplasm.

- Mitosis usually results in the production of two genetically identical cells.

- Sexual reproduction produces genetically variable progeny and allows for accelerated evolution. It includes meiosis, in which haploid sex cells are produced, and fertilization, the fusion of sex cells. Meiosis includes two cell divisions. In meiosis I, crossing over takes place and homologous chromosomes separate. In meiosis II, chromatids separate.

- The usual result of meiosis is the production of four haploid cells that are genetically variable. Genetic variation in meiosis is produced by crossing over and by the random distribution of maternal and paternal chromosomes.

- In animals, a diploid spermatogonium undergoes meiosis to produce four haploid sperm cells. A diploid oogonium undergoes meiosis to produce one large haploid ovum and one or more smaller polar bodies.

- In plants, a diploid microsporocyte in the stamen undergoes meiosis to produce four pollen grains, each with two haploid sperm cells. In the ovary, a diploid megasporocyte undergoes meiosis to produce eight haploid nuclei, one of which forms the egg.

Important Terms

prokaryote (p. 17)
eukaryote (p. 17)
eubacteria (p. 17)
archaea (p. 17)
nucleus (p. 17)
histone (p. 17)
chromatin (p. 17)
homologous pair (p. 19)
diploid (p. 19)
haploid (p. 19)
telomere (p. 20)
origin of replication (p. 20)
sister chromatid (p. 20)
cell cycle (p. 20)
checkpoint (p. 21)
interphase (p. 21)
M phase (p. 21)
mitosis (p. 21)
cytokinesis (p. 21)

prophase (p. 22)
prometaphase (p. 22)
metaphase (p. 22)
anaphase (p. 22)
telophase (p. 22)
meiosis (p. 25)
fertilization (p. 25)
prophase I (p. 26)
synapsis (p. 26)
bivalent (p. 26)
tetrad (p. 26)
crossing over (p.26)
metaphase I (p. 27)
anaphase I (p. 27)
telophase I (p. 27)
interkinesis (p. 27)
prophase II (p. 28)
metaphase II (p. 28)
anaphase II (p. 28)

telophase II (p. 28)
recombination (p. 28)
spermatogenesis (p. 31)
spermatogonium (p. 31)
primary spermatocyte (p. 31)
secondary spermatocyte (p. 31)
spermatid (p. 31)
oogenesis (p. 31)
oogonium (p. 31)
primary oocyte (p. 31)
secondary oocyte (p. 31)
first polar body (p. 31)
ovum (p. 31)
second polar body (p. 31)
microsporocyte (p. 33)
microspore (p. 33)
megasporocyte (p. 33)
megaspore (p. 33)

Answers to Concept Checks

1. Eubacteria and archaea are prokaryotes. They differ from eukaryotes in possessing no nucleus, a genome that usually consists of a single, circular chromosome, and a small amount of DNA.

2. b

3. A centromere, a pair of telomeres, and an origin of replication

4. a

5. d

6. d

7. d

Worked Problem

1. A student examines a thin section of an onion-root tip and records the number of cells that are in each stage of the cell cycle. She observes 94 cells in interphase, 14 cells in prophase, 3 cells in prometaphase, 3 cells in metaphase, 5 cells in anaphase, and 1 cell in telophase. If the complete cell cycle in an onion-root tip requires 22 hours, what is the average duration of each stage in the cycle? Assume that all cells are in the active cell cycle (not G_0).

• **Solution**

This problem is solved in two steps. First, we calculate the proportions of cells in each stage of the cell cycle, which correspond to the amount of time that an average cell spends in each stage. For example, if cells spend 90% of their time in interphase, then, at any given moment, 90% of the cells will be in interphase. The second step is to convert the proportions into lengths of time, which is done by multiplying the proportions by the total time of the cell cycle (22 hours).

Step 1. Calculate the proportion of cells at each stage.
The proportion of cells at each stage is equal to the number of cells found in that stage divided by the total number of cells examined:

Interphase $^{91}/_{120} = 0.783$

Prophase $^{14}/_{120} = 0.117$

Prometaphase $^{3}/_{120} = 0.025$

Metaphase $^{3}/_{120} = 0.025$

Anaphase $^{5}/_{120} = 0.042$

Telophase $^{1}/_{120} = 0.08$

We can check our calculations by making sure that the proportions sum to 1.0, which they do.

Step 2. Determine the average duration of each stage.
To determine the average duration of each stage, multiply the proportion of cells in each stage by the time required for the entire cell cycle:

Interphase 0.783×22 hours $= 17.23$ hours

Prophase 0.117×22 hours $= 2.57$ hours

Prometaphase 0.025×22 hours $= 0.55$ hour

Metaphase 0.025×22 hours $= 0.55$ hour

Anaphase 0.042×22 hours $= 0.92$ hour

Telophase 0.008×22 hours $= 0.18$ hour

Comprehension Questions

Section 2.1

*1. Give some genetic differences between prokaryotic and eukaryotic cells.

2. Why are the viruses that infect mammalian cells useful for studying the genetics of mammals?

Section 2.2

*3. List three fundamental events that must take place in cell reproduction.

4. Name three essential structural elements of a functional eukaryotic chromosome and describe their functions.

*5. Sketch and identify four different types of chromosomes based on the position of the centromere.

6. List the stages of interphase and the major events that take place in each stage.

*7. List the stages of mitosis and the major events that take place in each stage.

*8. What are the genetically important results of the cell cycle?

9. Why are the two cells produced by the cell cycle genetically identical?

Section 2.3

10. What are the stages of meiosis and what major events take place in each stage?

*11. What are the major results of meiosis?

12. What two processes unique to meiosis are responsible for genetic variation? At what point in meiosis do these processes take place?

*13. List similarities and differences between mitosis and meiosis. Which differences do you think are most important and why?

14. Outline the process of spermatogenesis in animals. Outline the process of oogenesis in animals.

15. Outline the process by which male gametes are produced in plants. Outline the process of female gamete formation in plants.

Application Questions and Problems

Section 2.2

16. A certain species has three pairs of chromosomes: an acrocentric pair, a metacentric pair, and a submetacentric pair. Draw a cell of this species as it would appear in metaphase of mitosis.

17. A biologist examines a series of cells and counts 160 cells in interphase, 20 cells in prophase, 6 cells in prometaphase, 2 cells in metaphase, 7 cells in anaphase, and 5 cells in telophase. If the complete cell cycle requires 24 hours, what is the average duration of the M phase in these cells? Of metaphase?

Section 2.3

*18. A cell in G_1 of interphase has 12 chromosomes. How many chromosomes and DNA molecules will be found per cell when this original cell progresses to the following stages?
  a. G_2 of interphase
  b. Metaphase I of meiosis
  c. Prophase of mitosis
  d. Anaphase I of meiosis
  e. Anaphase II of meiosis
  f. Prophase II of meiosis
  g. After cytokinesis following mitosis
  h. After cytokinesis following meiosis II

19. How are the events that take place in spermatogenesis and oogenesis similar? How are they different?

*20. All of the following cells, shown in various stages of mitosis and meiosis, come from the same rare species of plant. What is the diploid number of chromosomes in this plant? Give the names of each stage of mitosis or meiosis shown.

21. The amount of DNA per cell of a particular species is measured in cells found at various stages of meiosis, and the following amounts are obtained:

Amount of DNA per cell

_____ 3.7 pg   _____ 7.3 pg   _____ 14.6 pg

Match the amounts of DNA above with the corresponding stages of the cell cycle (*a* through *f*). You may use more than one stage for each amount of DNA.

Stage of meiosis

  a. G_1
  b. Prophase I
  c. G_2
  d. Following telophase II and cytokinesis
  e. Anaphase I
  f. Metaphase II

*22. Fill in the following table.

Event	Mitosis	Meiosis I	Meiosis II
Does crossing over take place?	_____	_____	_____
What separates in anaphase?	_____	_____	_____
What lines up on the metaphase plate?	_____	_____	_____
Does cell division usually take place?	_____	_____	_____
Do homologs pair?	_____	_____	_____
Is genetic variation produced?	_____	_____	_____

23. A cell has 8 chromosomes in G_1 of interphase. Draw a picture of this cell with its chromosomes at the following stages. Indicate how many DNA molecules are present at each stage.
  a. Metaphase of mitosis
  b. Anaphase of mitosis
  c. Anaphase II of meiosis

*24. The fruit fly *Drosophila melanogaster* has four pairs of chromosomes, whereas the house fly *Musca domestica* has six pairs of chromosomes. Other things being equal, in which species would you expect to see more genetic variation among the progeny of a cross? Explain your answer.

*25. A cell has two pairs of submetacentric chromosomes, which we will call chromosomes I_a, I_b, II_a, and II_b (chromosomes I_a and I_b are homologs, and chromosomes II_a and II_b are homologs). Allele M is located on the long arm of chromosome I_a, and allele m is located at the same position on chromosome I_b. Allele P is located on the short arm of chromosome I_a, and allele p is located at the same position on chromosome I_b. Allele R is located on chromosome II_a and allele r is located at the same position on chromosome II_b.

a. Draw these chromosomes, identifying genes M, m, P, p, R, and r, as they might appear in metaphase I of meiosis. Assume that there is no crossing over.

b. Taking into consideration the random separation of chromosomes in anaphase I, draw the chromosomes (with genes identified) present in all possible types of gametes that might result from this cell's undergoing meiosis. Assume that there is no crossing over.

26. A horse has 64 chromosomes and a donkey has 62 chromosomes. A cross between a female horse and a male donkey produces a mule, which is usually sterile. How many chromosomes does a mule have? Can you think of any reasons for the fact that most mules are sterile?

27. Normal somatic cells of horses have 64 chromosomes ($2n = 64$). How many chromosomes and DNA molecules will be present in the following types of horse cells?

Cell type	Number of chromosomes	Number of DNA Molecules
a. Spermatogonium	_____	_____
b. First polar body	_____	_____
c. Primary oocyte	_____	_____
d. Secondary spermatocyte	_____	_____

Challenge Questions

Section 2.3

28. From 80% to 90% of the most common human chromosome abnormalities arise because the chromosomes fail to divide properly in oogenesis. Can you think of a reason why failure of chromosome division might be more common in female gametogenesis than in male gametogenesis?

*29. Female bees are diploid, and male bees are haploid. The haploid males produce sperm and can successfully mate with diploid females. Fertilized eggs develop into females and unfertilized eggs develop into males. How do you think the process of sperm production in male bees differs from sperm production in other animals?

3 Basic Principles of Heredity

Red hair is caused by recessive mutations at the melanocortin 1 receptor gene. *Lady Lilith*, 1868, by Dante Charles Gabriel Rossetti. Oil on canvas. *[© Delaware Art Museum, Wilmington, USA/Samuel and Mary R. Bancroft Memorial/The Bridgeman Art Library.]*

The Genetics of Red Hair

Whether because of its exotic hue or its novelty, red hair has long been a subject of fascination for historians, poets, artists, and scientists. Greek historians made special note of the fact that Boudica, the Celtic queen who led a revolt against the Roman Empire, possessed a "great mass of red hair." Early Christian artists frequently portrayed Mary Magdalene as a striking red head (though there is no mention of her red hair in the Bible), and the famous artist Botticelli painted the goddess Venus as a red-haired beauty in his masterpiece *The Birth of Venus*. Queen Elizabeth I of England possessed curly red hair; during her reign, red hair was quite fashionable in London society.

The color of our hair is caused largely by a pigment called melanin that comes in two primary forms: eumelanin, which is black or brown, and pheomelanin, which is red or yellow. The color of a person's hair is determined by two factors: (1) the amount of melanin produced (more melanin causes darker hair; less melanin causes lighter hair) and (2) the relative amounts of eumelanin compared with pheomelanin (more eumelanin produces black or brown hair; more pheomelanin produces red or yellow hair). The color of our hair is not just an academic curiosity; melanin protects against the harmful effects of sunlight, and people with red hair are usually fair skinned and particularly susceptible to skin cancer.

The inheritance of red hair has long been a subject of scientific debate. In 1909, Charles and Gertrude Davenport speculated on the inheritance of hair color in humans. Charles Davenport was an early enthusiast of genetics, particularly of inheritance in humans, and was the first director of the Biological Laboratory in Cold Spring Harbor, New York. He later became a leading proponent of eugenics, a movement—now discredited—that advocated improvement of the human race through genetics. The Davenports' study was based on family histories sent in by untrained amateurs and was methodologically flawed, but their results suggested that red hair is recessive to black and brown, meaning that a person must inherit two copies of a red-hair gene—one from each parent—to have red hair. Subsequent research contradicted this initial conclusion, suggesting that red hair is inherited instead as a dominant trait and that a person will have red hair even if possessing only a single red-hair gene. Controversy over whether red hair color is dominant or recessive, or even dependent on combinations of several different genes, continued for many years.

In 1993, scientists who were investigating a gene that affects the color of fur in mice discovered that the gene encodes the melanocortin-1 receptor. This receptor, when

39

activated, increases the production of black eumelanin and decreases the production of red pheomelanin, resulting in black or brown fur. Shortly thereafter, the same melanocortin-1 receptor gene (*MC1R*) was located on human chromosome 16, cloned, and sequenced. When this gene is mutated in humans, red hair results. Most people with red hair carry two defective copies of the *MC1R* gene, which means that the trait is recessive (as originally proposed by the Davenports back in 1909). However, from 10% to 20% of red heads possess only a single mutant copy of *MC1R*, muddling the recessive interpretation of red hair (the people with a single mutant copy of the gene tend to have lighter red hair than those who harbor two mutant copies). The type and frequency of mutations at the *MC1R* gene vary widely among human populations, accounting for ethnic differences in the preponderance of red hair: Among those of African and Asian descent, mutations for red hair are uncommon, whereas almost 40% of the people from the northern part of the United Kingdom carry at least one mutant gene for red hair.

This chapter is about the principles of heredity: how genes—like the one for the melanocortin-1 receptor—are passed from generation to generation and how factors such as dominance influence that inheritance. The principles of heredity were first put forth by Gregor Mendel, and so we begin this chapter by examining Mendel's scientific achievements. We then turn to simple genetic crosses, those in which a single characteristic is examined. We will learn some techniques for predicting the outcome of genetic crosses and then turn to crosses in which two or more characteristics are examined. We will see how the principles applied to simple genetic crosses and the ratios of offspring that they produce serve as the key for understanding more complicated crosses. The chapter ends with a discussion of statistical tests for analyzing crosses.

Throughout this chapter, a number of concepts are interwoven: Mendel's principles of segregation and independent assortment, probability, and the behavior of chromosomes. These concepts might at first appear to be unrelated, but they are actually different views of the same phenomenon, because the genes that undergo segregation and independent assortment are located on chromosomes. The principal aim of this chapter is to examine these different views and to clarify their relations.

3.1 Gregor Mendel Discovered the Basic Principles of Heredity

In 1909, when the Davenports speculated about the inheritance of red hair, the basic principles of heredity were just becoming widely known among biologists. Surprisingly, these principles had been discovered some 44 years earlier by Johann Gregor Mendel (1822–1884).

Mendel was born in what is now part of the Czech Republic. Although his parents were simple farmers with little money, he was able to achieve a sound education and was admitted to the Augustinian monastery in Brno in September 1843. After graduating from seminary, Mendel was ordained a priest and appointed to a teaching position in a local school. He excelled at teaching, and the abbot of the monastery recommended him for further study at the University of Vienna, which he attended from 1851 to 1853. There, Mendel enrolled in the newly opened Physics Institute and took courses in mathematics, chemistry, entomology, paleontology, botany, and plant physiology. It was probably there that Mendel acquired knowledge of the scientific method, which he later applied so successfully to his genetics experiments. After 2 years of study in Vienna, Mendel returned to Brno, where he taught school and began his experimental work with pea plants. He conducted breeding experiments from 1856 to 1863 and presented his results publicly at meetings of the Brno Natural Science Society in 1865. Mendel's paper from these lectures was published in 1866. In spite of widespread interest in heredity, the effect of his research on the scientific community was minimal. At the time, no one seemed to have noticed that Mendel had discovered the basic principles of inheritance.

In 1868, Mendel was elected abbot of his monastery, and increasing administrative duties brought an end to his teaching and eventually to his genetics experiments. He died at the age of 61 on January 6, 1884, unrecognized for his contribution to genetics.

The significance of Mendel's discovery was unappreciated until 1900, when three botanists—Hugo de Vries, Erich von Tschermak, and Karl Correns—began independently conducting similar experiments with plants and arrived at conclusions similar to those of Mendel. Coming across Mendel's paper, they interpreted their results in accord with his principles and drew attention to his pioneering work.

Mendel's Success

Mendel's approach to the study of heredity was effective for several reasons. Foremost was his choice of experimental subject, the pea plant *Pisum sativum* (**Figure 3.1**), which

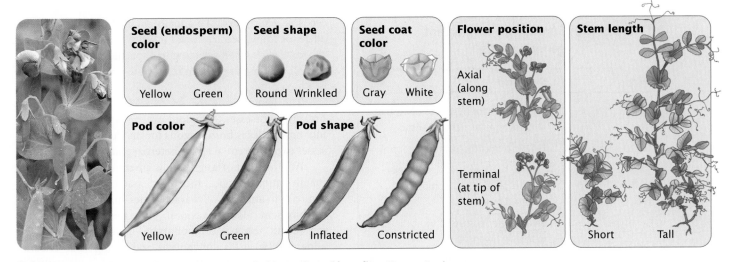

3.1 Mendel used the pea plant *Pisum sativum* in his studies of heredity. He examined seven characteristics that appeared in the seeds and in plants grown from the seeds. [*Photograph by Wally Eberhart/Visuals Unlimited.*]

offered clear advantages for genetic investigation. The plant is easy to cultivate, and Mendel had the monastery garden and greenhouse at his disposal. Compared with some other plants, peas grow relatively rapidly, completing an entire generation in a single growing season. By today's standards, one generation per year seems frightfully slow—fruit flies complete a generation in 2 weeks and bacteria in 20 minutes—but Mendel was under no pressure to publish quickly and was able to follow the inheritance of individual characteristics for several generations. Had he chosen to work on an organism with a longer generation time—horses, for example—he might never have discovered the basis of inheritance. Pea plants also produce many offspring—their seeds—which allowed Mendel to detect meaningful mathematical ratios in the traits that he observed in the progeny.

The large number of varieties of peas that were available to Mendel also was crucial, because these varieties differed in various traits and were genetically pure. Mendel was therefore able to begin with plants of variable, known genetic makeup.

Much of Mendel's success can be attributed to the seven characteristics that he chose for study (see Figure 3.1). He avoided characteristics that display a range of variation; instead, he focused his attention on those that exist in two easily differentiated forms, such as white versus gray seed coats, round versus wrinkled seeds, and inflated versus constricted pods.

Finally, Mendel was successful because he adopted an experimental approach and interpreted his results by using mathematics. Unlike many earlier investigators who just described the *results* of crosses, Mendel formulated *hypotheses* based on his initial observations and then conducted additional crosses to test his hypotheses. He kept careful records of the numbers of progeny possessing each

type of trait and computed ratios of the different types. He paid close attention to detail, was adept at seeing patterns in detail, and was patient and thorough, conducting his experiments for 10 years before attempting to write up his results.

Concepts

Gregor Mendel put forth the basic principles of inheritance, publishing his findings in 1866. The significance of his work did not become widely appreciated until 1900.

✔ Concept Check 1

Which of the following factors did not contribute to Mendel's success in his study of heredity?

a. His use of the pea plant

b. His study of plant chromosomes

c. His adoption of an experimental approach

d. His use of mathematics

Genetic Terminology

Before we examine Mendel's crosses and the conclusions that he drew from them, it will be helpful to review some terms commonly used in genetics (**Table 3.1**). The term *gene* is a word that Mendel never knew. It was not coined until 1909, when Danish geneticist Wilhelm Johannsen first used it. The definition of a gene varies with the context of its use, and so its definition will change as we explore different aspects of heredity. For our present use in the context of genetic crosses, we will define a **gene** as an inherited factor that determines a characteristic.

Table 3.1	Summary of important genetic terms
Term	**Definition**
Gene	A genetic factor (region of DNA) that helps determine a characteristic
Allele	One of two or more alternate forms of a gene
Locus	Specific place on a chromosome occupied by an allele
Genotype	Set of alleles possessed by an individual organism
Heterozygote	An individual organism possessing two different alleles at a locus
Homozygote	An individual organism possessing two of the same alleles at a locus
Phenotype or trait	The appearance or manifestation of a character
Character or characteristic	An attribute or feature

Genes frequently come in different versions called **alleles** (**Figure 3.2**). In Mendel's crosses, seed shape was determined by a gene that exists as two different alleles: one allele encodes round seeds and the other encodes wrinkled seeds. All alleles for any particular gene will be found at a specific place on a chromosome called the **locus** for that gene. (The plural of locus is loci; it's bad form in genetics—and incorrect—to speak of locuses.) Thus, there is a specific place—a locus—on a chromosome in pea plants where the shape of

seeds is determined. This locus might be occupied by an allele for round seeds or one for wrinkled seeds. We will use the term *allele* when referring to a specific version of a gene; we will use the term *gene* to refer more generally to any allele at a locus.

The **genotype** is the set of alleles that an individual organism possesses. A diploid organism that possesses two identical alleles is **homozygous** for that locus. One that possesses two different alleles is **heterozygous** for the locus.

Another important term is **phenotype**, which is the manifestation or appearance of a characteristic. A phenotype can refer to any type of characteristic—physical, physiological, biochemical, or behavioral. Thus, the condition of having round seeds is a phenotype, a body weight of 50 kilograms (50 kg) is a phenotype, and having sickle-cell anemia is a phenotype. In this book, the term *characteristic* or *character* refers to a general feature such as eye color; the term *trait* or *phenotype* refers to specific manifestations of that feature, such as blue or brown eyes.

A given phenotype arises from a genotype that develops within a particular environment. The genotype determines the potential for development; it sets certain limits, or boundaries, on that development. How the phenotype develops within those limits is determined by the effects of other genes and of environmental factors, and the balance between these influences varies from character to character. For some characters, the differences between phenotypes are determined largely by differences in genotype; in other words, the genetic limits for that phenotype are narrow. Seed shape in Mendel's peas is a good example of a characteristic for which the genetic limits are narrow and the phenotypic differences are largely genetic. For other characters, environmental differences are more important; in this case, the limits imposed by the genotype are broad. The height reached by an oak tree at maturity is a phenotype that is strongly influenced by environmental factors, such as the availability of water, sunlight, and nutrients. Nevertheless, the tree's genotype still imposes some limits on its height: an oak tree will never grow to be 300 meters (300 m) tall no matter how much sunlight, water, and fertilizer are provided. Thus, even the height of an oak tree is determined to some degree by genes. For many characteristics, both genes and environment are important in determining phenotypic differences.

An obvious but important concept is that only the genotype is inherited. Although the phenotype is determined, at least to some extent, by genotype, organisms do not transmit their phenotypes to the next generation. The distinction between genotype and phenotype is one of the most important principles of modern genetics. The next section describes Mendel's careful observation of phenotypes through several generations of breeding experiments. These experiments allowed him to deduce not only the genotypes of the individual plants, but also the rules governing their inheritance.

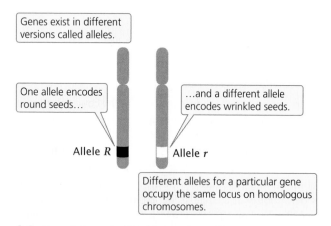

Genes exist in different versions called alleles.

One allele encodes round seeds...

...and a different allele encodes wrinkled seeds.

Allele *R*

Allele *r*

Different alleles for a particular gene occupy the same locus on homologous chromosomes.

3.2 At each locus, a diploid organism possesses two alleles located on different homologous chromosomes.

Concepts

Each phenotype results from a genotype developing within a specific environment. The genotype, not the phenotype, is inherited.

✔ **Concept Check 2**

Distinguish among the following terms: locus, allele, genotype.

3.2 Monohybrid Crosses Reveal the Principle of Segregation and the Concept of Dominance

Mendel started with 34 varieties of peas and spent 2 years selecting those varieties that he would use in his experiments. He verified that each variety was genetically pure (homozygous for each of the traits that he chose to study) by growing the plants for two generations and confirming that all offspring were the same as their parents. He then carried out a number of crosses between the different varieties. Although peas are normally self-fertilizing (each plant crosses with itself), Mendel conducted crosses between different plants by opening the buds before the anthers were fully developed, removing the anthers, and then dusting the stigma with pollen from a different plant (**Figure 3.3**).

Mendel began by studying **monohybrid crosses**—those between parents that differed in a single characteristic. In one experiment, Mendel crossed a pea plant homozygous for round seeds with one that was homozygous for wrinkled seeds (see Figure 3.3). This first generation of a cross is the **P (parental) generation**.

After crossing the two varieties in the P generation, Mendel observed the offspring that resulted from the cross. In regard to seed characteristics, such as seed shape, the phenotype develops as soon as the seed matures, because the seed traits are determined by the newly formed embryo within the seed. For characters associated with the plant itself, such as stem length, the phenotype doesn't develop until the plant grows from the seed; for these characters, Mendel had to wait until the following spring, plant the seeds, and then observe the phenotypes on the plants that germinated.

The offspring from the parents in the P generation are the F_1 (**filial 1**) **generation**. When Mendel examined the F_1 generation of this cross, he found that they expressed only one of the phenotypes present in the parental generation: all the F_1 seeds were round. Mendel carried out 60 such crosses and always obtained this result. He also conducted **reciprocal crosses**: in one cross, pollen (the male gamete) was taken from a plant with round seeds and, in its reciprocal cross,

Experiment

Question: When peas with two different traits—round and wrinkled seeds—are crossed, will their progeny exhibit one of those traits, both of those traits, or a "blended" intermediate trait?

Methods

Stigma

Anthers

♀ Flower

♂ Flower   ×

Cross

1 To cross different varieties of peas, Mendel removed the anthers from flowers to prevent self-fertilization…

2 …and dusted the stigma with pollen from a different plant.

3 The pollen fertilized ova, which developed into seeds.

4 The seeds grew into plants.

P generation   Homozygous round seeds   Homozygous wrinkled seeds

×

Cross

5 Mendel crossed two homozygous varieties of peas.

F_1 generation

×

Self-fertilize

6 All the F_1 seeds were round. Mendel allowed plants grown from these seeds to self-fertilize.

Results

F_2 generation

Fraction of progeny seeds

5474 round seeds   ¾ round

1850 wrinkled seeds   ¼ wrinkled

7 ¾ of F_2 seeds were round and ¼ were wrinkled, a 3 : 1 ratio.

Conclusion: The traits of the parent plants do not blend. Although F_1 plants display the phenotype of one parent, both traits are passed to F_2 progeny in a 3 : 1 ratio.

3.3 Mendel conducted monohybrid crosses.

pollen was taken from a plant with wrinkled seeds. Reciprocal crosses gave the same result: all the F_1 were round.

Mendel wasn't content with examining only the seeds arising from these monohybrid crosses. The following spring, he planted the F_1 seeds, cultivated the plants that germinated from them, and allowed the plants to self-fertilize, producing a second generation—the F_2 (**filial 2**) **generation**. Both of the traits from the P generation emerged in the F_2 generation; Mendel counted 5474 round seeds and 1850 wrinkled seeds in the F_2 (see Figure 3.3). He noticed that the number of the round and wrinkled seeds constituted approximately a 3 to 1 ratio; that is, about $^3/_4$ of the F_2 seeds were round and $^1/_4$ were wrinkled. Mendel conducted monohybrid crosses for all seven of the characteristics that he studied in pea plants and, in all of the crosses, he obtained the same result: all of the F_1 resembled only one of the two parents, but both parental traits emerged in the F_2 in an approximate ratio of 3 : 1.

What Monohybrid Crosses Reveal

Mendel drew several important conclusions from the results of his monohybrid crosses. First, he reasoned that, although the F_1 plants display the phenotype of only one parent, they must inherit genetic factors from both parents because they transmit both phenotypes to the F_2 generation. The presence of both round and wrinkled seeds in the F_2 could be explained only if the F_1 plants possessed both round and wrinkled genetic factors that they had inherited from the P generation. He concluded that each plant must therefore possess two genetic factors encoding a character.

The genetic factors (now called alleles) that Mendel discovered are, by convention, designated with letters; the allele for round seeds is usually represented by R, and the allele for wrinkled seeds by r. The plants in the P generation of Mendel's cross possessed two identical alleles: RR in the round-seeded parent and rr in the wrinkled-seeded parent (**Figure 3.4a**).

The second conclusion that Mendel drew from his monohybrid crosses was that the two alleles in each plant separate when gametes are formed, and one allele goes into each gamete. When two gametes (one from each parent) fuse to produce a zygote, the allele from the male parent unites with the allele from the female parent to produce the genotype of the offspring. Thus, Mendel's F_1 plants inherited an R allele from the round-seeded plant and an r allele from the wrinkled-seeded plant (**Figure 3.4b**). However, only the trait encoded by round allele (R) was *observed* in the F_1—all the F_1 progeny had round seeds. Those traits that appeared unchanged in the F_1 heterozygous offspring Mendel called **dominant**, and those traits that disappeared in the F_1 heterozygous offspring he called **recessive**. When dominant and recessive alleles are present together, the recessive allele is masked, or suppressed. The concept of dominance was the

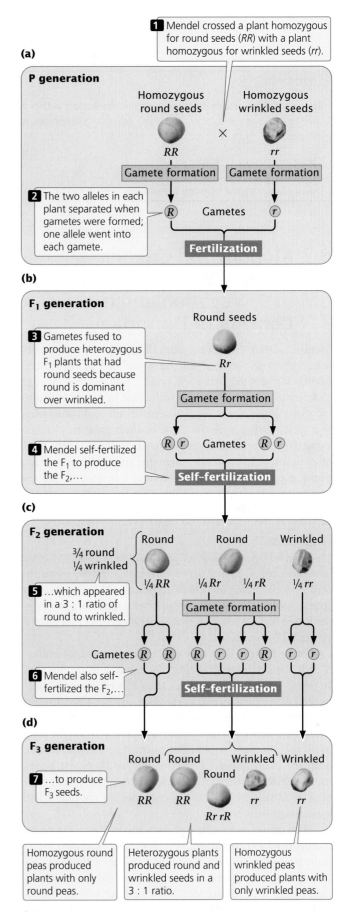

3.4 Mendel's monohybrid crosses revealed the principle of segregation and the concept of dominance.

third important conclusion that Mendel derived from his monohybrid crosses.

Mendel's fourth conclusion was that the two alleles of an individual plant separate with equal probability into the gametes. When plants of the F_1 (with genotype Rr) produced gametes, half of the gametes received the R allele for round seeds and half received the r allele for wrinkled seeds. The gametes then paired randomly to produce the following genotypes in equal proportions among the F_2: RR, Rr, rR, rr (**Figure 3.4c**). Because round (R) is dominant over wrinkled (r), there were three round progeny in the F_2 (RR, Rr, rR) for every one wrinkled progeny (rr) in the F_2. This 3 : 1 ratio of round to wrinkled progeny that Mendel observed in the F_2 could occur only if the two alleles of a genotype separated into the gametes with equal probability.

The conclusions that Mendel developed about inheritance from his monohybrid crosses have been further developed and formalized into the principle of segregation and the concept of dominance. The **principle of segregation** (Mendel's first law) states that each individual diploid organism possesses two alleles for any particular characteristic. These two alleles segregate (separate) when gametes are formed, and one allele goes into each gamete. Furthermore, the two alleles segregate into gametes in equal proportions. The **concept of dominance** states that, when two different alleles are present in a genotype, only the trait encoded by one of them—the "dominant" allele—is observed in the phenotype.

Mendel confirmed these principles by allowing his F_2 plants to self-fertilize and produce an F_3 generation. He found that the F_2 plants grown from the wrinkled seeds—those displaying the recessive trait (rr)—produced an F_3 in which all plants produced wrinkled seeds. Because his wrinkled-seeded plants were homozygous for wrinkled alleles (rr), they could pass on only wrinkled alleles to their progeny (**Figure 3.4d**).

The F_2 plants grown from round seeds—the dominant trait—fell into two types (see Figure 3.4c). On self-fertilization, about $2/3$ of the F_2 plants grown from round seeds produced both round and wrinkled seeds in the F_3 generation. These F_2 plants were heterozygous (Rr); so they produced $1/4$ RR (round), $1/2$ Rr (round), and $1/4$ rr (wrinkled) seeds, giving a 3 : 1 ratio of round to wrinkled in the F_3. About $1/3$ of the F_2 plants grown from round seeds were of the second type; they produced only the dominant round-seeded trait in the F_3. These F_2 plants were homozygous for the round allele (RR) and could thus produce only round offspring in the F_3 generation. Mendel planted the seeds obtained in the F_3 and carried these plants through three more rounds of self-fertilization. In each generation, $2/3$ of the round-seeded plants produced round and wrinkled offspring, whereas $1/3$ produced only round offspring. These results are entirely consistent with the principle of segregation.

Concepts

The principle of segregation states that each individual organism possesses two alleles that can encode a characteristic. These alleles segregate when gametes are formed, and one allele goes into each gamete. The concept of dominance states that, when the two alleles of a genotype are different, only the trait encoded by one of them—the "dominant" allele—is observed.

✔ Concept Check 3

How did Mendel know that each of his pea plants carried two alleles encoding a characteristic?

Connecting Concepts

Relating Genetic Crosses to Meiosis

We have now seen how the results of monohybrid crosses are explained by Mendel's principle of segregation. Many students find that they enjoy working genetic crosses but are frustrated by the abstract nature of the symbols. Perhaps you feel the same at this point. You may be asking, "What do these symbols really represent? What does the genotype RR mean in regard to the biology of the organism?" The answers to these questions lie in relating the abstract symbols of crosses to the structure and behavior of chromosomes, the repositories of genetic information (see Chapter 2).

In 1900, when Mendel's work was rediscovered and biologists began to apply his principles of heredity, the relation between genes and chromosomes was still unclear. The theory that genes are located on chromosomes (the **chromosome theory of heredity**) was developed in the early 1900s by Walter Sutton, then a graduate student at Columbia University. Through the careful study of meiosis in insects, Sutton documented the fact that each homologous pair of chromosomes consists of one maternal chromosome and one paternal chromosome. Showing that these pairs segregate independently into gametes in meiosis, he concluded that this process is the biological basis for Mendel's principles of heredity. German cytologist and embryologist Theodor Boveri came to similar conclusions at about the same time.

Sutton knew that diploid cells have two sets of chromosomes. Each chromosome has a pairing partner, its homologous chromosome. One chromosome of each homologous pair is inherited from the mother and the other is inherited from the father. Similarly, diploid cells possess two alleles at each locus, and these alleles constitute the genotype for that locus. The principle of segregation indicates that one allele of the genotype is inherited from each parent.

This similarity between the number of chromosomes and the number of alleles is not accidental—the two alleles of a genotype are located on homologous chromosomes. The symbols used in genetic crosses, such as R and r, are just shorthand notations for particular sequences of DNA in the chromosomes that encode particular phenotypes. The two alleles of a genotype are found on different but homologous chromosomes. In the S phase of meiotic interphase, each chromosome replicates, producing two copies of each allele, one on each chromatid (**Figure 3.5a**). The homologous

(a)

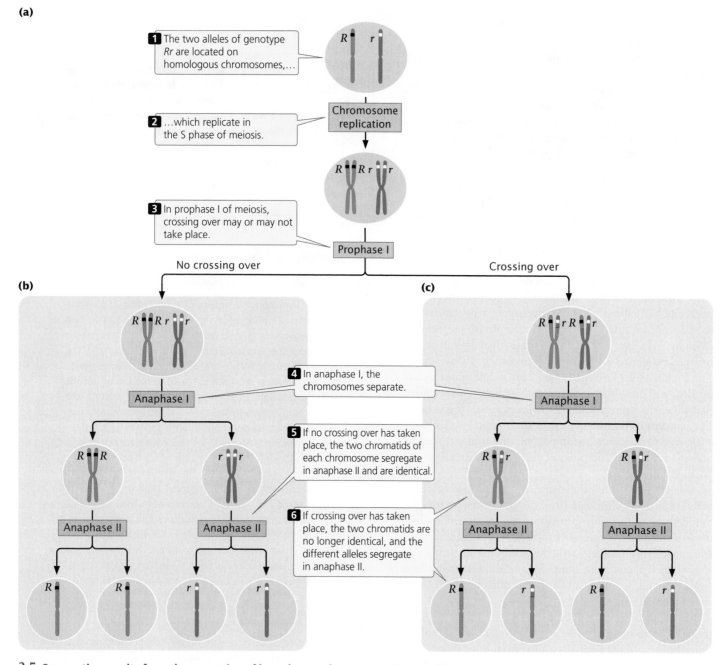

1 The two alleles of genotype *Rr* are located on homologous chromosomes,…

2 …which replicate in the S phase of meiosis.

Chromosome replication

3 In prophase I of meiosis, crossing over may or may not take place.

Prophase I

No crossing over

Crossing over

(b)

(c)

4 In anaphase I, the chromosomes separate.

Anaphase I

Anaphase I

5 If no crossing over has taken place, the two chromatids of each chromosome segregate in anaphase II and are identical.

6 If crossing over has taken place, the two chromatids are no longer identical, and the different alleles segregate in anaphase II.

Anaphase II   Anaphase II

Anaphase II   Anaphase II

3.5 Segregation results from the separation of homologous chromosomes in meiosis.

chromosomes segregate in anaphase I, thereby separating the two different alleles (**Figure 3.5b**). This chromosome segregation is the basis of the principle of segregation. In anaphase II of meiosis, the two chromatids of each replicated chromosome separate; so each gamete resulting from meiosis carries only a single allele at each locus, as Mendel's principle of segregation predicts.

If crossing over has taken place in prophase I of meiosis, then the two chromatids of each replicated chromosome are no longer identical, and the segregation of different alleles takes place at anaphase I and anaphase II (**Figure 3.5c**). However, Mendel didn't know anything about chromosomes; he formulated his principles of

heredity entirely on the basis of the results of the crosses that he carried out. Nevertheless, we should not forget that these principles work because they are based on the behavior of actual chromosomes in meiosis.

Predicting the Outcomes of Genetic Crosses

One of Mendel's goals in conducting his experiments on pea plants was to develop a way to predict the outcome of crosses between plants with different phenotypes. In this section, we will first learn a simple, shorthand method for predicting outcomes of genetic crosses (the Punnett square), and then we will learn how to use probability to predict the results of crosses.

The Punnett square   The Punnett square was developed by English geneticist Reginald C. Punnett in 1917. To illustrate the Punnett square, let's examine another cross that Mendel carried out. By crossing two varieties of peas that differed in height, Mendel established that tall (T) was dominant over short (t). He tested his theory concerning the inheritance of dominant traits by crossing an F_1 tall plant that was heterozygous (Tt) with the short homozygous parental variety (tt). This type of cross, between an F_1 genotype and either of the parental genotypes, is called a **backcross.**

To predict the types of offspring that result from this backcross, we first determine which gametes will be produced by each parent (**Figure 3.6a**). The principle of segregation tells us that the two alleles in each parent separate, and one allele passes to each gamete. All gametes from the homozygous tt short plant will receive a single short (t) allele. The tall plant in this cross is heterozygous (Tt); so 50% of its gametes will receive a tall allele (T) and the other 50% will receive a short allele (t).

A **Punnett square** is constructed by drawing a grid, putting the gametes produced by one parent along the upper edge and the gametes produced by the other parent down the left side (**Figure 3.6b**). Each cell (a block within the Punnett square) contains an allele from each of the corresponding gametes, generating the genotype of the progeny produced by fusion of those gametes. In the upper left-hand cell of the Punnett square in Figure 3.6b, a gamete containing T from the tall plant unites with a gamete containing t from the short plant, giving the genotype of the progeny (Tt). It is useful to write the phenotype expressed by each genotype; here the progeny will be tall, because the tall allele is dominant over the short allele. This process is repeated for all the cells in the Punnett square.

By simply counting, we can determine the types of progeny produced and their ratios. In Figure 3.6b, two cells contain tall (Tt) progeny and two cells contain short (tt) progeny; so the genotypic ratio expected for this cross is 2 Tt to 2 tt (a 1 : 1 ratio). Another way to express this result is to say that we expect $^1/_2$ of the progeny to have genotype Tt (and phenotype tall) and $^1/_2$ of the progeny to have genotype tt (and phenotype short). In this cross, the genotypic ratio and the phenotypic ratio are the same, but this outcome need not be the case. Try completing a Punnett square for the cross in which the F_1 round-seeded plants in Figure 3.4 undergo self-fertilization (you should obtain a phenotypic ratio of 3 round to 1 wrinkled and a genotypic ratio of 1 RR to 2 Rr to 1 rr).

3.6 The Punnett square can be used to determine the results of a genetic cross.

Concepts

The Punnett square is a shorthand method of predicting the genotypic and phenotypic ratios of progeny from a genetic cross.

✔ Concept Check 4

If an F_1 plant depicted in Figure 3.4 is backcrossed to the parent with round seeds, what proportion of the progeny will have wrinkled seeds? (Use a Punnett square.)

a. $^3/_4$ c. $^1/_4$

b. $^1/_2$ d. 0

Probability as a tool in genetics   Another method for determining the outcome of a genetic cross is to use the rules of probability, as Mendel did with his crosses. **Probability** expresses the likelihood of the occurrence of a particular event. It is the number of times that a particular event occurs, divided by the number of all possible outcomes. For example, a deck of 52 cards contains only one king of hearts. The probability of drawing one card from the deck at random and obtaining the king of hearts is $\frac{1}{52}$, because there is only one card that is the king of hearts (one event) and there are 52 cards that can be drawn from the deck (52 possible outcomes). The probability of drawing a card and obtaining an ace is $\frac{4}{52}$, because there are four cards that are aces (four events) and 52 cards (possible outcomes). Probability can be expressed either as a fraction ($\frac{4}{52}$ in this case) or as a decimal number (0.077 in this case).

The probability of a particular event may be determined by knowing something about *how* the event occurs or *how often* it occurs. We know, for example, that the probability of rolling a six-sided die and getting a four is $\frac{1}{6}$, because the die has six sides and any one side is equally likely to end up on top. So, in this case, understanding the nature of the event—the shape of the thrown die—allows us to determine the probability. In other cases, we determine the probability of an event by making a large number of observations. When a weather forecaster says that there is a 40% chance of rain on a particular day, this probability was obtained by observing a large number of days with similar atmospheric conditions and finding that it rains on 40% of those days. In this case, the probability has been determined empirically (by observation).

The multiplication rule   Two rules of probability are useful for predicting the ratios of offspring produced in genetic crosses. The first is the **multiplication rule**, which states that the probability of two or more independent events occurring together is calculated by multiplying their independent probabilities.

To illustrate the use of the multiplication rule, let's again consider the roll of a die. The probability of rolling one die and obtaining a four is $\frac{1}{6}$. To calculate the probability of rolling a die twice and obtaining 2 fours, we can apply the multiplication rule. The probability of obtaining a four on the first roll is $\frac{1}{6}$ and the probability of obtaining a four on the second roll is $\frac{1}{6}$; so the probability of rolling a four on both is $\frac{1}{6} \times \frac{1}{6} = \frac{1}{36}$ (**Figure 3.7a**). The key indicator for applying the multiplication rule is the word *and*; in the example just considered, we wanted to know the probability of obtaining a four on the first roll *and* a four on the second roll.

For the multiplication rule to be valid, the events whose joint probability is being calculated must be independent—the outcome of one event must not influence the outcome of the other. For example, the number that comes up on one roll of the die has no influence on the number that comes up

(a) The multiplication rule

(b) The addition rule

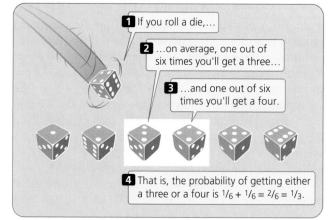

3.7 The multiplication and addition rules can be used to determine the probability of combinations of events.

on the other roll; so these events are independent. However, if we wanted to know the probability of being hit on the head with a hammer and going to the hospital on the same day, we could not simply multiply the probability of being hit on the head with a hammer by the probability of going to the hospital. The multiplication rule cannot be applied here, because the two events are not independent—being hit on the head with a hammer certainly influences the probability of going to the hospital.

The addition rule   The second rule of probability frequently used in genetics is the **addition rule**, which states that the probability of any one of two or more mutually exclusive events is calculated by adding the probabilities of these events. Let's look at this rule in concrete terms. To obtain the probability of throwing a die once and rolling *either* a three *or* a four, we would use the addition rule, adding the probability of obtaining a three ($\frac{1}{6}$) to the probability of obtaining a four (again, $\frac{1}{6}$), or $\frac{1}{6} + \frac{1}{6} = \frac{2}{6} = \frac{1}{3}$ (**Figure 3.7b**). The key indicators for applying the addition rule are the words *either* and *or*.

For the addition rule to be valid, the events whose probability is being calculated must be mutually exclusive, meaning that one event excludes the possibility of the occurrence of the other event. For example, you cannot throw a single die just once and obtain both a three and a four, because only one side of the die can be on top. These events are mutually exclusive.

Concepts

The multiplication rule states that the probability of two or more independent events occurring together is calculated by multiplying their independent probabilities. The addition rule states that the probability that any one of two or more mutually exclusive events occurring is calculated by adding their probabilities.

✔ Concept Check 5

If the probability of being blood-type A is $\frac{1}{8}$ and the probability of blood-type O is $\frac{1}{2}$, what is the probability of being either blood-type A or blood-type O?

a. $\frac{5}{8}$ c. $\frac{1}{8}$

b. $\frac{1}{2}$ d. $\frac{1}{16}$

The application of probability to genetic crosses   The multiplication and addition rules of probability can be used in place of the Punnett square to predict the ratios of progeny expected from a genetic cross. Let's first consider a cross between two pea plants heterozygous for the locus that determines height, $Tt \times Tt$. Half of the gametes produced by each plant have a T allele, and the other half have a t allele; so the probability for each type of gamete is $\frac{1}{2}$.

The gametes from the two parents can combine in four different ways to produce offspring. Using the multiplication rule, we can determine the probability of each possible type. To calculate the probability of obtaining TT progeny, for example, we multiply the probability of receiving a T allele from the first parent ($\frac{1}{2}$) times the probability of receiving a T allele from the second parent ($\frac{1}{2}$). The multiplication rule should be used here because we need the probability of

receiving a T allele from the first parent *and* a T allele from the second parent—two independent events. The four types of progeny from this cross and their associated probabilities are:

TT	(T gamete and T gamete)	$\frac{1}{2} \times \frac{1}{2} = \frac{1}{4}$	tall
Tt	(T gamete and t gamete)	$\frac{1}{2} \times \frac{1}{2} = \frac{1}{4}$	tall
tT	(t gamete and T gamete)	$\frac{1}{2} \times \frac{1}{2} = \frac{1}{4}$	tall
tt	(t gamete and t gamete)	$\frac{1}{2} \times \frac{1}{2} = \frac{1}{4}$	short

Notice that there are two ways for heterozygous progeny to be produced: a heterozygote can either receive a T allele from the first parent and a t allele from the second or receive a t allele from the first parent and a T allele from the second.

After determining the probabilities of obtaining each type of progeny, we can use the addition rule to determine the overall phenotypic ratios. Because of dominance, a tall plant can have genotype TT, Tt, or tT; so, using the addition rule, we find the probability of tall progeny to be $\frac{1}{4} + \frac{1}{4} + \frac{1}{4} = \frac{3}{4}$. Because only one genotype encodes short (tt), the probability of short progeny is simply $\frac{1}{4}$.

Two methods have now been introduced to solve genetic crosses: the Punnett square and the probability method. At this point, you may be asking, "Why bother with probability rules and calculations? The Punnett square is easier to understand and just as quick." For simple monohybrid crosses, the Punnett square is simpler than the probability method and is just as easy to use. However, for tackling more-complex crosses concerning genes at two or more loci, the probability method is both clearer and quicker than the Punnett square.

The Testcross

A useful tool for analyzing genetic crosses is the **testcross,** in which one individual of unknown genotype is crossed with another individual with a homozygous recessive genotype for the trait in question. Figure 3.6 illustrates a testcross (as well as a backcross). A testcross tests, or reveals, the genotype of the first individual.

Suppose you were given a tall pea plant with no information about its parents. Because tallness is a dominant trait in peas, your plant could be either homozygous (TT) or heterozygous (Tt), but you would not know which. You could determine its genotype by performing a testcross. If the plant were homozygous (TT), a testcross would produce all tall progeny ($TT \times tt \rightarrow$ all Tt); if the plant were heterozygous (Tt), the testcross would produce half tall progeny and half short progeny ($Tt \times tt \rightarrow \frac{1}{2} Tt$ and $\frac{1}{2} tt$). When a testcross is performed, any recessive allele in the unknown genotype is expressed in the progeny, because it will be paired with a recessive allele from the homozygous recessive parent.

Incomplete Dominance

We have seen that, in a cross between two individuals heterozygous for a dominant trait ($Aa \times Aa$), the offspring have a $\frac{3}{4}$ probability of exhibiting the dominant trait and a $\frac{1}{4}$ probability of exhibiting the recessive trait. We also examined the outcome of a cross between an individual heterozygous for a dominant trait and an individual homozygous for a recessive trait ($Aa \times aa$); in this case, $\frac{1}{2}$ of the offspring exhibit the dominant trait and $\frac{1}{2}$ exhibit the recessive trait. Later in the chapter, we will see how probabilities for such individual traits can be combined to determine the overall probability of offspring with combinations of two or more different traits. But, before exploring the inheritance of multiple traits, we must consider an additional phenotypic ratio that is obtained when dominance is lacking.

All of the seven characters in pea plants that Mendel chose to study extensively exhibited dominance and produced a 3 : 1 phenotypic ratio in the progeny. However, Mendel did realize that not all characters have traits that exhibit dominance. He conducted some crosses concerning the length of time that pea plants take to flower. When he crossed two homozygous varieties that differed in their flowering time by an average of 20 days, the length of time taken by the F_1 plants to flower was intermediate between those of the two parents. When the heterozygote has a phenotype intermediate between the phenotypes of the two homozygotes, the trait is said to display **incomplete dominance**.

Incomplete dominance is also exhibited in the fruit color of eggplants. When a homozygous plant that produces purple fruit (PP) is crossed with a homozygous plant that produces white fruit (pp), all the heterozygous F_1 (Pp) produce violet fruit (**Figure 3.8a**). When the F_1 are crossed with each other, $\frac{1}{4}$ of the F_2 are purple (PP), $\frac{1}{2}$ are violet (Pp), and $\frac{1}{4}$ are white (pp), as shown in **Figure 3.8b**. This 1 : 2 : 1 ratio is different from the 3 : 1 ratio that we would observe if eggplant fruit color exhibited dominance. When a trait displays incomplete dominance, the genotypic ratios and phenotypic ratios of the offspring are the *same*, because each genotype has its own phenotype. It is impossible to obtain eggplants that are pure breeding for violet fruit, because all plants with violet fruit are heterozygous.

Another example of incomplete dominance is feather color in chickens. A cross between a homozygous black chicken and a homozygous white chicken produces F_1 chickens that are gray. If these gray F_1 are intercrossed, they produce F_2 birds in a ratio of 1 black : 2 gray : 1 white. Leopard white spotting in horses is incompletely dominant over

(a)

P generation

Purple fruit   ×   White fruit

PP pp

Gametes P p

Fertilization

F_1 generation

Violet fruit   ×   Violet fruit

Pp Pp

Gametes P p P p

Fertilization

(b)

F_2 generation

	P	p
P	PP Purple	Pp Violet
p	Pp Violet	pp White

Conclusion: Genotypic ratio $1PP : 2Pp : 1pp$
Phenotypic ratio 1 purple : 2 violet : 1 white

3.8 Fruit color in eggplant is inherited as an incompletely dominant characteristic.

unspotted horses: LL horses are white with numerous dark spots, heterozygous Ll horses have fewer spots, and ll horses have no spots (**Figure 3.9**). The concept of dominance and some of its variations are discussed further in Chapter 4.

called the **wild type** because it is the allele usually found in the wild—is often symbolized by one or more letters and a plus sign (+). The letter or letters chosen are usually based on the mutant (unusual) phenotype. For example, the recessive allele for yellow eyes in the Oriental fruit fly is represented by *ye*, whereas the allele for wild-type eye color is represented by *ye*+. At times, the letters for the wild-type allele are dropped and the allele is represented simply by a plus sign.

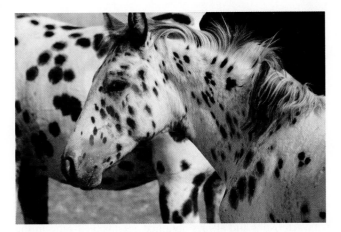

3.9 Leopard spotting in horses exhibits incomplete dominance. *[PhotoDisc.]*

Connecting Concepts

Ratios in Simple Crosses

Now that we have had some experience with genetic crosses, let's review the ratios that appear in the progeny of simple crosses, in which a single locus is under consideration. Understanding these ratios and the parental genotypes that produce them will allow you to work simple genetic crosses quickly, without resorting to the Punnett square. Later, we will use these ratios to work more-complicated crosses entailing several loci.

There are only four phenotypic ratios to understand (**Table 3.2**). The 3 : 1 ratio arises in a simple genetic cross when both of the parents are heterozygous for a dominant trait (*Aa* × *Aa*). The second phenotypic ratio is the 1 : 2 : 1 ratio, which arises in the progeny of crosses between two parents heterozygous for a character that exhibits incomplete dominance (*Aa* × *Aa*). The third phenotypic ratio is the 1 : 1 ratio, which results from the mating of a homozygous parent and a heterozygous parent. If the character exhibits dominance, the homozygous parent in this cross must carry two recessive alleles (*Aa* × *aa*) to obtain a 1 : 1 ratio, because a cross between a homozygous dominant parent and a heterozygous parent (*AA* × *Aa*) produces offspring displaying only the dominant trait. For a character with incomplete dominance, a 1 : 1 ratio results from a cross between the heterozygote and either homozygote (*Aa* × *aa* or *Aa* × *AA*).

✔ Concept Check 6

If an F_1 individual in Figure 3.8 is used in a testcross, what proportion of the progeny from this cross will be white?

a. All the progeny   c. $\frac{1}{4}$

b. $\frac{1}{2}$ d. 0

Genetic Symbols

As we have seen, genetic crosses are usually depicted with the use of symbols to designate the different alleles. There are a number of different ways in which alleles can be represented. Lowercase letters are traditionally used to designate recessive alleles, and uppercase letters are for dominant alleles. The common allele for a character—

Table 3.2	Phenotypic ratios for simple genetic crosses (crosses for a single locus)		
Ratio	**Genotypes of Parents**	**Genotypes of Progeny**	**Type of Dominance**
3 : 1	*Aa* × *Aa*	$\frac{3}{4}$ *A_* : $\frac{1}{4}$ *aa*	Dominance
1 : 2 : 1	*Aa* × *Aa*	$\frac{1}{4}$ *AA* : $\frac{1}{2}$ *Aa* : $\frac{1}{4}$ *aa*	Incomplete dominance
1 : 1	*Aa* × *aa*	$\frac{1}{2}$ *Aa* : $\frac{1}{2}$ *aa*	Dominance or incomplete dominance
	Aa × *AA*	$\frac{1}{2}$ *Aa* : $\frac{1}{2}$ *AA*	Incomplete dominance
Uniform progeny	*AA* × *AA*	All *AA*	Dominance or incomplete dominance
	aa × *aa*	All *aa*	Dominance or incomplete dominance
	AA × *aa*	All *Aa*	Dominance or incomplete dominance
	AA × *Aa*	All *A_*	Dominance

Note: A line in a genotype, such as *A_*, indicates that any allele is possible.

Table 3.3 **Genotypic ratios for simple genetic crosses (crosses for a single locus)**

Genotypic Ratio	Genotypes of Parents	Genotypes of Progeny
1 : 2 : 1	$Aa \times Aa$	$\frac{1}{4}\,AA : \frac{1}{2}\,Aa : \frac{1}{4}\,aa$
1 : 1	$Aa \times aa$	$\frac{1}{2}\,Aa : \frac{1}{2}\,aa$
	$Aa \times AA$	$\frac{1}{2}\,Aa : \frac{1}{2}\,AA$
Uniform progeny	$AA \times AA$	All AA
	$aa \times aa$	All aa
	$AA \times aa$	All Aa

The fourth phenotypic ratio is not really a ratio—all the off-spring have the same phenotype. Several combinations of parents can produce this outcome (see Table 3.2). A cross between any two homozygous parents—either between two of the same homozygotes ($AA \times AA$ and $aa \times aa$) or between two different homozygotes ($AA \times aa$)—produces progeny all having the same phenotype. Progeny of a single phenotype can also result from a cross between a homozygous dominant parent and a heterozygote ($AA \times Aa$).

If we are interested in the ratios of genotypes instead of phenotypes, there are only three outcomes to remember (**Table 3.3**): the 1 : 2 : 1 ratio, produced by a cross between two heterozygotes; the 1 : 1 ratio, produced by a cross between a heterozygote and a homozygote; and the uniform progeny produced by a cross between two homozygotes. These simple phenotypic and genotypic ratios and the parental genotypes that produce them provide the key to understanding crosses for a single locus and, as you will see in the next section, for multiple loci.

3.3 Dihybrid Crosses Reveal the Principle of Independent Assortment

We will now extend Mendel's principle of segregation to more-complex crosses entailing alleles at multiple loci. Understanding the nature of these crosses will require an additional principle, the principle of independent assortment.

Dihybrid Crosses

In addition to his work on monohybrid crosses, Mendel crossed varieties of peas that differed in *two* characteristics (a **dihybrid cross**). For example, he had one homozygous variety of pea with seeds that were round and yellow; another homozygous variety with seeds that were wrinkled and green. When he crossed the two varieties, the seeds of all the F_1 prog-

eny were round and yellow. He then self-fertilized the F_1 and obtained the following progeny in the F_2: 315 round, yellow seeds; 101 wrinkled, yellow seeds; 108 round, green seeds; and 32 wrinkled, green seeds. Mendel recognized that these traits appeared approximately in a 9 : 3 : 3 : 1 ratio; that is, $\frac{9}{16}$ of the progeny were round and yellow, $\frac{3}{16}$ were wrinkled and yellow, $\frac{3}{16}$ were round and green, and $\frac{1}{16}$ were wrinkled and green.

The Principle of Independent Assortment

Mendel carried out a number of dihybrid crosses for pairs of characteristics and always obtained a 9 : 3 : 3 : 1 ratio in the F_2. This ratio makes perfect sense in regard to segregation and dominance if we add a third principle, which Mendel recognized in his dihybrid crosses: the **principle of independent assortment** (Mendel's second law). This principle states that alleles at different loci separate independently of one another.

A common mistake is to think that the principle of segregation and the principle of independent assortment refer to two different processes. The principle of independent assortment is really an extension of the principle of segregation. The principle of segregation states that the two alleles of a locus separate when gametes are formed; the principle of independent assortment states that, when these two alleles separate, their separation is independent of the separation of alleles at *other* loci.

Let's see how the principle of independent assortment explains the results that Mendel obtained in his dihybrid cross. Each plant possesses two alleles encoding each characteristic, and so the parental plants must have had genotypes *RR YY* and *rr yy* (**Figure 3.10a**). The principle of segregation indicates that the alleles for each locus separate, and one allele for each locus passes to each gamete. The gametes produced by the round, yellow parent therefore contain alleles *RY*, whereas the gametes produced by the wrinkled, green parent contain alleles *ry*. These two types of gametes unite to produce the F_1, all with genotype *Rr Yy*. Because round is dominant over wrinkled and yellow is dominant over green, the phenotype of the F_1 will be round and yellow.

When Mendel self-fertilized the F_1 plants to produce the F_2, the alleles for each locus separated, with one allele going into each gamete. This event is where the principle of independent assortment becomes important. Each pair of alleles can separate in two ways: (1) *R* separates with *Y* and *r* separates with *y* to produce gametes *RY* and *ry* or (2) *R* separates with *y* and *r* separates with *Y* to produce gametes *Ry* and *rY*. The principle of independent assortment tells us that the alleles at each locus separate independently; thus, both kinds of separation occur equally and all four type of gametes (*RY*, *ry*, *Ry*, and *rY*) are produced in equal proportions (**Figure 3.10b**). When these four types of gametes are combined to produce the F_2 generation, the progeny consist of $\frac{9}{16}$ round and yellow, $\frac{3}{16}$ wrinkled and yellow, $\frac{3}{16}$ round and green, and $\frac{1}{16}$ wrinkled and green, resulting in a 9 : 3 : 3 : 1 phenotypic ratio (**Figure 3.10c**).

Relating the Principle of Independent Assortment to Meiosis

An important qualification of the principle of independent assortment is that it applies to characters encoded by loci located on different chromosomes because, like the principle of segregation, it is based wholly on the behavior of chromosomes during meiosis. Each pair of homologous chromosomes separates independently of all other pairs in anaphase I of meiosis (see Figure 2.13); so genes located on different pairs of homologs will assort independently. Genes that happen to be located on the same chromosome will travel together during anaphase I of meiosis and will arrive at the same destination—within the same gamete (unless crossing over takes place). Genes located on the same chromosome therefore do not assort independently (unless they are located sufficiently far apart that crossing over takes place every meiotic division, as will be discussed fully in Chapter 5).

Concepts

The principle of independent assortment states that genes encoding different characteristics separate independently of one another when gametes are formed, owing to the independent separation of homologous pairs of chromosomes in meiosis. Genes located close together on the same chromosome do not, however, assort independently.

✔ Concept Check 7

How are the principles of segregation and independent assortment related and how are they different?

Applying Probability and the Branch Diagram to Dihybrid Crosses

When the genes at two loci separate independently, a dihybrid cross can be understood as two monohybrid crosses. Let's examine Mendel's dihybrid cross ($Rr\ Yy \times Rr\ Yy$) by considering each characteristic separately (**Figure 3.11a**). If we consider only the shape of the seeds, the cross was $Rr \times Rr$, which yields a 3 : 1 phenotypic ratio ($\frac{3}{4}$ round and $\frac{1}{4}$ wrinkled progeny, see Table 3.2). Next consider the other characteristic, the color of the seed. The cross was $Yy \times Yy$, which produces a 3 : 1 phenotypic ratio ($\frac{3}{4}$ yellow and $\frac{1}{4}$ green progeny).

We can now combine these monohybrid ratios by using the multiplication rule to obtain the proportion of progeny with different combinations of seed shape and color. The proportion of progeny with round and yellow seeds is $\frac{3}{4}$ (the

Experiment

Question: Do alleles encoding different traits separate independently?

(a)

Methods

P generation

Round, yellow seeds × Wrinkled, green seeds

$RR\ YY$ $rr\ yy$

Gametes RY ry

Fertilization

(b)

F₁ generation   Round, yellow seeds

$Rr\ Yy$

Gametes RY ry Ry rY

Self-fertilization

(c)

Results

F₂ generation

	RY	ry	Ry	rY
RY	$RR\ YY$	$Rr\ Yy$	$RR\ Yy$	$Rr\ YY$
ry	$Rr\ Yy$	$rr\ yy$	$Rr\ yy$	$rr\ Yy$
Ry	$RR\ Yy$	$Rr\ yy$	$RR\ yy$	$Rr\ Yy$
rY	$Rr\ YY$	$rr\ Yy$	$Rr\ Yy$	$rr\ YY$

Phenotypic ratio
9 round, yellow : 3 round, green : 3 wrinkled, yellow : 1 wrinkled, green

Conclusion: The allele encoding color separated independently of the allele encoding seed shape, producing a 9 : 3 : 3 : 1 ratio in the F₂ progeny.

3.10 Mendel's dihybrid crosses revealed the principle of independent assortment.

3.11 A branch diagram can be used to determine the phenotypes and expected proportions of offspring from a dihybrid cross (*Rr Yy* × *Rr Yy*).

probability of round) × $\frac{3}{4}$ (the probability of yellow) = $\frac{9}{16}$. The proportion of progeny with round and green seeds is $\frac{3}{4} \times \frac{1}{4} = \frac{3}{16}$; the proportion of progeny with wrinkled and yellow seeds is $\frac{1}{4} \times \frac{3}{4} = \frac{3}{16}$; and the proportion of progeny with wrinkled and green seeds is $\frac{1}{4} \times \frac{1}{4} = \frac{1}{16}$.

Branch diagrams are a convenient way of organizing all the combinations of characteristics (**Figure 3.11b**). In the first column, list the proportions of the phenotypes for one character (here, $\frac{3}{4}$ round and $\frac{1}{4}$ wrinkled). In the second column, list the proportions of the phenotypes for the second character ($\frac{3}{4}$ yellow and $\frac{1}{4}$ green) twice, next to each of the phenotypes in the first column: put $\frac{3}{4}$ yellow and $\frac{1}{4}$ green next to the round phenotype and again next to the wrinkled phenotype. Draw lines between the phenotypes in

the first column and each of the phenotypes in the second column. Now follow each branch of the diagram, multiplying the probabilities for each trait along that branch. One branch leads from round to yellow, yielding round and yellow progeny. Another branch leads from round to green, yielding round and green progeny, and so forth. We calculate the probability of progeny with a particular combination of traits by using the multiplication rule: the probability of round ($\frac{3}{4}$) and yellow ($\frac{3}{4}$) seeds is $\frac{3}{4} \times \frac{3}{4} = \frac{9}{16}$. The advantage of the branch diagram is that it helps keep track of all the potential combinations of traits that may appear in the progeny. It can be used to determine phenotypic or genotypic ratios for any number of characteristics.

Using probability is much faster than using the Punnett square for crosses that include multiple loci. Genotypic and phenotypic ratios can be quickly worked out by combining, with the multiplication rule, the simple ratios in Tables 3.2 and 3.3. The probability method is particularly efficient if we need the probability of only a *particular* phenotype or genotype among the progeny of a cross. Suppose we needed to know the probability of obtaining the genotype *Rr yy* in the F_2 of the dihybrid cross in Figure 3.10. The probability of obtaining the *Rr* genotype in a cross of *Rr* × *Rr* is $\frac{1}{2}$ and that of obtaining *yy* progeny in a cross of *Yy* × *Yy* is $\frac{1}{4}$ (see Table 3.3). Using the multiplication rule, we find the probability of *Rr yy* to be $\frac{1}{2} \times \frac{1}{4} = \frac{1}{8}$.

To illustrate the advantage of the probability method, consider the cross *Aa Bb cc Dd Ee* × *Aa Bb Cc dd Ee*. Suppose we wanted to know the probability of obtaining offspring with the genotype *aa bb cc dd ee*. If we used a Punnett square to determine this probability, we might be working on the solution for months. However, we can quickly figure the probability of obtaining this one genotype by breaking this cross into a series of single-locus crosses:

Progeny cross	Genotype	Probability
Aa × *Aa*	*aa*	$\frac{1}{4}$
Bb × *Bb*	*bb*	$\frac{1}{4}$
cc × *Cc*	*cc*	$\frac{1}{2}$
Dd × *dd*	*dd*	$\frac{1}{2}$
Ee × *Ee*	*ee*	$\frac{1}{4}$

The probability of an offspring from this cross having genotype *aa bb cc dd ee* is now easily obtained by using the multiplication rule: $\frac{1}{4} \times \frac{1}{4} \times \frac{1}{2} \times \frac{1}{2} \times \frac{1}{4} = \frac{1}{256}$. This calculation assumes that genes at these five loci all assort independently.

Concepts

A cross including several characteristics can be worked by breaking the cross down into single-locus crosses and using the multiplication rule to determine the proportions of combinations of characteristics (provided the genes assort independently).

The Dihybrid Testcross

Let's practice using the branch diagram by determining the types and proportions of phenotypes in a dihybrid testcross between the round and yellow F$_1$ plants ($Rr\,Yy$) obtained by Mendel in his dihybrid cross and the wrinkled and green plants ($rr\,yy$), as shown in **Figure 3.12**. Break the cross down into a series of single-locus crosses. The cross $Rr \times rr$ yields $\frac{1}{2}$ round (Rr) progeny and $\frac{1}{2}$ wrinkled (rr) progeny. The cross $Yy \times yy$ yields $\frac{1}{2}$ yellow (Yy) progeny and $\frac{1}{2}$ green (yy) progeny. Using the multiplication rule, we find the proportion of round and yellow progeny to be $\frac{1}{2}$ (the probability of round) $\times \frac{1}{2}$ (the probability of yellow) $= \frac{1}{4}$. Four combinations of traits with the following proportions appear in the offspring: $\frac{1}{4}\,Rr\,Yy$, round yellow; $\frac{1}{4}\,Rr\,yy$, round green; $\frac{1}{4}\,rr\,Yy$, wrinkled yellow; and $\frac{1}{4}\,rr\,yy$, wrinkled green.

3.12 A branch diagram can be used to determine the phenotypes and expected proportions of offspring from a dihybrid testcross ($Rr\,Yy \times rr\,yy$).

Worked Problem

Not only are the principles of segregation and independent assortment important because they explain how heredity works, but they also provide the means for predicting the outcome of genetic crosses. This predictive power has made genetics a powerful tool in agriculture and other fields, and the ability to apply the principles of heredity is an important skill for all students of genetics. Practice with genetic problems is essential for mastering the basic principles of heredity—no amount of reading and memorization can substitute for the experience gained by deriving solutions to specific problems in genetics.

Students may have difficulty with genetics problems when they are unsure of where to begin or how to organize the problem and plan a solution. In genetics, every problem is different, and so no common series of steps can be applied to all genetics problems. Logic and common sense must be used to analyze a problem and arrive at a solution. Nevertheless, certain steps can facilitate the process, and solving the following problem will serve to illustrate these steps.

In mice, black coat color (B) is dominant over brown (b), and a solid pattern (S) is dominant over white spotted (s). Color and spotting are controlled by genes that assort independently. A homozygous black, spotted mouse is crossed with a homozygous brown, solid mouse. All the F$_1$ mice are black and solid. A testcross is then carried out by mating the F$_1$ mice with brown, spotted mice.

a. Give the genotypes of the parents and the F$_1$ mice.

b. Give the genotypes and phenotypes, along with their expected ratios, of the progeny expected from the testcross.

• Solution

Step 1: Determine the questions to be answered. What question or questions is the problem asking? Is it asking for genotypes, genotypic ratios, or phenotypic ratios? This problem asks you to provide the *genotypes* of the parents and the F$_1$, the *expected genotypes* and *phenotypes* of the progeny of the testcross, and their *expected proportions*.

Step 2: Write down the basic information given in the problem. This problem provides important information about the dominance relations of the characters and about the mice being crossed. Black is dominant over brown, and solid is dominant over white spotted. Furthermore, the genes for the two characters assort independently. In this problem, symbols are provided for the different alleles (B for black, b for brown, S for solid, and s for spotted); had these symbols not been provided, you would need to choose symbols to represent these alleles. It is useful to record these symbols at the beginning of the solution:

B—black S—solid
b—brown s—white spotted

Next, write out the crosses given in the problem.

P   Homozygous   ×   Homozygous
           black, spotted   brown, solid
                          ↓
F₁   Black, solid

Testcross   Black, solid   ×   Brown, spotted

Step 3: Write down any genetic information that can be determined from the phenotypes alone. From the phenotypes and the statement that they are homozygous, you know that the P-generation mice must be *BB ss* and *bb SS*. The F₁ mice are black and solid, both dominant traits, and so the F₁ mice must possess at least one black allele (*B*) and one solid allele (*S*). At this point, you cannot be certain about the other alleles; so represent the genotype of the F₁ as *B_ S_*. The brown, spotted mice in the testcross must be *bb ss*, because both brown and spotted are recessive traits that will be expressed only if two recessive alleles are present. Record these genotypes on the crosses that you wrote out in step 2:

P   Homozygous   ×   Homozygous
           black, spotted   brown, solid
              BB ss   *bb SS*
                          ↓
F₁   Black, solid
                          *B_ S_*
Testcross   Black, solid   ×   Brown, spotted
              *B_ S_*   *bb ss*

Step 4: Break the problem down into smaller parts. First, determine the genotype of the F₁. After this genotype has been determined, you can predict the results of the testcross and determine the genotypes and phenotypes of the progeny from the testcross. Second, because this cross includes two independently assorting loci, it can be conveniently broken down into two single-locus crosses: one for coat color and the other for spotting. Third, use a branch diagram to determine the proportion of progeny of the testcross with different combinations of the two traits.

Step 5: Work the different parts of the problem. Start by determining the genotype of the F₁ progeny. Mendel's first law indicates that the two alleles at a locus separate, one going into each gamete. Thus, the gametes produced by the black, spotted parent contain *B s* and the gametes produced by the brown, solid parent contain *b S*, which combine to produce F₁ progeny with the genotype *Bb Ss*:

P   Homozygous   ×   Homozygous
           black, spotted   brown, solid
              BB ss   *bb SS*
                          ↓
Gametes   (Bs)   (bS)

F₁   *Bb Ss*

Use the F₁ genotype to work the testcross (*Bb Ss* × *bb ss*), breaking it into two single-locus crosses. First, consider the cross for coat color: *Bb* × *bb*. Any cross between a heterozygote and a homozygous recessive genotype produces a 1 : 1 phenotypic ratio of progeny (see Table 3.2):

$$Bb \times bb$$
$$\downarrow$$
$$^1/_2 \; Bb \text{ black}$$
$$^1/_2 \; bb \text{ brown}$$

Next, do the cross for spotting: *Ss* × *ss*. This cross also is between a heterozygote and a homozygous recessive genotype and will produce $^1/_2$ solid (*Ss*) and $^1/_2$ spotted (*ss*) progeny (see Table 3.2).

$$Ss \times ss$$
$$\downarrow$$
$$^1/_2 \; Ss \text{ solid}$$
$$^1/_2 \; ss \text{ spotted}$$

Finally, determine the proportions of progeny with combinations of these characters by using the branch diagram.

$^1/_2$ *Bb* black
   ⟨ $^1/_2$ *Ss* solid → *Bb Ss* black, solid $^1/_2 × ^1/_2 = ^1/_4$
      $^1/_2$ *ss* spotted → *Bb ss* black, spotted $^1/_2 × ^1/_2 = ^1/_4$

$^1/_2$ *bb* brown
   ⟨ $^1/_2$ *Ss* solid → *bb Ss* brown, solid $^1/_2 × ^1/_2 = ^1/_4$
      $^1/_2$ *ss* spotted → *bb ss* brown, spotted $^1/_2 × ^1/_2 = ^1/_4$

Step 6: Check all work. As a last step, reread the problem, checking to see if your answers are consistent with the information provided. You have used the genotypes *BB ss* and *bb SS* in the P generation. Do these genotypes encode the phenotypes given in the problem? Are the F₁ progeny phenotypes consistent with the genotypes that you assigned? The answers are consistent with the information.

> **?** Now that we have stepped through a genetics problem together, try your hand at Problem 24 at the end of the chapter.

3.4 Observed Ratios of Progeny May Deviate from Expected Ratios by Chance

When two individual organisms of known genotype are crossed, we expect certain ratios of genotypes and phenotypes in the progeny; these expected ratios are based on the

Mendelian principles of segregation, independent assortment, and dominance. The ratios of genotypes and phenotypes *actually* observed among the progeny, however, may deviate from these expectations.

For example, in German cockroaches, brown body color (Y) is dominant over yellow body color (y). If we cross a brown, heterozygous cockroach (Yy) with a yellow cockroach (yy), we expect a 1 : 1 ratio of brown (Yy) and yellow (yy) progeny. Among 40 progeny, we would therefore expect to see 20 brown and 20 yellow offspring. However, the observed numbers might deviate from these expected values; we might in fact see 22 brown and 18 yellow progeny.

Chance plays a critical role in genetic crosses, just as it does in flipping a coin. When you flip a coin, you expect a 1 : 1 ratio—$^1/_2$ heads and $^1/_2$ tails. If you flip a coin 1000 times, the proportion of heads and tails obtained would probably be very close to that expected 1 : 1 ratio. However, if you flip the coin 10 times, the ratio of heads to tails might be quite different from 1 : 1. You could easily get 6 heads and 4 tails, or 3 heads and 7 tails, just by chance. You might even get 10 heads and 0 tails. The same thing happens in genetic crosses. We may expect 20 brown and 20 yellow cockroaches, but 22 brown and 18 yellow progeny *could* arise as a result of chance.

The Goodness-of-Fit Chi-Square Test

If you expected a 1 : 1 ratio of brown and yellow cockroaches but the cross produced 22 brown and 18 yellow, you probably wouldn't be too surprised even though it wasn't a perfect 1 : 1 ratio. In this case, it seems reasonable to assume that chance produced the deviation between the expected and the observed results. But, if you observed 25 brown and 15 yellow, would the ratio still be 1 : 1? Something other than chance might have caused the deviation. Perhaps the inheritance of this character is more complicated than was assumed or perhaps some of the yellow progeny died before they were counted. Clearly, we need some means of evaluating how likely it is that chance is responsible for the deviation between the observed and the expected numbers.

To evaluate the role of chance in producing deviations between observed and expected values, a statistical test called the **goodness-of-fit chi-square test** is used. This test provides information about how well the observed values fit expected values. Before we learn how to calculate the chi square, it is important to understand what this test does and does not indicate about a genetic cross.

The chi-square test cannot tell us whether a genetic cross has been correctly carried out, whether the results are correct, or whether we have chosen the correct genetic explanation for the results. What it does indicate is the *probability* that the difference between the observed and the expected values is due to chance. In other words, it indicates the likelihood that chance alone could produce the deviation between the expected and the observed values.

If we expected 20 brown and 20 yellow progeny from a genetic cross, the chi-square test gives the probability that we might observe 25 brown and 15 yellow progeny simply owing to chance deviations from the expected 20 : 20 ratio. This hypothesis, that chance alone is responsible for any deviations between observed and expected values, is sometimes called the null hypothesis. When the probability calculated from the chi-square test is high, we assume that chance alone produced the difference (the null hypothesis is true). When the probability is low, we assume that some factor other than chance—some significant factor—produced the deviation (the null hypothesis is false).

To use the goodness-of-fit chi-square test, we first determine the expected results. The chi-square test must always be applied to numbers of progeny, not to proportions or percentages. Let's consider a locus for coat color in domestic cats, for which black color (B) is dominant over gray (b). If we crossed two heterozygous black cats ($Bb \times Bb$), we would expect a 3 : 1 ratio of black and gray kittens. A series of such crosses yields a total of 50 kittens—30 black and 20 gray. These numbers are our *observed* values. We can obtain the expected numbers by multiplying the *expected* proportions by the total number of observed progeny. In this case, the expected number of black kittens is $^3/_4 \times 50 = 37.5$ and the expected number of gray kittens is $^1/_4 \times 50 = 12.5$. The chi-square (χ^2) value is calculated by using the following formula:

$$\chi^2 = \sum \frac{(\text{observed} - \text{expected})^2}{\text{expected}}$$

where $\sum$ means the sum. We calculate the sum of all the squared differences between observed and expected and divide by the expected values. To calculate the chi-square value for our black and gray kittens, we would first subtract the number of *expected* black kittens from the number of *observed* black kittens ($30 - 37.5 = -7.5$) and square this value: $-7.5^2 = 56.25$. We then divide this result by the expected number of black kittens, $56.25/37.5 = 1.5$. We repeat the calculations on the number of expected gray kittens: $(20 - 12.5)^2/12.5 = 4.5$. To obtain the overall chi-square value, we sum the (observed $-$ expected)^2/expected values: $1.5 + 4.5 = 6.0$.

The next step is to determine the probability associated with this calculated chi-square value, which is the probability that the deviation between the observed and the expected results could be due to chance. This step requires us to compare the calculated chi-square value (6.0) with theoretical values that have the same degrees of freedom in a chi-square table. The degrees of freedom represent the number of ways in which the expected classes are free to vary. For a goodness-of-fit chi-square test, the degrees of freedom are equal to $n - 1$, where n is the number of different expected phenotypes. In our example, there are two expected phenotypes (black and gray); so $n = 2$, and the degree of freedom equals $2 - 1 = 1$.

Now that we have our calculated chi-square value and have figured out the associated degrees of freedom, we are

					P				
df	0.995	0.975	0.9	0.5	0.1	0.05	0.025	0.01	0.005
1	0.000	0.000	0.016	0.455	2.706	3.841	5.024	6.635	7.879
2	0.010	0.051	0.211	1.386	4.605	5.991	7.378	9.210	10.597
3	0.072	0.216	0.584	2.366	6.251	7.815	9.348	11.345	12.838
4	0.207	0.484	1.064	3.357	7.779	9.488	11.143	13.277	14.860
5	0.412	0.831	1.610	4.351	9.236	11.070	12.832	15.086	16.750
6	0.676	1.237	2.204	5.348	10.645	12.592	14.449	16.812	18.548
7	0.989	1.690	2.833	6.346	12.017	14.067	16.013	18.475	20.278
8	1.344	2.180	3.490	7.344	13.362	15.507	17.535	20.090	21.955
9	1.735	2.700	4.168	8.343	14.684	16.919	19.023	21.666	23.589
10	2.156	3.247	4.865	9.342	15.987	18.307	20.483	23.209	25.188
11	2.603	3.816	5.578	10.341	17.275	19.675	21.920	24.725	26.757
12	3.074	4.404	6.304	11.340	18.549	21.026	23.337	26.217	28.300
13	3.565	5.009	7.042	12.340	19.812	22.362	24.736	27.688	29.819
14	4.075	5.629	7.790	13.339	21.064	23.685	26.119	29.141	31.319
15	4.601	6.262	8.547	14.339	22.307	24.996	27.488	30.578	32.801

Table 3.4 Critical values of the χ^2 distribution

P, probability; df, degrees of freedom.

ready to obtain the probability from a chi-square table (**Table 3.4**). The degrees of freedom are given in the left-hand column of the table and the probabilities are given at the top; within the body of the table are chi-square values associated with these probabilities. First, find the row for the appropriate degrees of freedom; for our example with 1 degree of freedom, it is the first row of the table. Find where our calculated chi-square value (6.0) lies among the theoretical values in this row. The theoretical chi-square values increase from left to right and the probabilities decrease from left to right. Our chi-square value of 6.0 falls between the value of 5.024, associated with a probability of 0.025, and the value of 6.635, associated with a probability of 0.01.

Thus, the probability associated with our chi-square value is less than 0.025 and greater than 0.01. So there is less than a 2.5% probability that the deviation that we observed between the expected and the observed numbers of black and gray kittens could be due to chance.

Most scientists use the 0.05 probability level as their cutoff value: if the probability of chance being responsible for the deviation is greater than or equal to 0.05, they accept that chance may be responsible for the deviation between the observed and the expected values. When the probability is less than 0.05, scientists assume that chance is not responsible and a significant difference exists. The expression *significant difference* means that some factor other than chance is responsible for the observed values being different from the expected values. In regard to the kittens, perhaps one of the genotypes had a greater mortality rate before the progeny

were counted or perhaps other genetic factors skewed the observed ratios.

In choosing 0.05 as the cutoff value, scientists have agreed to assume that chance is responsible for the deviations between observed and expected values unless there is strong evidence to the contrary. It is important to bear in mind that, even if we obtain a probability of, say, 0.01, there is still a 1% probability that the deviation between the observed and the expected numbers is due to nothing more than chance. Calculation of the chi-square value is illustrated in **Figure 3.13**.

Concepts

Differences between observed and expected ratios can arise by chance. The goodness-of-fit chi-square test can be used to evaluate whether deviations between observed and expected numbers are likely to be due to chance or to some other significant factor.

✔ Concept Check 8

A chi-square test comparing observed and expected progeny is carried out, and the probability associated with the calculated chi-square value is 0.72. What does this probability represent?

a. Probability that the correct results were obtained

b. Probability of obtaining the observed numbers

c. Probability that the difference between observed and expected numbers is significant

d. Probability that the difference between observed and expected numbers could be due to chance.

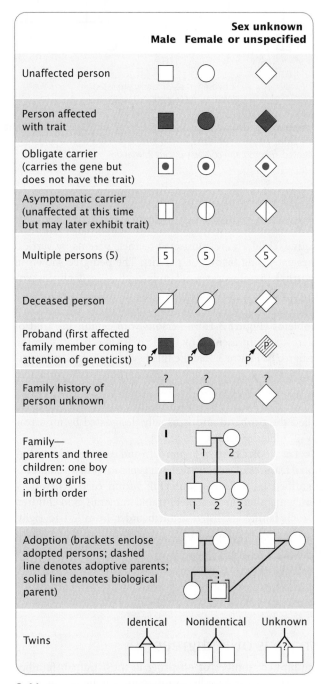

3.13 A chi-square test is used to determine the probability that the difference between observed and expected values is due to chance.

geneticists have been forced to develop special techniques that are uniquely suited to human biology and culture.

One technique used by geneticists to study human inheritance is the analysis of pedigrees. A **pedigree** is a pictorial representation of a family history, essentially a family tree that outlines the inheritance of one or more characteristics. When a particular characteristic or disease is observed in a person, a geneticist often studies the family of this affected person by drawing a pedigree.

The symbols commonly used in pedigrees are summarized in **Figure 3.14**. Males in a pedigree are represented by

3.5 Geneticists Often Use Pedigrees to Study the Inheritance of Human Characteristics

The study of human genetic characteristics presents some major obstacles. First, controlled matings are not possible. With other organisms, geneticists carry out specific crosses to test their hypotheses about inheritance. Unfortunately (for the geneticist at least), matings between humans are more frequently determined by romance, family expectations, and—occasionally—accident than they are by the requirements of the geneticist. Other obstacles are the long generation time and generally small family size. To overcome these obstacles,

3.14 Standard symbols are used in pedigrees.

(a)

Each generation in a pedigree is identified by a Roman numeral.

Within each generation, family members are identified by Arabic numerals.

Filled symbols represent family members with Waardenburg syndrome…

…and open symbols represent unaffected members.

Children in each family are listed left to right in birth order.

(b)

3.15 Waardenburg syndrome is (a) inherited as an autosomal dominant trait and (b) characterized by deafness, fair skin, visual problems, and a white forelock. The proband (P) is the person from whom this pedigree is initiated. *[Photograph courtesy of Guy Rowland.]*

squares, females by circles. A horizontal line drawn between two symbols representing a man and a woman indicates a mating; children are connected to their parents by vertical lines extending below the parents. The pedigree shown in **Figure 3.15a** illustrates a family with Waardenburg syndrome, an autosomal dominant type of deafness that may be accompanied by fair skin, a white forelock, and visual problems (**Figure 3.15b**). Persons who exhibit the trait of interest are represented by filled circles and squares; in the pedigree of Figure 3.15a, the filled symbols represent members of the family who have Waardenburg syndrome. Unaffected members are represented by open circles and squares. The person from whom the pedigree is initiated is called the **proband** and is usually designated by an arrow (IV-2 in Figure 3.15a).

Let's look closely at Figure 3.15 and consider some additional features of a pedigree. Each generation in a pedigree is identified by a Roman numeral; within each generation, family members are assigned Arabic numerals, and children in each family are listed in birth order from left to right. Person II-4, a man with Waardenburg syndrome, mated with II-5, an unaffected woman, and they produced five children. The oldest of their children is III-8, a male with Waardenburg syndrome, and the youngest is III-14, an unaffected female.

Analysis of Pedigrees

The limited number of offspring in most human families means that clear Mendelian ratios in a single pedigree are usually impossible to discern. Pedigree analysis requires a certain amount of genetic sleuthing, based on recognizing patterns associated with different modes of inheritance.

Recessive traits Recessive traits normally appear with equal frequency in both sexes and appear only when a person inherits two alleles for the trait, one from each parent. If the trait is uncommon, most parents of affected offspring are heterozygous and unaffected; consequently, the trait seems to skip generations (**Figure 3.16**). Frequently, a recessive allele may be passed for a number of generations without the trait appearing in a pedigree. Whenever both parents are heterozygous, approximately $\frac{1}{4}$ of the offspring are expected to

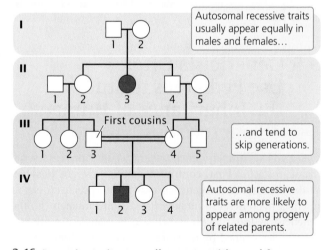

Autosomal recessive traits usually appear equally in males and females…

First cousins

…and tend to skip generations.

Autosomal recessive traits are more likely to appear among progeny of related parents.

3.16 Recessive traits normally appear with equal frequency in both sexes and seem to skip generations. The double line between III-3 and III-4 represents consanguinity (mating between related persons).

express the trait, but this ratio will not be obvious unless the family is large. In the rare event that both parents are affected by an autosomal recessive trait, all the offspring will be affected.

When a recessive trait is rare, persons from outside the family are usually homozygous for the normal allele. Thus, when an affected person mates with someone outside the family ($aa \times AA$), usually none of the children will display the trait, although all will be carriers (i.e., heterozygous). A recessive trait is more likely to appear in a pedigree when two people within the same family mate, because there is a greater chance of both parents carrying the same recessive allele. Mating between closely related people is called **consanguinity**. In the pedigree shown in Figure 3.16, persons III-3 and III-4 are first cousins, and both are heterozygous for the recessive allele; when they mate, $^{1}/_{4}$ of their children are expected to have the recessive trait.

Dominant traits Dominant traits appear in both sexes with equal frequency, and both sexes are capable of transmitting these traits to their offspring. Every person with a dominant trait must have inherited the allele from at least one parent; autosomal dominant traits therefore do not skip generations (**Figure 3.17**).

Sex-linked traits also have a distinctive pattern of inheritance. Characteristics of sex-linked traits will be considered in Chapter 4.

Concepts

Recessive traits appear in pedigrees with equal frequency in males and females. Affected children are commonly born to unaffected parents who are carriers of the gene for the trait, and the trait tends to skip generations. Recessive traits appear in pedigrees more frequently among the offspring of consanguine matings.

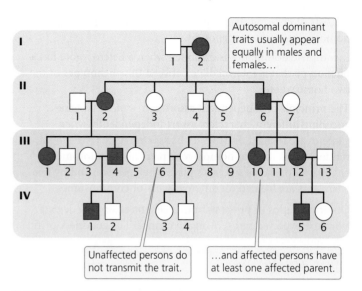

3.17 Dominant traits normally appear with equal frequency in both sexes and do not skip generations.

Dominant traits also appear in both sexes with equal frequency. An affected person has an affected parent (unless the person carries new mutations), and the trait does not skip generations. Unaffected persons do not transmit the trait.

✔ Concept Check 9

Recessive traits often appear in pedigrees in which there have been consanguine matings, because these traits

a. tend to skip generations.

b. appear only when both parents carry a copy of the gene for the trait, which is more likely when the parents are related.

c. usually arise in children born to parents who are unaffected.

d. appear equally in males and females.

Concepts Summary

- Gregor Mendel discovered the principles of heredity. His success can be attributed to his choice of the pea plant as an experimental organism, the use of characters with a few, easily distinguishable phenotypes, his experimental approach, the use of mathematics to interpret his results, and careful attention to detail.

- Genes are inherited factors that determine a character. Alternate forms of a gene are called alleles. The alleles are located at a specific place, a locus, on a chromosome, and the set of genes that an individual organism possesses is its genotype. Phenotype is the manifestation or appearance of a characteristic and may refer to a physical, biochemical, or behavioral characteristic. Only the genotype—not the phenotype—is inherited.

- The principle of segregation states that an individual organism possesses two alleles encoding a trait and that these two alleles separate in equal proportions when gametes are formed.

- The concept of dominance indicates that, when two different alleles are present in a heterozygote, only the trait of one of them, the "dominant" allele, is observed in the phenotype. The other allele is said to be "recessive."

- The two alleles of a genotype are located on homologous chromosomes. The separation of homologous chromosomes in anaphase I of meiosis brings about the segregation of alleles.

- Probability is the likelihood that a particular event will occur. The multiplication rule of probability states that the probability of two or more independent events occurring together is calculated by multiplying the probabilities of the independent events. The addition rule of probability states that the probability of any of two or more mutually exclusive events occurring is calculated by adding the probabilities of the events.

- A testcross reveals the genotype (homozygote or heterozygote) of an individual organism having a dominant trait and consists of crossing that individual with one having the homozygous recessive genotype.

- Incomplete dominance is exhibited when a heterozygote has a phenotype that is intermediate between the phenotypes of the two homozygotes.

- The principle of independent assortment states that genes encoding different characters assort independently when gametes are formed. Independent assortment is based on the random separation of homologous pairs of chromosomes in anaphase I of meiosis; it takes place when genes encoding two characters are located on different pairs of chromosomes.

- Observed ratios of progeny from a genetic cross may deviate from the expected ratios owing to chance. The goodness-of-fit chi-square test can be used to determine the probability that a difference between observed and expected numbers is due to chance.

- Pedigrees are often used to study the inheritance of traits in humans. Recessive traits typically appear with equal frequency in both sexes and tend to skip generations. They are more likely to appear in families with consanguinity (mating between closely related persons). Dominant traits usually appear equally in both sexes and do not skip generations. Unaffected people do not normally transmit an autosomal dominant trait to their offspring.

Important Terms

gene (p. 41)
allele (p. 42)
locus (p. 42)
genotype (p. 42)
homozygous (p. 42)
heterozygous (p. 42)
phenotype (p. 42)
monohybrid cross (p. 43)
P (parental) generation (p. 43)
F_1 (filial 1) generation (p. 43)
reciprocal crosses (p. 43)

F_2 (filial 2) generation (p. 44)
dominant (p. 44)
recessive (p. 44)
principle of segregation (Mendel's first
  law) (p. 45)
concept of dominance (p. 45)
chromosome theory of heredity (p. 45)
backcross (p. 47)
Punnett square (p. 47)
probability (p. 48)
multiplication rule (p. 48)

addition rule (p. 49)
testcross (p. 49)
incomplete dominance (p. 50)
wild type (p. 51)
dihybrid cross (p. 52)
principle of independent assortment
  (Mendel's second law) (p. 52)
goodness-of-fit chi-square test (p. 57)
pedigree (p. 59)
proband (p. 60)
consanguinity (p. 61)

Answers to Concept Checks

1. b

2. A locus is a place on a chromosome where genetic information encoding a trait is located. An allele is a copy of a gene that encodes a specific trait. A gene is an inherited factor that determines a trait.

3. Because the traits for both alleles appeared in the F_2 progeny

4. d

5. a

6. b

7. The principle of segregation and the principle of independent assortment both refer to the separation of alleles in anaphase I of meiosis. The principle of segregation says that these alleles separate, and the principle of independent assortment says that they separate independently of alleles at other loci.

8. d

9. b

Worked Problems

1. The following genotypes are crossed:

$$Aa\ Bb\ Cc\ Dd \times Aa\ Bb\ Cc\ Dd$$

Give the proportion of the progeny of this cross having each of the following genotypes: (a) *Aa Bb Cc Dd*, (b) *aa bb cc dd*, (c) *Aa Bb cc Dd*.

• Solution

This problem is easily worked if the cross is broken down into simple crosses and the multiplication rule is used to find the different combinations of genotypes:

Locus 1	$Aa \times Aa = \frac{1}{4}AA, \frac{1}{2}Aa, \frac{1}{4}aa$
Locus 2	$Bb \times Bb = \frac{1}{4}BB, \frac{1}{2}Bb, \frac{1}{4}bb$
Locus 3	$Cc \times Cc = \frac{1}{4}CC, \frac{1}{2}Cc, \frac{1}{4}cc$
Locus 4	$Dd \times Dd = \frac{1}{4}DD, \frac{1}{2}Dd, \frac{1}{4}dd$

To find the probability of any combination of genotypes, simply multiply the probabilities of the different genotypes:

a. $Aa\ Bb\ Cc\ Dd$ $\frac{1}{2}(Aa) \times \frac{1}{2}(Bb) \times \frac{1}{2}(Cc) \times \frac{1}{2}(Dd) = \frac{1}{16}$

b. $aa\ bb\ cc\ dd$ $\frac{1}{4}(aa) \times \frac{1}{4}(bb) \times \frac{1}{4}(cc) \times \frac{1}{4}(dd) = \frac{1}{256}$

c. $Aa\ Bb\ cc\ Dd$ $\frac{1}{2}(Aa) \times \frac{1}{2}(Bb) \times \frac{1}{4}(cc) \times \frac{1}{2}(Dd) = \frac{1}{32}$

2. In corn, purple kernels are dominant over yellow kernels, and full kernels are dominant over shrunken kernels. A corn plant having purple and full kernels is crossed with a plant having yellow and shrunken kernels, and the following progeny are obtained:

> purple, full   112
> purple, shrunken 103
> yellow, full   91
> yellow, shrunken 94

What are the most likely genotypes of the parents and progeny? Test your genetic hypothesis with a chi-square test.

• Solution

The best way to begin this problem is by breaking the cross down into simple crosses for a single characteristic (seed color or seed shape):

P   purple × yellow   full × shrunken

F_1 112 + 103 = 215 purple   112 + 91 = 203 full
      91 + 94 = 185 yellow   103 + 94 = 197 shrunken

Purple × yellow produces approximately $\frac{1}{2}$ purple and $\frac{1}{2}$ yellow. A 1 : 1 ratio is usually caused by a cross between a heterozygote and a homozygote. Because purple is dominant, the purple parent must be heterozygous (Pp) and the yellow parent must be homozygous (pp). The purple progeny produced by this cross will be heterozygous (Pp) and the yellow progeny must be homozygous (pp).

Now let's examine the other character. Full × shrunken produces $\frac{1}{2}$ full and $\frac{1}{2}$ shrunken, or a 1 : 1 ratio, and so these progeny phenotypes also are produced by a cross between a heterozygote (Ff) and a homozygote (ff); the full-kernel progeny will be heterozygous (Ff) and the shrunken-kernel progeny will be homozygous (ff).

Now combine the two crosses and use the multiplication rule to obtain the overall genotypes and the proportions of each genotype:

P   Purple, full   ×   Yellow, shrunken
        $Pp\ Ff$ $pp\ ff$

F_1 $Pp\ Ff = \frac{1}{2}$ purple × $\frac{1}{2}$ full $= \frac{1}{4}$ purple, full

    $Pp\ ff = \frac{1}{2}$ purple × $\frac{1}{2}$ shrunken $= \frac{1}{4}$ purple, shrunken

    $pp\ Ff = \frac{1}{2}$ yellow × $\frac{1}{2}$ full $= \frac{1}{4}$ yellow, full

    $pp\ ff = \frac{1}{2}$ yellow × $\frac{1}{2}$ shrunken $= \frac{1}{4}$ yellow, shrunken

Our genetic explanation predicts that, from this cross, we should see $\frac{1}{4}$ purple, full-kernel progeny; $\frac{1}{4}$ purple, shrunken-kernel progeny; $\frac{1}{4}$ yellow, full-kernel progeny; and $\frac{1}{4}$ yellow,

shrunken-kernel progeny. A total of 400 progeny were produced; so $\frac{1}{4} \times 400 = 100$ of each phenotype are expected. These observed numbers do not fit the expected numbers exactly. Could the difference between what we observe and what we expect be due to chance? If the probability is high that chance alone is responsible for the difference between observed and expected, we will assume that the progeny have been produced in the 1 : 1 : 1 : 1 ratio predicted by the cross. If the probability that the difference between observed and expected is due to chance is low, the progeny are not really in the predicted ratio and some other, *significant* factor must be responsible for the deviation.

The observed and expected numbers are:

Phenotype	Observed	Expected
purple, full	112	$\frac{1}{4} \times 400 = 100$
purple, shrunken	103	$\frac{1}{4} \times 400 = 100$
yellow, full	91	$\frac{1}{4} \times 400 = 100$
yellow, shrunken	94	$\frac{1}{4} \times 400 = 100$

To determine the probability that the difference between observed and expected is due to chance, we calculate a chi-square value with the formula $\chi^2 = \sum [(\text{observed} - \text{expected})^2 / \text{expected}]$:

$$\chi^2 = \frac{(112 - 100)^2}{100} + \frac{(103 - 100)^2}{100} + \frac{(91 - 100)^2}{100}$$
$$+ \frac{(94 - 100)^2}{100}$$
$$= \frac{12^2}{100} + \frac{3^2}{100} + \frac{9^2}{100} + \frac{6^2}{100}$$
$$= \frac{144}{100} + \frac{9}{100} + \frac{81}{100} + \frac{36}{100}$$
$$= 1.44 + 0.09 + 0.81 + 0.36 = 2.70$$

Now that we have the chi-square value, we must determine the probability that this chi-square value is due to chance. To obtain this probability, we first calculate the degrees of freedom, which for a goodness-of-fit chi-square test are $n - 1$, where n equals the number of expected phenotypic classes. In this case, there are four expected phenotypic classes; so the degrees of freedom equal $4 - 1 = 3$. We must now look up the chi-square value in a chi-square table (see Table 3.4). We select the row corresponding to 3 degrees of freedom and look along this row to find our calculated chi-square value. The calculated chi-square value of 2.7 lies between 2.366 (a probability of 0.5) and 6.251 (a probability of 0.1). The probability (P) associated with the calculated chi-square value is therefore $0.5 < P < 0.1$. This is the probability that the difference between what we observed and what we expect is due to chance, which in this case is relatively high, and so chance is likely responsible for the deviation. We can conclude that the progeny *do* appear in the 1 : 1 : 1 : 1 ratio predicted by our genetic explanation.

3. Joanna has "short fingers" (brachydactyly). She has two older brothers who are identical twins; both have short fingers. Joanna's two younger sisters have normal fingers. Joanna's mother has normal fingers, and her father has short fingers. Joanna's paternal grandmother (her father's mother) has short fingers; her paternal grandfather (her father's father), who is now deceased, had normal fingers. Both of Joanna's maternal grandparents (her mother's parents) have normal fingers. Joanna marries Tom, who has normal fingers; they adopt a son named Bill who has normal fingers. Bill's biological parents both have normal fingers. After adopting Bill, Joanna and Tom produce two children: an older daughter with short fingers and a younger son with normal fingers.

a. Using standard symbols and labels, draw a pedigree illustrating the inheritance of short fingers in Joanna's family.

b. What is the most likely mode of inheritance for short fingers in this family?

c. If Joanna and Tom have another biological child, what is the probability (based on your answer to part *b*) that this child will have short fingers?

• **Solution**

a. In the pedigree for the family, identify persons with the trait (short fingers) by filled circles (females) and filled squares (males). Connect Joanna's identical twin brothers to the line above by drawing diagonal lines that have a horizontal line between them. Enclose the adopted child of Joanna and Tom in brackets; connect him to his biological parents by drawing a diagonal line and to his adopted parents by a dashed line.

b. The most likely mode of inheritance for short fingers in this family is dominant. The trait appears equally in males and females and does not skip generations. When one parent has the trait, it appears in approximately half of that parent's sons and daughters, although the number of children in the families is small.

c. If having short fingers is dominant, Tom must be homozygous (*bb*) because he has normal fingers. Joanna must be heterozygous (*Bb*) because she and Tom have produced both short- and normal- fingered offspring. In a cross between a heterozygote and homozygote, half of the progeny are expected to be heterozygous and half homozygous (*Bb × bb*: $\frac{1}{2}$ *Bb*, $\frac{1}{2}$ *bb*); so the probability that Joanna's and Tom's next biological child will have short fingers is $\frac{1}{2}$.

Comprehension Questions

Section 3.1

*1. Why was Mendel's approach to the study of heredity so successful?

2. What is the difference between genotype and phenotype?

Section 3.2

*3. What is the principle of segregation? Why is it important?

4. How are Mendel's principles different from the concept of blending inheritance discussed in Chapter 1?

5. What is the concept of dominance? How does dominance differ from incomplete dominance?

6. What are the addition and multiplication rules of probability and when should they be used?

7. Give the genotypic ratios that may appear among the progeny of simple crosses and the genotypes of the parents that may give rise to each ratio.

*8. What is the chromosome theory of inheritance? Why was it important?

Section 3.3

*9. What is the principle of independent assortment? How is it related to the principle of segregation?

10. In which phases of mitosis and meiosis are the principles of segregation and independent assortment at work?

Section 3.4

11. How is the goodness-of-fit chi-square test used to analyze genetic crosses? What does the probability associated with a chi-square value indicate about the results of a cross?

Section 3.5

12. What features are exhibited by a pedigree of a recessive trait? What features if the trait is dominant?

Application Questions and Problems

Section 3.1

13. What characteristics of an organism would make it suitable for studies of the principles of inheritance? Can you name several organisms that have these characteristics?

Section 3.2

*14. In cucumbers, orange fruit color (*R*) is dominant over cream fruit color (*r*). A cucumber plant homozygous for orange fruits is crossed with a plant homozygous for cream fruits. The F_1 are intercrossed to produce the F_2.

a. Give the genotypes and phenotypes of the parents, the F_1, and the F_2.

b. Give the genotypes and phenotypes of the offspring of a backcross between the F_1 and the orange parent.

c. Give the genotypes and phenotypes of a backcross between the F_1 and the cream parent.

*15. In cats, blood-type A results from an allele (I^A) that is dominant over an allele (i^B) that produces blood-type B. There is no O blood type. The blood types of male and female cats that were mated and the blood types of their kittens follow. Give the most likely genotypes for the parents of each litter.

	Male parent	Female parent	Kittens
a.	A	B	4 kittens with type A, 3 kittens with type B
b.	B	B	6 kittens with type B
c.	B	A	8 kittens with type A
d.	A	A	7 kittens with type A, 2 kittens with type B
e.	A	A	10 kittens with type A
f.	A	B	4 kittens with type A, 1 kitten with type B

*16. In humans, alkaptonuria is a metabolic disorder in which affected persons produce black urine. Alkaptonuria results from an allele (a) that is recessive to the allele for normal metabolism (A). Sally has normal metabolism, but her brother has alkaptonuria. Sally's father has alkaptonuria, and her mother has normal metabolism.

a. Give the genotypes of Sally, her mother, her father, and her brother.

b. If Sally's parents have another child, what is the probability that this child will have alkaptonuria?

c. If Sally marries a man with alkaptonuria, what is the probability that their first child will have alkaptonuria?

*17. Hairlessness in American rat terriers is recessive to the presence of hair. Suppose that you have a rat terrier with hair. How can you determine whether this dog is homozygous or heterozygous for the hairy trait?

18. In snapdragons, red flower color (R) is incompletely dominant over white flower color (r); the heterozygotes produce pink flowers. A red snapdragon is crossed with a white snapdragon, and the F_1 are intercrossed to produce the F_2.

a. Give the genotypes and phenotypes of the F_1 and F_2, along with their expected proportions.

b. If the F_1 are backcrossed to the white parent, what will the genotypes and phenotypes of the offspring be?

c. If the F_1 are backcrossed to the red parent, what will the genotypes and phenotypes of the offspring be?

19. What is the probability of rolling one six-sided die and obtaining the following numbers?

a. 2   c. An even number

b. 1 or 2   d. Any number but a 6

*20. What is the probability of rolling two six-sided dice and obtaining the following numbers?

a. 2 and 3

b. 6 and 6

c. At least one 6

d. Two of the same number (two 1s, or two 2s, or two 3s, etc.)

e. An even number on both dice

f. An even number on at least one die

21. Phenylketonuria (PKU) is a disease that results from a recessive gene. Two normal parents produce a child with PKU.

a. What is the probability that a sperm from the father will contain the PKU allele?

b. What is the probability that an egg from the mother will contain the PKU allele?

c. What is the probability that their next child will have PKU?

d. What is the probability that their next child will be heterozygous for the PKU gene?

*22. In German cockroaches, curved wing (cv) is recessive to normal wing (cv^1). A homozygous cockroach having normal wings is crossed with a homozygous cockroach having curved wings. The F_1 are intercrossed to produce the F_2. Assume that the pair of chromosomes containing the locus for wing shape is metacentric. Draw this pair of chromosomes as it would appear in the parents, the F_1, and each class of F_2 progeny at metaphase I of meiosis. Assume that no crossing over takes place. At each stage, label a location for the alleles for wing shape (cv and cv^1) on the chromosomes.

*23. In guinea pigs, the allele for black fur (B) is dominant over the allele for brown (b) fur. A black guinea pig is crossed with a brown guinea pig, producing five F_1 black guinea pigs and six F_1 brown guinea pigs.

a. How many copies of the black allele (B) will be present in each cell from an F_1 black guinea pig at the following stages: G_1, G_2, metaphase of mitosis, metaphase I of meiosis, metaphase II of meiosis, and after the second cytokinesis following meiosis? Assume that no crossing over takes place.

b. How many copies of the brown allele (*b*) will be present in each cell from an F_1 brown guinea pig at the same stages as those listed in part *a*? Assume that no crossing over takes place.

Section 3.3

24. In watermelons, bitter fruit (*B*) is dominant over sweet fruit (*b*), and yellow spots (*S*) are dominant over no spots (*s*). The genes for these two characteristics assort independently. A homozygous plant that has bitter fruit and yellow spots is crossed with a homozygous plant that has sweet fruit and no spots. The F_1 are intercrossed to produce the F_2.
   a. What will be the phenotypic ratios in the F_2?
   b. If an F_1 plant is backcrossed with the bitter, yellow-spotted parent, what phenotypes and proportions are expected in the offspring?
   c. If an F_1 plant is backcrossed with the sweet, nonspotted parent, what phenotypes and proportions are expected in the offspring?

*25. The following two genotypes are crossed: *Aa Bb Cc dd Ee* × *Aa bb Cc Dd Ee*. What will the proportion of the following genotypes be among the progeny of this cross?
   a. *Aa Bb Cc Dd Ee*
   b. *Aa bb Cc dd ee*
   c. *aa bb cc dd ee*
   d. *AA BB CC DD EE*

26. In cucumbers, dull fruit (*D*) is dominant over glossy fruit (*d*), orange fruit (*R*) is dominant over cream fruit (*r*), and bitter cotyledons (*B*) are dominant over nonbitter cotyledons (*b*). The three characters are encoded by genes located on different pairs of chromosomes. A plant homozygous for dull, orange fruit and bitter cotyledons is crossed with a plant that has glossy, cream fruit and nonbitter cotyledons. The F_1 are intercrossed to produce the F_2.
   a. Give the phenotypes and their expected proportions in the F_2.
   b. An F_1 plant is crossed with a plant that has glossy, cream fruit and nonbitter cotyledons. Give the phenotypes and expected proportions among the progeny of this cross.

*27. Alleles *A* and *a* are located on a pair of metacentric chromosomes. Alleles *B* and *b* are located on a pair of acrocentric chromosomes. A cross is made between individuals having the following genotypes: *Aa Bb* × *aa bb*.
   a. Draw the chromosomes as they would appear in each type of gamete produced by the individuals of this cross.

b. For each type of progeny resulting from this cross, draw the chromosomes as they would appear in a cell at G_1, G_2, and metaphase of mitosis.

Section 3.4

*28. J. A. Moore investigated the inheritance of spotting patterns in leopard frogs (J. A. Moore. 1943. *Journal of Heredity* 34:3–7). The pipiens phenotype had the normal spots that give leopard frogs their name. In contrast, the burnsi phenotype lacked spots on its back. Moore carried out the following crosses, producing the progeny indicated.

Parent phenotypes	Progeny phenotypes
burnsi × burnsi	39 burnsi, 6 pipiens
burnsi × pipiens	23 burnsi, 33 pipiens
burnsi × pipiens	196 burnsi, 210 pipiens

   a. On the basis of these results, what is the most likely mode of inheritance of the burnsi phenotype?
   b. Give the most likely genotypes of the parent in each cross.
   c. Use a chi-square test to evaluate the fit of the observed numbers of progeny to the number expected on the basis of your proposed genotypes.

*29. In the California poppy, an allele for yellow flowers (*C*) is dominant over an allele for white flowers (*c*). At an independently assorting locus, an allele for entire petals (*F*) is dominant over an allele for fringed petals (*f*). A plant that is homozygous for yellow and entire petals is crossed with a plant that is white and fringed. A resulting F_1 plant is then crossed with a plant that is white and fringed, and the following progeny are produced: 54 yellow and entire; 58 yellow and fringed, 53 white and entire, and 10 white and fringed.
   a. Use a chi-square test to compare the observed numbers with those expected for the cross.
   b. What conclusion can you make from the results of the chi-square test?
   c. Suggest an explanation for the results.

Section 3.5

30. Many studies have suggested a strong genetic predisposition to migraine headaches, but the mode of inheritance is not clear. L. Russo and colleagues examined migraine headaches in several families, two of which are shown below (L. Russo et al. 2005. *American Journal of Human Genetics* 76:327–333). What is the most likely mode of inheritance for migraine headaches in these families? Explain your reasoning.

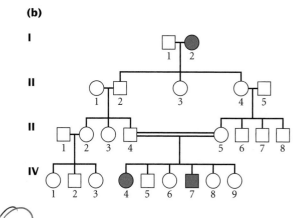

*31. For each of the following pedigrees, give the most likely mode of inheritance, assuming that the trait is rare. Carefully explain your reasoning.

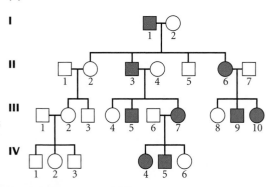

32. Ectodactyly is a rare condition in which the fingers are absent and the hand is split. This condition is usually inherited as an autosomal dominant trait. Ademar Freire-Maia reported the appearance of ectodactyly in a family in São Paulo, Brazil, whose pedigree is shown here. Is this pedigree consistent with autosomal dominant inheritance? If not, what mode of inheritance is most likely? Explain your reasoning.

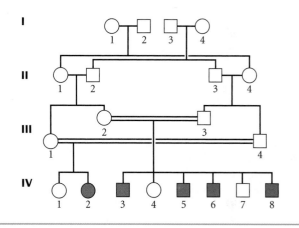

Challenge Questions

Section 3.2

*33. A geneticist discovers an obese mouse in his laboratory colony. He breeds this obese mouse with a normal mouse. All the F_1 mice from this cross are normal in size. When he interbreeds two F_1 mice, eight of the F_2 mice are normal in size and two are obese. The geneticist then intercrosses two of his obese mice, and he finds that all of the progeny from this cross are obese. These results lead the geneticist to conclude that obesity in mice results from a recessive allele.

A second geneticist at a different university also discovers an obese mouse in her laboratory colony. She carries out the same crosses as those done by the first geneticist and obtains the same results. She also concludes that obesity in mice results from a recessive allele. One day the two geneticists meet at a genetics conference, learn of

each other's experiments, and decide to exchange mice. They both find that, when they cross two obese mice from the different laboratories, all the offspring are normal; however, when they cross two obese mice from the same laboratory, all the offspring are obese. Explain their results.

34. Albinism in humans is a recessive trait (see the introduction to Chapter 1). A geneticist studies a series of families in which both parents are normal and at least one child has albinism. The geneticist reasons that both parents in these families must be heterozygotes and that albinism should appear in $\frac{1}{4}$ of the children of these families. To his surprise, the geneticist finds that the frequency of albinism among the children of these families is considerably greater than $\frac{1}{4}$. Can you think of an explanation for the higher-than-expected frequency of albinism among these families?

Extensions and Modifications of Basic Principles

Cuénot's Odd Yellow Mice

At the start of the twentieth century, Mendel's work on inheritance in pea plants became widely known (see Chapter 3), and a number of biologists set out to verify his conclusions by conducting crosses with other organisms. One of these biologists was Lucien Cuénot, a French scientist working at the University of Nancy. Cuénot experimented with coat colors in mice and was among the first to show that Mendel's principles applied to animals.

Cuénot observed that the coat colors of his mice followed the same patterns of inheritance that Mendel had observed in his pea plants. Cuénot found that, when he crossed pure-breeding gray mice with pure-breeding white mice, all of the F_1 progeny were gray, and interbreeding the F_1 produced a 3 : 1 ratio of gray and white mice in the F_2, as would be expected if gray were dominant over white. The results of Cuénot's breeding experiments perfectly fit Mendel's rules—with one exception. His crosses of yellow mice suggested that yellow coat color was dominant over gray, but he was never able to obtain true-breeding (homozygous) yellow mice. Whenever Cuénot crossed two yellow mice, he obtained yellow and gray mice in approximately a 3 : 1 ratio, suggesting that the yellow mice were heterozygous ($Yy \times Yy \rightarrow {}^{3}/_{4}\ Y\_$ and ${}^{1}/_{4}\ yy$). If yellow were indeed dominant over gray, some of the yellow progeny from this cross should have been homozygous for yellow (YY) and crossing two of these mice should have yielded all yellow offspring ($YY \times YY \rightarrow YY$). However, he never obtained all yellow progeny in his crosses. Cuénot was puzzled by these results, which failed to conform to Mendel's predications. He speculated that yellow gametes were incompatible with each other and would not fuse to form a zygote. Other biologists thought that additional factors might affect the inheritance of the yellow coat color, but the genetics of the yellow mice remained a mystery.

Yellow coat color in mice is caused by a recessive lethal gene, producing distorted phenotypic ratios in the progeny of two yellow mice. William Castle and Clarence Little discovered the lethal nature of the yellow gene in 1910. *[Reprinted with permission of Dr. Loan Phan and* In Vivo, *a publication of Columbia University Medical Center.]*

In 1910, William Ernest Castle and his student Clarence Little solved the mystery of Cuénot's unusual results. They carried out a large series of crosses between two yellow mice and showed that the progeny appeared, not in the 3 : 1 ratio that Cuénot thought he had observed but actually in a 2 : 1 ratio of yellow and nonyellow. Castle and Little recognized that the allele for yellow was lethal when homozygous (**Figure 4.1**), and thus all the yellow mice were heterozygous (Yy). A cross between two yellow heterozygous mice produces an initial genotypic ratio of ${}^{1}/_{4}\ YY$, ${}^{1}/_{4}\ Yy$, and ${}^{1}/_{4}\ yy$, but the homozygous YY mice die early in development and do not appear among the progeny, resulting in a 2 : 1 ratio of Yy (yellow) to yy (nonyellow) in offspring. Indeed, Castle and Little found that crosses of yellow × yellow mice resulted in smaller litters compared with litters of

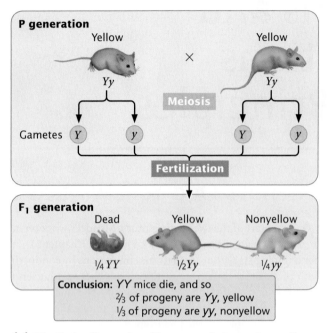

4.1 The 2 : 1 ratio produced by a cross between two yellow mice results from a lethal allele.

yellow × nonyellow mice. Because only mice homozygous for the Y allele die, the *yellow* allele is a recessive lethal. The *yellow* allele in mice is unusual in that it acts as a *recessive* allele in its effect on development but acts as a *dominant* allele in its effect on coat color.

Cuénot went on to make a number of other important contributions to genetics. He was the first to propose that more than two alleles could exist at a single locus, and he described how genes at different loci could interact in the determination of coat color in mice (aspects of inheritance that we will consider in this chapter). He observed that some types of cancer in mice display a hereditary predisposition; he also proposed, far ahead of his time, that genes might encode enzymes. Unfortunately, Cuénot's work brought him little recognition in his lifetime and was not well received by other French biologists, many of them openly hostile to the idea of Mendelian genetics. Cuénot's studies were interrupted by World War I, when foreign troops occupied his town and he was forced to abandon his laboratory at the university. He later returned to find his stocks of mice destroyed, and he never again took up genetic investigations.

In Chapter 3, we studied Mendel's principles of segregation and independent assortment and saw how these principles explain much about the nature of inheritance. After Mendel's principles were rediscovered in 1900, biologists began to conduct genetic studies on a wide array of different organisms. As they applied Mendel's principles more widely, exceptions were observed, and it became necessary to devise extensions to his basic principles of heredity. Like a number of other genetic phenomena, the lethal *yellow* gene discovered by Cuénot does not produce the ratios predicted by Mendel's principles of heredity. This lack of adherence to Mendel's rules doesn't mean that Mendel was wrong; rather, it demonstrates that Mendel's principles are not, by themselves, sufficient to explain the inheritance of all genetic characteristics. Our modern understanding of genetics has been greatly enriched by the discovery of a number of modifications and extensions of Mendel's basic principles, which are the focus of this chapter.

4.1 Sex Is Determined by a Number of Different Mechanisms

One of the first extensions of Mendel's principles is the inheritance of characteristics encoded by genes located on the sex chromosomes, which often differ in males and females (**Figure 4.2**). To understand the inheritance of sex-linked characteristics, we must first know how sex is determined—why some members of a species are male and others are female.

Sexual reproduction is the formation of offspring that are genetically distinct from their parents; most often, two parents contribute genes to their offspring and the genes are assorted into new combinations through meiosis. Among most eukaryotes, sexual reproduction consists of two

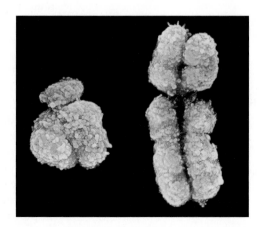

4.2 The sex chromosomes of males (Y, at the left) and females (X, at the right) differ in size and shape.
[*Biophoto Associates/Photo Researchers.*]

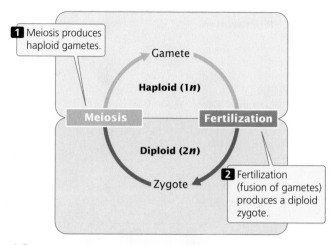

1 Meiosis produces haploid gametes.

Gamete

Haploid (1*n*)

Meiosis   Fertilization

Diploid (2*n*)

Zygote

2 Fertilization (fusion of gametes) produces a diploid zygote.

4.3 In most eukaryotic organisms, sexual reproduction consists of an alternation of haploid (1*n*) and diploid (2*n*) cells.

processes that lead to an alternation of haploid and diploid cells: meiosis produces haploid gametes (or spores in plants), and fertilization produces diploid zygotes (**Figure 4.3**).

The term **sex** refers to sexual phenotype. Most organisms have only two sexual phenotypes: male and female. The fundamental difference between males and females is gamete size: males produce small gametes; females produce relatively larger gametes (**Figure 4.4**).

The mechanism by which sex is established is termed **sex determination**. We define the sex of an individual organism in reference to its phenotype. Sometimes an individual organism has chromosomes or genes that are normally associated with one sex but a morphology corresponding to the opposite sex. For instance, the cells of female humans normally have two X chromosomes, and the cells of males have one X chromosome and one Y chromosome. A few rare persons have male anatomy, although their cells each contain two X chromosomes. Even though these people are genetically female, we refer to them as male because their sexual

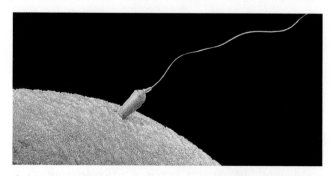

4.4 Male and female gametes (sperm and egg, respectively) differ in size. In this photograph, a human sperm (with flagellum) penetrates a human egg cell. *[Francis Leroy, Biocosmos/Science Photo Library/Photo Researchers.]*

phenotype is male. (As we will see later in the chapter, these XX males usually have a small piece of the Y chromosome, which is attached to another chromosome.)

Concepts

In sexual reproduction, parents contribute genes to produce an offspring that is genetically distinct from both parents. In most eukaryotes, sexual reproduction consists of meiosis, which produces haploid gametes (or spores), and fertilization, which produces a diploid zygote.

✔ Concept Check 1

What process causes the genetic variation seen in offspring produced by sexual reproduction?

Chromosomal Sex-Determining Systems

Sex in many organisms is determined by a pair of chromosomes, the **sex chromosomes**, which differ between males and females. The nonsex chromosomes, which are the same for males and females, are called **autosomes**. We think of sex in these organisms as being determined by the presence of the sex chromosomes, but, in fact, the individual genes located on the sex chromosomes are usually responsible for the sexual phenotypes.

XX-XO sex determination   In some insects, sex is determined by the XX-XO system. In this system, females have two X chromosomes (XX), and males possess a single X chromosome (XO). There is no O chromosome; the letter O signifies the absence of a sex chromosome.

In meiosis in females, the two X chromosomes pair and then separate, with one X chromosome entering each haploid egg. In males, the single X chromosome segregates in meiosis to half the sperm cells—the other half receive no sex chromosome. Because males produce two different types of gametes with respect to the sex chromosomes, they are said to be the **heterogametic sex**. Females, which produce gametes that are all the same with respect to the sex chromosomes, are the **homogametic sex**. In the XX-XO system, the sex of an individual organism is therefore determined by which type of male gamete fertilizes the egg. X-bearing sperm unite with X-bearing eggs to produce XX zygotes, which eventually develop as females. Sperm lacking an X chromosome unite with X-bearing eggs to produce XO zygotes, which develop into males.

XX-XY sex determination   In many species, the cells of males and females have the same number of chromosomes, but the cells of females have two X chromosomes (XX) and the cells of males have a single X chromosome and a smaller sex chromosome called the Y chromosome (XY). In humans

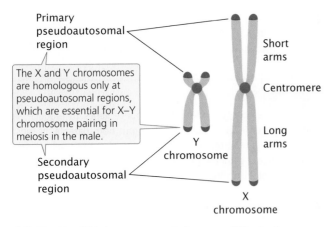

4.5 **The X and Y chromosomes in humans differ in size and genetic content.** They are homologous only at the pseudoautosomal regions.

and many other organisms, the Y chromosome is acrocentric (**Figure 4.5**), not Y shaped as is commonly assumed. In this type of sex-determining system, the male is the heterogametic sex—half of his gametes have an X chromosome and half have a Y chromosome. The female is the homogametic sex—all her egg cells contain a single X chromosome. A sperm containing a Y chromosome unites with an X-bearing egg to produce an XY male, whereas a sperm containing an X chromosome unites with an X-bearing egg to produce an XX female, which accounts for the 50 : 50 sex ratio observed in most organisms (**Figure 4.6**). Many organisms, including some plants, insects, and reptiles and all mammals (including humans), have the XX-XY sex-determining system. Other organisms have odd variations of the XX-XY system of sex determination, including the duck-billed platypus, in which females have five pairs of X chromosomes and males have five pairs of X and Y chromosomes.

Although the X and Y chromosomes are not generally homologous, they do pair and segregate into different cells in meiosis. They can pair because these chromosomes are homologous at small regions called the **pseudoautosomal regions** (see Figure 4.5), in which they carry the same genes. In humans, there are pseudoautosomal regions at both tips of the X and Y chromosomes.

ZZ-ZW sex determination   In this system, the female is heterogametic and the male is homogametic. To prevent confusion with the XX-XY system, the sex chromosomes in this system are labeled Z and W, but the chromosomes do not resemble Zs and Ws. Females in this system are ZW; after meiosis, half of the eggs have a Z chromosome and the other half have a W chromosome. Males are ZZ; all sperm contain a single Z chromosome. The ZZ-ZW system is found in birds, snakes, butterflies, some amphibians, and some fishes. It is also found in some isopods, commonly known as pill bugs or rolly-pollies.

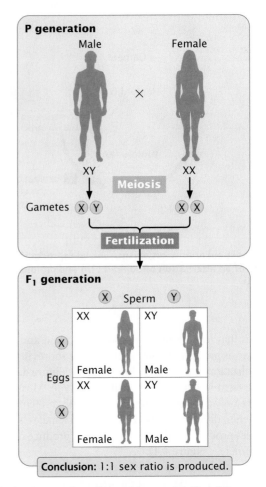

4.6 **Inheritance of sex in organisms with X and Y chromosomes results in equal numbers of male and female offspring.**

Genic Sex-Determining Systems

In some plants and protozoans, sex is genetically determined, but there are no obvious differences in the chromosomes of males and females: there are no sex chromosomes. These organisms have **genic sex determination**; genotypes at one or more loci determine the sex of an individual plant or protozoan.

It is important to understand that, even in chromosomal sex-determining systems, sex is actually determined by individual genes. For example, in mammals, a gene (*SRY*, discussed later in this chapter) located on the Y chromosome determines the male phenotype. In both genic sex determination and chromosomal sex determination, sex is controlled by individual genes; the difference is that, with chromosomal sex determination, the chromosomes that carry those genes look different in males and females.

Environmental Sex Determination

Genes have had a role in all of the examples of sex determination discussed thus far, but sex is determined fully or in part by environmental factors in a number of organisms. For example, environmental factors are important in determining sex in many reptiles. Although most snakes and lizards have sex chromosomes, the sexual phenotype of many turtles, crocodiles, and alligators is affected by temperature during embryonic development. In turtles, for example, warm temperatures produce females during certain times of the year, whereas cool temperatures produce males. In alligators, the reverse is true.

Now that we have surveyed some of the different ways that sex can be determined, we will examine one mechanism in detail: the XX-XY system. Both fruit flies and humans possess XX-XY sex determination but, as we will see, the way in which the X and Y chromosomes determine sex in these two organisms is quite different.

Concepts

In genic sex determination, sex is determined by genes at one or more loci, but there are no obvious differences in the chromosomes of males and females. In environmental sex determination, sex is determined fully or in part by environmental factors.

✔ Concept Check 2

How do chromosomal, genic, and environmental sex determination differ?

Sex Determination in *Drosophila melanogaster*

The fruit fly *Drosophila melanogaster* has eight chromosomes: three pairs of autosomes and one pair of sex chromosomes. Thus, it has two haploid sets of autosomes and two sex chromosomes, one set of autosomes and one sex chromosome inherited from each parent. Normally, females have two X chromosomes and males have an X chromosome and a Y chromosome. However, the presence of the Y chromosome does not determine maleness in *Drosophila*; instead, each fly's sex is determined by a balance between genes on the autosomes and genes on the X chromosome. This type of sex determination is called the **genic balance system**. In this system, a number of different genes influence sexual development. The X chromosome contains genes with female-producing effects, whereas the autosomes

Table 4.1	Chromosome complements and sexual phenotypes in *Drosophila*		
Sex-Chromosome Complement	Haploid Sets of Autosomes	X : A Ratio	Sexual Phenotype
XX	AA	1.0	Female
XY	AA	0.5	Male
XO	AA	0.5	Male
XXY	AA	1.0	Female
XXX	AA	1.5	Metafemale
XXXY	AA	1.5	Metafemale
XX	AAA	0.67	Intersex
XO	AAA	0.33	Metamale
XXXX	AAA	1.3	Metafemale

contain genes with male-producing effects. Consequently, a fly's sex is determined by the **X : A ratio**, the number of X chromosomes divided by the number of haploid sets of autosomal chromosomes.

An X : A ratio of 1.0 produces a female fly; an X : A ratio of 0.5 produces a male. If the X : A ratio is less than 0.5, a male phenotype is produced, but the fly is weak and sterile—such flies are sometimes called metamales. An X : A ratio between 1.0 and 0.5 produces an intersex fly, with a mixture of male and female characteristics. If the X : A ratio is greater than 1.0, a female phenotype is produced, but this fly (called a metafemale) has serious developmental problems and many never complete development. **Table 4.1** presents some different chromosome complements in *Drosophila* and their associated sexual phenotypes. Normal females have two X chromosomes and two sets of autosomes (XX, AA), and so their X : A ratio is 1.0. Males, on the other hand, normally have a single X and two sets of autosomes (XY, AA), and so their X : A ratio is 0.5. Flies with XXY sex chromosomes and two sets of autosomes (an X : A ratio of 1.0) develop as fully fertile females, in spite of the presence of a Y chromosome. Flies with only a single X and two sets of autosomes (XO, AA, for an X : A ratio of 0.5) develop as males, although they are sterile. These observations confirm that the Y chromosome does not determine sex in *Drosophila*.

Concepts

The sexual phenotype of a fruit fly is determined by the ratio of the number of X chromosomes to the number of haploid sets of autosomal chromosomes (the X : A ratio).

✔ Concept Check 3

What will be the sexual phenotype of a fruit fly with XXYYY sex chromosomes and two sets of autosomes?

a. Male   c. Intersex

b. Female   d. Metamale

Sex Determination in Humans

Humans, like *Drosophila*, have XX-XY sex determination, but, in humans, the presence of a gene (*SRY*) on the Y chromosome determines maleness. The phenotypes that result from abnormal numbers of sex chromosomes, which arise when the sex chromosomes do not segregate properly in meiosis or mitosis, illustrate the importance of the Y chromosome in human sex determination.

Turner syndrome   Persons who have **Turner syndrome** are female and often have underdeveloped secondary sex characteristics. This syndrome is seen in 1 of 3000 female births. Affected women are frequently short and have a low hairline, a relatively broad chest, and folds of skin on the neck. Their intelligence is usually normal. Most women who have Turner syndrome are sterile. In 1959, Charles Ford used new techniques to study human chromosomes and discovered that cells from a 14-year-old girl with Turner syndrome had only a single X chromosome; this chromosome complement is usually referred to as XO.

There are no known cases in which a person is missing both X chromosomes, an indication that at least one X chromosome is necessary for human development. Presumably, embryos missing both Xs are spontaneously aborted in the early stages of development.

Klinefelter syndrome   Persons who have **Klinefelter syndrome**, which occurs with a frequency of about 1 in 1000 male births, have cells with one or more Y chromosomes and multiple X chromosomes. The cells of most males having this condition are XXY, but cells of a few Klinefelter males are XXXY, XXXXY, or XXYY. Persons with this condition are male, frequently with small testes and reduced facial and pubic hair. They are often taller than normal and sterile; most have normal intelligence.

Poly-X females   In about 1 in 1000 female births, the infant's cells possess three X chromosomes, a condition often referred to as **triplo-X syndrome**. These persons have no distinctive features other than a tendency to be tall and thin. Although a few are sterile, many menstruate regularly and are fertile. The incidence of mental retardation among triple-X females is slightly greater than that in the general population, but most XXX females have normal intelligence. Much rarer are females whose cells contain four or five X chromosomes. These women usually have normal female anatomy but are mentally retarded and have a number of physical problems. The severity of mental retardation increases as the number of X chromosomes increases beyond three.

The male-determining gene in humans   The phenotypes associated with sex-chromosome anomalies show that the Y chromosome in humans and all other mammals is of paramount importance in producing a male phenotype. However, scientists discovered a few rare XX males whose cells apparently lack a Y chromosome. For many years, these males presented a real enigma: How could a male phenotype exist without a Y chromosome? Close examination eventually revealed a small part of the Y chromosome attached to another chromosome. This finding indicates that it is not the entire Y chromosome that determines maleness in humans; rather, it is a gene on the Y chromosome.

Early in development, all humans possess undifferentiated gonads and both male and female reproductive ducts. Then, about 6 weeks after fertilization, a gene on the Y chromosome becomes active. By an unknown mechanism, this gene causes the neutral gonads to develop into testes, which begin to secrete two hormones: testosterone and Mullerian-inhibiting substance. Testosterone induces the development of male characteristics, and Mullerian-inhibiting substance causes the degeneration of the female reproductive ducts. In the absence of this male-determining gene, the neutral gonads become ovaries, and female features develop.

The male-determining gene in humans, called the **sex-determining region Y** (*SRY*) **gene**, was discovered in 1990 (**Figure 4.7**). This gene is found in XX males and is missing from all XY females; it is also found on the Y chromosome of all mammals examined to date. Definitive proof that *SRY* is the male-determining gene came when scientists placed a copy of this gene into XX mice by means of genetic engineering. The XX mice that received this gene, although sterile, developed into anatomical males. Although *SRY* is the primary determinant of maleness in humans, other genes (some X linked, others Y linked, and still others autosomal) also play a role in fertility and the development of sex differences.

4.7 The *SRY* gene is on the Y chromosome and causes the development of male characteristics.

Concepts

The presence of the *SRY* gene on the Y chromosome causes a human embryo to develop as a male. In the absence of this gene, a human embryo develops as a female.

✔ Concept Check 4

In humans, what will be the phenotype of a person with XXXY sex chromosomes?

a. Klinefelter syndrome

b. Turner syndrome

c. Poly-X female

4.2 Sex-Linked Characteristics Are Determined by Genes on the Sex Chromosomes

In Chapter 3, we learned several basic principles of heredity that Mendel discovered from his crosses among pea plants. A major extension of these Mendelian principles is the pattern of inheritance exhibited by **sex-linked characteristics**, characteristics determined by genes located on the sex chromosomes. Genes on the X chromosome determine **X-linked characteristics**; those on the Y chromosome determine **Y-linked characteristics**. Because the Y chromosome of many organisms contains little genetic information, most sex-linked characteristics are X linked. Males and females differ in their sex chromosomes; so the pattern of inheritance for sex-linked characteristics differs from that exhibited by genes located on autosomal chromosomes.

X-Linked White Eyes in *Drosophila*

The first person to explain sex-linked inheritance was American biologist Thomas Hunt Morgan (**Figure 4.8**).

Morgan began his career as an embryologist, but the discovery of Mendel's principles inspired him to begin conducting genetic experiments, initially on mice and rats. In 1909, Morgan switched to *Drosophila melanogaster*; a year later, he discovered among the flies of his laboratory colony a single male that possessed white eyes, in stark contrast with the red eyes of normal fruit flies. This fly had a tremendous effect on the future of genetics and on Morgan's career as a biologist.

To explain the inheritance of the white-eyed characteristic in fruit flies, Morgan systematically carried out a series of genetic crosses. First, he crossed pure-breeding, red-eyed females with his white-eyed male, producing F_1 progeny of which all had red eyes (**Figure 4.9a**). (In fact, Morgan found 3 white-eyed males among the 1237 progeny, but he assumed that the white eyes were due to new mutations.) Morgan's results from this initial cross were consistent with Mendel's principles: a cross between a homozygous dominant individual and a homozygous recessive individual produces heterozygous offspring exhibiting the dominant trait. His results suggested that white eyes are a simple recessive trait. However, when Morgan crossed the F_1 flies with one another, he found that all the female F_2 flies possessed red eyes but that half the male F_2 flies had red eyes and the other half had white eyes. This finding was clearly not the expected result for a simple recessive trait, which should appear in $\frac{1}{4}$ of both male and female F_2 offspring.

To explain this unexpected result, Morgan proposed that the locus affecting eye color is on the X chromosome (i.e., eye color is X linked). He recognized that the eye-color alleles are present only on the X chromosome; no homologous allele is present on the Y chromosome. Because the cells of females possess two X chromosomes, females can be homozygous or heterozygous for the eye-color alleles. The cells of males, on the other hand, possess only a single X chromosome and can carry only a single eye-color allele. Males therefore cannot be either homozygous or heterozygous but are said to be **hemizygous** for X-linked loci.

(a)   **(b)**

4.8 Thomas Hunt Morgan's work with *Drosophila* **helped unravel many basic principles of genetics, including X-linked inheritance.** (a) Morgan. (b) The Fly Room, where Morgan and his students conducted genetic research. *[Part a: AP/Wide World Photos. Part b: American Philosophical Society.]*

Experiment

Question: Are white eyes in fruit flies inherited as an autosomal recessive trait?

Methods Perform reciprocal crosses.

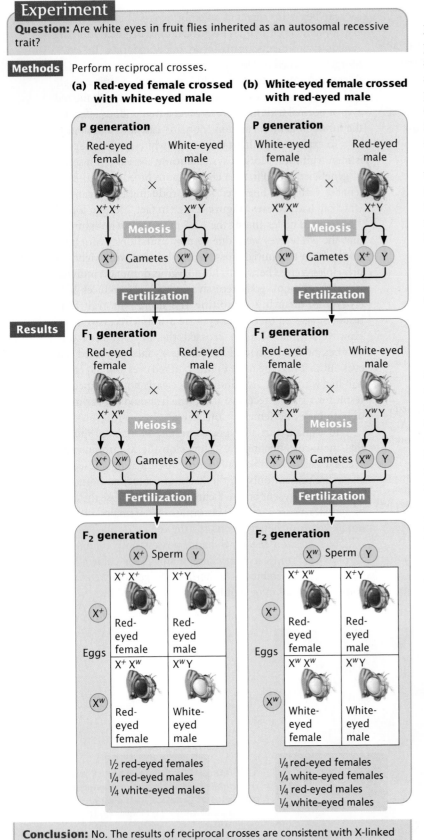

(a) Red-eyed female crossed with white-eyed male

(b) White-eyed female crossed with red-eyed male

Results

(a)
½ red-eyed females
¼ red-eyed males
¼ white-eyed males

(b)
¼ red-eyed females
¼ white-eyed females
¼ red-eyed males
¼ white-eyed males

Conclusion: No. The results of reciprocal crosses are consistent with X-linked inheritance.

4.9 Morgan's X-linked crosses for white eyes in fruit flies. (a) Original and F₁ crosses. (b) Reciprocal crosses.

To verify his hypothesis that the white-eye trait is X linked, Morgan conducted additional crosses. He predicted that a cross between a white-eyed female and a red-eyed male would produce all red-eyed females and all white-eyed males (**Figure 4.9b**). When Morgan performed this cross, the results were exactly as predicted. Note that this cross is the reciprocal of the original cross and that the two reciprocal crosses produced different results in the F_1 and F_2 generations. Morgan also crossed the F_1 heterozygous females with their white-eyed father, the red-eyed F_2 females with white-eyed males, and white-eyed females with white-eyed males. In all of these crosses, the results were consistent with Morgan's conclusion that white eyes is an X-linked characteristic.

Model Genetic Organism
The Fruit Fly *Drosophila melanogaster*

Drosophila melanogaster, a fruit fly (**Figure 4.10**), was among the first organisms used for genetic analysis and, today, it is one of the most widely used and best known genetically of all eukaryotic organisms. It has played an important role in studies of linkage, epistasis, chromosome genetics, development, behavior, and evolution. Because all organisms use a common genetic system, understanding a process such as replication or transcription in fruit flies helps us to understand these same processes in humans and other eukaryotes.

Drosophila is a genus of more than 1000 described species of small flies (about 1 to 2 mm in length) that frequently feed and reproduce on fruit, although they rarely cause damage and are not considered economic pests. The best known and most widely studied of the fruit flies is *D. melanogaster*, but genetic studies have also been extended to many other species of the genus.

D. melanogaster first began to appear in biological laboratories about 1900. After first taking up breeding experiments with mice and rats, as mentioned earlier, Thomas Hunt Morgan began using fruit flies in experimental studies of heredity at Columbia University.

Morgan's laboratory, located on the top floor of Schermerhorn Hall, became known as the Fly Room (see Figure 4.8b). To say that the Fly Room was unimpressive is an understatement. The cramped room, only about 16 by 23 feet, was filled with eight desks, each occupied by a student and his experiments. The primitive laboratory equipment consisted of little more than milk bottles for rearing the flies and hand-held lenses for observing their traits. Later, microscopes replaced the hand-held lenses, and crude incubators were added to maintain the fly cultures, but even these additions did little to increase the physical

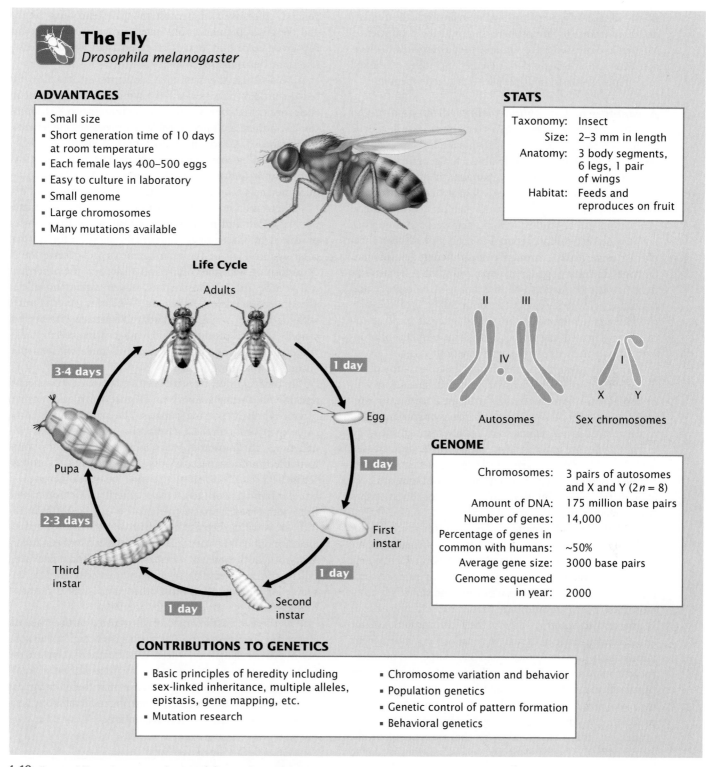

The Fly
Drosophila melanogaster

ADVANTAGES

- Small size
- Short generation time of 10 days at room temperature
- Each female lays 400–500 eggs
- Easy to culture in laboratory
- Small genome
- Large chromosomes
- Many mutations available

STATS

Taxonomy:	Insect
Size:	2–3 mm in length
Anatomy:	3 body segments, 6 legs, 1 pair of wings
Habitat:	Feeds and reproduces on fruit

Life Cycle

Adults

3-4 days

1 day

Egg

1 day

First instar

1 day

Second instar

1 day

Third instar

2-3 days

Pupa

II   III

IV

I

X   Y

Autosomes   Sex chromosomes

GENOME

Chromosomes:	3 pairs of autosomes and X and Y ($2n = 8$)
Amount of DNA:	175 million base pairs
Number of genes:	14,000
Percentage of genes in common with humans:	~50%
Average gene size:	3000 base pairs
Genome sequenced in year:	2000

CONTRIBUTIONS TO GENETICS

- Basic principles of heredity including sex-linked inheritance, multiple alleles, epistasis, gene mapping, etc.
- Mutation research
- Chromosome variation and behavior
- Population genetics
- Genetic control of pattern formation
- Behavioral genetics

4.10 *Drosophila melanogaster* is a model genetic organism.

sophistication of the laboratory. Morgan and his students were not tidy: cockroaches were abundant (living off spilled *Drosophila* food), dirty milk bottles filled the sink, ripe bananas—food for the flies—hung from the ceiling, and escaped fruit flies hovered everywhere.

In spite of its physical limitations, the Fly Room was the source of some of the most important research in the history of biology. There was daily excitement among the students, some of whom initially came to the laboratory as undergraduates. The close quarters facilitated informality and the free flow of ideas. Morgan and the Fly Room illustrate the tremendous importance of "atmosphere" in producing good science. Morgan and his students eventually used *Drosophila* to elucidate many basic principles of heredity, including sex-linked inheritance, epistasis, multiple alleles, and gene mapping.

Advantages of Drosophila melanogaster as a model genetic organism   *Drosophila*'s widespread use in genetic studies is no accident. The fruit fly has a number of characteristics that make it an ideal subject for genetic investigations. Compared with other organisms, it has a relatively short generation time; fruit flies will complete an entire generation in about 10 days at room temperature, and so several generations can be studied within a few weeks. Although *D. melanogaster* has a short generation time, it possesses a complex life cycle, passing through several different developmental stages, including egg, larva, pupa, and adult. A female fruit fly is capable of mating within 8 hours of emergence and typically begins to lay eggs after about 2 days. Fruit flies also produce a large number of offspring, laying as many as 400 to 500 eggs in a 10-day period. Thus, large numbers of progeny can be obtained from a single genetic cross.

Another advantage is that fruit flies are easy to culture in the laboratory. They are usually raised in small glass vials or bottles (milk bottles were originally used) with easily prepared, pastelike food consisting of bananas or corn meal and molasses. Males and females are readily distinguished and virgin females are easily isolated, facilitating genetic crosses. The flies are small, requiring little space—several hundred can be raised in a small half-pint bottle—but they are large enough for many mutations to be easily observed with a hand lens or a dissecting microscope.

Finally, *D. melanogaster* is an organism of choice for many geneticists because it has a relatively small genome consisting of 175 million base pairs of DNA, which is only about 5% of the human genome. It has four pairs of chromosomes: three pairs of autosomes and one pair of sex chromosomes. The X chromosome (designated chromosome 1) is large and acrocentric, whereas the Y chromosome is large and submetacentric, although it contains very little genetic information. Chromosomes 2 and 3 are large and metacentric; chromosome 4 is a very small acrocentric chromosome. In the salivary glands, the chromosomes are very large, making *Drosophila* an excellent subject for chromosome studies. In 2000, the complete genome of *D. melanogaster* was sequenced, followed in 2005 by the sequencing of the genome of *D. pseudoobscura*. *Drosophila* continues today to be one of the most versatile and powerful of all genetic model organisms. ∎

X-Linked Color Blindness in Humans

To further examine X-linked inheritance, let's consider another X-linked characteristic: red–green color blindness in humans. Mutations that produce defective color vision are generally recessive and, because the genes encoding the red and the green pigments are located on the X chromosome, red–green color blindness is inherited as an X-linked recessive characteristic.

We will use the symbol X^c to represent an allele for red–green color blindness and the symbol X^+ to represent an allele for normal color vision. Females possess two X chromosomes; so there are three possible genotypes among females: X^+X^+ and X^+X^c, which produce normal vision, and X^cX^c, which produces color blindness. Males have only a single X chromosome and two possible genotypes: X^+Y, which produces normal vision, and X^cY which produces color blindness.

If a woman homozygous for normal color vision mates with a color-blind man (**Figure 4.11a**), all of the gametes produced by the woman will contain an allele for normal color vision. Half of the man's gametes will receive the X chromosome with the color-blind allele, and the other half will receive the Y chromosome, which carries no alleles affecting color vision. When an X^c-bearing sperm unites with the X^+-bearing egg, a heterozygous female with normal vision (X^+X^c) is produced. When a Y-bearing sperm unites with the X-bearing egg, a hemizygous male with normal vision (X^+Y) is produced.

In the reciprocal cross between a color-blind woman and a man with normal color vision (**Figure 4.11b**), the woman produces only X^c-bearing gametes. The man produces some gametes that contain the X chromosome and others that contain the Y chromosome. Males inherit the X chromosome from their mothers; because both of the mother's X chromosomes bear the X^c allele in this case, all the male offspring will be color blind. In contrast, females inherit an X chromosome from both parents; thus all the female offspring of this reciprocal cross will be heterozygous with normal vision. Females are color blind only when color-blind alleles have been inherited from both parents, whereas a color-blind male need inherit a color-blind allele from his mother only; for this reason, color blindness and most other rare X-linked recessive characteristics are more common in males.

In these crosses for color blindness, notice that an affected woman passes the X-linked recessive trait to her sons but not to her daughters, whereas an affected man passes the trait to his grandsons through his daughters but never to his sons. X-linked recessive characteristics may therefore appear to alternate between the sexes, appearing in females one generation and in males the next generation.

Worked Problem

Now that we understand the pattern of X-linked inheritance, let's apply our knowledge to answer a specific question in regard to X-linked inheritance of color blindness in humans.

Betty has normal vision, but her mother is color blind. Bill is color blind. If Bill and Betty marry and have a child together, what is the probability that the child will be color blind?

(a) Normal female and color-blind male

(b) Reciprocal cross

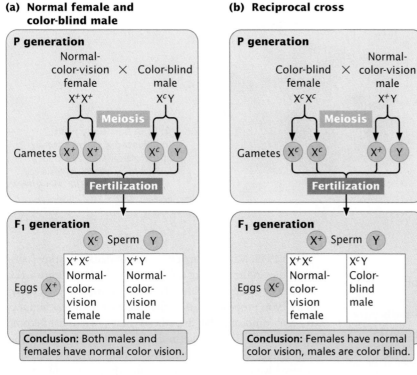

Conclusion: Both males and females have normal color vision.

Conclusion: Females have normal color vision, males are color blind.

4.11 Red–green color blindness is inherited as an X-linked recessive trait in humans.

• Solution

Because color blindness is an X-linked recessive characteristic, Betty's color-blind mother must be homozygous for the color-blind allele ($X^c X^c$). Females inherit one X chromosome from each of their parents; so Betty must have inherited a color-blind allele from her mother. Because Betty has normal color vision, she must have inherited an allele for normal vision (X^+) from her father; thus Betty is heterozygous ($X^+ X^c$). Bill is color blind. Because males are hemizygous for X-linked alleles, he must be ($X^c Y$). A mating between Betty and Bill is represented as:

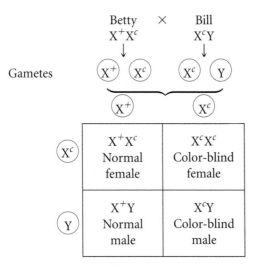

Thus, $\frac{1}{4}$ of the children are expected to be female with normal color vision, $\frac{1}{4}$ female with color blindness, $\frac{1}{4}$ male with normal color vision, and $\frac{1}{4}$ male with color blindness.

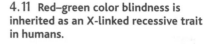

? Get some additional practice with X-linked inheritance by working Problem 12 at the end of this chapter.

Concepts

Characteristics determined by genes on the sex chromosomes are called sex-linked characteristics. Diploid females have two alleles at each X-linked locus, whereas diploid males possess a single allele at each X-linked locus. Females inherit X-linked alleles from both parents, but males inherit a single X-linked allele from their mothers.

✔ Concept Check 5

Hemophilia (reduced blood clotting) is an X-linked recessive disease in humans. A woman with hemophilia mates with a man who exhibits normal blood clotting. What is the probability that their child will have hemophilia?

(a)   **(b)**

4.12 A Barr body is an inactivated X chromosome. (a) Female cell with a Barr body (indicated by arrow). (b) Male cell without a Barr body. *[M. Abbey/Photo Researchers.]*

Symbols for X-Linked Genes

There are several different ways to record genotypes for X-linked traits. Sometimes the genotypes are recorded in the same fashion as for autosomal characteristics—the hemizygous males are simply given a single allele: the genotype of a female *Drosophila* with white eyes would be *ww*, and the genotype of a white-eyed hemizygous male would be *w*. Another method is to include the Y chromosome, designating it with a diagonal slash (/). With this method, the white-eyed female's genotype would still be *ww* and the white-eyed male's genotype would be *w/*. Perhaps the most useful method is to write the X and Y chromosomes in the genotype, designating the X-linked alleles with superscripts, as done in this chapter. With this method, a white-eyed female would be $X^w X^w$ and a white-eyed male $X^w Y$. Using Xs and Ys in the genotype has the advantage of reminding us that the genes are X linked and that the male must always have a single allele, inherited from the mother.

Dosage Compensation

The presence of different numbers of X chromosomes in males and females presents a special problem in development. Because females have two copies of every X-linked gene and males possess one copy, the amount of gene product (protein) from X-linked genes would differ in the two sexes: females would produce twice as much gene product as that produced by males. This difference could be highly detrimental because protein concentration plays a critical role in development. Animals overcome this potential problem through **dosage compensation**, which equalizes the amount of protein produced by X-linked genes in the two sexes. In fruit flies, dosage compensation is achieved by a doubling of the activity of the genes on the X chromosome of the male. In the worm *Caenorhabditis elegans*, it is achieved by a halving of the activity of genes on both of the X chromosomes in the female. Placental mammals use yet another mechanism of dosage compensation: genes on one of the X chromosomes in the female are inactivated.

In 1949, Murray Barr observed condensed, darkly staining bodies in the nuclei of cells from female cats (**Figure 4.12**); this darkly staining structure became known as a **Barr body**. Mary Lyon proposed in 1961 that the Barr body was an inactive X chromosome; her hypothesis (now proved) has become known as the **Lyon hypothesis**. She suggested that, within each female cell, one of the two X chromosomes becomes inactive; which X chromosome is inactivated is random. If a cell contains more than two X chromosomes, all but one of them is inactivated. The number of Barr bodies present in human cells with different complements of sex chromosomes is shown in **Table 4.2**.

As a result of X inactivation, females are functionally hemizygous at the cellular level for X-linked genes. In females that are heterozygous at an X-linked locus, approxi-

Table 4.2	Number of Barr bodies in human cells with different complements of sex chromosomes	
Sex Chromosomes	Syndrome	Number of Barr Bodies
XX	None	1
XY	None	0
XO	Turner	0
XXY	Klinefelter	1
XXYY	Klinefelter	1
XXXY	Klinefelter	2
XXXXY	Klinefelter	3
XXX	Triplo-X	2
XXXX	Poly-X female	3
XXXXX	Poly-X female	4

mately 50% of the cells will express one allele and 50% will express the other allele; thus, in heterozygous females, proteins encoded by both alleles are produced, although not within the same cell. This functional hemizygosity means that cells in females are not identical with respect to the expression of the genes on the X chromosome; females are mosaics for the expression of X-linked genes.

Random X inactivation takes place early in development—in humans, within the first few weeks of development. After an X chromosome has become inactive in a cell, it remains inactive and is inactive in all somatic cells that descend from the cell. Thus, neighboring cells tend to have the same X chromosome inactivated, producing a patchy pattern (mosaic) for the expression of an X-linked characteristic in heterozygous females.

This patchy distribution can be seen in tortoiseshell (**Figure 4.13**) and calico cats. Although many genes contribute to coat color and pattern in domestic cats, a single X-linked locus determines the presence of orange color. There are two possible alleles at this locus: X^+, which produces nonorange (usually black) fur, and X^o, which produces orange fur. Males are hemizygous and thus may be black (X^+Y) or orange (X^oY) but not black and orange. (Rare tortoiseshell males can arise from the presence of two X chromosomes, X^+X^oY.) Females may be black (X^+X^+), orange (X^oX^o), or tortoiseshell (X^+X^o), the tortoiseshell pattern arising from a patchy mixture of black and orange fur. Each orange patch is a clone of cells derived from an original cell in which the black allele is inactivated, and each black patch is a clone of cells derived from an original cell in which the orange allele is inactivated.

4.13 The patchy distribution of color on tortoiseshell cats results from the random inactivation of one X chromosome in females. [*Chanan Photography.*]

Lyon's hypothesis suggests that the presence of variable numbers of X chromosomes should not be detrimental in mammals, because any X chromosomes in excess of one X chromosome should be inactivated. However, persons with Turner syndrome (XO) differ from normal females, and those with Klinefelter syndrome (XXY) differ from normal males. These disorders probably arise because some X-linked genes escape inactivation.

Concepts

In mammals, dosage compensation ensures that the same amount of X-linked gene product will be produced in the cells of both males and females. All but one X chromosome are inactivated in each cell; which of the X chromosomes is inactivated is random and varies from cell to cell.

✔ **Concept Check 6**

How many Barr bodies will a male with XXXYY chromosomes have in each of his cells? What are these Barr bodies?

Y-Linked Characteristics

Y-linked traits exhibit a distinct pattern of inheritance and are present only in males, because only males possess a Y chromosome. All male offspring of a male with a Y-linked trait will display the trait, because every male inherits the Y chromosome from his father.

Use of Y-linked genetic markers DNA sequences in the Y chromosome undergo mutation with the passage of time and vary among individual males. Like Y-linked traits, these variants—called genetic markers—are passed from father to son and can be used to study male ancestry. Although the markers themselves do not encode any physical traits, they can be detected with the use of molecular methods. Much of the Y chromosome is nonfunctional; so mutations readily accumulate. Many of these mutations are unique; they arise only once and are passed down through the generations without undergoing recombination. Individual males possessing the same set of mutations are therefore related, and the distribution of these genetic markers on Y chromosomes provides clues about genetic relationships of present-day people.

Y-linked markers have been used to study the offspring of Thomas Jefferson, principal author of the Declaration of Independence and third president of the United States. In 1802, Jefferson was accused by a political enemy of fathering a child by his slave Sally Hemings, but the evidence was circumstantial. Hemings, who worked in the Jefferson household and accompanied Jefferson on a trip to Paris, had five children. Jefferson was accused of fathering the first child, Tom, but rumors about the paternity of the other children circulated as well. Hemings's last child, Eston, bore a striking resemblance to Jefferson, and her fourth child, Madison, testified late in life that Jefferson was the father of all of Hemings's children. Ancestors of Hemings's children maintained that they were descendants of the Jefferson line, but some Jefferson descendants refused to recognize their claim.

To resolve this long-standing controversy, geneticists examined markers from the Y chromosomes of male-line descendants of Hemings's first son (Thomas Woodson), her last son (Eston Hemings), and a paternal uncle of Thomas Jefferson with whom Jefferson had Y chromosomes in common. (Descendants of Jefferson's uncle were used because Jefferson himself had no verified male descendants.) Geneticists determined that Jefferson possessed a rare and distinctive set of genetic markers on his Y chromosome. The same markers were also found on the Y chromosomes of the male-line descendants of Eston Hemings. The probability of such a match arising by chance is less than 1%. (The markers were not found on the Y chromosomes of the descendants of Thomas Woodson.) Together with the circumstantial historical evidence, these matching markers strongly suggest that Jefferson fathered Eston Hemings but not Thomas Woodson.

Concepts

Y-linked characteristics exhibit a distinct pattern of inheritance: they are present only in males, and all male offspring of a male with a Y-linked trait inherit the trait.

Connecting Concepts

Recognizing Sex-Linked Inheritance

What features should we look for to identify a trait as sex linked? A common misconception is that any genetic characteristic in which the phenotypes of males and females differ must be sex linked. In fact, the expression of many *autosomal* characteristics differs between males and females. The genes that encode these characteristics are the same in both sexes, but their expression is influenced by sex hormones. The different sex hormones of males and females cause the same genes to generate different phenotypes in males and females.

Another misconception is that any characteristic that is found more frequently in one sex is sex linked. A number of autosomal traits are expressed more commonly in one sex than in the other. These traits are said to be sex influenced. Some autosomal traits are expressed in only one sex; these traits are said to be sex limited. Both sex-influenced and sex-limited characteristics will be discussed in more detail later in the chapter.

Several features of sex-linked characteristics make them easy to recognize. Y-linked traits are found only in males, but this fact does not guarantee that a trait is Y linked, because some autosomal characteristics are expressed only in males. A Y-linked trait is unique, however, in that all the male offspring of an affected male will express the father's phenotype, and a Y-linked trait can be inherited only from the father's side of the family. Thus, a Y-linked trait can be inherited only from the paternal grandfather (the father's father), never from the maternal grandfather (the mother's father).

X-linked characteristics also exhibit a distinctive pattern of inheritance. X linkage is a possible explanation when the results of reciprocal crosses differ. If a characteristic is X linked, a cross between an affected male and an unaffected female will not give the same results as a cross between an affected female and an unaffected male. For almost all autosomal characteristics, the results of reciprocal crosses are the same. We should not conclude, however, that, when the reciprocal crosses give different results, the characteristic is X linked. Other sex-associated forms of inheritance, discussed later in the chapter, also produce different results in reciprocal crosses. The key to recognizing X-linked inheritance is to remember that a male always inherits his X chromosome from his mother, not from his father. Thus, an X-linked characteristic is not passed directly from father to son; if a male clearly inherits a characteristic from his father—and the mother is not heterozygous—it cannot be X linked.

4.3 Dominance, Penetrance, and Lethal Alleles Modify Phenotypic Ratios

A number of factors potentially modify the phenotypic ratios presented in Chapter 3. These factors include different types of dominance, variable penetrance, and lethal alleles.

Dominance Is Interaction Between Genes at the Same Locus

One of Mendel's important contributions to the study of heredity is the concept of *dominance*—the idea that an individual organism possesses two different alleles for a characteristic, but the trait encoded by only one of the alleles is observed in the phenotype. With dominance, the heterozygote possesses the same phenotype as one of the homozygotes. When biologists began to apply Mendel's principles to organisms other than peas, it quickly became apparent that many characteristics do not exhibit this type of dominance. Indeed, Mendel himself was aware that dominance is not universal, because he observed that a pea plant heterozygous for long and short flowering times had a flowering time that was intermediate between those of its homozygous parents. This situation, in which the heterozygote is intermediate in phenotype between the two homozygotes, is termed *incomplete dominance*. As discussed in Chapter 3, a cross between two individuals

4.14 The type of dominance exhibited by a trait depends on how the phenotype of the heterozygote relates to the phenotypes of the homozygotes.

heterozygous for an incompletely dominant trait produces a 1 : 2 : 1 phenotypic ratio in the offspring.

Dominance can be understood in regard to how the phenotype of the heterozygote relates to the phenotypes of the homozygotes. In the example presented in **Figure 4.14**, flower color potentially ranges from red to white. One homozygous genotype, A^1A^1, encodes red flowers, and another, A^2A^2, encodes white flowers. Where the heterozygote falls in the range of phenotypes determines the type of dominance. If the heterozygote (A^1A^2) has flowers that are the same color as those of the A^1A^1 homozygote (red), then the A^1 allele is *completely dominant* over the A^2 allele; that is, red is dominant over white. If, on the other hand, the heterozygote has flowers that are the same color as the A^2A^2 homozygote (white), then the A^2 allele is completely dominant, and white is dominant over red. When the heterozygote falls in between the phenotypes of the two homozygotes, dominance is incomplete. With incomplete dominance, the heterozygote need not be exactly intermediate (pink in our example) between the two homozygotes; it might be a slightly lighter shade of red or a slightly pink shade of white. As long as the heterozygote's phenotype can be differentiated and falls within the range of the two homozygotes, dominance is incomplete. The important thing to remember about dominance is that it affects the phenotype that genes produce, but not the way in which genes are *inherited*.

Another type of interaction between alleles is **codominance**, in which the phenotype of the heterozygote is not intermediate between the phenotypes of the homozygotes; rather, the heterozygote simultaneously expresses the phenotypes of both homozygotes. An example of codominance is seen in the MN blood types.

The MN locus encodes one of the types of antigens on red blood cells. Unlike antigens foreign to the ABO and Rh blood groups (which also encode red-blood-cell antigens), foreign MN antigens do not elicit a strong immunological reaction, and therefore the MN blood types are not routinely considered in blood transfusions. At the MN locus, there are two alleles: the L^M allele, which encodes the M antigen; and the L^N allele, which encodes the N antigen. Homozygotes with genotype L^ML^M express the M antigen on their red blood cells and have the M blood type. Homozygotes with genotype L^NL^N express the N antigen and have the N blood type. Heterozygotes with genotype L^ML^N exhibit codominance and express both the M and the N antigens; they have blood-type MN. The differences between dominance, incomplete dominance, and codominance are summarized in **Table 4.3**.

Phenotypes can frequently be observed at several different levels, including the anatomical level, the physiological level, and the molecular level. The type of dominance exhibited by a character depends on the level of the phenotype examined. This dependency is seen in cystic fibrosis, one of the more common genetic disorders found in Caucasians and usually considered to be a recessive disease. People who have cystic fibrosis produce large quantities of thick, sticky mucus, which plugs up the airways of the lungs and clogs the ducts leading from the pancreas to the intestine, causing frequent respiratory infections and digestive problems. Even with medical treatment, patients with cystic fibrosis suffer chronic, life-threatening medical problems.

The gene responsible for cystic fibrosis resides on the long arm of chromosome 7. It encodes a protein termed *cystic fibrosis transmembrane conductance regulator*, abbreviated CFTR,

Table 4.3	Differences between dominance, incomplete dominance, and codominance
Type of Dominance	**Definition**
Dominance	Phenotype of the heterozygote is the same as the phenotype of one of the homozygotes.
Incomplete dominance	Phenotype of the heterozygote is intermediate (falls within the range) between the phenotypes of the two homozygotes.
Codominance	Phenotype of the heterozygote includes the phenotypes of both homozygotes.

which acts as a gate in the cell membrane and regulates the movement of chloride ions into and out of the cell. Patients with cystic fibrosis have a mutated, dysfunctional form of CFTR that causes the channel to stay closed, and so chloride ions build up in the cell. This buildup causes the formation of thick mucus and produces the symptoms of the disease.

Most people have two copies of the normal allele for CFTR and produce only functional CFTR protein. Those with cystic fibrosis possess two copies of the mutated CFTR allele and produce only the defective CFTR protein. Heterozygotes, having one normal and one defective CFTR allele, produce both functional and defective CFTR protein. Thus, at the molecular level, the alleles for normal and defective CFTR are codominant, because both alleles are expressed in the heterozygote. However, because one functional allele produces enough functional CFTR protein to allow normal chloride ion transport, the heterozygote exhibits no adverse effects, and the mutated CFTR allele appears to be recessive at the physiological level. The type of dominance expressed by an allele, as illustrated in this example, is a function of the phenotypic aspect of the allele that is observed.

In summary, several important characteristics of dominance should be emphasized. First, dominance is a result of interactions between genes at the same locus; in other words, dominance is *allelic* interaction. Second, dominance does not alter the way in which the genes are inherited; it only influences the way in which they are expressed as a phenotype. The allelic interaction that characterizes dominance is therefore interaction between the *products* of the genes. Finally, dominance is frequently "in the eye of the beholder," meaning that the classification of dominance depends on the level at which the phenotype is examined. As seen for cystic fibrosis, an allele may exhibit codominance at one level and be recessive at another level.

Concepts

Dominance entails interactions between genes at the same locus (allelic genes) and is an aspect of the phenotype; dominance does not affect the way in which genes are inherited. The type of dominance exhibited by a characteristic frequently depends on the level at which the phenotype is examined.

✔ Concept Check 7

How do complete dominance, incomplete dominance, and codominance differ?

Penetrance and Expressivity Describe How Genes Are Expressed As Phenotype

In the genetic crosses presented thus far, we have considered only the interactions of alleles and have assumed that every individual organism having a particular genotype expresses

4.15 Human polydactyly (extra digits) exhibits incomplete penetrance and variable expressivity. *[SPL/Photo Researchers.]*

the expected phenotype. We assumed, for example, that the genotype *Rr* always produces round seeds and that the genotype *rr* always produces wrinkled seeds. For some characters, however, such an assumption is incorrect: the genotype does not always produce the expected phenotype, a phenomenon termed **incomplete penetrance**.

Incomplete penetrance is seen in human polydactyly, the condition of having extra fingers and toes (**Figure 4.15**). There are several different forms of human polydactyly, but the trait is usually caused by a dominant allele. Occasionally, people possess the allele for polydactyly (as evidenced by the fact that their children inherit the polydactyly) but nevertheless have a normal number of fingers and toes. In these cases, the gene for polydactyly is not fully penetrant. **Penetrance** is defined as the percentage of individuals having a particular genotype that express the expected phenotype. For example, if we examined 42 people having an allele for polydactyly and found that only 38 of them were polydactylous, the penetrance would be $^{38}/_{42} = 0.90$ (90%).

A related concept is that of **expressivity**, the degree to which a character is expressed. In addition to incomplete penetrance, polydactyly exhibits variable expressivity. Some polydactylous persons possess extra fingers and toes that are fully functional, whereas others possess only a small tag of extra skin.

Incomplete penetrance and variable expressivity are due to the effects of other genes and to environmental factors that can alter or completely suppress the effect of a particular gene. For example, a gene may encode an enzyme that produces a particular phenotype only within a limited temperature range. At higher or lower temperatures, the enzyme does not function and the phenotype is not expressed; the allele encoding such an enzyme is therefore penetrant only within a particular temperature range. Many characters exhibit incomplete penetrance and variable expressivity; thus the mere presence of a gene does not guarantee its expression.

Concepts

Penetrance is the percentage of individuals having a particular genotype that express the associated phenotype. Expressivity is the degree to which a trait is expressed. Incomplete penetrance and variable expressivity result from the influence of other genes and environmental factors on the phenotype.

✔ **Concept Check 8**

Assume that long fingers are inherited as a recessive trait with 80% penetrance. Two people heterozygous for long fingers mate. What is the probability that their first child will have long fingers?

Lethal Alleles May Alter Phenotypic Ratios

As described in the introduction to this chapter, Lucien Cuénot reported the first case of a lethal allele, the allele for yellow coat color in mice (see Figure 4.1). A **lethal allele** causes death at an early stage of development—often before birth—and so some genotypes may not appear among the progeny.

Another example of a lethal allele, originally described by Erwin Baur in 1907, is found in snapdragons. The *aurea* strain in these plants has yellow leaves. When two plants with yellow leaves are crossed, $^2/_3$ of the progeny have yellow leaves and $^1/_3$ have green leaves. When green is crossed with green, all the progeny have green leaves; however, when yellow is crossed with green, $^1/_2$ of the progeny have green leaves and $^1/_2$ have yellow leaves, confirming that all yellow-leaved snapdragons are heterozygous. A 2 : 1 ratio is almost always produced by a recessive lethal allele; so observing this ratio among the progeny of a cross between individuals with the same phenotype is a strong clue that one of the alleles is lethal.

In this example, like that of yellow coat color in mice, the lethal allele is recessive because it causes death only in homozygotes. Unlike its effect on *survival*, the effect of the allele on *color* is dominant; in both mice and snapdragons, a single copy of the allele in the heterozygote produces a yellow color. It illustrates the point made earlier that the type of dominance depends on the aspect of the phenotype examined.

Lethal alleles also can be dominant; in this case, homozygotes and heterozygotes for the allele die. Truly dominant lethal alleles cannot be transmitted unless they are expressed after the onset of reproduction, as in Huntington disease.

Concepts

A lethal allele causes death, frequently at an early developmental stage, and so one or more genotypes are missing from the progeny of a cross. Lethal alleles modify the ratio of progeny resulting from a cross.

4.4 Multiple Alleles at a Locus Create a Greater Variety of Genotypes and Phenotypes Than Do Two Alleles

Most of the genetic systems that we have examined so far consist of two alleles. In Mendel's peas, for instance, one allele encoded round seeds and another encoded wrinkled seeds; in cats, one allele produced a black coat and another produced a gray coat. For some loci, more than two alleles are present within a group of individuals—the locus has **multiple alleles**. (Multiple alleles may also be referred to as an *allelic series*.) Although there may be more than two alleles present within a *group*, the genotype of each individual diploid organism still consists of only two alleles. The inheritance of characteristics encoded by multiple alleles is no different from the inheritance of characteristics encoded by two alleles, except that a greater variety of genotypes and phenotypes are possible.

The ABO Blood Group

Another multiple-allele system is at the locus for the ABO blood group. This locus determines your ABO blood type and, like the MN locus, encodes antigens on red blood cells. The three common alleles for the ABO blood group locus are: I^A, which encodes the A antigen; I^B, which encodes the B antigen; and i, which encodes no antigen (O). We can represent the dominance relations among the ABO alleles as follows: $I^A > i, I^B > i, I^A = I^B$. Both the I^A and the I^B alleles are dominant over i and are codominant with each other; the AB phenotype is due to the presence of an I^A allele and an I^B allele, which results in the production of A and B antigens on red blood cells. A person with genotype ii produces neither antigen and has blood type O. The six common genotypes at this locus and their phenotypes are shown in **Figure 4.16a**.

Antibodies are produced against any foreign antigens (see Figure 4.16a). For instance, a person having blood-type A produces B antibodies, because the B antigen is foreign. A person having blood-type B produces A antibodies, and someone having blood-type AB produces neither A nor B antibodies, because neither A nor B antigen is foreign. A person having blood-type O possesses no A or B antigens; consequently, that person produces both A antibodies and B antibodies. The presence of antibodies against foreign ABO antigens means that successful blood transfusions are possible only between persons with certain compatible blood types (**Figure 4.16b**).

The inheritance of alleles at the ABO locus is illustrated by a paternity suit against the famous movie actor Charlie Chaplin. In 1941, Chaplin met a young actress named Joan Barry, with whom he had an affair. The affair ended in February 1942 but, 20 months later, Barry gave

(b)

(a)

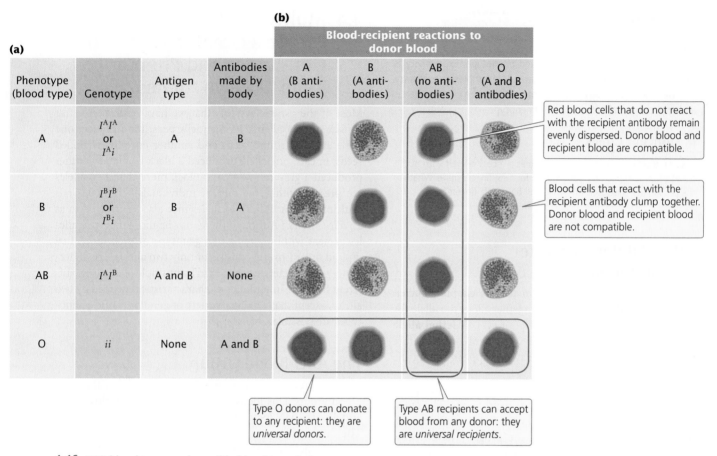

Phenotype (blood type)	Genotype	Antigen type	Antibodies made by body	Blood-recipient reactions to donor blood			
				A (B antibodies)	B (A antibodies)	AB (no antibodies)	O (A and B antibodies)
A	$I^A I^A$ or $I^A i$	A	B				
B	$I^B I^B$ or $I^B i$	B	A				
AB	$I^A I^B$	A and B	None				
O	ii	None	A and B				

Red blood cells that do not react with the recipient antibody remain evenly dispersed. Donor blood and recipient blood are compatible.

Blood cells that react with the recipient antibody clump together. Donor blood and recipient blood are not compatible.

Type O donors can donate to any recipient: they are *universal donors*.

Type AB recipients can accept blood from any donor: they are *universal recipients*.

4.16 ABO blood types and possible blood transfusions.

birth to a baby girl and claimed that Chaplin was the father. Barry then sued for child support. At this time, blood typing had just come into widespread use, and Chaplin's attorneys had Chaplin, Barry, and the child blood typed. Barry had blood-type A, her child had blood-type B, and Chaplin had blood-type O. Could Chaplin have been the father of Barry's child?

Your answer should be no. Joan Barry had blood-type A, which can be produced by either genotype $I^A I^A$ or genotype $I^A i$. Her baby possessed blood-type B, which can be produced by either genotype $I^B I^B$ or genotype $I^B i$. The baby could not have inherited the I^B allele from Barry (Barry could not carry an I^B allele if she were blood-type A); therefore the baby must have inherited the i allele from her. Barry must have had genotype $I^A i$, and the baby must have had genotype $I^B i$. Because the baby girl inherited her i allele from Barry, she must have inherited the I^B allele from her father. Having blood-type O, produced only by genotype ii, Chaplin could not have been the father of Barry's child. Although blood types can be used to exclude the possibility of paternity (as in this case), they cannot prove that a person is the

parent of a child, because many different people have the same blood type.

In the course of the trial to settle the paternity suit against Chaplin, three pathologists came to the witness stand and declared that it was genetically impossible for Chaplin to have fathered the child. Nevertheless, the jury ruled that Chaplin was the father and ordered him to pay child support and Barry's legal expenses.

Concepts

More than two alleles (multiple alleles) may be present within a group of individuals, although each individual diploid organism still has only two alleles at that locus.

✔ **Concept Check 9**

What blood types are possible among the children of a cross between a man who is blood-type A and a woman of blood-type B?

4.5 Gene Interaction Takes Place When Genes at Multiple Loci Determine a Single Phenotype

In the dihybrid crosses that we examined in Chapter 3, each locus had an independent effect on the phenotype. When Mendel crossed a homozygous round and yellow plant (*RR YY*) with a homozygous wrinkled and green plant (*rr yy*) and then self-fertilized the F₁, he obtained F₂ progeny in the following proportions:

$\frac{9}{16}$	*R_ Y_*	round, yellow
$\frac{3}{16}$	*R_ yy*	round, green
$\frac{3}{16}$	*rr Y_*	wrinkled, yellow
$\frac{1}{16}$	*rr yy*	wrinkled, green

In this example, the genes showed two kinds of independence. First, the genes at each locus are independent in their *assortment* in meiosis, which is what produces the 9 : 3 : 3 : 1 ratio of phenotypes in the progeny, in accord with Mendel's principle of independent assortment. Second, the genes are independent in their *phenotypic expression*; the *R* and *r* alleles affect only the shape of the seed and have no influence on the color of the endosperm; the *Y* and *y* alleles affect only color and have no influence on the shape of the seed.

Frequently, genes exhibit independent assortment but do not act independently in their phenotypic expression; instead, the effects of genes at one locus depend on the presence of genes at other loci. This type of interaction between the effects of genes at different loci (genes that are not allelic) is termed **gene interaction**. With gene interaction, the products of genes at different loci combine to produce new phenotypes that are not predictable from the single-locus effects alone. In our consideration of gene interaction, we'll focus primarily on interaction between the effects of genes at two loci, although interactions among genes at three, four, or more loci are common.

Concepts

In gene interaction, genes at different loci contribute to the determination of a single phenotypic characteristic.

Gene Interaction That Produces Novel Phenotypes

Let's first examine gene interaction in which genes at two loci interact to produce a single characteristic. Fruit color in the pepper *Capsicum annuum* is determined in this way. Certain types of pepper plants produce fruits in one of four colors: red, peach, orange (sometimes called yellow), and cream (or white). If a homozygous plant with red peppers is crossed with a homozygous plant with cream peppers, all

(a)

(b)

4.17 Gene interaction in which two loci determine a single characteristic, fruit color, in the pepper *Capsicum annuum*.

the F₁ plants have red peppers (**Figure 4.17a**). When the F₁ are crossed with one another, the F₂ are in a ratio of 9 red : 3 peach : 3 orange : 1 cream (**Figure 4.17b**). This dihybrid ratio (see Chapter 3) is produced by a cross between two plants that are both heterozygous for two loci ($Y^+y\ C^+c \times Y^+y\ C^+c$). In this example, the *Y* locus and the *C* locus interact to produce a single phenotype—the color of the pepper:

Genotype	Phenotype
$Y^+\_ C^+\_$	red
$Y^+\_ cc$	peach
$yy\ C^+\_$	orange
$yy\ cc$	cream

To illustrate how Mendel's rules of heredity can be used to understand the inheritance of characteristics determined by gene interaction, let's consider a testcross between an F_1 plant from the cross in Figure 4.17 ($Y^+y\ C^+c$) and a plant with cream peppers ($yy\ cc$). As outlined in Chapter 3 for independent loci, we can work this cross by breaking it down into two simple crosses. At the first locus, the heterozygote Y^+y is crossed with the homozygote yy; this cross produces $\frac{1}{2}\ Y^+y$ and $\frac{1}{2}\ yy$ progeny. Similarly, at the second locus, the heterozygous genotype C^+c is crossed with the homozygous genotype cc, producing $\frac{1}{2}\ C^+c$ and $\frac{1}{2}\ cc$ progeny. In accord with Mendel's principle of independent assortment, these single-locus ratios can be combined by using the multiplication rule: the probability of obtaining the genotype $Y^+y\ C^+c$ is the probability of Y^+y ($\frac{1}{2}$) multiplied by the probability of C^+c ($\frac{1}{2}$), or $\frac{1}{4}$. The probability of each progeny genotype resulting from the testcross is:

Progeny genotype	Probability at each locus	Overall probability	Phenotype
$Y^+y\ C^+c$	$\frac{1}{2} \times \frac{1}{2} =$	$\frac{1}{4}$	red peppers
$Y^+y\ cc$	$\frac{1}{2} \times \frac{1}{2} =$	$\frac{1}{4}$	peach peppers
$yy\ C^+c$	$\frac{1}{2} \times \frac{1}{2} =$	$\frac{1}{4}$	orange peppers
$yy\ cc$	$\frac{1}{2} \times \frac{1}{2} =$	$\frac{1}{4}$	cream peppers

When you work problems with gene interaction, it is especially important to determine the probabilities of single-locus genotypes and to multiply the probabilities of *genotypes*, not phenotypes, because the phenotypes cannot be determined without considering the effects of the genotypes at all the contributing loci.

Gene Interaction with Epistasis

Sometimes the effect of gene interaction is that one gene masks (hides) the effect of another gene at a different locus, a phenomenon known as **epistasis**. This phenomenon is similar to dominance, except that dominance entails the masking of genes at the *same* locus (allelic genes). In epistasis, the gene that does the masking is called an **epistatic gene**; the gene whose effect is masked is a **hypostatic gene**. Epistatic genes may be recessive or dominant in their effects.

Recessive epistasis   Recessive epistasis is seen in the genes that determine coat color in Labrador retrievers. These dogs may be black, brown, or yellow; their different coat colors are determined by interactions between genes at two loci (although a number of other loci also help to determine coat color). One locus determines the type of pigment produced by the skin cells: a dominant allele *B* encodes black pigment, whereas a recessive allele *b* encodes brown pigment. Alleles at a second locus affect the *deposition* of the pigment in the shaft of the hair; allele *E* allows dark pigment (black or brown) to be deposited, whereas a recessive allele *e* prevents the deposi-

tion of dark pigment, causing the hair to be yellow. The presence of genotype *ee* at the second locus therefore masks the expression of the black and brown alleles at the first locus. The genotypes that determine coat color and their phenotypes are:

Genotype	Phenotype
$B\_\ E\_$	black
$bb\ E\_$	brown (frequently called chocolate)
$B\_\ ee$	yellow
$bb\ ee$	yellow

If we cross a black Labrador homozygous for the dominant alleles with a yellow Labrador homozygous for the recessive alleles and then intercross the F_1, we obtain progeny in the F_2 in a 9 : 3 : 4 ratio:

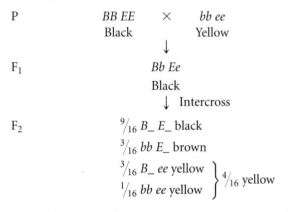

Notice that yellow dogs can carry alleles for either black or brown pigment, but these alleles are not expressed in their coat color.

In this example of gene interaction, allele *e* is epistatic to *B* and *b*, because *e* masks the expression of the alleles for black and brown pigments, and alleles *B* and *b* are hypostatic to *e*. In this case, *e* is a recessive epistatic allele, because two copies of *e* must be present to mask the expression of the black and brown pigments.

Another example of an epistatic gene is the Bombay phenotype, which suppresses the expression of alleles at the ABO locus. In most people, a dominant allele (*H*) encodes an enzyme that makes H, a molecule necessary for the production of antigens. People with the Bombay phenotype are homozygous for a recessive mutation (*h*) that encodes a defective enzyme. The defective enzyme is incapable of making H and, because H is not produced, no ABO antigens are synthesized. People with genotype *hh*, who would normally have A, B, or AB blood types, do not produce antigens and therefore express an O phenotype. In this example, the alleles at the ABO locus are hypostatic to the recessive *h* allele.

Dominant epistasis   Dominant epistasis is seen in the interaction of two loci that determine fruit color in summer squash, which is commonly found in one of three colors: yellow, white, or green. When a homozygous plant that produces white squash is crossed with a homozygous plant that

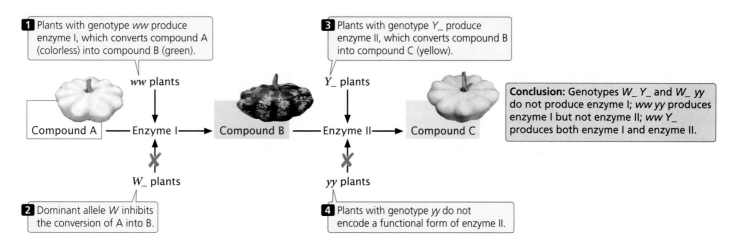

1. Plants with genotype *ww* produce enzyme I, which converts compound A (colorless) into compound B (green).

3. Plants with genotype *Y_* produce enzyme II, which converts compound B into compound C (yellow).

ww plants

*Y_* plants

Conclusion: Genotypes *W_ Y_* and *W_ yy* do not produce enzyme I; *ww yy* produces enzyme I but not enzyme II; *ww Y_* produces both enzyme I and enzyme II.

Compound A ── Enzyme I ⟶ Compound B ── Enzyme II ⟶ Compound C

*W_* plants

yy plants

2. Dominant allele *W* inhibits the conversion of A into B.

4. Plants with genotype *yy* do not encode a functional form of enzyme II.

4.18 Yellow pigment in summer squash is produced in a two-step pathway.

produces green squash and the F_1 plants are crossed with each other, the following results are obtained:

P   Plants with   Plants with
          white squash   ×   green squash
                              ↓
F_1 Plants with
                        white squash
                              ↓ Intercross
F_2 $^{12}/_{16}$ plants with white squash
              $^{3}/_{16}$ plants with yellow squash
              $^{1}/_{16}$ plants with green squash

How can gene interaction explain these results?

In the F_2, $^{12}/_{16}$, or $^{3}/_{4}$, of the plants produce white squash and $^{3}/_{16} + ^{1}/_{16} = ^{4}/_{16} = ^{1}/_{4}$ of the plants produce squash having color. This outcome is the familiar 3 : 1 ratio produced by a cross between two heterozygotes, which suggests that a dominant allele at one locus inhibits the production of pigment, resulting in white progeny. If we use the symbol *W* to represent the dominant allele that inhibits pigment production, the genotype *W_* inhibits pigment production and produces white squash, whereas *ww* allows pigment and results in colored squash.

Among those *ww* F_2 plants with pigmented fruit, we observe $^{3}/_{16}$ yellow and $^{1}/_{16}$ green (a 3 : 1 ratio). In this outcome, a second locus determines the type of pigment produced in the squash, with yellow (*Y_*) dominant over green (*yy*). This locus is expressed only in *ww* plants, which lack the dominant inhibitory allele *W*. We can assign the genotype *ww Y_* to plants that produce yellow squash and the genotype *ww yy* to plants that produce green squash. The genotypes and their associated phenotypes are:

*W_ Y_*	white squash
W_ yy	white squash
*ww Y_*	yellow squash
ww yy	green squash

Allele *W* is epistatic to *Y* and *y*: it suppresses the expression of these pigment-producing genes. Allele *W* is a dominant epistatic allele because, in contrast with *e* in Labrador retriever coat color, a single copy of the allele is sufficient to inhibit pigment production.

Yellow pigment in the squash is most likely produced in a two-step biochemical pathway (**Figure 4.18**). A colorless (white) compound (designated A in Figure 4.18) is converted by enzyme I into green compound B, which is then converted into compound C by enzyme II. Compound C is the yellow pigment in the fruit. Plants with the genotype *ww* produce enzyme I and may be green or yellow, depending on whether enzyme II is present. When allele *Y* is present at a second locus, enzyme II is produced and compound B is converted into compound C, producing a yellow fruit. When two copies of allele *y*, which does not encode a functional form of enzyme II, are present, squash remain green. The presence of *W* at the first locus inhibits the conversion of compound A into compound B; plants with genotype *W_* do not make compound B and their fruit remains white, regardless of which alleles are present at the second locus.

Concepts

Epistasis is the masking of the expression of one gene by another gene at a different locus. The epistatic gene does the masking; the hypostatic gene is masked. Epistatic genes can be dominant or recessive.

✔ Concept Check 10

A number of all-white cats are crossed and they produce the following types of progeny: $^{12}/_{16}$ all-white, $^{3}/_{16}$ black, and $^{1}/_{16}$ gray. Give the genotypes of the progeny. Which gene is epistatic?

Connecting Concepts

Interpreting Ratios Produced by Gene Interaction

A number of modified ratios that result from gene interaction are shown in **Table 4.4**. Each of these examples represents a modification of the basic $9:3:3:1$ dihybrid ratio. In interpreting the genetic basis of modified ratios, we should keep several points in mind. First, the inheritance of the genes producing these characteristics is no different from the inheritance of genes encoding simple genetic characters. Mendel's principles of segregation and independent assortment still apply; each individual possesses two alleles at each locus, which separate in meiosis, and genes at the different loci assort independently. The only difference is in how the *products* of the genotypes interact to produce the phenotype. Thus, we cannot consider the expression of genes at each locus separately; instead, we must take into consideration how the genes at different loci interact.

A second point is that, in the examples that we have considered, the phenotypic proportions were always in sixteenths because, in all the crosses, pairs of alleles segregated at two independently assorting loci. The probability of inheriting one of the two alleles at a locus is $\frac{1}{2}$. Because there are two loci, each with two alleles, the probability of inheriting any particular combination of genes is $\left(\frac{1}{2}\right)^4 = \frac{1}{16}$. For a trihybrid cross, the progeny proportions should be in sixty-fourths, because $\left(\frac{1}{2}\right)^6 = \frac{1}{64}$. In general, the progeny proportions should be in fractions of $\left(\frac{1}{2}\right)^{2n}$, where n equals the number of loci with two alleles segregating in the cross.

Crosses rarely produce exactly 16 progeny; therefore, modifications of a dihybrid ratio are not always obvious. Modified dihybrid ratios are more easily seen if the number of individuals of each phenotype is expressed in sixteenths:

$$\frac{x}{16} = \frac{\text{number of progeny with a phenotype}}{\text{total number of progeny}}$$

where $x/16$ equals the proportion of progeny with a particular phenotype. If we solve for x (the proportion of the particular phenotype in sixteenths), we have:

$$x = \frac{\text{number of progeny with a phenotype} \times 16}{\text{total number of progeny}}$$

For example, suppose we cross two homozygotes, interbreed the F_1, and obtain 63 red, 21 brown, and 28 white F_2 individuals. Using the preceding formula, we find the phenotypic ratio in the F_2 to be: red $= (63 \times 16)/112 = 9$; brown $= (21 \times 16)/112 = 3$; and white $= (28 \times 16)/112 = 4$. The phenotypic ratio is $9:3:4$.

A final point to consider is how to assign genotypes to the phenotypes in modified ratios that result from gene interaction. Don't try to *memorize* the genotypes associated with all the modified ratios in Table 4.4. Instead, practice relating modified ratios to known ratios, such as the $9:3:3:1$ dihybrid ratio. Suppose we obtain $\frac{15}{16}$ green progeny and $\frac{1}{16}$ white progeny in a cross between two plants. If we compare this $15:1$ ratio with the standard $9:3:3:1$ dihybrid ratio, we see that $\frac{9}{16} + \frac{3}{16} + \frac{3}{16}$ equals $\frac{15}{16}$. All the genotypes associated with these proportions in the dihybrid cross ($A\_ B\_$, $A\_ bb$, and $aa B\_$) must give the same phenotype, the green progeny. Genotype $aa\,bb$ makes up $\frac{1}{16}$ of the progeny in a dihybrid cross, the white progeny in this cross.

In assigning genotypes to phenotypes in modified ratios, students sometimes become confused about which letters to assign to which phenotype. Suppose we obtain the following phenotypic ratio: $\frac{9}{16}$ black : $\frac{3}{16}$ brown : $\frac{4}{16}$ white. Which genotype do we assign to the brown progeny, $A\_ bb$ or $aa B\_$? Either answer is correct, because the letters are just arbitrary symbols for the genetic information. The important thing to realize about this ratio is that the brown phenotype arises when two recessive alleles are present at one locus.

| Table 4.4 | Modified dihybrid phenotypic ratios due to gene interaction | | | | | |

Ratio*	Genotype				Type of Interaction	Example
	$A\_ B\_$	$A\_ bb$	$aa B\_$	$aa\,bb$		
$9:3:3:1$	9	3	3	1	None	Seed shape and endosperm color in peas
$9:3:4$	9	3	4		Recessive epistasis	Coat color in Labrador retrievers
$12:3:1$	12		3	1	Dominant epistasis	Color in squash
$9:7$	9	7			Duplicate recessive epistasis	—
$9:6:1$	9	6		1	Duplicate interaction	—
$15:1$	15			1	Duplicate dominant epistasis	—
$13:3$	13		3		Dominant and recessive epistasis	—

*Each ratio is produced by a dihybrid cross ($Aa\,Bb \times Aa\,Bb$). Shaded bars represent combinations of genotypes that give the same phenotype.

Worked Problem

A homozygous strain of yellow corn is crossed with a homozygous strain of purple corn. The F_1 are intercrossed, producing an ear of corn with 119 purple kernels and 89 yellow kernels (the progeny). What is the genotype of the yellow kernels?

• Solution

We should first consider whether the cross between yellow and purple strains might be a monohybrid cross for a simple dominant trait, which would produce a 3 : 1 ratio in the F_2 ($Aa \times Aa \rightarrow {}^3/_4 \, A\_$ and ${}^1/_4 \, aa$). Under this hypothesis, we would expect 156 purple progeny and 52 yellow progeny:

Phenotype	Genotype	Observed number	Expected number
purple	$A\_$	119	${}^3/_4 \times 208 = 156$
yellow	aa	89	${}^1/_4 \times 208 = 52$
Total		208	

We see that the expected numbers do not closely fit the observed numbers. If we performed a chi-square test (see Chapter 3), we would obtain a calculated chi-square value of 35.08, which has a probability much less than 0.05, indicating that it is extremely unlikely that, when we expect a 3 : 1 ratio, we would obtain 119 purple progeny and 89 yellow progeny. Therefore, we can reject the hypothesis that these results were produced by a monohybrid cross.

Another possible hypothesis is that the observed F_2 progeny are in a 1 : 1 ratio. However, we learned in Chapter 3 that a 1 : 1 ratio is produced by a cross between a heterozygote and a homozygote ($Aa \times aa$) and, from the information given, the cross was not between a heterozygote and a homozygote, because both original parental strains were homozygous. Furthermore, a chi-square test comparing the observed numbers with an expected 1 : 1 ratio yields a calculated chi-square value of 4.32, which has a probability of less than 0.05.

Next, we should look to see if the results can be explained by a dihybrid cross ($Aa\,Bb \times Aa\,Bb$). A dihybrid cross results in phenotypic proportions that are in sixteenths. We can apply the formula given earlier in the chapter to determine the number of sixteenths for each phenotype:

$$x = \frac{\text{number of progeny with a phenotype} \times 16}{\text{total number of progeny}}$$

$$x_{(purple)} = \frac{119 \times 16}{208} = 9.15$$

$$x_{(yellow)} = \frac{89 \times 16}{208} = 6.85$$

Thus, purple and yellow appear in an approximate ratio of 9 : 7. We can test this hypothesis with a chi-square test:

Phenotype	Genotype	Observed number	Expected number
purple	?	119	${}^9/_{16} \times 208 = 117$
yellow	?	89	${}^7/_{16} \times 208 = 91$
Total		208	

$$x^2 = \sum \frac{(\text{observed} - \text{expected})^2}{\text{expected}}$$

$$= \frac{(119 - 117)^2}{117} + \frac{(89 - 91)^2}{91}$$

$$= 0.034 + 0.44 = 0.078$$

Degree of freedom $= n - 1 = 2 - 1 = 1$

$$P > 0.05$$

The probability associated with the chi-square value is greater than 0.05, indicating that there is a good fit between the observed results and a 9 : 7 ratio.

We now need to determine how a dihybrid cross can produce a 9 : 7 ratio and what genotypes correspond to the two phenotypes. A dihybrid cross without epistasis produces a 9 : 3 : 3 : 1 ratio:

$$Aa\,Bb \times Aa\,Bb$$
$$\downarrow$$
$$A\_ \, B\_ \quad {}^9/_{16}$$
$$A\_ \, bb \quad {}^3/_{16}$$
$$aa \, B\_ \quad {}^3/_{16}$$
$$aa \, bb \quad {}^1/_{16}$$

Because ${}^9/_{16}$ of the progeny from the corn cross are purple, purple must be produced by genotypes $A\_ \, B\_$; in other words, individual kernels that have at least one dominant allele at the first locus and at least one dominant allele at the second locus are purple. The proportions of all the other genotypes ($A\_ \, bb$, $aa \, B\_$, and $aa \, bb$) sum to ${}^7/_{16}$, which is the proportion of the progeny in the corn cross that are yellow, and so any individual kernel that does not have a dominant allele at both the first and the second locus is yellow.

> **?** Now test your understanding of epistasis by working Problem 27 at the end of the chapter.

Complementation: Determining Whether Mutations Are at the Same Locus or at Different Loci

How do we know whether different mutations that affect a characteristic occur at the same locus (are allelic) or at different loci? In fruit flies, for example, *white* is an X-linked mutation that produces white eyes instead of the red eyes found in wild-type flies; *apricot* is an X-linked recessive mutation that produces light orange-colored eyes. Do the *white* and *apricot* mutations occur at the same locus or at different loci? We can use the complementation test to answer this question.

To carry out a **complementation test** on recessive mutations, parents that are homozygous for different mutations are crossed, producing offspring that are heterozygous. If the mutations are allelic (occur at the same locus), then the heterozygous offspring have only mutant alleles (*a b*) and exhibit a mutant phenotype:

If, on the other hand, the mutations occur at different loci, each of the homozygous parents possesses wild-type genes at the other locus ($aa\ b^+b^+$ and $a^+a^+\ bb$); so the heterozygous offspring inherit a mutant allele and a wild-type allele at each locus. In this case, the mutations complement each other and the heterozygous offspring have the wild-type phenotype:

Complementation has occurred if an individual possessing two mutant genes has a wild-type phenotype and is an indicator that the mutations are nonallelic genes.

When the complementation test is applied to *white* and *apricot* mutations, all of the heterozygous offspring have light-colored eyes, demonstrating that white eyes and apricot eyes are produced by mutations that occur at the same locus and are allelic.

Concepts

A complementation test is used to determine whether two mutations occur at the same locus (are allelic) or occur at different loci.

4.6 Sex Influences the Inheritance and Expression of Genes in a Variety of Ways

In Section 4.2, we considered characteristics encoded by genes located on the sex chromosomes (sex-linked traits) and how their inheritance differs from the inheritance of traits encoded by autosomal genes. X-linked traits, for example, are passed from father to daughter, but never from father to son, and Y-linked traits are passed from father to all sons. Now, we will examine additional influences of sex, including the effect of the sex of an individual on the expression of genes on autosomal chromosomes, on characteristics determined by genes located in the cytoplasm, and on characteristics for which the genotype of only the maternal parent determines the phenotype of the offspring. Finally, we'll look at situations in which the expression of genes on autosomal chromosomes is affected by the sex of the parent from whom they are inherited.

Sex-Influenced and Sex-Limited Characteristics

Sex-influenced characteristics are determined by autosomal genes and are inherited according to Mendel's principles, but they are expressed differently in males and females. In this case, a particular trait is more readily expressed in one sex; in other words, the trait has higher penetrance in one of the sexes.

For example, the presence of a beard on some goats is determined by an autosomal gene (B^b) that is dominant in males and recessive in females. In males, a single allele is required for the expression of this trait: both the homozygote (B^bB^b) and the heterozygote (B^bB^+) have beards, whereas the B^+B^+ male is beardless. In contrast, females require two alleles in order for this trait to be expressed: the homozygote B^bB^b has a beard, whereas the heterozygote (B^bB^+) and the other homozygote (B^+B^+) are beardless. The key to understanding the expression of the bearded gene is to look at the heterozygote. In males (for which the presence of a beard is dominant), the heterozygous genotype produces a beard but, in females (for which the presence of a beard is recessive and its absence is dominant), the heterozygous genotype produces a goat without a beard.

An extreme form of sex-influenced inheritance, a **sex-limited characteristic** is encoded by autosomal genes that are expressed in only one sex; the trait has zero penetrance in the other sex. In domestic chickens, some males display a plumage pattern called cock feathering. Other males and all females display a pattern called hen feathering. Cock feathering is an autosomal recessive trait that is sex limited to males. Because the trait is autosomal, the genotypes of males and females are the same, but the phenotypes produced by these genotypes are different in males and females. Only homozygous recessive males will have the cock-feathering phenotype.

Concepts

Sex-influenced characteristics are encoded by autosomal genes that are more readily expressed in one sex. Sex-limited characteristics are encoded by autosomal genes whose expression is limited to one sex.

✔ **Concept Check 11**

How do sex-influenced and sex-limited traits differ from sex-linked traits?

Cytoplasmic Inheritance

Mendel's principles of segregation and independent assortment are based on the assumption that genes are located on chromosomes in the nucleus of the cell. For most genetic characteristics, this assumption is valid, and Mendel's principles allow us to predict the types of offspring that will be produced in a genetic cross. However, not all the genetic material of a cell is found in the nucleus; some characteristics are encoded by genes located in the cytoplasm. These characteristics exhibit **cytoplasmic inheritance**.

A few organelles, notably chloroplasts and mitochondria, contain DNA. Each human mitochondrion contains about 15,000 nucleotides of DNA, encoding 37 genes. Compared with that of nuclear DNA, which contains some 3 billion nucleotides encoding perhaps 25,000 genes, the amount of mitochondrial DNA (mtDNA) is very small; nevertheless, mitochondrial and chloroplast genes encode some important characteristics.

Cytoplasmic inheritance differs from the inheritance of characteristics encoded by nuclear genes in several important respects. A zygote inherits nuclear genes from both parents; but, typically, all its cytoplasmic organelles, and thus all its cytoplasmic genes, come from only one of the gametes, usually the egg. A sperm generally contributes only a set of nuclear genes from the male parent. In a few organisms, cytoplasmic genes are inherited from the male parent or from both parents; however, for most organisms, all the cytoplasm is inherited from the egg. In this case, cytoplasmically inherited traits are present in both males and females and are passed from mother to offspring, never from father to offspring. Reciprocal crosses, therefore, give different results when cytoplasmic genes encode a trait.

Cytoplasmically inherited characteristics frequently exhibit extensive phenotypic variation, because no mechanism analogous to mitosis or meiosis ensures that cytoplasmic genes are evenly distributed in cell division. Thus, different cells and individual offspring will contain various proportions of cytoplasmic genes.

Consider mitochondrial genes. Most cells contain thousands of mitochondria, and each mitochondrion contains

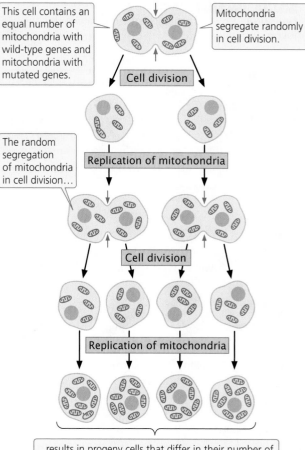

This cell contains an equal number of mitochondria with wild-type genes and mitochondria with mutated genes.

Mitochondria segregate randomly in cell division.

Cell division

Replication of mitochondria

The random segregation of mitochondria in cell division…

Cell division

Replication of mitochondria

…results in progeny cells that differ in their number of mitochondria with wild-type and mutated genes.

4.19 Cytoplasmically inherited characteristics frequently exhibit extensive phenotypical variation because cells and individual offspring contain various proportions of cytoplasmic genes. Mitochondria that have wild-type mitochondrial DNA are shown in red; those having mutant mtDNA are shown in blue.

from 2 to 10 copies of mtDNA. Suppose that half of the mitochondria in a cell contain a normal wild-type copy of mtDNA and the other half contain a mutated copy (**Figure 4.19**). In cell division, the mitochondria segregate into progeny cells at random. Just by chance, one cell may receive mostly mutated mtDNA and another cell may receive mostly wild-type mtDNA. In this way, different progeny from the same mother and even cells within an individual offspring may vary in their phenotypes. Traits encoded by chloroplast DNA (cpDNA) are similarly variable.

In 1909, cytoplasmic inheritance was recognized by Carl Correns as one of the first exceptions to Mendel's principles. Correns, one of the biologists who rediscovered Mendel's work, studied the inheritance of leaf variegation in the four-o'clock plant, *Mirabilis jalapa*. Correns found that the leaves and shoots of one variety of four-o'clock were variegated, displaying a mixture of green and white splotches. He also noted that some branches of the variegated strain had all-green leaves; other branches had all-white leaves. Each

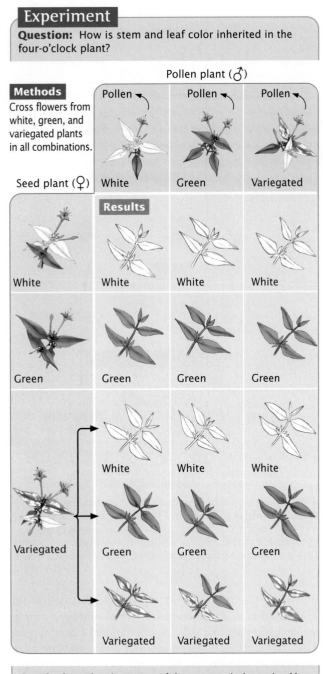

Experiment

Question: How is stem and leaf color inherited in the four-o'clock plant?

Pollen plant (♂)

Methods
Cross flowers from white, green, and variegated plants in all combinations.

Seed plant (♀)	Pollen ← White	Pollen ← Green	Pollen ← Variegated

Results

White	White	White	White
Green	Green	Green	Green
Variegated	White	White	White
	Green	Green	Green
	Variegated	Variegated	Variegated

Conclusion: The phenotype of the progeny is determined by the phenotype of the branch from which the seed originated, not from the branch on which the pollen originated. Stem and leaf color exhibits cytoplasmic inheritance.

4.20 Crosses for leaf type in four-o'clocks illustrate cytoplasmic inheritance.

branch produced flowers; so Correns was able to cross flowers from variegated, green, and white branches in all combinations (**Figure 4.20**). The seeds from green branches always gave rise to green progeny, no matter whether the pollen was from a green, white, or variegated branch. Similarly, flowers on white branches always produced white progeny. Flowers on the variegated branches gave rise to green, white, and variegated progeny, in no particular ratio.

Correns's crosses demonstrated cytoplasmic inheritance of variegation in the four-o'clocks. The phenotypes of the offspring were determined entirely by the maternal parent, never by the paternal parent (the source of the pollen). Furthermore, the production of all three phenotypes by flowers on variegated branches is consistent with cytoplasmic inheritance. Variegation in these plants is caused by a defective gene in the cpDNA, which results in a failure to produce the green pigment chlorophyll. Cells from green branches contain normal chloroplasts only, cells from white branches contain abnormal chloroplasts only, and cells from variegated branches contain a mixture of normal and abnormal chloroplasts. In the flowers from variegated branches, the random segregation of chloroplasts in the course of oogenesis produces some egg cells with normal cpDNA, which develop into green progeny; other egg cells with only abnormal cpDNA develop into white progeny; and, finally, still other egg cells with a mixture of normal and abnormal cpDNA develop into variegated progeny.

A number of human diseases (mostly rare) that exhibit cytoplasmic inheritance have been identified. These disorders arise from mutations in mtDNA, most of which occur in genes encoding components of the electron-transport chain, which generates most of the ATP (adenosine triphosphate) in aerobic cellular respiration. One such disease is Leber hereditary optic neuropathy (LHON). Patients who have this disorder experience rapid loss of vision in both eyes, resulting from the death of cells in the optic nerve. This loss of vision typically occurs in early adulthood (usually between the ages of 20 and 24), but it can occur any time after adolescence. There is much clinical variability in the severity of the disease, even within the same family. Leber hereditary optic neuropathy exhibits maternal inheritance: the trait is always passed from mother to child.

Genetic Maternal Effect

A genetic phenomenon that is sometimes confused with cytoplasmic inheritance is **genetic maternal effect**, in which the phenotype of the offspring is determined by the genotype of the mother. In cytoplasmic inheritance, the genes for a characteristic are inherited from only one parent, usually the mother. In genetic maternal effect, the genes are inherited from both parents, but the offspring's phenotype is determined not by its own genotype but by the genotype of its mother.

Genetic maternal effect frequently arises when substances present in the cytoplasm of an egg (encoded by the mother's nuclear genes) are pivotal in early development. An excellent example is the shell coiling of the snail *Limnaea peregra* (**Figure 4.21**). In most snails of this species, the shell coils to the right, which is termed dextral coiling. However, some snails possess a left-coiling shell, exhibiting sinistral coiling. The direction of coiling is determined by a pair of alleles; the allele for dextral (s^+) is dominant over the allele for sinistral (s). However, the direction of coiling is deter-

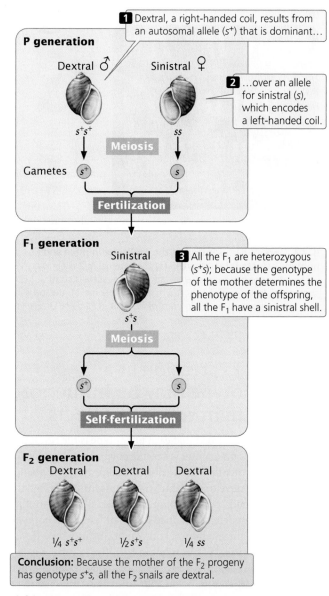

P generation

1 Dextral, a right-handed coil, results from an autosomal allele (s^+) that is dominant…

Dextral ♂   Sinistral ♀

2 …over an allele for sinistral (s), which encodes a left-handed coil.

s^+s^+ ss

Meiosis

Gametes s^+ s

Fertilization

F₁ generation

Sinistral

3 All the F₁ are heterozygous (s^+s); because the genotype of the mother determines the phenotype of the offspring, all the F₁ have a sinistral shell.

s^+s

Meiosis

s^+ s

Self-fertilization

F₂ generation

Dextral   Dextral   Dextral

¼ s^+s^+ ½ s^+s ¼ ss

Conclusion: Because the mother of the F₂ progeny has genotype s^+s, all the F₂ snails are dextral.

4.21 In genetic maternal effect, the genotype of the maternal parent determines the phenotype of the offspring. The shell coiling of a snail is a trait that exhibits genetic maternal effect.

dextral coiled because the genotype of their mother (s^+s) encodes a right-coiling shell and determines their phenotype. With genetic maternal effect, the phenotype of the progeny is not necessarily the same as the phenotype of the mother, because the progeny's phenotype is determined by the mother's *genotype*, not her phenotype. Neither the male parent's nor the offspring's own genotype has any role in the offspring's phenotype. However, a male does influence the phenotype of the F₂ generation: by contributing to the genotypes of his daughters, he affects the phenotypes of their offspring. Genes that exhibit genetic maternal effect are therefore transmitted through males to future generations. In contrast, genes that exhibit cytoplasmic inheritance are always transmitted through only one of the sexes (usually the female).

Concepts

Characteristics exhibiting cytoplasmic inheritance are encoded by genes in the cytoplasm and are usually inherited from one parent, most commonly the mother. In genetic maternal effect, the genotype of the mother determines the phenotype of the offspring.

Genomic Imprinting

A basic tenet of Mendelian genetics is that the parental origin of a gene does not affect its expression and, therefore, reciprocal crosses give identical results. However, the expression of some genes is significantly affected by their parental origin. This phenomenon, the differential expression of genetic material depending on whether it is inherited from the male or female parent, is called **genomic imprinting**.

A gene that exhibits genomic imprinting in both mice and humans is *Igf2*, which encodes a protein called insulin-like growth factor II (Igf-II). Offspring inherit one *Igf2* allele from their mother and one from their father. The paternal copy of *Igf2* is actively expressed in the fetus and placenta, but the maternal copy is completely silent (**Figure 4.22**). Both male and female offspring possess *Igf2* genes; the key to whether the gene is expressed is the sex of the parent transmitting the gene. In the present example, the gene is expressed only when it is transmitted by a male parent. In other genomically imprinted traits, the trait is expressed only when the gene is transmitted by the female parent.

Genomic imprinting is brought about through differential methylation of DNA—the addition of methyl (CH_3) groups to DNA nucleotides. In mammals, methylation is erased in the germ cells each generation and then reestablished during gamete formation, with different levels of methylation occuring in sperm and eggs, which then causes the differential expression of male and female alleles in the offspring.

mined not by that snail's own genotype, but by the genotype of its *mother*. The direction of coiling is affected by the way in which the cytoplasm divides soon after fertilization, which in turn is determined by a substance produced by the mother and passed to the offspring in the cytoplasm of the egg.

If a male homozygous for dextral alleles (s^+s^+) is crossed with a female homozygous for sinistral alleles (ss), all of the F₁ are heterozygous (s^+s) and have a sinistral shell, because the genotype of the mother (ss) encodes sinistral coiling (Figure 4.21). If these F₁ snails are self-fertilized, the genotypic ratio of the F₂ is 1 s^+s^+ : 2 s^+s : 1 ss.

Notice that that the phenotype of all the F₂ snails is dextral coiled, regardless of their genotypes. The F₂ offspring are

(a)

Paternal allele — *Igf2*

Maternal allele — *Igf2*

The paternal allele is *active* and its protein product stimulates fetal growth. — *Igf2*   *Igf2* — The maternal allele is *silent*. The absence of its protein product does not further stimulate fetal growth.

The size of the fetus is determined by the combined effects of both alleles.

(b)

Human chromosome 11 — *Igf2*

4.22 Genomic imprinting of the *Igf2* gene in mice and humans affects fetal growth. (a) The paternal *Igf2* allele is active in the fetus and placenta, whereas the maternal allele is silent. (b) The human *Igf2* locus is on the short arm of chromosome 11; the locus in mice is on chromosome 7. [*Courtesy of Dr. Thomas Ried and Dr. Evelin Schrock.*]

Genomic imprinting is just one form of a phenomenon known as **epigenetics**, in which reversible changes to DNA influence the expression of traits. Some of the ways in which sex interacts with heredity are summarized in **Table 4.5**.

Concepts

In genomic imprinting, the expression of a gene is influenced by the sex of the parent that transmits the gene to the offspring. Epigenetic marks are reversible changes to DNA that do not alter the base sequence but may affect how a gene is expressed.

Table 4.5	Sex influences on heredity
Genetic Phenomenon	**Phenotype determined by**
Sex-linked characteristic	Genes located on the sex chromosome
Sex-influenced characteristic	Genes on autosomal chromosomes that are more readily expressed in one sex
Sex-limited characteristic	Autosomal genes whose expression is limited to one sex
Genetic maternal effect	Nuclear genotype of the maternal parent
Cytoplasmic inheritance	Cytoplasmic genes, which are usually inherited entirely from only one parent
Genomic imprinting	Genes whose expression is affected by the sex of the transmitting parent

4.7 The Expression of a Genotype May Be Influenced by Environmental Effects

In Chapter 3, we learned that each phenotype is the result of a genotype developing within a specific environment; the genotype sets the potential for development, but how the phenotype actually develops within the limits imposed by the genotype depends on environmental effects. Stated another way, each genotype may produce several different phenotypes, depending on the environmental conditions in which development takes place. For example, a fruit fly homozygous for the vestigial mutation (*vg vg*) develops reduced wings when raised at a temperature below 29°C, but the same genotype develops much longer wings when raised at 31°C. The range of phenotypes (in this case, wing length) produced by a genotype in different environments is called the **norm of reaction**.

For most of the characteristics discussed so far, the effect of the environment on the phenotype has been slight. Mendel's peas with genotype *yy*, for example, developed green endosperm regardless of the environment in which they were raised. Similarly, persons with genotype $I^A I^A$ have the A antigen on their red blood cells regardless of their diet, socioeconomic status, or family environment. For other phenotypes, however, environmental effects play a more important role.

Environmental Effects on Gene Expression

The expression of some genotypes critically depends on the presence of a specific environment. For example, the *himalayan* allele in rabbits produces dark fur at the extremities of the body—on the nose, ears, and feet

Reared at 20°C or less   Reared at temperatures above 30°C

4.23 The expression of some genotypes depends on specific environments. The expression of a temperature-sensitive allele, *himalayan*, is shown in rabbits reared at different temperatures.

(**Figure 4.23**). The dark pigment develops, however, only when the rabbit is reared at a temperature of 25°C or less; if a Himalayan rabbit is reared at 30°C, no dark patches develop. The expression of the *himalayan* allele is thus temperature dependent; an enzyme necessary for the production of dark pigment is inactivated at higher temperatures. The pigment is restricted to the nose, feet, and ears of a Himalayan rabbit because the animal's core body temperature is normally above 25°C and the enzyme is functional only in the cells of the relatively cool extremities. The *himalayan* allele is an example of a **temperature-sensitive allele**, an allele whose product is functional only at certain temperatures.

Environmental factors also play an important role in the expression of a number of human genetic diseases. Glucose-6-phosphate dehydrogenase is an enzyme that helps to supply energy to the cell. In humans, there are a number of genetic variants of glucose-6-phosphate dehydrogenase, some of which destroy red blood cells when the body is stressed by infection or by the ingestion of certain drugs or foods. The symptoms of the genetic disease, called glucose-6-phosphate dehydrogenase deficiency, appear only in the presence of these specific environmental factors.

These examples illustrate the point that genes and their products do not act in isolation; rather, they frequently interact with environmental factors. Occasionally, environmental factors alone can produce a phenotype that is the same as the phenotype produced by a genotype; this phenotype is called a **phenocopy**. In fruit flies, for example, the autosomal recessive mutation *eyeless* produces greatly reduced eyes. The eyeless phenotype can also be produced by exposing the larvae of normal flies to sodium metaborate.

Concepts

The expression of many genes is modified by the environment. The range of phenotypes produced by a genotype in different environments is called the norm of reaction. A phenocopy is a trait produced by environmental effects that mimics the phenotype produced by a genotype.

The Inheritance of Continuous Characteristics

So far, we've dealt primarily with characteristics that have only a few distinct phenotypes. In Mendel's peas, for example, the seeds were either smooth or wrinkled, yellow or green; the coats of dogs were black, brown, or yellow; blood types were of four distinct types, A, B, AB, or O. Such characteristics, which have a few easily distinguished phenotypes, are called **discontinuous characteristics**.

Not all characteristics exhibit discontinuous phenotypes. Human height is an example of such a characteristic; people do not come in just a few distinct heights but, rather, display a continuum of heights. Indeed, there are so many possible phenotypes of human height that we must use a measurement to describe a person's height. Characteristics that exhibit a continuous distribution of phenotypes are termed **continuous characteristics**. Because such characteristics have many possible phenotypes and must be described in quantitative terms, continuous characteristics are also called **quantitative characteristics**.

Continuous characteristics frequently arise because genes at many loci interact to produce the phenotypes. When a single locus with two alleles encodes a characteristic, there are three genotypes possible: AA, Aa, and aa. With two loci, each with two alleles, there are $3^2 = 9$ genotypes possible. The number of genotypes encoding a characteristic is 3^n, where n equals the number of loci with two alleles that influence the characteristic. For example, when a characteristic is determined by eight loci, each with two alleles, there are $3^8 = 6561$ different genotypes possible for this characteristic. If each genotype produces a different phenotype, many phenotypes will be possible. The slight differences between the phenotypes will be indistinguishable, and the characteristic will appear continuous. Characteristics encoded by genes at many loci are called **polygenic characteristics**.

The converse of polygeny is **pleiotropy**, in which one gene affects multiple characteristics. Many genes exhibit pleiotropy. Phenylketonuria, mentioned earlier, results from a recessive allele; persons homozygous for this allele, if untreated, exhibit mental retardation, blue eyes, and light skin color.

Frequently, the phenotypes of continuous characteristics are also influenced by environmental factors. Each genotype is capable of producing a range of phenotypes: it has a broad norm of reaction. In this situation, the particular phenotype that results depends on both the genotype and the environmental conditions in which the genotype develops. For example, only three genotypes may encode a characteristic, but, because each genotype has a broad norm of reaction, the phenotype of the characteristic exhibits a continuous distribution. Many continuous characteristics are both polygenic and influenced by environmental factors; such characteristics are called **multifactorial characteristics** because many factors help determine the phenotype.

The inheritance of continuous characteristics may appear to be complex, but the alleles at each locus follow Mendel's principles and are inherited in the same way as alleles encoding simple, discontinuous characteristics. However, because many genes participate, because environmental factors influence the phenotype, and because the phenotypes do not sort out into a few distinct types, we cannot observe the distinct ratios that have allowed us to interpret the genetic basis of discontinuous characteristics. To analyze continuous characteristics, we must employ special statistical tools, as will be discussed in Chapter 16.

Concepts

Discontinuous characteristics exhibit a few distinct phenotypes; continuous characteristics exhibit a range of phenotypes. A continuous characteristic is frequently produced when genes at many loci and environmental factors combine to determine a phenotype.

✔ **Concept Check 12**

What is the difference between polygeny and pleiotropy?

Concepts Summary

- Sexual reproduction is the production of offspring that are genetically distinct from their parents. Most organisms have two sexual phenotypes—males and females. Males produce small gametes; females produce large gametes.

- The mechanism by which sex is specified is termed sex determination. Sex may be determined by differences in specific chromosomes, genotypes, or environment.

- Sex chromosomes differ in number and appearance between males and females. The homogametic sex produces gametes that are all identical with regard to sex chromosomes; the heterogametic sex produces gametes that differ in their sex-chromosome composition.

- In the XX-XO system, females possess two X chromosomes, and males possess a single X chromosome. In the XX-XY system, females possess two X chromosomes, and males possess a single X and a single Y chromosome. In the ZZ-ZW system of sex determination, males possess two Z chromosomes and females possess a Z and a W chromosome.

- In some organisms, environmental factors determine sex.

- In *Drosophila melanogaster*, sex is determined by a balance between genes on the X chromosomes and genes on the autosomes, the X : A ratio.

- In humans, sex is ultimately determined by the presence or absence of the *SRY* gene located on the Y chromosome.

- Sex-linked characteristics are determined by genes on the sex chromosomes; X-linked characteristics are encoded by genes on the X chromosome, and Y-linked characteristics are encoded by genes on the Y chromosome.

- A female inherits X-linked alleles from both parents; a male inherits X-linked alleles from his female parent only.

- The fruit fly *Drosophila melanogaster* has a number of characteristics that make it an ideal model organism for genetic studies, including a short generation time, large numbers of progeny, small size, ease of rearing, and a small genome.

- Dosage compensation equalizes the amount of protein produced by X-linked genes in males and females. In placental mammals, one of the two X chromosomes in females normally becomes inactivated. Which X chromosome is inactivated is random and varies from cell to cell.

- Y-linked characteristics are found only in males and are passed from father to all sons.

- Dominance always refers to genes at the same locus (allelic genes). Dominance is complete when a heterozygote has the same phenotype as a homozygote, is incomplete when the heterozygote has a phenotype intermediate between those of two parental homozygotes, and is codominant when the heterozygote exhibits traits of both parental homozygotes.

- Penetrance is the percentage of individuals having a particular genotype that exhibit the expected phenotype. Expressivity is the degree to which a character is expressed.

- Lethal alleles cause the death of an individual possessing them, usually at an early stage of development, and may alter phenotypic ratios.

- Multiple alleles refer to the presence of more than two alleles at a locus within a group. Their presence increases the number of genotypes and phenotypes possible.

- Gene interaction refers to the interaction between genes at different loci to produce a single phenotype. An epistatic gene at one locus suppresses or masks the expression of hypostatic genes at other loci. Gene interaction frequently produces phenotypic ratios that are modifications of dihybrid ratios.

- Sex-influenced characteristics are encoded by autosomal genes that are expressed more readily in one sex. Sex-limited characteristics are encoded by autosomal genes expressed in only one sex.

- In cytoplasmic inheritance, the genes for the characteristic are found in the organelles and are usually inherited from a single (usually maternal) parent. Genetic maternal effect is present when an offspring inherits genes from both parents, but the nuclear genes of the mother determine the offspring's phenotype.

- Genomic imprinting refers to characteristics encoded by autosomal genes whose expression is affected by the sex of the parent transmitting the genes.

- Phenotypes are often modified by environmental effects. A phenocopy is a phenotype produced by an environmental effect that mimics a phenotype produced by a genotype.

- Continuous characteristics are those that exhibit a wide range of phenotypes; they are frequently produced by the combined effects of many genes and environmental effects.

Important Terms

sex (p. 71)
sex determination (p. 71)
sex chromosome (p. 71)
autosome (p. 71)
heterogametic sex (p. 71)
homogametic sex (p. 71)
pseudoautosomal region (p. 72)
genic sex determination (p. 72)
genic balance system (p. 73)
X : A ratio (p. 73)
Turner syndrome (p. 74)
Klinefelter syndrome (p. 74)
triplo-X syndrome (p. 74)
sex-determining region Y (*SRY*)
    gene (p. 74)
sex-linked characteristic (p. 75)
X-linked characteristic (p. 75)

Y-linked characteristic (p. 75)
hemizygosity (p. 75)
dosage compensation (p. 80)
Barr body (p. 80)
Lyon hypothesis (p. 80)
codominance (p. 83)
incomplete penetrance (p. 84)
penetrance (p. 84)
expressivity (p. 84)
lethal allele (p. 85)
multiple alleles (p. 85)
gene interaction (p. 87)
epistasis (p. 88)
epistatic gene (p. 88)
hypostatic gene (p. 88)
complementation test (p. 92)
complementation (p. 92)

sex-influenced characteristic (p. 92)
sex-limited characteristic (p. 92)
cytoplasmic inheritance (p. 93)
genetic maternal effect (p. 94)
genomic imprinting (p. 95)
epigenetics (p. 96)
norm of reaction (p. 96)
temperature-sensitive allele (p. 97)
phenocopy (p. 97)
discontinuous characteristic (p. 97)
continuous characteristic (p. 97)
quantitative characteristic (p. 97)
polygenic characteristic (p. 97)
pleiotropy (p. 97)
multifactorial characteristic (p. 98)

Answers to Concept Checks

1. Meiosis

2. In chromosomal sex determination, males and females have chromosomes that are distinguishable. In genic sex determination, sex is determined by genes but the chromosomes of males and females are indistinguishable. In environmental sex determination, sex is determined by environmental effects.

3. b

4. a

5. All male offspring will have hemophilia, and all female offspring will not have hemophilia; so the overall probability of hemophilia in the offspring is $^1/_2$.

6. Two Barr bodies. A Barr body is an inactivated X chromosome.

7. With complete dominance, the heterozygote expresses the same phenotype as that of one of the homozygotes. With incomplete dominance, the heterozygote has a phenotype that is

intermediate between the two homozygotes. And, with codominance, the heterozygote has a phenotype that simultaneously expresses the phenotypes of both homozygotes.

8. The cross is $Ll \times Ll$, where l is an allele for long fingers and L is an allele for normal fingers. The probability that the child will possess the genotype for long fingers (ll) is $^1/_4$, or 0.25. The trait has a penetrance of 80%, which indicates that a person with the genotype for long fingers has a probability of 0.8 of actually having long fingers. The probability that the child will have long fingers is found by multiplying the probability of the genotype by the probability that a person with that genotype will express the trait: $0.25 \times 0.8 = 0.2$.

9. People with blood-type A can be I^AI^A or I^Ai. People with blood-type B can be either I^BI^B or I^Bi. The types of matings possible between a man with blood-type A and a woman with blood-type B, along with the offspring that each mating would produce, are as follows:

Possible matings	Offspring
$I^A I^A \times I^B I^B$	$I^A I^B$
$I^A I^A \times I^B i$	$I^A I^B, I^A i$
$I^A i \times I^B I^B$	$I^A I^B, I^B i$
$I^A i \times I^B i$	$I^A I^B, I^A i, I^B i, ii$

Thus, the offspring could have blood-types AB, A, B, or O.

10. The 12 all-white : 3 black : 1 gray ratio is a modification of the 9 : 3 : 3 : 1 ratio produced in a cross between two double heterozygotes:

$$Ww\ Gg \times Ww\ Gg$$
$$\downarrow$$

$9/16$ $W\_\ G\_$ all-white

$3/16$ $W\_\ gg$ all-white

$3/16$ $ww\ G\_$ black

$1/16$ $ww\ gg$ gray

Therefore, the all-white cats have a dominant epistatic allele (W) and are genotype $W\_\ G\_$ and $W\_\ gg$, the black cats lack the epistatic allele (W) and have a dominant allele for black ($ww\ G\_$), and the gray cats lack the epistatic W and are recessive for gray ($ww\ gg$). The allele for all-white (W) is a dominant epistatic gene.

11. Both sex-influenced and sex-limited traits are encoded by autosomal genes whose expression is affected by the sex of the individual who possesses the gene. Sex-linked traits are encoded by genes on the sex chromosomes.

12. Polygeny refers to the influence of multiple genes on the expression of a single characteristic. Pleiotropy refers to the effect of a single gene on the expression of multiple characteristics.

Worked Problems

1. In *Drosophila melanogaster*, forked bristles are caused by an allele (X^f) that is X linked and recessive to an allele for normal bristles (X^I). Brown eyes are caused by an allele (b) that is autosomal and recessive to an allele for red eyes (b^+). A female fly that is homozygous for normal bristles and red eyes mates with a male fly that has forked bristles and brown eyes. The F_1 are intercrossed to produce the F_2. What will the phenotypes and proportions of the F_2 flies be from this cross?

• Solution

This problem is best worked by breaking the cross down into two separate crosses, one for the X-linked genes that determine the type of bristles and one for the autosomal genes that determine eye color.

Let's begin with the autosomal characteristics. A female fly that is homozygous for red eyes (b^+b^+) is crossed with a male with brown eyes. Because brown eyes are recessive, the male fly must be homozygous for the brown-eyed allele (bb). All of the offspring of this cross will be heterozygous (b^+b) and will have red eyes:

P b^+b^+ × bb
          Red eyes   Brown eyes
              ↓   ↓

Gametes b^+ b

F_1 b^+b
                 Red eyes

The F_1 are then intercrossed to produce the F_2. Whenever two individual organisms heterozygous for an autosomal recessive characteristic are crossed, $3/4$ of the offspring will have the

dominant trait and will have the recessive trait; thus, of the F_2 flies will have red eyes and $1/4$ will have brown eyes:

F_1 b^+b × b^+b
          Red eyes   Red eyes
              ↓   ↓

Gametes b^+ b b^+ b

F_2 $1/4\ b^+b^+$ red

               $1/2\ b^+b$ red

               $1/4\ bb$ brown

               $3/4$ red, $1/4$ brown

Next, we work out the results for the X-linked characteristic. A female that is homozygous for normal bristles (X^+X^+) is crossed with a male that has forked bristles (X^fY). The female F_1 from this cross are heterozygous (X^+X^f), receiving an X chromosome with a normal-bristle allele from their mother (X^+) and an X chromosome with a forked-bristle allele (X^f) from their father. The male F_1 are hemizygous (X^+Y), receiving an X chromosome with a normal-bristle allele from their mother (X^+) and a Y chromosome from their father:

P X^+X^+ × X^fY
          Normal   Forked
          bristles   bristles
              ↓   ↓

Gametes X^+ X^f Y

F_1 $1/2\ X^+X^f$ normal bristle

               $1/2\ X^+Y$ normal bristle

When these F_1 are intercrossed, $\frac{1}{2}$ of the F_2 will be normal-bristle females, $\frac{1}{4}$ will be normal-bristle males, and $\frac{1}{4}$ will be forked-bristle males:

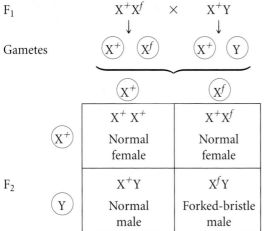

F_1 $\qquad$ X^+X^f $\qquad \times \qquad$ X^+Y

Gametes

	X^+	X^f
X^+	X^+X^+ Normal female	X^+X^f Normal female
Y	X^+Y Normal male	X^fY Forked-bristle male

F_2

$\frac{1}{2}$ normal female, $\frac{1}{4}$ normal male, $\frac{1}{4}$ forked-bristle male

To obtain the phenotypic ratio in the F_2, we now combine these two crosses by using the multiplication rule of probability and the branch diagram:

Eye color	Bristle and sex	F_2 phenotype	Probability
red ($\frac{3}{4}$)	normal female ($\frac{1}{2}$)	red normal female	$\frac{3}{4} \times \frac{1}{2} = \frac{3}{8}$ $= \frac{6}{16}$
	normal male ($\frac{1}{4}$)	red normal male	$\frac{3}{4} \times \frac{1}{4} = \frac{3}{16}$
	forked-bristled male ($\frac{1}{4}$)	red forked male	$\frac{3}{4} \times \frac{1}{4} = \frac{3}{16}$
brown ($\frac{1}{4}$)	normal female ($\frac{1}{2}$)	brown normal female	$\frac{1}{4} \times \frac{1}{2} = \frac{1}{8}$ $= \frac{2}{16}$
	normal male ($\frac{1}{4}$)	brown normal male	$\frac{1}{4} \times \frac{1}{4} = \frac{1}{16}$
	forked-bristled male ($\frac{1}{4}$)	brown forked male	$\frac{1}{4} \times \frac{1}{4} = \frac{1}{16}$

2. The type of plumage found in mallard ducks is determined by three alleles at a single locus: M^R, which encodes restricted plumage; M, which encodes mallard plumage; and m^d, which encodes dusky plumage. The restricted phenotype is dominant over mallard and dusky; mallard is dominant over dusky ($M^R > M > m^d$). Give the expected phenotypes and proportions of offspring produced by the following crosses.

a. $M^RM \times m^dm^d$ $\qquad$ **c.** $M^Rm^d \times M^RM$

b. $M^Rm^d \times Mm^d$ $\qquad$ **d.** $M^RM \times Mm^d$

• Solution

We can determine the phenotypes and proportions of offspring by (1) determining the types of gametes produced by each parent

and (2) combining the gametes of the two parents with the use of a Punnett square.

a. Parents $\qquad$ $M^RM \qquad \times \qquad m^dm^d$

Gametes $\qquad$ $M^R \quad M \qquad\qquad m^d$

	M^R	M
m^d	M^Rm^d Restricted	Mm^d Mallard

$\frac{1}{2}$ restricted, $\frac{1}{2}$ mallard

b. Parents $\qquad$ $M^Rm^d \qquad \times \qquad Mm^d$

Gametes $\qquad$ $M^R \quad m^d \qquad\qquad M \quad m^d$

	M	m^d
M^R	M^RM Restricted	M^Rm^d Restricted
m^d	Mm^d Mallard	m^dm^d Dusky

$\frac{1}{2}$ restricted, $\frac{1}{4}$ mallard, $\frac{1}{4}$ dusky

c. Parents $\qquad$ $M^Rm^d \qquad \times \qquad M^RM$

Gametes $\qquad$ $M^R \quad m^d \qquad\qquad M^R \quad M$

	M^R	M
M^R	M^RM^R Restricted	M^RM Restricted
m^d	M^Rm^d Restricted	Mm^d Mallard

$\frac{3}{4}$ restricted, $\frac{1}{4}$ mallard

d. Parents $\qquad$ $M^RM \qquad \times \qquad Mm^d$

Gametes $\qquad$ $M^R \quad M \qquad\qquad M \quad m^d$

	M	m^d
M^R	M^RM Restricted	M^Rm^d Restricted
M	MM Mallard	Mm^d Mallard

$\frac{1}{2}$ restricted, $\frac{1}{2}$ mallard

3. In some sheep, the presence of horns is produced by an autosomal allele that is dominant in males and recessive in females. A horned female is crossed with a hornless male. One of the resulting F_1 females is crossed with a hornless male. What proportion of the male and female progeny from this cross will have horns?

• **Solution**

The presence of horns in these sheep is an example of a sex-influenced characteristic. Because the phenotypes associated with the genotypes differ for the two sexes, let's begin this problem by writing out the genotypes and phenotypes for each sex. We will let H represent the allele that encodes horns and H^+ represent the allele that encodes hornless. In males, the allele for horns is dominant over the allele for hornless, which means that males homozygous (HH) and heterozygous (H^+H) for this gene are horned. Only males homozygous for the recessive hornless allele (H^+H^+) will be hornless. In females, the allele for horns is recessive, which means that only females homozygous for this allele (HH) will be horned; females heterozygous (H^+H) and homozygous (H^+H^+) for the hornless allele will be hornless. The following table summarizes genotypes and associated phenotypes:

Genotype	Male phenotype	Female phenotype
HH	horned	horned
HH^+	horned	hornless
H^+H^+	hornless	hornless

In the problem, a horned female is crossed with a hornless male. From the preceding table, we see that a horned female must be homozygous for the allele for horns (HH) and a hornless male must be homozygous for the allele for hornless (H^+H^+); so all the F_1 will be heterozygous; the F_1 males will be horned and the F_1 females will be hornless, as shown in the following diagram:

$$P \qquad\qquad H^+H^+ \quad \times \quad HH$$
$$\downarrow$$
$$F_1 \qquad\qquad H^+H$$

Horned males and hornless females

A heterozygous hornless F_1 female (H^+H) is then crossed with a hornless male (H^+H^+):

$$H^+H \qquad \times \qquad H^+H^+$$

Hornless female   Hornless male

$$\downarrow$$

	Males	Females
$\frac{1}{2} H^+H^+$	hornless	hornless
$\frac{1}{2} H^+H$	horned	hornless

Therefore, $\frac{1}{2}$ of the male progeny will be horned, but none of the female progeny will be horned.

Comprehension Questions

Section 4.1

1. How does sex determination in the XX-XY system differ from sex determination in the ZZ-ZW system?

2. What is meant by genic sex determination?

3. How does sex determination in *Drosophila* differ from sex determination in humans?

Section 4.2

*4. What characteristics are exhibited by an X-linked trait?

5. Explain why tortoiseshell cats are almost always female and why they have a patchy distribution of orange and black fur.

Section 4.3

*6. How do incomplete dominance and codominance differ?

*7. What is incomplete penetrance and what causes it?

Section 4.5

8. What is gene interaction? What is the difference between an epistatic gene and a hypostatic gene?

*9. What is a complementation test and what is it used for?

Section 4.6

*10. What characteristics are exhibited by a cytoplasmically inherited trait?

Application Questions and Problems

Section 4.1

*11. What is the sexual phenotype of fruit flies having the following chromosomes?

	Sex chromosomes	Autosomal chromosomes
a.	XX	all normal
b.	XY	all normal
c.	XO	all normal
d.	XXY	all normal
e.	XXYY	all normal
f.	XXX	all normal
g.	XXX	four haploid sets
h.	X	three haploid sets
i.	XY	three haploid sets

Section 4.2

*12. Joe has classic hemophilia, an X-linked recessive disease. Could Joe have inherited the gene for this disease from the following persons?

	Yes	No
a. His mother's mother	_____	_____
b. His mother's father	_____	_____
c. His father's mother	_____	_____
d. His father's father	_____	_____

*13. In *Drosophila*, yellow body is due to an X-linked gene that is recessive to the gene for gray body.

   a. A homozygous gray female is crossed with a yellow male. The F_1 are intercrossed to produce F_2. Give the genotypes and phenotypes, along with the expected proportions, of the F_1 and F_2 progeny.

   b. A yellow female is crossed with a gray male. The F_1 are intercrossed to produce the F_2. Give the genotypes and phenotypes, along with the expected proportions, of the F_1 and F_2 progeny.

*14. Red–green color blindness in humans is due to an X-linked recessive gene. Both John and Cathy have normal color vision. After 10 years of marriage to John, Cathy gave birth to a color-blind daughter. John filed for divorce, claiming that he is not the father of the child. Is John justified in his claim of nonpaternity? Explain why. If Cathy had given birth to a color-blind son, would John be justified in claiming nonpaternity?

*15. The following pedigree illustrates the inheritance of Nance–Horan syndrome, a rare genetic condition in which affected persons have cataracts and abnormally shaped teeth.

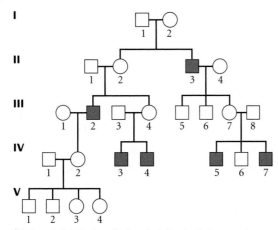

(Pedigree after D. Stambollan, R. A. Lewis, K. Buetow, A. Bond, and R. Nussbaum. 1990. *American Journal of Human Genetics* 47:15.)

   a. On the basis of this pedigree, what do you think is the most likely mode of inheritance for Nance–Horan syndrome?

   b. If couple III-7 and III-8 have another child, what is the probability that the child will have Nance–Horan syndrome?

   c. If III-2 and III-7 were to mate, what is the probability that one of their children would have Nance–Horan syndrome?

*16. Bob has XXY chromosomes (Klinefelter syndrome) and is color blind. His mother and father have normal color vision, but his maternal grandfather is color blind. Assume that Bob's chromosome abnormality arose from nondisjunction in meiosis. In which parent and in which meiotic division did nondisjunction occur? Explain your answer.

17. The Talmud, an ancient book of Jewish civil and religious laws, states that, if a woman bears two sons who die of bleeding after circumcision (removal of the foreskin from the penis), any additional sons that she has should not be circumcised. (The bleeding is most likely due to the X-linked disorder hemophilia.) Furthermore, the Talmud states that the sons of her sisters must not be circumcised, whereas the sons of her brothers should. Is this religious law consistent with sound genetic principles? Explain your answer.

18. Craniofrontonasal syndrome (CFNS) is a birth defect in which premature fusion of the cranial sutures leads to abnormal head shape, widely spaced eyes, nasal clefts, and various other skeletal abnormalities. George Feldman and his colleagues, looked at several families in which CFNS occurred and recorded the results shown in the following table (G. J. Feldman. 1997. *Human Molecular Genetics* 6:1937–1941).

Family number	Parents		Offspring			
			Normal		CFNS	
	Father	Mother	Male	Female	Male	Female
1	normal	CFNS	1	0	2	1
5	normal	CFNS	0	2	1	2
6	normal	CFNS	0	0	1	2
8	normal	CFNS	1	1	1	0
10a	CFNS	normal	3	0	0	2
10b	normal	CFNS	1	1	2	0
12	CFNS	normal	0	0	0	1
13a	normal	CFNS	0	1	2	1
13b	CFNS	normal	0	0	0	2
7b	CFNS	normal	0	0	0	2

   a. On the basis of the families given, what is the most likely mode of inheritance for CFNS?

   b. Give the most likely genotypes of the parents in families numbered 1 and 10a.

*19. How many Barr bodies would you expect to see in a human cell containing the following chromosomes?

a. XX   d. XXY   g. XYY
b. XY   e. XXYY   h. XXX
c. XO   f. XXXY   i. XXXX

*20. Miniature wings in *Drosophila melanogaster* result from an X-linked gene (X^m) that is recessive to an allele for long wings (X^+). Sepia eyes are produced by an autosomal gene (s) that is recessive to an allele for red eyes (s^+).

a. A female fly that has miniature wings and sepia eyes is crossed with a male that has normal wings and is homozygous for red eyes. The F_1 are intercrossed to produce the F_2. Give the phenotypes and their proportions expected in the F_1 and F_2 flies from this cross.

b. A female fly that is homozygous for normal wings and has sepia eyes is crossed with a male that has miniature wings and is homozygous for red eyes. The F_1 are intercrossed to produce the F_2. Give the phenotypes and their proportions expected in the F_1 and F_2 flies from this cross.

Section 4.3

*21. Palomino horses have a golden yellow coat, chestnut horses have a brown coat, and cremello horses have a coat that is almost white. A series of crosses between the three different types of horses produced the following offspring:

Cross	Offspring
palomino × palomino	13 palomino, 6 chestnut, 5 cremello
chestnut × chestnut	16 chestnut
cremello × cremello	13 cremello
palomino × chestnut	8 palomino, 9 chestnut
palomino × cremello	11 palomino, 11 cremello
chestnut × cremello	23 palomino

a. Explain the inheritance of the palomino, chestnut, and cremello phenotypes in horses.

b. Assign symbols for the alleles that determine these phenotypes, and list the genotypes of all parents and offspring given in the preceding table.

*22. The L^M and L^N alleles at the MN blood group locus exhibit codominance. Give the expected genotypes and phenotypes and their ratios in progeny resulting from the following crosses.

a. $L^M L^M \times L^M L^N$ d. $L^M L^N \times L^N L^N$
b. $L^N L^N \times L^N L^N$ e. $L^M L^M \times L^N L^N$
c. $L^M L^N \times L^M L^N$

*23. When a Chinese hamster with white spots is crossed with another hamster that has no spots, approximately $1/2$ of the offspring have white spots and $1/2$ have no spots. When two

hamsters with white spots are crossed, $2/3$ of the offspring possess white spots and $1/3$ have no spots.

a. What is the genetic basis of white spotting in Chinese hamsters?

b. How might you go about producing Chinese hamsters that breed true for white spotting?

24. As discussed in the introduction to this chapter, Cuénot studied the genetic basis of yellow coat color in mice. He carried out a number of crosses between two yellow mice and obtained what he thought was a 3 : 1 ratio of yellow to gray mice in the progeny. The following table gives Cuénot's actual results, along with the results of a much larger series of crosses carried out by Castle and Little (W. E. Castle and C. C. Little. 1910. *Science* 32:868–870).

Progeny Resulting from Crosses of Yellow × Yellow Mice

Investigators	Yellow progeny	Nonyellow progeny	Total progeny
Cuénot	263	100	363
Castle and Little	800	435	1235
Both combined	1063	535	1598

a. Using a chi-square test, determine whether Cuénot's results are significantly different from the 3 : 1 ratio that he thought he observed. Are they different from a 2 : 1 ratio?

b. Determine whether Castle and Little's results are significantly different from a 3 : 1 ratio. Are they different from a 2 : 1 ratio?

c. Combine the results of Castle and Cuénot and determine whether they are significantly different from a 3 : 1 ratio and a 2 : 1 ratio.

d. Offer an explanation for the different ratios that Cuénot and Castle obtained.

Section 4.4

25. In this chapter, we considered Joan Barry's paternity suit against Charlie Chaplin and how, on the basis of blood types, Chaplin could not have been the father of her child.

a. What blood types are possible for the father of Barry's child?

b. If Chaplin had possessed one of these blood types, would that prove that he fathered Barry's child?

Section 4.5

*26. In chickens, comb shape is determined by alleles at two loci (R, r and P, p). A walnut comb is produced when at least one dominant allele R is present at one locus and at least one dominant allele P is present at a second locus (genotype $R\_ P\_$). A rose comb is produced when at least one dominant allele is present at the first locus and two recessive alleles are present at the second locus (genotype $R\_ pp$). A pea comb is produced when two recessive alleles are present at the first locus and at least one dominant allele is present at the second (genotype $rr\ P\_$). If two recessive alleles are present at the first and at the second locus ($rr\ pp$), a single comb is produced. Progeny with what types of combs and in what proportions will result from the following crosses?

a. $RR\ PP \times rr\ pp$

b. $Rr\ Pp \times rr\ pp$

c. $Rr\ Pp \times Rr\ Pp$

d. $Rr\ pp \times Rr\ pp$

e. $Rr\ pp \times rr\ Pp$

f. $Rr\ pp \times rr\ pp$

27. Tatuo Aida investigated the genetic basis of color variation in the Medaka (*Aplocheilus latipes*), a small fish that occurs naturally in Japan (T. Aida. 1921. *Genetics* 6:554–573). Aida found that genes at two loci (B, b and R, r) determine the color of the fish: fish with a dominant allele at both loci ($B\_R\_$) are brown, fish with a dominant allele at the B locus only ($B\_\ rr$) are blue, fish with a dominant allele at the R locus only ($bb\ R\_$) are red, and fish with recessive alleles at both loci ($bb\ rr$) are white. Aida crossed a homozygous brown fish with a homozygous white fish. He then back-crossed the F_1 with the homozygous white parent and obtained 228 brown fish, 230 blue fish, 237 red fish, and 222 white fish.

a. Give the genotypes of the backcross progeny.

b. Use a chi-square test to compare the observed numbers of backcross progeny with the number expected. What conclusion can you make from your chi-square results?

c. What results would you expect for a cross between a homozygous red fish and a white fish?

d. What results would you expect if you crossed a homozygous red fish with a homozygous blue fish and then backcrossed the F_1 with a homozygous red parental fish?

28. E. W. Lindstrom crossed two corn plants with green seedlings and obtained the following progeny: 3583 green seedlings, 853 virescent-white seedlings, and 260 yellow seedlings (E. W. Lindstrom. 1921. *Genetics* 6:91–110).

a. Give the genotypes for the green, virescent-white, and yellow progeny.

b. Provide an explanation for how color is determined in these seedlings.

c. Does epistasis occur among the genes that determine color in the maize seedlings? If so, which gene is epistatic and which is hypostatic.

*29. A summer-squash plant that produces disc-shaped fruit is crossed with a summer-squash plant that produces long fruit. All the F_1 have disc-shaped fruit. When the F_1 are

intercrossed, F_2 progeny are produced in the following ratio: $9/16$ disc-shaped fruit : $6/16$ spherical fruit : $1/16$ long fruit. Give the genotypes of the F_2 progeny.

30. Some sweet-pea plants have purple flowers and other plants have white flowers. A homozygous variety of pea that has purple flowers is crossed with a homozygous variety that has white flowers. All the F_1 have purple flowers. When these F_1 are self-fertilized, the F_2 appear in a ratio of $9/16$ purple to $7/16$ white.

a. Give genotypes for the purple and white flowers in these crosses.

b. Draw a hypothetical biochemical pathway to explain the production of purple and white flowers in sweet peas.

Section 4.6

31. Shell coiling of the snail *Limnaea peregra* results from a genetic maternal effect. An autosomal allele for a right-handed shell (s^+), called dextral, is dominant over the allele for a left-handed shell (s), called sinistral. A pet snail called Martha is sinistral and reproduces only as a female (the snails are hermaphroditic). Indicate which of the following statements are true and which are false. Explain your reasoning in each case.

a. Martha's genotype *must* be ss.

b. Martha's genotype *cannot* be s^+s^+.

c. All the offspring produced by Martha *must* be sinistral.

d. At least some of the offspring produced by Martha *must* be sinistral.

e. Martha's mother *must* have been sinistral.

f. All of Martha's brothers *must* be sinistral.

Section 4.7

32. Which of the following statements is an example of a phenocopy? Explain your reasoning.

a. Phenylketonuria results from a recessive mutation that causes light skin as well as mental retardation.

b. Human height is influenced by genes at many different loci.

c. Dwarf plants and mottled leaves in tomatoes are caused by separate genes that are linked.

d. Vestigial wings in *Drosophila* are produced by a recessive mutation. This trait is also produced by high temperature during development.

e. Intelligence in humans is influenced by both genetic and environmental factors.

Challenge Question

Section 4.2

33. A geneticist discovers a male mouse with greatly enlarged testes in his laboratory colony. He suspects that this trait

results from a new mutation that is either Y linked or autosomal dominant. How could he determine whether the trait is autosomal dominant or Y linked?

5

Linkage, Recombination, and Eukaryotic Gene Mapping

Alfred Henry Sturtevant, an early geneticist, developed the first genetic map. *[Institute Archives, California Institute of Technology.]*

Alfred Sturtevant and the First Genetic Map

In 1909, Thomas Hunt Morgan taught the introduction to zoology class at Columbia University. Seated in the lecture hall were sophomore Alfred Henry Sturtevant and freshman Calvin Bridges. Sturtevant and Bridges were excited by Morgan's teaching style and intrigued by his interest in biological problems. They asked Morgan if they could work in his laboratory and, the following year, both young men were given desks in the "Fly Room," Morgan's research laboratory where the study of *Drosophila* genetics was in its infancy (see pp. 75–76 in Chapter 4). Sturtevant, Bridges, and Morgan's other research students virtually lived in the laboratory, raising fruit flies, designing experiments, and discussing their results.

In the course of their research, Morgan and his students observed that some pairs of genes did not segregate randomly according to Mendel's principle of independent assortment but instead tended to be inherited together. Morgan suggested that possibly the genes were located on the same chromosome and thus traveled together during meiosis. He further proposed that closely linked genes—those that are rarely shuffled by recombination—lie close together on the same chromosome, whereas loosely linked genes—those more frequently shuffled by recombination—lie farther apart.

One day in 1911, Sturtevant and Morgan were discussing independent assortment when, suddenly, Sturtevant had a flash of inspiration: variation in the strength of linkage indicated how genes are positioned along a chromosome, providing a way of mapping genes. Sturtevant went home and, neglecting his undergraduate homework, spent most of the night working out the first genetic map (**Figure 5.1**). Sturtevant's first chromosome map was remarkably accurate, and it established the basic methodology used today for mapping genes.

Sturtevant went on to become a leading geneticist. His research included gene mapping and basic mechanisms of inheritance in *Drosophila*, cytology, embryology, and evolution. Sturtevant's career was deeply influenced by his early years in the Fly Room, where Morgan's unique personality and the close quarters combined to stimulate intellectual excitement and the free exchange of ideas.

Sturtevant's symbols:	**B C**		**P R**		**M**
X-chromosome locations:	**00 10**		**30.7 33.7**		**57.6**
Modern symbols:	*y w*		*v m*		*r*

| Yellow | White | | Vermilion | Miniature | | Rudimentary |
| body | eyes | | eyes | wings | | wings |

5.1   Sturtevant's map included five genes on the X chromosome of *Drosophila*. The genes are yellow body (*y*), white eyes (*w*), vermilion eyes (*v*), miniature wings (*m*), and rudimentary wings (*r*).

This chapter explores the inheritance of genes located on the same chromosome. These linked genes do not strictly obey Mendel's principle of independent assortment; rather, they tend to be inherited together. This tendency requires a new approach to understanding their inheritance and predicting the types of offspring produced. A critical piece of information necessary for predicting the results of these crosses is the arrangement of the genes on the chromosomes; thus, it will be necessary to think about the relation between genes and chromosomes. A key to understanding the inheritance of linked genes is to make the conceptual connection between the genotypes in a cross and the behavior of chromosomes in meiosis.

We will begin our exploration of linkage by first comparing the inheritance of two linked genes with the inheritance of two genes that assort independently. We will then examine how crossing over breaks up linked genes. This knowledge of linkage and recombination will be used for predicting the results of genetic crosses in which genes are linked and for mapping genes. Later in the chapter, we will focus on physical methods of determining the chromosomal locations of genes.

5.1 Linked Genes Do Not Assort Independently

Chapter 3 introduced Mendel's principles of segregation and independent assortment. Let's take a moment to review these two important concepts. The principle of segregation states that each individual diploid organism possesses two alleles at a locus that separate in meiosis, with one allele going into each gamete. The principle of independent assortment provides additional information about the process of segregation: it tells us that, in the process of separation, the two alleles at a locus act independently of alleles at other loci.

The independent separation of alleles results in *recombination*, the sorting of alleles into new combinations. Consider a cross between individuals homozygous for two different pairs of alleles: *AA BB* × *aa bb*. The first parent, *AA BB*, produces gametes with alleles *A B*, and the second parent, *aa bb*, produces gametes with the alleles *a b*, resulting

in F₁ progeny with genotype *Aa Bb* (**Figure 5.2**). Recombination means that, when one of the F₁ progeny reproduces, the combination of alleles in its gametes may differ from the combinations in the gametes from its parents. In other words, the F₁ may produce gametes with alleles *A b* or *a B* in addition to gametes with *A B* or *a b*.

Mendel derived his principles of segregation and independent assortment by observing the progeny of genetic crosses, but he had no idea of what biological processes produced these phenomena. In 1903, Walter Sutton proposed a biological basis for Mendel's principles, called the chromosome theory of heredity (see Chapter 3). This theory holds that genes are found on chromosomes. Let's restate Mendel's two principles in relation to the chromosome theory of heredity. The principle of segregation states that a diploid organism possesses two alleles for a trait, each of which is

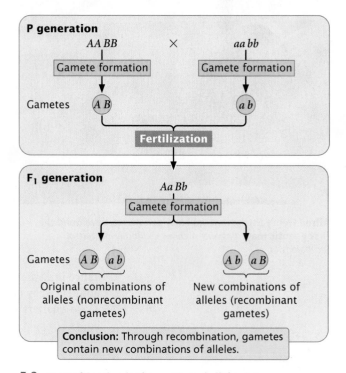

5.2   Recombination is the sorting of alleles into new combinations.

located at the same position, or locus, on each of the two homologous chromosomes. These chromosomes segregate in meiosis, with each gamete receiving one homolog. The principle of independent assortment states that, in meiosis, each pair of homologous chromosomes assorts independently of other homologous pairs. With this new perspective, it is easy to see that the number of chromosomes in most organisms is limited and that there are certain to be more genes than chromosomes; so some genes must be present on the same chromosome and should not assort independently. Genes located close together on the same chromosome are called **linked genes** and belong to the same **linkage group**. Linked genes travel together during meiosis, eventually arriving at the same destination (the same gamete), and are not expected to assort independently.

All of the characteristics examined by Mendel in peas did display independent assortment and, after the rediscovery of Mendel's work, the first genetic characteristics studied in other organisms also seemed to assort independently. How could genes be carried on a limited number of chromosomes and yet assort independently?

The apparent inconsistency between the principle of independent assortment and the chromosome theory of heredity soon disappeared as biologists began finding genetic characteristics that did not assort independently. One of the first cases was reported in sweet peas by William Bateson, Edith Rebecca Saunders, and Reginald C. Punnett in 1905. They crossed a homozygous strain of peas having purple flowers and long pollen grains with a homozygous strain having red flowers and round pollen grains. All the F_1 had purple flowers and long pollen grains, indicating that purple was dominant over red and long was dominant over round. When they intercrossed the F_1, the resulting F_2 progeny did not appear in the 9 : 3 : 3 : 1 ratio expected with independent assortment (**Figure 5.3**). An excess of F_2 plants had purple flowers and long pollen or red flowers and round pollen (the parental phenotypes). Although Bateson, Saunders, and Punnett were unable to explain these results, we now know that the two loci that they examined lie close together on the same chromosome and therefore do not assort independently.

5.2 Linked Genes Segregate Together and Crossing Over Produces Recombination Between Them

Genes that are close together on the same chromosome usually segregate as a unit and are therefore inherited together. However, genes occasionally switch from one homologous chromosome to the other through the process of crossing over (see Chapter 2), as illustrated in **Figure 5.4**. Crossing over results in recombination; it breaks up the associations

Experiment

Question: Do the genes for flower color and pollen shape in sweet peas assort independently?

Methods Cross two strains homozygous for two different traits.

P generation
Homozygous strains

Purple flowers, long pollen   ×   Red flowers, round pollen

Pollen

Fertilization

F₁ generation

Purple flowers, long pollen

Self-fertilization

Results   **F₂ generation**

| 284 Purple flowers, long pollen | 21 Purple flowers, round pollen | 21 Red flowers, long pollen | 55 Red flowers, round pollen |

Conclusion: F_2 progeny do not appear in the 9 : 3 : 3 : 1 ratio expected with independent assortment.

5.3 **Nonindependent assortment of flower color and pollen shape in sweet peas.**

of genes that are close together on the same chromosome. Linkage and crossing over can be seen as processes that have opposite effects: linkage keeps particular genes together, and crossing over mixes them up. In Chapter 4 we considered a number of exceptions and extensions to Mendel's principles of heredity. The concept of linked genes adds a further complication to interpretations of the results of genetic crosses. However, with an understanding of how linkage affects

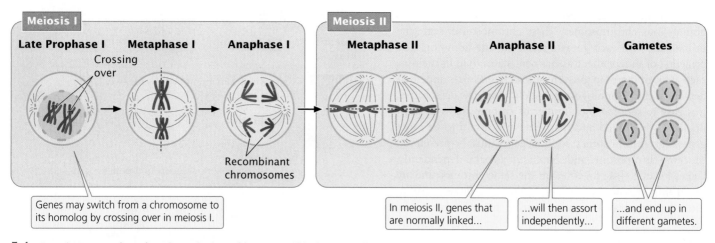

5.4 Crossing over takes place in meiosis and is responsible for recombination.

heredity, we can analyze crosses for linked genes and successfully predict the types of progeny that will be produced.

Notation for Crosses with Linkage

In analyzing crosses with linked genes, we must know not only the genotypes of the individuals crossed, but also the arrangement of the genes on the chromosomes. To keep track of this arrangement, we introduce a new system of notation for presenting crosses with linked genes. Consider a cross between an individual homozygous for dominant alleles at two linked loci and another individual homozygous for recessive alleles at those loci ($AA\ BB \times aa\ bb$). For linked genes, it's necessary to write out the specific alleles as they are arranged on each of the homologous chromosomes:

$$\frac{A\quad B}{A\quad B} \times \frac{a\quad b}{a\quad b}$$

In this notation, each line represents one of the two homologous chromosomes. Inheriting one chromosome from each parent, the F_1 progeny will have the following genotype:

$$\frac{A\quad B}{a\quad b}$$

Here, the importance of designating the alleles on each chromosome is clear. One chromosome has the two dominant alleles A and B, whereas the homologous chromosome has the two recessive alleles a and b. The notation can be simplified by drawing only a single line, with the understanding that genes located on the same side of the line lie on the same chromosome:

$$\frac{A\quad B}{a\quad b}$$

This notation can be simplified further by separating the alleles on each chromosome with a slash: AB/ab.

Remember that the two alleles at a locus are always located on different homologous chromosomes and therefore must lie on opposite sides of the line. Consequently, we would *never* write the genotypes as

$$\frac{A\quad a}{B\quad b}$$

because the alleles A and a can *never* be on the same chromosome.

It is also important to always keep the same order of the genes on both sides of the line; thus, we should *never* write

$$\frac{A\quad B}{b\quad a}$$

because it would imply that alleles A and b are allelic (at the same locus).

Complete Linkage Compared with Independent Assortment

We will first consider what happens to genes that exhibit complete linkage, meaning that they are located very close together on the same chromosome and do not exhibit crossing over. Genes are rarely completely linked but, by assuming that no crossing over occurs, we can see the effect of linkage more clearly. We will then consider what happens when genes assort independently. Finally, we will consider the results obtained if the genes are linked but exhibit some crossing over.

A testcross reveals the effects of linkage. For example, if a heterozygous individual is test-crossed with a homozygous recessive individual ($Aa\ Bb \times aa\ bb$), the alleles that are present in the gametes contributed by the heterozygous parent will be expressed in the phenotype of the offspring, because the homozygous parent could not contribute dominant alleles that might mask them. Consequently, traits that appear in the progeny reveal which alleles were transmitted by the heterozygous parent.

Consider a pair of linked genes in tomato plants. One pair affects the type of leaf: an allele for mottled leaves (m) is recessive to an allele that produces normal leaves (M). Nearby on the same chromosome is another locus that determines the height of the plant: an allele for dwarf (d) is recessive to an allele for tall (D).

Testing for linkage can be done with a testcross, which requires a plant heterozygous for both characteristics. A geneticist might produce this heterozygous plant by crossing a variety of tomato that is homozygous for normal leaves and tall height with a variety that is homozygous for mottled leaves and dwarf height:

$$\text{P} \qquad \frac{M \quad D}{M \quad D} \times \frac{m \quad d}{m \quad d}$$
$$\downarrow$$
$$\text{F}_1 \qquad \frac{M \quad D}{m \quad d}$$

The geneticist would then use these F_1 heterozygotes in a testcross, crossing them with plants homozygous for mottled leaves and dwarf height:

$$\frac{M \quad D}{m \quad d} \times \frac{m \quad d}{m \quad d}$$

The results of this testcross are diagrammed in **Figure 5.5a**. The heterozygote produces two types of gametes: some with the $\underline{M \quad D}$ chromosome and others with the $\underline{m \quad d}$ chromosome. Because no crossing over occurs, these gametes are the only types produced by the heterozygote. Notice that these gametes contain only combinations of alleles that were present in the original parents: either the allele for normal leaves together with the allele for tall height (M and D) or the allele for mottled leaves together with the allele for dwarf height (m and d). Gametes that contain only original combinations of alleles present in the parents are **nonrecombinant gametes**, or *parental* gametes.

The homozygous parent in the testcross produces only one type of gamete; it contains chromosome $\underline{m \quad d}$ and pairs with one of the two gametes generated by the heterozygous parent (see Figure 5.5a). Two types of progeny result: half have normal leaves and are tall:

$$\frac{M \quad D}{m \quad d}$$

and half have mottled leaves and are dwarf:

$$\frac{m \quad d}{m \quad d}$$

These progeny display the original combinations of traits present in the P generation and are **nonrecombinant progeny**, or *parental* progeny. No new combinations of the two traits, such as normal leaves with dwarf or mottled leaves with tall, appear in the offspring, because the genes affecting the two traits are completely linked and are inherited together. New combinations of traits could arise only if the physical connection between M and D or between m and d were broken.

These results are distinctly different from the results that are expected when genes assort independently (**Figure 5.5b**). If the M and D loci assorted independently, the heterozygous plant ($Mm\ Dd$) would produce four types of gametes: two nonrecombinant gametes containing the original combinations of alleles ($M\ D$ and $m\ d$) and two gametes containing new combinations of alleles ($M\ d$ and $m\ D$). Gametes with new combinations of alleles are called **recombinant gametes**. With independent assortment, nonrecombinant and recombinant gametes are produced in equal proportions. These four types of gametes join with the single type of gamete produced by the homozygous parent of the testcross to produce four kinds of progeny in equal proportions (see Figure 5.5b). The progeny with new combinations of traits formed from recombinant gametes are termed **recombinant progeny**.

In summary, a testcross in which one of the plants is heterozygous for two completely linked genes yields two types of progeny, each type displaying one of the original combinations of traits present in the P generation. Independent assortment, in contrast, produces progeny in a 1 : 1 : 1 : 1 ratio. That is, there are four types of progeny—two types of recombinant progeny and two types of nonrecombinant progeny in equal proportions.

Crossing Over with Linked Genes

Usually, there is some crossing over between genes that lie on the same chromosome, producing new combinations of traits. Genes that exhibit crossing over are incompletely linked. Let's see how it takes place.

Theory   The effect of crossing over on the inheritance of two linked genes is shown in **Figure 5.6**. Crossing over, which takes place in prophase I of meiosis, is the exchange of genetic material between nonsister chromatids (see Figures 2.12 and 2.14). After a single crossover has taken place, the two chromatids that did not participate in crossing over are unchanged; gametes that receive these chromatids are nonrecombinants. The other two chromatids, which did participate in crossing over, now contain new combinations of alleles; gametes that receive these chromatids are recombinants. For each meiosis in which a single crossover takes place, then, two nonrecombinant gametes and two recombinant gametes will be produced. This result is the same as that produced by independent assortment (see Figure 5.5b); so, when crossing over between two loci takes place in every meiosis, it is impossible to determine whether the genes are on the same chromosome and crossing over took place or whether the genes are on different chromosomes.

For closely linked genes, crossing over does not take place in every meiosis. In meioses in which there is no

(a) If genes are completely linked (no crossing over)

Normal leaves, tall × Mottled leaves, dwarf

M D / m d × m d / m d

Gamete formation   Gamete formation

½ (M D) ½ (m d)   (m d)
Nonrecombinant gametes

Fertilization

Normal leaves, tall   Mottled leaves, dwarf

½ M D / m d   ½ m d / m d

All nonrecombinant progeny

Conclusion: With complete linkage, only nonrecombinant progeny are produced.

(b) If genes are unlinked (assort independently)

Normal leaves, tall × Mottled leaves, dwarf

Mm Dd × mm dd

Gamete formation   Gamete formation

¼ (M D) ¼ (m d)   ¼ (M d) ¼ (m D)   (m d)
Nonrecombinant gametes   Recombinant gametes

Fertilization

Normal leaves, tall   Mottled leaves, dwarf   Normal leaves, dwarf   Mottled leaves, tall

¼ Mm Dd   ¼ mm dd   ¼ Mm dd   ¼ mm Dd
Nonrecombinant progeny   Recombinant progeny

Conclusion: With independent assortment, half the progeny are recombinant and half the progeny are not.

5.5 A testcross reveals the effects of linkage. Results of a testcross for two loci in tomatoes that determine leaf type and plant height.

crossing over, only nonrecombinant gametes are produced. In meioses in which there is a single crossover, half the gametes are recombinants and half are nonrecombinants (because a single crossover affects only two of the four chromatids); so the total percentage of recombinant gametes is always half the percentage of meioses in which

crossing over takes place. Even if crossing over between two genes takes place in every meiosis, only 50% of the resulting gametes will be recombinants. Thus, the frequency of recombinant gametes is always half the frequency of crossing over, and the maximum proportion of recombinant gametes is 50%.

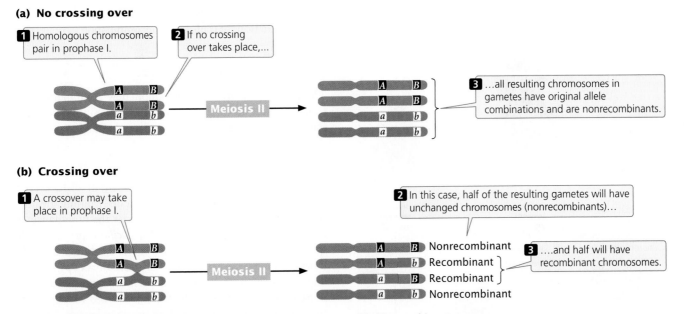

5.6 **A single crossover produces half nonrecombinant gametes and half recombinant gametes.**

Concepts

Linkage between genes causes them to be inherited together and reduces recombination; crossing over breaks up the associations of such genes. In a testcross for two linked genes, each crossover produces two recombinant gametes and two nonrecombinants. The frequency of recombinant gametes is half the frequency of crossing over, and the maximum frequency of recombinant gametes is 50%.

✔ Concept Check 1

For single crossovers, the frequency of recombinant gametes is half the frequency of crossing over because

a. a test cross between a homozygote and heterozygote produces $\frac{1}{2}$ heterozygous and $\frac{1}{2}$ homozygous progeny.

b. the frequency of recombination is always 50%.

c. each crossover takes place between only two of the four chromatids of a homologous pair.

d. crossovers occur in about 50% of meioses.

Application   Let's apply what we have learned about linkage and recombination to a cross between tomato plants that differ in the genes that encode leaf type and plant height. Assume now that these genes are linked and that some crossing over takes place between them. Suppose a geneticist carried out the testcross outlined earlier:

$$\frac{M \quad D}{m \quad d} \times \frac{m \quad d}{m \quad d}$$

When crossing over takes place between the genes for leaf type and height, two of the four gametes produced will be recombinants. When there is no crossing over, all four result-ing gametes will be nonrecombinants. Thus, over all meioses, the majority of gametes will be nonrecombinants. These gametes then unite with gametes produced by the homozygous recessive parent, which contain only the recessive alleles, resulting in mostly nonrecombinant progeny and a few recombinant progeny (**Figure 5.7**). In this cross, we see that 55 of the testcross progeny have normal leaves and are tall and 53 have mottled leaves and are dwarf. These plants are the nonrecombinant progeny, containing the original combinations of traits that were present in the parents. Of the 123 progeny, 15 have new combinations of traits that were not seen in the parents: 8 are normal leaved and dwarf, and 7 are mottle leaved and tall. These plants are the recombinant progeny.

The results of a cross such as the one illustrated in Figure 5.7 reveal several things. A testcross for two independently assorting genes is expected to produce a 1 : 1 : 1 : 1 phenotypic ratio in the progeny. The progeny of this cross clearly do not exhibit such a ratio; so we might suspect that the genes are not assorting independently. When linked genes undergo some crossing over, the result is mostly nonrecombinant progeny and fewer recombinant progeny. This result is what we observe among the progeny of the testcross illustrated in Figure 5.7; so we conclude that the two genes show evidence of linkage with some crossing over.

Calculating Recombination Frequency

The percentage of recombinant progeny produced in a cross is called the **recombination frequency**, which is calculated as follows:

$$\text{recombinant frequency} = \frac{\text{number of recombinant progeny}}{\text{total number of progeny}} \times 100\%$$

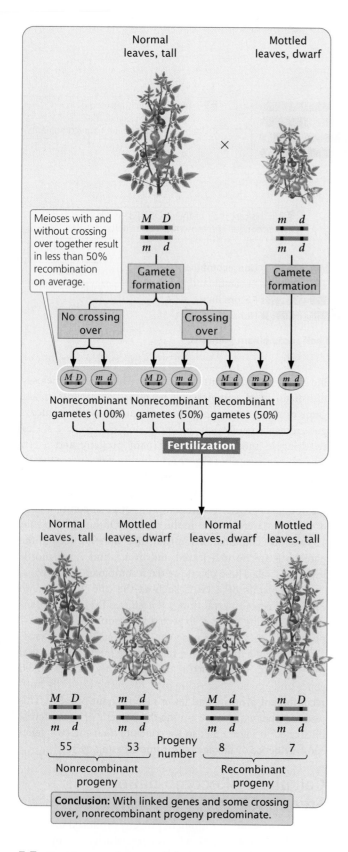

Meioses with and without crossing over together result in less than 50% recombination on average.

Normal leaves, tall × Mottled leaves, dwarf

$M\ D$
$m\ d$

$m\ d$
$m\ d$

Gamete formation   Gamete formation

No crossing over   Crossing over

$M\ D$ $m\ d$ $M\ D$ $m\ d$ $M\ d$ $m\ D$ $m\ d$

Nonrecombinant gametes (100%)   Nonrecombinant gametes (50%)   Recombinant gametes (50%)

Fertilization

Normal leaves, tall   Mottled leaves, dwarf   Normal leaves, dwarf   Mottled leaves, tall

$M\ D$
$m\ d$

$m\ d$
$m\ d$

$M\ d$
$m\ d$

$m\ D$
$m\ d$

55   53   Progeny number   8   7

Nonrecombinant progeny   Recombinant progeny

Conclusion: With linked genes and some crossing over, nonrecombinant progeny predominate.

5.7 Crossing over between linked genes produces nonrecombinant and recombinant offspring.
In this testcross, genes are linked and there is some crossing over.

In the testcross shown in Figure 5.7, 15 progeny exhibit new combinations of traits; so the recombination frequency is:

$$\frac{8+7}{55+53+8+7} \times 100\% = \frac{15}{123} \times 100\% = 12.2\%$$

Thus, 12.2% of the progeny exhibit new combinations of traits resulting from crossing over. The recombination frequency can also be expressed as a decimal fraction (0.122).

Coupling and Repulsion

In crosses for linked genes, the arrangement of alleles on the homologous chromosomes is critical in determining the outcome of the cross. For example, consider the inheritance of two genes in the Australian blowfly, *Lucilia cuprina*. In this species, one locus determines the color of the thorax: a purple thorax (p) is recessive to the normal green thorax (p^+). A second locus determines the color of the puparium: a black puparium (b) is recessive to the normal brown puparium (b^+). These loci are located close together on the chromosome. Suppose we test cross a fly that is heterozygous at both loci with a fly that is homozygous recessive at both. Because these genes are linked, there are two possible arrangements on the chromosomes of the heterozygous progeny fly. The dominant alleles for green thorax (p^+) and brown puparium (b^+) might reside on the same chromosome, and the recessive alleles for purple thorax (p) and black puparium (b) might reside on the other homologous chromosome:

$$\frac{p^+ \quad b^+}{p \quad b}$$

This arrangement, in which wild-type alleles are found on one chromosome and mutant alleles are found on the other chromosome, is referred to as the **coupling** or **cis configuration**. Alternatively, one chromosome might bear the alleles for green thorax (p^+) and black puparium (b), and the other chromosome would carry the alleles for purple thorax (p) and brown puparium (b^+):

$$\frac{p^+ \quad b}{p \quad b^+}$$

This arrangement, in which each chromosome contains one wild-type and one mutant allele, is called the **repulsion** or **trans configuration**. Whether the alleles in the heterozygous parent are in coupling or repulsion determines which phenotypes will be most common among the progeny of a testcross.

When the alleles are in the coupling configuration, the most numerous progeny types are those with green thorax and brown puparium and those with purple thorax and black puparium (**Figure 5.8a**); but, when the alleles of the heterozygous parent are in repulsion, the most numerous progeny types are those with green thorax and black puparium and those with purple thorax and brown puparium (**Figure 5.8b**). Notice that the genotypes of the parents in Figure 5.8a and b are the same ($p^+p\ b^+b \times pp\ bb$) and that the dramatic dif-

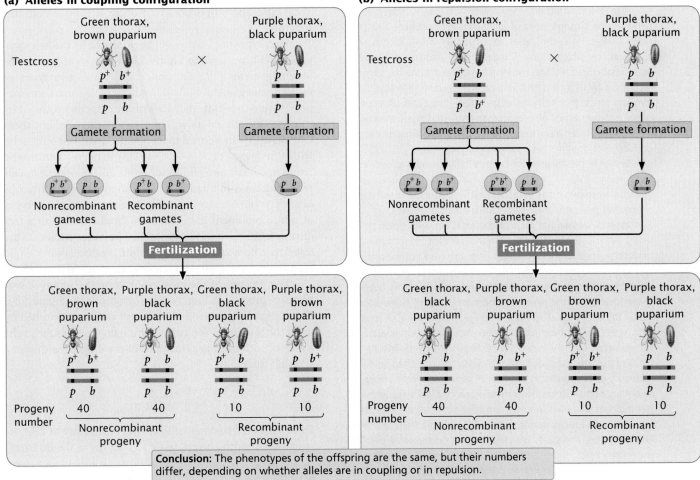

5.8 The arrangement (coupling or repulsion) of linked genes on a chromosome affects the results of a testcross. Linked loci in the Australian blowfly. *Luciliá cuprina*, determine the color of the thorax and that of the puparium.

ference in the phenotypic ratios of the progeny in the two crosses results entirely from the configuration—coupling or repulsion—of the chromosomes. It is essential to know the arrangement of the alleles on the chromosomes to accurately predict the outcome of crosses in which genes are linked.

Concepts

In a cross, the arrangement of linked alleles on the chromosomes is critical for determining the outcome. When two wild-type alleles are on one homologous chromosome and two mutant alleles are on the other, they are in the coupling configuration; when each chromosome contains one wild-type allele and one mutant allele, the alleles are in repulsion.

✔ Concept Check 2

The following testcross produces the progeny shown: *Aa Bb* × *aa bb* → 10 *Aa Bb*, 40 *Aa bb*, 40 *aa Bb*, 10 *aa bb*. What is the percent recombination between the *A* and *B* loci? Were the genes in the *Aa Bb* parent in coupling or in repulsion?

Connecting Concepts

Relating Independent Assortment, Linkage, and Crossing Over

We have now considered three situations concerning genes at different loci. First, the genes may be located on different chromosomes; in this case, they exhibit independent assortment and combine randomly when gametes are formed. An individual heterozygous at two loci (*Aa Bb*) produces four types of gametes (*A B*, *a b*, *A b*, and *a B*) in equal proportions: two types of nonrecombinants and two types of recombinants.

Second, the genes may be completely linked—meaning that they're on the same chromosome and lie so close together that crossing over between them is rare. In this case, the genes do not recombine. An individual heterozygous for two closely linked genes in the coupling configuration

$$\frac{A \quad B}{a \quad b}$$

produces only the nonrecombinant gametes containing alleles *A B* or *a b*. The alleles do not assort into new combinations such as *A b* or *a B*.

The third situation, incomplete linkage, is intermediate between the two extremes of independent assortment and complete linkage. Here, the genes are physically linked on the same chromosome, which prevents independent assortment. However, occasional crossovers break up the linkage and allow the genes to recombine. With incomplete linkage, an individual heterozygous at two loci produces four types of gametes—two types of recombinants and two types of nonrecombinants—but the nonrecombinants are produced more frequently than the recombinants because crossing over does not take place in every meiosis.

Earlier in the chapter, the term recombination was defined as the sorting of alleles into new combinations. We can now distinguish between two types of recombination that differ in the mechanism that generates these new combinations of alleles.

Interchromosomal recombination is between genes on *different* chromosomes. It arises from independent assortment—the random segregation of chromosomes in anaphase I of meiosis. This is the kind of recombination that Mendel discovered while studying dihybrid crosses. **Intrachromosomal recombination** is between genes located on the *same* chromosome. It arises from crossing over—the exchange of genetic material in prophase I of meiosis. Both types of recombination produce new allele combinations in the gametes; so they cannot be distinguished by examining the types of gametes produced. Nevertheless, they can often be distinguished by the *frequencies* of types of gametes: interchromosomal recombination produces 50% nonrecombinant gametes and 50% recombinant gametes, whereas intrachromosomal recombination frequently produces fewer than 50% recombinant gametes. However, when the genes are very far apart on the same chromosome, they assort independently, as if they were on different chromosomes. In this case, intrachromosomal recombination also produces 50% recombinant gametes. Intrachromosomal recombination of genes that lie far apart on the same chromosome and interchromosomal recombination are genetically indistinguishable.

Predicting the Outcomes of Crosses with Linked Genes

Knowing the arrangement of alleles on a chromosome allows us to predict the types of progeny that will result from a cross entailing linked genes and to determine which of these types will be the most numerous. Determining the *proportions* of the types of offspring requires an additional piece of information—the recombination frequency. The recombination frequency provides us with information about how often the alleles in the gametes appear in new combinations and allows us to predict the proportions of offspring phenotypes that will result from a specific cross with linked genes.

In cucumbers, smooth fruit (t) is recessive to warty fruit (T) and glossy fruit (d) is recessive to dull fruit (D). Geneticists have determined that these two genes exhibit a recombination frequency of 16%. Suppose we cross a plant homozygous for warty and dull fruit with a plant homozygous for smooth and glossy fruit and then carry out a testcross by using the F_1:

$$\frac{T \quad D}{t \quad d} \times \frac{t \quad d}{t \quad d}$$

What types and proportions of progeny will result from this testcross?

Four types of gametes will be produced by the heterozygous parent, as shown in **Figure 5.9**: two types of nonrecombinant gametes ($\underline{T \quad D}$ and $\underline{t \quad d}$) and two types of recombinant gametes ($\underline{T \quad d}$ and $\underline{t \quad D}$). The recombination frequency tells us that 16% of the gametes produced by the heterozygous parent will be recombinants. Because there are two types of recombinant gametes, each should arise with a frequency of $^{16\%}/_2 = 8\%$ This frequency can also be represented as a probability of 0.08. All the other gametes will be nonrecombinants; so they should arise with a frequency of $100\% - 16\% = 84\%$. Because there are two types of nonrecombinant gametes, each should arise with a frequency of $^{84\%}/_2 = 42\%$ (or 0.42). The other parent in the testcross is homozygous and therefore produces only a single type of gamete ($\underline{t \quad d}$) with a frequency of 100% (or 1.00).

The progeny of the cross result from the union of two gametes, producing four types of progeny (see Figure 5.9). The expected proportion of each type can be determined by using the multiplication rule, multiplying together the probability of each uniting gamete. Testcross progeny with warty and dull fruit

$$\frac{T \quad D}{t \quad d}$$

appear with a frequency of 0.42 (the probability of inheriting a gamete with chromosome $\underline{T \quad D}$ from the heterozygous parent) × 1.00 (the probability of inheriting a gamete with chromosome $\underline{t \quad d}$ from the recessive parent) = 0.42. The proportions of the other types of F_2 progeny can be calculated in a similar manner (see Figure 5.9). This method can be used for predicting the outcome of any cross with linked genes for which the recombination frequency is known.

Testing for Independent Assortment

In some crosses, the genes are obviously linked because there are clearly more nonrecombinant progeny than recombinant progeny. In other crosses, the difference between independent assortment and linkage is not so obvious. For example, suppose we did a testcross for two pairs of genes, such as *Aa Bb* × *aa bb*, and observed the following numbers of progeny: 54 *Aa Bb*, 56 *aa bb*, 42 *Aa bb*, and 48 *aa Bb*. Is this outcome the 1 : 1 : 1 : 1 ratio we would expect if *A* and *B* assorted independently? Not exactly, but it's pretty close. Perhaps these genes assorted independently and chance produced the slight deviations between the observed numbers and the expected 1 : 1 : 1 : 1 ratio. Alternatively, the genes might be linked, with considerable crossing over taking place between them, and so the number of nonrecombinants is only slightly greater than the number of recombinants. How do we distinguish between the role of chance and the role of linkage in producing deviations from the results expected with independent assortment?

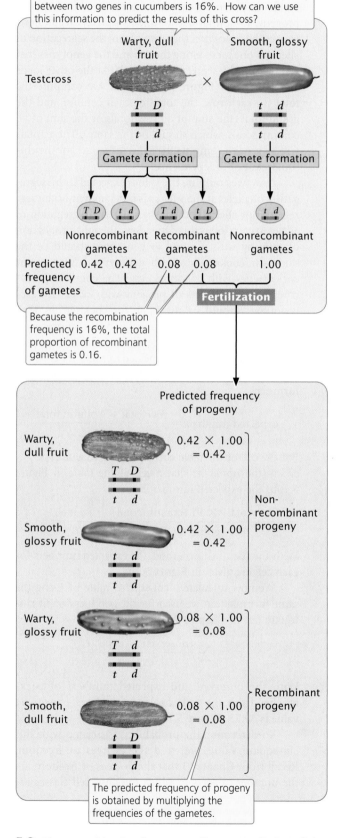

Geneticists have determined that the recombination frequency between two genes in cucumbers is 16%. How can we use this information to predict the results of this cross?

Because the recombination frequency is 16%, the total proportion of recombinant gametes is 0.16.

The predicted frequency of progeny is obtained by multiplying the frequencies of the gametes.

5.9 The recombination frequency allows a prediction of the proportions of offspring expected for a cross entailing linked genes.

We encountered a similar problem in crosses in which genes were unlinked—the problem of distinguishing between deviations due to chance and those due to other factors. We addressed this problem (in Chapter 3) with the goodness-of-fit chi-square test, which helps us evaluate the likelihood that chance alone is responsible for deviations between the numbers of progeny that we observed and the numbers that we expected by applying the principles of inheritance. Here, we are interested in a different question: Is the inheritance of alleles at one locus independent of the inheritance of alleles at a second locus? If the answer to this question is yes, then the genes are assorting independently; if the answer is no, then the genes are probably linked.

A possible way to test for independent assortment is to calculate the expected probability of each progeny type, assuming independent assortment, and then use the goodness-of-fit chi-square test to evaluate whether the observed numbers deviate significantly from the expected numbers. With independent assortment, we expect $1/4$ of each phenotype: $1/4$ Aa Bb, $1/4$ aa bb, $1/4$ Aa bb, and $1/4$ aa Bb. This expected probability of each genotype is based on the multiplication rule of probability, which we considered in Chapter 3. For example, if the probability of Aa is $1/2$ and the probability of Bb is $1/2$, then the probability of Aa Bb is $1/2 \times 1/2 = 1/4$. In this calculation, we are making two assumptions: (1) the probability of each single-locus genotype is $1/2$ and (2) genotypes at the two loci are inherited independently ($1/2 \times 1/2 = 1/4$).

One problem with this approach is that a significant chi-square test can result from a violation of either assumption. If the genes are linked, then the inheritances of genotypes at the two loci are not independent (assumption 2), and we will get a significant deviation between observed and expected numbers. But we can also get a significant deviation if the probability of each single-locus genotype is not $1/2$ (assumption 1), even when the genotypes are assorting independently. We may obtain a significant deviation, for example, if individuals with one genotype have a lower probability of surviving or the penetrance of a genotype is not 100%. We could test both assumptions by conducting a series of chi-square tests, first testing the inheritance of genotypes at each locus separately (assumption 1) and then testing for independent assortment (assumption 2). However, a faster method is to test for independence in genotypes with a *chi-square test of independence*.

The chi-square test of independence allows us to evaluate whether the segregation of alleles at one locus is independent of the segregation of alleles at another locus, without making any assumption about the probability of single-locus genotypes. To illustrate this analysis, we will examine results from a cross between German cockroaches, in which yellow body (y) is recessive to brown body (y^+) and curved wings (cv) are recessive to straight wings (cv^+). A testcross (y^+y $cv^+cv \times yy$ $cvcv$) produced the progeny shown in **Figure 5.10a**.

(a)

Brown body, straight wings × Yellow body, curved wings

$y^+y \; cv^+cv$ $yy \; cvcv$

1 A testcross is carried out between cockroaches differing in two characteristics.

Cross

63 $y^+y \; cv^+cv$ brown body, straight wings
28 $y^+y \; cvcv$ brown body, curved wings
33 $yy \; cv^+cv$ yellow body, straight wings
77 $yy \; cvcv$ yellow body, curved wings

(b) Contingency table

2 To test for independent assortment of alleles encoding the two traits, a table is constructed...

3 ...with genotypes for one locus along the top...

Segregation of y^+ and y

Segregation of cv^+ and cv

	y^+y	yy	Row totals
cv^+cv	63	33	96
$cv\,cv$	28	77	105
Column totals	91	110	201

Grand total

4 ...and genotypes for the other locus along the left side.

5 Numbers of each genotype are placed in the table cells, and the row totals, column totals, and grand total are computed.

(c)

Genotype	Number observed	Number expected $\left(\dfrac{\text{row total} \times \text{column total}}{\text{grand total}}\right)$
$y^+y \; cv^+cv$	63	$\dfrac{96 \times 91}{201} = 43.46$
$y^+y \; cvcv$	28	$\dfrac{105 \times 91}{201} = 47.54$
$yy \; cv^+cv$	33	$\dfrac{96 \times 110}{201} = 52.46$
$yy \; cvcv$	77	$\dfrac{105 \times 110}{201} = 57.46$

6 The expected numbers of progeny, assuming independent assortment, are calculated.

(d)

7 A chi-square value is calculated.

$$\chi^2 = \sum \frac{(\text{observed} - \text{expected})^2}{\text{expected}}$$

$$\chi^2 = \frac{(63 - 43.46)^2}{43.46} + \frac{(28 - 47.54)^2}{47.54} + \frac{(33 - 52.54)^2}{52.54} + \frac{(77 - 57.46)^2}{57.46}$$

$$\chi^2 = 8.79 + 8.03 + 7.27 + 6.64$$

$$\chi^2 = 30.73$$

(e)

8 The probability is less than 0.005, indicating that the *difference* between numbers of observed and expected progeny is probably not due to chance.

$df = (\text{number of rows} - 1) \times (\text{number of columns} - 1)$
$df = (2 - 1) \times (2 - 1) = 1 \times 1 = 1$
$P < 0.005$

Conclusion: The genes for body color and type of wing are not assorting independently and must be linked.

To carry out the chi-square test of independence, we first construct a table of the observed numbers of progeny, somewhat like a Punnett square, except, here, we put the genotypes that result from the segregation of alleles at one locus along the top and the genotypes that result from the segregation of alleles at the other locus along the side (**Figure 5.10b**). Next, we compute the total for each row, the total for each column, and the grand total (the sum of all row totals or the sum of all column totals, which should be the same). These totals will be used to compute the expected values for the chi-square test of independence.

Now, we compute the values expected if the segregation of alleles at the y locus is independent of the segregation of alleles at the cv locus. If the segregation of alleles at each locus is independent, then the proportion of progeny with y^+y and yy genotypes should be the same for cockroaches with genotype cv^+cv and for cockroaches with genotype $cvcv$. The converse is also true; the proportions of progeny with cv^+cv and $cvcv$ genotypes should be the same for cockroaches with genotype y^+y and for cockroaches with genotype yy. With the assumption that the alleles at the two loci segregate independently, the expected number for each cell of the table can be computed by using the following formula:

$$\text{expected number} = \frac{\text{row total} \times \text{column total}}{\text{grand total}}$$

For the cell of the table corresponding to genotype y^+y cv^+cv (the upper-left-hand cell of the table in Figure 5.10b) the expected number is:

$$\frac{96 \,(\text{row total}) \times 91 \,(\text{column total})}{201 \,(\text{grand total})} = \frac{8736}{201} = 43.46$$

With the use of this method, the expected numbers for each cell are given in **Figure 5.10c**.

We now calculate a chi-square value by using the same formula that we used for the goodness-of-fit chi-square test in Chapter 3:

$$x^2 = \sum \frac{(\text{observed} - \text{expected})^2}{\text{expected}}$$

With the observed and expected numbers of cockroaches from the testcross, the calculated chi-square value is 30.73 (**Figure 5.10d**).

To determine the probability associated with this chi-square value, we need the degrees of freedom. Recall from Chapter 3 that the degrees of freedom are the number of ways in which the observed classes are

5.10 A chi-square test of independence can be used to determine if genes at two loci are assorting independently.

free to vary from the expected values. In general, for the chi-square test of independence, the degrees of freedom equal the number of rows in the table minus 1 multiplied by the number of columns in the table minus 1 (**Figure 5.10e**), or

$$df = (\text{number of rows} - 1) \times (\text{number of columns} - 1)$$

In our example, there were two rows and two columns, and so the degrees of freedom are:

$$df = (2 - 1) \times (2 - 1) = 1 \times 1 = 1$$

Therefore, our calculated chi-square value is 30.73, with 1 degree of freedom. We can use Table 3.4 to find the associated probability. Looking at Table 3.4, we find our calculated chi-square value is larger than the largest chi-square value given for 1 degree of freedom, which has a probability of 0.005. Thus, our calculated chi-square value has a probability less than 0.005. This very small probability indicates that the genotypes are not in the proportions that we would expect if independent assortment were taking place. Our conclusion, then, is that these genes are not assorting independently and must be linked. As is the case for the goodness-of-fit chi-square test, geneticists generally consider that any chi-square value for the test of independence with a probability less than 0.05 is significantly different from the expected values and is therefore evidence that the genes are not assorting independently.

Gene Mapping with Recombination Frequencies

Morgan and his students developed the idea that physical distances between genes on a chromosome are related to the rates of recombination. They hypothesized that crossover events occur more or less at random up and down the chromosome and that two genes that lie far apart are more likely to undergo a crossover than are two genes that lie close together. They proposed that recombination frequencies could provide a convenient way to determine the order of genes along a chromosome and would give estimates of the relative distances between the genes. Chromosome maps calculated by using the genetic phenomenon of recombination are called **genetic maps**. In contrast, chromosome maps calculated by using physical distances along the chromosome (often expressed as numbers of base pairs) are called **physical maps**.

Distances on genetic maps are measured in **map units** (abbreviated m.u.); one map unit equals 1% recombination. Map units are also called **centiMorgans** (cM), in honor of Thomas Hunt Morgan; 100 centiMorgans equals one **Morgan**. Genetic distances measured with recombination rates are approximately additive: if the distance from gene A to gene B is 5 m.u., the distance from gene B to gene C is 10 m.u., and the distance from gene A to gene C is 15 m.u., then gene B must be located between genes A and C. On the basis of the map distances just given, we can draw a simple genetic map for genes A, B, and C, as shown here:

We could just as plausibly draw this map with C on the left and A on the right:

Both maps are correct and equivalent because, with information about the relative positions of only three genes, the most that we can determine is which gene lies in the middle. If we obtained distances to an additional gene, then we could position A and C relative to that gene. An additional gene D, examined through genetic crosses, might yield the following recombination frequencies:

Gene pair	Recombination frequency (%)
A and D	8
B and D	13
C and D	23

Notice that C and D exhibit the greatest amount of recombination; therefore, C and D must be farthest apart, with genes A and B between them. Using the recombination frequencies and remembering that 1 m.u. = 1% recombination, we can now add D to our map:

By doing a series of crosses between pairs of genes, we can construct genetic maps showing the linkage arrangements of a number of genes.

Two points should be emphasized about constructing chromosome maps from recombination frequencies. First, recall that we cannot distinguish between genes on different chromosomes and genes located far apart on the same chromosome. If genes exhibit 50% recombination, the most that can be said about them is that they belong to different groups of linked genes (different linkage groups), either on different chromosomes or far apart on the same chromosome.

The second point is that a testcross for two genes that are relatively far apart on the same chromosome tends to underestimate the true physical distance, because the cross does not reveal double crossovers that might take place between the two genes (**Figure 5.11**). A double crossover arises when two separate crossover events take place between two loci. (For now, we will consider only double crossovers that occur between two of the four chromatids of a homologus pair—a two-strand double crossover. Double crossovers entailing three and four chromatids will be considered later.) Whereas a single crossover produces combinations of alleles that were

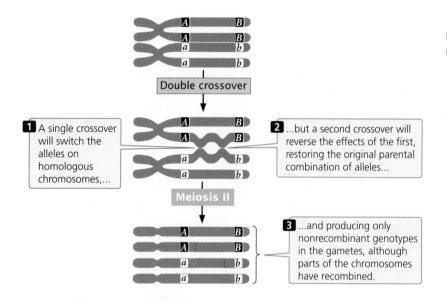

1 A single crossover will switch the alleles on homologous chromosomes,...

2 ...but a second crossover will reverse the effects of the first, restoring the original parental combination of alleles...

3 ...and producing only nonrecombinant genotypes in the gametes, although parts of the chromosomes have recombined.

5.11 **A two-strand double crossover between two linked genes produces only nonrecombinant gametes.**

not present on the original parental chromosomes, a second crossover between the same two genes reverses the effects of the first, thus restoring the original parental combination of alleles (see Figure 5.11). Two-strand double crossovers produce only nonrecombinant gametes, and so we cannot distinguish between the progeny produced by double crossovers and the progeny produced when there is no crossing over at all. As we shall see in the next section, we can detect double crossovers if we examine a third gene that lies between the two crossovers. Because double crossovers between two genes go undetected, map distances will be underestimated whenever double crossovers take place. Double crossovers are more frequent between genes that are far apart; therefore genetic maps based on short distances are usually more accurate than those based on longer distances.

Concepts

A genetic map provides the order of the genes on a chromosome and the approximate distances from one gene to another based on recombination frequencies. In genetic maps, 1% recombination equals 1 map unit, or 1 centiMorgan. Double crossovers between two genes go undetected; so map distances between distant genes tend to underestimate genetic distances.

✔ Concept Check 3

How does a genetic map differ from a physical map?

Constructing a Genetic Map with Two-Point Testcrosses

Genetic maps can be constructed by conducting a series of testcrosses between pairs of genes and examining the recombination frequencies between them. A testcross between two genes is called a **two-point testcross** or a two-point cross for short. Suppose that we carried out a series of two-point crosses for four genes, *a*, *b*, *c*, and *d*, and obtained the following recombination frequencies:

Gene loci in testcross	Recombination frequency (%)
a and *b*	50
a and *c*	50
a and *d*	50
b and *c*	20
b and *d*	10
c and *d*	28

We can begin constructing a genetic map for these genes by considering the recombination frequencies for each pair of genes. The recombination frequency between *a* and *b* is 50%, which is the recombination frequency expected with independent assortment. Therefore, genes *a* and *b* may either be on different chromosomes or be very far apart on the same chromosome; so we will place them in different linkage groups with the understanding that they may or may not be on the same chromosome:

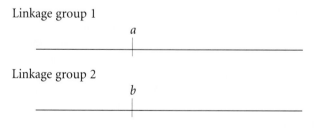

The recombination frequency between *a* and *c* is 50%, indicating that they, too, are in different linkage groups. The recombination frequency between *b* and *c* is 20%; so these genes are linked and separated by 20 map units:

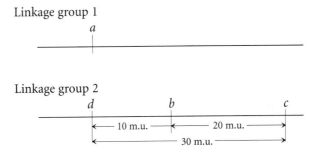

The recombination frequency between *a* and *d* is 50%, indicating that these genes belong to different linkage groups, whereas genes *b* and *d* are linked, with a recombination frequency of 10%. To decide whether gene *d* is 10 m.u. to the left or to the right of gene *b*, we must consult the *c*-to-*d* distance. If gene *d* is 10 m.u. to the left of gene *b*, then the distance between *d* and *c* should be 20 m.u. + 10 m.u. = 30 m.u. This distance will be only approximate because any double crossovers between the two genes will be missed and the map distance will be underestimated. If, on the other hand, gene *d* lies to the right of gene *b*, then the distance between gene *d* and gene *c* will be much shorter, approximately 20 m.u. − 10 m.u. = 10 m.u.

By examining the recombination frequency between *c* and *d*, we can distinguish between these two possibilities. The recombination frequency between *c* and *d* is 28%; so gene *d* must lie to the left of gene *b*. Notice that the sum of the recombination between *d* and *b* (10%) and between *b* and *c* (20%) is greater than the recombination between *d* and *c* (28%). As already discussed, this discrepancy arises because double crossovers between the two outer genes go undetected, causing an underestimation of the true map distance. The genetic map of these genes is now complete:

5.3 A Three-Point Testcross Can Be Used to Map Three Linked Genes

Genetic maps can be constructed from a series of testcrosses for pairs of genes, but this approach is not particularly efficient, because numerous two-point crosses must be carried out to establish the order of the genes and because double crossovers are missed. A more efficient mapping technique is a testcross for three genes—a **three-point testcross**, or three-point cross. With a three-point cross, the order of the three genes can be established in a single set of progeny and some double crossovers can usually be detected, providing more accurate map distances.

Consider what happens when crossing over takes place among three hypothetical linked genes. **Figure 5.12** illustrates a pair of homologous chromosomes from an individual that is heterozygous at three loci (*Aa Bb Cc*). Notice that the genes are in the coupling configuration; that is, all the dominant alleles are on one chromosome (*A   B   C*) and all the recessive alleles are on the other chromosome

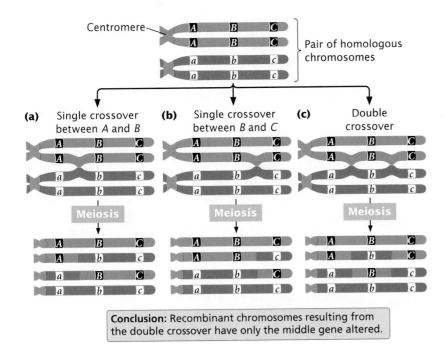

Conclusion: Recombinant chromosomes resulting from the double crossover have only the middle gene altered.

5.12 Three types of crossovers can take place among three linked loci.

(a b c). Three types of crossover events can take place between these three genes: two types of single crossovers (see Figure 5.12a and b) and a double crossover (see Figure 5.12c). In each type of crossover, two of the resulting chromosomes are recombinants and two are nonrecombinants.

Notice that, in the recombinant chromosomes resulting from the double crossover, the outer two alleles are the same as in the nonrecombinants, but the middle allele is different. This result provides us with an important clue about the order of the genes. In progeny that result from a double crossover, only the middle allele should differ from the alleles present in the nonrecombinant progeny.

Constructing a Genetic Map with the Three-Point Testcross

To examine gene mapping with a three-point testcross, we will consider three recessive mutations in the fruit fly *Drosophila melanogaster*. In this species, scarlet eyes (st) are recessive to red eyes (st^+), ebony body color (e) is recessive to gray body color (e^+), and spineless (ss)—that is, the presence of small bristles—is recessive to normal bristles (ss^+). All three mutations are linked and located on the third chromosome.

We will refer to these three loci as st, e, and ss, but keep in mind that either the recessive alleles (st, e, and ss) or the dominant alleles (st^+, e^+, and ss^+) may be present at each locus. So, when we say that there are 10 m.u. between st and ss, we mean that there are 10 m.u. between the loci at which these mutations occur; we could just as easily say that there are 10 m.u. between st^+ and ss^+.

To map these genes, we need to determine their order on the chromosome and the genetic distances between them. First, we must set up a three-point testcross, a cross between a fly heterozygous at all three loci and a fly homozygous for recessive alleles at all three loci. To produce flies heterozygous for all three loci, we might cross a stock of flies that are homozygous for normal alleles at all three loci with flies that are homozygous for recessive alleles at all three loci:

$$\text{P} \quad \frac{st^+ \quad e^+ \quad ss^+}{st^+ \quad e^+ \quad ss^+} \times \frac{st \quad e \quad ss}{st \quad e \quad ss}$$

$$\downarrow$$

$$\text{F}_1 \quad \frac{st^+ \quad e^+ \quad ss^+}{st \quad e \quad ss}$$

The order of the genes has been arbitrarily assigned because, at this point, we do not know which is the middle gene. Additionally, the alleles in these heterozygotes are in coupling configuration (because all the wild-type dominant alleles were inherited from one parent and all the recessive mutations from the other parent), although the testcross can also be done with alleles in repulsion.

In the three-point testcross, we cross the F_1 heterozygotes with flies that are homozygous for all three recessive mutations. In many organisms, it makes no difference whether the heterozygous parent in the testcross is male or

female (provided that the genes are autosomal) but, in *Drosophila*, no crossing over takes place in males. Because crossing over in the heterozygous parent is essential for determining recombination frequencies, the heterozygous flies in our testcross must be female. So we mate female F_1 flies that are heterozygous for all three traits with male flies that are homozygous for all the recessive traits:

$$\frac{st^+ \quad e^+ \quad ss^+}{st \quad e \quad ss} \text{ Female} \times \frac{st \quad e \quad ss}{st \quad e \quad ss} \text{ Male}$$

The progeny produced from this cross are listed in **Figure 5.13**. For each locus, two classes of progeny are produced: progeny that are heterozygous, displaying the dominant trait, and progeny that are homozygous, displaying the recessive trait. With two classes of progeny possible for each of the three loci, there will be $2^3 = 8$ classes of phenotypes possible in the progeny. In this example, all eight phenotypic classes are present but, in some three-point crosses, one or more of the phenotypes may be missing if the number of progeny is limited. Nevertheless, the absence of a particular class can provide important information about which combination of traits is least frequent and, ultimately, about the order of the genes, as we will see.

To map the genes, we need information about where and how often crossing over has taken place. In the homozygous recessive parent, the two alleles at each locus are the same, and so crossing over will have no effect on the types of gametes produced; with or without crossing over, all gametes from this parent have a chromosome with three recessive alleles (st e ss). In contrast, the heterozygous parent has different alleles on its two chromosomes, and so crossing over can be detected. The information that we need for mapping, therefore, comes entirely from the gametes produced by the heterozygous parent. Because chromosomes contributed by the homozygous parent carry only recessive alleles, whatever alleles are present on the chromosome contributed by the heterozygous parent will be expressed in the progeny.

As a shortcut, we often do not write out the complete genotypes of the testcross progeny, listing instead only the alleles expressed in the phenotype, which are the alleles inherited from the heterozygous parent. This convention is used in the discussion that follows.

Concepts

To map genes, information about the location and number of crossovers in the gametes that produced the progeny of a cross is needed. An efficient way to obtain this information is to use a three-point testcross, in which an individual heterozygous at three linked loci is crossed with an individual that is homozygous recessive at the three loci.

✔ Concept Check 4

Write the genotypes of all recombinant and nonrecombinant progeny expected from the following three-point cross:

$$\frac{m^+ \quad p^+ \quad s^+}{m \quad p \quad s} \times \frac{m \quad p \quad s}{m \quad p \quad s}$$

First, determine which progeny are the nonrecombinants; they will be the two most-numerous classes of progeny. (Even if crossing over takes place in every meiosis, the nonrecombinants will constitute at least 50% of the progeny.) Among the progeny of the testcross in Figure 5.13, the most numerous are those with all three dominant traits (st^+ e^+ ss^+) and those with all three recessive traits (st e ss).

Next, identify the double-crossover progeny. These progeny should always have the two least-numerous phenotypes, because the probability of a double crossover is always less than the probability of a single crossover. The least-common progeny among those listed in Figure 5.13 are progeny with spineless bristles (st^+ e^+ ss) and progeny with scarlet eyes and ebony body (st e ss^+); so they are the double-crossover progeny.

Three orders of genes on the chromosome are possible: the eye-color locus could be in the middle (e st ss), the body-color locus could be in the middle (st e ss), or the bristle locus could be in the middle (st ss e). To determine which gene is in the middle, we can draw the chromosomes of the heterozygous parent with all three possible gene orders and then see if a double crossover produces the combination of genes observed in the double-crossover progeny. The three possible gene orders and the types of progeny produced by their double crossovers are:

5.13   The results of a three-point testcross can be used to map linked genes. In this three-point testcross of *Drosophila melanogaster*, the recessive mutations scarlet eyes (*st*), ebony body color (*e*), and spineless bristles (*ss*) are at three linked loci. The order of the loci has been designated arbitrarily, as has the sex of the progeny flies.

The only gene order that produces chromosomes with the set of alleles observed in the least-numerous progeny or double crossovers (st^+ e^+ ss and st e ss^+ in Figure 5.13) is the one in which the *ss* locus for bristles lies in the middle (gene-order 3). Therefore, this order (st ss e) must be the correct sequence of genes on the chromosome.

With a little practice, we can quickly determine which locus is in the middle without writing out all the gene orders. The phenotypes of the progeny are expressions of the alleles inherited from the heterozygous parent. Recall that, when we looked at the results of double crossovers (see Figure 5.12), only the alleles at the middle locus differed from the nonrecombinants. If we compare the nonrecombinant progeny with double-crossover progeny, they should differ only in alleles of the middle locus.

Determining the gene order   The first task in mapping the genes is to determine their order on the chromosome. In Figure 5.13, we arbitrarily listed the loci in the order *st, e, ss*, but we had no way of knowing which of the three loci was between the other two. We can now identify the middle locus by examining the double-crossover progeny.

Let's compare the alleles in the double-crossover progeny $\dfrac{st^+ \quad e^+ \quad ss}{st^+ \quad e^+ \quad ss^+}$ with those in the nonrecombinant progeny. We see that both have an allele for red eyes (st^+) and both have an allele for gray body (e^+), but the nonrecombinants have an allele for normal bristles (ss^+), whereas the double crossovers have an allele for spineless bristles (ss). Because the bristle locus is the only one that differs, it must lie in the middle. We would obtain the same results if we compared the other class of double-crossover progeny ($st \quad e \quad ss^+$) with other nonrecombinant progeny ($st \quad e \quad ss$). Again, the only locus that differs is the one for bristles. Don't forget that the nonrecombinants and the double crossovers should differ only at one locus; if they differ at two loci, the wrong classes of progeny are being compared.

Concepts

To determine the middle locus in a three-point cross, compare the double-crossover progeny with the nonrecombinant progeny. The double crossovers will be the two least-common classes of phenotypes; the nonrecombinants will be the two most-common classes of phenotypes. The double-crossover progeny should have the same alleles as the nonrecombinant types at two loci and different alleles at the locus in the middle.

✔ Concept Check 5

A three-point test cross is carried out between three linked genes. The resulting nonrecombinant progeny are $s^+ r^+ c^+$ and $s r c$ and the double-crossover progeny are $s r c^+$ and $s^+ r^+ c$. Which is the middle locus?

Determining the locations of crossovers When we know the correct order of the loci on the chromosome, we should rewrite the phenotypes of the testcross progeny in Figure 5.13 with the alleles in the correct order so that we can determine where crossovers have taken place (**Figure 5.14**).

Among the eight classes of progeny, we have already identified two classes as nonrecombinants ($\underline{st^+ \quad ss^+ \quad e^+}$ and $\underline{st \quad ss \quad e}$) and two classes as double crossovers ($\underline{st^+ \quad ss \quad e^+}$ and $\underline{st \quad ss^+ \quad e}$). The other four classes include progeny that resulted from a chromosome that underwent a single crossover: two underwent single crossovers between st and ss, and two underwent single crossovers between ss and e.

To determine where the crossovers took place in these progeny, compare the alleles found in the single-crossover progeny with those found in the nonrecombinants, just as we did for the double crossovers. For example, consider progeny with chromosome $\underline{st^+ \quad ss \quad e}$. The first allele ($st^+$) came from the nonrecombinant chromosome $\underline{st^+ \quad ss^+ \quad e^+}$ and the other two alleles (ss and e) must have come from the other nonrecombinant chromosome $\underline{st \quad ss \quad e}$ through crossing over:

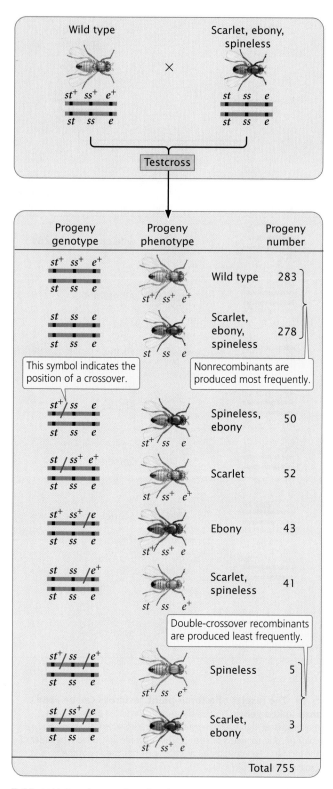

5.14 Writing the results of a three-point testcross with the loci in the correct order allows the locations of crossovers to be determined. These results are from the testcross illustrated in Figure 5.13, with the loci shown in the correct order. The location of a crossover is indicated by a slash (/). The sex of the progeny flies has been designated arbitrarily.

This same crossover also produces the st ss^+ e^+ progeny.

This method can also be used to determine the location of crossing over in the other two types of single-crossover progeny. Crossing between ss and e produces st^+ ss^+ e and st ss e^+ chromosomes:

We now know the locations of all the crossovers. Their locations are marked with a slash in Figure 5.14.

Calculating the recombination frequencies

Next, we can determine the map distances, which are based on the frequencies of recombination. We calculate recombination frequency by adding up all of the recombinant progeny, dividing this number by the total number of progeny from the cross, and multiplying the number obtained by 100%. To determine the map distances accurately, we must include all crossovers (both single and double) that take place between two genes.

Recombinant progeny that possess a chromosome that underwent crossing over between the eye-color locus (st) and the bristle locus (ss) include the single crossovers (st^+ / ss e and st / ss^+ e^+) and the two double crossovers (st^+ / ss / e^+ and st / ss^+ / e); see Figure 5.14. There are a total of 755 progeny; so the recombination frequency between ss and st is:

st–ss recombination frequency =
$$\frac{(50 + 52 + 5 + 3)}{755} \times 100\% = 14.6\%$$

The distance between the st and ss loci can be expressed as 14.6 m.u.

The map distance between the bristle locus (ss) and the body locus (e) is determined in the same manner. The recombinant progeny that possess a crossover between ss and e are the single crossovers st^+ ss^+ / e and st ss / e^+ and the double crossovers st^+ / ss / e^+ and st / ss^+ / e . The recombination frequency is:

ss–e recombination frequency =
$$\frac{(43 + 41 + 5 + 3)}{755} \times 100\% = 12.2\%$$

Thus, the map distance between ss and e is 12.2 m.u.

Finally, calculate the map distance between the outer two loci, st and e. This map distance can be obtained by summing the map distances between st and ss and between ss and

e (14.6 m.u. + 12.2 m.u. = 26.8 m.u.). We can now use the map distances to draw a map of the three genes on the chromosome:

A genetic map of *D. melanogaster* is illustrated in **Figure 5.15.**

Interference and coefficient of coincidence

Map distances give us information not only about the distances that separate genes, but also about the proportions of recombinant and nonrecombinant gametes that will be produced in a cross. For example, knowing that genes st and ss on the third chromosome of *D. melanogaster* are separated by a distance of 14.6 m.u. tells us that 14.6% of the gametes produced by a fly heterozygous at these two loci will be recombinants. Similarly, 12.2% of the gametes from a fly heterozygous for ss and e will be recombinants.

Theoretically, we should be able to calculate the proportion of double-recombinant gametes by using the multiplication rule of probability (see Chapter 3), which states that the probability of two independent events occurring together is calculated by multiplying the probabilities of the independent events. Applying this principle, we should find that the proportion (probability) of gametes with double crossovers between st and e is equal to the probability of recombination between st and ss multiplied by the probability of recombination between ss and e, or $0.146 \times 0.122 = 0.0178$. Multiplying this probability by the total number of progeny gives us the *expected* number of double-crossover progeny from the cross: $0.0178 \times 755 = 13.4$. Only 8 double crossovers—considerably fewer than the 13 expected—were observed in the progeny of the cross (see Figure 5.14).

This phenomenon is common in eukaryotic organisms. The calculation assumes that each crossover event is independent and that the occurrence of one crossover does not influence the occurrence of another. But crossovers are frequently *not* independent events: the occurrence of one crossover tends to inhibit additional crossovers in the same region of the chromosome, and so double crossovers are less frequent than expected.

The degree to which one crossover interferes with additional crossovers in the same region is termed the **interference**. To calculate the interference, we first determine the **coefficient of coincidence**, which is the ratio of observed double crossovers to expected double crossovers:

coefficient of coincidence =
$$\frac{\text{number of observed double crossovers}}{\text{number of expected double crossovers}}$$

5.15 *Drosophila melanogaster* **has four linkage groups corresponding to its four pairs of chromosomes.** Distances between genes within a linkage group are in map units.

For the loci that we mapped on the third chromosome of *D. melanogaster* (see Figure 5.14), we find that the

coefficient of coincidence =

$$\frac{5 + 3}{0.146 \times 0.122 \times 755} = \frac{8}{13.4} = 0.6$$

which indicates that we are actually observing only 60% of the double crossovers that we expected on the basis of the single-crossover frequencies. The interference is calculated as

interference = 1 − coefficient of coincidence

So the interference for our three-point cross is:

interference = 1 − 0.6 = 0.4

This value of interference tells us that 40% of the double-crossover progeny expected will not be observed, because of interference. When interference is complete and no double-crossover progeny are observed, the coefficient of coincidence is 0 and the interference is 1.

Sometimes a crossover *increases* the probability of another crossover taking place nearby and we see *more* double-crossover progeny than expected. In this case, the coefficient of coincidence is greater than 1 and the interference is negative.

Concepts

The coefficient of coincidence equals the number of double crossovers observed divided by the number of double crossovers expected on the basis of the single-crossover frequencies. The interference equals 1 − the coefficient of coincidence; it indicates the degree to which one crossover interferes with additional crossovers.

✔ Concept Check 6

In analyzing the results of a three-point testcross, a student determines that the interference is −0.23. What does this negative interference value indicate?

a. Fewer double crossovers took place than expected on the basis of single-crossover frequencies.

b. More double crossovers took place than expected on the basis of single-crossover frequencies.

c. Fewer single crossovers took place than expected.

d. A crossover in one region interferes with additional crossovers in the same region.

Connecting Concepts

Stepping Through the Three-Point Cross

We have now examined the three-point cross in considerable detail and have seen how the information derived from the cross can be used to map a series of three linked genes. Let's briefly review the steps required to map genes from a three-point cross.

1. **Write out the phenotypes and numbers of progeny produced in the three-point cross.** The progeny phenotypes will be easier to interpret if you use allelic symbols for the traits (such as $st^+ e^+ ss$).

2. **Write out the genotypes of the original parents used to produce the triply heterozygous individual** in the testcross and, if known, the arrangement (coupling or repulsion) of the alleles on their chromosomes.

3. **Determine which phenotypic classes among the progeny are the nonrecombinants and which are the double crossovers.** The nonrecombinants will be the two most-common phenotypes; double crossovers will be the two least-common phenotypes.

4. **Determine which locus lies in the middle.** Compare the alleles present in the double crossovers with those present in the nonrecombinants; each class of double crossovers should be like one of the nonrecombinants for two loci and should differ for one locus. The locus that differs is the middle one.

5. **Rewrite the phenotypes with genes in correct order.**

6. **Determine where crossovers must have taken place to give rise to the progeny phenotypes.** To do so, compare each phenotype with the phenotype of the nonrecombinant progeny.

7. **Determine the recombination frequencies.** Add the numbers of the progeny that possess a chromosome with a crossover between a pair of loci. Add the double crossovers to this number. Divide this sum by the total number of progeny from the cross, and multiply by 100%; the result is the recombination frequency between the loci, which is the same as the map distance.

8. **Draw a map of the three loci.** Indicate which locus lies in the middle, and indicate the distances between them.

9. **Determine the coefficient of coincidence and the interference.** The coefficient of coincidence is the number of observed double-crossover progeny divided by the number of expected double-crossover progeny. The expected number can be obtained by multiplying the product of the two single-recombination probabilities by the total number of progeny in the cross.

Worked Problem

In *D. melanogaster*, cherub wings (*ch*), black body (*b*), and cinnabar eyes (*cn*) result from recessive alleles that are all located on chromosome 2. A homozygous wild-type fly was mated with a cherub, black, and cinnabar fly, and the resulting F_1 females were test-crossed with cherub, black, and cinnabar males. The following progeny were produced from the testcross:

ch	b^+	*cn*	105
ch^+	b^+	cn^+	750
ch^+	*b*	*cn*	40
ch^+	b^+	*cn*	4
ch	*b*	*cn*	753
ch	b^+	cn^+	41
ch^+	*b*	cn^+	102
ch	*b*	cn^+	5
Total			1800

a. Determine the linear order of the genes on the chromosome (which gene is in the middle).

b. Calculate the recombinant distances between the three loci.

c. Determine the coefficient of coincidence and the interference for these three loci.

• Solution

a. We can represent the crosses in this problem as follows:

$$\text{P} \quad \frac{ch^+ \ b^+ \ cn^+}{ch^+ \ b^+ \ cn^+} \times \frac{ch \ b \ cn}{ch \ b \ cn}$$

$$\downarrow$$

$$\text{F}_1 \quad \frac{ch^+ \ b^+ \ cn^+}{ch \ b \ cn}$$

$$\text{Testcross} \quad \frac{ch^+ \ b^+ \ cn^+}{ch \ b \ cn} \times \frac{ch \ b \ cn}{ch \ b \ cn}$$

Note that we do not know, at this point, the order of the genes; we have arbitrarily put *b* in the middle.

The next step is to determine which of the testcross progeny are nonrecombinants and which are double crossovers. The nonrecombinants should be the most-frequent phenotype; so they must be the progeny with phenotypes encoded by ch^+ b^+ cn^+ and ch b cn. These genotypes are consistent with the genotypes of the parents, given earlier. The double crossovers are the least-frequent phenotypes and are encoded by ch^+ b^+ cn and ch b cn^+. We can determine the gene order by comparing the alleles present in the double crossovers with those present in the nonrecombinants. The double-crossover progeny should be like one of the nonrecombinants at two loci and unlike it at one locus; the allele that differs should be in the middle. Compare the double-crossover progeny ch b cn^+ with the nonrecombinant ch b cn. Both have cherub wings (ch) and black body (b), but the double-crossover progeny have wild-type eyes (cn^+), whereas the nonrecombinants have cinnabar eyes (cn). The locus that determines cinnabar eyes must be in the middle.

b. To calculate the recombination frequencies among the genes, we first write the phenotypes of the progeny with the genes encoding them in the correct order. We have already identified the nonrecombinant and double-crossover progeny; so the other four progeny types must have resulted from single crossovers. To determine *where* single crossovers took place, we compare the alleles found in the single-crossover progeny with those in the nonrecombinants. Crossing over must have taken place where the alleles switch from those found in one nonrecombinant to those found in the other nonrecombinant. The locations of the crossovers are indicated with a slash:

ch	cn	/	b^+	105	single crossover
ch^+	cn^+		b^+	750	nonrecombinant
ch^+ /	cn		b	40	single crossover
ch^+ /	cn	/	b^+	4	double crossover
ch	cn		b	753	nonrecombinant
ch /	cn^+		b^+	41	single crossover
ch^+	cn^+ /		b	102	single crossover
ch /	cn^+ /		b	5	double crossover
Total				1800	

Next, we determine the recombination frequencies and draw a genetic map:

ch–cn recombination frequency =

$$\frac{40 + 4 + 41 + 5}{1800} \times 100\% = 5\%$$

cn–b recombination frequency =

$$\frac{105 + 4 + 102 + 5}{1800} \times 100\% = 12\%$$

ch–b map distance = 5% + 12% = 17%

c. The coefficient of coincidence is the number of observed double crossovers divided by the number of expected double crossovers. The number of expected double crossovers is obtained by multiplying the probability of a crossover between ch and cn (0.05) × the probability of a crossover between cn and b (0.12) × the total number of progeny in the cross (1800):

coefficient of coincidence =

$$\frac{4 + 5}{0.05 + 0.12 \times 1800} = 0.84$$

Finally, the interference is equal to 1 – the coefficient of coincidence:

$$\text{interference} = 1 - 0.83 = 0.17$$

> **?** To increase your skill with three-point crosses, try working Problem 18 at the end of this chapter.

Effect of Multiple Crossovers

So far, we have examined the effects of double crossovers taking place between only two of the four chromatids of a homologous pair. These crossovers are called two-strand crossovers. Double crossovers including three and even four of the chromatids of a homologous pair also may take place (**Figure 5.16**). If we examine only the alleles at loci on either side of both crossover events, two-strand double crossovers result in no new combinations of alleles, and no recombinant gametes are produced (see Figure 5.16). Three-strand double crossovers result in two of the four gametes being recombinant, and four-strand double crossovers result in all four gametes being recombinant. Thus, two-strand double crossovers produce 0% recombination, three-strand double crossovers produce 50% recombination, and four-strand double crossovers produce 100% recombination. The overall result is that all types of double crossovers, taken together, produce an average of 50% recombinant progeny.

As we have seen, two-strand double crossover cause alleles on either side of the crossovers to remain the same and produce no recombinant progeny. Three-strand and four-strand crossovers produce recombinant progeny, but these progeny are the same types produced by single crossovers. Consequently, some multiple crossovers go undetected when the progeny of a genetic cross are observed. Therefore, map distances based on recombination rates will underestimate

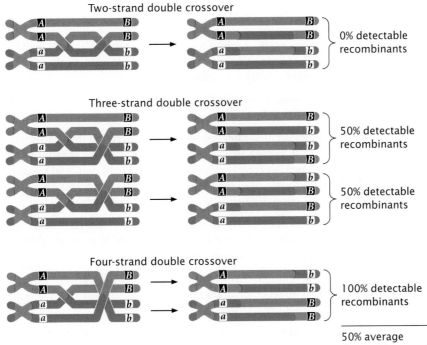

Two-strand double crossover

0% detectable recombinants

Three-strand double crossover

50% detectable recombinants

50% detectable recombinants

Four-strand double crossover

100% detectable recombinants

50% average detectable recombinants

5.16 Results of two-, three-, and four-strand double crossovers on recombination between two genes.

the true physical distances between genes, because some multiple crossovers are not detected among the progeny of a cross. When genes are very close together, multiple crossovers are unlikely, and the distances based on recombination rates accurately correspond to the physical distances on the chromosome. But, as the distance between genes increases, more multiple crossovers are likely, and the discrepancy between genetic distances (based on recombination rates) and physical distances increases. To correct for this discrepancy, geneticists have developed mathematical **mapping functions**, which relate recombination frequencies to actual physical distances between genes (**Figure 5.17**). Most of these functions are based on the Poisson distribution, which predicts the probability of multiple rare events. With the use of such mapping functions, map distances based on recombination rates can be more accurately estimated.

Mapping with Molecular Markers

For many years, gene mapping was limited in most organisms by the availability of **genetic markers**—that is, variable genes with easily observable phenotypes whose inheritance could be studied. Traditional genetic markers include genes that encode easily observable characteristics such as flower color, seed shape, blood types, and biochemical differences. The paucity of these types of characteristics in many organisms limited mapping efforts.

In the 1980s, new molecular techniques made it possible to examine variations in DNA itself, providing an almost unlimited number of genetic markers that can be used for creating genetic maps and studying linkage relations. The earliest of these molecular markers consisted of restriction

fragment length polymorphisms (RFLPs), which are variations in DNA sequence detected by cutting the DNA with restriction enzymes (see Chapter 14). Later, methods were developed for detecting variable numbers of short DNA sequences repeated in tandem, called microsatellites. More recently, DNA sequencing allows the direct detection of individual variations in the DNA nucleotides, called single nucleotide polymorphisms (SNPs; see Chapter 14). All of these methods have expanded the availability of genetic markers and greatly facilitated the creation of genetic maps.

Gene mapping with molecular markers is done essentially in the same manner as mapping performed with traditional phenotypic markers: the cosegregation of two or more markers is studied, and map distances are based on the rates of recombination between markers. These methods and their use in mapping are presented in more detail in Chapter 14.

5.17 Percent recombination underestimates the true physical distance between genes at higher map distances.

Concepts Summary

- Linked genes do not assort independently. In a testcross for two completely linked genes (no crossing over), only nonrecombinant progeny are produced. When two genes assort independently, recombinant progeny and nonrecombinant progeny are produced in equal proportions. When two genes are linked with some crossing over between them, more nonrecombinant progeny than recombinant progeny are produced.

- Recombination frequency is calculated by summing the number of recombinant progeny, dividing by the total number of progeny produced in the cross, and multiplying by 100%. The recombination frequency is half the frequency of crossing over, and the maximum frequency of recombinant gametes is 50%.

- Coupling and repulsion refer to the arrangement of alleles on a chromosome. Whether genes are in coupling configuration or in repulsion determines which combination of phenotypes will be most frequent in the progeny of a testcross.

- Interchromosomal recombination takes place among genes located on different chromosomes through the random

segregation of chromosomes in meiosis. Intrachromosomal recombination takes place among genes located on the same chromosome through crossing over.

- A chi-square test of independence can be used to determine if genes are linked.

- Recombination rates can be used to determine the relative order of genes and distances between them on a chromosome. One percent recombination equals one map unit. Maps based on recombination rates are called genetic maps; maps based on physical distances are called physical maps.

- Genetic maps can be constructed by examining recombination rates from a series of two-point crosses or by examining the progeny of a three-point testcross.

- Some multiple crossovers go undetected; thus, genetic maps based on recombination rates underestimate the true physical distances between genes.

- Molecular techniques that allow the detection of variable differences in DNA sequence have greatly facilitated gene mapping.

Important Terms

linked genes (p. 109)
linkage group (p. 109)
nonrecombinant (parental) gamete (p. 111)
nonrecombinant (parental) progeny (p. 111)
recombinant gamete (p. 111)
recombinant progeny (p. 111)

recombination frequency (p. 113)
coupling (cis) configuration (p. 114)
repulsion (trans) configuration (p. 114)
interchromosomal recombination (p. 116)
intrachromosomal recombination (p. 116)
genetic map (p. 119)
physical map (p. 119)
map unit (m.u.) (p. 119)

centiMorgan (p. 119)
Morgan (p. 119)
two-point testcross (p. 120)
three-point testcross (p. 121)
interference (p. 125)
coefficient of coincidence (p. 125)
mapping function (p. 129)
genetic marker (p. 129)

Answers to Concept Checks

1. c

2. 20%, in repulsion

3. Genetic maps are based on rates of recombination; physical maps are based on physical distances.

4. $\dfrac{m^+ p^+ s^+}{m\ p\ s}\ \dfrac{m^+ p\ s}{m\ p\ s}\ \dfrac{m\ p^+ s^+}{m\ p\ s}\ \dfrac{m^+ p^+ s}{m\ p\ s}\ \dfrac{m\ p\ s^+}{m\ p\ s}\ \dfrac{m^+ p\ s^+}{m\ p\ s}\ \dfrac{m\ p^+ s}{m\ p\ s}\ \dfrac{m\ p\ s}{m\ p\ s}$

5. The c locus

6. b

Worked Problems

1. In guinea pigs, white coat (w) is recessive to black coat (W) and wavy hair (v) is recessive to straight hair (V). A breeder crosses a guinea pig that is homozygous for white coat and wavy hair with a guinea pig that is black with straight hair. The F_1 are then crossed with guinea pigs having white coats and wavy hair in a series of testcrosses. The following progeny are produced from these testcrosses:

black, straight	30
black, wavy	10
white, straight	12
white, wavy	<u>31</u>
Total	83

a. Are the genes that determine coat color and hair type assorting independently? Carry out chi-square tests to test your hypothesis.

b. If the genes are not assorting independently, what is the recombination frequency between them?

· Solution

a. Assuming independent assortment, outline the crosses conducted by the breeder:

P $ww\ vv \times WW\ VV$
                        $\downarrow$
F_1 $Ww\ Vv$
                        $\downarrow$
Testcross $Ww\ Vv \times ww\ vv$
                        $\downarrow$

$Ww\ Vv$ $\frac{1}{4}$ black, straight
$Ww\ vv$ $\frac{1}{4}$ black, wavy
$ww\ Vv$ $\frac{1}{4}$ white, straight
$ww\ vv$ $\frac{1}{4}$ white, wavy

Because a total of 83 progeny were produced in the testcrosses, we expect $\frac{1}{4} \times 83 = 20.75$ of each. The observed numbers of progeny from the testcross (30, 10, 12, 31) do not appear to fit the expected numbers (20.75, 20.75, 20.75, 20.75) well; so independent assortment may not have taken place.

To test the hypothesis, carry out a chi-square test of independence. Construct a table, with the genotypes of the first locus along the top and the genotypes of the second locus along the side. Compute the totals for the rows and columns and the grand total.

	Ww	ww	Row totals
Vv	30	12	42
vv	10	31	41
Column totals	40	43	83 ← Grand total

The expected value for each cell of the table is calculated with the formula:

$$\text{expected number} = \frac{\text{row total} \times \text{column total}}{\text{grand total}}$$

Using this formula, we find the expected values (given in parentheses) to be:

	Ww	ww	Row totals
Vv	30 (20.24)	12 (21.76)	42
vv	10 (19.76)	31 (21.24)	41
Column totals	40	43	83 ← Grand total

Using these observed and expected numbers, we find the calculated chi-square value to be:

$$\chi^2 = \sum \frac{(\text{observed} - \text{expected})^2}{\text{expected}}$$

$$= \frac{(30-20.24)^2}{20.24} + \frac{(10-19.76)^2}{19.76} + \frac{(12-21.76)^2}{21.76} + \frac{(31-21.24)^2}{21.24}$$

$$= 4.71 + 4.82 + 4.38 + 4.48$$

$$= 18.39$$

The degrees of freedom for the chi-square test of independence are $df = (\text{number of rows} - 1) \times (\text{number of columns} - 1)$. There are two rows and two columns, so the degrees of freedom are:

$$df = (2 - 1) \times (2 - 1) = 1 \times 1 = 1$$

In Table 3.4, the probability associated with a chi-square value of 18.39 and 1 degree of freedom is less than 0.005, indicating that chance is very unlikely to be responsible for the differences between the observed numbers and the numbers expected with independent assortment. The genes for coat color and hair type have therefore not assorted independently.

b. To determine the recombination frequencies, identify the recombinant progeny. Using the notation for linked genes, write the crosses:

P $\dfrac{w\ \ v}{w\ \ v} \times \dfrac{W\ \ V}{W\ \ V}$
                        $\downarrow$
F_1 $\dfrac{W\ \ V}{w\ \ v}$
                        $\downarrow$
Testcross $\dfrac{W\ \ V}{w\ \ v} \times \dfrac{w\ \ v}{w\ \ v}$
                        $\downarrow$

$\dfrac{W\ \ V}{w\ \ v}$ 30 black, straight (nonrecombinant progeny)

$\dfrac{w\ \ v}{w\ \ v}$ 31 white, wavy (nonrecombinant progeny)

$\dfrac{W\ \ v}{w\ \ v}$ 10 black, wavy (recombinant progeny)

$\dfrac{w\ \ V}{w\ \ v}$ 12 white, straight (recombinant progeny)

The recombination frequency is:

$$\frac{\text{number of recombinant progeny}}{\text{total number progeny}} \times 100\%$$

or

$$\text{recombinant frequency} = \frac{10 + 12}{30 + 31 + 10 + 12} \times 100\%$$

$$= \frac{22}{83} \times 100\% = 26.5$$

2. A series of two-point crosses were carried out among seven loci (*a*, *b*, *c*, *d*, *e*, *f*, and *g*), producing the following recombination frequencies. Using these recombination frequencies, map the seven loci, showing their linkage groups, the order of the loci in each linkage group, and the distances between the loci of each group:

Loci	Recombination frequency (%)	Loci	Recombination frequency (%)
a and *b*	10	*c* and *d*	50
a and *c*	50	*c* and *e*	8
a and *d*	14	*c* and *f*	50
a and *e*	50	*c* and *g*	12
a and *f*	50	*d* and *e*	50
a and *g*	50	*d* and *f*	50
b and *c*	50	*d* and *g*	50
b and *d*	4	*e* and *f*	50
b and *e*	50	*e* and *g*	18
b and *f*	50	*f* and *g*	50
b and *g*	50		

• Solution

To work this problem, remember that 1% recombination equals 1 map unit and a recombination frequency of 50% means that genes at the two loci are assorting independently (located in different linkage groups).

The recombination frequency between *a* and *b* is 10%; so these two loci are in the same linkage group, approximately 10 m.u. apart.

The recombination frequency between *a* and *c* is 50%; so *c* must lie in a second linkage group.

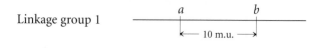

The recombination frequency between *a* and *d* is 14%; so *d* is located in linkage group 1. Is locus *d* 14 m.u. to the right or to the left of gene *a*? If *d* is 14 m.u. to the left of *a*, then the *b*-to-*d* distance should be 10 m.u. + 14 m.u. = 24 m.u. On the other hand, if *d* is to the right of *a*, then the distance between *b* and *d* should be 14 m.u. − 10 m.u. = 4 m.u. The *b*–*d* recombination frequency is 4%; so *d* is 14 m.u. to the right of *a*. The updated map is:

The recombination frequencies between each of loci *a*, *b*, and *d*, and locus *e* are all 50%; so *e* is not in linkage group 1 with *a*, *b*, and *d*. The recombination frequency between *e* and *c* is 8 m.u.; so *e* is in linkage group 2:

There is 50% recombination between *f* and all the other genes; so *f* must belong to a third linkage group:

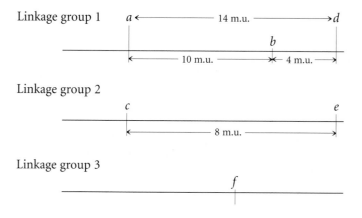

Finally, position locus *g* with respect to the other genes. The recombination frequencies between *g* and loci *a*, *b*, and *d* are all 50%; so *g* is not in linkage group 1. The recombination frequency between *g* and *c* is 12 m.u.; so *g* is a part of linkage group 2. To determine whether *g* is 12 m.u. to the right or left of *c*, consult the *g*–*e* recombination frequency. Because this recombination frequency is 18%, *g* must lie to the left of *c*:

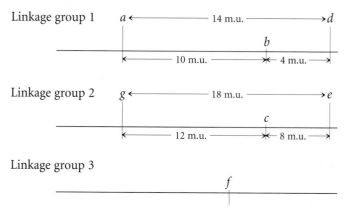

Note that the g-to-e distance (18 m.u.) is shorter than the sum of the g-to-c (12 m.u.) and c-to-e distances (8 m.u.), because of undetectable double crossovers between g and e.

3. Ebony body color (e), rough eyes (ro), and brevis bristles (bv) are three recessive mutations that occur in fruit flies. The loci for these mutations have been mapped and are separated by the following map distances:

The interference between these genes is 0.4.

A fly with ebony body, rough eyes, and brevis bristles is crossed with a fly that is homozygous for the wild-type traits. The resulting F_1 females are test-crossed with males that have ebony body, rough eyes, and brevis bristles; 1800 progeny are produced. Give the expected numbers of phenotypes in the progeny of the testcross.

• **Solution**

The crosses are:

$$
P \qquad \frac{e^+ \quad ro^+ \quad bv^+}{e^+ \quad ro^+ \quad bv^+} \times \frac{e \quad ro \quad bv}{e \quad ro \quad bv}
$$

$$
\downarrow
$$

$$
F_1 \qquad \frac{e^+ \quad ro^+ \quad bv^+}{e \quad ro \quad bv}
$$

$$
\downarrow
$$

$$
\text{Testcross} \qquad \frac{e^+ \quad ro^+ \quad bv^+}{e \quad ro \quad bv} \times \frac{e \quad ro \quad bv}{e \quad ro \quad bv}
$$

In this case, we know that ro is the middle locus because the genes have been mapped. Eight classes of progeny will be produced from this cross:

e^+	ro^+	bv^+	nonrecombinant
e	ro	bv	nonrecombinant
e^+ / ro	bv		single crossover between e and ro
e / ro^+	bv^+		single crossover between e and ro
e^+	ro^+ / bv		single crossover between ro and bv
e	ro / bv^+		single crossover between ro and bv
e^+ / ro / bv^+			double crossover
e / ro^+ / bv			double crossover

To determine the numbers of each type, use the map distances, starting with the double crossovers. The expected number of double crossovers is equal to the product of the single-crossover probabilities:

expected number of double crossovers =
$$0.20 \times 0.12 \times 1800 = 43.2$$

However, some interference occurs; so the observed number of double crossovers will be less than the expected. The interference is 1 − coefficient of coincidence; so the coefficient of coincidence is:

$$\text{coefficient of coincidence} = 1 - \text{interference}$$

The interference is given as 0.4; so the coefficient of coincidence equals $1 - 0.4 = 0.6$. Recall that the coefficient of coincidence is:

$$\text{coefficient of coincidence} = \frac{\text{number of observed double crossovers}}{\text{number of expected double crossovers}}$$

Rearranging this equation, we obtain:

$$\text{number of observed double crossovers} = \text{coefficient of coincidence} \times \text{number of expected double crossover}$$

$$\text{number of observed double crossovers} = 0.6 \times 43.2 = 26$$

A total of 26 double crossovers should be observed. Because there are two classes of double crossovers (e^+ / ro / bv^+ and e / ro^+ / bv), we expect to observe 13 of each.

Next, we determine the number of single-crossover progeny. The genetic map indicates that the distance between e and ro is 20 m.u.; so 360 progeny (20% of 1800) are expected to have resulted from recombination between these two loci. Some of them will be single-crossover progeny and some will be double-crossover progeny. We have already determined that the number of double-crossover progeny is 26; so the number of progeny resulting from a single crossover between e and ro is $360 - 26 = 334$, which will be divided equally between the two single-crossover phenotypes (e / ro^+ / bv^+ and e^+ / ro / bv).

The distance between ro and bv is 12 m.u.; so the number of progeny resulting from recombination between these two genes is $0.12 \times 1800 = 216$. Again, some of these recombinants will be single-crossover progeny and some will be double-crossover progeny. To determine the number of progeny resulting from a single crossover, subtract the double crossovers: $216 - 26 = 190$. These single-crossover progeny will be divided between the two single-crossover phenotypes (e^+ / ro^+ / bv and e / ro / bv^+); so there will be $\frac{190}{2} = 95$ of each of these phenotypes. The remaining progeny will be nonrecombinants, and they can be obtained by subtraction: $1800 - 26 - 334 - 190 = 1250$; there are two nonrecombinants (e^+ ro^+ bv^+ and e ro bv); so there will be $\frac{1250}{2} = 625$ of each. The numbers of the various phenotypes are listed here:

e^+	ro^+	bv^+	625	nonrecombinant
e	ro	bv	625	nonrecombinant
e^+ / ro	bv		167	single crossover between e and ro
e / ro^+	bv^+		167	single crossover between e and ro
e^+	ro^+ / bv		95	single crossover between ro and bv
e	ro / bv^+		95	single crossover between ro and bv
e^+ / ro / bv^+			13	double crossover
e / ro^+ / bv			13	double crossover
Total			1800	

Comprehension Questions

Section .5.1

*1. What does the term recombination mean? What are two causes of recombination?

Section 5.2

2. What effect does crossing over have on linkage?

3. Why is the frequency of recombinant gametes always half the frequency of crossing over?

*4. What is the difference between genes in coupling configuration and genes in repulsion? What effect does the arrangement of linked genes (whether they are in coupling configuration or in repulsion) have on the results of a cross?

5. What is the difference between a genetic map and a physical map?

Section 5.3

6. Explain how to determine which of three linked loci is the middle locus from the progeny of a three-point testcross.

*7. What does the interference tell us about the effect of one crossover on another?

Application Questions and Problems

Section 5.2

*8. In the snail *Cepaea nemoralis*, an autosomal allele causing a banded shell (B^B) is recessive to the allele for an unbanded shell (B^O). Genes at a different locus determine the background color of the shell; here, yellow (C^Y) is recessive to brown (C^{Bw}). A banded, yellow snail is crossed with a homozygous brown, unbanded snail. The F_1 are then crossed with banded, yellow snails (a testcross).

   a. What will be the results of the testcross if the loci that control banding and color are linked with no crossing over?

   b. What will be the results of the testcross if the loci assort independently?

   c. What will be the results of the testcross if the loci are linked and 20 m.u. apart?

*9. A geneticist discovers a new mutation in *Drosophila melanogaster* that causes the flies to shake and quiver. She calls this mutation spastic (*sps*) and determines that spastic is due to an autosomal recessive gene. She wants to determine if the gene encoding spastic is linked to the recessive gene for vestigial wings (*vg*). She crosses a fly homozygous for spastic and vestigial traits with a fly homozygous for the wild-type traits and then uses the resulting F_1 females in a testcross. She obtains the following flies from this testcross.

vg^+	sps^+	230
vg	sps	224
vg	sps^+	97
vg^+	sps	99
Total		650

Are the genes that cause vestigial wings and the spastic mutation linked? Do a chi-square test of independence to determine if the genes have assorted independently.

10. In cucumbers, heart-shaped leaves (*hl*) are recessive to normal leaves (*Hl*) and having numerous fruit spines (*ns*) is recessive to having few fruit spines (*Ns*). The genes for leaf shape and for number of spines are located on the same chromosome; findings from mapping experiments indicate that they are 32.6 m.u. apart. A cucumber plant having heart-shaped leaves and numerous spines is crossed with a plant that is homozygous for normal leaves and few spines. The F_1 are crossed with plants that have heart-shaped leaves and numerous spines. What phenotypes and proportions are expected in the progeny of this cross?

*11. In tomatoes, tall (*D*) is dominant over dwarf (*d*) and smooth fruit (*P*) is dominant over pubescent fruit (*p*), which is covered with fine hairs. A farmer has two tall and smooth tomato plants, which we will call plant A and plant B. The farmer crosses plants A and B with the same dwarf and pubescent plant and obtains the following numbers of progeny:

	Progeny of	
	Plant A	**Plant B**
Dd Pp	122	2
Dd pp	6	82
dd Pp	4	82
dd pp	124	4

   a. What are the genotypes of plant A and plant B?

   b. Are the loci that determine height of the plant and pubescence linked? If so, what is the map distance between them?

   c. Explain why different proportions of progeny are produced when plant A and plant B are crossed with the same dwarf pubescent plant.

12. Daniel McDonald and Nancy Peer determined that eyespot (a clear spot in the center of the eye) in flour beetles is caused by an X-linked gene (es) that is recessive to the allele for the absence of eyespot (es^+). They conducted a series of crosses to determine the distance between the gene for eyespot and a dominant X-linked gene for stripped (St), which acted as a recessive lethal (is lethal when homozygous in females or hemizygous in males). The following cross was carried out (D. J. McDonald and N. J. Peer. 1961. *Journal of Heredity* 52:261–264).

$$\female \ \frac{es^+ \ St}{es \ St^+} \times \frac{es \ St^+}{Y} \ \male$$

$$\downarrow$$

$$\frac{es^+ \quad St}{es \quad St^+} \quad 1630$$

$$\frac{es \quad St^+}{es \quad St^+} \quad 1665$$

$$\frac{es \quad St^+}{es \quad St^+} \quad 935$$

$$\frac{es^+ \quad St^+}{es \quad St^+} \quad 1005$$

$$\frac{es \quad St^+}{Y} \quad 1661$$

$$\frac{es^+ \quad St^+}{Y} \quad 1024$$

a. Which progeny are the recombinants and which progeny are the nonrecombinants?

b. Calculate the recombination frequency between *es* and *St*.

c. Are some potential genotypes missing among the progeny of the cross? If so, which ones and why?

13. In German cockroaches, bulging eyes (bu) are recessive to normal eyes (bu^+) and curved wings (cv) are recessive to straight wings (cv^+). Both traits are encoded by autosomal genes that are linked. A cockroach has genotype $bu^+ bu \ cv^+ cv$ and the genes are in repulsion. Which of the following sets of genes will be found in the most-common gametes produced by this cockroach?

a. $bu^+ \ cv^+$

b. $bu \ cv$

c. $bu^+ \ bu$

d. $cv^+ \ cv$

e. $bu \ cv^+$

Explain your answer.

*14. In *Drosophila melanogaster*, ebony body (e) and rough eyes (ro) are encoded by autosomal recessive genes found on chromosome 3; they are separated by 20 m.u. The gene that encodes forked bristles (f) is X-linked recessive and assorts independently of e and ro. Give the phenotypes of progeny and their expected proportions when a female of each of the following genotypes is test-crossed with a male.

a. $\dfrac{e^+ \ ro^+}{e \ ro} \quad \dfrac{f^+}{f}$

b. $\dfrac{e^+ \ ro}{e \ ro^+} \quad \dfrac{f^+}{f}$

*15. A series of two-point crosses were carried out among seven loci (a, b, c, d, e, f, and g), producing the following recombination frequencies. Map the seven loci, showing their linkage groups, the order of the loci in each linkage group, and the distances between the loci of each group.

Loci	Percent recombination	Loci	Percent recombination
a and b	50	c and d	50
a and c	50	c and e	26
a and d	12	c and f	50
a and e	50	c and g	50
a and f	50	d and e	50
a and g	4	d and f	50
b and c	10	d and g	8
b and d	50	e and f	50
b and e	18	e and g	50
b and f	50	f and g	50
b and g	50		

16. R. W. Allard and W. M. Clement determined recombination rates for a series of genes in lima beans (R. W. Allard and W. M. Clement. 1959. *Journal of Heredity* 50:63–67). The following table lists paired recombination rates for eight of the loci (D, Wl, R, S, L_1, Ms, C, and G) that they mapped. On the basis of these data, draw a series of genetic maps for the different linkage groups of the genes, indicating the distances between the genes. Keep in mind that these rates are estimates of the true recombination rates and that some error is associated with each estimate. An asterisk beside a recombination frequency indicates that the recombination frequency is significantly different from 50%.

Recombination Rates (%) among Seven Loci in Lima Beans

	Wl	R	S	L_1	Ms	C	G
D	2.1*	39.3*	52.4	48.1	53.1	51.4	49.8
Wl		38.0*	47.3	47.7	48.8	50.3	50.4
R			51.9	52.7	54.6	49.3	52.6
S				26.9*	54.9	52.0	48.0
L_1					48.2	45.3	50.4
Ms						14.7*	43.1
C							52.0

*Significantly different from 50%.

Section 5.3

17. Raymond Popp studied linkage among genes for pink eye (p), shaker-1 (sh-1), and hemoglobin (Hb) in mice (R. A. Popp. 1962. *Journal of Heredity* 53:73–80). He performed a series of test crosses, in which mice heterozygous for pink eye, shaker-1, and hemoglobin 1 and 2 were crossed with

mice that were homozygous for pink eye, shaker-1 and hemoglobin 2.

$$\frac{P\ Sh{-}1\ Hb^1}{p\ sh{-}1\ Hb^2} \times \frac{P\ sh{-}1\ Hb^2}{p\ sh{-}1\ Hb^2}$$

The following progeny were produced.

Progeny genotype	Number
$\dfrac{p\ sh{-}1\ Hb^2}{p\ sh{-}1\ Hb^2}$	274
$\dfrac{P\ Sh{-}1\ Hb^1}{p\ sh{-}1\ Hb^2}$	320
$\dfrac{P\ sh{-}1\ Hb^2}{p\ sh{-}1\ Hb^2}$	57
$\dfrac{p\ Sh{-}1\ Hb^1}{p\ sh{-}1\ Hb^2}$	45
$\dfrac{p\ Sh{-}1\ Hb^2}{p\ sh{-}1\ Hb^2}$	6
$\dfrac{p\ sh{-}1\ Hb^1}{p\ sh{-}1\ Hb^2}$	5
$\dfrac{p\ Sh{-}1\ Hb^2}{p\ sh{-}1\ Hb^2}$	0
$\dfrac{P\ sh{-}1\ Hb^1}{p\ sh{-}1\ Hb^2}$	1
Total	708

a. Determine the order of these genes on the chromosome.

b. Calculate the map distances between the genes.

c. Determine the coefficient of coincidence and the interference among these genes.

*18. Waxy endosperm (*wx*), shrunken endosperm (*sh*), and yellow seedling (*v*) are encoded by three recessive genes in corn that are linked on chromosome 5. A corn plant homozygous for all three recessive alleles is crossed with a plant homozygous for all the dominant alleles. The resulting F_1 are then crossed with a plant homozygous for the recessive alleles in a three-point testcross. The progeny of the testcross are:

wx	*sh*	*V*	87
Wx	*Sh*	*v*	94
Wx	*Sh*	*V*	3,479
wx	*sh*	*v*	3,478
Wx	*sh*	*V*	1,515
wx	*Sh*	*v*	1,531
wx	*Sh*	*V*	292
Wx	*sh*	*v*	280
Total			10,756

a. Determine the order of these genes on the chromosome.

b. Calculate the map distances between the genes.

c. Determine the coefficient of coincidence and the interference among these genes.

19. Priscilla Lane and Margaret Green studied the linkage relations of three genes affecting coat color in mice: mahogany (*mg*), agouti (*a*), and ragged (*Rg*). They carried out a series of three-point crosses, mating mice that were heterozygous at all three loci with mice that were homozygous for the recessive alleles at these loci (P W. Lane and M. C. Green. 1960. *Journal of Heredity* 51:228–230). The following table lists the results of the test crosses.

Phenotype			Number
a	*Rg*	+	1
+	+	*mg*	1
a	+	+	15
+	*Rg*	*mg*	9
+	+	+	16
a	*Ra*	*mg*	36
a	+	*mg*	76
+	*Ra*	+	69
Total			213

a. Determine the order of the loci that encode mahogany, agouti, and ragged on the chromosome, the map distances between them, and the interference and coefficient of coincidence for these genes.

b. Draw a picture of the two chromosomes in the triply heterozygous mice used in the testcrosses, indicating which of the alleles are present on each chromosome.

*20. In *Drosophila melanogaster*, black body (*b*) is recessive to gray body (b^+), purple eyes (*pr*) are recessive to red eyes (pr^+), and vestigial wings (*vg*) are recessive to normal wings (vg^+). The loci encoding these traits are linked, with the following map distances:

The interference among these genes is 0.5. A fly with a black body, purple eyes, and vestigial wings is crossed with a fly homozygous for a gray body, red eyes, and normal wings. The female progeny are then crossed with males that have a black body, purple eyes, and vestigial wings. If 1000 progeny are produced from this testcross, what will be the phenotypes and proportions of the progeny?

Challenge Question

Section 5.3

21. Transferrin is a blood protein that is encoded by the transferrin locus (Trf). In house mice the two alleles at this locus (Trf^a and Trf^b) are codominant and encode three types of transferrin:

Genotype	Phenotype
Trf^a/Trf^a	Trf-a
Trf^a/Trf^b	Trf-ab
Trf^b/Trf^b	Trf-b

The dilution locus, found on the same chromosome, determines whether the color of a mouse is diluted or full; an allele for dilution (d) is recessive to an allele for full color (d^+):

Genotype	Phenotype
d^+d^+	d^+ (full color)
d^+d	d^+ (full color)
dd	d (dilution)

Donald Shreffler conducted a series of crosses to determine the map distance between the tranferrin locus and the dilution locus (D. C. Shreffler. 1963 *Journal of Heredity* 54:127–129). The table at right presents a series of crosses carried out by Shreffler and the progeny resulting from these crosses.

a. Calculate the recombinant frequency between the *Trf* and the *d* loci by using the pooled data from all the crosses.

b. Which crosses represent recombination in male gamete formation and which crosses represent recombination in female gamete formation?

c. On the basis of your answer to part *b*, calculate the frequency of recombination among male parents and female parents separately.

d. Are the rates of recombination in males and females the same? If not, what might produce the difference?

			Progeny phenotypes				
			d^+	d^+	d	d	
Cross	♂	♀	Trf-ab	Trf-b	Trf-ab	Trf-b	Total
1	$\dfrac{d^+\ Trf^a}{d\ Trf^b}$	$\times\ \dfrac{d\ Trf^b}{d\ Trf^b}$	32	3	6	21	62
2	$\dfrac{d\ Trf^b}{d\ Trf^b}$	$\times\ \dfrac{d^+\ Trf^a}{d\ Trf^b}$	16	0	2	20	38
3	$\dfrac{d^+\ Trf^a}{d\ Trf^b}$	$\times\ \dfrac{d\ Trf^b}{d\ Trf^b}$	35	9	4	30	78
4	$\dfrac{d\ Trf^b}{d\ Trf^b}$	$\times\ \dfrac{d^+\ Trf^a}{d\ Trf^b}$	21	3	2	19	45
5	$\dfrac{d^+\ Trf^b}{d\ Trf^a}$	$\times\ \dfrac{d\ Trf^b}{d\ Trf^b}$	8	29	22	5	64
6	$\dfrac{d\ Trf^b}{d\ Trf^b}$	$\times\ \dfrac{d^+\ Trf^b}{d\ Trf^a}$	4	14	11	0	29

6 Bacterial and Viral Genetic Systems

Past human migrations can be charted by examining the present-day genetic diversity of the bacterium *Helicobacter pylori*, which resides in the human stomach and causes peptic ulcers. This bacterium has been transported throughout the world in the guts of its human hosts. *[©2004 Gwendolyn Knight Lawrence/ Artists Rights Society (ARS), New York/The Phillips Collection, Washington, D.C.]*

Gutsy Travelers

Peptic ulcers are tissue-damaging sores of the stomach and upper intestinal tract that affect 25 million Americans and, in serious cases, lead to life-threatening blood loss. For many years, peptic ulcers were attributed to stress and spicy foods, and ulcers were treated by encouraging changes in diet and life style, as well as by drugs that limited acid production by the stomach. Although these treatments often brought short-term relief, peptic ulcers in many patients returned and proved to be a recurring problem.

In 1982, physicians Barry Marshall and Robin Warren made a startling proposal. They suggested that most peptic ulcers are actually caused by a bacterium, *Helicobacter pylori* (**Figure 6.1**), which is able to tolerate the acidic environment of the stomach. Treatment for ulcers has now changed from an adjustment in life style to the administration of antibiotics, which has proved to be effective in eliminating the presence of *H. pylori* and permanently curing the disease. For their discovery, Marshall and Warren were awarded the Nobel Prize in medicine or physiology in 2005.

Interestingly, about half of the world's population is infected with *H. pylori*, but only a few people suffer from peptic ulcers. Thus, infection alone cannot be responsible for peptic ulcers, and other factors, including stress, diet, genetic differences among *H. pylori* strains, and even the genetic constitution of the host, are thought to play important roles in whether peptic ulcers arise.

Most people become infected with *H. pylori* in infancy or early childhood and remain infected for life. The source of the infection is usually other family members. A strain of *H. pylori* from one family can be differentiated from a strain from another family and so, like a surname, provides an accurate record of familial connections. Geneticists are now using this property of *H. pylori* to trace historical migrations of human populations.

In 2003, a team of geneticists led by Mark Achtman at the Max Planck Institute for Infection Biology in Berlin examined DNA sequences in eight genes of *H. pylori* bacteria collected from humans throughout the world. They observed that the bacterial sequences clustered into four major groups—two from Africa, one from east Asia, and one from Europe—which corresponded well to the human populations from which the bacteria were isolated. Further analysis revealed affinities among human populations within and between the groups. For example, bacteria from the Maoris (a group of native people from New Zealand) were closely related to those from Polynesia, corroborating other evidence

6.1 *Helicobacter pylori* **is the bacterium that causes peptic ulcers.** *[Veronika Burmeister/Visuals Unlimited.]*

that Maoris originated from South Pacific islanders who migrated to New Zealand several thousand years ago. Similarly, bacteria from Native Americans clustered with the East Asian bacterial strains, concurring with the Asian origin of Native Americans. The genes of the bacteria also fit together with more recent human migrations: African strains were found in high frequency among African Americans in Louisiana and Tennessee, and European strains of the bacteria were found among people in Singapore, South Africa, and North America. The results of these studies reveal that *H. pylori* travels the world in the guts of its human hosts and can be used to help resolve the details of past human migrations.

Just as scientists are recognizing the value of *H. pylori* in studies of human evolution and history, the bacteria seem to be disappearing from human guts, particularly those of people in developed countries. In developing countries, from 70% to 100% of children are infected with *H. pylori*, but fewer than 10% of children in the United States have the bacteria. The cause for this bacterial decline is unknown—widespread use of antibiotics and better hygiene are suspected—but it has led to a dramatic decrease in the incidence of peptic ulcers and stomach cancer (which is also associated with the presence of the bacteria) in developed countries.

I n this chapter, we will examine some of the mechanisms by which bacteria like *H. pylori* exchange and recombine their genes. Since the 1940s, the genetic systems of bacteria and viruses have contributed to the discovery of many important concepts in genetics. The study of molecular genetics initially focused almost entirely on their genes; today, bacteria and viruses are still essential tools for probing the nature of genes in more-complex organisms, in part because they possess a number of characteristics that make them suitable for genetic studies (**Table 6.1**).

The genetic systems of bacteria and viruses are also studied because these organisms play important roles in human society. As illustrated by *H. pylori*, many bacteria are an important part of human ecology. They have been harnessed to produce a number of economically important substances, and they are of immense medical significance, causing many human diseases. In this chapter, we focus on several unique aspects of bacterial and viral genetic systems. Important processes of gene transfer and recombination, like those that contribute to the genetic structure of *H. pylori*, will be described, and we will see how these processes can be used to map bacterial and viral genes.

Table 6.1	Advantages of using bacteria and viruses for genetic studies

1. Reproduction is rapid.

2. Many progeny are produced.

3. Haploid genome allows all mutations to be expressed directly.

4. Asexual reproduction simplifies the isolation of genetically pure strains.

5. Growth in the laboratory is easy and requires little space.

6. Genomes are small.

7. Techniques are available for isolating and manipulating their genes.

8. They have medical importance.

9. They can be genetically engineered to produce substances of commercial value.

6.1 Genetic Analysis of Bacteria Requires Special Approaches and Methods

Heredity in bacteria is fundamentally similar to heredity in more-complex organisms, but the bacterial haploid genome and the small size of bacteria (which makes observation of their phenotypes difficult) require different approaches and methods. First, we will consider how bacteria are studied and, then, we will examine several processes that transfer genes from one bacterium to another.

Techniques for the Study of Bacteria

Microbiologists have defined the nutritional needs of a number of bacteria and developed culture media for growing them in the laboratory. Culture media typically contain a carbon source, essential elements such as nitrogen and phos-

6.2 Bacteria can be grown (a) in liquid medium or (b) on solid medium.

phorus, certain vitamins, and other required ions and nutrients. Wild-type (*prototrophic*) bacteria can use these simple ingredients to synthesize all the compounds that they need for growth and reproduction. A medium that contains only the nutrients required by prototrophic bacteria is termed **minimal medium**. Mutant strains called *auxotrophs* lack one or more enzymes necessary for metabolizing nutrients or synthesizing essential molecules and will grow only on medium supplemented with one or more nutrients. For example, auxotrophic strains that are unable to synthesize the amino acid leucine will not grow on minimal medium but *will* grow on medium to which leucine has been added. **Complete medium** contains all the substances, such as the amino acid leucine, required by bacteria for growth and reproduction.

Cultures of bacteria are often grown in test tubes that contain sterile liquid medium (**Figure 6.2a**). A few bacteria are added to a tube, and they grow and divide until all the nutrients are used up or—more commonly—until the concentration of their waste products becomes toxic. Bacteria are also grown in petri plates (**Figure 6.2b**). Growth medium suspended in agar is poured into the bottom half of the petri plate, providing a solid, gel-like base for bacterial growth. In a process called plating, a dilute solution of bacteria is spread

over the surface of an agar-filled petri plate. As each bacterium grows and divides, it gives rise to a visible clump of genetically identical cells (a **colony**). Genetically pure strains of the bacteria can be isolated by collecting bacteria from a single colony and transferring them to a new test tube or petri plate. The chief advantage of growing bacteria on a petri plate is that it allows one to isolate and count bacteria, which individually are too small to see without a microscope.

Because individual bacteria are too small to be seen directly, it is often easier to study phenotypes that affect the appearance of the colony (**Figure 6.3**) or can be detected by simple chemical tests. Auxotrophs are commonly studied phenotypes. Suppose we want to detect auxotrophs that cannot synthesize leucine (*leu⁻* mutants). We first spread the bacteria on a petri plate containing medium that includes leucine; both prototrophs that have the *leu⁺* allele and auxotrophs that have *leu⁻* alleles will grow on it (**Figure 6.4**). Next, using a technique called replica plating, we transfer a few cells from each of the colonies on the original plate to two new replica plates: one plate contains medium to which leucine has been added; the other plate contains selective medium—that is, a medium in this case lacking leucine. The *leu⁺* bacteria will grow on both media, but the *leu⁻* mutants will grow only on the medium supplemented by leucine,

6.3 Bacteria have a variety of phenotypes. (a) *Serratia marcescens* with color variation. (b) *Bacillus cereus*. [Part a: Dr. E. Bottone/Peter Arnold. Part b: Biophoto Associates/Photo Researchers.]

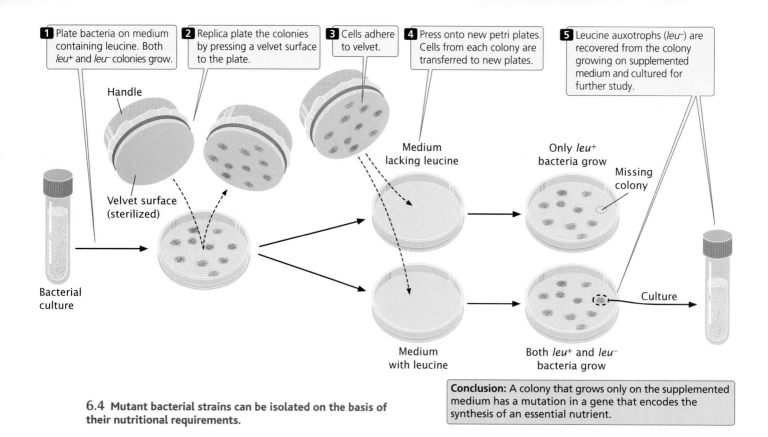

1 Plate bacteria on medium containing leucine. Both *leu+* and *leu–* colonies grow.

2 Replica plate the colonies by pressing a velvet surface to the plate.

3 Cells adhere to velvet.

4 Press onto new petri plates. Cells from each colony are transferred to new plates.

5 Leucine auxotrophs (*leu–*) are recovered from the colony growing on supplemented medium and cultured for further study.

Handle

Velvet surface (sterilized)

Bacterial culture

Medium lacking leucine

Only *leu+* bacteria grow

Missing colony

Medium with leucine

Both *leu+* and *leu–* bacteria grow

Culture

6.4 Mutant bacterial strains can be isolated on the basis of their nutritional requirements.

Conclusion: A colony that grows only on the supplemented medium has a mutation in a gene that encodes the synthesis of an essential nutrient.

because they cannot synthesize their own leucine. Any colony that grows on medium that contains leucine but not on medium that lacks leucine consists of *leu–* bacteria. The auxotrophs that grow on the supplemented medium can then be cultured for further study.

The Bacterial Genome

Bacteria are unicellular organisms that lack a nuclear membrane. Most bacterial genomes consist of a circular chromosome that contains a single DNA molecule several million base pairs (bp) in length (**Figure 6.5**). For example, the genome of *E. coli* has approximately 4.6 million base pairs of DNA. However, some bacteria (such as *Vibrio cholerae*, which causes cholera) contain multiple chromosomes, and a few even have linear chromosomes.

Plasmids

In addition to having a chromosome, many bacteria possess **plasmids**—small, circular DNA molecules (**Figure 6.6**).

6.5 Most bacterial cells possess a single, circular chromosome, shown here emerging from a ruptured bacterial cell. *[David L. Nelson and Michael M. Cox, Lehninger Principles of Biochemistry, 4th ed. (New York: Worth Publishers, 2004), from Huntington Potter and David Dressler, Harvard Medical School, Department of Neurobiology.]*

6.6 Many bacteria contain plasmids—small, circular molecules of DNA. *[Professor Stanley N. Cohen/Photo Researchers.]*

1 Replication in a plasmid begins at the origin of replication, the *ori* site.

2 Strands separate and replication takes place in both directions,...

3 ...eventually producing two circular DNA molecules.

Origin of replication (*ori* site)

Newly synthesized DNA

Strand separation

Replication

Separation of daughter plasmids

Double-stranded DNA

Strands separate at *oriV*

New strand

Old strand

6.7 A plasmid replicates independently of its bacterial chromosome. Replication begins at the origin of replication (*ori*) and continues around the circle. In this diagram, replication is taking place in both directions; in some plasmids, replication is in one direction only.

Some plasmids are present in many copies per cell, whereas others are present in only one or two copies. In general, plasmids carry genes that are not essential to bacterial function but that may play an important role in the life cycle and growth of their bacterial hosts. Some plasmids promote mating between bacteria; others produce compounds that kill other bacteria. Of great importance, plasmids are used extensively in genetic engineering (see Chapter 14), and some of them play a role in the spread of antibiotic resistance among bacteria.

Most plasmids are circular and several thousand base pairs in length, although plasmids consisting of several hundred thousand base pairs also have been found. Possessing its own origin of replication, a plasmid replicates independently of the bacterial chromosome. Replication proceeds from the origin in one or two directions until the entire plasmid is copied. In **Figure 6.7**, the origin of replication is *ori*. A few plasmids have multiple replication origins.

Episomes are plasmids that are capable of freely replicating and able to integrate into the bacterial chromosomes. The **F** (fertility) **factor** of *E. coli* (**Figure 6.8**) is an episome that controls mating and gene exchange between *E. coli* cells, as will be discussed shortly.

Concepts

Bacteria can be studied in the laboratory by growing them on defined liquid or solid medium. A typical bacterial genome consists of a single circular chromosome that contains several million base pairs. Some bacterial genes may be present on plasmids, which are small, circular DNA molecules that replicate independently of the bacterial chromosome.

✔ Concept Check 1

Which is true of plasmids?

a. They are composed of RNA.

b. They normally exist outside of bacterial cells.

c. They possess only a single strand of DNA.

d. They replicate independently of the bacterial chromosome.

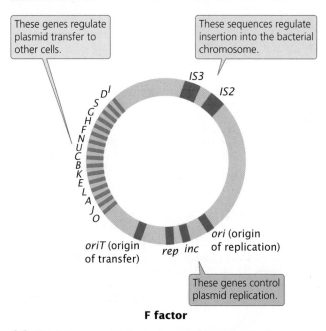

These genes regulate plasmid transfer to other cells.

These sequences regulate insertion into the bacterial chromosome.

IS3

IS2

oriT (origin of transfer)

rep *inc*

ori (origin of replication)

These genes control plasmid replication.

F factor

6.8 The F factor, a circular episome of *E. coli*, contains a number of genes that regulate transfer into the bacterial cell, replication, and insertion into the bacterial chromosome. Replication is initiated at *ori*. Insertion sequences *IS3* and *IS2* control both insertion into the bacterial chromosome and excision from it.

Gene Transfer in Bacteria

Bacteria exchange genetic material by three different mechanisms, all entailing some type of DNA transfer and recombination between the transferred DNA and the bacterial chromosome.

1. **Conjugation** takes place when genetic material passes directly from one bacterium to another (**Figure 6.9a**). In conjugation, two bacteria lie close together and a connection forms between them. A plasmid or a part

of the bacterial chromosome passes from one cell (the donor) to the other (the recipient). Subsequent to conjugation, crossing over may take place between homologous sequences in the transferred DNA and the chromosome of the recipient cell. In conjugation, DNA is transferred only from donor to recipient, with no reciprocal exchange of genetic material.

2. **Transformation** takes place when a bacterium takes up DNA from the medium in which it is growing

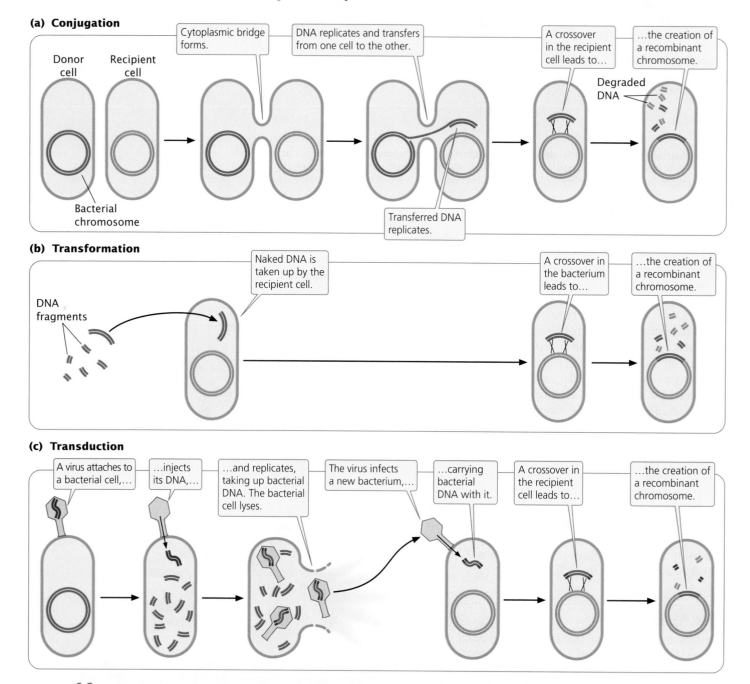

6.9 Conjugation, transformation, and transduction are three processes of gene transfer in bacteria. For the transferred DNA to be stably inherited, all three processes require the transferred DNA to undergo recombination with the bacterial chromosome.

(**Figure 6.9b**). After transformation, recombination may take place between the introduced genes and those of the bacterial chromosome.

3. **Transduction** takes place when bacterial viruses (bacteriophages) carry DNA from one bacterium to another (**Figure 6.9c**). Inside the bacterium, the newly introduced DNA may undergo recombination with the bacterial chromosome.

Not all bacterial species exhibit all three types of genetic transfer. Conjugation takes place more frequently in some species than in others. Transformation takes place to a limited extent in many species of bacteria, but laboratory techniques increase the rate of DNA uptake. Most bacteriophages have a limited host range; so transduction is normally between bacteria of the same or closely related species only.

These processes of genetic exchange in bacteria differ from diploid eukaryotic sexual reproduction in two important ways. First, DNA exchange and reproduction are not coupled in bacteria. Second, donated genetic material that is not recombined into the host DNA is usually degraded, and so the recipient cell remains haploid. Each type of genetic transfer can be used to map genes, as will be discussed in the following sections.

Concepts

DNA may be transferred between bacterial cells through conjugation, transformation, or transduction. Each type of genetic transfer consists of a one-way movement of genetic information to the recipient cell, sometimes followed by recombination. These processes are not connected to cellular reproduction in bacteria.

✔ Concept Check 2

Which process of DNA transfer in bacteria requires a virus?

a. Conjugation   c. Transformation

b. Transduction   d. All of the above

Conjugation

In 1946, Joshua Lederberg and Edward Tatum demonstrated that bacteria can transfer and recombine genetic information, paving the way for the use of bacteria in genetic studies. In the course of their research, Lederberg and Tatum studied auxotrophic strains of *E. coli*. The Y10 strain required the amino acids threonine (and was genotypically *thr⁻*) and leucine (*leu⁻*) and the vitamin thiamine (*thi⁻*) for growth but did not require the vitamin biotin (*bio⁺*) or the amino acids phenylalanine (*phe⁺*) and cysteine (*cys⁺*); the genotype of this strain can be written as *thr⁻ leu⁻ thi⁻ bio⁺ phe⁺ cys⁺*. The Y24 strain required biotin, phenylalanine, and cysteine in its medium, but it did not require threonine, leucine, or thiamine; its genotype was *thr⁺ leu⁺ thi⁺ bio⁻ phe⁻ cys⁻*. In one experiment, Lederberg and Tatum mixed Y10 and Y24 bacteria together and plated them on minimal medium (**Figure 6.10**). Each strain was also plated separately on minimal medium.

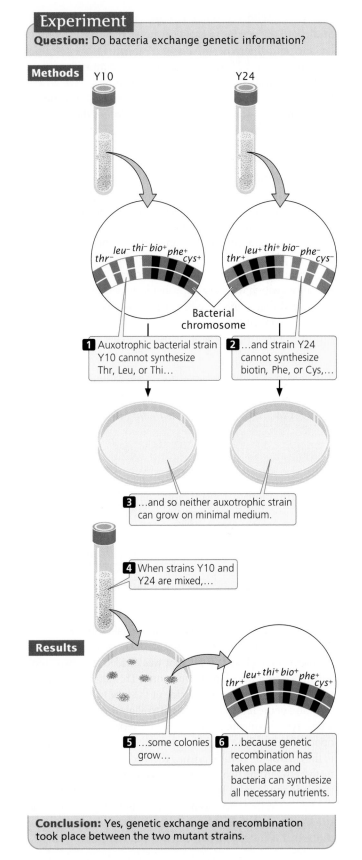

Experiment

Question: Do bacteria exchange genetic information?

Methods

1 Auxotrophic bacterial strain Y10 cannot synthesize Thr, Leu, or Thi…

2 …and strain Y24 cannot synthesize biotin, Phe, or Cys,…

3 …and so neither auxotrophic strain can grow on minimal medium.

4 When strains Y10 and Y24 are mixed,…

Results

5 …some colonies grow…

6 …because genetic recombination has taken place and bacteria can synthesize all necessary nutrients.

Conclusion: Yes, genetic exchange and recombination took place between the two mutant strains.

6.10 Lederberg and Tatum's experiment demonstrated that bacteria undergo genetic exchange.

Alone, neither Y10 nor Y24 grew on minimal medium. Strain Y10 was unable to grow, because it required threonine, leucine, and thiamine, which were absent in the minimal medium; strain Y24 was unable to grow, because it required biotin, phenylalanine, and cysteine, which also were absent from the minimal medium. When Lederberg and Tatum mixed the two strains, however, a few colonies did grow on the minimal medium. These prototrophic bacteria must have had genotype $thr^+ leu^+ thi^+ bio^+ phe^+ cys^+$. Where had they come from?

If mutations were responsible for the prototrophic colonies, then some colonies should also have grown on the plates containing Y10 or Y24 alone, but no bacteria grew on these plates. Multiple simultaneous mutations ($thr^- \rightarrow thr^+$, $leu^- \rightarrow leu^+$, and $thi^- \rightarrow thi^+$ in strain Y10 or $bio^- \rightarrow bio^+$, $phe^- \rightarrow phe^+$, and $cys^- \rightarrow cys^+$ in strain Y24) would have been required for either strain to become prototrophic by mutation, which was very improbable. Lederberg and Tatum concluded that some type of genetic transfer and recombination had taken place:

Auxotrophic strain

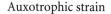
Y10 $\overline{thr^- \ leu^- \ thi^- \ bio^+ \ phe^+ \ cys^+}$

Y24 $\overline{thr^+ \ leu^+ \ thi^+ \ bio^- \ phe^- \ cys^-}$

↓

$\overline{thr^- \ leu^- \ thi^- \quad \ \ bio^+ \ phe^+ \ cys^+}$
$\overline{thr^+ \ leu^+ \ thi^+ \quad \ \ bio^- \ phe^- \ cys^-}$

↓

$\overline{thr^- \ leu^- \ thi^- \ bio^- \ phe^- \ cys^-}$

Prototrophic strain $\overline{thr^+ \ leu^+ \ thi^+ \ bio^+ \ phe^+ \ cys^+}$

What they did not know was *how* it had taken place.

To study this problem, Bernard Davis constructed a U-shaped tube (**Figure 6.11**) that was divided into two compartments by a filter having fine pores. This filter allowed liquid medium to pass from one side of the tube to the other, but the pores of the filter were too small to allow the passage of bacteria. Two auxotrophic strains of bacteria were placed on opposite sides of the filter, and suction was applied alternately to the ends of the U-tube, causing the medium to flow back and forth between the two compartments. Despite hours of incubation in the U-tube, bacteria plated out on minimal medium did not grow; there had been no genetic exchange between the strains. The exchange of bacterial genes clearly required direct contact, or conjugation, between the bacterial cells.

F^+ and F^- cells In most bacteria, conjugation depends on a fertility (F) factor that is present in the donor cell and

Experiment

Question: How did the genetic exchange seen in Lederberg and Tatum's experiment take place?

Methods

Results

Conclusion: Genetic exchange requires direct contact between bacterial cells.

6.11 Davis's U-tube experiment.

absent in the recipient cell. Cells that contain F factor are referred to as F^+, and cells lacking F factor are F^-.

The F factor contains an origin of replication and a number of genes required for conjugation (see Figure 6.8). For example, some of these genes encode sex **pili** (singular, pilus), slender extensions of the cell membrane. A cell containing F factor produces the sex pili, one of which makes contact with a receptor on an F^- cell (**Figure 6.12**) and pulls the two cells together. DNA is then transferred from the F^+ cell to the F^- cell. Conjugation can take place only between a cell that possesses F factor and a cell that lacks F factor.

In most cases, the only genes transferred during conjugation between an F^+ and F^- cell are those on the F factor (**Figure 6.13a and b**). Transfer is initiated when one of the DNA strands on the F factor is nicked at an origin (*oriT*). One end of the nicked DNA separates from the circle and passes into the recipient cell (**Figure 6.13c**). Replication takes

6.12 A sex pilus connects F⁺ and F⁻ cells during bacterial conjugation. *E. coli* cells in conjugation. [Dr. Dennis Kunkel/Phototake.]

place on the nicked strand, proceeding around the circular plasmid in the F⁺ cell and replacing the transferred strand (**Figure 6.13d**). Because the plasmid in the F⁺ cell is always nicked at the *oriT* (origin of transfer) site, this site always enters the recipient cell first, followed by the rest of the plasmid. Thus, the transfer of genetic material has a defined direction. Inside the recipient cell, the single strand is replicated, producing a circular, double-stranded copy of the F plasmid (**Figure 6.13e**). If the entire F factor is transferred to the recipient F⁻ cell, that cell becomes an F⁺ cell.

Hfr cells Conjugation transfers genetic material in the F plasmid from F⁺ to F⁻ cells but does not account for the transfer of chromosomal genes observed by Lederberg and Tatum. In Hfr (high-frequency recombination) strains, the F

factor is integrated into the bacterial chromosome (**Figure 6.14**). Hfr cells behave as F⁺ cells, forming sex pili and undergoing conjugation with F⁻ cells.

In conjugation between Hfr and F⁻ cells (**Figure 6.15a**), the integrated F factor is nicked, and the end of the nicked strand moves into the F⁻ cell (**Figure 6.15b**), just as it does in conjugation between F⁺ and F⁻ cells. Because, in an Hfr cell, the F factor has been integrated into the bacterial chromosome, the chromosome follows it into the recipient cell. How much of the bacterial chromosome is transferred depends on the length of time that the two cells remain in conjugation.

Inside the recipient cell, the donor DNA strand is replicated (**Figure 6.15c**), and crossing over between it and the original chromosome of the F⁻ cell (**Figure 6.15d**) may take place. This gene transfer between Hfr and F⁻ cells is how the recombinant prototrophic cells observed by Lederberg and Tatum were produced. After crossing over has taken place in the recipient cell, the donated chromosome is degraded, and the recombinant recipient chromosome remains (**Figure 6.15e**), to be replicated and passed on to later generations of bacterial cells by binary fission (cell division).

In a mating of Hfr × F⁻, the F⁻ cell almost never becomes F⁺ or Hfr, because the F factor is nicked in the middle in the initiation of strand transfer, placing part of the F factor at the beginning and part at the end of the strand to be transferred. To become F⁺ or Hfr, the recipient cell must receive the entire F factor, requiring the entire bacterial chromosome to be transferred. This event happens rarely, because most conjugating cells break apart before the entire chromosome has been transferred.

The F plasmid in F⁺ cells integrates into the bacterial chromosome, causing an F⁺ cell to become Hfr, at a

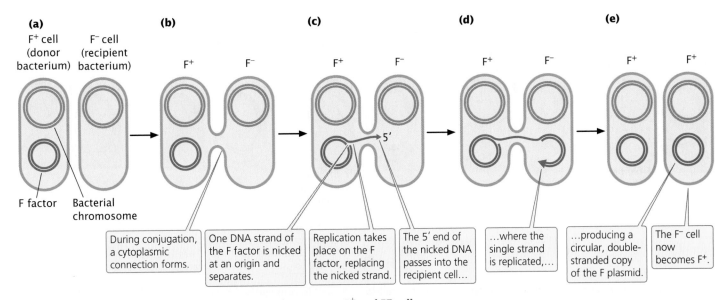

(a)
F⁺ cell (donor bacterium)   F⁻ cell (recipient bacterium)

F factor   Bacterial chromosome

During conjugation, a cytoplasmic connection forms.

(b)
F⁺   F⁻

One DNA strand of the F factor is nicked at an origin and separates.

(c)
F⁺   F⁻

5′

Replication takes place on the F factor, replacing the nicked strand.

The 5′ end of the nicked DNA passes into the recipient cell...

(d)
F⁺   F⁻

...where the single strand is replicated,...

(e)
F⁺   F⁺

...producing a circular, double-stranded copy of the F plasmid.

The F⁻ cell now becomes F⁺.

6.13 The F factor is transferred during conjugation between an F⁺ and F⁻ cell.

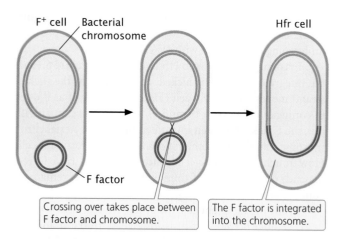

F⁺ cell   Bacterial chromosome

Hfr cell

F factor

Crossing over takes place between F factor and chromosome.

The F factor is integrated into the chromosome.

6.14 The F factor is integrated into the bacterial chromosome in an Hfr cell.

frequency of only about $\frac{1}{10,000}$. This low frequency accounts for the low rate of recombination observed by Lederberg and Tatum in their F⁺ cells. The F factor is excised from the bacterial chromosome at a similarly low rate, causing a few Hfr cells to become F⁺.

F' cells   When an F factor does excise from the bacterial chromosome, a small amount of the bacterial chromosome may be removed with it, and these chromosomal genes will then be carried with the F plasmid (**Figure 6.16**). Cells containing an F plasmid with some bacterial genes are called F prime (F'). For example, if an F factor integrates into a chromosome adjacent to the *lac* genes (genes that enable a cell to metabolize the sugar lactose), the F factor may pick up *lac* genes when it excises, becoming F'*lac*. F' cells can conjugate with F⁻ cells, given that F' cells possess the F plasmid with all the genetic information necessary for conjugation and gene

transfer. Characteristics of different mating types of *E. coli* (cells with different types of F) are summarized in **Table 6.2**.

During conjugation between an F'*lac* cell and an F⁻ cell, the F plasmid is transferred to the F⁻ cell, which means that any genes on the F plasmid, including those from the bacterial chromosome, may be transferred to F⁻ recipient cells. This process is called sexduction. It produces partial diploids, or *merozygotes*, which are cells with two copies of some genes, one on the bacterial chromosome and one on the newly introduced F plasmid. The outcomes of conjugation between different mating types of *E. coli* are summarized in **Table 6.3**.

Concepts

Conjugation in *E. coli* is controlled by an episome called the F factor. Cells containing F (F⁺ cells) are donors during gene transfer; cells without F (F⁻ cells) are recipients. Hfr cells possess F integrated into the bacterial chromosome; they donate DNA to F⁻ cells at a high frequency. F' cells contain a copy of F with some bacterial genes.

✔ Concept Check 3

Conjugation between an F⁺ and an F⁻ cell usually results in

a. two F⁺ cells.   c. an F⁺ and an F⁻ cell.

b. two F⁻ cells.   d. an Hfr cell and an F⁺ cell.

Mapping bacterial genes with the use of interrupted conjugation   The transfer of DNA that takes place during conjugation between Hfr and F⁻ cells allows bacterial genes to be mapped. In conjugation, the chromosome of the Hfr cell is transferred to the F⁻ cell. Transfer of the entire *E. coli* chromosome requires about 100 minutes; if conjugation is interrupted before 100 minutes have elapsed, only part of the chromosome will pass into the F⁻

(a) Hfr cell   F⁻ cell   **(b)**   **(c)**   **(d)**   **(e)** Hfr cell   F⁻ cell

Bacterial chromosome

F factor

In conjugation, F is nicked and the 5' end moves into the F⁻ cell.

The transferred strand is replicated,…

Hfr chromosome (F factor plus bacterial genes)

…and crossing over takes place between the donated Hfr chromosome and the original chromosome of the F⁻ cell.

Crossing over may lead to the recombination of alleles (bright blue in place of black segment).

The linear chromosome is degraded.

6.15 Bacterial genes may be transferred from an Hfr cell to an F⁻ cell in conjugation. In an Hfr cell, the F factor has been integrated into the bacterial chromosome.

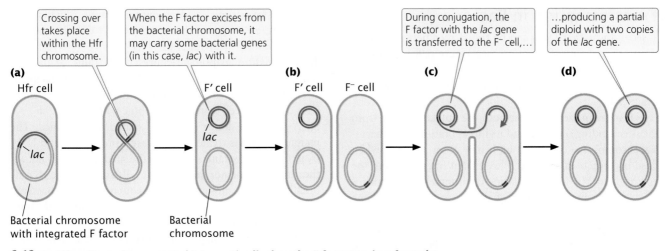

6.16 An Hfr cell may be converted into an F′ cell when the F factor excises from the bacterial chromosome and carries bacterial genes with it. Conjugation produces a partial diploid.

Table 6.2	Characteristics of *E. coli* cells with different types of F factor	
Type	**F Factor Characteristics**	**Role in Conjugation**
F⁺	Present as separate circular DNA	Donor
F⁻	Absent	Recipient
Hfr	Present, integrated into bacterial chromosome	High-frequency donor
F′	Present as separate circular DNA, carrying some bacterial genes	Donor

cell and have an opportunity to recombine with the recipient chromosome. Chromosome transfer always begins within the integrated F factor and proceeds in a continuous direction; so genes are transferred according to their sequence on the chromosome. The time required for

Table 6.3	Results of conjugation between cells with different F factors
Conjugating	**Cell Types Present after Conjugation**
F⁺ × F⁻	Two F⁺ cells (F⁻ cell becomes F⁺)
Hfr × F⁻	One Hfr cell and one F⁻ (no change)*
F′ × F⁻	Two F′ cells (F⁻ cell becomes F′)

*Rarely, the F⁻ cell becomes F⁺ in an Hfr × F⁻ conjugation if the entire chromosome is transferred during conjugation.

individual genes to be transferred indicates their relative positions on the chromosome. In most genetic maps, distances are expressed as percent recombination; but, in bacterial maps constructed with interrupted conjugation, the basic unit of distance is a minute.

Concepts

Conjugation can be used to map bacterial genes by mixing Hfr and F⁻ cells that differ in genotype and interrupting conjugation at regular intervals. The amount of time required for individual genes to be transferred from the Hfr to the F⁻ cells indicates the relative positions of the genes on the bacterial chromosome.

Natural Gene Transfer and Antibiotic Resistance

Many pathogenic bacteria have developed resistance to antibiotics, particularly in environments where antibiotics are routinely used, such as hospitals and fish farms. (Massive amounts of antibiotics are often used in aquaculture to prevent infection in the fish and enhance their growth.) The continual presence of antibiotics in these environments creates selection for resistant bacteria, which reduces the effectiveness of antibiotic treatment for medically important infections.

Antibiotic resistance in bacteria frequently results from the action of genes located on *R plasmids*, small circular plasmids that can be transferred by conjugation. R plasmids have evolved in the past 60 years (since the beginning of widespread use of antibiotics), and some convey resistance to several antibiotics simultaneously. Ironic but plausible sources of some of the resistance genes found in R plasmids are the microbes that produce antibiotics in the first place.

Transformation in Bacteria

A second way that DNA can be transferred between bacteria is through transformation (see Figure 6.9b). Transformation played an important role in the initial identification of DNA as the genetic material, which will be discussed in Chapter 8.

Transformation requires both the uptake of DNA from the surrounding medium and its incorporation into a bacterial chromosome or a plasmid. It may occur naturally when dead bacteria break up and release DNA fragments into the environment. In soil and marine environments, this means may be an important route of genetic exchange for some bacteria.

Cells that take up DNA through their envelopes are said to be **competent**. Some species of bacteria take up DNA more easily than do others; competence is influenced by growth stage, the concentration of available DNA, and environmental challenges. The DNA taken up by a competent cell need not be bacterial: virtually any type of DNA (bacterial or otherwise) can be taken up by competent cells under the appropriate conditions.

As a DNA fragment enters the cell in the course of transformation (**Figure 6.17**), one of the strands is hydrolyzed, whereas the other strand moves across the membrane and may pair with a homologous region and become integrated into the bacterial chromosome. This integration requires two crossover events, after which the remaining single-stranded DNA is degraded by bacterial enzymes.

Bacterial geneticists have developed techniques to increase the frequency of transformation in the laboratory to introduce particular DNA fragments into cells. They have developed strains of bacteria that are more competent than wild-type cells. Treatment with calcium chloride, heat shock, or an electrical field makes bacterial membranes more porous and permeable to DNA, and the efficiency of transformation can also be increased by using high concentrations of DNA. These techniques make it possible to transform bacteria such as *E. coli*, which are not naturally competent.

Transformation, like conjugation, is used to map bacterial genes, especially in those species that do not undergo conjugation or transduction (see Figure 6.9a and c). Transformation mapping requires two strains of bacteria that differ in several genetic traits; for example, the recipient strain might be $a^- \ b^- \ c^-$ (auxotrophic for three nutrients), with the donor cell being prototrophic with alleles $a^+ \ b^+ \ c^+$. DNA from the donor strain is isolated and purified. The recipient strain is treated to increase competency, and DNA from the donor strain is added to the medium. Fragments of the donor DNA enter the recipient cells and undergo recombination with homologous DNA sequences on the bacterial chromosome. Cells that receive genetic material through transformation are called **transformants**.

Genes can be mapped by observing the rate at which two or more genes are transferred together, or **cotransformed**, in transformation. When the DNA is fragmented during isolation, genes that are physically close on the chromosome are more likely to be present on the same DNA fragment and transferred together, as shown for genes a^+ and b^+ in **Figure 6.18**. Genes that are far apart are unlikely to be present on the same DNA fragment and will rarely be transferred together. Inside the cell, DNA becomes incorporated into the bacterial chromosome through recombination. If two genes are close together on the same fragment, any two crossovers are likely to take place on either side of the two genes, allowing both to become part of the recipient chromosome. If the two genes are far apart, there may be one crossover between them, allowing one gene but not the other to recombine with the bacterial chromosome. Thus, two genes are more likely to be transferred together when they are close together on the chromosome, and genes located far apart are rarely cotransformed. Therefore, the frequency of cotransformation can be used to map bacterial genes. If genes *a* and *b* are frequently cotransformed, and genes *b* and *c* are frequently cotransformed, but genes *a* and *c* are rarely cotransformed, then gene *b* must be between *a* and *c*—the gene order is *a b c*.

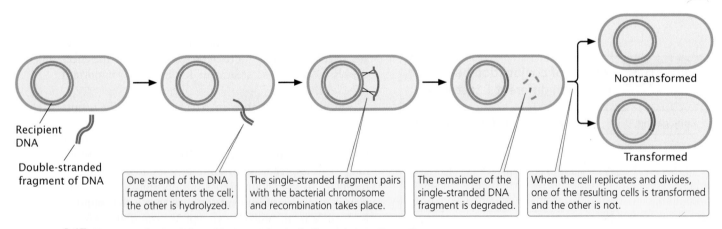

Recipient
DNA

Double-stranded
fragment of DNA

One strand of the DNA fragment enters the cell; the other is hydrolyzed.

The single-stranded fragment pairs with the bacterial chromosome and recombination takes place.

The remainder of the single-stranded DNA fragment is degraded.

When the cell replicates and divides, one of the resulting cells is transformed and the other is not.

Nontransformed

Transformed

6.17 Genes can be transferred between bacteria through transformation.

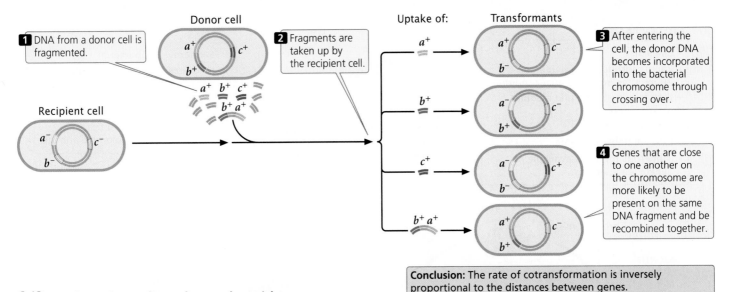

6.18 **Transformation can be used to map bacterial genes.**

Concepts

Genes can be mapped in bacteria by taking advantage of transformation—the ability of cells to take up DNA from the environment and incorporate it into their chromosomes through crossing over. The relative rate at which pairs of genes are cotransformed indicates the distance between them: the higher the rate of cotransformation, the closer the genes are on the bacterial chromosome.

✔ Concept Check 4

DNA from a bacterial strain with genotype $his^- \ leu^- \ thr^-$ is transformed with DNA from a strain that is $his^+ \ leu^+ \ thr^+$. A few $leu^+ \ thr^+$ cells and a few $his^+ \ thr^+$ cells are found, but no $his^+ \ leu^+$ cells are observed. Which genes are farthest apart?

Bacterial Genome Sequences

Genetic maps serve as a foundation for more-detailed information provided by DNA sequencing. Geneticists have now determined the complete nucleotide sequence of a number of bacterial genomes, and many additional microbial sequencing projects are underway. The size and content of bacterial genomes is discussed in Chapter 14.

One practical application of bacterial DNA sequencing is the identification and tracing of the sources of bacterial contamination and infection. This use of DNA sequences is illustrated by the study of anthrax-causing bacteria that were used in bioterrorism in the United States in 2001. Anthrax is caused by long-lasting spores of the bacterium *Bacillus anthracis* and was the cause of 18 deaths shortly after the terrorist attacks on the World Trade Center and the Pentagon in the United States on September 11, 2001. The source of the anthrax was traced to letters sent to U.S. senators and people in the news media.

To determine if the anthrax spores from the contaminated letters and the bacteria that infected the 18 victims came from the same source, investigators turned to DNA typing. They examined the variable number of tandem repeats (VNTRs, also called microsatellites), which are short DNA sequences that are repeated different numbers of times in different bacterial strains (see Chapter 14). This analysis showed that all of the spores found in the letters and bacteria found in the victims were related and probably originated from a single source, although the person or persons responsible for this act of bioterrorism have never been conclusively identified. The entire genome of *Bacillus anthracis* was sequenced in 2003 and now provides a much larger set of variable DNA sequences that can be used to effectively trace the origins of future disease outbreaks.

Model Genetic Organism
The Bacterium *Escherichia coli*

The most widely studied prokaryotic organism and one of the best genetically characterized of all species is the bacterium *Escherichia coli* (**Figure 6.19**). Although some strains of *E. coli* are toxic and cause disease, most are benign and reside naturally in the intestinal tracts of humans and other warm-blooded animals.

Advantages of E. coli as a model genetic organism *Escherichia coli* is one of the true workhorses of genetics; its twofold advantage is rapid reproduction and small size. Under optimal conditions, this organism can reproduce every 20 minutes and, in a mere 7 hours, a single bacterial cell can give rise to more than 2 million descendants. One of the values of rapid reproduction is that enormous numbers

Bacterium
Escherichia coli

ADVANTAGES

- Small size
- Rapid reproduction, dividing every 20 minutes under optimal conditions
- Easy to culture in liquid medium or on petri plates
- Small genome
- Many mutants available
- Numerous methods available for genetic engineering

STATS

Taxonomy:	Eubacteria
Size:	1–2 μm in length
Anatomy:	Single cell surrounded by cell wall with nucleoid region
Habitat:	Intestinal tract of warm-blooded animals

Life Cycle

Chromosome

Bacterial chromosome replicates

Asexual reproduction

Chromosomes separate

Cell division

Genetic exchange

Conjugation

F factor

F⁺

F⁻

Transfer of genetic information

F⁺

F⁺

Chromosome

GENOME

Chromosomes:	1 circular chromosome
Amount of DNA:	4.64 million base pairs
Number of genes:	4300
Percentage of genes in common with humans:	8%
Average gene size:	1000 base pairs
Genome sequenced in:	1997

CONTRIBUTIONS TO GENETICS

- Gene regulation
- Molecular biology and biochemistry of genetic processes, such as replication, transcription, translation, recombination
- Gene structure and organization in bacteria
- Workhorse of recombinant DNA
- Gene mutations

6.19 *Escherichia coli* is a model genetic organism.

of cells can be grown quickly, and so even very rare mutations will appear in a short period. Consequently, numerous mutations in *E. coli*, affecting everything from colony appearance to drug resistance, have been isolated and characterized.

Escherichia coli is easy to culture in the laboratory in liquid medium (see Figure 6.2a) or on solid medium within petri plates (see Figure 6.2b). In liquid culture, *E. coli* cells will grow to a concentration of a billion cells per milliliter, and trillions of bacterial cells can be easily grown in a single test tube. When *E. coli* cells are diluted and spread onto the solid medium of a petri dish, individual bacteria reproduce asexually, giving rise to a concentrated clump of 10 million to 100 million genetically identical cells, called a colony. This colony formation makes it easy to isolate genetically pure strains of the bacteria.

The E. coli genome The *E. coli* genome is on a single chromosome and—compared with those of humans, mice, plants, and other multicellular organisms—is relatively small, consisting of 4,638,858 base pairs. The information within the *E. coli* chromosome is compact, having little non-coding DNA between and within the genes and having few sequences for which there is more than one copy. The *E. coli* genome contains an estimated 4300 genes, more than half of which have no known function. These "orphan genes" may play important roles in adapting to unusual environments, coordinating metabolic pathways, organizing the chromosome, or communicating with other bacterial cells. The haploid genome of *E. coli* makes it easy to isolate mutations, because there are no dominant genes at the same locus to suppress and mask recessive mutations.

The E. coli life cycle Wild-type *E. coli* is prototrophic and can grow on minimal medium that contains only glucose and some inorganic salts. Under most conditions, *E. coli* divides about once an hour, although, in a richer medium containing sugars and amino acids, it will divide every 20 minutes. It normally reproduces through simple binary fission, in which the single chromosome of a bacterium replicates and migrates to opposite sides of the cell, followed by cell division, giving rise to two identical daughter cells. Mating between bacteria, called conjugation, is controlled by fertility genes normally located on the F plasmid (see pp. 145–149). As stated earlier, in conjugation, one bacterium donates genetic material to another bacterium, followed by genetic recombination that integrates new alleles into the bacterial chromosome. Genetic material can also be exchanged between strains of *E. coli* through transformation and transduction (see Figure 6.9).

Genetic techniques with E. coli *Escherichia coli* is used in a number of experimental systems in which fundamental genetic processes are studied in detail. For example, in vitro translation systems contain within a test tube all the components necessary to translate the genetic information of a messenger RNA into a polypeptide chain. Similarly, in vitro systems based on components from *E. coli* cells allow transcription, replication, gene expression, and many other important genetic functions to be studied and analyzed under controlled laboratory conditions.

Escherichia coli is also used widely in genetic engineering (recombinant DNA; see Chapter 14). Plasmids have been isolated from *E. coli* and genetically modified to create effective vectors for transferring genes into bacteria and eukaryotic cells. Often, new genetic constructs (DNA sequences created in the laboratory) are assembled and cloned in *E. coli* before transfer to other organisms. Methods have been developed to introduce specific mutations within *E. coli* genes, and so genetic analysis no longer depends on the isolation of randomly occurring mutations. New DNA sequences produced by recombinant DNA can be introduced by transformation into special strains of *E. coli* that are particularly efficient (competent) at taking up DNA.

Because of its powerful advantages as a model genetic organism, *E. coli* has played a leading role in many fundamental discoveries in genetics, including elucidation of the genetic code, probing the nature of replication, and working out the basic mechanisms of gene regulation. ■

6.2 Viruses Are Simple Replicating Systems Amenable to Genetic Analysis

All organisms—plants, animals, fungi, and bacteria—are infected by viruses. A **virus** is a simple replicating structure made up of nucleic acid surrounded by a protein coat (see Figure 2.3). Viruses come in a great variety of shapes and sizes (**Figure 6.20**). Some have DNA as their genetic material, whereas others have RNA; the nucleic acid may be double stranded or single stranded, linear or circular. Not surprisingly, viruses reproduce in a number of different ways.

Bacteriophages (phages) have played a central role in genetic research since the late 1940s. They are ideal for many types of genetic research because they have small and easily manageable genomes, reproduce rapidly, and produce large numbers of progeny. Bacteriophages have two alternative life cycles: the lytic and the lysogenic cycles. In the lytic cycle, a

(a)

(b)

6.20 Viruses come in different structures and sizes. (a) T4 bacteriophage (bright orange). (b) Influenza A virus (green structures). *[Left: Biozentrum, University of Basel/Photo Researchers. Right: Eye of Science/Photo Researchers.]*

phage attaches to a receptor on the bacterial cell wall and injects its DNA into the cell (**Figure 6.21**). Inside the host cell, the phage DNA is replicated, transcribed, and translated, producing more phage DNA and phage proteins. New phage particles are assembled from these components. The phages then produce an enzyme that breaks open the host cell, releasing the new phages. **Virulent phages** reproduce strictly through the lytic cycle and always kill their host cells.

Temperate phages can undergo either the lytic or the lysogenic cycle. The lysogenic cycle begins like the lytic cycle (see Figure 6.21) but, inside the cell, the phage DNA integrates into the bacterial chromosome, where it remains as an inactive **prophage**. The prophage is replicated along with the bacterial DNA and is passed on when the bacterium divides. Certain stimuli can cause the prophage to dissociate from the bacterial chromosome and enter into the lytic cycle, producing new phage particles and lysing the cell.

Techniques for the Study of Bacteriophages

Viruses reproduce only within host cells; so bacteriophages must be cultured in bacterial cells. To do so, phages and bacteria are mixed together and plated on solid medium on a petri plate. A high concentration of bacteria is used so that the colonies grow into one another and produce a continuous layer of bacteria, or "lawn," on the agar. An individual phage infects a single bacterial cell and goes through its lytic cycle. Many new phages are released from the lysed cell and infect additional cells; the cycle is then repeated. The bacteria grow on solid medium; so the diffusion of the phages is restricted, and only nearby cells are infected. After several rounds of phage reproduction, a clear patch of lysed cells, or **plaque**, appears on the plate (**Figure 6.22**). Each plaque represents a single phage that multiplied and lysed many cells. Plating a known volume of a dilute solution of phages on a bacterial lawn and counting the number of plaques that appear can be used to determine the original concentration of phage in the solution.

Concepts

Viral genomes may be DNA or RNA, circular or linear, and double or single stranded. Bacteriophages are used in many types of genetic research.

✔ Concept Check 5

In which bacteriophage life cycle does the phage DNA become incorporated into the bacterial chromosome?

a. Lytic

b. Lysogenic

c. Both lytic and lysogenic

d. Neither lytic or lysogenic

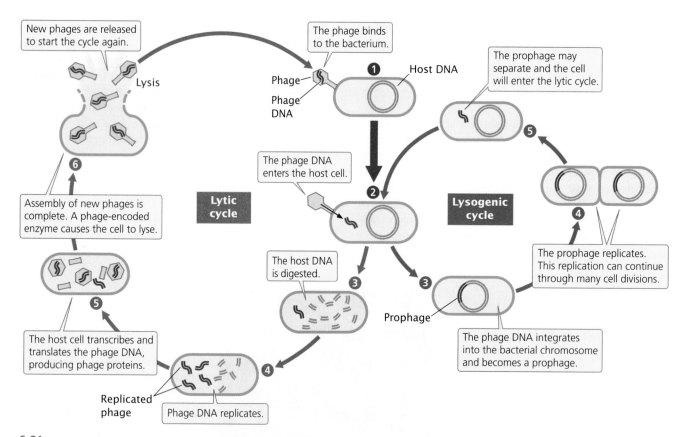

6.21 Bacteriophages have two alternative life cycles: lytic and lysogenic.

6.22 Plaques are clear patches of lysed cells on a lawn of bacteria. [Carolina Biological/Visuals Unlimited.]

Transduction: Using Phages to Map Bacterial Genes

In the discussion of bacterial genetics, three mechanisms of gene transfer were identified: conjugation, transformation, and transduction (see Figure 6.9). Let's take a closer look at transduction, in which genes are transferred between bacteria by viruses. In **generalized transduction**, any gene may be transferred. In **specialized transduction**, only a few genes are transferred.

Generalized transduction Joshua Lederberg and Norton Zinder discovered generalized transduction in 1952. They were trying to produce recombination in the bacterium *Salmonella typhimurium* by conjugation. They mixed a strain of *S. typhimurium* that was phe^+ trp^+ tyr^+ met^- his^- with a strain that was phe^- trp^- tyr^- met^+ his^+ (**Figure 6.23**) and plated them on minimal medium. A few prototrophic recombinants (phe^+ trp^+ tyr^+ met^+ his^+) appeared, suggesting that conjugation had taken place. However, when they tested the two strains in a U-shaped tube similar to the one used by Davis, some phe^+ trp^+ tyr^+ met^+ his^+ prototrophs were obtained on one side of the tube (compare Figure 6.23 with Figure 6.11). This apparatus separated the two strains by a filter with pores too small for the passage of bacteria; so how were genes being transferred between bacteria in the absence of conjugation? The results of subsequent studies revealed that the agent of transfer was a bacteriophage.

In the lytic cycle of phage reproduction, the bacterial chromosome is broken into random fragments (**Figure 6.24**). For some types of bacteriophage, a piece of the bacterial chromosome instead of phage DNA occasionally gets packaged into a phage coat; these phage particles are called **transducing phages**. The transducing phage infects a new cell, releasing the bacterial DNA, and the introduced genes may then become integrated into the bacterial chromosome by a double crossover. Bacterial genes can, by this process, be moved from one bacterial strain to another, producing recombinant bacteria called **transductants**.

Experiment

Question: Does genetic exchange between bacteria always require cell-to-cell contact?

Methods

$phe^+ trp^+ tyr^+ met^- his^-$ $phe^- trp^- tyr^- met^+ his^+$ $phe^+ trp^+ tyr^+ met^- his^-$ $phe^- trp^- tyr^- met^+ his^+$

1 Two auxotrophic strains of *Salmonella typhimurium* were mixed…

4 When the two strains were placed in a Davis U-tube,…

Filter

2 …and plated on minimal medium.

5 …which separated the strains by a filter with pores too small for the bacteria to pass through,…

Results

Prototrophic colonies   No colonies   Prototrophic colonies

$phe^+ trp^+ tyr^+ met^+ his^+$ $phe^+ trp^+ tyr^+ met^+ his^+$

3 Some prototrophic colonies were obtained.

6 …prototrophic colonies were obtained from only one side of the tube.

Conclusion: Genetic exchange did not take place through conjugation. A phage was later shown to be the agent of transfer.

6.23 The Lederberg and Zinder experiment.

6.24 Genes can be transferred from one bacterium to another through generalized transduction.

Not all phages are capable of transduction, a rare event that requires (1) that the phage degrade the bacterial chromosome; (2) that the process of packaging DNA into the phage protein not be specific for phage DNA; and (3) that the bacterial genes transferred by the virus recombine with the chromosome in the recipient cell.

Because of the limited size of a phage particle, only about 1% of the bacterial chromosome can be transduced. Only genes located close together on the bacterial chromosome will be transferred together, or **cotransduced**. The overall rate of transduction ranges from only about 1 in 100,000 to 1 in 1,000,000. Because the chance of a cell being transduced by two separate phages is exceedingly small, any cotransduced genes are usually located close together on the bacterial chromosome. Thus, rates of cotransduction, like rates of cotransformation, give an indication of the physical distances between genes on a bacterial chromosome.

To map genes by using transduction, two bacterial strains with different alleles at several loci are used. The donor strain is infected with phages (**Figure 6.25**), which reproduce within the cell. When the phages have lysed the donor cells, a suspension of the progeny phages is mixed with a recipient strain of bacteria, which is then plated on several different kinds of media to determine the phenotypes of the transducing progeny phages.

Concepts

In transduction, bacterial genes become packaged into a viral coat, are transferred to another bacterium by the virus, and become incorporated into the bacterial chromosome by crossing over. Bacterial genes can be mapped with the use of generalized transduction.

✔ Concept Check 6

In gene mapping experiments using generalized transduction, bacterial genes that are cotransduced are

a. far apart on the bacterial chromosome.

b. on different bacterial chromosomes.

c. close together on the bacterial chromosome.

d. on a plasmid.

Connecting Concepts

Three Methods for Mapping Bacterial Genes

Three methods of mapping bacterial genes have now been outlined: (1) interrupted conjugation; (2) transformation; and (3) transduction. These methods have important similarities and differences.

Mapping with interrupted conjugation is based on the time required for genes to be transferred from one bacterium to another by means of cell-to-cell contact. The key to this technique is that the bacterial chromosome itself is transferred, and the order of genes and the time required for their transfer provide information about the positions of the genes on the chromosome. In contrast with other mapping methods, the distance between genes is measured not in recombination frequencies but in units of time required for genes to be transferred. Here, the basic unit of conjugation mapping is a minute.

In gene mapping with transformation, DNA from the donor strain is isolated, broken up, and mixed with the recipient strain. Some fragments pass into the recipient cells, where the transformed DNA may recombine with the bacterial chromosome. The unit of transfer here is a random fragment of the chromosome. Loci that are close together on the donor chromosome tend to be on the same DNA fragment; so the rates of cotransformation provide information about the relative positions of genes on the chromosome.

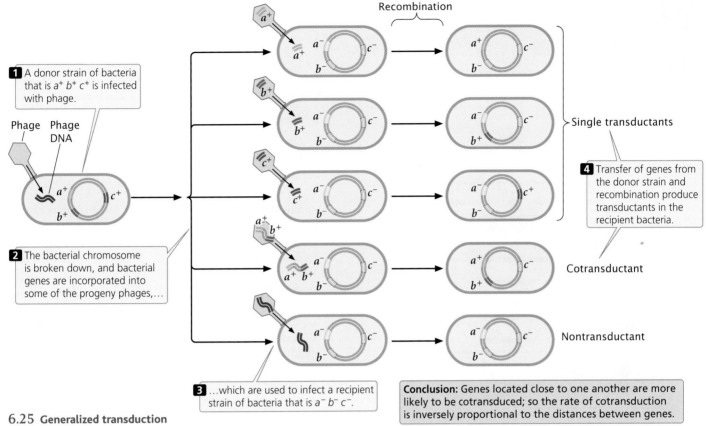

1 A donor strain of bacteria that is a^+ b^+ c^+ is infected with phage.

Phage   Phage DNA

2 The bacterial chromosome is broken down, and bacterial genes are incorporated into some of the progeny phages,...

Recombination

Single transductants

4 Transfer of genes from the donor strain and recombination produce transductants in the recipient bacteria.

Cotransductant

Nontransductant

3 ...which are used to infect a recipient strain of bacteria that is a^- b^- c^-.

Conclusion: Genes located close to one another are more likely to be cotransduced; so the rate of cotransduction is inversely proportional to the distances between genes.

6.25 Generalized transduction can be used to map genes.

Transduction mapping also relies on the transfer of genes between bacteria that differ in two or more traits, but, here, the vehicle of gene transfer is a bacteriophage. In a number of respects, transduction mapping is similar to transformation mapping. Small fragments of DNA are carried by the phage from donor to recipient bacteria, and the rates of cotransduction, like the rates of cotransformation, provide information about the relative distances between the genes.

All of the methods use a common strategy for mapping bacterial genes. The movement of genes from donor to recipient is detected by using strains that differ in two or more traits, and the transfer of one gene relative to the transfer of others is examined. Additionally, all three methods rely on recombination between the transferred DNA and the bacterial chromosome. In mapping with interrupted conjugation, the relative order and timing of gene transfer provide the information necessary to map the genes; in transformation and transduction, the rate of cotransfer provides this information.

In conclusion, the same basic strategies are used for mapping with interrupted conjugation, transformation, and transduction. The methods differ principally in their mechanisms of transfer: in conjugation mapping, DNA is transferred though contact between bacteria; in transformation, DNA is transferred as small naked fragments; and, in transduction, DNA is transferred by bacteriophages.

Gene Mapping in Phages

Mapping genes in the bacteriophages themselves requires the application of the same principles as those applied to mapping genes in eukaryotic organisms (see Chapter 5). Crosses are made between viruses that differ in two or more genes, and recombinant progeny phages are identified and counted. The proportion of recombinant progeny is then used to estimate the distances between the genes and their linear order on the chromosome.

In 1949, Alfred Hershey and Raquel Rotman examined rates of recombination in the T2 bacteriophage, which has single-stranded DNA. They studied recombination between genes in two strains that differed in plaque appearance and host range (the bacterial strains that the phages could infect). One strain was able to infect and lyse type B *E. coli* cells but not B/2 cells (wild type with normal host range, h^+) and produced an abnormal plaque that was large with distinct borders (r^-). The other strain was able to infect and lyse *both* B *and* B/2 cells (mutant host range, h^-) and produced wild-type plaques that were small with fuzzy borders (r^+).

Hershey and Rotman crossed the h^+ r^- and h^- r^+ strains of T2 by infecting type B *E. coli* cells with a mixture of the two strains. They used a high concentration of phages so that most cells could be simultaneously infected by both

Experiment

Question: How can we determine the position of a gene on a phage chromosome?

Method

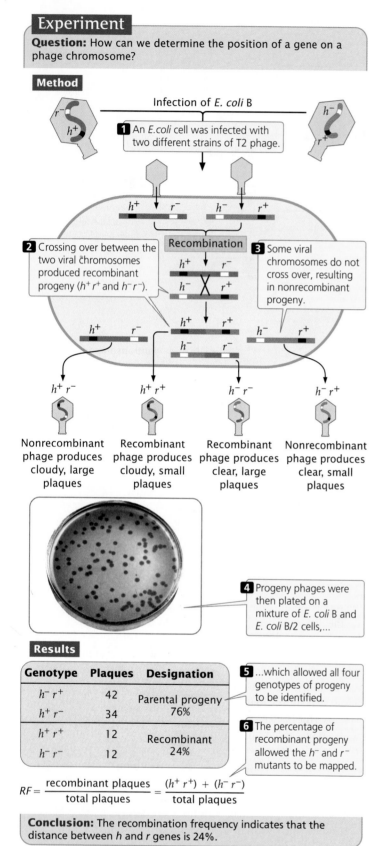

Infection of *E. coli* B

1 An *E.coli* cell was infected with two different strains of T2 phage.

Recombination

2 Crossing over between the two viral chromosomes produced recombinant progeny ($h^+ r^+$ and $h^- r^-$).

3 Some viral chromosomes do not cross over, resulting in nonrecombinant progeny.

$h^+ r^-$ — Nonrecombinant phage produces cloudy, large plaques

$h^+ r^+$ — Recombinant phage produces cloudy, small plaques

$h^- r^-$ — Recombinant phage produces clear, large plaques

$h^- r^+$ — Nonrecombinant phage produces clear, small plaques

4 Progeny phages were then plated on a mixture of *E. coli* B and *E. coli* B/2 cells,...

5 ...which allowed all four genotypes of progeny to be identified.

6 The percentage of recombinant progeny allowed the h^- and r^- mutants to be mapped.

Results

Genotype	Plaques	Designation
$h^- r^+$	42	Parental progeny 76%
$h^+ r^-$	34	
$h^+ r^+$	12	Recombinant 24%
$h^- r^-$	12	

$$RF = \frac{\text{recombinant plaques}}{\text{total plaques}} = \frac{(h^+ r^+) + (h^- r^-)}{\text{total plaques}}$$

Conclusion: The recombination frequency indicates that the distance between *h* and *r* genes is 24%.

6.26 Hershey and Rotman developed a technique for mapping viral genes. [*Photograph from G. S. Stent*, Molecular Biology of Bacterial Viruses. © 1963 by W. H. Freeman and Company.]

Table 6.4	Progeny phages produced from $h^- r^+ \times h^+ r^-$
Phenotype	**Genotype**
Clear and small	$h^- r^+$
Cloudy and large	$h^+ r^-$
Cloudy and small	$h^+ r^+$
Clear and large	$h^- r^-$

strains (**Figure 6.26**). Homologous recombination occasionally took place between the chromosomes of the different strains, producing $h^+ r^+$ and $h^- r^-$ chromosomes, which were then packaged into new phage particles. When the cells lysed, the recombinant phages were released, along with the nonrecombinant $h^+ r^-$ phages and $h^- r^+$ phages.

Hershey and Rotman diluted and plated the progeny phages on a bacterial lawn that consisted of a *mixture* of B and B/2 cells. Phages carrying the h^+ allele (which conferred the ability to infect only B cells) produced a cloudy plaque because the B/2 cells did not lyse. Phages carrying the h^- allele produced a clear plaque because all the cells within the plaque were lysed. The r^+ phages produced small plaques, whereas the r^- phages produced large plaques. The genotypes of these progeny phages could therefore be determined by the appearance of the plaque (see Figure 6.26 and **Table 6.4**).

In this type of phage cross, the recombination frequency (*RF*) between the two genes can be calculated by using the following formula:

$$RF = \frac{\text{recombinant plaques}}{\text{total plaques}}$$

In Hershey and Rotman's cross, the recombinant plaques were $h^+ r^+$ and $h^- r^-$; so the recombination frequency was

$$RF = \frac{(h^+ r^+) + (h^- r^-)}{\text{total plaques}}$$

Recombination frequencies can be used to determine the distances between genes and their order on the phage chromosome, just as recombination frequencies are used to map genes in eukaryotes. In the 1950s and 1960s, Seymour Benzer used this method of analyzing recombination frequencies to map the location of thousands of *rII* mutations in T4 bacteriophage, providing the first detailed look at the structure of an individual gene.

Concepts

To map phage genes, bacterial cells are infected with viruses that differ in two or more genes. Recombinant plaques are counted, and rates of recombination are used to determine the linear order of the genes on the chromosome and the distances between them.

(a)

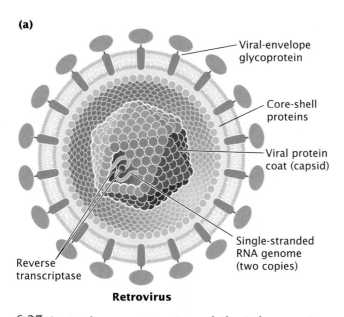

- Viral-envelope glycoprotein
- Core-shell proteins
- Viral protein coat (capsid)
- Single-stranded RNA genome (two copies)
- Reverse transcriptase

Retrovirus

6.27 A retrovirus uses reverse transcription to incorporate its RNA into the host DNA. (a) Structure of a typical retrovirus. Two copies of the single-stranded RNA genome and the reverse transcriptase enzyme are shown enclosed within a protein capsid. The capsid is surrounded by a viral envelope that is studded with viral glycoproteins. (b) The retrovirus life cycle.

RNA Viruses

Viral genomes may be encoded in either DNA or RNA, as stated earlier. RNA is the genetic material of some medically important human viruses, including those that cause influenza, common colds, polio, and AIDS. Almost all viruses that infect plants have RNA genomes. The medical and economic importance of RNA viruses has encouraged their study.

RNA viruses capable of integrating into the genomes of their hosts, much as temperate phages insert themselves into bacterial chromosomes, are called **retroviruses** (**Figure 6.27a**). Because the retroviral genome is RNA, whereas that of the host is DNA, a retrovirus must produce **reverse transcriptase**, an enzyme that synthesizes complementary DNA (cDNA) from either an RNA or a DNA template. A retrovirus uses reverse transcriptase to make a double-stranded DNA copy from its single-stranded RNA genome. The DNA copy then integrates into the host chromosome to form a **provirus**, which is replicated by host enzymes when the host chromosome is duplicated (**Figure 6.27b**).

When conditions are appropriate, the provirus undergoes transcription to produce numerous copies of the original RNA genome. This RNA encodes viral proteins and serves as genomic RNA for new viral particles. As these viruses escape the cell, they collect patches of the cell membrane to use as their envelopes.

All known retroviral genomes have in common three genes: *gag*, *pol*, and *env*, each encoding a precursor protein that is cleaved into two or more functional proteins. The *gag*

(b)

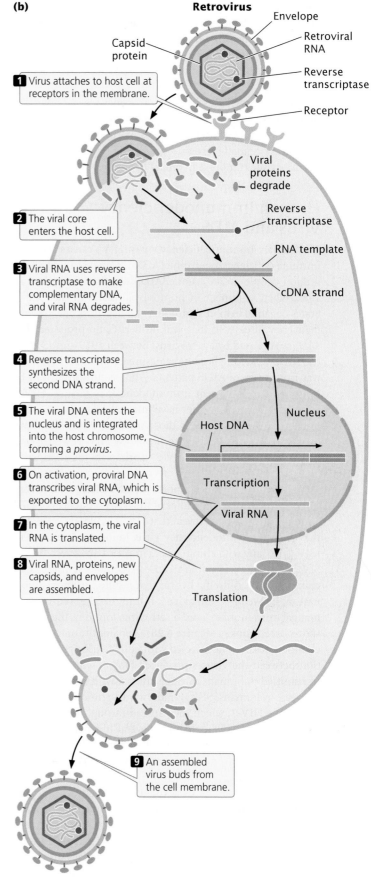

Retrovirus

- Envelope
- Capsid protein
- Retroviral RNA
- Reverse transcriptase
- Receptor

1 Virus attaches to host cell at receptors in the membrane.

Viral proteins degrade

2 The viral core enters the host cell.

Reverse transcriptase

RNA template

3 Viral RNA uses reverse transcriptase to make complementary DNA, and viral RNA degrades.

cDNA strand

4 Reverse transcriptase synthesizes the second DNA strand.

5 The viral DNA enters the nucleus and is integrated into the host chromosome, forming a *provirus*.

Nucleus

Host DNA

6 On activation, proviral DNA transcribes viral RNA, which is exported to the cytoplasm.

Transcription

Viral RNA

7 In the cytoplasm, the viral RNA is translated.

8 Viral RNA, proteins, new capsids, and envelopes are assembled.

Translation

9 An assembled virus buds from the cell membrane.

gene encodes the three or four proteins that make up the viral capsid. The *pol* gene encodes reverse transcriptase and an enzyme, called **integrase**, that inserts the viral DNA into the host chromosome. The *env* gene codes for the glycoproteins that appear on the viral envelope that surrounds the viral capsid.

Some retroviruses contain **oncogenes** (see Chapter 15) that may stimulate cell division and cause the formation of tumors. The first retrovirus to be isolated, the Rous sarcoma virus, was originally recognized by its ability to produce connective-tissue tumors (sarcomas) in chickens.

Human Immunodeficiency Virus and AIDS

The human immunodeficiency virus (HIV) causes acquired immune deficiency syndrome (AIDS), a disease that killed 2 million people in 2007 alone. AIDS was first recognized in 1982, when a number of homosexual males in the United States began to exhibit symptoms of a new immune-system-deficiency disease. In that year, Robert Gallo proposed that AIDS was caused by a retrovirus. Between 1983 and 1984, as the AIDS epidemic became widespread, the HIV retrovirus was isolated from AIDS patients. AIDS is now known to be caused by two different immunodeficiency viruses, HIV-1 and HIV-2, which together have infected more than 65 million people worldwide. Of those infected, 25 million have died. Most cases of AIDS are caused by HIV-1, which now has a global distribution; HIV-2 is primarily found in western Africa.

HIV illustrates the importance of genetic recombination in viral evolution. Studies of the DNA sequences of HIV and other retroviruses reveal that HIV-1 is closely related to the simian immunodeficiency virus found in chimpanzees (SIV_{cpz}). Many wild chimpanzees in Africa are infected with SIV_{cpz}, although it doesn't cause AIDS-like symptoms in chimps. SIV_{cpz} is itself a hybrid that resulted from recombination between a retrovirus found in the red-capped mangabey (a monkey) and a retrovirus found in the greater spot-nosed monkey (**Figure 6.28**). Apparently, one or more chimpanzees became infected with both viruses; recombination between the viruses produced SIV_{cpz}, which was then transmitted to humans through contact with infected chimpanzees. In humans, SIV_{cpz} underwent significant evolution to become HIV-1, which then spread throughout the world to produce the AIDS epidemic. Several independent transfers of SIV_{cpz} to humans gave rise to different strains of HIV-1. HIV-2 evolved from a different retrovirus, SIV_{sm}, found in sooty mangabeys.

HIV is transmitted by sexual contact between humans and through any type of blood-to-blood contact, such as that caused by the sharing of dirty needles by drug addicts. Until screening tests could identify HIV-infected blood, transfusions and clotting factors used by hemophiliacs also were sources of infection.

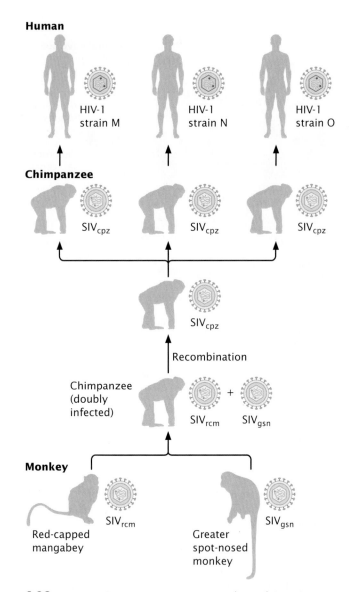

6.28 HIV-1 evolved from a similar virus (SIV_{cpz}) found in chimpanzees and was transmitted to humans. SIV_{cpz} arose from recombination taking place between retroviruses in red-capped mangabeys and greater spot-nosed monkeys.

HIV principally attacks a class of blood cells called helper T lymphocytes or, simply, helper T cells (**Figure 6.29**). HIV enters a helper T cell, undergoes reverse transcription, and integrates into the chromosome. The virus reproduces rapidly, destroying the T cell as new virus particles escape from the cell. Because helper T cells are central to immune function and are destroyed in the infection, AIDS patients have a diminished immune response; most AIDS patients die of secondary infections that develop because they have lost the ability to fight off pathogens.

The HIV genome is 9749 nucleotides long and carries *gag*, *pol*, *env*, and six other genes that regulate the life cycle of the virus. HIV's reverse transcriptase is very error prone, giv-

6.29 HIV principally attacks T lymphocytes. Electron micrograph showing a T cell infected with HIV, visible as small circles with dark centers. *[Courtesy of Dr. Hans Gelderblom.]*

ing the virus a high mutation rate and allowing it to evolve rapidly, even within a single host. This rapid evolution makes the development of an effective vaccine against HIV particularly difficult. Genetic variation within the human population also affects the virus. To date, more than 10 loci in humans that affect HIV infection and the progression of AIDS have been identified.

Concepts

A retrovirus is an RNA virus that integrates into its host's chromosome by making a DNA copy of its RNA genome through the process of reverse transcription. Human immunodeficiency virus, the causative agent of AIDS, is a retrovirus. It evolved from related retroviruses found in other primates.

✔ Concept Check 7

What enzyme is used by a retrovirus to make a DNA copy of its genome?

Concepts Summary

- Bacteria and viruses are well suited to genetic studies: they are small, have a small haploid genome, undergo rapid reproduction, and produce large numbers of progeny through asexual reproduction.

- The bacterial genome normally consists of a single, circular molecule of double-stranded DNA. Plasmids are small pieces of bacterial DNA that can replicate independently of the large chromosome.

- DNA may be transferred between bacteria by means of conjugation, transformation, and transduction.

- Conjugation is the union of two bacterial cells and the transfer of genetic material between them. It is controlled by an episome called F. The rate at which individual genes are transferred during conjugation provides information about the order of the genes and the distances between them on the bacterial chromosome.

- Bacteria take up DNA from the environment through the process of transformation. Frequencies of the cotransformation of genes provide information about the physical distances between chromosomal genes.

- The bacterium *Escherichia coli* is an important model genetic organism that has the advantages of small size, rapid reproduction, and a small genome.

- Viruses are replicating structures with DNA or RNA genomes that may be double stranded or single stranded, linear or circular.

- Bacterial genes become incorporated into phage coats and are transferred to other bacteria by phages through the process of transformation. Rates of cotransduction can be used to map bacterial genes.

- Phage genes can be mapped by infecting bacterial cells with two different phage strains and counting the number of recombinant plaques produced by the progeny phages.

- A number of viruses have RNA genomes. Retroviruses encode a reverse transcriptase enzyme used to make a DNA copy of the viral genome, which then integrates into the host genome as a provirus. HIV is a retrovirus that is the causative agent for AIDS.

Important Terms

minimal medium (p. 141)
complete medium (p. 141)
colony (p. 141)
plasmid (p. 142)
episome (p. 143)

F factor (p. 143)
conjugation (p. 144)
transformation (p. 144)
transduction (p. 145)
pili (singular, pilus) (p. 146)

competent cell (p. 150)
transformant (p. 150)
cotransformation (p. 150)
virus (p. 153)
virulent phage (p. 154)

temperate phage (p. 154)
prophage (p. 154)
plaque (p. 154)
generalized transduction (p. 155)
specialized transduction (p. 155)

transducing phage (p. 155)
transductant (p. 155)
cotransduction (p. 156)
retrovirus (p. 159)

reverse transcriptase (p. 159)
provirus (p. 159)
integrase (p. 160)
oncogene (p. 160)

Answers to Concept Checks

1. d

2. b

3. a

4. *his* and *leu*

5. b

6. c

7. Reverse transcriptase

Worked Problems

1. DNA from a strain of bacteria with genotype $a^+ \, b^+ \, c^+ \, d^+ \, e^+$ was isolated and used to transform a strain of bacteria that was $a^- \, b^- \, c^- \, d^- \, e^-$. The transformed cells were tested for the presence of donated genes. The following genes were cotransformed:

a^+ and d^+ c^+ and d^+
b^+ and e^+ c^+ and e^+

What is the order of genes *a*, *b*, *c*, *d*, and *e* on the bacterial chromosome?

• **Solution**

The rate at which genes are cotransformed is inversely proportional to the distance between them: genes that are close together are frequently cotransformed, whereas genes that are far apart are rarely cotransformed. In this transformation experiment, gene c^+ is cotransformed with both genes e^+ and d^+, but genes e^+ and d^+ are not cotransformed; therefore the *c* locus must be between the *d* and *e* loci:

Gene e^+ is also cotransformed with gene b^+; so the *e* and *b* loci must be located close together. Locus *b* could be on either side of locus *e*. To determine whether locus *b* is on the same side of *e* as locus *c*, we look to see whether genes b^+ and c^+ are cotransformed. They are not; so locus *b* must be on the opposite side of *e* from *c*:

Gene a^+ is cotransformed with gene d^+; so they must be located close together. If locus *a* were located on the same side of *d* as locus *c*, then genes a^+ and c^+ would be cotransformed. Because these genes display no cotransformation, locus *a* must be on the opposite side of locus *d*:

2. Consider three genes in *E. coli*: *thr*⁺ (the ability to synthesize threonine), *ara*⁺ (the ability to metabolize arabinose), and *leu*⁺ (the ability to synthesize leucine). All three of these genes are close together on the *E. coli* chromosome. Phages are grown in a *thr*⁺ *ara*⁺ *leu*⁺ strain of bacteria (the donor strain). The phage lysate is collected and used to infect a strain of bacteria that is *thr*⁻ *ara*⁻ *leu*⁻. The recipient bacteria are then tested on medium lacking leucine. Bacteria that grow and form colonies on this medium (*leu*⁺ transductants) are then replica plated onto medium lacking threonine and onto medium lacking arabinose to see which are *thr*⁺ and which are *ara*⁺.

Another group of recipient bacteria are tested on medium lacking threonine. Bacteria that grow and form colonies on this medium (*thr*⁺ transductants) are then replica plated onto medium lacking leucine and onto medium lacking arabinose to see which are *ara*⁺ and which are *leu*⁺. Results from these experiments are as follows:

Selected marker	Cells with cotransduced genes (%)
leu⁺	3 *thr*⁺
	76 *ara*⁺
thr⁺	3 *leu*⁺
	0 *ara*⁺

How are the loci arranged on the chromosome?

• **Solution**

Notice that, when we select for *leu*⁺ (the top half of the table), most of the selected cells are also *ara*⁺. This finding indicates that the *leu* and *ara* genes are located close together, because they are usually cotransduced. In contrast, *thr*⁺ is only rarely cotransduced with *leu*⁺, indicating that *leu* and *thr* are much farther apart. On the basis of these observations, we know that *leu* and *ara* are closer together than are *leu* and *thr*, but we don't yet know the order of three genes—whether *thr* is on the same side of *ara* as *leu* or on the opposite side, as shown here:

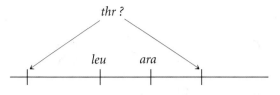

We can determine the position of *thr* with respect to the other two genes by looking at the cotransduction frequencies when *thr⁺* is selected (the bottom half of the preceding table). Notice that,

although the cotransduction frequency for *thr* and *leu* also is 3%, no *thr⁺ ara⁺* cotransductants are observed. This finding indicates that *thr* is closer to *leu* than to *ara*, and therefore *thr* must be to the left of *leu*, as shown here:

Comprehension Questions

Section 6.1

1. Briefly explain the differences between F⁺, F⁻, Hfr, and F′ cells.

*2. What types of matings are possible between F⁺, F⁻, Hfr, and F′ cells? What outcomes do these matings produce? What is the role of F factor in conjugation?

*3. Explain how interrupted conjugation, transformation, and transduction can be used to map bacterial genes. How are these methods similar and how are they different?

Section 6.2

*4. List some of the characteristics that make bacteria and viruses ideal organisms for many types of genetic studies.

5. What types of genomes do viruses have?

6. Briefly describe the differences between the lytic cycle of virulent phages and the lysogenic cycle of temperate phages.

7. Briefly explain how genes in phages are mapped.

8. Briefly describe the genetic structure of a typical retrovirus.

9. What are the evolutionary origins of HIV-1 and HIV-2?

Application Questions and Problems

Section 6.1

*10. John Smith is a pig farmer. For the past 5 years, Smith has been adding vitamins and low doses of antibiotics to his pig food; he says that these supplements enhance the growth of the pigs. Within the past year, however, several of his pigs died from infections of common bacteria, which failed to respond to large doses of antibiotics. Can you offer an explanation for the increased rate of mortality due to infection in Smith's pigs? What advice might you offer Smith to prevent this problem in the future?

11. Austin Taylor and Edward Adelberg isolated some new
DATA strains of Hfr cells that they then used to map several genes
ANALYSIS in *E. coli* by using interrupted conjugation (A. L. Taylor and E. A. Adelberg. 1960. *Genetics* 45:1233–1243). In one experiment, they mixed cells of Hfr strain AB-312, which were *xyl⁺ mtl⁺ mal⁺ met⁺* and sensitive to phage T6, with F⁻ strain AB-531, which was *xyl⁻ mtl⁻ mal⁻ met⁻* and resistant to phage T6. The cells were allowed to undergo conjugation. At regular intervals, the researchers removed a sample of cells and interrupted conjugation by killing the Hfr cells with phage T6. The F⁻ cells, which were resistant to phage T6, survived and were then tested for the presence of genes transferred from the Hfr strain. The results of this

experiment are shown in the accompanying graph. On the basis of these data, give the order of the *xyl, mtl, mal,* and *met* genes on the bacterial chromosome and indicate the minimum distances between them.

12. DNA from a strain of *Bacillus subtilis* with genotype *a⁺ b⁺ c⁺ d⁺ e⁺* is used to transform a strain with genotype *a⁻ b⁻*

c^- d^- e^-. Pairs of genes are checked for cotransformation and the following results are obtained:

Pair of genes	Cotransformation	Pair of genes	Cotransformation
a^+ and b^+	no	b^+ and d^+	no
a^+ and c^+	no	b^+ and e^+	yes
a^+ and d^+	yes	c^+ and d^+	no
a^+ and e^+	yes	c^+ and e^+	yes
b^+ and c^+	yes	d^+ and e^+	no

On the basis of these results, what is the order of the genes on the bacterial chromosome?

13. DNA from a bacterial strain that is his^+ leu^+ lac^+ is used to transform a strain that is his^- leu^- lac^-. The following percentages of cells were transformed:

Donor strain	Recipient strain	Genotype of transformed cells	Percentage
his^+ leu^+ lac^+	his^- leu^- lac^-	his^+ leu^+ lac^+	0.02
		his^+ leu^+ lac^-	0.00
		his^+ leu^- lac^+	2.00
		his^+ leu^- lac^-	4.00
		his^- leu^+ lac^+	0.10
		his^- leu^- lac^+	3.00
		his^- leu^+ lac^-	1.50

a. What conclusions can you make about that order of these three genes on the chromosome?

b. Which two genes are closest?

14. Rollin Hotchkiss and Julius Marmur studied transformation in the bacterium *Streptococcus pneumoniae* (R. D. Hotchkiss and J. Marmur. 1954. *Proceedings of the National Academy of Sciences of the United States of America* 40:55–60). They examined four mutations in this bacterium: penicillin resistance (*P*), streptomycin resistance (*S*), sulfanilamide resistance (*F*), and the ability to utilize mannitol (*M*). They extracted DNA from strains of bacteria with different combinations of different mutations and used this DNA to transform wild-type bacterial cells (P^+ S^+ F^+ M^+). The results from one of their transformation experiments are shown here.

Donor DNA	Recipient DNA	Transformants	Percentage of all cells
M S F	M^+ S^+ F^+	M^+ S F^+	4.0
		M^+ S^+ F	4.0
		M S^+ F^+	2.6
		M S F^+	0.41
		M^+ S F	0.22
		M S^+ F	0.0058
		M S F	0.0071

a. Hotchkiss and Marmur noted that the percentage of cotransformation was higher than would be expected on a random basis. For example, the results show that the 2.6% of the cells were transformed into *M* and 4% were transformed into *S*. If the *M* and *S* traits were inherited independently, the expected probability of cotransformation of *M* and *S* (*M S*) would be $0.026 \times 0.04 = 0.001$, or 0.1%. However, they observed 0.41% *M S* cotransformants, four times more than they expected. What accounts for the relatively high frequency of cotransformation of the traits they observed?

b. On the basis of the results, what conclusion can you make about the order of the *M*, *S*, and *F* genes on the bacterial chromosome?

c. Why is the rate of cotransformation for all three genes (*M S F*) almost the same as the cotransformation of *M F* alone?

Section 6.2

15. Two mutations that affect plaque morphology in phages (a^- and b^-) have been isolated. Phages carrying both mutations (a^- b^-) are mixed with wild-type phages (a^+ b^+) and added to a culture of bacterial cells. Subsequent to infection and lysis, samples of the phage lysate are collected and cultured on bacterial cells. The following numbers of plaques are observed:

Plaque phenotype	Number
a^+ b^+	2043
a^+ b^-	320
a^- b^+	357
a^- b^-	2134

What is the frequency of recombination between the *a* and *b* genes?

16. T. Miyake and M. Demerec examined proline-requiring mutations in the bacterium *Salmonella typhimurium* (T. Miyake and M. Demerec. 1960. *Genetics* 45:755–762). On the basis of complementation studies, they found four proline auxotrophs: *proA*, *proB*, *proC*, and *proD*. To determine if *proA*, *proB*, *proC*, and *proD* loci were located close together on the bacterial chromosome, they conducted a transduction experiment. Bacterial strains that were $proC^+$ and had mutations at *proA*, *proB*, or *proD*, were used as donors. The donors were infected with bacteriophages, and progeny phages were allowed to infect recipient bacteria with genotype $proC^-$ $proA^+$ $proB^+$ $proD^+$. The bacteria were then plated on a selective medium that allowed only $proC^+$ bacteria to grow. The following results were obtained:

Donor genotype	Transductant genotype	Number
$proC^+$ $proA^-$ $proB^+$ $proD^+$	$proC^+$ $proA^+$ $proB^+$ $proD^+$	2765
	$proC^+$ $proA^-$ $proB^+$ $proD^+$	3
$proC^+$ $proA^+$ $proB^-$ $proD^+$	$proC^+$ $proA^+$ $proB^+$ $proD^+$	1838
	$proC^+$ $proA^+$ $proB^-$ $proD^+$	2
$proC^+$ $proA^+$ $proB^+$ $proD^-$	$proC^+$ $proA^+$ $proB^+$ $proD^+$	1166
	$proC^+$ $proA^+$ $proB^+$ $proD^-$	0

a. Why are there no *proC⁻* genotypes among the transductants?

b. Which genotypes represent single transductants and which represent cotransductants?

c. Is there evidence that *proA*, *proB*, and *proD* are located close to *proC*? Explain your answer.

*17. A geneticist isolates two mutations in a bacteriophage. One mutation causes clear plaques (*c*), and the other produces minute plaques (*m*). Previous mapping experiments have established that the genes responsible for these two mutations are 8 m.u. apart. The geneticist mixes phages with genotype $c^+ m^+$ and genotype $c^- m^-$ and uses the mixture to infect bacterial cells. She collects the progeny phages and cultures a sample of them on plated bacteria. A total of 1000 plaques are observed. What numbers of the different types of plaques $(c^+ m^+, c^- m^-, c^+ m^-, c^- m^+)$ should she expect to see?

*18. *E. coli* cells are simultaneously infected with two strains of phage λ. One strain has a mutant host range, is temperature sensitive, and produces clear plaques (genotype is *h st c*); another strain carries the wild-type alleles (genotype is $h^+ st^+ c^+$). Progeny phages are collected from the lysed cells and are plated on bacteria. The genotypes of the progeny phages are given here:

Progeny phage genotype	Number of plaques
$h^+ c^+ st^+$	321
h c st	338
$h^+ c\ st$	26
$h\ c^+ st^+$	30
$h^+ c\ st^+$	106
$h\ c^+ st$	110
$h^+ c^+ st$	5
$h\ c\ st^+$	6

a. Determine the order of the three genes on the phage chromosome.

b. Determine the map distances between the genes.

c. Determine the coefficient of coincidence and the interference (see pp. 120–125 in Chapter 5).

19. A donor strain of bacteria with genes $a^+ b^+ c^+$ is infected with phages to map the donor chromosome with generalized transduction. The phage lysate from the bacterial cells is collected and used to infect a second strain of bacteria that are $a^- b^- c^-$. Bacteria with the a^+ gene are selected, and the percentage of cells with cotransduced b^+ and c^+ genes are recorded.

Donor	Recipient	Selected gene	Cells with cotransduced gene (%)
$a^+ b^+ c^+$	$a^- b^- c^-$	a^+	25 b^+
		a^+	3 c^+

Is the *b* or *c* gene closer to *a*? Explain your reasoning.

Challenge Question

Section 6.1

20. A group of genetics students mix two auxotrophic strains of bacteria: one is $leu^+ trp^+ his^- met^-$ and the other is $leu^- trp^- his^+ met^+$. After mixing the two strains, they plate the bacteria on minimal medium and observe a few pro-totrophic colonies ($leu^+ trp^+ his^+ met^+$). They assume that some gene transfer has taken place between the two strains. How can they determine whether the transfer of genes is due to conjugation, transduction, or transformation?

7

Chromosome Variation

Down syndrome is caused by the presence of three copies of one or more genes located on chromosome 21. [Stockbyte.]

Trisomy 21 and the Down-Syndrome Critical Region

In 1866, John Langdon Down, physician and medical superintendent of the Earlswood Asylum in Surrey, England, noticed a remarkable resemblance among a number of his mentally retarded patients: all of them possessed a broad, flat face, a thick tongue, a small nose, and oval-shaped eyes. Their features were so similar, in fact, that he felt that they might easily be mistaken as children of the same family. Down did not understand the cause of their retardation, but his original description faithfully records the physical characteristics of this most common genetic form of mental retardation. In his honor, the disorder is today known as Down syndrome.

As early as the 1930s, geneticists suggested that Down syndrome might be due to a chromosome abnormality, but not until 1959 did researchers firmly establish the cause of Down syndrome: most people with the disorder have three copies of chromosome 21, a condition known as trisomy 21. In a few rare cases, people having the disorder are trisomic for smaller parts of chromosome 21. By comparing the parts of chromosome 21 that these people have in common, geneticists have established that a specific segment—called the Down-syndrome critical region or DSCR—probably contains one or more genes responsible for the features of Down syndrome. Sequencing of the human genome has established that the DSCR consists of a little more than 5 million base pairs and only 33 genes.

Research by several groups in 2006 was a source of insight into the roles of two genes in the DSCR. Mice that have deficiencies in the regulatory pathway controlled by two genes, called *DSCR1* and *DYRK1A*, found in the DSCR exhibit many of the features seen in Down syndrome. Mice genetically engineered to express *DYRK1A* and *DSCR1* at high levels have abnormal heart development, similar to congenital heart problems seen in many people with Down syndrome.

167

In spite of this exciting finding, the genetics of Down syndrome appears to be more complex than formerly thought. Geneticist Lisa Olson and her colleagues had conducted an earlier study on mice to determine if genes within the DSCR are solely responsible for Down syndrome. Mouse breeders have developed several strains of mice that are trisomic for most of the genes found on human chromosome 21 (the equivalent mouse genes are actually found on mouse chromosome 16). These mice display many of the same anatomical features found in people with Down syndrome, as well as altered behavior, and they are considered an animal model for Down syndrome. Olson and her colleagues carefully created mice that were trisomic for genes in the DSCR but possessed the normal two copies of other genes found on human chromosome 21. If one or more of the genes in the DSCR are indeed responsible for Down syndrome, these mice should exhibit the features of Down syndrome. Surprisingly, the engineered mice exhibited none of the anatomical features of Down syndrome, demonstrating that three copies of genes in the DSCR are not the sole cause of the features of Down syndrome, at least in mice. Another study examined a human gene called *APP*, which lies outside of the DSCR. This gene appears to be responsible for at least some of the Alzheimer-like features observed in older Down-syndrome people.

Taken together, findings from these studies suggest that Down syndrome is not due to a single gene but is instead caused by complex interactions among multiple genes that are affected when an extra copy of chromosome 21 is present. Research on Down syndrome illustrates the principle that chromosome abnormalities often affect many genes that interact in complex ways.

Most species have a characteristic number of chromosomes, each with a distinct size and structure, and all the tissues of an organism (except for gametes) generally have the same set of chromosomes. Nevertheless, variations in chromosome number—such as the extra chromosome 21 that leads to Down syndrome—do periodically arise. Variations may also arise in chromosome structure: individual chromosomes may lose or gain parts and the order of genes within a chromosome may become altered. These variations in the number and structure of chromosomes are termed **chromosome mutations**, and they frequently play an important role in evolution.

We begin this chapter by briefly reviewing some basic concepts of chromosome structure, which we learned in Chapter 2. We then consider the different types of chromosome mutations, their definitions, features, phenotypic effects, and influence on evolution.

7.1 Chromosome Mutations Include Rearrangements, Aneuploids, and Polyploids

Before we consider the different types of chromosome mutations, their effects, and how they arise, we will review the basics of chromosome structure.

Chromosome Morphology

Each functional chromosome has a centromere, where spindle fibers attach, and two telomeres that stabilize the chromosome (see Figure 2.5). Chromosomes are classified into four basic types:

1. **Metacentric.** The centromere is located approximately in the middle, and so the chromosome has two arms of equal length.

2. **Submetacentric.** The centromere is displaced toward one end, creating a long arm and a short arm. (On human chromosomes, the short arm is designated by the letter p and the long arm by the letter q.)

3. **Acrocentric.** The centromere is near one end, producing a long arm and a knob, or satellite, at the other end.

4. **Telocentric.** The centromere is at or very near the end of the chromosome (see Figure 2.6).

The complete set of chromosomes possessed by an organism is called its *karyotype* and is usually presented as a picture of metaphase chromosomes lined up in descending order of their size (**Figure 7.1**). Karyotypes are prepared from actively dividing cells, such as white blood cells, bone-marrow cells, or cells from meristematic tissues of plants. After treatment with a chemical (such as colchicine) that prevents them from entering anaphase, the cells are chemically preserved, spread on a microscope slide, stained, and photographed. The photograph is then enlarged, and the individual chromosomes are cut out and arranged in a karyotype. For human chromosomes, karyotypes are often routinely prepared by automated machines, which scan a slide with a video camera attached to a microscope, looking for chromosome spreads. When a spread has been located, the camera takes a picture of the

7.1 A human karyotype consists of 46 chromosomes.
A karyotype for a male is shown here; a karyotype for a female would have two X chromosomes. *[ISM/Phototake.]*

(a)

(b)

(c)

chromosomes, the image is digitized, and the chromosomes are sorted and arranged electronically by a computer.

Preparation and staining techniques help to distinguish among chromosomes of similar size and shape. For instance, the staining of chromosomes with a special dye called Giemsa reveals G bands, which distinguish areas of DNA that are rich in adenine–thymine (A–T) base pairs (**Figure 7.2a**; see Chapter 8). Q bands (**Figure 7.2b**) are revealed by staining chromosomes with quinacrine mustard and viewing the chromosomes under ultraviolet light; variation in the brightness of Q bands results from differences in the relative amounts of cytosine–guanine (C–G) and adenine–thymine base pairs. Other techniques reveal C bands (**Figure 7.2c**), which are regions of DNA occupied by centromeric heterochromatin, and R bands (**Figure 7.2d**), which are rich in cytosine–guanine base pairs.

Types of Chromosome Mutations

Chromosome mutations can be grouped into three basic categories: chromosome rearrangements, aneuploids, and polyploids (**Figure 7.3**). Chromosome rearrangements alter the *structure* of chromosomes; for example, a piece of a chromosome might be duplicated, deleted, or inverted. In aneuploidy, the *number* of chromosomes is altered: one or more individual chromosomes are added or deleted. In polyploidy, one or more complete *sets* of chromosomes are added. Some organisms (such as yeast) possess a single chromosome set (1n) for most of their life cycles and are referred to as haploid, whereas others possess two chromosome sets and are referred to as diploid (2n). A polyploid is any organism that has more than two sets of chromosomes (3n, 4n, 5n, or more).

(d)

7.2 Chromosome banding is revealed by special staining techniques. (a) G banding. (b) Q banding. (c) C banding. (d) R banding. *[Part a: Leonard Lessin/Peter Arnold. Parts b and c: University of Washington Pathology Department. Part d: Dr. Ram Verma/Phototake.]*

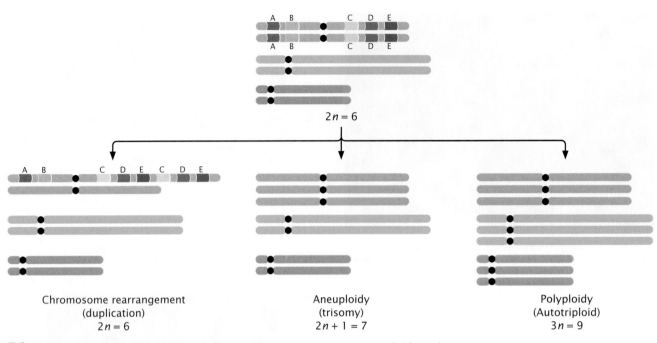

$2n = 6$

Chromosome rearrangement
(duplication)
$2n = 6$

Aneuploidy
(trisomy)
$2n + 1 = 7$

Polyploidy
(Autotriploid)
$3n = 9$

7.3 Chromosome mutations consist of chromosome rearrangements, aneuploids, and polyploids. Duplications, trisomy, and autotriploids are examples of each category of mutation.

7.2 Chromosome Rearrangements Alter Chromosome Structure

Chromosome rearrangements are mutations that change the structure of individual chromosomes. The four basic types of rearrangements are duplications, deletions, inversions, and translocations (**Figure 7.4**).

Duplications

A **chromosome duplication** is a mutation in which part of the chromosome has been doubled (see Figure 7.4a). Consider a chromosome with segments AB•CDEFG, in which • represents the centromere. A duplication might include the EF segments, giving rise to a chromosome with segments AB•CDEF<u>EF</u>G. This type of duplication, in which the duplicated region is immediately adjacent to the

(a) Duplication

A B C D E F G
Original chromosome

In a chromosome duplication, a segment of the chromosome is duplicated.

A B C D E F E F G
Rearranged chromosome

(b) Deletion

A B C D E F G

In a chromosome deletion, a segment of the chromosome is deleted.

A B C D G

(c) Inversion

A B C D E F G

In a chromosome inversion, a segment of the chromosome is turned 180°.

A B C Ⅎ Ǝ D G

(d) Translocation

A B C D E F G

M N O P Q R S

In a translocation, a segment of a chromosome moves from one chromosome to a nonhomologous chromosome or to another place on the same chromosome (the latter not shown here).

A B C D Q R G

M N O P E F S

7.4 The four basic types of chromosome rearrangements are duplication, deletion, inversion, and translocation.

original segment, is called a **tandem duplication**. If the duplicated segment is located some distance from the original segment, either on the same chromosome or on a different one, the chromosome rearrangement is called a **displaced duplication**. An example of a displaced duplication would be AB•CDEFG<u>EF</u>. A duplication can be either in the same orientation as that of the original sequence, as in the two preceding examples, or inverted: AB•CDEFF<u>EG</u>. When the duplication is inverted, it is called a **reverse duplication**.

An individual homozygous for a duplication carries the duplication (the mutated sequence) on both homologous chromosomes, and an individual heterozygous for a duplication has one unmutated chromosome and one chromosome with the duplication. In the heterozygotes (**Figure 7.5a**), problems arise in chromosome pairing at prophase I of meiosis, because the two chromosomes are not homologous throughout their length. The pairing and synapsis of homologous regions require that one or both chromosomes loop and twist so that these regions are able to line up (**Figure 7.5b**). The appearance of this characteristic loop structure in meiosis is one way to detect duplications.

Duplications may have major effects on the phenotype. Among fruit flies (*Drosophila melanogaster*), for example, a fly having a *Bar* mutation has a reduced number of facets in the eye, making the eye smaller and bar shaped instead of

(a) Wild type female B^+B^+

Bar region

(b) Heterozygous *Bar* female B^+B

(c) Homozygous *Bar* female BB

(d) Heterozygous *double Bar* female B^+B^D

7.6 The Bar phenotype in *Drosophila melanogaster* results from an X-linked duplication. (a) Wild-type fruit flies have normal-size eyes. (b) Flies heterozygous and (c) homozygous for the *Bar* mutation have smaller, bar-shaped eyes. (d) Flies with *double Bar* have three copies of the duplication and much smaller bar-shaped eyes.

(a)

Normal chromosome

A   B   C   D   E   F   G

One chromosome has a duplication (E and F).

Chromosome with duplication

A   B   C   D   E   F   E   F   G

Alignment in prophase I of meiosis

(b)

A   B   C   D   E   F   G

A   B   C   D   E   F   G

E   F

The duplicated EF region must loop out to allow the homologous sequences of the chromosomes to align.

7.5 In an individual heterozygous for a duplication, the duplicated chromosome loops out during pairing in prophase I.

oval (**Figure 7.6**). The *Bar* mutation results from a small duplication on the X chromosome that is inherited as an incompletely dominant, X-linked trait: heterozygous female flies have somewhat smaller eyes (the number of facets is reduced; see Figure 7.6b), whereas, in homozygous female and hemizygous male flies, the number of facets is greatly reduced (see Figure 7.6c). Occasionally, a fly carries three copies of the *Bar* duplication on its X chromosome; for flies carrying such mutations, which are termed *double Bar*, the number of facets is extremely reduced (see Figure 7.6d). The *Bar* mutation arises from unequal crossing over, a duplication-generating process (**Figure 7.7**; see also Figure 13.14).

How does a chromosome duplication alter the phenotype? After all, gene sequences are not altered by duplications, and no genetic information is missing; the only change is the presence of additional copies of normal sequences. The answer to this question is not well understood, but the effects are most likely due to imbalances in the amounts of gene products (abnormal gene dosage). The amount of a particular protein synthesized by a cell is often

7.7 Unequal crossing over produces *Bar* and *double-Bar* mutations.

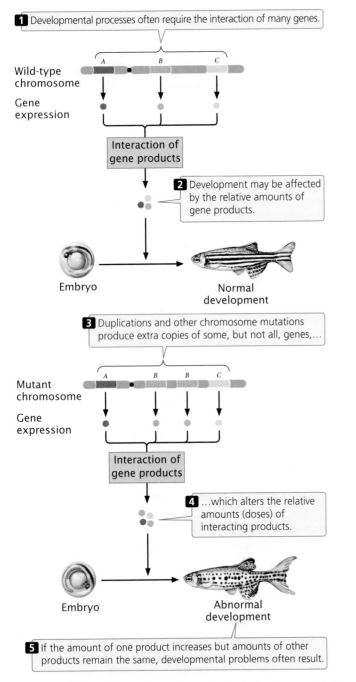

7.8 Unbalanced gene dosage leads to developmental abnormalities.

directly related to the number of copies of its corresponding gene: an individual organism with three functional copies of a gene often produces 1.5 times as much of the protein encoded by that gene as that produced by an individual with two copies. Because developmental processes require the interaction of many proteins, they may critically depend on the relative amounts of the proteins. If the amount of one protein increases while the amounts of others remain constant, problems can result (**Figure 7.8**). Although duplications can have severe consequences when the precise balance of a gene product is critical to cell function, duplications have arisen frequently throughout the evolution of many eukaryotic organisms and are a source of new genes that may provide novel functions. For example, humans have a series of genes that encode different globin chains, some of which function as an oxygen carrier during adult stages and others that function during embryonic and fetal development. All of these globin genes arose from an original ancestral gene that underwent a series of duplications. Human phenotypes associated with some duplications are summarized in **Table 7.1**.

Concepts

A chromosome duplication is a mutation that doubles part of a chromosome. In individuals heterozygous for a chromosome duplication, the duplicated region of the chromosome loops out when homologous chromosomes pair in prophase I of meiosis. Duplications often have major effects on the phenotype, possibly by altering gene dosage.

✔ Concept Check 1

Chromosome duplications often result in abnormal phenotypes because

a. developmental processes depend on the relative amounts of proteins encoded by different genes.

b. extra copies of the genes within the duplicated region do not pair in meiosis.

c. the chromosome is more likely to break when it loops in meiosis.

d. extra DNA must be replicated, which slows down cell division.

Table 7.1	Effects of some human chromosome rearrangements			
Type of Rearrangement	**Chromosome**	**Disorder**		**Symptoms**
Duplication	4, short arm	—		Small head, short neck, low hairline, growth and mental retardation
Duplication	4, long arm	—		Small head, sloping forehead, hand abnormalities
Duplication	7, long arm	—		Delayed development, asymmetry of the head, fuzzy scalp, small nose, low-set ears
Duplication	9, short arm	—		Characteristic face, variable mental retardation, high and broad forehead, hand abnormalities
Deletion	5, short arm	*Cri-du-chat* syndrome		Small head, distinctive cry, widely spaced eyes, round face, mental retardation
Deletion	4, short arm	Wolf–Hirschhorn syndrome		Small head with high forehead, wide nose, cleft lip and palate, severe mental retardation
Deletion	4, long arm	—		Small head, from mild to moderate mental retardation, cleft lip and palate, hand and foot abnormalities
Deletion	7, long arm	Williams–Beuren syndrome		Facial features, heart defects, mental impairment
Deletion	15, long arm	Prader–Willi syndrome		Feeding difficulty at early age, but becoming obese after 1 year of age, from mild to moderate mental retardation
Deletion	18, short arm	—		Round face, large low-set ears, from mild to moderate mental retardation
Deletion	18, long arm	—		Distinctive mouth shape, small hands, small head, mental retardation

Deletions

A second type of chromosome rearrangement is a **chromosome deletion**, the loss of a chromosome segment (see Figure 7.4b). A chromosome with segments AB•CDEFG that undergoes a deletion of segment EF would generate the mutated chromosome AB•CDG.

A large deletion can be easily detected because the chromosome is noticeably shortened. In individuals heterozygous for deletions, the normal chromosome must loop during the pairing of homologs in prophase I of meiosis (**Figure 7.9**) to allow the homologous regions of the two chromosomes to align and undergo synapsis. This looping out generates a structure that looks very much like that seen for individuals heterozygous for duplications.

The phenotypic consequences of a deletion depend on which genes are located in the deleted region. If the deletion includes the centromere, the chromosome will not segregate in meiosis or mitosis

The heterozygote has one normal chromosome…

…and one chromosome with a deletion.

Formation of deletion loop during pairing of homologs in prophase I

In prophase I, the normal chromosome must loop out for the homologous sequences of the chromosomes to align.

Appearance of homologous chromosomes during pairing

7.9 In an individual heterozygous for a deletion, the normal chromosome loops out during chromosome pairing in prophase I.

7.10 The Notch phenotype is produced by a chromosome deletion that includes the *Notch* gene. (Left) Normal wing veination. (Right) Wing veination produced by *Notch* mutation. *[Spyros Artavanis-Tsakonas, Kenji Matsuno, and Mark E. Fortini.]*

and will usually be lost. Many deletions are lethal in the homozygous state because all copies of any essential genes located in the deleted region are missing. Even individuals heterozygous for a deletion may have multiple defects for three reasons.

First, the heterozygous condition may produce imbalances in the amounts of gene products, similar to the imbalances produced by extra gene copies. Second, recessive mutations on the homologous chromosome lacking the deletion may be expressed when the wild-type allele has been deleted (and is no longer present to mask the recessive allele's expression). The expression of a recessive mutation is referred to as **pseudodominance**, and it is an indication that one of the homologous chromosomes has a deletion. Third, some genes must be present in two copies for normal function. When a single copy of a gene is not sufficient to produce a wild-type phenotype, it is said to be a **haploinsufficient gene.** *Notch* is a series of X-linked wing mutations in *Drosophila* that often result from chromosome deletions. *Notch* deletions behave as dominant mutations: when heterozygous for the *Notch* deletion, a fly has wings that are notched at the tips and along the edges (**Figure 7.10**). The *Notch* locus is therefore haploinsufficient. Females that are homozygous for a *Notch* deletion (or males that are hemizygous) die early in embryonic development. The *Notch* gene encodes a receptor that normally transmits signals received from outside the cell to the cell's interior and is important in fly development. The deletion acts as a recessive lethal because loss of all copies of the *Notch* gene prevents normal development.

Concepts

A chromosomal deletion is a mutation in which a part of a chromosome is lost. In individuals heterozygous for a deletion, the normal chromosome loops out during prophase I of meiosis. Deletions cause recessive genes on the homologous chromosome to be expressed and may cause imbalances in gene products.

✔ **Concept Check 2**

What is pseudodominance and how is it produced by a chromosome deletion?

Inversions

A third type of chromosome rearrangement is a **chromosome inversion**, in which a chromosome segment is inverted—turned 180 degrees (see Figure 7.4c). If a chromosome originally had segments AB•CDEFG, then chromosome AB•CFEDG represents an inversion that includes segments DEF. For an inversion to take place, the chromosome must break in two places. Inversions that do not include the centromere, such as AB•CFEDG, are termed **paracentric inversions** (*para* meaning "next to"), whereas inversions that include the centromere, such as ADC•BEFG, are termed **pericentric inversions** (*peri* meaning "around").

Individual organisms with inversions have neither lost nor gained any genetic material; just the gene order has been altered. Nevertheless, these mutations often have pronounced phenotypic effects. An inversion may break a gene into two parts, with one part moving to a new location and destroying the function of that gene. Even when the chromosome breaks are between genes, phenotypic effects may arise from the inverted gene order in an inversion. Many genes are regulated in a position-dependent manner; if their positions are altered by an inversion, they may be expressed at inappropriate times or in inappropriate tissues, an outcome referred to as a **position effect**.

When an individual is homozygous for a particular inversion, no special problems arise in meiosis, and the two homologous chromosomes can pair and separate normally. When an individual is heterozygous for an inversion, however, the gene order of the two homologs differs, and the homologous sequences can align and pair only if the two chromosomes form an inversion loop (**Figure 7.11**).

Individuals heterozygous for inversions also exhibit reduced recombination among genes located in the inverted region. The frequency of crossing over within the inversion is not actually diminished but, when crossing over does take place, the result is abnormal gametes that result in nonviable offspring, and thus no recombinant progeny are observed. Let's see why this result occurs.

Figure 7.12 illustrates the results of crossing over within a paracentric inversion. The individual is heterozygous for an inversion (see Figure 7.12a), with one wild-type, unmutated chromosome (AB•CDEFG) and one inverted chromosome (AB•EDCFG). In prophase I of meiosis, an inversion loop forms, allowing the homologous sequences to pair up (see Figure 7.12b). If a single crossover takes place in the inverted region (between segments C and

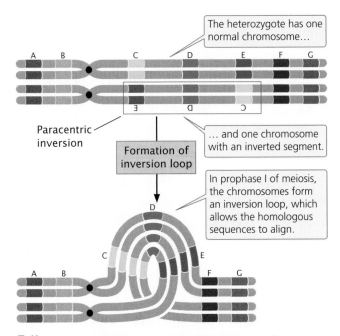

Paracentric inversion

Formation of inversion loop

The heterozygote has one normal chromosome…

… and one chromosome with an inverted segment.

In prophase I of meiosis, the chromosomes form an inversion loop, which allows the homologous sequences to align.

7.11 In an individual heterozygous for a paracentric inversion, the chromosomes form an inversion loop during pairing in prophase I.

D in Figure 7.12), an unusual structure results (see Figure 7.12c). The two outer chromatids, which did not participate in crossing over, contain original, nonrecombinant gene sequences. The two inner chromatids, which did cross over, are highly abnormal: each has two copies of some genes and no copies of others. Furthermore, one of the four chromatids now has two centromeres and is said to be a **dicentric chromatid**; the other lacks a centromere and is an **acentric chromatid**.

In anaphase I of meiosis, the centromeres are pulled toward opposite poles and the two homologous chromosomes separate. This action stretches the dicentric chromatid across the center of the nucleus, forming a structure called a **dicentric bridge** (see Figure 7.12d). Eventually, the dicentric bridge breaks, as the two centromeres are pulled farther apart. Spindle fibers do not attach to the acentric fragment, and so this fragment does not segregate into a nucleus in meiosis and is usually lost.

In the second division of meiosis, the chromatids separate and four gametes are produced (see Figure 7.12e). Two of the gametes contain the original, nonrecombinant chromosomes (AB•CDEFG and AB•EDCFG). The other two gametes contain recombinant chromosomes that are missing some genes; these gametes will not produce viable offspring. Thus, no recombinant progeny result when crossing over takes place within a paracentric inversion. The key is to recognize that crossing over still takes place, but, when it does so, the resulting recombinant gametes are not viable; so no recombinant progeny are observed.

7.12 In a heterozygous individual, a single crossover within a paracentric inversion leads to abnormal gametes.

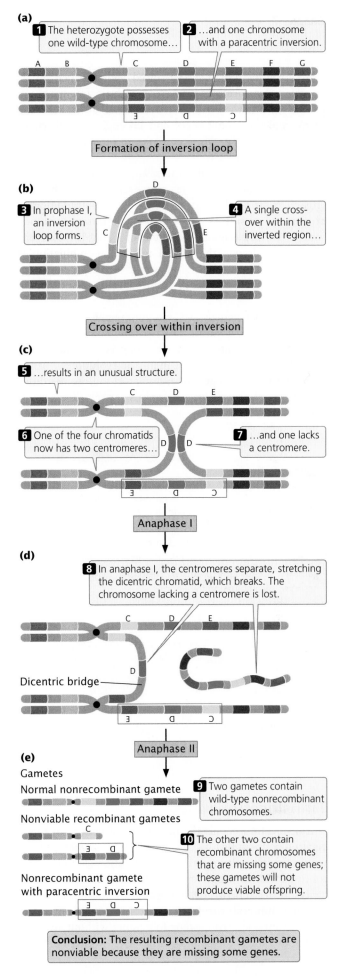

(a)

1 The heterozygote possesses one wild-type chromosome…

2 …and one chromosome with a paracentric inversion.

Formation of inversion loop

(b)

3 In prophase I, an inversion loop forms.

4 A single cross-over within the inverted region…

Crossing over within inversion

(c)

5 …results in an unusual structure.

6 One of the four chromatids now has two centromeres…

7 …and one lacks a centromere.

Anaphase I

(d)

8 In anaphase I, the centromeres separate, stretching the dicentric chromatid, which breaks. The chromosome lacking a centromere is lost.

Dicentric bridge

Anaphase II

(e)

Gametes

Normal nonrecombinant gamete

9 Two gametes contain wild-type nonrecombinant chromosomes.

Nonviable recombinant gametes

10 The other two contain recombinant chromosomes that are missing some genes; these gametes will not produce viable offspring.

Nonrecombinant gamete with paracentric inversion

Conclusion: The resulting recombinant gametes are nonviable because they are missing some genes.

Recombination is also reduced within a pericentric inversion. No dicentric bridges or acentric fragments are produced, but the recombinant chromosomes have too many copies of some genes and no copies of others; so gametes that receive the recombinant chromosomes cannot produce viable progeny.

Figure 7.12 illustrates the results of single crossovers within inversions. Double crossovers, in which both crossovers are on the same two strands (two-strand double crossovers), result in functional recombinant chromosomes. (Try drawing out the results of a double crossover.) Thus, even though the overall rate of recombination is reduced within an inversion, some viable recombinant progeny may still be produced through two-strand double crossovers.

Inversion heterozygotes are common in many organisms, including a number of plants, some species of *Drosophila*, mosquitoes, and grasshoppers. Inversions may have played an important role in human evolution: G-banding patterns reveal that several human chromosomes differ from those of chimpanzees by only a pericentric inversion (**Figure 7.13**).

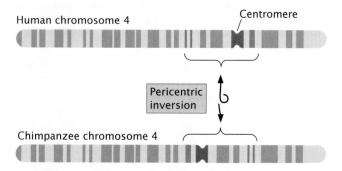

7.13 Chromosome 4 differs in humans and chimpanzees in a pericentric inversion.

Concepts

In an inversion, a segment of a chromosome is inverted. Inversions cause breaks in some genes and may move others to new locations. In heterozygotes for a chromosome inversion, the homologous chromosomes form a loop in prophase I of meiosis. When crossing over takes place within the inverted region, nonviable gametes are usually produced, resulting in a depression in observed recombination frequencies.

✔ Concept Check 3

A dicentric chromosome is produced when crossing over takes place in an individual heterozygous for which type of chromosome rearrangement?

a. Duplication

b. Deletion

c. Paracentric inversion

d. Pericentric inversion

Translocations

A **translocation** entails the movement of genetic material between nonhomologous chromosomes (see Figure 7.4d) or within the same chromosome. Translocation should not be confused with crossing over, in which there is an exchange of genetic material between *homologous* chromosomes.

In a **nonreciprocal translocation**, genetic material moves from one chromosome to another without any reciprocal exchange. Consider the following two nonhomolo-gous chromosomes: AB•CDEFG and MN•OPQRS. If chromosome segment EF moves from the first chromosome to the second without any transfer of segments from the second chromosome to the first, a nonreciprocal translocation has taken place, producing chromosomes AB•CDG and MN•OPEFQRS. More commonly, there is a two-way exchange of segments between the chromosomes, resulting in a **reciprocal translocation**. A reciprocal translocation between chromosomes AB•CDEFG and MN•OPQRS might give rise to chromosomes AB•CDQRG and MN•OPEFS.

Translocations can affect a phenotype in several ways. First, they can create new linkage relations that affect gene expression (a position effect): genes translocated to new locations may come under the control of different regulatory sequences or other genes that affect their expression—an example is found in Burkitt lymphoma, to be discussed in Chapter 15.

Second, the chromosomal breaks that bring about translocations may take place within a gene and disrupt its function. Molecular geneticists have used these types of effects to map human genes. Neurofibromatosis is a genetic disease characterized by numerous fibrous tumors of the skin and nervous tissue; it results from an autosomal dominant mutation. Linkage studies first placed the locus for neurofibromatosis on chromosome 17. Geneticists later identified two patients with neurofibromatosis who possessed a translocation affecting chromosome 17. These patients were assumed to have developed neurofibromatosis because one of the chromosome breaks that occurred in the translocation disrupted a particular gene that, when mutated, causes neurofibromatosis. DNA from the regions around the breaks was sequenced and eventually led to the identification of the gene responsible for neurofibromatosis.

Deletions frequently accompany translocations. In a **Robertsonian translocation**, for example, the long arms of two acrocentric chromosomes become joined to a common centromere through a translocation, generating a metacentric chromosome with two long arms and another chromosome

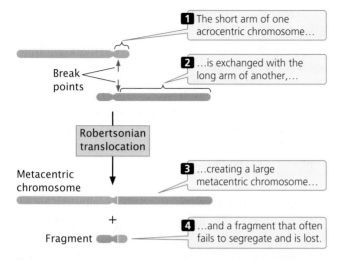

1 The short arm of one acrocentric chromosome…

2 …is exchanged with the long arm of another,…

Break points

Robertsonian translocation

Metacentric chromosome

3 …creating a large metacentric chromosome…

+

Fragment

4 …and a fragment that often fails to segregate and is lost.

7.14 In a Robertsonian translocation, the short arm of one acrocentric chromosome is exchanged with the long arm of another.

Let's consider what happens in an individual heterozygous for a reciprocal translocation. Suppose that the original chromosomes were AB•CDEFG and MN•OPQRS (designated N_1 and N_2, respectively) and that a reciprocal translocation takes place, producing chromosomes AB•CDQRS and MN•OP<u>EFG</u> (designated T_1 and T_2, respectively). An individual heterozygous for this translocation would possess one normal copy of each chromosome and one translocated copy (**Figure 7.15a**). Each of these chromosomes contains segments that are homologous to two other chromosomes. When the homologous sequences pair in prophase I of meiosis, crosslike configurations consisting of all four chromosomes (**Figure 7.15b**) form. Whether viable or nonviable gametes can be produced depends on how the chromosomes in these crosslike configurations separate. Only about half of the gametes from an individual heterozygous for a reciprocal translocation are expected to be functional, and so these individuals frequently exhibit reduced fertility.

with two very short arms (**Figure 7.14**). The smaller chromosome often fails to segregate, leading to an overall reduction in chromosome number. As we will see, Robertsonian translocations are the cause of some cases of Down syndrome, a chromosome disorder discussed in the introduction to the chapter.

The effects of a translocation on chromosome segregation in meiosis depend on the nature of the translocation.

Concepts

In translocations, parts of chromosomes move to other, nonhomologous chromosomes or to other regions of the same chromosome. Translocations may affect the phenotype by causing genes to move to new locations, where they come under the influence of new regulatory sequences, or by breaking genes and disrupting their function.

(a)

1 An individual heterozygous for this translocation possesses one normal copy of each chromosome (N_1 and N_2)…

2 …and one translocated copy (T_1 and T_2).

(b)

3 Because each chromosome has sections that are homologous to two other chromosomes, a crosslike configuration forms in prophase I of meiosis.

7.15 In an individual heterozygous for a reciprocal translocation, crosslike structures form in homologous pairing.

Fragile Sites

Chromosomes of cells grown in culture sometimes develop constrictions or gaps at particular locations called **fragile sites** (**Figure 7.16**), because they are prone to breakage under certain conditions. A number of fragile sites have been identified on human chromosomes. One of the most intensively studied is a fragile site on the human X chromosome, a site associated with mental retardation known as the fragile-X syndrome. Exhibiting X-linked inheritance and arising with a frequency of about 1 in 1250 male births, fragile-X syndrome has been shown to result from an increase in the number of repeats of a CGG trinucleotide (see Chapter 13). However, other common fragile sites do not consist of trinucleotide repeats, and their nature is still incompletely understood.

7.3 Aneuploidy Is an Increase or Decrease in the Number of Individual Chromosomes

In addition to chromosome rearrangements, chromosome mutations include changes in the number of chromosomes. Variations in chromosome number can be classified into two basic types: **aneuploidy**, which is a change in the number of individual chromosomes, and **polyploidy**, which is a change in the number of chromosome sets.

Aneuploidy can arise in several ways. First, a chromosome may be lost in the course of mitosis or meiosis if, for example, its centromere is deleted. Loss of the centromere

7.16 Fragile sites are chromosomal regions susceptible to breakage under certain conditions. Shown here is a fragile site on human chromosome X. *[University of Wisconsin Cytogenic Services Laboratory.]*

prevents the spindle fibers from attaching; so the chromosome fails to move to the spindle pole and does not become incorporated into a nucleus after cell division. Second, the small chromosome generated by a Robertsonian translocation may be lost in mitosis or meiosis. Third, aneuploids may arise through **nondisjunction**, the failure of homologous chromosomes or sister chromatids to separate in meiosis or mitosis. Nondisjunction leads to some gametes or cells that contain an extra chromosome and others that are missing a chromosome (**Figure 7.17**).

Types of Aneuploidy

We will consider four types of common aneuploid conditions in diploid individuals: nullisomy, monosomy, trisomy, and tetrasomy.

1. **Nullisomy** is the loss of both members of a homologous pair of chromosomes. It is represented as $2n - 2$, where n refers to the haploid number of chromosomes. Thus, among humans, who normally possess $2n = 46$ chromosomes, a nullisomic person has 44 chromosomes.

2. **Monosomy** is the loss of a single chromosome, represented as $2n - 1$. A monosomic person has 45 chromosomes.

3. **Trisomy** is the gain of a single chromosome, represented as $2n + 1$. A trisomic person has 47 chromosomes. The gain of a chromosome means that there are three homologous copies of one chromosome. Most cases of Down syndrome, discussed in the introduction to the chapter, result from trisomy of chromosome 21.

4. **Tetrasomy** is the gain of two homologous chromosomes, represented as $2n + 2$. A tetrasomic person has 48 chromosomes. Tetrasomy is not the gain of *any* two extra chromosomes, but rather the gain of two homologous chromosomes; so there will be four homologous copies of a particular chromosome.

More than one aneuploid mutation may occur in the same individual organism. An individual that has an extra copy of two different (nonhomologous) chromosomes is referred to as being double trisomic and represented as $2n + 1 + 1$. Similarly, a double monosomic has two fewer nonhomologous chromosomes ($2n - 1 - 1$), and a double tetrasomic has two extra pairs of homologous chromosomes ($2n + 2 + 2$).

Effects of Aneuploidy

Aneuploidy usually alters the phenotype drastically. In most animals and many plants, aneuploid mutations are lethal. Because aneuploidy affects the number of gene copies but not their nucleotide sequences, the effects of aneuploidy are most likely due to abnormal gene dosage. Aneuploidy alters the dosage for some, but not all, genes, disrupting the relative

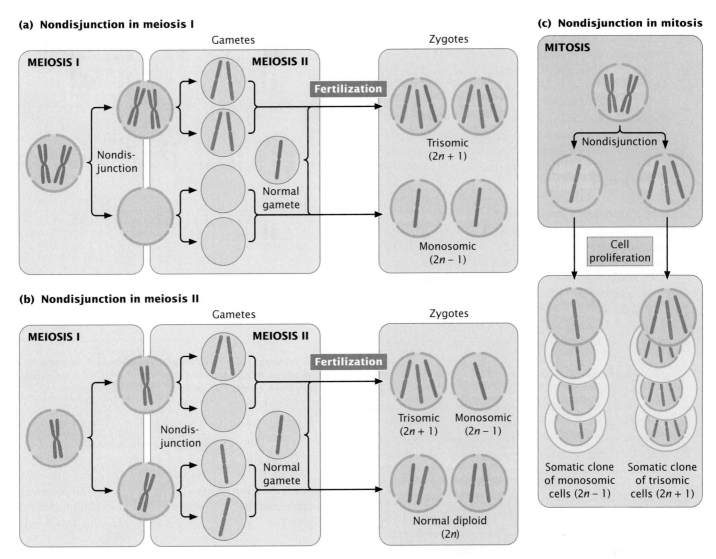

7.17 Aneuploids can be produced through nondisjunction in meiosis I, meiosis II, and mitosis. The gametes that result from meioses with nondisjunction combine with a gamete (with blue chromosome) that results from normal meiosis to produce the zygotes. (a) Nondisjunction in meiosis I. (b) Nondisjunction in meiosis II. (c) Nondisjunction in mitosis.

concentrations of gene products and often interfering with normal development.

A major exception to the relation between gene number and protein dosage pertains to genes on the mammalian X chromosome. In mammals, X-chromosome inactivation ensures that males (who have a single X chromosome) and females (who have two X chromosomes) receive the same functional dosage for X-linked genes (see pp. 80–81 in Chapter 4 for further discussion of X-chromosome inactivation). Extra X chromosomes in mammals are inactivated; so we might expect that aneuploidy of the sex chromosomes would be less detrimental in these animals. Indeed, it is the case for mice and humans, for whom aneuploids of the sex chromosomes are the most common form of aneuploidy seen in living organisms. Y-chromosome aneuploids are probably common because there is so little information on the Y chromosome.

Concepts

Aneuploidy, the loss or gain of one or more individual chromosomes, may arise from the loss of a chromosome subsequent to translocation or from nondisjunction in meiosis or mitosis. It disrupts gene dosage and often has severe phenotypic effects.

✔ Concept Check 5

A diploid organism has $2n = 36$ chromosomes. How many chromosomes will be found in a trisomic member of this species?

Aneuploidy in Humans

For unknown reasons, an incredibly high percentage of all human embryos that are conceived possess chromosome abnormalities. Findings from studies of women who are attempting pregnancy suggest that more than 30% of all conceptions spontaneously abort (miscarry), usually so early in development that the mother is not even aware of her pregnancy. Chromosome defects are present in at least 50% of spontaneously aborted human fetuses, with aneuploidy accounting for most of them. This rate of chromosome abnormality in humans is higher than in other organisms that have been studied; in mice, for example, aneuploidy is found in no more than 2% of fertilized eggs. Aneuploidy in humans usually produces such serious developmental problems that spontaneous abortion results. Only about 2% of all fetuses with a chromosome defect survive to birth.

Sex-chromosome aneuploids The most common aneuploidy seen in living humans has to do with the sex chromosomes. As is true of all mammals, aneuploidy of the human sex chromosomes is better tolerated than aneuploidy of autosomal chromosomes. Both Turner syndrome and Klinefelter syndrome (see Chapter 4) result from aneuploidy of the sex chromosomes.

Autosomal aneuploids Autosomal aneuploids resulting in live births are less common than sex-chromosome aneuploids in humans, probably because there is no mechanism of dosage compensation for autosomal chromosomes. Most autosomal aneuploids are spontaneously aborted, with the exception of aneuploids of some of the small autosomes such as chromosome 21. Because these chromosomes are small and carry fewer genes, the presence of extra copies is less detrimental than it is for larger chromosomes. For example, the most common autosomal aneuploidy in humans is **trisomy 21**, also called **Down syndrome** (discussed in the introduction to the chapter). The number of genes on different human chromosomes is not precisely known at the present time, but DNA sequence data indicate that chromosome 21 has fewer genes than any other autosome, with only about 230 genes of a total of 20,000 to 25,000 for the entire genome.

The incidence of Down syndrome in the United States is similar to that of the world, about 1 in 700 human births, although the incidence increases among children born to older mothers. Approximately 92% of those who have Down syndrome have three full copies of chromosome 21 (and therefore a total of 47 chromosomes), a condition termed **primary Down syndrome** (**Figure 7.18**). Primary Down syndrome usually arises from spontaneous nondisjunction in egg formation: about 75% of the nondisjunction events that cause Down syndrome are maternal in origin, most arising in meiosis I. Most children with Down syndrome are born to normal parents, and the failure of the chromosomes to divide has little hereditary tendency. A couple who has conceived one child with primary Down syndrome has only a slightly higher risk of conceiving a second child with Down syndrome (compared with other couples of similar age who have not had any Down-syndrome children). Similarly, the

7.18 Primary Down syndrome is caused by the presence of three copies of chromosome 21. Karyotype of a person who has primary Down syndrome. [*L. Willatt, East Anglian Regional Genetics Service/Science Photo Library/Photo Researchers.*]

couple's relatives are not more likely to have a child with primary Down syndrome.

About 4% of people with Down syndrome have 46 chromosomes, but an extra copy of part of chromosome 21 is attached to another chromosome through a translocation (**Figure 7.19**). This condition is termed **familial Down syn-**

7.19 The translocation of chromosome 21 onto another chromosome results in familial Down syndrome. Here, the long arm of chromosome 21 is attached to chromosome 15. This karyotype is from a translocation carrier, who is phenotypically normal but is at increased risk for producing children with Down syndrome. [*Dr. Dorothy Warburton, HICCC, Columbia University.*]

drome because it has a tendency to run in families. The phenotypic characteristics of familial Down syndrome are the same as those for primary Down syndrome.

Familial Down syndrome arises in offspring whose parents are carriers of chromosomes that have undergone a Robertsonian translocation, most commonly between chromosome 21 and chromosome 14: the long arm of 21 and the short arm of 14 exchange places. This exchange produces a chromosome that includes the long arms of chromosomes 14 and 21, and a very small chromosome that consists of the short arms of chromosomes 21 and 14. The small chromosome is generally lost after several cell divisions. Although exchange between chromosomes 21 and 14 is the most-common cause of familial Down syndrome, the condition can also be caused by translocations between 21 and other chromosomes such as 15 (illustrated in Figure 7.19).

Persons with the translocation, called **translocation carriers**, do not have Down syndrome. Although they possess only 45 chromosomes, their phenotypes are normal because they have two copies of the long arms of chromosomes 14 and 21, and apparently the short arms of these chromosomes (which are lost) carry no essential genetic information. Although translocation carriers are completely healthy, they have an increased chance of producing children with Down syndrome (**Figure 7.20**).

Few autosomal aneuploids besides trisomy 21 result in human live births. **Trisomy 18**, also known as **Edward syndrome**, arises with a frequency of approximately 1 in 8000 live births. Babies with Edward syndrome are severely retarded and have low-set ears, a short neck, deformed feet, clenched fingers, heart problems, and other disabilities. Few live for more than a year after birth. **Trisomy 13** has a frequency of about 1 in 15,000 live births and produces features that are collectively known as **Patau syndrome**. Characteristics of this condition include severe mental retardation, a small head, sloping forehead, small eyes, cleft lip and palate, extra fingers and toes, and numerous other problems. About half of children with trisomy 13 die within the first month of life, and 95% die by the age of 3. Rarer still is **trisomy 8**, which arises with a frequency ranging from about 1 in 25,000 to 1 in 50,000 live births. This aneuploid is characterized by mental retardation, contracted fingers and toes, low-set malformed ears, and a prominent forehead. Many who have this condition have normal life expectancy.

Aneuploidy and maternal age   Most cases of Down syndrome and other types of aneuploidy in humans arise from maternal nondisjunction, and the frequency of aneuploidy correlates with maternal age (**Figure 7.21**). Why maternal age is associated with nondisjunction is not known for certain, but

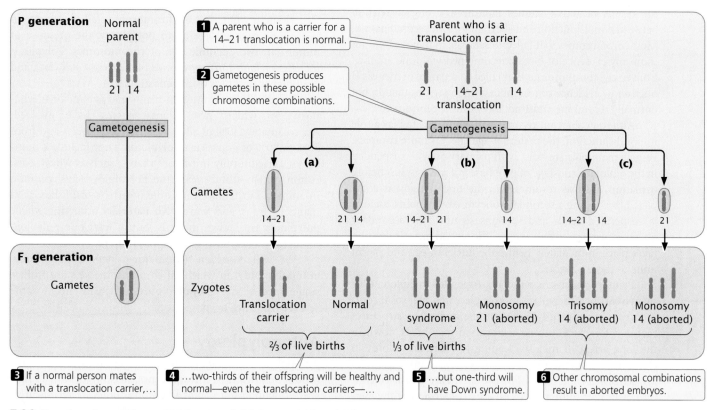

7.20 Translocation carriers are at increased risk for producing children with Down syndrome.

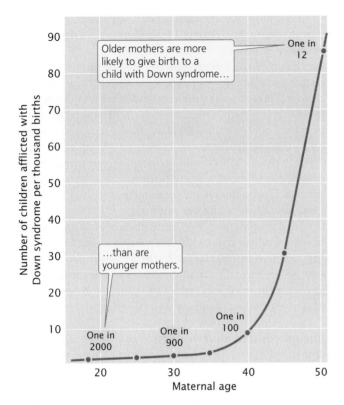

7.21 The incidence of primary Down syndrome and other aneuploids increases with maternal age.

the results of recent studies indicate a strong correlation between nondisjunction and aberrant meiotic recombination. Most chromosomes that failed to separate in meiosis I do not show any evidence of having recombined with one another. Conversely, chromosomes that failed to separate in meiosis II often show evidence of recombination in regions that do not normally recombine, most notably near the centromere.

Although aberrant recombination appears to play a role in nondisjunction, the maternal-age effect is more complex. Female mammals are born with primary oocytes suspended in the diplotene substage of prophase I of meiosis. Just before ovulation, meiosis resumes and the first division is completed, producing a secondary oocyte. At this point, meiosis is suspended again and remains so until the secondary oocyte is penetrated by a sperm. The second meiotic division takes place immediately before the nuclei of egg and sperm unite to form a zygote.

Primary oocytes may remain suspended in diplotene for many years before ovulation takes place and meiosis recommences. Components of the spindle and other structures required for chromosome segregation may break down in the long arrest of meiosis, leading to more aneuploidy in children born to older mothers. According to this theory, no age effect is seen in males, because sperm are produced continuously after puberty with no long suspension of the meiotic divisions.

Aneuploidy and cancer Many tumor cells have extra or missing chromosomes or both; some types of tumors are consistently associated with specific chromosome mutations, including aneuploidy and chromosome rearrangements. The role of chromosome mutations in cancer will be explored in Chapter 15.

Concepts

In humans, sex-chromosome aneuploids are more common than are autosomal aneuploids. X-chromosome inactivation prevents problems of gene dosage for X-linked genes. Down syndrome results from three functional copies of chromosome 21, either through trisomy (primary Down syndrome) or a Robertsonian translocation (familial Down syndrome).

✔ Concept Check 6

Briefly explain why, in humans and mammals, sex-chromosome aneuploids are more common than autosomal aneuploids?

7.4 Polyploidy Is the Presence of More Than Two Sets of Chromosomes

Most eukaryotic organisms are diploid (2*n*) for most of their life cycles, possessing two sets of chromosomes. Occasionally, whole sets of chromosomes fail to separate in meiosis or mitosis, leading to polyploidy, the presence of more than two genomic sets of chromosomes. Polyploids include *triploids* (3*n*), *tetraploids* (4*n*), *pentaploids* (5*n*), and even higher numbers of chromosome sets.

Polyploidy is common in plants and is a major mechanism by which new plant species have evolved. Approximately 40% of all flowering-plant species and from 70% to 80% of grasses are polyploids. They include a number of agriculturally important plants, such as wheat, oats, cotton, potatoes, and sugar cane. Polyploidy is less common in animals but is found in some invertebrates, fishes, salamanders, frogs, and lizards. No naturally occurring, viable polyploids are known in birds, but at least one polyploid mammal—a rat in Argentina—has been reported.

We will consider two major types of polyploidy: **autopolyploidy**, in which all chromosome sets are from a single species; and **allopolyploidy**, in which chromosome sets are from two or more species.

Autopolyploidy

Autopolyploidy occurs when accidents of meiosis or mitosis produce extra sets of chromosomes, all derived from a single species. Nondisjunction of all chromosomes in mitosis in an early 2*n* embryo, for example, doubles the chromosome

(a) Autopolyploidy through mitosis

(b) Autopolyploidy through meiosis

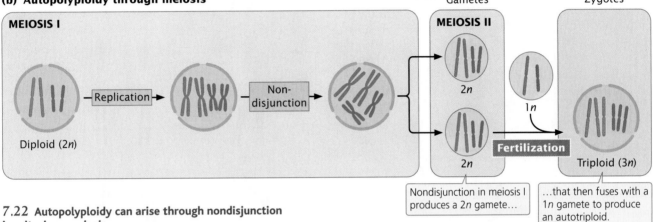

7.22 **Autopolyploidy can arise through nondisjunction in mitosis or meiosis.**

number and produces an autotetraploid (4n) (**Figure 7.22a**). An autotriploid (3n) may arise when nondisjunction in meiosis produces a diploid gamete that then fuses with a normal haploid gamete to produce a triploid zygote (**Figure 7.22b**). Alternatively, triploids may arise from a cross between an autotetraploid that produces 2n gametes and a diploid that produces 1n gametes.

Because all the chromosome sets in autopolyploids are from the same species, they are homologous and attempt to align in prophase I of meiosis, which usually results in sterility. Consider meiosis in an autotriploid (**Figure 7.23**). In meiosis in a diploid cell, two chromosome homologs pair and align, but, in autotriploids, three homologs are present. One of the three homologs may fail to align with the other two, and this unaligned chromosome will segregate randomly (see Figure 7.23a). Which gamete gets the extra chromosome will be determined by chance and will differ for each homologous group of chromosomes. The resulting gametes will have two copies of some chromosomes and one copy of others. Even if all three chromosomes do align, two chromosomes must segregate to one gamete and one chromosome to the other (see Figure 7.23b). Occasionally, the presence of a third chromosome interferes with normal alignment, and all three chromosomes segregate to the same gamete (see Figure 7.23c).

No matter how the three homologous chromosomes align, their random segregation will create **unbalanced gametes**, with various numbers of chromosomes. A gamete produced by meiosis in such an autotriploid might receive, say, two copies of chromosome 1, one copy of chromosome 2, three copies of chromosome 3, and no copies of chromosome 4. When the unbalanced gamete fuses with a normal gamete (or with another unbalanced gamete), the resulting zygote has different numbers of the four types of chromosomes. This difference in number creates unbalanced gene dosage in the zygote, which is often lethal. For this reason, triploids do not usually produce viable offspring.

In even-numbered autopolyploids, such as autotetraploids, the homologous chromosomes can theoretically form pairs and divide equally. However, this event rarely happens; so these types of autotetraploids also produce unbalanced gametes.

The sterility that usually accompanies autopolyploidy has been exploited in agriculture. Wild diploid bananas (2n = 22), for example, produce seeds that are hard and inedible, but triploid bananas (3n = 33) are sterile, and produce no seeds—they are the bananas sold commercially. Similarly, seedless triploid watermelons have been created and are now widely sold.

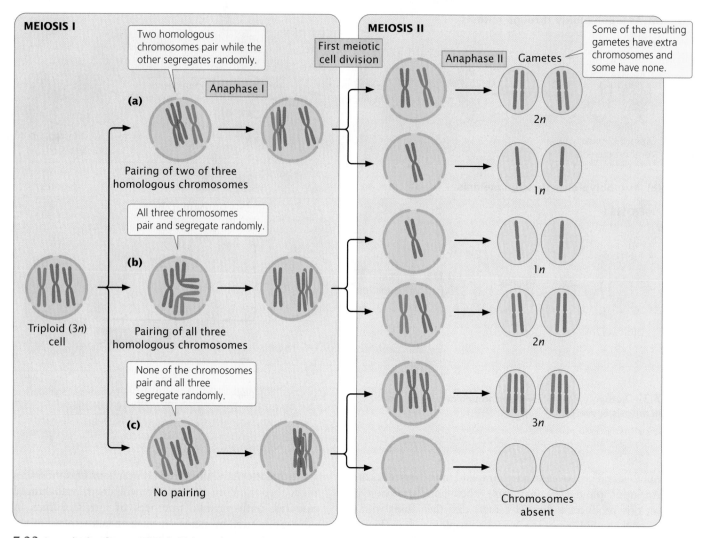

MEIOSIS I

Two homologous chromosomes pair while the other segregates randomly.

First meiotic cell division

MEIOSIS II

Anaphase II   Gametes

Some of the resulting gametes have extra chromosomes and some have none.

Anaphase I

(a)

Pairing of two of three homologous chromosomes

2n

1n

All three chromosomes pair and segregate randomly.

(b)

Triploid (3n) cell

Pairing of all three homologous chromosomes

1n

2n

None of the chromosomes pair and all three segregate randomly.

(c)

No pairing

3n

Chromosomes absent

7.23 In meiosis of an autotriploid, homologous chromosomes can pair or not pair in three ways. This example illustrates the pairing and segregation of a single homologous set of chromosomes.

Allopolyploidy

Allopolyploidy arises from hybridization between two species; the resulting polyploid carries chromosome sets derived from two or more species. **Figure 7.24** shows how allopolyploidy can arise from two species that are sufficiently related that hybridization occurs between them. Species I (AABBCC, $2n = 6$) produces haploid gametes with chromosomes ABC, and species II (GGHHII, $2n = 6$) produces gametes with chromosomes GHI. If gametes from species I and II fuse, a hybrid with six chromosomes (ABCGHI) is created. The hybrid has the same chromosome number as that of both diploid species; so the hybrid is considered diploid. However, because the hybrid chromosomes are not homologous, they will not pair and segregate properly in meiosis; so this hybrid is functionally haploid and sterile.

The sterile hybrid is unable to produce viable gametes through meiosis, but it may be able to perpetuate itself through mitosis (asexual reproduction). On rare occasions, nondisjunction takes place in a mitotic division, which leads to a doubling of chromosome number and an allotetraploid with chromosomes AABBCCGGHHII. This type of allopolyploid, consisting of two combined diploid genomes, is sometimes called an **amphidiploid**. Although the chromosome number has doubled compared with what was present in each of the parental species, the amphidiploid is functionally diploid: every chromosome has one and only one homologous partner, which is exactly what meiosis requires for proper segregation. The amphidiploid can now undergo normal meiosis to produce balanced gametes having six chromosomes.

George Karpechenko created polyploids experimentally in the 1920s. Today, as well as in the early twentieth century, cabbage (*Brassica oleracea*, $2n = 18$) and radishes (*Raphanus sativa*, $2n = 18$) are agriculturally important plants, but only

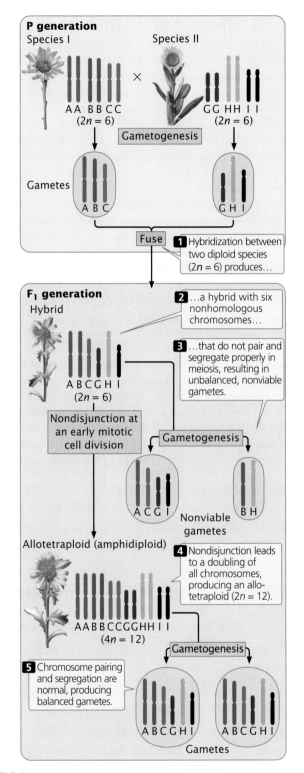

7.24 Most allopolyploids arise from hybridization between two species followed by chromosome doubling.

the leaves of the cabbage and the roots of the radish are normally consumed. Karpechenko wanted to produce a plant that had cabbage leaves and radish roots so that no part of the plant would go to waste. Because both cabbage and

radish possess 18 chromosomes, Karpechenko was able to successfully cross them, producing a hybrid with $2n = 18$, but, unfortunately, the hybrid was sterile. After several crosses, Karpechenko noticed that one of his hybrid plants produced a few seeds. When planted, these seeds grew into plants that were viable and fertile. Analysis of their chromosomes revealed that the plants were allotetraploids, with $2n = 36$ chromosomes. To Karpechencko's great disappointment, however, the new plants possessed the roots of a cabbage and the leaves of a radish.

Worked Problem

Species I has $2n = 14$ and species II has $2n = 20$. Give all possible chromosome numbers that may be found in the following individuals.

a. An autotriploid of species I

b. An autotetraploid of species II

c. An allotriploid formed from species I and species II

d. An allotetraploid formed from species I and species II

• Solution

The haploid number of chromosomes (n) for species I is 7 and for species II is 10.

a. A triploid individual is $3n$. A common mistake is to assume that $3n$ means three times as many chromosomes as in a normal individual, but remember that normal individuals are $2n$. Because n for species I is 7 and all genomes of an autopolyploid are from the same species, $3n = 3 \times 7 = 21$.

b. A autotetraploid is $4n$ with all genomes from the same species. The n for species II is 10, so $4n = 4 \times 10 = 40$.

c. A triploid individual is $3n$. By definition, an allopolyploid must have genomes from two different species. An allotriploid could have $1n$ from species I and $2n$ from species II or $(1 \times 7) + (2 \times 10) = 27$. Alternatively, it might have $2n$ from species I and $1n$ from species II, or $(2 \times 7) + (1 \times 10) = 24$. Thus, the number of chromosomes in an allotriploid could be 24 or 27.

d. A tetraploid is $4n$. By definition, an allotetraploid must have genomes from at least two different species. An allotetraploid could have $3n$ from species I and $1n$ from species II or $(3 \times 7) + (1 \times 10) = 31$; or $2n$ from species I and $2n$ from species II or $(2 \times 7) + (2 \times 10) = 34$; or $1n$ from species I and $3n$ from species II or $(1 \times 7) + (3 \times 10) = 37$. Thus, the number of chromosomes could be 31, 34, or 37.

> **?** For additional practice, try Problem 23 at the end of this chapter.

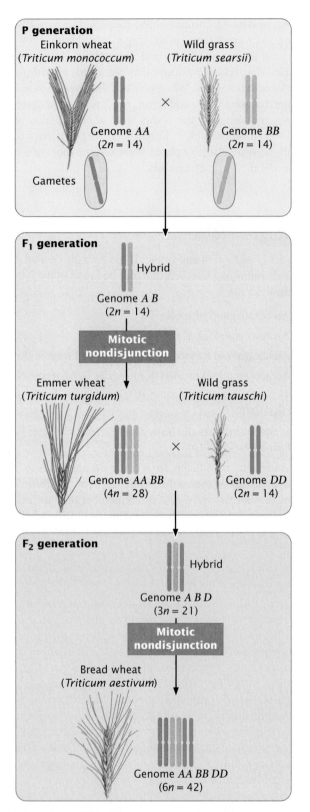

7.25 Modern bread wheat, *Triticum aestivum*, is a hexaploid with genes derived from three different species. Two diploid species, *T. monococcum* ($n = 14$) and probably *T. searsii* ($n = 14$), originally crossed to produce a diploid hybrid ($2n = 14$) that underwent chromosome doubling to create *T. turgidum* ($4n = 28$). A cross between *T. turgidum* and *T. tauschi* ($2n = 14$) produced a triploid hybrid ($3n = 21$) that then underwent chromosome doubling to produce *T. aestivum*, which is a hexaploid ($6n = 42$).

The Significance of Polyploidy

In many organisms, cell volume is correlated with nuclear volume, which, in turn, is determined by genome size. Thus, the increase in chromosome number in polyploidy is often associated with an increase in cell size, and many polyploids are physically larger than diploids. Breeders have used this effect to produce plants with larger leaves, flowers, fruits, and seeds. The hexaploid ($6n = 42$) genome of wheat probably contains chromosomes derived from three different wild species (**Figure 7.25**). Many other cultivated plants also are polyploid (**Table 7.2**).

Polyploidy is less common in animals than in plants for several reasons. As discussed, allopolyploids require hybridization between different species, which happens less frequently in animals than in plants. Animal behavior often prevents interbreeding among species, and the complexity of animal development causes most interspecific hybrids to be nonviable. Many of the polyploid animals that do arise are in groups that reproduce through parthenogenesis (a type of reproduction in which the animal develops from an unfertilized egg). Thus, asexual reproduction may facilitate the development of polyploids, perhaps because the perpetuation of hybrids through asexual reproduction provides greater opportunities for nondisjunction than does sexual reproduction. Only a few human polyploid babies have been reported, and most died within a few days of birth. Polyploidy—usually triploidy—is seen in about 10% of all spontaneously aborted human fetuses. Different types of chromosome mutations are summarized in **Table 7.3**.

Table 7.2	Examples of polyploid crop plants		
Plant	Type of Polyploidy	Ploidy	Chromosome Number
Potato	Autopolyploid	$4n$	48
Banana	Autopolyploid	$3n$	33
Peanut	Autopolyploid	$4n$	40
Sweet potato	Autopolyploid	$6n$	90
Tobacco	Allopolyploid	$4n$	48
Cotton	Allopolyploid	$4n$	52
Wheat	Allopolyploid	$6n$	42
Oats	Allopolyploid	$6n$	42
Sugar cane	Allopolyploid	$8n$	80
Strawberry	Allopolyploid	$8n$	56

Source: After F. C. Elliot, *Plant Breeding and Cytogenetics* (New York: McGraw-Hill, 1958).

Table 7.3	Different types of chromosome mutations
Chromosome Mutation	**Definition**
Chromosome rearrangement	Change in chromosome structure
Chromosome duplication	Duplication of a chromosome segment
Chromosome deletion	Deletion of a chromosome segment
Inversion	Chromosome segment inverted 180 degrees
Paracentric inversion	Inversion that does not include the centromere in the inverted region
Pericentric inversion	Inversion that includes the centromere in the inverted region
Translocation	Movement of a chromosome segment to a nonhomologous chromosome or to another region of the same chromosome
Nonreciprocal translocation	Movement of a chromosome segment to a nonhomologous chromosome or to another region of the same chromosome without reciprocal exchange
Reciprocal translocation	Exchange between segments of nonhomologous chromosomes or between regions of the same chromosome
Aneuploidy	Change in number of individual chromosomes
Nullisomy	Loss of both members of a homologous pair
Monosomy	Loss of one member of a homologous pair
Trisomy	Gain of one chromosome, resulting in three homologous chromosomes
Tetrasomy	Gain of two homologous chromosomes, resulting in four homologous chromosomes
Polyploidy	Addition of entire chromosome sets
Autopolyploidy	Polyploidy in which extra chromosome sets are derived from the same species
Allopolyploidy	Polyploidy in which extra chromosome sets are derived from two or more species

Concepts

Polyploidy is the presence of extra chromosome sets: autopolyploids possess extra chromosome sets from the same species; allopolyploids possess extra chromosome sets from two or more species. Problems in chromosome pairing and segregation often lead to sterility in autopolyploids, but many allopolyploids are fertile.

✔ **Concept Check 7**

Species A has $2n = 16$ chromosomes and species B has $2n = 14$. How many chromosomes would be found in an allotriploid of these two species?

a. 21 or 24   c. 22 or 23

b. 42 or 48   d. 45

7.5 Chromosome Variation Plays an Important Role in Evolution

Chromosome variations are potentially important in evolution and, within a number of different groups of organisms, have clearly played a significant role in past evolution. Chromosome duplications provide one way in which new genes may evolve. In many cases, existing copies of a gene are not free to vary, because they encode a product that is essential to development or function. However, after a chromosome undergoes duplication, extra copies of genes within the duplicated region are present. The original copy can provide the essential function while an extra copy from the duplication is free to undergo mutation and change. Over evolutionary time, the extra copy may acquire enough mutations to assume a new function that benefits the organism.

Inversions also can play important evolutionary roles by suppressing recombination among a set of genes. As we have seen, crossing over within an inversion in an individual heterozygous for a pericentric or paracentric inversion leads to unbalanced gametes and no recombinant progeny. This suppression of recombination allows particular sets of coadapted alleles that function well together to remain intact, unshuffled by recombination.

Polyploidy, particularly allopolyploidy, often gives rise to new species and has been particularly important in the evolution of flowering plants. Occasional genome doubling through polyploidy has been a major contributor to evolutionary success in animal groups. For example, *Saccharomyces cerevisiae* (yeast) is a tetraploid, having undergone whole-genome duplication about 100 million years ago. The vertebrate genome has duplicated twice, once in the common ancestor to jawed vertebrates and again in the ancestor of fishes. Certain groups of vertebrates, such as some frogs and some fishes, have undergone additional polyploidy. Cereal plants have undergone several genome duplication events.

Concepts Summary

- Three basic types of chromosome mutations are: (1) chromosome rearrangements, which are changes in the structure of chromosomes; (2) aneuploidy, which is an increase or decrease in chromosome number; and (3) polyploidy, which is the presence of extra chromosome sets.
- Chromosome rearrangements include duplications, deletions, inversions, and translocations.
- In individuals heterozygous for a duplication, the duplicated region will form a loop when homologous chromosomes pair in meiosis. Duplications often have pronounced effects on the phenotype owing to unbalanced gene dosage.
- In individuals heterozygous for a deletion, one of the chromosomes will loop out during pairing in meiosis. Deletions may cause recessive alleles to be expressed.
- Pericentric inversions include the centromere; paracentric inversions do not. In individuals heterozygous for an inversion, the homologous chromosomes form inversion loops in meiosis, with reduced recombination taking place within the inverted region.
- In translocation heterozygotes, the chromosomes form crosslike structures in meiosis.

- Fragile sites are constrictions or gaps that appear at particular regions on the chromosomes of cells grown in culture and are prone to breakage under certain conditions.
- Nullisomy is the loss of two homologous chromosomes; monosomy is the loss of one homologous chromosome; trisomy is the addition of one homologous chromosome; tetrasomy is the addition of two homologous chromosomes.
- Aneuploidy usually causes drastic phenotypic effects because it leads to unbalanced gene dosage.
- Primary Down syndrome is caused by the presence of three full copies of chromosome 21, whereas familial Down syndrome is caused by the presence of two normal copies of chromosome 21 and a third copy that is attached to another chromosome through a translocation.
- All the chromosomes in an autopolyploid derive from one species; chromosomes in an allopolyploid come from two or more species.
- Chromosome variations have played an important role in the evolution of many groups of organisms.

Important Terms

chromosome mutation (p. 168)
metacentric chromosome (p. 168)
submetacentric chromosome (p. 168)
acrocentric chromosome (p. 168)
telocentric chromosome (p. 168)
chromosome rearrangement (p. 170)
chromosome duplication (p. 170)
tandem duplication (p. 171)
displaced duplication (p. 171)
reverse duplication (p. 171)
chromosome deletion (p. 173)
pseudodominance (p. 174)
haploinsufficient gene (p. 174)
chromosome inversion (p. 174)
paracentric inversion (p. 174)

pericentric inversion (p. 174)
position effect (p. 174)
dicentric chromatid (p. 175)
acentric chromatid (p. 175)
dicentric bridge (p. 175)
translocation (p. 176)
nonreciprocal translocation (p. 176)
reciprocal translocation (p. 176)
Robertsonian translocation (p. 176)
fragile site (p. 178)
aneuploidy (p. 178)
polyploidy (p. 178)
nondisjunction (p. 178)
nullisomy (p. 178)
monosomy (p. 178)

trisomy (p. 178)
tetrasomy (p. 178)
Down syndrome (trisomy 21) (p. 180)
primary Down syndrome (p. 180)
familial Down syndrome (p. 180)
translocation carrier (p. 181)
Edward syndrome (trisomy 18) (p. 181)
Patau syndrome (trisomy 13) (p. 181)
trisomy 8 (p. 181)
autopolyploidy (p. 182)
allopolyploidy (p. 182)
unbalanced gametes (p. 183)
amphidiploid (p. 184)

Answers to Concept Checks

1. a

2. Pseudodominance is the expression of a recessive mutation. It is produced when the wild-type allele in a heterozygous individual is absent due to a deletion on one chromosome.

3. c

4. b

5. 37

6. Dosage compensation prevents the expression of additional copies of X-linked genes in mammals, and there is little information on the Y chromosome; so extra copies of the X and Y chromosomes do not have major effects on development. In contrast, there is no mechanism of dosage compensation for autosomes, and so extra copies of autosomal genes are expressed, upsetting development and causing the spontaneous abortion of aneuploid embryos.

7. c

Worked Problems

1. A chromosome has the following segments, where • represents the centromere.

$$\text{A B C D E • F G}$$

What types of chromosome mutations are required to change this chromosome into each of the following chromosomes? (In some cases, more than one chromosome mutation may be required.)

a. A B E • F G

b. A E D C B • F G

c. A B A B C D E • F G

d. A F • E D C B G

e. A B C D E E D C • F G

• Solution

The types of chromosome mutations are identified by comparing the mutated chromosome with the original, wild-type chromosome.

a. The mutated chromosome (A B E • F G) is missing segment C D; so this mutation is a deletion.

b. The mutated chromosome (A E D C B • F G) has one and only one copy of all the gene segments, but segment B C D E has been inverted 180 degrees. Because the centromere has not changed location and is not in the inverted region, this chromosome mutation is a paracentric inversion.

c. The mutated chromosome (A B A B C D E • F G) is longer than normal, and we see that segment A B has been duplicated. This mutation is a tandem duplication.

d. The mutated chromosome (A F • E D C B G) is normal length, but the gene order and the location of the centromere have changed; this mutation is therefore a pericentric inversion of region (B C D E • F).

e. The mutated chromosome (A B C D E E D C • F G) contains a duplication (C D E) that is also inverted; so this chromosome has undergone a duplication and a paracentric inversion.

2. Species I is diploid ($2n = 4$) with chromosomes AABB; related species II is diploid ($2n = 6$) with chromosomes MMNNOO. Give the chromosomes that would be found in individuals with the following chromosome mutations.

a. Autotriploidy in species I

b. Allotetraploidy including species I and II

c. Monosomy in species I

d. Trisomy in species II for chromosome M

e. Tetrasomy in species I for chromosome A

f. Allotriploidy including species I and II

g. Nullisomy in species II for chromosome N

• Solution

To work this problem, we should first determine the haploid genome complement for each species. For species I, $n = 2$ with chromosomes AB and, for species II, $n = 3$ with chromosomes MNO.

a. An autotriploid is $3n$, with all the chromosomes coming from a single species; so an autotriploid of species I would have chromosomes AAABBB ($3n = 6$).

b. An allotetraploid is $4n$, with the chromosomes coming from more than one species. An allotetraploid could consist of $2n$ from species I and $2n$ from species II, giving the allotetraploid ($4n = 2 + 2 + 3 + 3 = 10$) chromosomes AABBMMNNOO. An allotetraploid could also possess $3n$ from species I and $1n$ from species II ($4n = 2 + 2 + 2 + 3 = 9$; AAABBBMNO) or $1n$ from species I and $3n$ from species II ($4n = 2 + 3 + 3 + 3 = 11$; ABMMMNNNOOO).

c. A monosomic is missing a single chromosome; so a monosomic for species I would be $2n - 1 = 4 - 1 = 3$. The monosomy might include either of the two chromosome pairs, with chromosomes ABB or AAB.

d. Trisomy requires an extra chromosome; so a trisomic of species II for chromosome M would be $2n + 1 = 6 + 1 = 7$ (MMMNNOO).

e. A tetrasomic has two extra homologous chromosomes; so a tetrasomic of species I for chromosome A would be $2n + 2 = 4 + 2 = 6$ (AAAABB).

f. An allotriploid is $3n$ with the chromosomes coming from two different species; so an allotriploid could be $3n = 2 + 2 + 3 = 7$ (AABBMNO) or $3n = 2 + 3 + 3 = 8$ (ABMMNNOO).

g. A nullisomic is missing both chromosomes of a homologous pair; so a nullisomic of species II for chromosome N would be $2n - 2 = 6 - 2 = 4$ (MMOO).

Comprehension Questions

Section 7.1

*1. List the different types of chromosome mutations and define each one.

Section 7.2

*2. Why do extra copies of genes sometimes cause drastic phenotypic effects?

3. Draw a pair of chromosomes as they would appear during synapsis in prophase I of meiosis in an individual heterozygous for a chromosome duplication.

*4. What is the difference between a paracentric and a pericentric inversion?

5. How do inversions cause phenotypic effects?

6. Explain why recombination is suppressed in individuals heterozygous for paracentric inversions.

*7. How do translocations produce phenotypic effects?

Section 7.3

8. List four major types of aneuploidy.

*9. What is the difference between primary Down syndrome and familial Down syndrome? How does each type arise?

Section 7.4

*10. What is the difference between autopolyploidy and allopolyploidy? How does each arise?

11. Explain why autopolyploids are usually sterile, whereas allopolyploids are often fertile.

Application Questions and Problems

Section 7.1

*12. Which types of chromosome mutations
   a. increase the amount of genetic material in a particular chromosome?
   b. increase the amount of genetic material in all chromosomes?
   c. decrease the amount of genetic material in a particular chromosome?
   d. change the position of DNA sequences in a single chromosome without changing the amount of genetic material?
   e. move DNA from one chromosome to a nonhomologous chromosome?

Section 7.2

*13. A chromosome has the following segments, where • represents the centromere:

$$A B • C D E F G$$

What types of chromosome mutations are required to change this chromosome into each of the following chromosomes? (In some cases, more than one chromosome mutation may be required.)
   a. A B A B • C D E F G
   b. A B • C D E A B F G
   c. A B • C F E D G
   d. A • C D E F G
   e. A B • C D E
   f. A B • E D C F G
   g. C • B A D E F G
   h. A B • C F E D F E D G
   i. A B • C D E F C D F E G

14. The following diagram represents two nonhomologous chromosomes:

$$A B • C D E F G$$
$$R S • T U V W X$$

What type of chromosome mutation would produce each of the following chromosomes?
   a. A B • C D
      R S • T U V W X E F G
   b. A U V B • C D E F G
      R S • T W X
   c. A B • T U V F G
      R S • C D E W X
   d. A B • C W G
      R S • T U V D E F X

15. The green-nose fly normally has six chromosomes, two metacentric and four acrocentric. A geneticist examines the chromosomes of an odd-looking green-nose fly and discovers that it has only five chromosomes; three of them are metacentric and two are acrocentric. Explain how this change in chromosome number might have taken place.

*16. A wild-type chromosome has the following segments:

$$A B C • D E F G H I$$

An individual is heterozygous for the following chromosome mutations. For each mutation, sketch how the wild-type and mutated chromosomes would pair in prophase I of meiosis, showing all chromosome strands.
   a. A B C • D E F D E F G H I
   b. A B C • D H I
   c. A B C • D G F E H I
   d. A B E D • C F G H I

17. As discussed in this chapter, crossing over within a pericentric inversion produces chromosomes that have extra copies of some genes and no copies of other genes. The fertilization of gametes containing such duplication/ deficient chromosomes often results in children with syndromes characterized by developmental delay, mental retardation, abnormal development of organ systems, and early death. Using a special two-color FISH (fluorescence in

situ hybridization) analysis that revealed the presence of crossing over within pericentric inversions, Maarit Jaarola and colleagues examined individual sperm cells of a male who was heterozygous for a pericentric inversion on chromosome 8 and determined that crossing over took place within the pericentric inversion in 26% of the meiotic divisions (M. Jaarola, R. H. Martin, and T. Ashley. 1998. *American Journal of Human Genetics* 63:218–224).

Assume that you are a genetic counselor and that a couple seeks genetic counseling from you. Both the man and the woman are phenotypically normal, but the woman is heterozygous for a pericentric inversion on chromosome 8. The man is karyotypically normal. What is the probability that this couple will produce a child with a debilitating syndrome as the result of crossing over within the pericentric inversion?

Section 7.3

18. Red–green color blindness is a human X-linked recessive disorder. A young man with a 47,XXY karyotype (Klinefelter syndrome) is color blind. His 46,XY brother also is color blind. Both parents have normal color vision. Where did the nondisjunction occur that gave rise to the young man with Klinefelter syndrome?

*19. Bill and Betty have had two children with Down syndrome. Bill's brother has Down syndrome and his sister has two children with Down syndrome. On the basis of these observations, which of the following statements is most likely correct? Explain your reasoning.

   a. Bill has 47 chromosomes.

   b. Betty has 47 chromosomes.

   c. Bill and Betty's children each have 47 chromosomes.

   d. Bill's sister has 45 chromosomes.

   e. Bill has 46 chromosomes.

   f. Betty has 45 chromosomes.

   g. Bill's brother has 45 chromosomes.

*20. In mammals, sex-chromosome aneuploids are more common than autosomal aneuploids but, in fish, sex-chromosome aneuploids and autosomal aneuploids are found with equal frequency. Offer an explanation for these differences in mammals and fish.

Section 7.4

*21. Species I has $2n = 16$ chromosomes. How many chromosomes will be found per cell in each of the following mutants in this species?

   a. Monosomic

   b. Autotriploid

   c. Autotetraploid

   d. Trisomic

   e. Double monosomic

   f. Nullisomic

   g. Autopentaploid

   h. Tetrasomic

22. Species I is diploid ($2n = 8$) with chromosomes AABBCCDD; related species II is diploid ($2n = 8$) with chromosomes MMNNOOPP. What types of chromosome mutations do individual organisms with the following sets of chromosomes have?

   a. AAABBCCDD

   b. MMNNOOOOPP

   c. AABBCDD

   d. AAABBBCCCDDD

   e. AAABBCCDDD

   f. AABBDD

   g. AABBCCDDMMNNOOPP

   h. AABBCCDDMNOP

23. Species I has $2n = 8$ chromosomes and species II has $2n = 14$ chromosomes. What would be the expected chromosome numbers in individual organisms with the following chromosome mutations? Give all possible answers.

   a. Allotriploidy including species I and II

   b. Autotetraploidy in species II

   c. Trisomy in species I

   d. Monosomy in species II

   e. Tetrasomy in species I

   f. Allotetraploidy including species I and II

24. *Nicotiana glutinosa* ($2n = 24$) and *N. tabacum* ($2n = 48$) are two closely related plants that can be intercrossed, but the F1 hybrid plants that result are usually sterile. In 1925, Roy Clausen and Thomas Goodspeed crossed *N. glutinosa* and *N. tabacum,* and obtained one fertile F_1 plant (R. E. Clausen and T. H. Goodspeed. 1925 *Genetics* 10:278–284). They were able to self-pollinate the flowers of this plant to produce an F_2 generation. Surprisingly, the F_2 plants were fully fertile and produced viable seed. When Clausen and Goodspeed examined the chromosomes of the F_2 plants, they observed 36 pairs of chromosomes in metaphase I and 36 individual chromosomes in metaphase II. Explain the origin of the F_2 plants obtained by Clausen and Goodspeed and the numbers of chromosomes observed.

25. What would be the chromosome number of progeny resulting from the following crosses in wheat (see Figure 7.25)? What type of polyploid (allotriploid, allotetraploid, etc.) would result from each cross?

   a. Einkorn wheat and Emmer wheat

   b. Bread wheat and Emmer wheat

   c. Einkorn wheat and bread wheat

26. Karl and Hally Sax crossed *Aegilops cylindrical* ($2n = 28$), a wild grass found in the Mediterranean region, with *Triticum vulgare* ($2n = 42$), a type of wheat (K. Sax and H. J. Sax. 1924. *Genetics* 9:454–464). The resulting F_1 plants from this cross had 35 chromosomes. Examination of metaphase I in the F_1 plants revealed the presence of 7 pairs of chromosomes (bivalents) and 21 unpaired chromosomes (univalents).

   a. If the unpaired chromosomes segregate randomly, what possible chromosome numbers will appear in the gametes of the F_1 plants?

   b. What does the appearance of the bivalents in the F_1 hybrids suggest about the origin of *Triticum vulgare* wheat?

Challenge Questions

Section 7.3

27. Red–green color blindness is a human X-linked recessive disorder. Jill has normal color vision, but her father is color blind. Jill marries Tom, who also has normal color vision.

    Jill and Tom have a daughter who has Turner syndrome and is color blind.

    a. How did the daughter inherit color blindness?

    b. Did the daughter inherit her X chromosome from Jill or from Tom?

28. Mules result from a cross between a horse ($2n = 64$) and a donkey ($2n = 62$), have 63 chromosomes and are almost always sterile. However, in the summer of 1985, a female mule named Krause who was pastured with a male donkey gave birth to a newborn foal (O. A. Ryder et al. 1985. *Journal of Heredity* 76:379–381). Blood tests established that the male foal, appropriately named Blue Moon, was the offspring of Krause and that Krause was indeed a mule. Both Blue Moon and Krause were fathered by the same donkey (see the illustration). The foal, like his mother, had 63 chromosomes—half of them horse chromosomes and the other half donkey chromosomes. Analyses of genetic markers showed that, remarkably, Blue Moon seemed to have inherited a complete set of horse chromosomes from his mother, instead of a random mixture of horse and donkey chromosomes that would be expected with normal meiosis. Thus, Blue Moon and Krause were not only mother and son, but also brother and sister.

a. With the use of a diagram, show how, if Blue Moon inherited only horse chromosomes from his mother, Blue Moon and Krause are both mother and son as well as brother and sister.

b. Although rare, additional cases of fertile mules giving births to offspring have been reported. In these cases, when a female mule mates with a male horse, the offspring is horselike in appearance but, when a female mule mates with a male donkey, the offspring is mulelike in appearance. Is this observation consistent with the idea that the offspring of fertile female mules inherit only a set of horse chromosomes from their mule mothers? Explain your reasoning.

c. Can you suggest a possible mechanism for how the offspring of fertile female mules might pass on a complete set of horse chromosomes to their offspring?

Section 7.5

29. Humans and many other complex organisms are diploid, possessing two sets of genes, one inherited from the mother and one from the father. However, a number of eukaryotic organisms spend most of their life cycles in a haploid state. Many of these eukaryotes, such as *Neurospora* and yeast, still undergo meiosis and sexual reproduction, but most of the cells that make up the organism are haploid.

    Considering that haploid organisms are fully capable of sexual reproduction and generating genetic variation, why are most complex eukaryotes diploid? In other words, what might be the evolutionary advantage of existing in a diploid state instead of a haploid state? And why might a few organisms, such as *Neurospora* and yeast, exist as haploids?

8 DNA: The Chemical Nature of the Gene

The remarkable stability of DNA facilitates the extraction and analysis of DNA from ancient remains, including Neanderthal bones that are more than 30,000 years old. *[John Reader/Photo Researchers.]*

Neanderthal's DNA

DNA, with its double-stranded spiral, is among the most elegant of all biological molecules. But the double helix is not just a beautiful structure; it also gives DNA incredible stability and permanence, providing geneticists with a unique window to the past.

In 1856, a group of men working a limestone quarry in the Neander Valley of Germany discovered a small cave containing a number of bones. The workers assumed that the bones were those of a cave bear, but a local schoolteacher recognized them as human, although they were clearly unlike any human bones the teacher had ever seen. The bones appeared to be those of a large person with great muscular strength, a low forehead, a large nose with broad nostrils, and massive protruding brows. Experts confirmed that the bones belonged to an extinct human, who became known as Neanderthal. In the next 100 years, similar fossils were discovered in Spain, Belgium, France, Croatia, and the Middle East.

Research has now revealed that Neanderthals roamed Europe and western Asia for at least 200,000 years, disappearing abruptly 30,000 to 40,000 years ago. During the last years of this period, Neanderthals coexisted with the direct ancestors of modern humans, the Cro-Magnons. The fate of the Neanderthals—why they disappeared—has captured the imagination of scientists and laypersons alike. Did Cro-Magnons, migrating out of Africa with a superior technology, cause the demise of the Neanderthal people, either through competition or perhaps through deliberate extermination? Or did the Neanderthals interbreed with Cro-Magnons, their genes becoming assimilated into the larger gene pool of modern humans? Support for the latter hypothesis came from the discovery of fossils that appeared to be transitional between Neanderthals and Cro-Magnons. Unfortunately, the meager fossil record of Neanderthals and Cro-Magnons did not allow a definitive resolution of these questions.

Whether Neanderthals interbred with the ancestors of modern humans became a testable question in 1997, when scientists extracted DNA from a bone of the original Neanderthal specimen collected from the Neander Valley more than 140 years earlier. This Neanderthal bone was at least 30,000 years old. Using a technique called the polymerase chain reaction (see Chapter 14), the scientists amplified 378 nucleotides of the Neanderthal's mitochondrial DNA a millionfold. They then determined the base sequence of this amplified DNA, and compared it with mitochondrial sequences from living humans.

The mitochondrial DNA from the Neanderthal bone differed markedly from mitochondrial DNAs that were isolated from numerous humans throughout the world. DNA from many different Neanderthal and Cro-Magnon specimens has provided clear evidence that Neanderthals were genetically distinct from both present-day humans and ancient Cro-Magnons. There is no Neanderthal mitochondrial DNA in the present-day human gene pool or in DNA extracted from fossils of early Cro-Magnons, suggesting that interbreeding between the two forms did not take place. Recently, *nuclear* DNA has been extracted from Neanderthal bones, and scientists are currently attempting to sequence the entire Neanderthal genome, which will provide a more definitive test of possible interbreeding between early humans and Neanderthals.

That Neanderthal's DNA persists and reveals its genetic affinities, even 40,000 years later, is testimony to the remarkable stability of the double helix. This chapter focuses on how DNA was identified as the source of genetic information and how this elegant molecule encodes the genetic instructions. We begin by considering the basic requirements of the genetic material and the history of our understanding of DNA—how its relation to genes was uncovered and its structure determined. The history of DNA illustrates several important points about the nature of scientific research. As with so many important scientific advances, the structure of DNA and its role as the genetic material were not discovered by any single person but were gradually revealed over a period of almost 100 years, thanks to the work of many investigators. Our understanding of the relation between DNA and genes was enormously enhanced in 1953, when James Watson and Francis Crick proposed a three-dimensional structure for DNA that brilliantly illuminated its role in genetics. As illustrated by Watson and Crick's discovery, major scientific advances are often achieved, not through the collection of new data, but through the interpretation of old data in new ways.

After reviewing the history of DNA, we will examine DNA structure. The structure of DNA is important in its own right, but the key genetic concept is the relation between the structure and the function of DNA—how its structure allows it to serve as the genetic material.

8.1 Genetic Material Possesses Several Key Characteristics

Life is characterized by tremendous diversity, but the coding instructions of all living organisms are written in the same genetic language—that of nucleic acids. Surprisingly, the idea that genes are made of nucleic acids was not widely accepted until after 1950. This skepticism was due in part to a lack of knowledge about the structure of deoxyribonucleic acid (DNA). Until the structure of DNA was understood, how DNA could store and transmit genetic information was unclear. Even before nucleic acids were identified as the genetic material, biologists recognized that, whatever the nature of the genetic material, it must possess three important characteristics.

1. **Genetic material must contain complex information.** First and foremost, the genetic material must be capable of storing large amounts of information—instructions for all the traits and functions of an organism. This information must have the capacity to vary, because different species and even individual members of a species differ in their genetic makeup. At the same time, the genetic material must be stable, because most alterations to the genetic instructions (mutations) are likely to be detrimental.

2. **Genetic material must replicate faithfully.** A second necessary feature is that genetic material must have the capacity to be copied accurately. Every organism begins life as a single cell, which must undergo billions of cell divisions to produce a complex, multicellular creature like yourself. At each cell division, the genetic instructions must be transmitted to descendant cells with great accuracy. When organisms reproduce and pass genes on to their progeny, the coding instructions must be copied with fidelity.

3. **Genetic material must encode the phenotype.** The genetic material (the genotype) must have the capacity to "code for" (determine) traits (the phenotype). The product of a gene is often a protein; so there must be a mechanism for genetic instructions to be translated into the amino acid sequence of a protein.

Concepts

The genetic material must be capable of carrying large amounts of information, replicating faithfully, and translating its coding instructions into phenotypes.

✔ **Concept Check 1**

Why was the discovery of the structure of DNA so important for understanding genetics?

8.2 All Genetic Information Is Encoded in the Structure of DNA

Although our understanding of how DNA encodes genetic information is relatively recent, the study of DNA structure stretches back more than 100 years.

Early Studies of DNA

In 1868, Johann Friedrich Miescher graduated from medical school in Switzerland. Influenced by an uncle who believed that the key to understanding disease lay in the chemistry of tissues, Miescher traveled to Tübingen, Germany, to study under Ernst Felix Hoppe-Seyler, an early leader in the emerging field of biochemistry. Under Hoppe-Seyler's direction, Miescher turned his attention to the chemistry of pus, a substance of clear medical importance. Pus contains white blood cells with large nuclei. Miescher determined that the nucleus contained a novel substance that was slightly acidic and high in phosphorus. This material consisted of DNA and protein.

By 1887, researchers had concluded that the physical basis of heredity lies in the nucleus. Chromatin was shown to consist of nucleic acid and proteins, but which of these substances is actually the genetic information was not clear. In the late 1800s, further work on the chemistry of DNA was carried out by Albrecht Kossel, who determined that DNA contains four nitrogenous bases: adenine, cytosine, guanine, and thymine (abbreviated A, C, G, and T).

Phoebus Aaron Levene later showed that DNA consists of a large number of linked, repeating units called **nucleotides**; each nucleotide contains a sugar, a phosphate, and a base.

Nucleotide

Levene incorrectly proposed that DNA consists of a series of four-nucleotide units, each unit containing all four bases—

Source of DNA	A	T	G	C	Ratio A/T	Ratio G/C	Ratio (A + G)/(T + C)
E. coli	26.0	23.9	24.9	25.2	1.09	0.99	1.04
Yeast	31.3	32.9	18.7	17.1	0.95	1.09	1.00
Sea urchin	32.8	32.1	17.7	18.4	1.02	0.96	1.00
Rat	28.6	28.4	21.4	21.5	1.01	1.00	1.00
Human	30.3	30.3	19.5	19.9	1.00	0.98	0.99

Table 8.1   Base composition of DNA from different sources and ratios of bases

adenine, guanine, cytosine, and thymine—in a fixed sequence. This concept, known as the tetranucleotide theory, implied that the structure of DNA is not variable enough to be the genetic material. The tetranucleotide theory contributed to the idea that protein is the genetic material because, with its 20 different amino acids, protein structure could be highly variable.

As additional studies of the chemistry of DNA were completed in the 1940s and 1950s, this notion of DNA as a simple, invariant molecule began to change. Erwin Chargaff and his colleagues carefully measured the amounts of the four bases in DNA from a variety of organisms and found that DNA from different organisms varies greatly in base composition. This finding disproved the tetranucleotide theory. They discovered that, within each species, there is some regularity in the ratios of the bases: the amount of adenine is always equal to the amount of thymine (A = T), and the amount of guanine is always equal to the amount of cytosine (G = C; **Table 8.1**). These findings became known as **Chargaff's rules**.

Concepts

Details of the structure of DNA were worked out by a number of scientists. At first, DNA was interpreted as being too regular in structure to carry genetic information but, by the 1940s, DNA from different organisms was shown to vary in its base composition.

DNA As the Source of Genetic Information

While chemists were working out the structure of DNA, biologists were attempting to identify the source of genetic information. Mendel identified the basic rules of heredity in 1866, but he had no idea about the physical nature of hereditary information. By the early 1900s, biologists had concluded that genes reside on chromosomes, which were known to contain both DNA and protein. Two sets of experiments, one conducted on bacteria and the other on viruses, provided pivotal evidence that DNA, rather than protein, was the genetic material.

The discovery of the transforming principle   The first clue that DNA was the carrier of hereditary information came with the demonstration that DNA was responsible for a phenomenon called *transformation*. The phenomenon was first observed in 1928 by Fred Griffith, an English physician whose special interest was the bacterium that causes pneumonia, *Streptococcus pneumoniae*. Griffith had succeeded in isolating several different strains of *S. pneumoniae* (type I, II, III, and so forth). In the virulent (disease-causing) forms of a strain, each bacterium is surrounded by a polysaccharide coat, which makes the bacterial colony appear smooth when grown on an agar plate; these forms are referred to as S, for

smooth. Griffith found that these virulent forms occasionally mutated to nonvirulent forms, which lack a polysaccharide coat and produce a rough-appearing colony; these forms are referred to as R, for rough.

Griffith observed that small amounts of living type IIIS bacteria injected into mice caused the mice to develop pneumonia and die; on autopsy, he found large amounts of type IIIS bacteria in the blood of the mice (**Figure 8.1a**). When Griffith injected type IIR bacteria into mice, the mice lived, and no bacteria were recovered from their blood (**Figure 8.1b**). Griffith knew that boiling killed all the bacteria and destroyed their virulence; when he injected large amounts of heat-killed type IIIS bacteria into mice, the mice lived and no type IIIS bacteria were recovered from their blood (**Figure 8.1c**).

The results of these experiments were not unusual. However, Griffith got a surprise when he infected his mice with a small amount of living type IIR bacteria along with a large amount of heat-killed type IIIS bacteria. Because both the type IIR bacteria and the heat-killed type IIIS bacteria were nonvirulent, he expected these mice to live. Surprisingly, 5 days after the injections, the mice became infected with pneumonia and died (**Figure 8.1d**). When Griffith examined blood from the hearts of these mice, he observed live type IIIS bacteria. Furthermore, these bacteria retained their type IIIS characteristics through several generations; so the infectivity was heritable.

Griffith finally concluded that the type IIR bacteria had somehow been transformed, acquiring the genetic virulence of the dead type IIIS bacteria. This transformation had produced a permanent, genetic change in the bacteria. Although Griffith didn't understand the nature of transformation, he theorized that some substance in the polysaccharide coat of the dead bacteria might be responsible. He called this substance the **transforming principle**.

Identification of the transforming principle   At the time of Griffith's report, Oswald Avery was a microbiologist at the Rockefeller Institute. At first Avery was skeptical but, after other microbiologists successfully repeated Griffith's experiments with other bacteria, Avery set out to identify the nature of the transforming substance.

After 10 years of research, Avery, Colin MacLeod, and Maclyn McCarty succeeded in isolating and purifying the transforming substance. They showed that it had a chemical composition closely matching that of DNA and quite different from that of proteins. Enzymes such as trypsin and chymotrypsin, known to break down proteins, had no effect on the transforming substance. Ribonuclease, an enzyme that destroys RNA, also had no effect. Enzymes capable of destroying DNA, however, eliminated the biological activity of the transforming substance (**Figure 8.2**). Avery, MacLeod, and McCarty showed that purified transforming substance precipitated at about the same rate as purified DNA and that it absorbed ultraviolet light at the same wavelengths as DNA. These results, published in 1944, provided compelling

Experiment

Question: Can an extract from dead bacterial cells genetically transform living cells?

Methods

(a)   (b)   (c)   (d)

| Type IIIS (virulent) bacteria are injected into a mouse. | Type IIR (nonvirulent) bacteria are injected into a mouse. | Heat-killed type IIIS bacteria are injected into a mouse. | A mixture of type IIR bacteria and heat-killed type IIIS bacteria are injected into a mouse. |

Results

Mouse dies   Mouse lives   Mouse lives   Mouse dies

Autopsy

| Type IIIS (virulent) bacteria recovered | No bacteria recovered | No bacteria recovered | Type IIIS (virulent) bacteria recovered |

Conclusion: A substance in the heat-killed virulent bacteria genetically transformed the type IIR bacteria into live, virulent type IIIS bacteria.

8.1 Griffith's experiments demonstrated transformation in bacteria.

Experiment

Question: What is the chemical nature of the transforming substance?

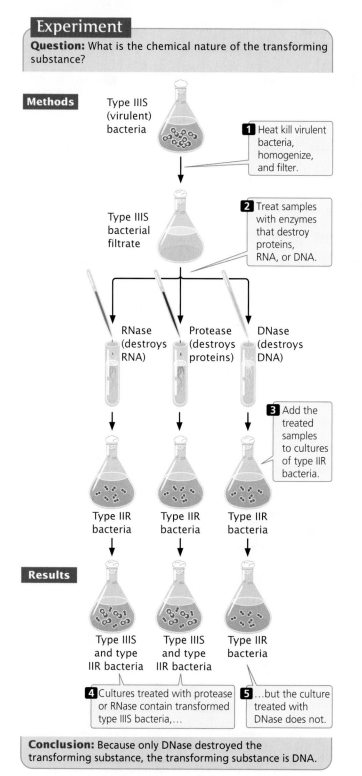

Methods

Type IIIS (virulent) bacteria

1 Heat kill virulent bacteria, homogenize, and filter.

Type IIIS bacterial filtrate

2 Treat samples with enzymes that destroy proteins, RNA, or DNA.

RNase (destroys RNA)

Protease (destroys proteins)

DNase (destroys DNA)

3 Add the treated samples to cultures of type IIR bacteria.

Type IIR bacteria

Type IIR bacteria

Type IIR bacteria

Results

Type IIIS and type IIR bacteria

Type IIIS and type IIR bacteria

Type IIR bacteria

4 Cultures treated with protease or RNase contain transformed type IIIS bacteria,...

5 ...but the culture treated with DNase does not.

Conclusion: Because only DNase destroyed the transforming substance, the transforming substance is DNA.

8.2 Avery, MacLeod, and McCarty's experiment revealed the nature of the transforming principle.

evidence that the transforming principle—and therefore genetic information—resides in DNA. Many biologists refused to accept the idea, however, still preferring the hypothesis that the genetic material is protein.

Concepts

The process of transformation indicates that some substance—the transforming principle—is capable of genetically altering bacteria. Avery, MacLeod, and McCarty demonstrated that the transforming principle is DNA, providing the first evidence that DNA is the genetic material.

✔ Concept Check 2

If Avery, MacLeod, and McCarty had found that samples of heat-killed bacteria treated with RNase and DNase transformed bacteria but that samples treated with protease did not, what conclusion would they have made?

a. Protease carries out transformation.

b. RNA and DNA are the genetic materials.

c. Protein is the genetic material.

d. RNase and DNase are necessary for transformation.

The Hershey–Chase experiment   A second piece of evidence implicating DNA as the genetic material resulted from a study of the T2 virus conducted by Alfred Hershey and Martha Chase. The T2 virus is a *bacteriophage* (phage) that infects the bacterium *Escherichia coli* (**Figure 8.3a**). As stated in Chapter 6, a phage reproduces by attaching to the outer wall of a bacterial cell and injecting its DNA into the cell, where it replicates and directs the cell to synthesize phage protein. The phage DNA becomes encapsulated within the proteins, producing progeny phages that lyse (break open) the cell and escape (**Figure 8.3b**).

At the time of the Hershey–Chase study (their paper was published in 1952), biologists did not understand exactly how phages reproduce. What they did know was that the T2 phage is approximately 50% protein and 50% nucleic acid, that a phage infects a cell by first attaching to the cell wall, and that progeny phages are ultimately produced within the cell. Because the progeny carry the same traits as the infecting phage, genetic material from the infecting phage must be transmitted to the progeny, but how this takes place was unknown.

Hershey and Chase designed a series of experiments to determine whether the phage *protein* or the phage *DNA* is transmitted in phage reproduction. To follow the fate of protein and DNA, they used radioactive forms, or **isotopes**, of phosphorus and sulfur. A radioactive isotope can be used as a tracer to identify the location of a specific molecule, because any molecule containing the isotope will be radioactive and therefore easily detected. DNA contains phosphorus but not sulfur; so Hershey and Chase used ^{32}P to follow phage DNA during reproduction. Protein contains sulfur but not phosphorus; so they used ^{35}S to follow the protein.

Hershey and Chase grew one batch of *E. coli* in a medium containing ^{32}P and infected the bacteria with T2 phage so that all the new phages would have DNA labeled with ^{32}P (**Figure 8.4**). They grew a second batch of *E. coli* in

(a)

Phage genome is DNA.

All other parts of the bacteriophage are protein.

(b)

Phage

E. coli

Phage chromosome

Bacterial chromosome

1 Phage attaches to E. coli and injects its chromosome.

2 Bacterial chromosome breaks down and the phage chromosome replicates.

3 Expression of phage genes produces phage structural components.

4 Progeny phage particles assemble.

5 Bacterial wall lyses, releasing progeny phages.

Experiment

Question: Which part of the phage—its DNA or its protein—serves as the genetic material and is transmitted to phage progeny?

Methods   Protein — DNA

T2 phage   E. coli

1 Infect E. coli grown in medium containing ^{35}S.

1 Infect E. coli grown in medium containing ^{32}P.

^{35}S

^{32}P

2 ^{35}S is taken up in phage protein, which contains S but not P.

2 ^{32}P is taken up in phage DNA, which contains P but not S.

3 Phages with ^{35}S infect unlabeled E. coli.

3 Phages with ^{32}P infect unlabeled E. coli.

4 Shear off protein coats in blender…

5 …and separate protein from cells by centrifuging.

Results

^{35}S

6 After centrifugation, ^{35}S is recovered in the fluid containing the virus coats.

6 After centrifugation, infected bacteria form a pellet containing ^{32}P in the bottom of the tube.

^{32}P

^{32}P

Phage reproduction

7 No radioactivity is detected, indicating that protein has not been transmitted to the progeny phages.

7 The progeny phages are radioactive, indicating that DNA has been transmitted to progeny phages.

Conclusion: DNA—not protein—is the genetic material in bacteriophages.

8.3 T2 is a bacteriophage that infects E. coli. (a) T2 phage. (b) Its life cycle. [Part a: © Lee D. Simon/Photo Researchers.]

8.4 Hershey and Chase demonstrated that DNA carries the genetic information in bacteriophages.

a medium containing ^{35}S and infected these bacteria with T2 phage so that all these new phages would have protein labeled with ^{35}S. Hershey and Chase then infected separate batches of unlabeled *E. coli* with the ^{35}S- and ^{32}P-labeled phages. After allowing time for the phages to infect the cells, they placed the *E. coli* cells in a blender and sheared off the then-empty protein coats (ghosts) from the cell walls. They separated out the protein coats and cultured the infected bacterial cells.

When phages labeled with ^{35}S infected the bacteria, most of the radioactivity was detected in the protein ghosts and little was detected in the cells. Furthermore, when new phages emerged from the cell, they contained almost no ^{35}S (see Figure 8.4). This result indicated that, although the protein component of a phage is necessary for infection, it does not enter the cell and is not transmitted to progeny phages.

In contrast, when Hershey and Chase infected bacteria with ^{32}P-labeled phages and removed the protein ghosts, the bacteria were still radioactive. Most significantly, after the cells lysed and new progeny phages emerged, many of these phages emitted radioactivity from ^{32}P, demonstrating that DNA from the infecting phages had been passed on to the progeny (see Figure 8.4). These results confirmed that DNA, not protein, is the genetic material of phages.

Concepts

Using radioactive isotopes, Hershey and Chase traced the movement of DNA and protein during phage infection. They demonstrated that DNA, not protein, enters the bacterial cell during phage reproduction and that only DNA is passed on to progeny phages.

✔ Concept Check 3

Could Hershey and Chase have used a radioactive isotope of carbon instead of ^{32}P? Why or why not?

Watson and Crick's Discovery of the Three-Dimensional Structure of DNA

The experiments on the nature of the genetic material set the stage for one of the most important advances in the history of biology—the discovery of the three-dimensional structure of DNA by James Watson and Francis Crick in 1953. Before Watson and Crick's breakthrough, much of the basic chemistry of DNA had already been determined by Miescher, Kossel, Levene, Chargaff, and others, who had established that DNA consists of nucleotides and that each nucleotide contains a sugar, a base, and a phosphate group. However, how the nucleotides fit together in the three-dimensional structure of the molecule was not at all clear.

In 1947, William Ashbury began studying the three-dimensional structure of DNA by using a technique called **X-ray diffraction** (**Figure 8.5**), in which X-rays beamed at a molecule are reflected in specific patterns that reveal aspects of the structure of the molecule. But his diffraction pictures did not provide enough resolution to reveal the structure. A research group at King's College in London, led by Maurice Wilkins and Rosalind Franklin, also used X-ray diffraction to study DNA and obtained strikingly better pictures of the molecule. Wilkins and Franklin, however, were unable to develop a complete structure of the molecule; their progress was impeded by the personal discord that existed between them.

Watson and Crick investigated the structure of DNA, not by collecting new data but by using all available information about the chemistry of DNA to construct molecular models (**Figure 8.6**). By applying the laws of structural chemistry, they were able to limit the number of possible structures that DNA could assume. They tested various structures by building models made of wire and metal plates. With their models, they were able to see whether a structure was compatible with chemical principles and with the X-ray images.

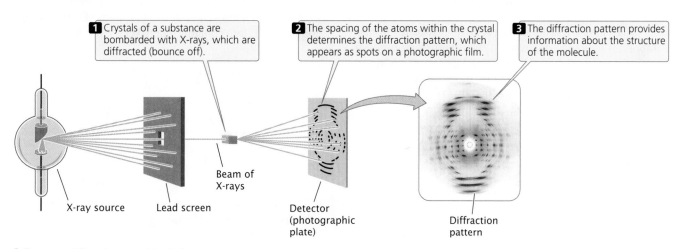

1 Crystals of a substance are bombarded with X-rays, which are diffracted (bounce off).

2 The spacing of the atoms within the crystal determines the diffraction pattern, which appears as spots on a photographic film.

3 The diffraction pattern provides information about the structure of the molecule.

X-ray source   Lead screen   Beam of X-rays   Detector (photographic plate)   Diffraction pattern

8.5 X-ray diffraction provides information about the structures of molecules. [*Photograph from M. H. F. Wilkins, Department of Biophysics, King's College, University of London.*]

8.6 James Watson (left) and Francis Crick (right) provided a three-dimensional model of the structure of DNA. [A. Barrington Brown/Science Photo Library/Photo Researchers.]

The key to solving the structure came when Watson recognized that an adenine base could bond with a thymine base and that a guanine base could bond with a cytosine base; these pairings accounted for the base ratios that Chargaff had discovered earlier. The model developed by Watson and Crick showed that DNA consists of two strands of nucleotides wound around each other to form a right-handed helix, with the sugars and phosphates on the outside and the bases in the interior. They published an electrifying description of their model in *Nature* in 1953. At the same time, Wilkins and Franklin published their X-ray diffraction data, which demonstrated experimentally the theory that DNA was helical in structure.

Many have called the solving of DNA's structure the most important biological discovery of the twentieth century. For their discovery, Watson and Crick, along with Maurice Wilkins, were awarded a Nobel Prize in 1962. Rosalind Franklin had died of cancer in 1957 and thus could not be considered a candidate for the shared prize.

Concepts

By collecting existing information about the chemistry of DNA and building molecular models, Watson and Crick were able to discover the three-dimensional structure of the DNA molecule.

8.3 DNA Consists of Two Complementary and Antiparallel Nucleotide Strands That Form a Double Helix

DNA, though relatively simple in structure, has an elegance and beauty unsurpassed by other large molecules. It is useful to consider the structure of DNA at three levels of increasing complexity, known as the primary, secondary, and tertiary structures of DNA. The primary structure of DNA refers to its nucleotide structure and how the nucleotides are joined together. The secondary structure refers to DNA's stable three-dimensional configuration, the helical structure worked out by Watson and Crick. Later, we will consider DNA's tertiary structures, which are the complex packing arrangements of double-stranded DNA in chromosomes.

The Primary Structure of DNA

The primary structure of DNA consists of a string of nucleotides joined together by phosphodiester linkages.

Nucleotides   DNA is typically a very long molecule and is therefore termed a macromolecule. For example, within each human chromosome is a single DNA molecule that, if stretched out straight, would be several centimeters in length. In spite of its large size, DNA has a quite simple structure: it is a polymer—that is, a chain made up of many repeating units linked together. The repeating units of DNA are *nucleotides*, each comprising three parts: (1) a sugar, (2) a phosphate, and (3) a nitrogen-containing base.

The sugars of nucleic acids—called pentose sugars—have five carbon atoms, numbered $1'$, $2'$, $3'$, and so forth (**Figure 8.7**) The sugars of DNA and RNA are slightly different in structure. RNA's sugar, called **ribose**, has a hydroxyl group (–OH) attached to the $2'$-carbon atom, whereas DNA's sugar, or **deoxyribose**, has a hydrogen atom (–H) at this position and therefore contains one oxygen atom fewer overall.

Ribose   **Deoxyribose**

8.7 A nucleotide contains either a ribose sugar (in RNA) or a deoxyribose sugar (in DNA). The carbon atoms are assigned primed numbers.

Purine
(basic structure)

Pyrimidine
(basic structure)

Adenine (A)   **Guanine (G)**

Cytosine (C)   **Thymine (T)**   **Uracil (U)**
(present in DNA)   (present in RNA)

8.8 A nucleotide contains either a purine or a pyrimidine base. The atoms of the rings in the bases are assigned unprimed numbers.

This difference gives rise to the names ribonucleic acid (RNA) and *deoxy*ribonucleic acid (DNA). This minor chemical difference is recognized by all the cellular enzymes that interact with DNA or RNA, thus yielding specific functions for each nucleic acid. Furthermore, the additional oxygen atom in the RNA nucleotide makes it more reactive and less chemically stable than DNA. For this reason, DNA is better suited to serve as the long-term repository of genetic information.

The second component of a nucleotide is its **nitrogenous base**, which may be of two types—a **purine** or a **pyrimidine** (**Figure 8.8**). Each purine consists of a six-sided ring attached to a five-sided ring, whereas each pyrimidine consists of a six-sided ring only. Both DNA and RNA contain two purines, **adenine** and **guanine** (A and G), which differ in the positions of their double bonds and in the groups attached to the six-sided ring. Three pyrimidines are common in nucleic acids: **cytosine** (C), **thymine** (T), and **uracil** (U). Cytosine is present in both DNA and RNA; however, thymine is restricted to DNA, and uracil is found only in RNA. The three pyrimidines differ in the groups or atoms attached to the carbon atoms of the ring and in the number

of double bonds in the ring. In a nucleotide, the nitrogenous base always forms a covalent bond with the 1′-carbon atom of the sugar (see Figure 8.7). A deoxyribose or a ribose sugar and a base together are referred to as a **nucleoside**.

The third component of a nucleotide is the **phosphate group**, which consists of a phosphorus atom bonded to four oxygen atoms (**Figure 8.9**). Phosphate groups are found in every nucleotide and frequently carry a negative charge, which makes DNA acidic. The phosphate group is always bonded to the 5′-carbon atom of the sugar (see Figure 8.7) in a nucleotide.

The DNA nucleotides are properly known as **deoxyribonucleotides** or deoxyribonucleoside 5′-monophosphates. Because there are four types of bases, there are four different kinds of DNA nucleotides (**Figure 8.10**). The equivalent RNA nucleotides are termed **ribonucleotides** or ribonucleoside 5′-monophosphates. RNA molecules sometimes contain additional rare bases, which are modified forms of the four common bases. These modified bases will be discussed in more detail when we examine the function of RNA molecules in Chapter 10.

Phosphate

8.9 A nucleotide contains a phosphate group.

Concepts

The primary structure of DNA consists of a string of nucleotides. Each nucleotide consists of a five-carbon sugar, a phosphate, and a base. There are two types of DNA bases: purines (adenine and guanine) and pyrimidines (thymine and cytosine).

Deoxyadenosine
5′-monophosphate
(dAMP)

Deoxyguanosine
5′-monophosphate
(dGMP)

Deoxythymidine
5′-monophosphate
(dTMP)

Deoxycytidine
5′-monophosphate
(dCMP)

8.10 There are four types of DNA nucleotides.

✔ **Concept Check 4**

How do the sugars of RNA and DNA differ?

a. RNA has a six-carbon sugar; DNA has a five-carbon sugar.

b. The sugar of RNA has a hydroxyl group that is not found in the sugar of DNA.

c. RNA contains uracil; DNA contains thymine.

d. DNA's sugar has a phosphorus atom; RNA's sugar does not.

Polynucleotide strands   DNA is made up of many nucleotides connected by covalent bonds, which join the 5′-phosphate group of one nucleotide to the 3′-carbon atom of the next nucleotide (**Figure 8.11**). These bonds, called **phosphodiester linkages**, are strong covalent bonds; a series of nucleotides linked in this way constitutes a **polynucleotide strand**. The backbone of the polynucleotide strand is composed of alternating sugars and phosphates; the bases project away from the long axis of the strand. The negative charges of the phosphate groups are frequently neutralized by the association of positive charges on proteins, metals, or other molecules.

An important characteristic of the polynucleotide strand is its direction, or polarity. At one end of the strand, a free (meaning that it's unattached on one side) phosphate group is attached to the 5′-carbon atom of the sugar in the nucleotide. This end of the strand is therefore referred to as the **5′ end**. The other end of the strand, referred to as the **3′ end**, has a free OH group attached to the 3′-carbon atom of the sugar.

RNA nucleotides also are connected by phosphodiester linkages to form similar polynucleotide strands (see Figure 8.11).

Concepts

The nucleotides of DNA are joined in polynucleotide strands by phosphodiester bonds that connect the 3′-carbon atom of one nucleotide to the 5′-phosphate group of the next. Each polynucleotide strand has polarity, with a 5′ direction and a 3′ direction.

Secondary Structures of DNA

The secondary structure of DNA refers to its three-dimensional configuration—its fundamental helical structure. DNA's secondary structure can assume a variety of configurations, depending on its base sequence and the conditions in which it is placed.

The double helix   A fundamental characteristic of DNA's secondary structure is that it consists of two polynucleotide strands wound around each other—it's a *double* helix. The sugar–phosphate linkages are on the outside of the helix, and the bases are stacked in the interior of the molecule (see Figure 8.11). The two polynucleotide strands run in opposite directions—they are **antiparallel**, which means that the 5′ end of one strand is opposite the 3′ end of the other strand.

The strands are held together by two types of molecular forces. Hydrogen bonds link the bases on opposite strands (see Figure 8.11). These bonds are relatively weak compared with the covalent phosphodiester bonds that connect the sugar and phosphate groups of adjoining nucleotides on the same strand. As we will see, several important functions of DNA require the separation of its two nucleotide strands, and this separation can be readily accomplished because of the relative ease of breaking and reestablishing the hydrogen bonds.

The nature of the hydrogen bond imposes a limitation on the types of bases that can pair. Adenine normally pairs only with thymine through two hydrogen bonds, and cytosine normally pairs only with guanine through three hydrogen bonds (see Figure 8.11). Because three hydrogen bonds form between C and G and only two hydrogen bonds form between A and T, C–G pairing is stronger than A–T pairing. The specificity of the base pairing means that, wherever there is an A on one strand, there must be a T in the corresponding position on the other strand, and, wherever there is a G on one strand, a C must be on the other. The two polynucleotide strands of a DNA molecule are therefore not identical but are **complementary DNA strands**.

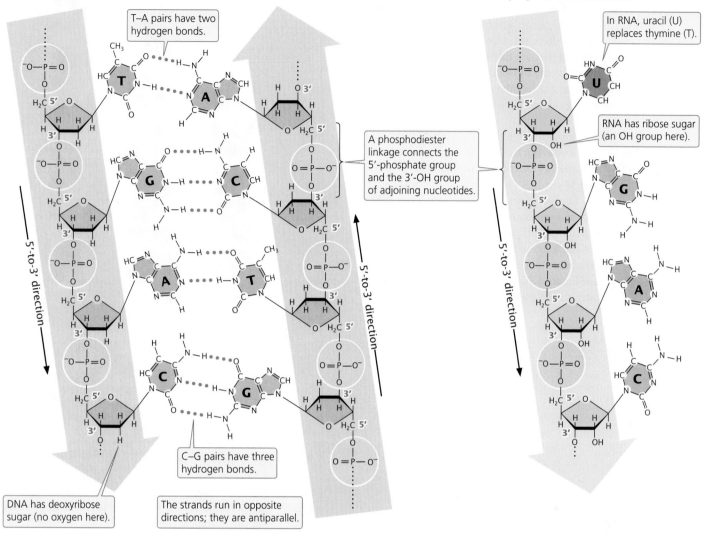

8.11 DNA and RNA consist of polynucleotide strands.

The second force that holds the two DNA strands together is the interaction between the stacked base pairs. These stacking interactions contribute to the stability of the DNA molecule but do not require that any particular base follow another. Thus, the base sequence of the DNA molecule is free to vary, allowing DNA to carry genetic information.

Concepts

DNA consists of two polynucleotide strands. The sugar–phosphate groups of each polynucleotide strand are on the outside of the molecule, and the bases are in the interior. Hydrogen bonding joins the bases of the two strands: guanine pairs with cytosine, and adenine pairs with thymine. The two polynucleotide strands of a DNA molecule are complementary and antiparallel.

✔ Concept Check 5

The antiparallel nature of DNA refers to

a. its charged phosphate groups.

b. the pairing of bases on one strand with bases on the other strand.

c. the formation of hydrogen bonds between bases from opposite strands.

d. the opposite direction of the two strands of nucleotides.

Different secondary structures As we have seen, DNA normally consists of two polynucleotide strands that are antiparallel and complementary (exceptions are single-stranded DNA molecules in a few viruses). The precise three-dimensional shape of the molecule can vary, however, depending on the conditions in which the DNA is placed and, in some cases, on the base sequence itself.

The three-dimensional structure of DNA described by Watson and Crick is termed the **B-DNA** structure (**Figure 8.12**). This structure exists when plenty of water surrounds the molecule and there is no unusual base sequence in the DNA—conditions that are likely to be present in cells. The B-DNA structure is the most stable configuration for a random sequence of nucleotides under physiological conditions, and most evidence suggests that it is the predominate structure in the cell.

B-DNA is an alpha helix, meaning that it has a right-handed, or clockwise, spiral. There are approximately 10 base pairs (bp) per 360-degree rotation of the helix; so each base pair is twisted 36 degrees relative to the adjacent bases (see Figure 8.12b). The base pairs are 0.34 nanometer (nm) apart; so each complete rotation of the molecule encompasses 3.4 nm. The diameter of the helix is 2 nm, and the bases are perpendicular to the long axis of the DNA molecule. A space-filling model shows that B-DNA has a slim and elongated structure (see Figure 8.12a). The spiraling of the nucleotide strands creates major and minor grooves in the helix, fea-

(a) A form   (b) B form   (c) Z form

8.13 DNA can assume several different secondary structures. [After J. M. Berg, J. L. Tymoczko, and L. Stryer, Biochemistry, 6th ed. (New York: W. H. Freeman and Company, 2002), pp. 785, 787.]

tures that are important for the binding of some proteins that regulate the expression of genetic information (see Chapter 12).

Another secondary structure that DNA can assume is the **A-DNA** structure, which exists if less water is present. Like B-DNA, A-DNA is an alpha (right-handed) helix (**Figure 8.13a**), but it is shorter and wider than B-DNA (**Figure 8.13b**) and its bases are tilted away from the main axis of the molecule. There is little evidence that A-DNA exists under physiological conditions.

A radically different secondary structure, called **Z-DNA** (**Figure 8.13c**), forms a left-handed helix. In this form, the sugar–phosphate backbone zigzags back and forth, giving rise to its name. A Z-DNA structure can result when DNA is placed in a high-salt solution. It can arise under physiological conditions if the molecule contains particular base sequences, such as stretches of alternating C and G nucleotides. Recently, researchers have found that Z-DNA-specific antibodies bind to regions of the DNA that are being transcribed into RNA, suggesting that Z-DNA may play some role in gene expression.

Concepts

DNA can assume different secondary structures, depending on the conditions in which it is placed and on its base sequence. B-DNA is thought to be the most common configuration in the cell.

✔ Concept Check 6

How does Z-DNA differ from B-DNA?

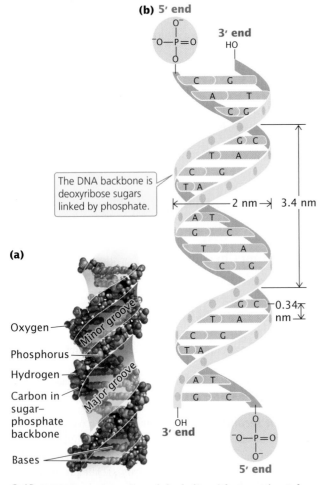

8.12 B-DNA consists of an alpha helix with approximately 10 bases per turn. (a) Space-filling model of B-DNA showing major and minor grooves. (b) Diagrammatic representation.

(a) Major information pathways

(b) Special information pathways

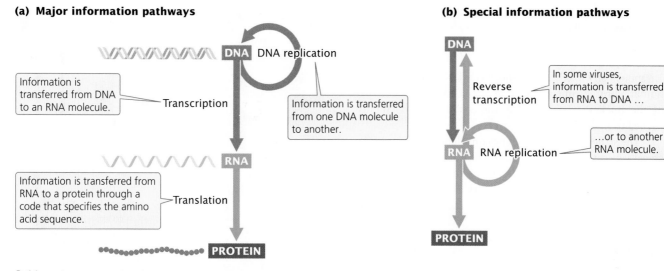

8.14 Pathways of information transfer within the cell.

Connecting Concepts

Genetic Implications of DNA Structure

Watson and Crick's great contribution was their elucidation of the genotype's chemical structure, making it possible for geneticists to begin to examine genes directly, instead of looking only at the phenotypic consequences of gene action. The determination of the structure of DNA led to the birth of molecular genetics—the study of the chemical and molecular nature of genetic information.

Watson and Crick's structure did more than just create the potential for molecular genetic studies; it was an immediate source of insight into key genetic processes. At the beginning of this chapter, three fundamental properties of the genetic material were identified. First, it must be capable of carrying large amounts of information; so it must vary in structure. Watson and Crick's model suggested that genetic instructions are encoded in the base sequence, the only variable part of the molecule.

The second necessary property of genetic material is its ability to replicate faithfully. The complementary polynucleotide strands of DNA make this replication possible. Watson and Crick proposed that, in replication, the two polynucleotide strands unzip, breaking the weak hydrogen bonds between the two strands, and each strand serves as a template on which a new strand is synthesized. The specificity of the base pairing means that only one possible sequence of bases—the complementary sequence—can be synthesized from each template. Newly replicated double-stranded DNA molecules will therefore be identical with the original double-stranded DNA molecule (see Chapter 9 on DNA replication).

The third essential property of genetic material is the ability to translate its instructions into the phenotype. DNA expresses its genetic instructions by first transferring its information to an RNA molecule, in a process termed **transcription** (see Chapter 10). The term *transcription* is appropriate because, although the information is transferred from DNA to RNA, the information remains in the language of nucleic acids. The RNA molecule then transfers the genetic information to a protein by specifying its amino acid sequence. This process is termed **translation** (see Chapter 11) because the information must be *translated* from the language of nucleotides into the language of amino acids.

We can now identify three major pathways of information flow in the cell (**Figure 8.14a**): in **replication**, information passes from one DNA molecule to other DNA molecules; in transcription, information passes from DNA to RNA; and, in translation, information passes from RNA to protein. This concept of information flow was formalized by Francis Crick in a concept that he called the **central dogma** of molecular biology. The central dogma states that genetic information passes from DNA to protein in a one-way information pathway. We now realize, however, that the central dogma is an oversimplification. In addition to the three general information pathways of replication, transcription, and translation, other transfers may take place in certain organisms or under special circumstances, including the transfer of information from RNA to DNA (in **reverse transcription**) and the transfer of information from RNA to RNA (in **RNA replication**; **Figure 8.14b**). Retroviruses (see Chapter 6) and some transposable elements (see Chapter 13) utilize reverse transcription;. some RNA viruses use RNA replication.

8.4 Large Amounts of DNA Are Packed into a Cell

The packaging of tremendous amounts of genetic information into the small space within a cell has been called the ultimate storage problem. Consider the chromosome of the bacterium *E. coli*, a single molecule of DNA with approximately 4.6 million base pairs. Stretched out straight, this DNA would be about 1000 times as long as the cell within which it resides (**Figure 8.15**). Human cells contain more than 6 billion base pairs of DNA, which would measure some

E. coli bacterium

Bacterial chromosome

8.15 The DNA in *E. coli* is about 1000 times as long as the cell itself.

1.8 meters stretched end to end. Even DNA in the smallest human chromosome would stretch 14,000 times the length of the nucleus. Clearly, DNA molecules must be tightly packed to fit into such small spaces.

The structure of DNA can be considered at three hierarchical levels: the primary structure of DNA is its nucleotide sequence; the secondary structure is the double-stranded helix; and the tertiary structure refers to higher-order folding that allows DNA to be packed into the confined space of a cell.

Concepts

Chromosomal DNA exists in the form of very long molecules, which must be tightly packed to fit into the small confines of a cell.

One type of DNA tertiary structure is **supercoiling**, which takes place when the DNA helix is subjected to strain by being overwound or underwound. The lowest energy state for B-DNA is when it has approximately 10 bp per turn of its helix. In this **relaxed state**, a stretch of 100 bp of DNA would assume about 10 complete turns (**Figure 8.16a**). If energy is used to add or remove any turns, strain is placed on the molecule, causing the helix to supercoil, or twist, on itself (**Figure 8.16b and c**). Molecules that are overrotated exhibit **positive supercoiling** (see Figure 8.16b). Underrotated molecules exhibit **negative supercoiling** (see Figure 8.16c). Supercoiling is a partial solution to the cell's DNA packing problem because supercoiled DNA occupies less space than relaxed DNA.

Supercoiling takes place when the strain of overrotating or underrotating cannot be compensated by the turning of the ends of the double helix, which is the case if the DNA is circular—that is, there are no free ends. If the chains *can* turn freely, their ends will simply turn as extra rotations are added or removed, and the molecule will spontaneously revert to the relaxed state. Both bacterial and eukaryotic DNA usually folds into loops stabilized by proteins (which prevent free rotation of the ends), and supercoiling takes place within the loops.

Supercoiling relies on **topoisomerases**, enzymes that add or remove rotations from the DNA helix by temporarily breaking the nucleotide strands, rotating the ends around each other, and then rejoining the broken ends. Thus topoisomerases can both induce and relieve supercoiling.

Most DNA found in cells is negatively supercoiled, which has two advantages over nonsupercoiled DNA. First, supercoiling makes the separation of the two strands of DNA easier during replication and transcription. Negatively supercoiled DNA is underrotated; so separation of the two

(a)

Relaxed circular DNA

A coiled telephone cord is like relaxed circular DNA.

(b) Add two turns (overrotate)

(c) Remove two turns (underrotate)

Positive supercoil

Negative supercoil

| Positive supercoiling occurs when DNA is overrotated; the helix twists on itself. | Negative supercoiling occurs when DNA is underrotated; the helix twists on itself in the opposite direction. | If you turn the receiver when you hang up, you induce a negative supercoil in the cord. |

8.16 Supercoiled DNA is overwound or underwound, causing it to twist on itself. Electron micrographs are of relaxed DNA (top) and supercoiled DNA (bottom). [Dr. Gopal Murti/Phototake.]

(b) Twisted loops of DNA

Proteins

8.17 Bacterial DNA is highly folded into a series of twisted loops. *[Part a: Dr. Gopal Murti/Photo Researchers.]*

strands during replication and transcription is more rapid and requires less energy. Second, supercoiled DNA can be packed into a smaller space than can relaxed DNA.

Concepts

Overrotation or underrotation of a DNA double helix places strain on the molecule, causing it to supercoil. Supercoiling is controlled by topoisomerase enzymes. Most cellular DNA is negatively supercoiled, which eases the separation of nucleotide strands during replication and transcription and allows DNA to be packed into small spaces.

✔ Concept Check 7

A DNA molecule 300 bp long has 20 complete rotations. This DNA molecule is

a. positively supercoiled.

b. negatively supercoiled.

c. relaxed.

8.5 A Bacterial Chromosome Consists of a Single Circular DNA Molecule

Most bacterial genomes consist of a single circular DNA molecule, although linear DNA molecules have been found in a few species. In circular bacterial chromosomes, the DNA does not exist in an open, relaxed circle; the 3 million to 4 million base pairs of DNA found in a typical bacterial genome would be much too large to fit into a bacterial cell (see Figure 8.15). Bacterial DNA is not attached to histone proteins as is eukaryotic DNA (discussed later in the chapter), but bacterial DNA is complexed to a number of proteins that help compact it.

When a bacterial cell is viewed with the electron microscope, its DNA frequently appears as a distinct clump, the **nucleoid**, which is confined to a definite region of the cytoplasm. If a bacterial cell is broken open gently, its DNA spills out in a series of twisted loops (**Figure 8.17a**). The ends of the loops are most likely held in place by proteins (**Figure 8.17b**). Many bacteria contain additional DNA in the form of small circular molecules called plasmids, which replicate independently of the chromosome (see Chapter 6).

Concepts

A typical bacterial chromosome consists of a large, circular molecule of DNA that is a series of twisted loops. Bacterial DNA appears as a distinct clump, the nucleoid, within the bacterial cell.

✔ Concept Check 8

How does bacterial DNA differ from eukaryotic DNA?

8.6 Eukaryotic Chromosomes Are DNA Complexed to Histone Proteins

Individual eukaryotic chromosomes contain enormous amounts of DNA. Like bacterial chromosomes, each eukaryotic chromosome consists of a single, extremely long molecule of DNA. For all of this DNA to fit into the nucleus, tremendous packing and folding are required, the extent of which must change in the course of the cell cycle. The chromosomes are in an elongated, relatively uncondensed state during interphase of the cell cycle (see pp. 21–22 in Chapter 2), but the term *relatively* is an important qualification here. Although the DNA of interphase chromosomes is less tightly packed than the DNA of mitotic chromosomes, it is still highly condensed; it's just *less* condensed. In the course of the cell cycle, the level of DNA packing changes: chromosomes progress from a highly packed state to a state of extreme condensation. DNA packing also changes locally in replication and transcription, when the two nucleotide strands must unwind so that particular base sequences are exposed. Thus, the packing of eukaryotic DNA (its tertiary chromosomal structure) is not static but changes regularly in response to cellular processes.

Chromatin Structure

Eukaryotic DNA is closely associated with proteins, creating *chromatin.* The two basic types of chromatin are **euchromatin**, which undergoes the normal process of condensation and decondensation in the cell cycle, and **heterochromatin**, which remains in a highly condensed state throughout the cell cycle, even during interphase. Euchromatin constitutes the majority of the chromosomal material and is where most transcription takes place. All chromosomes have heterochromatin at the centromeres and telomeres. Heterochromatin is also present at other specific places on some chromosomes, along the entire inactive X chromosome in female mammals (see pp. 80–81 in Chapter 4), and throughout most of the Y chromosome in males. Most, but not all, heterochromatin appears to be largely devoid of transcription.

The most abundant proteins in chromatin are the *histones,* which are small, positively charged proteins of five major types: H1, H2A, H2B, H3, and H4 (**Table 8.2**). All histones have a high percentage of arginine and lysine, positively charged amino acids that give the histones a net positive charge. The positive charges attract the negative charges on the phosphates of DNA; this attraction holds the DNA in contact with the histones.

A heterogeneous assortment of **nonhistone chromosomal proteins** make up about half of the protein mass of the chromosome. **Chromosomal scaffold proteins** (**Figure 8.18**), one class of nonhistone chromosomal protein, may help fold and pack the chromosome. Other structural proteins make up the kinetochore, cap the chromosome ends by attaching to telomeres, and constitute the molecular motors

8.18 Scaffold proteins play a role in the folding and packing of chromosomes. *[Professor U. Laemmli/Photo Researchers.]*

that move chromosomes in mitosis and meiosis. Some types of nonhistone chromosomal proteins have roles in genetic processes. They are components of the replication machinery (DNA polymerases, helicases, primases; see Chapter 9) and proteins that carry out and regulate transcription (RNA polymerases; see Chapter 10). The highly organized structure of chromatin is best viewed from several levels. In the next sections, we will examine these levels of chromatin organization.

Concepts

Chromatin, which consists of DNA complexed to proteins, is the material that makes up eukaryotic chromosomes. The most abundant of these proteins are the five types of positively charged histone proteins: H1, H2A, H2B, H3, and H4.

✔ Concept Check 9

Neutralizing their positive charges would have which effect on the histone proteins?

a. They would bind the DNA tighter.

b. They would separate from the DNA.

c. They would no longer be attracted to each other.

d. They would cause supercoiling of the DNA.

Table 8.2	Characteristics of histone proteins	
Histone Protein	Molecular Weight	Number of Amino Acids
H1	21,130	223
H2A	13,960	129
H2B	13,774	125
H3	15,273	135
H4	11,236	102

Note: The sizes of H1, H2A, and H2B histones vary somewhat from species to species. The values given are for bovine histones.

Source: Data are from A. L. Lehninger, D. L. Nelson, and M. M. Cox, *Principles of Biochemistry,* 3d ed. (New York: Worth Publishers, 1993), p. 924.

The nucleosome Chromatin has a highly complex structure with several levels of organization (**Figure 8.19**). The simplest level is the double-helical structure of DNA. At a more complex level, the DNA molecule is associated with proteins and is highly folded to produce a chromosome.

When chromatin is isolated from the nucleus of a cell and viewed with an electron microscope, it frequently looks like beads on a string (**Figure 8.20a**). If a small amount of nuclease is added to this structure, the enzyme cleaves the "string" between the "beads," leaving individual beads attached to about 200 bp of DNA (**Figure 8.20b**). If more

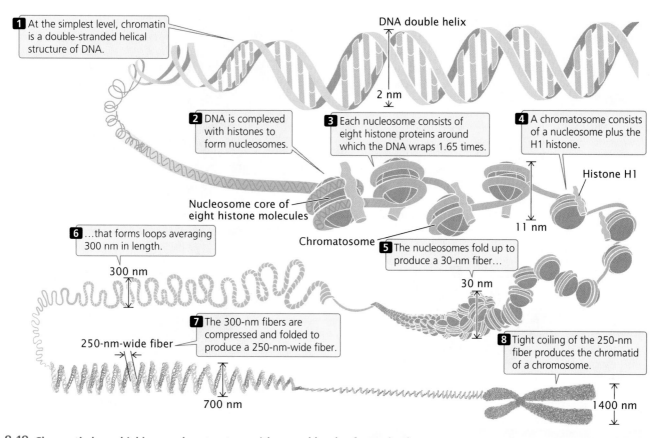

1 At the simplest level, chromatin is a double-stranded helical structure of DNA.

DNA double helix

2 nm

2 DNA is complexed with histones to form nucleosomes.

3 Each nucleosome consists of eight histone proteins around which the DNA wraps 1.65 times.

4 A chromatosome consists of a nucleosome plus the H1 histone.

Histone H1

Nucleosome core of eight histone molecules

Chromatosome

11 nm

6 …that forms loops averaging 300 nm in length.

5 The nucleosomes fold up to produce a 30-nm fiber…

30 nm

300 nm

7 The 300-nm fibers are compressed and folded to produce a 250-nm-wide fiber.

250-nm-wide fiber

8 Tight coiling of the 250-nm fiber produces the chromatid of a chromosome.

700 nm

1400 nm

8.19 Chromatin has a highly complex structure with several levels of organization.

nuclease is added, the enzyme chews up all of the DNA between the beads and leaves a core of proteins attached to a fragment of DNA (**Figure 8.20c**). Such experiments demonstrated that chromatin is not a random association of proteins and DNA but has a fundamental repeating structure.

The repeating core of protein and DNA produced by digestion with nuclease enzymes is the simplest level of chromatin structure, the **nucleosome** (see Figure 8.19). The nucleosome is a core particle consisting of DNA wrapped about two times around an octamer of eight histone proteins (two copies each of H2A, H2B, H3, and H4), much like thread wound around a spool (**Figure 8.20d**). The DNA in direct contact with the histone octamer is between 145 and 147 bp in length.

Each of the histone proteins that make up the nucleosome core particle has a flexible "tail," containing from 11 to 37 amino acids, that extends out from the nucleosome. Positively charged amino acids in the tails of the histones interact with the negative charges of the phosphates on the DNA, and the tails of one nucleosome may interact with neighboring nucleosomes. Chemical modifications of these histone tails bring about changes in chromatin structure that are necessary for gene expression.

The fifth type of histone, H1, is not a part of the core particle but plays an important role in the nucleosome structure. H1 binds to 20 to 22 bp of DNA where the DNA joins and leaves the octamer (see Figure 8.19) and helps to lock the DNA into place, acting as a clamp around the nucleosome octamer.

Together, the core particle and its associated H1 histone are called the **chromatosome** (see Figure 8.19), the next level of chromatin organization. Each chromatosome encompasses about 167 bp of DNA. Chromatosomes are located at regular intervals along the DNA molecule and are separated from one another by **linker DNA**, which varies in size among cell types; in most cells, linker DNA comprises from about 30 to 40 bp. Nonhistone chromosomal proteins may be associated with this linker DNA, and a few also appear to bind directly to the core particle.

Higher-order chromatin structure When chromatin is in a condensed form, adjacent nucleosomes are not separated by space equal to the length of the linker DNA; rather, nucleosomes fold on themselves to form a dense, tightly packed structure (see Figure 8.19) that makes up a fiber with a diameter of about 30 nm (**Figure 8.21a**). Two different models have been proposed for the 30-nm fiber: a solenoid

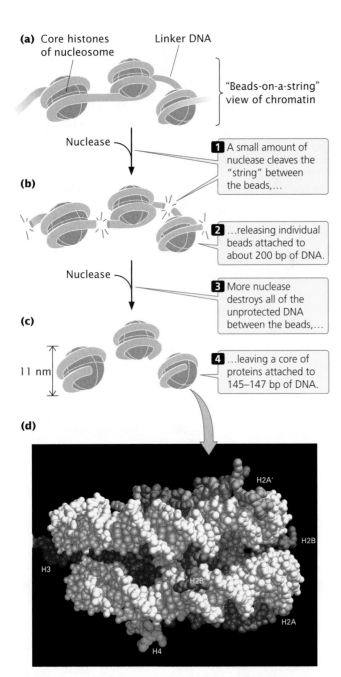

(a) Core histones of nucleosome   Linker DNA

"Beads-on-a-string" view of chromatin

Nuclease →

1 A small amount of nuclease cleaves the "string" between the beads,…

(b)

2 …releasing individual beads attached to about 200 bp of DNA.

Nuclease →

3 More nuclease destroys all of the unprotected DNA between the beads,…

(c)

11 nm

4 …leaving a core of proteins attached to 145–147 bp of DNA.

(d)

H2A'

H2B

H3

H2B'

H2A

H4

8.20 The nucleosome is the fundamental repeating unit of chromatin. The space-filling model shows that the nucleosome core particle consists of two copies each of H2A, H2B, H3, and H4, around which DNA (white) coils. *[Part d: From K. Luger et al., 1997, Nature 389:251; courtesy of T. H. Richmond.]*

model, in which a linear array of nucleosomes are coiled, and a helix model, in which nucleosomes are arranged in a zigzag ribbon that twists or supercoils. Recent evidence supports the helix model (**Figure 8.21b**).

The next-higher level of chromatin structure is a series of loops of 30-nm fibers, each anchored at its base by proteins in the nuclear scaffold (see Figure 8.19). On average, each loop encompasses some 20,000 to 100,000 bp of DNA

(a)

(b)   Individual nucleosomes

30-nm fiber

8.21 Adjacent nucleosomes pack together to form a 30-nm fiber. (a) Electron micrograph of nucleosomes. (b) One model of how nucleosomes associate to form the helical fiber. *[Part a: Jan Bednar, Rachel A. Horowitz, Sergei A. Grigoryev, Lenny M. Carruthers, Jeffrey C. Hansen, Abraham J. Koster, and Christopher L. Woodcock. Nucleosomes, linker DNA, and linker histone form a unique structural motif that directs the higher-order folding and compaction of chromatin. PNAS 1998; 95:14173–14178. Copyright 2004 National Academy of Sciences, U.S.A.]*

and is about 300 nm in length, but the individual loops vary considerably. The 300-nm loops are packed and folded to produce a 250-nm-wide fiber. Tight helical coiling of the 250-nm fiber, in turn, produces the structure that appears in metaphase—an individual chromatid approximately 700 nm in width.

Concepts

The nucleosome consists of a core particle of eight histone proteins and DNA that wraps around the core. Chromatosomes, which are nucleosomes bound to an H1 histone, are separated by linker DNA. Nucleosomes fold to form a 30-nm chromatin fiber, which appears as a series of loops that pack to create a 250-nm-wide fiber. Helical coiling of the 250-nm fiber produces a chromatid.

Centromere Structure

The centromere is a constricted region of the chromosome to which spindle fibers attach and is essential for proper chromosome movement in mitosis and meiosis (see Chapter 2).

The first centromeres to be isolated and studied at the molecular level came from yeast, which has small linear chromosomes. When molecular biologists attached DNA sequences from yeast centromeres to plasmids, the plasmids behaved in mitosis as if they were eukaryotic chromosomes. This finding indicated that the DNA sequences from yeast, called **centromeric sequences**, are functional centromeres that allow segregation to take place. Centromeric sequences are the binding sites for the kinetochore, to which spindle fibers attach.

The centromeres of different organisms exhibit considerable variation in centromeric sequences. Some organisms have chromosomes with diffuse centromeres, and spindle fibers attach along the entire length of each chromosome. Most have chromosomes with localized centromeres; in these organisms, spindle fibers attach at a specific place on the chromosome, but there can also be secondary constrictions at places that do not have centromeric functions. In *Drosophila*, *Arabidopsis*, and humans, centromeres span hundreds of thousands of base pairs. Most of the centromere is made up of short sequences of DNA that are repeated thousands of times in tandem.

Concepts

The centromere is a region of the chromosome to which spindle fibers attach. Centromeres display considerable variation in structure. In addition to their role in chromosome movement, centromeres help control the cell cycle by inhibiting anaphase until chromosomes are attached to spindle fibers from both poles.

Telomere Structure

Telomeres are the natural ends of a chromosome (see p. 20 in Chapter 2). Pioneering work by Hermann Muller (on fruit flies) and Barbara McClintock (on corn) showed that chromosome breaks produce unstable ends that have a tendency to stick together and enable the chromosome to be degraded. Because attachment and degradation don't happen to the ends of a chromosome that has telomeres, each telomere must serve as a cap that stabilizes the chromosome, much like the plastic tips on the ends of a shoelace that prevent the lace from unraveling. Telomeres also provide a means of replicating the ends of the chromosome, which will be discussed in Chapter 9.

Telomeres have now been isolated from protozoans, plants, humans, and other organisms; most are similar in structure. These **telomeric sequences** usually consist of a series of cytosine nucleotides followed by several adenine or thymine nucleotides or both, taking the form $5'-C_n(A \text{ or } T)_m-3'$, where n is 2 or more and m is from 1 to 4. For example, the repeating unit in human telomeres is CCCTAA, which may be repeated from 250 to 1500 times. The sequence is always oriented with the string of Cs and Gs toward the end of the chromosome, as shown here:

$$\begin{array}{ccc}
\text{end of} & \leftarrow 5'\text{-CCCTAA} \rightarrow & \text{toward} \\
\text{chromosome} & 3'\text{-GGGATT} & \text{centromere}
\end{array}$$

The G-rich strand often protrudes beyond the complementary C-rich strand at the end of the chromosome (**Figure 8.22a**). Special POT (protection of telomere) proteins bind to the G-rich single-stranded sequence, protecting the

(a)

5′ CCCTAACCCTAA 3′
3′ GGGATTGGGATTGGGATTGGGATTGGGATT 5′

DNA sequence at
end of chromosome

(b)

t-loop

G-rich single-strand
overhang
5′

AATCCCAATCCC
TTAGGGTTAGGG 3′

3′
5′

8.22 DNA at the ends of eukaryotic chromosomes consists of telomeric sequences.
(a) The G-rich strand at the telomere is longer than the C-rich strand. (b) In mammalian cells, the G-rich strand folds over and pairs with a short stretch of DNA to form a t-loop.

telomere from degradation and preventing the ends of chromosomes from sticking together. In mammalian cells, the single-stranded overhang may fold over and pair with a short stretch of DNA to form a structure called the t-loop, which also functions in protecting the telomere from degradation (**Figure 8.22b**).

Concepts

A telomere is the stabilizing end of a chromosome. At the end of each telomere are many short telomeric sequences.

✔ Concept Check 10

Which is a characteristic of DNA sequences at the telomeres?

a. They consist of cytosine and adenine nucleotides.

b. They consist of repeated sequences.

c. One strand protrudes beyond the other, creating some single-standed DNA at the end.

d. All of the above.

8.7 Eukaryotic DNA Contains Several Classes of Sequence Variation

Eukaryotic organisms differ dramatically in the amount of DNA per cell, a quantity termed an organism's **C value** (**Table 8.3**). Each cell of a fruit fly, for example, contains 35 times the amount of DNA found in a cell of the bacterium *E. coli*. In general, eukaryotic cells contain more DNA than prokaryotic cells do, but variability in the C values of different eukaryotes is huge. Human cells contain more than 10 times the amount of DNA found in *Drosophila* cells, whereas

Table 8.3	Genome sizes of various organisms
Organism	**Approximate Genome Size (bp)**
Lambda (λ) bacteriophage	50,000
Escherichia coli (bacterium)	4,640,000
Saccharomyces cerevisiae (yeast)	12,000,000
Arabidopsis thaliana (plant)	125,000,000
Drosophila melanogaster (insect)	170,000,000
Homo sapiens (human)	3,200,000,000
Zea mays (corn)	4,500,000,000
Amphiuma (salamander)	765,000,000,000

some salamander cells contain 20 times as much DNA as that in human cells. Clearly, these differences in C value cannot be explained simply by differences in organismal complexity. So, what is all this extra DNA in eukaryotic cells doing? We do not yet have a complete answer to this question, but eukaryotic DNA sequences reveal a complexity that is absent from prokaryotic DNA.

Types of DNA Sequences in Eukaryotes

Eukaryotic DNA consists of at least three types of sequences: unique-sequence DNA, moderately repetitive DNA, and highly repetitive DNA. **Unique-sequence DNA** consists of sequences that are present only once or, at most, a few times in the genome. This DNA includes sequences that encode proteins, as well as a great deal of DNA whose function is unknown. Genes that are present in a single copy constitute from roughly 25% to 50% of the protein-encoding genes in most multicellular eukaryotes. Other genes within unique-sequence DNA are present in several similar, but not identical, copies that arose through the duplication of an existing gene and are referred to as a **gene family**. Most gene families include just a few member genes, but some, such as those that encode immunoglobulin proteins in vertebrates, contain hundreds of members. The genes that encode β-like globins are another example of a gene family. In humans, there are seven β-globin genes, clustered together on chromosome 11. The polypeptides encoded by these genes join with α-globin polypeptides to form hemoglobin molecules, which transport oxygen in the blood.

Other sequences exist in many copies and are called **repetitive DNA**. A major class of repetitive DNA is called **moderately repetitive DNA**, which typically consists of sequences from 150 to 300 bp in length (although they may be longer) that are repeated many thousands of times. Some of these sequences perform important functions for the cell; for example, the genes for ribosomal RNAs (rRNAs) and transfer RNAs (tRNAs) make up a part of the moderately repetitive DNA. However, much of the moderately repetitive DNA has no known function in the cell. Moderately repetitive DNA itself is of two types of repeats. **Tandem repeat sequences** appear one after another and tend to be clustered at a few locations on the chromosomes. **Interspersed repeat sequences** are scattered throughout the genome. An example of an interspersed repeat is the *Alu* sequence, which consists of about 200 bp. The *Alu* sequence is present more than a million times in the human genome and apparently has no ceullular function. Short repeats, such as the *Alu* sequences, are called **SINEs** (**short interspersed elements**). Longer interspersed repeats consisting of several thousand base pairs are called **LINEs** (**long interspersed elements**). Most interspersed repeats are transposable elements, sequences that can multiply and move (see Chapter 13).

The other major class of repetitive DNA is **highly repetitive DNA**. These short sequences, often less than 10 bp in length, are present in hundreds of thousands to millions of copies that are repeated in tandem and clustered in certain regions of the chromosome, especially at centromeres and telomeres. Highly repetitive DNA is sometimes called satellite DNA, because its percentages of the four bases differ from those of other DNA sequences and, therefore, it separates as a satellite fraction when centrifuged at high speeds. Highly repetitive DNA is rarely transcribed into RNA. Although these sequences may contribute to centromere and telomere function, most highly repetitive DNA has no known function.

Direct sequencing of eukaryotic genomes also tell us a lot about how genetic information is organized within chromosomes. We now know that the density of genes varies greatly among and within chromosomes. For example, human chromosome 19 has a high density of genes, with about 26 genes per million base pairs. Chromosome 13, on the other hand, has only about 6.5 genes per million base pairs. Gene density can also vary within different regions of the same chromosome: some parts of the long arm of chro-mosome 13 have only 3 genes per million base pairs, whereas other parts have almost 30 genes per million base pairs. And the short arm of chromosome 13 contains almost no genes, consisting entirely of heterochromatin.

Concepts

Eukaryotic DNA comprises three major classes: unique-sequence DNA, moderately repetitive DNA, and highly repetitive DNA. Unique-sequence DNA consists of sequences that exist in one or only a few copies; moderately repetitive DNA consists of sequences that may be several hundred base pairs in length and is present in thousands to hundreds of thousands of copies. Highly repetitive DNA consists of very short sequences repeated in tandem and present in hundreds of thousands to millions of copies. The density of genes varies greatly among and even within chromosomes.

✔ Concept Check 11

Most of the genes that encode proteins are found in

a. unique-sequence DNA.
b. moderately repetitive DNA.
c. highly repetitive DNA.
d. all of the above.

Concepts Summary

- Genetic material must contain complex information, be replicated accurately, and have the capacity to be translated into the phenotype.

- Evidence that DNA is the source of genetic information came from the finding by Avery, MacLeod, and McCarty that transformation depended on DNA and from the demonstration by Hershey and Chase that viral DNA is passed on to progeny phages.

- James Watson and Francis Crick proposed a new model for the three-dimensional structure of DNA in 1953.

- A DNA nucleotide consists of a deoxyribose sugar, a phosphate group, and a nitrogenous base. RNA consists of a ribose sugar, a phosphate group, and a nitrogenous base.

- The bases of a DNA nucleotide are of two types: purines (adenine and guanine) and pyrimidines (cytosine and thymine). RNA contains the pyrimidine uracil instead of thymine.

- Nucleotides are joined together by phosphodiester linkages in a polynucleotide strand. Each polynucleotide strand has a free phosphate group at its 5′ end and a free hydroxyl group at its 3′ end.

- DNA consists of two nucleotide strands that wind around each other to form a double helix. The sugars and phosphates lie on the outside of the helix, and the bases are stacked in the interior. The two strands are joined together by hydrogen bonding between bases in each strand. The two strands are antiparallel and complementary.

- DNA molecules can form a number of different secondary structures, depending on the conditions in which the DNA is placed and on its base sequence.

- The structure of DNA has several important genetic implications. Genetic information resides in the base sequence of DNA, which ultimately specifies the amino acid sequence of proteins. Complementarity of the bases on DNA's two strands allows genetic information to be replicated.

- The central dogma of molecular biology proposes that information flows in a one-way direction, from DNA to RNA to protein. Exceptions to the central dogma are now known.

- Chromosomes contain very long DNA molecules that are tightly packed. Supercoiling results from strain produced when rotations are added to a relaxed DNA molecule or removed from it.

- A bacterial chromosome consists of a single, circular DNA molecule that is bound to proteins and exists as a series of large loops.

- Each eukaryotic chromosome contains a single, long linear DNA molecule that is bound to histone and nonhistone chromosomal proteins. Euchromatin undergoes the normal cycle of decondensation and condensation in the cell cycle. Heterochromatin remains highly condensed throughout the cell cycle.

- The nucleosome is a core of eight histone proteins and the DNA that wraps around the core.

- A nucleosome is folded into a 30-nm fiber that forms a series of 300-nm-long loops; these loops are anchored at their bases by proteins associated with the nuclear scaffold. The 300-nm loops are condensed to form a fiber that is itself tightly coiled to produce a chromatid.

- Centromeres are chromosomal regions where spindle fibers attach; chromosomes without centromeres are usually lost in the course of cell division. Telomeres stabilize the ends of chromosomes.

- Eukaryotic DNA exhibits three classes of sequences. Unique-sequence DNA exists in very few copies. Moderately repetitive DNA consists of moderately long sequences that are repeated from hundreds to thousands of times. Highly repetitive DNA consists of very short sequences that are repeated in tandem from many thousands to millions of times.

Important Terms

nucleotide (p. 195)
Chargaff's rules (p. 195)
transforming principle (p. 196)
isotope (p. 197)
X-ray diffraction (p. 199)
ribose (p. 200)
deoxyribose (p. 200)
nitrogenous base (p. 201)
purine (p. 201)
pyrimidine (p. 201)
adenine (A) (p. 201)
guanine (G) (p. 201)
cytosine (C) (p. 201)
thymine (T) (p. 201)
uracil (U) (p. 201)
nucleoside (p. 201)
phosphate group (p. 201)
deoxyribonucleotide (p. 201)
ribonucleotide (p. 201)
phosphodiester linkage (p. 202)

polynucleotide strand (p. 202)
5′ end (p. 202)
3′ end (p. 202)
antiparallel (p. 202)
complementary DNA strands (p. 202)
B-DNA (p. 204)
A-DNA (p. 204)
Z-DNA (p. 204)
transcription (p. 205)
translation (p. 205)
replication (p. 205)
central dogma (p. 205)
reverse transcription (p. 205)
RNA replication (p. 205)
supercoiling (p. 206)
relaxed state of DNA (p. 206)
positive supercoiling (p. 206)
negative supercoiling (p. 206)
topoisomerase (p. 206)
nucleoid (p. 207)

euchromatin (p. 208)
heterochromatin (p. 208)
nonhistone chromosomal protein (p. 208)
chromosomal scaffold protein (p. 208)
nucleosome (p. 209)
chromatosome (p. 209)
linker DNA (p. 209)
centromeric sequence (p. 211)
telomeric sequence (p. 211)
C value (p. 212)
unique-sequence DNA (p. 212)
gene family (p. 212)
repetitive DNA (p. 212)
moderately repetitive DNA (p. 212)
tandem repeat sequence (p. 212)
interspersed repeat sequence (p. 212)
short interspersed element (SINE) (p. 212)
long interspersed element (LINE) (p. 212)
highly repetitive DNA (p. 213)

Answers to Concept Checks

1. Without knowledge of the structure of DNA, an understanding of how genetic information was encoded or expressed was impossible to achieve.

2. c

3. No, because carbon is found in both protein and nucleic acid.

4. b

5. d

6. Z-DNA has a left-handed helix; B-DNA has a right-handed helix. The sugar–phosphate backbone of Z-DNA zigzags back and forth, whereas the sugar–phosphate backbone of B-DNA forms a smooth continuous ribbon.

7. b

8. Bacterial DNA is not complexed to histone proteins and is circular.

9. b

10. d

11. a

Worked Problems

1. The percentage of cytosine in a double-stranded DNA molecule is 40%. What is the percentage of thymine?

• Solution

In double-stranded DNA, A pairs with T, whereas G pairs with C; so the percentage of A equals the percentage of T, and the percentage of G equals the percentage of C. If C = 40%, then G also must be 40%. The total percentage of C + G is therefore 40% + 40% = 80%. All the remaining bases must be either A or T; so the total percentage of A + T = 100% − 80% = 20%; because the percentage of A equals the percentage of T, the percentage of T is 20%/2 = 10%.

2. Which of the following relations will be true for the percentage of bases in double-stranded DNA?

a. C + T = A + G

b. $\dfrac{C}{A} = \dfrac{T}{G}$

• Solution

An easy way to determine whether the relations are true is to arbitrarily assign percentages to the bases, remembering that, in double-stranded DNA, A = T and G = C. For example, if the percentages of A and T are each 30%, then the percentages of G and C are each 20%. We can substitute these values into the equations to see if the relations are true.

a. 20 + 30 = 30 + 20. This relation is true.

b. $^{20}/_{30} \neq {}^{30}/_{20}$. This relation is not true.

3. A diploid plant cell contains 2 billion base pairs of DNA.

a. How many nucleosomes are present in the cell?

b. Give the numbers of molecules of each type of histone protein associated with the genomic DNA.

• Solution

Each nucleosome encompasses about 200 bp of DNA: from 145 to 147 bp of DNA wrapped around the histone core, from 20 to 22 bp of DNA associated with the H1 protein, and another 30 to 40 bp of linker DNA.

a. To determine how many nucleosomes are present in the cell, we simply divide the total number of base pairs of DNA (2×10^9 bp) by the number of base pairs per nucleosome:

$$\frac{2 \times 10^9 \text{ nucleotides}}{2 \times 10^2 \text{ nucleotides per nucleosome}} = 1 \times 10^7 \text{ nucleosomes}$$

Thus, there are approximately 10 million nucleosomes in the cell.

b. Each nucleosome includes two molecules each of H2A, H2B, H3, and H4 histones. Therefore, there are 2×10^7 molecules each of H2A, H2B, H3, and H4 histones. Each nucleosome has associated with it one copy of the H1 histone; so there are 1×10^7 molecules of H1.

Comprehension Questions

Section 8.1

*1. What three general characteristics must the genetic material possess?

Section 8.2

2. What is transformation? How did Avery and his colleagues demonstrate that the transforming principle is DNA?

*3. How did Hershey and Chase show that DNA is passed to new phages in phage reproduction?

Section 8.3

*4. Draw and identify the three parts of a DNA nucleotide.

5. How does an RNA nucleotide differ from a DNA nucleotide?

*6. Draw a short segment of a single DNA polynucleotide strand, including at least three nucleotides. Indicate the polarity of the strand by identifying the 5′ end and the 3′ end.

7. What are some of the important genetic implications of the DNA structure?

*8. What are the major transfers of genetic information?

Section 8.4

*9. How does supercoiling arise? What is the difference between positive and negative supercoiling?

10. What functions does supercoiling serve for the cell?

Section 8.6

*11. Describe the composition and structure of the nucleosome. How do core particles differ from chromatosomes?

12. Describe in steps how the double helix of DNA, which is 2 nm in width, gives rise to a chromosome that is 700 nm in width.

*13. Describe the function and molecular structure of a telomere.

14. What is the difference between euchromatin and heterochromatin?

Section 8.7

*15. Describe the different types of DNA sequences that exist in eukaryotes.

Application Questions and Problems

Section 8.2

16. A student mixes some heat-killed type IIS *Streptococcus pneumoniae* bacteria with live type IIR bacteria and injects the mixture into a mouse. The mouse develops pneumonia and dies. The student recovers some type IIS bacteria from the dead mouse. It is the only experiment conducted by the student. Has the student demonstrated that transformation has taken place? What other explanations might explain the presence of the type IIS bacteria in the dead mouse?

Section 8.3

17. DNA molecules of different size are often separated with the use of a technique called electrophoresis (see Chapter 14). With this technique, DNA molecules are placed in a gel, an electrical current is applied to the gel, and the DNA molecules migrate toward the positive (+) pole of the current. What aspect of its structure causes a DNA molecule to migrate toward the positive pole?

18. What aspects of its structure contribute to the stability of the DNA molecule? Why is RNA less stable than DNA?

*19. Edwin Chargaff collected data on the proportions of nucleotide bases from the DNA of a variety of different organisms and tissues (E. Chargaff, in *The Nucleic Acids: Chemistry and Biology*, vol. 1, E. Chargaff and J. N. Davidson, Eds. New York: Academic Press, 1955). Data from the DNA of several organisms analyzed by Chargaff are shown below.

Organism and tissue	Percent			
	A	G	C	T
Sheep thymus	29.3	21.4	21.0	28.3
Pig liver	29.4	20.5	20.5	29.7
Human thymus	30.9	19.9	19.8	29.4
Rat bone marrow	28.6	21.4	20.4	28.4
Hen erythrocytes	28.8	20.5	21.5	29.2
Yeast	31.7	18.3	17.4	32.6
E. coli	26.0	24.9	25.2	23.9
Human sperm	30.9	19.1	18.4	31.6
Salmon sperm	29.7	20.8	20.4	29.1
Herring sperm	27.8	22.1	20.7	27.5

a. For each organism, compute the ratio of (A + G)/(T + C) and the ratio of (A + T)/(C + G).

b. Are these ratios constant or do they vary among the organisms? Explain why.

c. Is the (A + G)/(T + C) ratio different for the sperm samples? Would you expect it to be? Why or why not?

20. Boris Magasanik collected data on the amounts of the bases of RNA isolated from a number of sources (shown below), expressed relative to a value of 10 for adenine (B. Magasanik, in *The Nucleic Acids: Chemistry and Biology*, vol. 1, E Chargaff and J. N. Davidson, Eds. New York: Academic Press, 1955).

Organism and tissue	Percent			
	A	G	C	U
Rat liver nuclei	10	14.8	14.3	12.9
Rabbit liver nuclei	10	13.6	13.1	14.0
Cat brain	10	14.7	12.0	9.5
Carp muscle	10	21.0	19.0	11.0
Yeast	10	12.0	8.0	9.8

a. For each organism, compute the ratio of (A + G)/(U + C).

b. How do these ratios compare with the (A + G)/(T + C) ratio found in DNA (see Problem 19)? Explain.

*21. Which of the following relations will be found in the percentages of bases of a double-stranded DNA molecule?

a. A + T = G + C

b. A + T = T + C

c. A + C = G + T

d. $\dfrac{A + T}{C + G}$

e. $\dfrac{A + G}{C + T} = 1.0$

f. $\dfrac{A}{C} = \dfrac{G}{T}$

g. $\dfrac{A}{G} = \dfrac{T}{C}$

h. $\dfrac{A}{T} = \dfrac{G}{C}$

*22. If a double-stranded DNA molecule is 15% thymine, what are the percentages of all the other bases?

23. Heinz Shuster collected the following data on the base composition of ribgrass virus (H. Shuster, in *The Nucleic Acids: Chemistry and Biology*, vol. 3, E. Chargaff and J. N. Davidson, Eds. New York: Academic Press, 1955). On the basis of this information, is the hereditary information of the ribgrass virus RNA or DNA? Is it likely to be single stranded or double stranded?

	Percent				
	A	G	C	T	U
Ribgrass virus	29.3	25.8	18.0	0.0	27.0

24. For entertainment on a Friday night, a genetics professor proposed that his children diagram a polynucleotide strand of DNA. Having learned about DNA in preschool, his

5-year-old daughter was able to draw a polynucleotide strand, but she made a few mistakes. The daughter's diagram (represented here) contained at least 10 mistakes.

a. Make a list of all the mistakes in the structure of this DNA polynucleotide strand.

b. Draw the correct structure for the polynucleotide strand.

*25. Chapter 1 considered the theory of the inheritance of acquired characteristics and noted that this theory is no longer accepted. Is the central dogma consistent with the theory of the inheritance of acquired characteristics? Why or why not?

Section 8.6

*26. Compare and contrast prokaryotic and eukaryotic chromosomes. How are they alike and how do they differ?

*27. A diploid human cell contains approximately 6.4 billion base pairs of DNA.

   a. How many nucleosomes are present in such a cell? (Assume that the linker DNA encompasses 40 bp.)

   b. How many histone proteins are complexed to this DNA?

Challenge Questions

Section 8.1

*28. Suppose that an automated, unmanned probe is sent into deep space to search for extraterrestrial life. After wandering for many light-years among the far reaches of the universe, this probe arrives on a distant planet and detects life. The chemical composition of life on this planet is completely different from that of life on Earth, and its genetic material is not composed of nucleic acids. What predictions can you make about the chemical properties of the genetic material on this planet?

Section 8.2

29. How might ^{32}P and ^{35}S be used to demonstrate that the transforming principle is DNA? Briefly outline an experiment that would show that DNA rather than protein is the transforming principle.

Section 8.3

*30. Researchers have proposed that early life on Earth used RNA as its source of genetic information and that DNA eventually replaced RNA as the source of genetic information. What aspects of DNA structure might make it better suited than RNA to be the genetic material?

9 DNA Replication and Recombination

Molecular model of DNA polymerase η, a translesion polymerase that is able to bypass distortions in DNA structure but often makes errors in DNA synthesis, resulting in mutations.
[J. Trincao et al., Molecular Cell 8:417, 2001. Research Collaboratory for Structural Bioinformatics. H. M. Berman et al., The Protein Data Bank Nucleic Acids Research, 28:235–242, 2000. http://www.pdb.org/.]

Preventing Train Wrecks in Replication

David had a tough childhood. It was bad enough that his fair skin was densely covered with frecklelike spots, but what he really hated were the long-sleeved shirts, long pants, and large straw hat that his mother forced him to wear, even in the middle of summer, when the other kids were in shorts and tee shirts. But, as David grew older, he came to understand that his mother had not been unreasonable, because David suffers from a rare genetic disease called xeroderma pigmentosum, characterized by acute sensitivity to sunlight and a predisposition to skin cancer triggered by exposure to the sun. Xeroderma pigmentosum is an autosomal recessive disease that arises from a defect in one of several genes that encode DNA synthesis and repair enzymes.

DNA polymerases—the enzymes that synthesize DNA—are beautiful and efficient molecular machines. Some of them operate at incredibly high speed, synthesizing DNA at a rate of more than 1000 nucleotides per second, with less than one error per billion nucleotides. To achieve this speed and accuracy, these DNA polymerases, like a high-speed train, require a very smooth track. If the DNA template is damaged or blocked—by, for example, distortions of structure induced by ultraviolet (UV) light—the replication machinery comes to a grinding halt, resulting in gaps in the DNA, with disastrous consequences for the cell.

To overcome this problem, cells have evolved specialized, slower DNA polymerases that are able to bypass distortions that normally block the high-speed, high-fidelity polymerases that are the usual workhorses of replication. But the use of these special "translesion" polymerases comes at a price: they often make mistakes in those sections of DNA that they synthesize. However, most of the errors are corrected by DNA repair mechanisms, and the errors produced by the low-fidelity polymerases are not likely to be as detrimental as the gaps in DNA left by failure to bypass the lesion.

The importance of low-fidelity polymerases is revealed by people with xeroderma pigmentosum. About 20% of those having the disease have a defect in the *POLH* gene, which encodes DNA polymerase η, one of the specialized (low-fidelity) DNA polymerases with the ability to bypass distortions in the DNA template. One such distortion is the presence of bonds between adjacent thymine bases on the same DNA strand; two thymine bases bonded together are called a thymine dimer, which is produced by UV radiation. Because UV radiation is present in sunlight, exposure to the sun causes thymine dimers to form. In most people, thymine dimers are bypassed by specialized polymerases such as DNA polymerase η. Most of the errors that are caused by DNA polymerase η as it bypasses the lesion are later repaired by other mechanisms. However, DNA polymerase η is defective in some people who have xeroderma pigmentosum, and pyrimidine dimers are not bypassed in the normal manner, leading to numerous mutations that eventually produce skin cancer.

The synthesis of DNA is a complex process, fundamental to cell function and health, in which dozens of proteins, enzymes, and DNA structures take part in the copying of DNA. A single defective component, such as DNA polymerase η, can disrupt the whole process and result in severe disease symptoms.

This chapter focuses on DNA replication, the process by which a cell doubles its DNA before division. We begin with the basic mechanism of replication that emerged from the Watson and Crick structure of DNA. We then examine several different modes of replication, the requirements of replication, and the universal direction of DNA synthesis. We examine the enzymes and proteins that participate in this process and conclude the chapter by considering the molecular details of recombination, which is closely related to replication and is essential for the segregation of homologous chromosomes, for the production of genetic variation, and for DNA repair.

9.1 Genetic Information Must Be Accurately Copied Every Time a Cell Divides

In a schoolyard game, a verbal message, such as "John's brown dog ran away from home," is whispered to a child, who runs to a second child and repeats the message. The message is relayed from child to child around the schoolyard until it returns to the original sender. Inevitably, the last child returns with an amazingly transformed message, such as "Joe Brown has a pig living under his porch." The more children playing the game, the more garbled the message becomes. This game illustrates an important principle: errors arise whenever information is copied; the more times it is copied, the greater the number of errors.

A complex, multicellular organism faces a problem analogous to that of the children in the schoolyard game: how to faithfully transmit genetic instructions each time its cells divide. The solution to this problem is central to replication. A huge amount of genetic information and an enormous number of cell divisions are required to produce a multicellular adult organism; even a low rate of error during copying would be catastrophic. A single-celled human zygote contains 6.4 billion base pairs of DNA. If a copying error was made only once per million base pairs, 6400 mistakes would be made every time a cell divided—errors that would be compounded at each of the millions of cell divisions that take place in human development.

Not only must the copying of DNA be astoundingly accurate, it must also take place at breakneck speed. The

single, circular chromosome of *E. coli* contains about 4.6 million base pairs. At a rate of more than 1000 nucleotides per minute, replication of the entire chromosome would require almost 3 days. Yet, as already stated, these bacteria are capable of dividing every 20 minutes. *E. coli* actually replicates its DNA at a rate of 1000 nucleotides per second, with less than one error in a billion nucleotides. How is this extraordinarily accurate and rapid process accomplished?

9.2 All DNA Replication Takes Place in a Semiconservative Manner

From the three-dimensional structure of DNA proposed by Watson and Crick in 1953 (see Figure 8.12), several important genetic implications were immediately apparent. The complementary nature of the two nucleotide strands in a DNA molecule suggested that, during replication, each strand can serve as a template for the synthesis of a new strand. The specificity of base pairing (adenine with thymine; guanine with cytosine) implied that only one sequence of bases can be specified by each template, and so two DNA molecules built on the pair of templates will be identical with the original. This process is called **semiconservative replication**, because each of the original nucleotide strands remains intact (conserved), despite no longer being combined in the same molecule; the original DNA molecule is half (semi) conserved during replication.

Initially, three alternative models were proposed for DNA replication. In conservative replication (**Figure 9.1a**), the entire double-stranded DNA molecule serves as a template for a whole new molecule of DNA, and the original DNA molecule is *fully* conserved during replication. In dispersive replication (**Figure 9.1b**), both nucleotide strands break down (disperse) into fragments, which serve as templates for the synthesis of new DNA fragments, and then somehow reassemble into two complete DNA molecules. In this model, each resulting DNA molecule is interspersed with fragments of old and new DNA; none of the original molecule is conserved. Semiconservative replication (**Figure 9.1c**) is intermediate between these two models; the two nucleotide strands unwind and each serves as a template for a new DNA molecule.

These three models allow different predictions to be made about the distribution of original DNA and newly synthesized DNA after replication. With conservative replication, after one round of replication, 50% of the molecules would consist entirely of the original DNA and 50% would consist entirely of new DNA. After a second round of replication,

(a) Conservative replication   (b) Dispersive replication   (c) Semiconservative replication

Original DNA

First replication

Second replication

9.1 **Three proposed models of replication are conservative replication, dispersive replication, and semiconservative replication.**

25% of the molecules would consist entirely of the original DNA and 75% would consist entirely of new DNA. With each additional round of replication, the proportion of molecules with new DNA would increase, although the number of molecules with the original DNA would remain constant. Dispersive replication would always produce hybrid molecules, containing some original and some new DNA, but the proportion of new DNA within the molecules would increase with each replication event. In contrast, with semiconservative replication, one round of replication would produce two hybrid molecules, each consisting of half original DNA and half new DNA. After a second round of replication, half the molecules would be hybrid, and the other half would consist of new DNA only. Additional rounds of replication would produce more and more molecules consisting entirely of new DNA, and a few hybrid molecules would persist.

Meselson and Stahl's Experiment

To determine which of the three models of replication applied to *E. coli* cells, Matthew Meselson and Franklin Stahl needed a way to distinguish old and new DNA. They did so by using two isotopes of nitrogen, ^{14}N (the common form) and ^{15}N (a rare, heavy form). Meselson and Stahl grew a culture of *E. coli* in a medium that contained ^{15}N as the sole nitrogen source; after many generations, all the *E. coli* cells had ^{15}N incorporated into the purine and pyrimidine bases of DNA (see Figure 8.8). Meselson and Stahl took a sample of these bacteria, switched the rest of the bacteria to a medium that contained only ^{14}N, and then took additional samples of bacteria over the next few cellular generations. In each sample, the bacterial DNA that was synthesized before the change in medium contained ^{15}N and was relatively heavy, whereas any DNA synthesized after the switch contained ^{14}N and was relatively light.

Meselson and Stahl distinguished between the heavy ^{15}N-laden DNA and the light ^{14}N-containing DNA with the use of **equilibrium density gradient centrifugation** (**Figure 9.2**). In this technique, a centrifuge tube is filled with

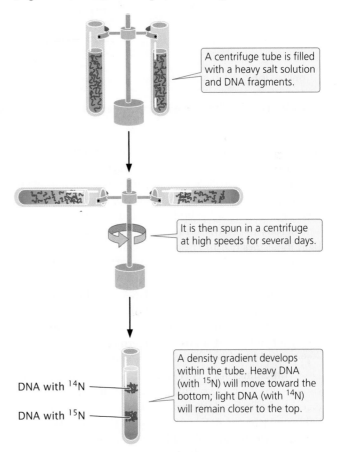

A centrifuge tube is filled with a heavy salt solution and DNA fragments.

It is then spun in a centrifuge at high speeds for several days.

DNA with ^{14}N

DNA with ^{15}N

A density gradient develops within the tube. Heavy DNA (with ^{15}N) will move toward the bottom; light DNA (with ^{14}N) will remain closer to the top.

9.2 **Meselson and Stahl used equilibrium density gradient centrifugation to distinguish between heavy, ^{15}N-laden DNA and lighter, ^{14}N-laden DNA.**

Experiment

Question: Which model of DNA replication—conservative, dispersive, or semiconservative—applies to *E. coli*?

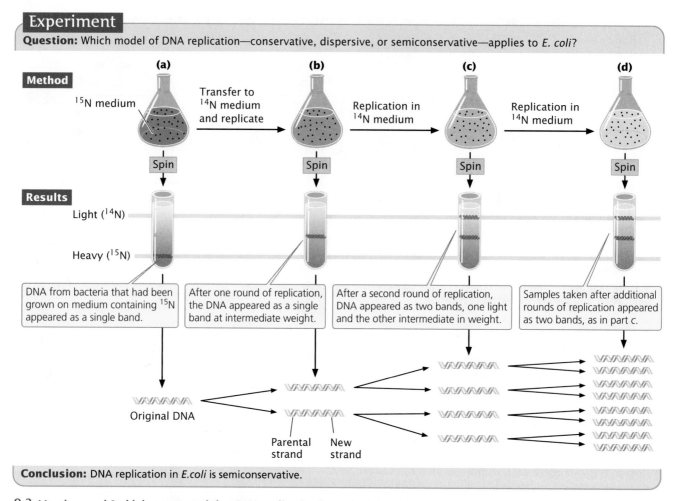

Conclusion: DNA replication in *E.coli* is semiconservative.

9.3 Meselson and Stahl demonstrated that DNA replication is semiconservative.

a heavy salt solution and a substance of which the density is to be measured—in this case, DNA fragments. The tube is then spun in a centrifuge at high speeds. After several days of spinning, a gradient of density develops within the tube, with high density at the bottom and low density at the top. The density of the DNA fragments matches that of the salt: light molecules rise and heavy molecules sink.

Meselson and Stahl found that DNA from bacteria grown only on medium containing ^{15}N produced a single band at the position expected of DNA containing only ^{15}N (**Figure 9.3a**). DNA from bacteria transferred to the medium with ^{14}N and allowed one round of replication also produced a single band but at a position intermediate between that expected of DNA containing only ^{15}N and that expected of DNA containing only ^{14}N (**Figure 9.3b**). This result is inconsistent with the conservative replication model, which predicts one heavy band (the original DNA molecules) and one light band (the new DNA molecules). A single band of intermediate density is predicted by both the semiconservative and the dispersive models.

To distinguish between these two models, Meselson and Stahl grew the bacteria in medium containing ^{14}N for a

second generation. After a second round of replication in medium with ^{14}N, two bands of equal intensity appeared, one in the intermediate position and the other at the position expected of DNA containing only ^{14}N (**Figure 9.3c**). All samples taken after additional rounds of replication produced two bands, and the band representing light DNA became progressively stronger (**Figure 9.3d**). Meselson and Stahl's results were exactly as expected for semiconservative replication and are incompatible with those predicated for both conservative and dispersive replication.

Concepts

Replication is semiconservative: each DNA strand serves as a template for the synthesis of a new DNA molecule. Meselson and Stahl convincingly demonstrated that replication in *E. coli* is semiconservative.

✔ **Concept Check 1**

How many bands of DNA would be expected in Meselson and Stahl's experiment after two rounds of *conservative* replication?

(a)

Origin of replication

Replication fork

Newly synthesized DNA

Replication bubble

1 Double-stranded DNA unwinds at the replication origin,…

2 …producing single-stranded templates for the synthesis of new DNA. A replication bubble forms, usually having a replication fork at each end.

3 The forks proceed around the circle.

4 Eventually two circular DNA molecules are produced.

Conclusion: The products of theta replication are two circular DNA molecules.

(b)

Replication fork

Origin of replication

Replication bubble

9.4 Theta replication is a type of replication common in *E. coli* and other organisms possessing circular DNA. *[Electron micrographs from Bernard Hirt, L'Institut Suisse de Recherche Expérimentale sur le Cancer.]*

Modes of Replication

After Meselson and Stahl's work, investigators confirmed that other organisms also use semiconservative replication. No evidence was found for conservative or dispersive replication. There are, however, several different ways in which semiconservative replication can take place, differing principally in the nature of the template DNA—whether it is linear or circular.

Individual units of replication are called **replicons**, each of which contains a **replication origin**. Replication starts at the origin and continues until the entire replicon has been replicated. Bacterial chromosomes have a single replication origin, whereas eukaryotic chromosomes contain many.

A common type of replication that takes place in circular DNA, such as that found in *E. coli* and other bacteria, is called **theta replication** (**Figure 9.4**) because it generates a structure that resembles the Greek letter theta (θ). In theta replication, double-stranded DNA begins to unwind at the replication origin, producing single-stranded nucleotide strands that then serve as templates on which new DNA can be synthesized. The unwinding of the double helix generates a loop, termed a **replication bubble**. Unwinding may be at one or both ends of the bubble, making it progressively larger. DNA replication on both of the template strands is simultaneous with unwinding. The point of unwinding, where the two single nucleotide strands separate from the double-stranded DNA helix, is called a **replication fork**.

If there are two replication forks, one at each end of the replication bubble, the forks proceed outward in both directions in a process called **bidirectional replication**, simultaneously unwinding and replicating the DNA until they eventually meet. If a single replication fork is present, it proceeds around the entire circle to produce two complete circular DNA molecules, each consisting of one old and one new nucleotide strand.

Circular DNA molecules that undergo theta replication have a single origin of replication. Because of the limited size of these DNA molecules, replication starting from one origin can traverse the entire chromosome in a reasonable amount of time. The large linear chromosomes in eukaryotic cells, however, contain far too much DNA to be replicated speedily from a single origin. Replication takes place on eukaryotic chromosomes simultaneously from thousands of origins.

Typical eukaryotic replicons are from 20,000 to 300,000 base pairs in length. At each replication origin, the DNA unwinds and produces a replication bubble. Replication takes place on both strands at each end of the bubble, with the two replication forks spreading outward. Eventually, the replication forks of adjacent replicons run into each other, and the replicons fuse to form long stretches of newly synthesized DNA (**Figure 9.5**). Replication and fusion of all the replicons leads to two identical DNA molecules. Important features of theta replication and linear eukaryotic replication are summarized in **Table 9.1**.

Requirements of Replication

Although the process of replication includes many components, they can be combined into three major groups:

1. a template consisting of single-stranded DNA,
2. raw materials (substrates) to be assembled into a new nucleotide strand, and
3. enzymes and other proteins that "read" the template and assemble the substrates into a DNA molecule.

Because of the semiconservative nature of DNA replication, a double-stranded DNA molecule must unwind to expose the bases that act as a template for the assembly of new polynucleotide strands, which are complementary and antiparallel to the template strands. The raw materials from which new DNA molecules are synthesized are deoxyribonucleoside triphosphates (dNTPs), each consisting of a deoxyribose sugar and a base (a nucleoside) attached to three phosphate groups (**Figure 9.6a**). In DNA synthesis, nucleotides are added to the 3'-hydroxyl (3'-OH) group of the growing nucleotide strand (**Figure 9.6b**). The 3'-OH group of the last nucleotide on the strand attacks the 5'-phosphate group of the incoming dNTP. Two phosphate groups are cleaved from the incoming dNTP, and a phosphodiester bond is created between the two nucleotides.

DNA synthesis does not happen spontaneously. Rather, it requires a host of enzymes and proteins that function in a coordinated manner. We will examine this complex array of proteins and enzymes as we consider the replication process in more detail.

1 Each chromosome contains numerous origins.

Origin 1   Origin 2   Origin 3

2 At each origin, the DNA unwinds, producing a replication bubble.

3 DNA synthesis takes place on both strands at each end of the bubble as the replication forks proceed outward.

4 Eventually, the forks of adjacent bubbles run into each other and the segments of DNA fuse,…

5 …producing two identical linear DNA molecules.

Newly synthesized DNA

Conclusion: The products of eukaryotic DNA replication are two linear DNA molecules.

9.5 Linear DNA replication takes place in eukaryotic chromosomes.

Table 9.1   Characteristics of theta and linear eukaryotic replication

Replication Model	DNA Template	Breakage of Nucleotide Strand	Number of Replicons	Unidirectional or Bidirectional	Products
Theta	Circular	No	1	Unidirectional or bidirectional	Two circular molecules
Linear eukaryotic	Linear	No	Many	Bidirectional	Two linear molecules

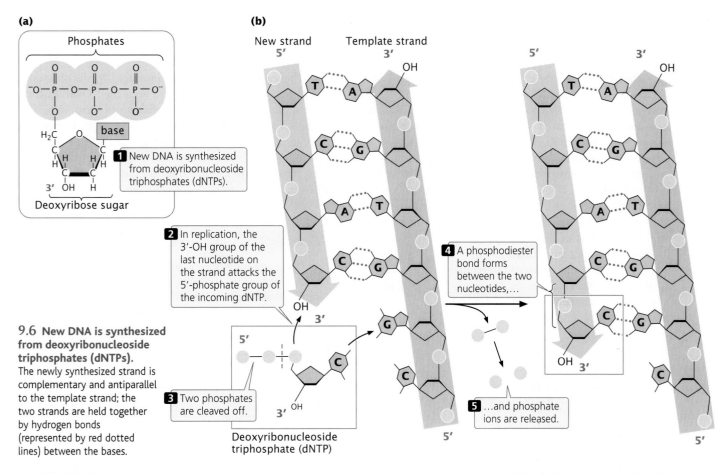

(a)

Phosphates

base

Deoxyribose sugar

1 New DNA is synthesized from deoxyribonucleoside triphosphates (dNTPs).

9.6 New DNA is synthesized from deoxyribonucleoside triphosphates (dNTPs). The newly synthesized strand is complementary and antiparallel to the template strand; the two strands are held together by hydrogen bonds (represented by red dotted lines) between the bases.

(b)

New strand 5′   Template strand 3′

2 In replication, the 3′-OH group of the last nucleotide on the strand attacks the 5′-phosphate group of the incoming dNTP.

3 Two phosphates are cleaved off.

Deoxyribonucleoside triphosphate (dNTP)

4 A phosphodiester bond forms between the two nucleotides,...

5 ...and phosphate ions are released.

Concepts

DNA synthesis requires a single-stranded DNA template, deoxyribonucleoside triphosphates, a growing nucleotide strand, and a group of enzymes and proteins.

Direction of Replication

In DNA synthesis, new nucleotides are joined one at a time to the 3′ end of the newly synthesized strand. **DNA polymerases**, the enzymes that synthesize DNA, can add nucleotides only to the 3′ end of the growing strand (not the 5′ end), and so new DNA strands always elongate in the same 5′-to-3′ direction (5′ → 3′). Because the two single-stranded

DNA templates are antiparallel and strand elongation is always 5′ → 3′, if synthesis on one template proceeds from, say, right to left, then synthesis on the other template must proceed in the opposite direction, from left to right (**Figure 9.7**). As DNA unwinds during replication, the antiparallel nature of the two DNA strands means that one template is exposed in the 5′ → 3′ direction and the other template is exposed in the 3′ → 5′ direction (see Figure 9.7); so how can synthesis take place simultaneously on both strands at the fork?

As the DNA unwinds, the template strand that is exposed in the 3′ → 5′ direction (the lower strand in Figures 9.7 and 9.8) allows the new strand to be synthesized continuously, in the 5′ → 3′ direction. This new strand, which undergoes **continuous replication**, is called the **leading strand**.

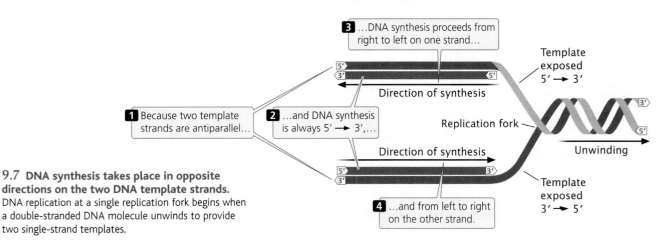

3 ...DNA synthesis proceeds from right to left on one strand...

Template exposed 5′ → 3′

Direction of synthesis

Replication fork

1 Because two template strands are antiparallel...

2 ...and DNA synthesis is always 5′ → 3′,...

Direction of synthesis

Unwinding

9.7 DNA synthesis takes place in opposite directions on the two DNA template strands. DNA replication at a single replication fork begins when a double-stranded DNA molecule unwinds to provide two single-strand templates.

4 ...and from left to right on the other strand.

Template exposed 3′ → 5′

The other template strand is exposed in the $5' \rightarrow 3'$ direction (the upper strand in Figures 9.7 and 9.8). After a short length of the DNA has been unwound, synthesis must proceed $5' \rightarrow 3'$; that is, in the direction *opposite* that of unwinding (**Figure 9.8**). Because only a short length of DNA needs to be unwound before synthesis on this strand gets started, the replication machinery soon runs out of template. By that time, more DNA has unwound, providing new template at the $5'$ end of the new strand. DNA synthesis must start anew at the replication fork and proceed in the direction opposite that of the movement of the fork until it runs into the previously replicated segment of DNA. This process is repeated again and again, and so synthesis of this strand is in short, discontinuous bursts. The newly made strand that undergoes **discontinuous replication** is called the **lagging strand**.

The short lengths of DNA produced by discontinuous replication of the lagging strand are called **Okazaki fragments,** after Reiji Okazaki, who discovered them. In bacterial cells, each Okazaki fragment ranges in length from about 1000 to 2000 nucleotides; in eukaryotic cells, they are about 100 to 200 nucleotides long. Okazaki fragments on the lagging strand are linked together to create a continuous new DNA molecule.

Concepts

All DNA synthesis is $5' \rightarrow 3'$, meaning that new nucleotides are always added to the $3'$ end of the growing nucleotide strand. At each replication fork, synthesis of the leading strand proceeds continuously and that of the lagging strand proceeds discontinuously.

✔ **Concept Check 2**

Discontinuous replication is a result of which property of DNA?

a. Complementary bases   c. Antiparallel nucleotide strands

b. Charged phosphate group   d. Five-carbon sugar

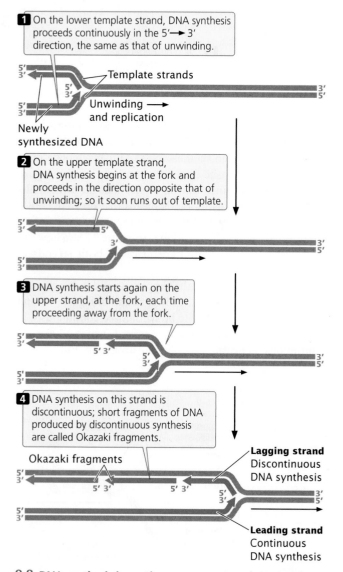

1 On the lower template strand, DNA synthesis proceeds continuously in the $5' \rightarrow 3'$ direction, the same as that of unwinding.

Template strands

Unwinding and replication

Newly synthesized DNA

2 On the upper template strand, DNA synthesis begins at the fork and proceeds in the direction opposite that of unwinding; so it soon runs out of template.

3 DNA synthesis starts again on the upper strand, at the fork, each time proceeding away from the fork.

4 DNA synthesis on this strand is discontinuous; short fragments of DNA produced by discontinuous synthesis are called Okazaki fragments.

Okazaki fragments

Lagging strand Discontinuous DNA synthesis

Leading strand Continuous DNA synthesis

9.8 DNA synthesis is continuous on one template strand of DNA and discontinuous on the other.

9.3 The Replication of DNA Requires a Large Number of Enzymes and Proteins

Replication takes place in four stages: initiation, unwinding, elongation, and termination.

Bacterial DNA Replication

The following discussion of the process of replication will focus on bacterial systems, where replication has been most thoroughly studied and is best understood. Although many aspects of replication in eukaryotic cells are similar to those in prokaryotic cells, there are some important differences. We will compare bacterial and eukaryotic replication later in the chapter.

Initiation   The circular chromosome of *E. coli* has a single replication origin (*oriC*). The minimal sequence required for *oriC* to function consists of 245 bp that contain several critical sites. An **initiator protein** (known as DnaA in *E. coli*) binds to *oriC* and causes a short section of DNA to unwind. This unwinding allows helicase and other single-strand-binding proteins to attach to the polynucleotide strand (**Figure 9.9**).

Unwinding   Because DNA synthesis requires a single-stranded template and because double-stranded DNA must be unwound before DNA synthesis can take place,

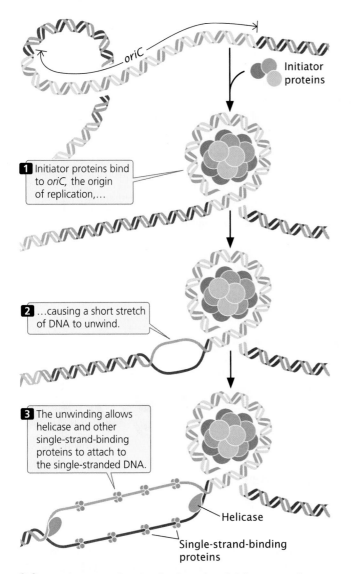

9.9 *E. coli* DNA replication begins when initiator proteins bind to *oriC*, the origin of replication.

1 Initiator proteins bind to *oriC,* the origin of replication,...

2 ...causing a short stretch of DNA to unwind.

3 The unwinding allows helicase and other single-strand-binding proteins to attach to the single-stranded DNA.

Initiator proteins

Helicase

Single-strand-binding proteins

the cell relies on several proteins and enzymes to accomplish the unwinding. A **DNA helicase** breaks the hydrogen bonds that exist between the bases of the two nucleotide strands of a DNA molecule. Helicase cannot *initiate* the unwinding of double-stranded DNA; the initiator protein first separates DNA strands at the origin, providing a short stretch of single-stranded DNA to which a helicase binds. Helicase binds to the lagging-strand template at each replication fork and moves in the $5' \rightarrow 3'$ direction along this strand, thus also moving the replication fork (**Figure 9.10**).

After DNA has been unwound by helicase, the single-stranded nucleotide chains have a tendency to form hydrogen bonds and reanneal (stick back together). Secondary structures also may form between complementary nucleotides on the same strand. To stabilize the single-stranded DNA long enough for replication to take place, **single-strand-binding proteins** (SSBs) attach tightly to the exposed single-stranded DNA (see Figure 9.10). Unlike many DNA-binding proteins, SSBs are indifferent to base sequence: they will bind to any single-stranded DNA. Single-strand-binding proteins form tetramers (groups of four); each tetramer covers from 35 to 65 nucleotides.

Another protein essential for the unwinding process is the enzyme **DNA gyrase**, a topoisomerase. As discussed in Chapter 8, topoisomerases control the supercoiling of DNA. In replication, DNA gyrase reduces the torsional strain (torque) that builds up ahead of the replication fork as a result of unwinding (see Figure 9.10). It reduces torque by making a double-stranded break in one segment of the DNA helix, passing another segment of the helix through the break, and then resealing the broken ends of the DNA.

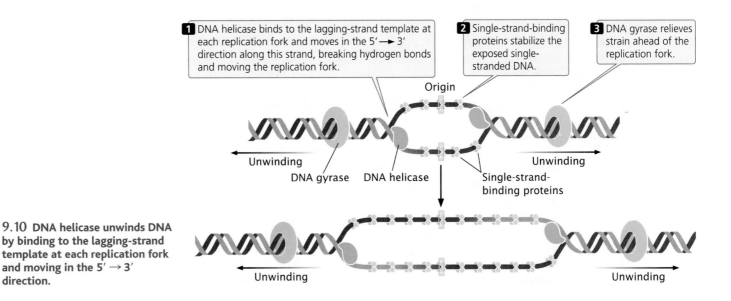

1 DNA helicase binds to the lagging-strand template at each replication fork and moves in the $5' \rightarrow 3'$ direction along this strand, breaking hydrogen bonds and moving the replication fork.

2 Single-strand-binding proteins stabilize the exposed single-stranded DNA.

3 DNA gyrase relieves strain ahead of the replication fork.

Origin

Unwinding

Unwinding

DNA gyrase   DNA helicase   Single-strand-binding proteins

Unwinding

Unwinding

9.10 DNA helicase unwinds DNA by binding to the lagging-strand template at each replication fork and moving in the $5' \rightarrow 3'$ direction.

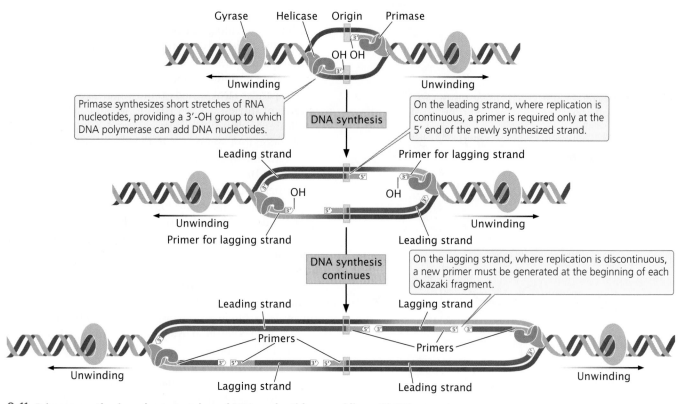

9.11 Primase synthesizes short stretches of RNA nucleotides, providing a 3′-OH group to which DNA polymerase can add DNA nucleotides.

This action removes a twist in the DNA and reduces the supercoiling.

Concepts

Replication is initiated at a replication origin, where an initiator protein binds and causes a short stretch of DNA to unwind. DNA helicase breaks hydrogen bonds at a replication fork, and single-strand-binding proteins stabilize the separated strands. DNA gyrase reduces the torsional strain that develops as the two strands of double-helical DNA unwind.

✔ Concept Check 3

Place the following components in the order in which they are first used in the course of replication: helicase, single-strand-binding protein, DNA gyrase, initiator protein.

Primers   All DNA polymerases require a nucleotide with a 3′-OH group to which a new nucleotide can be added. Because of this requirement, DNA polymerases cannot initiate DNA synthesis on a bare template; rather, they require a primer—an existing 3′-OH group—to get started. How, then, does DNA synthesis begin?

An enzyme called **primase** synthesizes short stretches of nucleotides, or **primers**, to get DNA replication started.

Primase synthesizes a short stretch of RNA nucleotides (about 10–12 nucleotides long), which provides a 3′-OH group to which DNA polymerase can attach DNA nucleotides. (Because primase is an RNA polymerase, it does not require a 3′-OH group to which nucleotides can be added.) All DNA molecules initially have short RNA primers embedded within them; these primers are later removed and replaced by DNA nucleotides.

On the leading strand, where DNA synthesis is continuous, a primer is required only at the 5′ end of the newly synthesized strand. On the lagging strand, where replication is discontinuous, a new primer must be generated at the beginning of each Okazaki fragment (**Figure 9.11**). Primase forms a complex with helicase at the replication fork and moves along the template of the lagging strand. The single primer on the leading strand is probably synthesized by the primase–helicase complex on the template of the lagging strand of the *other* replication fork, at the opposite end of the replication bubble.

Concepts

Primase synthesizes a short stretch of RNA nucleotides (primers), which provides a 3′-OH group for the attachment of DNA nucleotides to start DNA synthesis.

Primers are synthesized where on the lagging strand?

a. Only at the 5′ end of the newly synthesized strand

b. Only at the 3′ end of the newly synthesized strand

c. At the beginning of every Okazaki fragment

d. At multiple places within an Okazaki fragment

Elongation   After DNA is unwound and a primer has been added, DNA polymerases elongate the polynucleotide strand by catalyzing DNA polymerization. The best-studied polymerases are those of *E. coli*, which has at least five different DNA polymerases. Two of them, DNA polymerase I and DNA polymerase III, carry out DNA synthesis in replication; the other three have specialized functions in DNA repair (**Table 9.2**).

DNA polymerase III is a large multiprotein complex that acts as the main workhorse of replication. DNA polymerase III synthesizes nucleotide strands by adding new nucleotides to the 3′ end of a growing DNA molecule. This enzyme has two enzymatic activities (see Table 9.2). Its 5′ → 3′ polymerase activity allows it to add new nucleotides in the 5′ → 3′ direction. Its 3′ → 5′ exonuclease activity allows it to remove nucleotides in the 3′ → 5′ direction, enabling it to correct errors. If a nucleotide having an incorrect base is inserted into the growing DNA molecule, DNA polymerase III uses its 3′ → 5′ exonuclease activity to back up and remove the incorrect nucleotide. It then resumes its 5′ → 3′ polymerase activity. These two functions together allow DNA polymerase III to efficiently and accurately synthesize new DNA molecules.

The first *E. coli* polymerase to be discovered, **DNA polymerase I**, also has 5′ → 3′ polymerase and 3′ → 5′ exonuclease activities (see Table 9.2), permitting the enzyme to synthesize DNA and to correct errors. Unlike DNA polymerase III, however, DNA polymerase I also possesses 5′ → 3′ exonuclease activity, which is used to remove the primers laid down by primase and to replace them with DNA nucleotides by synthesizing in a 5′ → 3′ direction. The removal and replacement of primers appear to constitute the main function of DNA polymerase I. DNA polymerases II, IV, and V function in DNA repair.

Despite their differences, all of *E. coli*'s DNA polymerases

1. synthesize any sequence specified by the template strand;

2. synthesize in the 5′ → 3′ direction by adding nucleotides to a 3′-OH group;

3. use dNTPs to synthesize new DNA;

4. require a primer to initiate synthesis;

5. catalyze the formation of a phosphodiester bond by joining the 5′-phosphate group of the incoming nucleotide to the 3′-OH group of the preceding nucleotide on the growing strand, cleaving off two phosphates in the process;

6. produce newly synthesized strands that are complementary and antiparallel to the template strands; and

7. are associated with a number of other proteins.

Concepts

DNA polymerases synthesize DNA in the 5′ → 3′ direction by adding new nucleotides to the 3′ end of a growing nucleotide strand.

DNA ligase   After DNA polymerase III attaches a DNA nucleotide to the 3′-OH group on the last nucleotide of the RNA primer, each new DNA nucleotide then provides the 3′-OH group needed for the next DNA nucleotide to be added. This process continues as long as template is available (**Figure 9.12a**). DNA polymerase I follows DNA polymerase III and, using its 5′ → 3′ exonuclease activity, removes the RNA primer. It then uses its 5′ → 3′ polymerase activity to replace the RNA nucleotides with DNA nucleotides. DNA

Table 9.2   Characteristics of DNA Polymerases in *E. coli*

DNA Polymerase	5′ → 3′ Polymerization	3′ → 5′ Exonuclease	5′ → 3′ Exonuclease	Function
I	Yes	Yes	Yes	Removes and replaces primers
II	Yes	Yes	No	DNA repair; restarts replication after damaged DNA halts synthesis
III	Yes	Yes	No	Elongates DNA
IV	Yes	No	No	DNA repair
V	Yes	No	No	DNA repair; translesion DNA synthesis

(a) Template strand

DNA nucleotides have been added to the primer by DNA polymerase III.

RNA primer added by primase

DNA polymerase I

(b)

DNA polymerase I replaces the RNA nucleotides of the primer with DNA nucleotides.

RNA nucleotide   DNA dNTP

(c)

Nick

After the last nucleotide of the RNA primer has been replaced, a nick remains in the sugar–phosphate backbone of the strand.

(d)

DNA ligase

DNA ligase seals this nick with a phosphodiester bond between the 5′-P group of the initial nucleotide added by DNA polymerase III and the 3′-OH group of the final nucleotide added by DNA polymerase I.

9.12 DNA ligase seals the nick left by DNA polymerase I in the sugar–phosphate backbone.

polymerase I attaches the first nucleotide to the OH group at the 3′ end of the preceding Okazaki fragment and then continues, in the 5′ → 3′ direction along the nucleotide strand, removing and replacing, one at a time, the RNA nucleotides of the primer (**Figure 9.12b**).

After polymerase I has replaced the last nucleotide of the RNA primer with a DNA nucleotide, a nick remains in the sugar–phosphate backbone of the new DNA strand. The

Table 9.3	Components required for replication in bacterial cells
Component	**Function**
Initiator protein	Binds to origin and separates strands of DNA to initiate replication
DNA helicase	Unwinds DNA at replication fork
Single-strand-binding proteins	Attach to single-stranded DNA and prevent secondary structures from forming
DNA gyrase	Moves ahead of the replication fork, making and resealing breaks in the double-helical DNA to release the torque that builds up as a result of unwinding at the replication fork
DNA primase	Synthesizes a short RNA primer to provide a 3′-OH group for the attachment of DNA nucleotides
DNA polymerase III	Elongates a new nucleotide strand from the 3′-OH group provided by the primer
DNA polymerase I	Removes RNA primers and replaces them with DNA
DNA ligase backbone	Joins Okazaki fragments by sealing nicks in the sugar–phosphate of newly synthesized DNA

3′-OH group of the last nucleotide to have been added by DNA polymerase I is not attached to the 5′-phosphate group of the first nucleotide added by DNA polymerase III (**Figure 9.12c**). This nick is sealed by the enzyme **DNA ligase**, which catalyzes the formation of a phosphodiester bond without adding another nucleotide to the strand (**Figure 9.12d**). Some of the major enzymes and proteins required for replication are summarized in **Table 9.3**.

Concepts

After primers have been removed and replaced, the nick in the sugar–phosphate linkage is sealed by DNA ligase.

✔ Concept Check 5

Which bacterial enzyme removes the primers?

a. Primase

b. DNA polymerase I

c. DNA polymerase II

d. Ligase

Replication fork Now that the major enzymatic components of elongation—DNA polymerases, helicase, primase, and ligase—have been introduced, let's consider how these components interact at the replication fork. Because the synthesis of both strands takes place simultaneously, two units of DNA polymerase III must be present at the replication fork, one for each strand. In one model of the replication process, the two units of DNA polymerase III are connected (**Figure 9.13**); the lagging-strand template loops around so that it is in position for 5′ → 3′ replication. In this way, the DNA polymerase III complex is able to carry out 5′ → 3′ replication simultaneously on both templates, even though they run in opposite directions. After about 1000 bp of new DNA has been synthesized, DNA polymerase III releases the lagging-strand template, and a new loop forms (see Figure 9.13). Primase synthesizes a new primer on the lagging strand and DNA polymerase III then synthesizes a new Okazaki fragment.

In summary, each active replication fork requires five basic components:

1. helicase to unwind the DNA,

2. single-strand-binding proteins to keep the nucleotide strands separate long enough to allow replication,

3. the topoisomerase gyrase to remove strain ahead of the replication fork,

4. primase to synthesize primers with a 3′-OH group at the beginning of each DNA fragment, and

5. DNA polymerase to synthesize the leading and lagging nucleotide strands.

Termination In some DNA molecules, replication is terminated whenever two replication forks meet. In others, specific termination sequences block further replication. A termination protein, called Tus in *E. coli*, binds to these sequences. Tus blocks the movement of helicase, thus stalling the replication fork and preventing further DNA replication.

The fidelity of DNA replication Overall, the error rate in replication is less than one mistake per billion nucleotides. How is this incredible accuracy achieved?

DNA polymerases are very particular in pairing nucleotides with their complements on the template strand. Errors in nucleotide selection by DNA polymerase arise only about once per 100,000 nucleotides. Most of the errors that do arise in nucleotide selection are corrected in a second process called **proofreading**. When a DNA polymerase inserts an incorrect nucleotide into the growing strand, the 3′-OH group of the mispaired nucleotide is not correctly positioned in the active site of the DNA polymerase for accepting the next nucleotide. The incorrect positioning stalls the polymerization reaction, and the 3′ → 5′ exonuclease activity of DNA polymerase removes

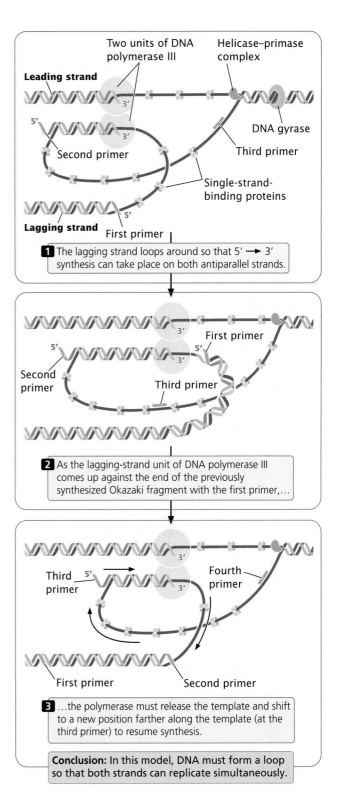

9.13 In one model of DNA replication in *E. coli*, the two units of DNA polymerase III are connected. The lagging-strand template forms a loop so that replication can take place on the two antiparallel DNA strands. Components of the replication machinery at the replication fork are shown at the top.

the incorrectly paired nucleotide. DNA polymerase then inserts the correct nucleotide. Together, proofreading and nucleotide selection result in an error rate of only one in 10 million nucleotides.

A third process, called **mismatch repair** (discussed further in Chapter 13), corrects errors after replication is complete. Any incorrectly paired nucleotides remaining after replication produce a deformity in the secondary structure of the DNA; the deformity is recognized by enzymes that excise an incorrectly paired nucleotide and use the original nucleotide strand as a template to replace the incorrect nucleotide.

In summary, the high level of accuracy in DNA replication is produced by a series of processes, each process catching errors missed by the preceding ones.

Concepts

Replication is extremely accurate, with less than one error per billion nucleotides. This accuracy is due to the processes of nucleotide selection, proofreading, and mismatch repair.

Connecting Concepts

The Basic Rules of Replication

Bacterial replication requires a number of enzymes (see Table 9.3), proteins, and DNA sequences that function together to synthesize a new DNA molecule. These components are important, but we must not become so immersed in the details of the process that we lose sight of the general principles of replication.

1. Replication is always semiconservative.
2. Replication begins at sequences called origins.
3. DNA synthesis is initiated by short segments of RNA called primers.
4. The elongation of DNA strands is always in the $5' \rightarrow 3'$ direction.
5. New DNA is synthesized from dNTPs; in the polymerization of DNA, two phosphate groups are cleaved from a dNTP and the resulting nucleotide is added to the 3'-OH group of the growing nucleotide strand.
6. Replication is continuous on the leading strand and discontinuous on the lagging strand.
7. New nucleotide strands are complementary and antiparallel to their template strands.
8. Replication takes place at very high rates and is astonishingly accurate, thanks to precise nucleotide selection, proofreading, and repair mechanisms.

Eukaryotic DNA Replication

Although eukaryotic replication resembles bacterial replication in many respects, replication in eukaryotic cells presents several additional challenges. First, the much greater size of eukaryotic genomes requires that replication be initiated at multiple origins. Second, eukaryotic chromosomes are linear, whereas prokaryotic chromosomes are circular. Third, the DNA template is associated with histone proteins in the form of nucleosomes, and nucleosome assembly must immediately follow DNA replication.

Eukaryotic origins The origins of replication of different eukaryotic organisms vary greatly in sequence, although they usually contain numerous A–T base pairs. In yeast, origins consist of 100 to 120 bp of DNA. A multiprotein complex, the origin-recognition complex (ORC), binds to origins and unwinds the DNA in this region. Interestingly, ORCs also function in regulating transcription.

Concepts

Eukaryotic DNA contains many origins of replication. At each origin, a multiprotein origin-recognition complex binds to initiate the unwinding of the DNA.

✔ Concept Check 6

In comparison with prokaryotes, what are some differences in the genome structure of eukaryotic cells that affect how replication takes place?

The licensing of DNA replication Eukaryotic cells utilize thousands of origins, and so the entire genome can be replicated in a timely manner. The use of multiple origins, however, creates a special problem in the timing of replication: the entire genome must be precisely replicated once and only once in each cell cycle so that no genes are left unreplicated and no genes are replicated more than once. How does a cell ensure that replication is initiated at thousands of origins only once per cell cycle?

The precise replication of DNA is accomplished by the separation of the initiation of replication into two distinct steps. In the first step, the origins are licensed, meaning that they are approved for replication. This step is early in the cell cycle when a **replication licensing factor** attaches to an origin. In the second step, the replication machinery initiates replication at each *licensed* origin. The key is that the replication machinery functions only at licensed origins. As the replication forks move away from the origin, the licensing factor is removed, leaving the origin in an unlicensed state, where replication cannot be initiated again until the license is renewed. To ensure that replication takes place only once per cell cycle, the licensing factor is active only after the cell has completed mitosis and before the replication is initiated.

Unwinding Several different helicases that separate double-stranded DNA have been isolated from eukaryotic cells, as

Table 9.4	DNA polymerases in eukaryotic cells			
DNA Polymerase	5′ → 3′ Polymerase Activity	3′ → 5′ Exonuclease Activity		Cellular Function
α (alpha)	Yes	No		Initiation of nuclear DNA synthesis and DNA repair; has primase activity
β (beta)	Yes	No		DNA repair and recombination of nuclear DNA
γ (gamma)	Yes	Yes		Replication and repair of mitochondrial DNA
δ (delta)	Yes	Yes		Lagging-strand synthesis of nuclear DNA, DNA repair, and translesion DNA synthesis
ε (epsilon)	Yes	Yes		Leading-strand synthesis
ζ (zeta)	Yes	No		Translesion DNA synthesis
η (eta)	Yes	No		Translesion DNA synthesis
θ (theta)	Yes	No		DNA repair
ι (iota)	Yes	No		Translesion DNA synthesis
κ (kappa)	Yes	No		Translesion DNA synthesis
λ (lambda)	Yes	No		DNA repair
μ (mu)	Yes	No		DNA repair
σ (sigma)	Yes	No		Nuclear DNA replication (possibly), DNA repair, and sister-chromatid cohesion
φ (phi)	Yes	No		Translesion DNA synthesis
Rev1	Yes	No		DNA repair

have single-strand-binding proteins and topoisomerases (which have a function equivalent to the DNA gyrase in bacterial cells). These enzymes and proteins are assumed to function in unwinding eukaryotic DNA in much the same way as their bacterial counterparts do.

Eukaryotic DNA polymerases Some significant differences in the processes of bacterial and eukaryotic replication are in the number and functions of DNA polymerases. Eukaryotic cells contain a number of different DNA polymerases that function in replication, recombination, and DNA repair (**Table 9.4**). **DNA polymerase α**, which contains primase activity, initiates nuclear DNA synthesis by synthesizing an RNA primer, followed by a short string of DNA nucleotides. After DNA polymerase α has laid down from 30 to 40 nucleotides, **DNA polymerase δ** completes replication on the lagging strand. Similar in structure and function to DNA polymerase δ, **DNA polymerase ε** replicates the leading strand. **DNA polymerase β** does not participate in replication but is associated with the repair and recombination of nuclear DNA. **DNA polymerase γ** replicates mitochondrial DNA; a γ-like polymerase also replicates chloroplast DNA. Other DNA polymerases (see Table

9.4) play a role in DNA repair, including the error-prone translesion DNA polymerases that were described at the beginning of the chapter.

Concepts

There are a large number of different DNA polymerases in eukaryotic cells. DNA polymerases α, δ, and ε carry out replication on the leading and lagging strands. Other DNA polymerases carry out DNA repair.

Replication at the Ends of Chromosomes

A fundamental difference between eukaryotic and bacterial replication arises because eukaryotic chromosomes are linear and thus have ends. The 3′-OH group needed for replication by DNA polymerases is provided at the initiation of replication by RNA primers that are synthesized by primase. This solution is temporary, because, eventually, the primers

(a) Circular DNA

OH Primer
3′ 5′ 3′

Replication around the circle provides a 3′-OH group in front of the primer; nucleotides can be added to the 3′-OH group when the primer is replaced.

(b) Linear DNA

Telomeres

1 In linear DNA with multiple origins of replication, elongation of DNA in adjacent replicons provides a 3′-OH group for replacement of each primer.

Lagging strand

Origin Primer

Leading strand

Unwinding

2 Primers at the ends of chromosomes cannot be replaced, because there is no adjacent 3′-OH to which DNA nucleotides can be attached.

3′OH

Primer

3 When the primer at the end of a chromosome is removed,…

5′
Primer

4 …there is no 3′-OH group to which DNA nucleotides can be attached, producing a gap.

Gap left by removal of primer

Conclusion: In the absence of special mechanisms, DNA replication would leave gaps owing to the removal of primers at the ends of chromosomes.

9.14 DNA synthesis at the ends of circular and linear chromosomes must differ.

must be removed and replaced by DNA nucleotides. In a circular DNA molecule, elongation around the circle eventually provides a 3′-OH group immediately in front of the primer (**Figure 9.14a**). After the primer has been removed, the replacement DNA nucleotides can be added to this 3′-OH group.

The end-of-chromosome problem In linear chromosomes with multiple origins, the elongation of DNA in

adjacent replicons also provides a 3′-OH group preceding each primer (**Figure 9.14b**). At the very end of a linear chromosome, however, there is no adjacent stretch of replicated DNA to provide this crucial 3′-OH group. When the primer at the end of the chromosome has been removed, it cannot be replaced with DNA nucleotides, which produces a gap at the end of the chromosome, suggesting that the chromosome should become progressively shorter with each round of replication. The chromosome would be shortened with each successive generation of an organism, leading to the eventual elimination of the entire telomere, destabilization of the chromosome, and cell death. But chromosomes don't normally become shorter each generation and destabilize; so how are the ends of linear chromosomes replicated?

Telomeres and telomerase The ends of chromosomes—the telomeres—possess several unique features, one of which is the presence of many copies of a short repeated sequence. In the protozoan *Tetrahymena*, this telomeric repeat is CCCCAA, with the G-rich strand typically protruding beyond the C-rich strand (**Figure 9.15a**):

end of   ←   5′-CCCCAA →   toward
chromosome   3′-GGGGTTGGGGTT   centromere

The single-stranded protruding end of the telomere can be extended by **telomerase**, an enzyme with both a protein and an RNA component (also known as a ribonucleoprotein). The RNA part of the enzyme contains from 15 to 22 nucleotides that are complementary to the sequence on the G-rich strand. This sequence pairs with the overhanging 3′ end of the DNA (**Figure 9.15b**) and provides a template for the synthesis of additional DNA copies of the repeats. DNA nucleotides are added to the 3′ end of the strand one at a time (**Figure 9.15c**) and, after several nucleotides have been added, the RNA template moves down the DNA and more nucleotides are added to the 3′ end (**Figure 9.15d**). Usually, from 14 to 16 nucleotides are added to the 3′ end of the G-rich strand.

In this way, the telomerase can extend the 3′ end of the chromosome without the use of a complementary DNA template (**Figure 9.15e**). How the complementary C-rich strand is synthesized (**Figure 9.15f**) is not yet clear. It may be synthesized by conventional replication, with DNA polymerase α synthesizing an RNA primer on the 5′ end of the extended (G-rich) template. The removal of this primer once again leaves a gap at the 5′ end of the chromosome, but this gap does not matter, because the end of the chromosome is extended at each replication by telomerase; so, the chromosome does not become shorter overall.

Telomerase is present in single-celled organisms, germ cells, early embryonic cells, and certain proliferative somatic cells (such as bone-marrow cells and cells lining the intestine), all of which must undergo continuous cell division. Most somatic cells have little or no telomerase activity, and chromosomes in these cells progressively shorten with each cell division. These cells are capable of only a limited number of divisions; when the telomeres have shortened beyond a critical point, a chromosome becomes unstable, has a tendency to undergo rearrangements, and is degraded. These events lead to cell death.

Concepts

The ends of eukaryotic chromosomes are replicated by an RNA–protein enzyme called telomerase. This enzyme adds extra nucleotides to the G-rich DNA strand of the telomere.

✔ Concept Check 7

What would be the result if an organism's telomerase were mutated and nonfunctional?

a. No DNA replication would take place.

b. The DNA polymerase enzyme would stall at the telomere.

c. Chromosomes would shorten each generation.

d. RNA primers could not be removed.

Telomerase, aging, and disease   The shortening of telomeres may contribute to the process of aging. The telomeres of genetically engineered mice that lack a functional telomerase gene (and therefore do not express telomerase in somatic or germ cells) undergo progressive shortening in successive generations. After several generations, these mice show some signs of premature aging, such as graying, hair loss, and delayed wound healing. Through genetic engineering, it is also possible to create somatic cells that express telomerase. In these cells, telomeres do not shorten, cell aging is inhibited, and the cells will divide indefinitely. Although these observations suggest that telomere length is associated with aging, the precise role of telomeres in human aging is uncertain and controversial.

Some diseases are associated with abnormalities of telomere replication. People with Werner syndrome, an autosomal recessive disease, show signs of premature aging that begins in adolescence or early adulthood, including wrinkled skin, graying of the hair, baldness, cataracts, and muscle atrophy. They often develop cancer, osteoporosis, heart and artery disease, and other ailments typically associated with aging. The causative gene, called *WRN*, has been mapped to human chromosome 8 and normally encodes a RecQ helicase enzyme. This enzyme is necessary for the

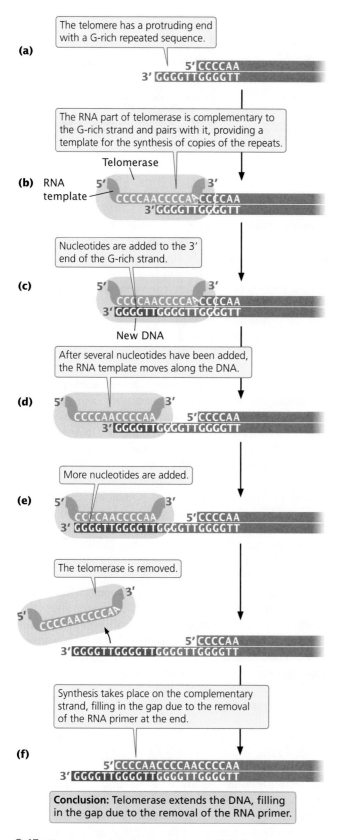

(a) The telomere has a protruding end with a G-rich repeated sequence.

5′ CCCCAA
3′ GGGGTTGGGGTT

(b) The RNA part of telomerase is complementary to the G-rich strand and pairs with it, providing a template for the synthesis of copies of the repeats.

Telomerase

RNA template

5′ CCCCAACCCCAACCCCAA 3′
3′ GGGGTTGGGGTT

(c) Nucleotides are added to the 3′ end of the G-rich strand.

5′ CCCCAACCCCAACCCCAA 3′
3′ GGGGTTGGGGTTGGGGTT

New DNA

(d) After several nucleotides have been added, the RNA template moves along the DNA.

5′ CCCCAACCCCAA 3′   5′ CCCCAA
3′ GGGGTTGGGGTTGGGGTT

(e) More nucleotides are added.

5′ CCCCAACCCCAA 3′   5′ CCCCAA
3′ GGGGTTGGGGTTGGGGTTGGGGTT

(f) The telomerase is removed.

3′
5′ CCCCAACCCCAA

5′ CCCCAA
3′ GGGGTTGGGGTTGGGGTTGGGGTT

Synthesis takes place on the complementary strand, filling in the gap due to the removal of the RNA primer at the end.

5′ CCCCAACCCCAACCCCAA
3′ GGGGTTGGGGTTGGGGTTGGGGTT

Conclusion: Telomerase extends the DNA, filling in the gap due to the removal of the RNA primer.

9.15 The enzyme telomerase is responsible for the replication of chromosome ends.

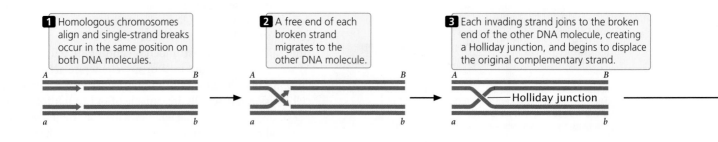

1 Homologous chromosomes align and single-strand breaks occur in the same position on both DNA molecules.

2 A free end of each broken strand migrates to the other DNA molecule.

3 Each invading strand joins to the broken end of the other DNA molecule, creating a Holliday junction, and begins to displace the original complementary strand.

Holliday junction

efficient replication of telomeres. In people with Werner syndrome, this helicase is defective and, consequently, the telomeres shorten prematurely.

Telomerase also appears to play a role in cancer. Cancer tumor cells have the capacity to divide indefinitely, and the telomerase enzyme is expressed in 90% of all cancers. As will be considered in Chapter 15, cancer is a complex, multistep process that usually requires mutations in at least several genes. Telomerase activation alone does not lead to cancerous growth in most cells, but it does appear to be required, along with other mutations, for cancer to develop.

Replication in Archaea

The process of replication in archaebacteria has a number of features in common with replication in eukaryotic cells; many of the proteins taking part are more similar to those in eukaryotic cells than to those in eubacteria. Like eubacteria, some archaebacteria have a single replication origin, but the archaean *Sulfolobus solfataricus* has two origins of replication, similar to the multiple origins seen in eukaryotic genomes. The replication origins of archaebacteria do not contain the typical sequences recognized by bacterial initiator proteins; instead, they have sequences that are similar to those found in eukaryotic origins. The initiator proteins of archaebacteria also are more similar to those of eukaryotes than those of eubacteria. These similarities in replication between archaeal and eukaryotic cells reinforce the conclusion that the archaea are more closely related to eukaryotic cells than to the prokaryotic eubacteria.

9.4 Recombination Takes Place Through the Breakage, Alignment, and Repair of DNA Strands

Recombination is the exchange of genetic information between DNA molecules; when the exchange is between homologous DNA molecules, it is called **homologous recombination**. This process takes place in crossing over, in which homologous regions of chromosomes are exchanged (see Figure 5.6) and genes are shuffled into new combinations. Recombination is an extremely important genetic process because it increases genetic variation. Rates of recombination provide important information about linkage relations among genes, which is used to create genetic maps (see Figures 5.13 and 5.14). Recombination is also essential for some types of DNA repair (Chapter 13).

Homologous recombination is a remarkable process: a nucleotide strand of one chromosome aligns precisely with a nucleotide strand of the homologous chromosome, breaks arise in corresponding regions of different DNA molecules, parts of the molecules precisely change place, and then the pieces are correctly joined. In this complicated series of events, no genetic information is lost or gained. Although the precise molecular mechanism of homologous recombination is still poorly known, the exchange is probably accomplished through the pairing of complementary bases. A single-stranded DNA molecule of one chromosome pairs with a single-stranded DNA molecule of another, forming **heteroduplex DNA**.

In meiosis, homologous recombination (crossing over) could theoretically take place before, during, or after DNA synthesis. Cytological, biochemical, and genetic evidence indicates that it takes place in prophase I of meiosis, whereas DNA replication takes place earlier, in interphase. Thus, crossing over must entail the breaking and rejoining of chromatids when homologous chromosomes are at the four-strand stage (see Figure 5.6).

Homologous recombination may take place through several different pathways. One pathway is initiated by a single-strand break in each of two DNA molecules and includes the formation of a special structure called the **Holliday junction** (**Figure 9.16**). In this model, double-stranded DNA molecules from two homologous chromosomes align precisely. A single-strand break in one of the DNA molecules provides a free end that invades and joins to the free end of the other DNA molecule. Strand invasion and joining take place on both DNA molecules, creating two heteroduplex DNAs, each consisting of one original

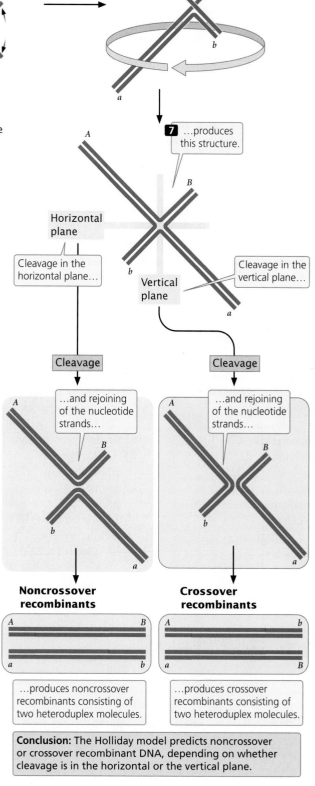

strand plus one new strand from the other DNA molecule. The point at which nucleotide strands pass from one DNA molecule to the other is the Holliday junction (see Figure 9.16). The junction moves along the molecules in a process called branch migration. The exchange of nucleotide strands and branch migration produce a structure termed the Holliday intermediate, which can be cleaved in one of two ways. Cleavage may be in the horizontal plane, followed by rejoining of the strands, producing noncrossover recombinants, in which the genes on either end of the molecules are identical with those originally present (gene *A* with gene *B*, and gene *a* with gene *b*). Cleavage in the vertical plane, followed by rejoining, produces crossover recombinants, in which the genes on either end of the molecules are different from those originally present (gene *A* with gene *b*, and gene *a* with gene *B*).

Concepts

Homologous recombination requires the formation of heteroduplex DNA consisting of one nucleotide strand from each of two homologous chromosomes. In the Holliday model, homologous recombination is accomplished through a single-strand break in the DNA, strand displacement, and branch migration.

✔ Concept Check 8

Why is recombination important?

9.16 The Holliday model of homologous recombination. In this model, recombination takes place through a single-strand break in each DNA duplex, strand displacement, branch migration, and resolution of a single Holliday junction.

Concepts Summary

- Replication is semiconservative: DNA's two nucleotide strands separate, and each serves as a template on which a new strand is synthesized.

- All DNA synthesis is in the $5' \rightarrow 3'$ direction. Because the two nucleotide strands of DNA are antiparallel, replication takes place continuously on one strand (the leading strand) and discontinuously on the other (the lagging strand).

- Replication begins when an initiator protein binds to a replication origin and unwinds a short stretch of DNA to which DNA helicase attaches. DNA helicase unwinds the DNA at the replication fork, single-strand-binding proteins bind to single nucleotide strands to prevent secondary structures, and DNA gyrase (a topoisomerase) removes the strain ahead of the replication fork that is generated by unwinding.

- During replication, primase synthesizes short primers of RNA nucleotides, providing a 3'-OH group to which DNA polymerase can add DNA nucleotides.

- DNA polymerase adds new nucleotides to the 3' end of a growing polynucleotide strand. Bacteria have two DNA polymerases that have primary roles in replication: DNA polymerase III, which synthesizes new DNA on the leading and lagging strands, and DNA polymerase I, which removes and replaces primers.

- DNA ligase seals the nicks that remain in the sugar–phosphate backbones when the RNA primers are replaced by DNA nucleotides.

- Several mechanisms ensure the high rate of accuracy in replication, including precise nucleotide selection, proofreading, and mismatch repair.

- Precise replication at multiple origins in eukaryotes is ensured by a licensing factor that must attach to an origin before replication can begin.

- The ends of linear eukaryotic DNA molecules are replicated by the enzyme telomerase.

- Homologous recombination takes place through breaks in nucleotide strands, alignment of homologous DNA segments, and rejoining of the strands.

Important Terms

semiconservative replication (p. 220)
equilibrium density gradient
 centrifugation (p. 221)
replicon (p. 223)
replication origin (p. 223)
theta replication (p. 223)
replication bubble (p. 223)
replication fork (p. 223)
bidirectional replication (p. 223)
DNA polymerase (p. 225)
continuous replication (p. 225)
leading strand (p. 225)
discontinuous replication (p. 226)

lagging strand (p. 226)
Okazaki fragment (p. 226)
initiator protein (p. 226)
DNA helicase (p. 227)
single-strand-binding protein (SSB)
 (p. 227)
DNA gyrase (p. 227)
primase (p. 228)
primer (p. 228)
DNA polymerase III (p. 229)
DNA polymerase I (p. 229)
DNA ligase (p. 230)

proofreading (p. 231)
mismatch repair (p. 232)
replication licensing factor (p. 232)
DNA polymerase α (p. 233)
DNA polymerase δ (p. 233)
DNA polymerase ε (p. 233)
DNA polymerase β (p. 233)
DNA polymerase γ (p. 233)
telomerase (p. 234)
homologous recombination (p. 236)
heteroduplex DNA (p. 236)
Holliday junction (p. 236)

Answers to Concept Checks

1. Two bands

2. c

3. Initiator protein, helicase, single-strand-binding protein, DNA gyrase

4. c

5. b

6. The size of eukaryotic genomes, the linear structure of eukaryotic chromosomes, and the association of DNA with histone proteins

7. c

8. Recombination is important for generating genetic variation.

Worked Problems

1. The following diagram represents the template strands of a replication bubble in a DNA molecule. Draw in the newly synthesized strands and identify the leading and lagging strands.

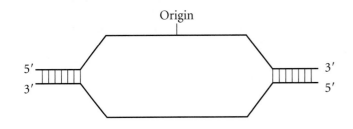

• **Solution**

To determine the leading and lagging strands, first note which end of each template strand is 5′ and which end is 3′. With a pencil, draw in the strands being synthesized on these templates, and identify their 5′ and 3′ ends, recalling that the newly synthesized strands must be antiparallel to the templates.

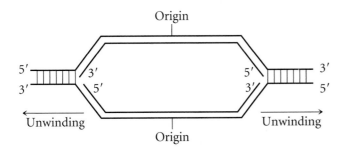

Next, determine the direction of replication for each new strand, which must be 5′ → 3′. You might draw arrows on the new strands to indicate the direction of replication. After you have established the direction of replication for each strand, look at each fork and determine whether the direction of replication for a strand is the same as the direction of unwinding. The strand on which replication is in the same direction as unwinding is the leading strand. The strand on which replication is in the direction opposite that of unwinding is the lagging strand. Make sure that you have one leading strand and one lagging strand for each fork.

2. Consider the experiment conducted by Meselson and Stahl in which they used ^{14}N and ^{15}N in cultures of *E. coli* and equilibrium density gradient centrifugation. Draw pictures to represent the bands produced by bacterial DNA in the density-gradient tube before the switch to medium containing ^{14}N and after one, two, and three rounds of replication after the switch to the medium containing ^{14}N. Use a separate set of drawings to show the bands that would appear if replication were (a) semiconservative; (b) conservative; (c) dispersive.

• **Solution**

DNA labeled with ^{15}N will be denser than DNA labeled with ^{14}N; therefore ^{15}N-labeled DNA will sink lower in the density-gradient tube. Before the switch to medium containing ^{14}N, all DNA in the bacteria will contain ^{15}N and will produce a single band in the lower end of the tube.

a. With semiconservative replication, the two strands separate, and each serves as a template on which a new strand is synthesized. After one round of replication, the original template strand of each molecule will contain ^{15}N and the new strand of each molecule will contain ^{14}N; so a single band will appear in the density gradient halfway between the positions expected of DNA with ^{15}N and of DNA with ^{14}N. In the next round of replication, the two strands again separate and serve as templates for new strands. Each of the new strands contains only ^{14}N, thus some DNA molecules will contain one strand with the original ^{15}N and one strand with new ^{14}N, whereas the other molecules will contain two strands with ^{14}N. This labeling will produce two bands, one at the intermediate position and one at a higher position in the tube. Additional rounds of replication should produce increasing amounts of DNA that contains only ^{14}N; so the higher band will get darker.

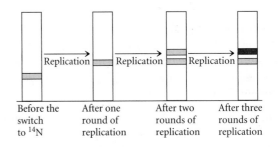

| Before the switch to ^{14}N | After one round of replication | After two rounds of replication | After three rounds of replication |

b. With conservative replication, the entire molecule serves as a template. After one round of replication, some molecules will consist entirely of ^{15}N, and others will consist entirely of ^{14}N; so two bands should be present. Subsequent rounds of replication will increase the fraction of DNA consisting entirely of new ^{14}N; thus the upper band will get darker. However, the original DNA with ^{15}N will remain, and so two bands will be present.

Before the switch to ¹⁴N | After one round of replication | After two rounds of replication | After three rounds of replication

c. In dispersive replication, both nucleotide strands break down into fragments that serve as templates for the synthesis of new DNA. The fragments then reassemble into DNA molecules. After one round of replication, all DNA should contain approximately half ^{15}N and half ^{14}N, producing a single band that is halfway between the positions expected of DNA labeled with ^{15}N and of DNA labeled with ^{14}N. With further rounds of replication, the proportion of ^{14}N in each molecule increases; so a single hybrid band remains, but its position in the density gradient will move upward. The band is also expected to get darker as the total amount of DNA increases.

Before the switch to ¹⁴N | After one round of replication | After two rounds of replication | After three rounds of replication

3. The *E. coli* chromosome contains 4.6 million base pairs of DNA. If synthesis at each replication fork takes place at a rate of 1000 nucleotides per second, how long will it take to completely replicate the *E. coli* chromosome with theta replication?

• Solution

Bacterial chromosomes contain a single origin of replication, and theta replication usually employs two replication forks, which proceed around the chromosome in opposite directions. Thus, the overall rate of replication for the whole chromosome is 2000 nucleotides per second. With a total of 4.6 million base pairs of DNA, the entire chromosome will be replicated in:

$$4,600,00 \text{ bp} \times \frac{1 \text{ second}}{2000 \text{ bp}} = 2300 \text{ seconds} \times \frac{1 \text{ minute}}{60 \text{ seconds}}$$

$$= 38.33 \text{ minutes}$$

At the beginning of this chapter, *E. coli* was said to be capable of dividing every 20 minutes. How is this rate possible if it takes almost twice as long to replicate its genome? The answer is that a second round of replication begins before the first round has finished. Thus, when an *E. coli* cell divides, the chromosomes that are passed on to the daughter cells are already partly replicated. In contrast, a eukaryotic cell replicates its entire genome once, and only once, in each cell cycle.

Comprehension Questions

Section 9.2

1. What is semiconservative replication?

*2. How did Meselson and Stahl demonstrate that replication in *E. coli* takes place in a semiconservative manner?

*3. Draw a molecule of DNA undergoing replication. On your drawing, identify (1) origin, (2) polarity (5′ and 3′ ends) of all template strands and newly synthesized strands, (3) leading and lagging strands, (4) Okazaki fragments, and (5) location of primers.

4. Draw a molecule of DNA undergoing eukaryotic linear replication. On your drawing, identify (1) origin, (2) polarity (5′ and 3′ ends) of all template and newly synthesized strands, (3) leading and lagging strands, (4) Okazaki fragments, and (5) location of primers.

5. What are three major requirements of replication?

*6. What substrates are used in the DNA synthesis reaction?

Section 9.3

7. List the different proteins and enzymes taking part in bacterial replication. Give the function of each in the replication process.

8. What similarities and differences exist in the enzymatic activities of DNA polymerases I, II, and III? What is the function of each type of DNA polymerase in bacterial cells?

*9. Why is primase required for replication?

10. What three mechanisms ensure the accuracy of replication in bacteria?

11. How does replication licensing ensure that DNA is replicated only once at each origin per cell cycle?

*12. In what ways is eukaryotic replication similar to bacterial replication, and in what ways is it different?

13. What is the end-of-chromosome problem for replication? Why, in the absence of telomerase, do the ends of chromosomes get progressively shorter each time the DNA is replicated?

14. Outline in words and pictures how telomeres at the ends of eukaryotic chromosomes are replicated.

Application Questions and Problems

Section 9.2

*15. Suppose a future scientist explores a distant planet and discovers a novel form of double-stranded nucleic acid. When this nucleic acid is exposed to DNA polymerases from *E. coli,* replication takes place continuously on both strands. What conclusion can you make about the structure of this novel nucleic acid?

16. A line of mouse cells is grown for many generations in a medium with ^{15}N. Cells in G_1 are then switched to a new medium that contains ^{14}N. Draw a pair of homologous chromosomes from these cells at the following stages, showing the two strands of DNA molecules found in the chromosomes. Use different colors to represent strands with ^{14}N and ^{15}N.

   a. Cells in G_1, before switching to medium with ^{14}N

   b. Cells in G_2, after switching to medium with ^{14}N

   c. Cells in anaphase of mitosis, after switching to medium with ^{14}N

   d. Cells in metaphase I of meiosis, after switching to medium with ^{14}N

   e. Cells in anaphase II of meiosis, after switching to medium with ^{14}N

17. A bacterium synthesizes DNA at each replication fork at a rate of 1000 nucleotides per second. If this bacterium completely replicates its circular chromosome by theta replication in 30 minutes, how many base pairs of DNA will its chromosome contain?

Section 9.3

*18. The following diagram represents a DNA molecule that is undergoing replication. Draw in the strands of newly synthesized DNA and identify the following items:

   a. Polarity of newly synthesized strands

   b. Leading and lagging strands

   c. Okazaki fragments

   d. RNA primers

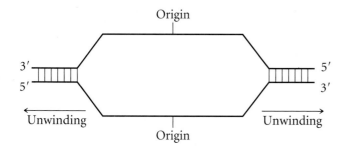

*19. What would be the effect on DNA replication of mutations that destroyed each of the following activities in DNA polymerase I?

   a. $3' \rightarrow 5'$ exonuclease activity

   b. $5' \rightarrow 3'$ exonuclease activity

   c. $5' \rightarrow 3'$ polymerase activity

20. How would DNA replication be affected in a cell that is lacking topoisomerase?

21. DNA polymerases are not able to prime replication, yet primase and other RNA polymerases can. Some geneticists have speculated that the inability of DNA polymerase to prime replication is due to its proofreading function. This hypothesis argues that proofreading is essential for faithful transmission of genetic information and that, because DNA polymerases have evolved the ability to proofread, they cannot prime DNA synthesis. Explain why proofreading and priming functions in the same enzyme might be incompatible?

22. A number of scientists who study ways to treat cancer have become interested in telomerase. Why would they be interested in telomerase? How might cancer-drug therapies that target telomerase work?

23. The enzyme telomerase is part protein and part RNA. What would be the most likely effect of a large deletion in the gene that encodes the RNA part of telomerase? How would the function of telomerase be affected?

24. Dyskeratosis congenita (DKC) is a rare genetic disorder characterized by abnormal fingernails and skin pigmentation, the formation of white patches on the tongue and cheek, and progressive failure of the bone marrow. An autosomal dominant form of DKC results from mutations in the gene that encodes the RNA component of telomerase. Tom Vulliamy and his colleagues examined 15 families with autosomal dominant DKC (T. Vulliamy et al. 2004. *Nature Genetics* 36:447–449). They observed that the median age of onset of DKC in parents was 37 years, whereas the median age of onset in the children of affected parents was 14.5 years. Thus, DKC in these families arose at progressively younger ages in successive generations, a phenomenon known as anticipation. The researchers measured telomere length of members of these families; the measurements are given in the table on page 242. Telomere length normally shortens with age, and so telomere length was adjusted for age. Note that the age-adjusted telomere length of all members of these families is negative, indicating that their telomeres are shorter than normal. For age-adjusted telomere length, the more negative the number, the shorter the telomere.

Age-Adjusted Telomere Length in Children and Their Parents in Families with DKC	
Parent telomere length	Child telomere length
−4.7	−6.1
	−6.6
	−6.0
−3.9	−0.6
−1.4	−2.2
−5.2	−5.4
−2.2	−3.6
−4.4	−2.0
−4.3	−6.8
−5.0	−3.8
−5.3	−6.4
−0.6	−2.5
−1.3	−5.1
	−3.9
−4.2	−5.9

a. How do the telomere lengths of parents compare with telomere length of their children?

b. Explain why the telomeres of people with DKC are shorter than normal and why DKC arises at an earlier age in subsequent generations.

Challenge Questions

Section 9.3

25. A conditional mutation expresses its mutant phenotype only under certain conditions (the restrictive conditions) and expresses the normal phenotype under other conditions (the permissive conditions). One type of conditional mutation is a temperature-sensitive mutation, which expresses the mutant phenotype only at certain temperatures.

    Strains of *E. coli* have been isolated that contain temperature-sensitive mutations in the genes encoding different components of the replication machinery. In each of these strains, the protein produced by the mutated gene is nonfunctional under the restrictive conditions. These strains are grown under permissive conditions and then abruptly switched to the restrictive condition. After one round of replication under the restrictive condition, the DNA from each strain is isolated and analyzed. What characteristics would you expect to see in the DNA isolated from each strain with a temperature-sensitive mutation in its gene that encodes in the following?

    a. DNA ligase
    b. DNA polymerase I
    c. DNA polymerase III
    d. Primase
    e. Initiator protein

26. DNA topoisomerases play important roles in DNA replication and supercoiling (see Chapter 8). These enzymes are also the targets for certain anticancer drugs. Eric Nelson and his colleagues studied m-AMSA, one of the anticancer compounds that acts on topisomerase enzymes (E. M. Nelson, K. M. Tewey, and L. F. Liu. 1984. *Proceedings of the National Academy of Sciences* 81:1361–1365). They found that m-AMSA stabilizes an intermediate produced in the course of the topoisomerase's action. The intermediate consisted of the topoisomerase bound to the broken ends of the DNA. Breaks in DNA that are produced by anticancer compounds such as m-AMSA inhibit the replication of the cellular DNA and thus stop cancer cells from proliferating. Propose a mechanism for how m-AMSA and other anticancer agents that target topoisomerase enzymes taking part in replication might lead to DNA breaks and chromosome rearrangements.

10 From DNA to Proteins: Transcription and RNA Processing

Molecular image of the hammerhead ribozyme (in blue) bound to RNA (in orange). Ribozymes are catalytic RNA molecules that may have been the first carriers of genetic information. *[Kenneth Eward/Photo Researchers.]*

RNA in the Primeval World

Life requires two basic functions. First, living organisms must be able to store and faithfully transmit genetic information during reproduction. Second, they must have the ability to catalyze chemical transformations—to fire the reactions that drive life processes. A long-held belief was that the functions of information storage and chemical transformation are handled by two entirely different types of molecules. Genetic information is stored in nucleic acids. The catalysis of chemical transformations was held to be the exclusive domain of certain proteins that serve as biological catalysts or enzymes, making reactions take place rapidly within the cell. This biochemical dichotomy—nucleic acid for information, proteins for catalysts—revealed a dilemma in our understanding of the early stages in the evolution of life. Which came first: proteins or nucleic acids? If nucleic acids carry the coding instructions for proteins, how could proteins be generated without them? Because nucleic acids are unable to copy themselves, how could they be generated without proteins? If DNA and proteins each require the other, how could life begin?

This apparent paradox disappeared in 1981 when Thomas Cech and his colleagues discovered that RNA can serve as a biological catalyst. They found that RNA from the protozoan *Tetrahymena thermophila* can excise 400 nucleotides from its RNA in the absence of any protein. Other examples of catalytic RNAs have now been discovered in different types of cells. Called **ribozymes**, these catalytic RNA molecules can cut out parts of their own sequences, connect some RNA molecules together, replicate others, and even catalyze the formation of peptide bonds between amino acids. The discovery of ribozymes complements other evidence suggesting that the original genetic material was RNA.

Ribozymes that were self-replicating probably first arose between 3.5 billion and 4 billion years ago and may have begun the evolution of life on Earth. Early life was an RNA world, with RNA molecules serving both as carriers of genetic information and as catalysts that drove the chemical reactions needed to sustain and perpetuate life. These catalytic RNAs may have acquired the ability to synthesize protein-based enzymes, which are more efficient catalysts. With enzymes taking over more and more of the catalytic functions, RNA probably became relegated to the role of information storage and transfer. DNA, with its chemical stability and faithful replication, eventually replaced RNA as the primary carrier of genetic information. In modern cells, RNA still plays a vital role in both DNA replication and protein synthesis.

The central dogma, first conceptualized by Francis Crick (see Chapter 8), provides a molecular explanation for how genotype encodes phenotype—that information in DNA (genotype) is first transferred to RNA and then to protein (phenotype). This chapter is about the first steps in this pathway of information transfer—the synthesis of an RNA molecule through the process of transcription and the processing of that RNA. We begin with a brief review of RNA structure and a discussion of the different classes of RNA. We then consider the process of transcription, which synthesizes an RNA molecule from a DNA template. Finally, we examine the function and processing of RNA.

10.1 RNA, Consisting of a Single Strand of Ribonucleotides, Participates in a Variety of Cellular Functions

Before we begin our study of transcription, let's review the structure of RNA and consider the different types of RNA molecules.

The Structure of RNA

RNA, like DNA, is a polymer consisting of nucleotides joined together by phosphodiester bonds (see Chapter 8 for a discussion of DNA structure). However, there are several important differences in the structures of DNA and RNA. Whereas DNA nucleotides contain deoxyribose sugars, RNA nucleotides have ribose sugars (**Figure 10.1a**). With a free hydroxyl group on the 2′-carbon atom of the ribose sugar, RNA is degraded rapidly under alkaline conditions. The deoxyribose sugar of DNA lacks this free hydroxyl group; so DNA is a more stable molecule. Another important difference is that thymine, one of the two pyrimidines found in DNA, is replaced by uracil in RNA.

A final difference in the structures of DNA and RNA is that RNA is usually single stranded, consisting of a single polynucleotide strand (**Figure 10.1b**), whereas DNA normally consists of two polynucleotide strands joined by hydrogen bonding between complementary bases. Although RNA is usually single stranded, short complementary

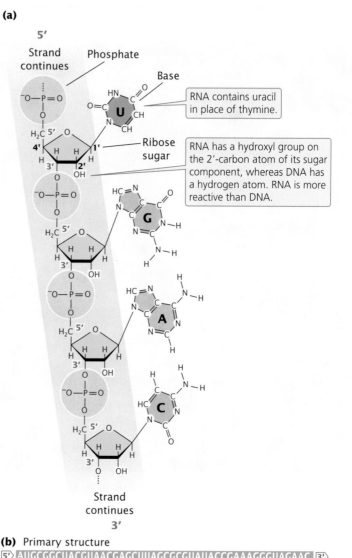

(a)

RNA contains uracil in place of thymine.

RNA has a hydroxyl group on the 2′-carbon atom of its sugar component, whereas DNA has a hydrogen atom. RNA is more reactive than DNA.

(b) Primary structure

5′ AUGCGGCUACGUAACGAGCUUAGCGCGUAUACCGAAAGGGUAGAAC 3′

An RNA molecule folds to form secondary structures... Folding

...owing to hydrogen bonding between complementary bases on the same strand.

Secondary structure

10.1 RNA has a primary and a secondary structure.

Table 10.1	The structures of DNA and RNA compared	
Characteristic	DNA	RNA
Composed of nucleotides	Yes	Yes
Type of sugar	Deoxyribose	Ribose
Presence of 2'-OH group	No	Yes
Bases	A, G, C, T	A, G, C, U
Nucleotides joined by phosphodiester bonds	Yes	Yes
Double or single stranded	Usually double	Usually single
Secondary structure	Double helix	Many types
Stability	Stable	Easily degraded

regions within a nucleotide strand can pair and form secondary structures (see Figure 10.1b). Exceptions to the rule that RNA is usually single stranded are found in a few RNA viruses that have double-stranded RNA genomes. Similarities and differences in DNA and RNA structures are summarized in **Table 10.1**.

Classes of RNA

RNA molecules perform a variety of functions in the cell. **Ribosomal RNA** (rRNA) along with ribosomal protein subunits make up the ribosome, the site of protein assembly. We'll take a more detailed look at the ribosome later in the chapter. **Messenger RNA** (mRNA) carries the coding instructions for polypeptide chains from DNA to the ribosome. After attaching to a ribosome, an mRNA molecule specifies the sequence of the amino acids in a polypeptide chain and provides a template for joining amino acids. Large precursor molecules, which are termed **pre-messenger RNAs** (pre-mRNAs), are the immediate products of transcription in eukaryotic cells. Pre-mRNAs (also called primary transcripts) are modified extensively before becoming mRNA and exiting the nucleus for translation into protein. Bacterial cells do not possess pre-mRNA; in these cells, transcription takes place concurrently with translation.

Transfer RNA (tRNA) serves as the link between the coding sequence of nucleotides in the mRNA and the amino acid sequence of a polypeptide chain. Each tRNA attaches to one particular type of amino acid and helps to incorporate that amino acid into a polypeptide chain.

Additional classes of RNA molecules are found in the nuclei of eukaryotic cells. **Small nuclear RNAs** (snRNAs) combine with small protein subunits to form **small nuclear ribonucleoproteins** (snRNPs, affectionately known as "snurps"). Some snRNAs participate in the processing of RNA, converting pre-mRNA into mRNA. **Small nucleolar RNAs** (snoRNAs) take part in the processing of rRNA. A class of very small and abundant RNA molecules, termed **microRNAs** (miRNAs) and **small interfering RNAs** (siRNAs), are found in eukaryotic cells and carry out RNA interference (RNAi), a process in which these small RNA molecules help trigger the degradation of mRNA or inhibit their translation into protein. The different classes of RNA molecules are summarized in **Table 10.2**.

Table 10.2	Location and functions of different classes of RNA molecules		
Class of RNA	Cell Type	Location of Function in Eukaryotic Cells*	Function
Ribosomal RNA (rRNA)	Bacterial and eukaryotic	Cytoplasm	Structural and functional components of the ribosome
Messenger RNA (mRNA)	Bacterial and eukaryotic	Nucleus and cytoplasm	Carries genetic code for proteins
Transfer RNA (tRNA)	Bacterial and eukaryotic	Cytoplasm	Helps incorporate amino acids into polypeptide chain
Small nuclear RNA (snRNA)	Eukaryotic	Nucleus	Processing of pre-mRNA
Small nucleolar RNA (snoRNA)	Eukaryotic	Nucleus	Processing and assembly of rRNA
MicroRNA (miRNA)	Eukaryotic	Cytoplasm	Inhibits translation of mRNA
Small interfering RNA (siRNA)	Eukaryotic	Cytoplasm	Triggers degradation of other RNA molecules

*All eukaryotic RNAs are transcribed in the nucleus.

Concepts

RNA differs from DNA in that RNA possesses a hydroxyl group on the 2′-carbon atom of its sugar, contains uracil instead of thymine, and is normally single stranded. Several classes of RNA exist within bacterial and eukaryotic cells.

✔ Concept Check 1

Which class of RNA is correctly paired with its function?

a. Small nuclear RNA (snRNA): processes rRNA

b. Transfer RNA (tRNA): attaches to an amino acid

c. MicroRNA (miRNA): carries information for the amino acid sequence of a protein

d. Ribosomal RNA (rRNA): carries out RNA interference

10.2 Transcription Is the Synthesis of an RNA Molecule from a DNA Template

All cellular RNAs are synthesized from DNA templates through the process of transcription (**Figure 10.2**). Transcription is in many ways similar to the process of replication, but a fundamental difference relates to the length of the template used. In replication, all the nucleotides in the DNA template are copied, but, in transcription, only small parts of the DNA molecule—usually a single gene or, at most, a few genes—are transcribed into RNA. Because not all gene products are needed at the same time or in the same cell, the constant transcription of all of a cell's genes would be highly inefficient. Furthermore, much of the DNA does not encode a functional product, and transcription of such sequences would be pointless. Transcription is, in fact, a highly selective process: individual genes are transcribed only as their products are needed. But this selectivity imposes a fundamental problem on the cell—the problem of how to recognize individual genes and transcribe them at the proper time and place.

Like replication, transcription requires three major components:

1. a DNA template;

2. the raw materials (substrates) needed to build a new RNA molecule; and

3. the transcription apparatus, consisting of the proteins necessary to catalyze the synthesis of RNA.

The Template for Transcription

In 1970, Oscar Miller, Jr., Barbara Hamkalo, and Charles Thomas used electron microscopy to examine cellular contents and demonstrate that RNA is transcribed from a DNA template. They saw within the cell Christmas-tree-like structures: thin central fibers (the trunk of the tree), to which were attached strings (the branches) with granules (**Figure 10.3**). The addition of deoxyribonuclease (an enzyme that degrades DNA) caused the central fibers to disappear, indicating that the "tree trunks" were DNA molecules. Ribonuclease (an enzyme that degrades RNA) removed the granular strings, indicating that the branches were RNA. Their conclusion was that each "Christmas tree" represented a gene undergoing transcription. The transcription of each gene begins at the top of the tree; there, little of the DNA has been transcribed and the RNA branches are short. As the transcription apparatus proceeds down the tree, transcribing more of the template, the RNA molecules lengthen, producing the long branches at the bottom.

The transcribed strand   The template for RNA synthesis, as for DNA synthesis, is a single strand of the DNA

10.2 All cellular types of RNA are transcribed from DNA.

10.3 Under the electron microscope, DNA molecules undergoing transcription exhibit Christmas-tree-like structures. The trunk of each "Christmas tree" (a transcription unit) represents a DNA molecule; the tree branches (granular strings attached to the DNA) are RNA molecules that have been transcribed from the DNA. As the transcription apparatus proceeds down the DNA, transcribing more of the template, the RNA molecules become longer and longer. *[Dr. Thomas Broker/Phototake.]*

double helix. Unlike replication, however, the transcription of a gene takes place on only one of the two nucleotide strands of DNA (**Figure 10.4**). The nucleotide strand used for transcription is termed the **template strand**. The other strand, called the **nontemplate strand**, is not ordinarily transcribed. Thus, within a gene, only one of the nucleotide strands is normally transcribed into RNA (there are some exceptions to this rule). Although only one strand within a single *gene* is normally transcribed, different genes may be transcribed from different strands, as illustrated in **Figure 10.5**.

During transcription, an RNA molecule that is complementary and antiparallel to the DNA template strand is synthesized (see Figure 10.4). The RNA transcript has the same polarity and base sequence as that of the nontemplate strand, with the exception that RNA contains U rather than T.

Concepts

Within a single gene, only one of the two DNA strands, the template strand, is usually transcribed into RNA.

✔ **Concept Check 2**

What is the difference between the template strand and nontemplate strand?

The transcription unit A transcription unit is a stretch of DNA that encodes an RNA molecule and the sequences necessary for its transcription. How does the complex of enzymes and proteins that performs transcription—the transcription apparatus—recognize a transcription unit? How does it know which DNA strand to read and where to start and stop? This information is encoded by the DNA sequence.

Included within a **transcription unit** are three critical regions: a promoter, an RNA-coding sequence, and a terminator (**Figure 10.6**). The **promoter** is a DNA sequence that the transcription apparatus recognizes and binds. It indicates which of the two DNA strands is to be read as the template and the direction of transcription. The promoter also determines the transcription start site, the first nucleotide that will be transcribed into RNA. In most transcription units, the promoter is located next to the transcription start site but is not, itself, transcribed.

The second critical region of the transcription unit is the **RNA-coding region**, a sequence of DNA nucleotides that is copied into an RNA molecule. The third component of the transcription unit is the **terminator**, a sequence of nucleotides that signals where transcription is to end. Terminators are usually part of the coding sequence; that is, transcription stops only after the terminator has been copied into RNA.

Molecular biologists often use the terms *upstream* and *downstream* to refer to the direction of transcription and the location of nucleotide sequences surrounding the RNA-coding

10.4 RNA molecules are synthesized that are complementary and antiparallel to one of the two nucleotide strands of DNA, the template strand.

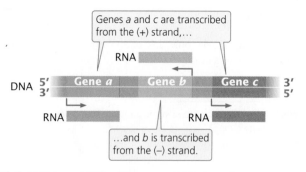

Genes *a* and *c* are transcribed from the (+) strand,...

...and *b* is transcribed from the (–) strand.

10.5 RNA is transcribed from one DNA strand. In most organisms, each gene is transcribed from a single DNA strand, but different genes may be transcribed from one or the other of the two DNA strands.

sequence. The transcription apparatus is said to move downstream as transcription takes place: it binds to the promoter (which is usually upstream of the start site) and moves toward the terminator (which is downstream of the start site).

When DNA sequences are written out, often the sequence of only one of the two strands is listed. Molecular biologists typically write the sequence of the nontemplate strand, because it will be the same as the sequence of the RNA transcribed from the template (with the exception that U in RNA replaces T in DNA). By convention, the sequence on the nontemplate strand is written with the 5′ end on the left and the 3′ end on the right. The first nucleotide transcribed (the transcription start site) is numbered +1; nucleotides downstream of the start site are assigned positive numbers, and nucleotides upstream of the start site are assigned negative numbers. So, nucleotide +34 would be 34 nucleotides downstream of the start site, whereas nucleotide –75 would be 75 nucleotides upstream of the start site. There is no nucleotide assigned 0.

Concepts

A transcription unit is a piece of DNA that encodes an RNA molecule and the sequences necessary for its proper transcription. Each transcription unit includes a promoter, an RNA-coding region, and a terminator.

The Substrate for Transcription

RNA is synthesized from **ribonucleoside triphosphates** (rNTPs; **Figure 10.7**). In synthesis, nucleotides are added one at a time to the 3′-OH group of the growing RNA molecule. Two phosphate groups are cleaved from the incoming ribonucleoside triphosphate; the remaining phosphate group participates in a phosphodiester bond that connects the nucleotide to the growing RNA molecule. The overall chemical reaction for the addition of each nucleotide is:

$$RNA_n + rNTP \rightarrow RNA_{n+1} + PP_i$$

where PP_i represents pyrophosphate. Nucleotides are always added to the 3′ end of the RNA molecule, and the direction of transcription is therefore 5′ $\rightarrow$ 3′ (**Figure 10.8**), the same as the direction of DNA synthesis in replication. The synthesis of RNA is complementary and antiparallel to one of the DNA strands (the template strand). Unlike DNA synthesis, RNA synthesis does not require a primer.

Concepts

RNA is synthesized from ribonucleoside triphosphates. Transcription is 5′ $\rightarrow$ 3′: each new nucleotide is joined to the 3′-OH group of the last nucleotide added to the growing RNA molecule.

The Transcription Apparatus

Recall that DNA replication requires a number of different enzymes and proteins. Transcription might initially appear to be quite different because a single enzyme—**RNA polymerase**—carries out all the required steps of transcription but, on closer inspection, the processes are actually similar. The action of RNA polymerase is enhanced by a number of accessory proteins that join and leave the polymerase at different stages of the process. Each accessory protein is responsible for providing or regulating a special function. Thus, transcription, like replication, requires an array of proteins.

Bacterial RNA polymerase   Bacterial cells typically possess only one type of RNA polymerase, which catalyzes the synthesis of all classes of bacterial RNA: mRNA, tRNA, and rRNA. Bacterial RNA polymerase is a large, multimeric enzyme (meaning that it consists of several polypeptide chains).

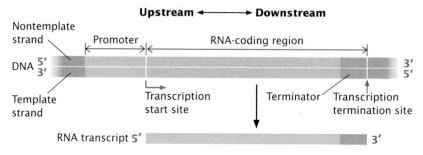

10.6 A transcription unit includes a promoter, a region that encodes RNA, and a terminator.

Triphosphate

Sugar

10.7 Ribonucleoside triphosphates are substrates used in RNA synthesis.

Table 10.3	Eukaryotic RNA polymerases
Type	**Transcribes**
RNA polymerase I	Large rRNAs
RNA polymerase II	Pre-mRNA, some snRNAs, snoRNAs, some miRNAs
RNA polymerase III	tRNAs, small rRNAs, some snRNAs, some miRNAs
RNA polymerase IV	Some siRNAs in plants

At the heart of most bacterial RNA polymerases are five subunits (individual polypeptide chains) that make up the **core enzyme**. This enzyme catalyzes the elongation of the RNA molecule by the addition of RNA nucleotides. Other functional subunits join and leave the core enzyme at particular stages of the transcription process. The **sigma (σ) factor** controls the binding of RNA polymerase to the promoter. Without sigma, RNA polymerase will initiate transcription at a random point along the DNA. After sigma has associated with the core enzyme (forming a **holoenzyme**), RNA polymerase binds stably only to the promoter region and initiates transcription at the proper start site. Sigma is required only for promoter binding and initiation; when a few RNA nucleotides have been joined together, sigma usually detaches from the core enzyme. Many bacteria have multiple types of sigma factors; each type of sigma initiates the binding of RNA polymerase to a particular set of promoters.

Eukaryotic RNA polymerases Most eukaryotic cells possess three distinct types of RNA polymerase, each of which is responsible for transcribing a different class of RNA: **RNA polymerase I** transcribes rRNA; **RNA polymerase II** transcribes pre-mRNAs, snoRNAs, some miRNAs, and some snRNAs; and **RNA polymerase III** transcribes small RNA molecules—specifically tRNAs, small rRNA, some miRNAs, and some snRNAs (Table 10.3). A fourth RNA polymerase, named **RNA polymerase IV**, has been found in plants. It functions in the nucleus and transcribes siRNAs that play a role in DNA methylation and chromatin structure.

All eukaryotic polymerases are large, multimeric enzymes, typically consisting of more than a dozen subunits. Some subunits are common to all RNA polymerases, whereas others are limited to one of the polymerases. As in bacterial cells, a number of accessory proteins bind to the core enzyme and affect its function.

Concepts

Bacterial cells possess a single type of RNA polymerase, consisting of a core enzyme and other subunits that participate in various stages of transcription. Eukaryotic cells possess three distinct types of RNA polymerase: RNA polymerase I transcribes rRNA; RNA polymerase II transcribes pre-mRNA, snoRNAs, and some snRNAs; and RNA polymerase III transcribes tRNAs, small rRNAs, and some snRNAs.

✔ **Concept Check 3**

What is the function of the sigma factor?

The Process of Bacterial Transcription

Now that we've considered some of the major components of transcription, we're ready to take a detailed look at the process. Transcription can be conveniently divided into three stages:

1. initiation, in which the transcription apparatus assembles on the promoter and begins the synthesis of RNA;

2. elongation, in which DNA is threaded through RNA polymerase, the polymerase unwinding the DNA and

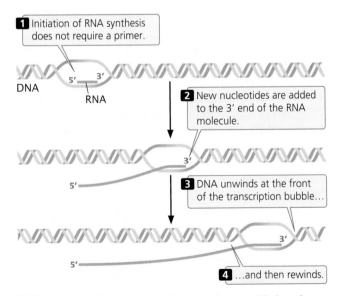

1 Initiation of RNA synthesis does not require a primer.

2 New nucleotides are added to the 3′ end of the RNA molecule.

3 DNA unwinds at the front of the transcription bubble…

4 …and then rewinds.

10.8 In transcription, nucleotides are always added to the 3′ end of the RNA molecule.

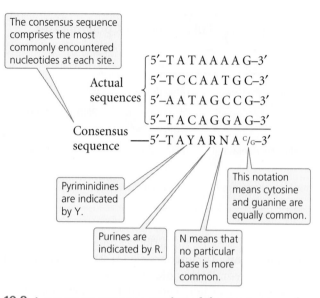

The consensus sequence comprises the most commonly encountered nucleotides at each site.

Actual sequences
5′–T A T A A A A G–3′
5′–T C C A A T G C–3′
5′–A A T A G C C G–3′
5′–T A C A G G A G–3′

Consensus sequence —5′–T A Y A R N A ᶜ/ɢ–3′

Pyrimidines are indicated by Y.

This notation means cytosine and guanine are equally common.

Purines are indicated by R.

N means that no particular base is more common.

10.9 A consensus sequence consists of the most commonly encountered bases at each position in a group of related sequences.

the frequency of transcription for a particular gene. Promoters also have different affinities for RNA polymerase. Even within a single promoter, the affinity can vary with the passage of time, depending on the promoter's interaction with RNA polymerase and a number of other factors.

Essential information for the transcription unit—where it will start transcribing, which strand is to be read, and in what direction the RNA polymerase will move—is imbedded in the nucleotide sequence of the promoter. Promoters are DNA sequences that are recognized by the transcription apparatus and are required for transcription to take place. In bacterial cells, promoters are usually adjacent to an RNA-coding sequence.

An examination of many promoters in *E. coli* and other bacteria reveals a general feature: although most of the nucleotides within the promoters vary in sequence, short stretches of nucleotides are common to many. Furthermore, the spacing and location of these nucleotides relative to the transcription start site are similar in most promoters. These short stretches of common nucleotides are called **consensus sequences**; "consensus sequence" refers to sequences that possess considerable similarity, or consensus (**Figure 10.9**). The presence of consensus in a set of nucleotides usually implies that the sequence is associated with an important function.

The most commonly encountered *consensus* sequence, found in almost all bacterial promoters, is centered about 10 bp upsteam of the start site. Called the **–10 consensus sequence** or, sometimes, the Pribnow box, its consensus sequence is

$$\text{5′–T A T A A T–3′}$$
$$\text{3′–A T A T T A–5′}$$

and is often written simply as TATAAT (**Figure 10.10**). Remember that TATAAT is just the *consensus* sequence—representing the most commonly encountered nucleotides at each of these positions. In most prokaryotic promoters, the actual sequence is not TATAAT.

Another consensus sequence common to most bacterial promoters is TTGACA, which lies approximately 35 nucleotides upstream of the start site and is termed the **–35 consensus sequence** (see Figure 10.10). The nucleotides on either side of the –10 and –35 consensus sequences and those between them vary greatly from promoter to promoter,

adding new nucleotides, one at a time, to the 3′ end of the growing RNA strand; and

3. termination, the recognition of the end of the transcription unit and the separation of the RNA molecule from the DNA template.

We will examine each of these steps in bacterial cells, where the process is best understood.

Initiation   Initiation comprises all the steps necessary to begin RNA synthesis, including (1) promoter recognition, (2) formation of the transcription bubble, (3) creation of the first bonds between rNTPs, and (4) escape of the transcription apparatus from the promoter.

Transcription initiation requires that the transcription apparatus recognize and bind to the promoter. At this step, the selectivity of transcription is enforced; the binding of RNA polymerase to the promoter determines which parts of the DNA template are to be transcribed and how often. Different genes are transcribed with different frequencies, and promoter binding is primarily responsible for determining

10.10 In bacterial promoters, consensus sequences are found upstream of the start site, approximately at positions –10 and –35.

suggesting that they are not very important in promoter recognition.

The sigma factor associates with the core enzyme (**Figure 10.11a**) to form a holoenzyme, which binds to the −35 and −10 consensus sequences in the DNA promoter (**Figure 10.11b**). Although it binds only the nucleotides of consensus sequences, the enzyme extends from −50 to +20 when bound to the promoter. The holoenzyme initially binds weakly to the promoter but then undergoes a change in structure that allows it to bind more tightly and unwind the double-stranded DNA (**Figure 10.11c**). Unwinding begins within the −10 consensus sequence and extends

downstream for about 14 nucleotides, including the start site (from nucleotides −12 to +2).

Concepts

A promoter is a DNA sequence adjacent to a gene and required for transcription. Promoters contain short consensus sequences that are important in the initiation of transcription.

After the holoenzyme has attached to the promoter, RNA polymerase is positioned over the start site for transcription (at position +1) and has unwound the DNA to

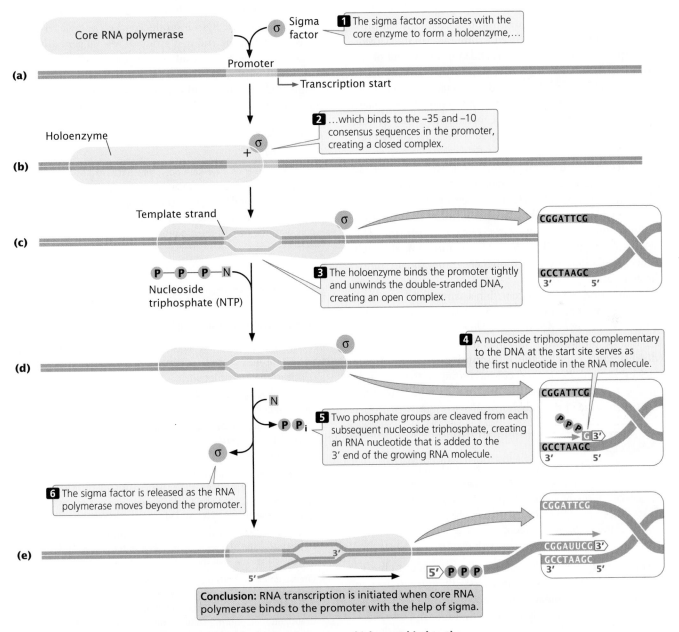

10.11 **Transcription in bacteria is catalyzed by RNA polymerase, which must bind to the sigma factor to initiate transcription.**

produce a single-stranded template. The orientation and spacing of consensus sequences on a DNA strand determine which strand will be the template for transcription and thereby determine the direction of transcription.

The position of the start site is determined not by the sequences located there but by the location of the consensus sequences, which positions RNA polymerase so that the enzyme's active site is aligned for the initiation of transcription at +1. If the consensus sequences are artificially moved upstream or downstream, the location of the starting point of transcription correspondingly changes.

To begin the synthesis of an RNA molecule, RNA polymerase pairs the base on a ribonucleoside triphosphate with its complementary base at the start site on the DNA template strand (**Figure 10.11d**). No primer is required to initiate the synthesis of the 5′ end of the RNA molecule. Two of the three phosphate groups are cleaved from the ribonucleoside triphosphate as the nucleotide is added to the 3′ end of the growing RNA molecule. However, because the 5′ end of the first ribonucleoside triphosphate does not take part in the formation of a phosphodiester bond, all three of its phosphate groups remain. An RNA molecule therefore possesses, at least initially, three phosphate groups at its 5′ end (**Figure 10.11e**).

Elongation   At the end of initiation, RNA polymerase undergoes a change in conformation (shape) and is thereafter no longer able to bind to the consensus sequences in the promoter. This change allows the polymerase to escape from the promoter and begin transcribing downstream. The sigma subunit is usually released after initiation, although some populations of RNA polymerase may retain sigma throughout elongation.

Transcription takes place within a short stretch of about 18 nucleotides of unwound DNA called the transcription bubble. As it moves downstream along the template, RNA polymerase progressively unwinds the DNA at the leading (downstream) edge of the transcription bubble, joining nucleotides to the RNA molecule according to the sequence on the template, and rewinds the DNA at the trailing (upstream) edge of the bubble.

Concepts

Transcription is initiated at the start site, which, in bacterial cells, is set by the binding of RNA polymerase to the consensus sequences of the promoter. No primer is required. Transcription takes place within the transcription bubble. DNA is unwound ahead of the bubble and rewound behind it.

Termination   RNA polymerase adds nucleotides to the 3′ end of the growing RNA molecule until it transcribes a terminator. Most terminators are found upstream of the site at which termination actually takes place. Transcription therefore does not suddenly stop when polymerase reaches a terminator, as does a car stopping at a stop sign. Rather, transcription stops after the terminator has been transcribed, like a car that stops only after running over a speed bump. At the terminator, several overlapping events are needed to bring an end to transcription: RNA polymerase must stop synthesizing RNA, the RNA molecule must be released from RNA polymerase, the newly made RNA molecule must dissociate fully from the DNA, and RNA polymerase must detach from the DNA template.

Bacterial cells possess two major types of terminators. **Rho-dependent terminators** are able to cause the termination of transcription only in the presence of an ancillary protein called the **rho factor**. **Rho-independent terminators** are able to cause the end of transcription in the absence of rho.

In bacteria, a group of genes is often transcribed into a single RNA molecule, which is termed a **polycistronic RNA**. Thus, polycistronic RNA is produced when a single terminator is present at the end of a group of several genes that are transcribed together, instead of each gene having its own terminator. Typically, eukaryotic genes are each transcribed and terminated separately, and so polycistronic mRNA is uncommon in eukaryotes.

Concepts

Transcription ends after RNA polymerase transcribes a terminator. Bacterial cells possess two types of terminator: a rho-independent terminator, which RNA polymerase can recognize by itself; and a rho-dependent terminator, which RNA polymerase can recognize only with the help of the rho protein.

Connecting Concepts

The Basic Rules of Transcription

Before we examine how RNA molecules are modified after transcription, let's pause to summarize some of the general principles of bacterial transcription.

1. Transcription is a selective process; only certain parts of the DNA are transcribed.

2. RNA is transcribed from single-stranded DNA. Within a gene, only one of the two DNA strands—the template strand—is normally copied into RNA.

3. Ribonucleoside triphosphates are used as the substrates in RNA synthesis. Two phosphate groups are cleaved from a ribonucleoside triphosphate, and the resulting nucleotide is joined to the 3′-OH group of the growing RNA strand.

4. RNA molecules are antiparallel and complementary to the DNA template strand. Transcription is always in the 5′ → 3′ direction, meaning that the RNA molecule grows at the 3′ end.

5. Transcription depends on RNA polymerase—a complex, multimeric enzyme. RNA polymerase consists of a core enzyme, which is capable of synthesizing RNA, and other subunits that may join transiently to perform additional functions. A sigma factor enables the core enzyme of RNA polymerase to bind to a promoter and initiate transcription.

6. Promoters contain short sequences crucial in the binding of RNA polymerase to DNA.

7. RNA polymerase binds to DNA at a promoter, begins transcribing at the start site of the gene, and ends transcription after a terminator has been transcribed.

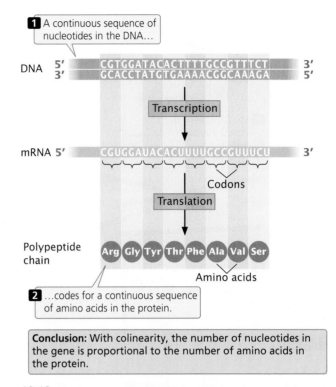

1 A continuous sequence of nucleotides in the DNA...

DNA

5′ CGTGGATACACTTTTGCCGTTTCT 3′
3′ GCACCTATGTGAAAACGGCAAAGA 5′

Transcription

mRNA 5′ CGUGGAUACACUUUUGCCGUUUCU 3′

Codons

Translation

Polypeptide chain

Arg Gly Tyr Thr Phe Ala Val Ser

Amino acids

2 ...codes for a continuous sequence of amino acids in the protein.

Conclusion: With colinearity, the number of nucleotides in the gene is proportional to the number of amino acids in the protein.

10.12 **The concept of colinearity suggests that a continuous sequence of nucleotides in DNA encodes a continuous sequence of amino acids in a protein.** As illustrated here, a codon specifies each amino acid.

10.3 Many Genes Have Complex Structures

What is a gene? As noted in Chapter 3, the definition of a gene changes as we explore different aspects of heredity. A gene was defined there as an inherited factor that determines a characteristic. This definition may have seemed vague, because it says only what a gene does but nothing about what a gene is. Nevertheless, this definition was appropriate for our purposes at the time, because our focus was on how genes influence the inheritance of traits. We did not have to consider the physical nature of the gene in learning the rules of inheritance.

Knowing something about the chemical structure of DNA and the process of transcription now enables us to be more precise about what a gene is. Chapter 8 described how genetic information is encoded in the base sequence of DNA; so a gene consists of a set of DNA nucleotides. But how many nucleotides are encompassed in a gene and how is the information in these nucleotides organized? In 1902, Archibald Garrod suggested, correctly, that genes encode proteins. Proteins are made of amino acids; so a gene contains the nucleotides that specify the amino acids of a protein. We could, then, define a gene as a set of nucleotides that specifies the amino acid sequence of a protein, which indeed was, for many years, the working definition of a gene. As geneticists learned more about the structure of genes, however, it became clear that this concept of a gene was an oversimplification.

Gene Organization

Early work on gene structure was carried out largely through the examination of mutations in bacteria and viruses. This research led Francis Crick in 1958 to propose that genes and proteins are **colinear**—that there is a direct correspondence between the nucleotide sequence of DNA and the amino acid sequence of a protein (**Figure 10.12**). The concept of colinearity suggests that the number of nucleotides in a gene should be proportional to the number of amino acids in the protein encoded by that gene. In a general sense, this concept is true for genes found in bacterial cells and many viruses, although these genes are slightly longer than would be expected if colinearity were strictly applied (the mRNAs encoded by the genes contain sequences at their ends that do not specify amino acids). At first, eukaryotic genes and proteins also were generally assumed to be colinear, but there were hints that eukaryotic gene structure is fundamentally different. Eukaryotic cells contain far more DNA than is required to encode proteins. Furthermore, many large RNA molecules observed in the nucleus were absent from the cytoplasm, suggesting that nuclear RNAs undergo some type of change before they are exported to the cytoplasm.

Most geneticists were nevertheless surprised by the announcement in the 1970s that four coding sequences in a gene from a eukaryotic virus were interrupted by nucleotides that did not specify amino acids. This discovery was made when the viral DNA was hybridized with the mRNA transcribed from it, and the hybridized structure was examined with the use of an electron microscope (**Figure 10.13**). The DNA was clearly much longer than the mRNA, because regions of DNA looped out from the hybridized molecules. These regions contained nucleotides in the DNA that were absent from the coding nucleotides in the mRNA. Many other examples of interrupted genes were subsequently discovered; it quickly became apparent that most eukaryotic genes consist of stretches of coding and noncoding nucleotides.

Concepts

When a continuous sequence of nucleotides in DNA encodes a continuous sequence of amino acids in a protein, the two are said to be colinear. In eukaryotes, not all genes are colinear with the proteins that they encode.

Question: Is the coding sequence in a gene always continuous?

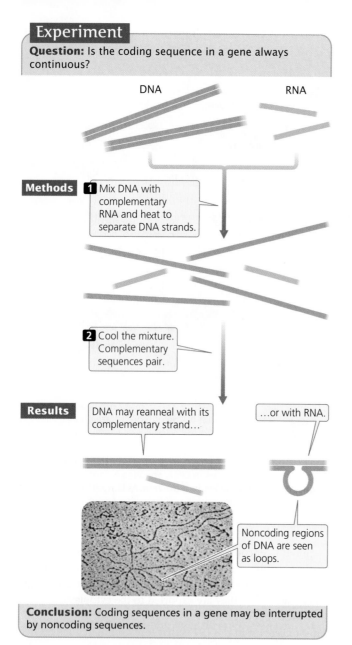

DNA

RNA

Methods **1** Mix DNA with complementary RNA and heat to separate DNA strands.

2 Cool the mixture. Complementary sequences pair.

Results DNA may reanneal with its complementary strand…

…or with RNA.

Noncoding regions of DNA are seen as loops.

Conclusion: Coding sequences in a gene may be interrupted by noncoding sequences.

10.13 The noncolinearity of eukaryotic genes was discovered by hybridizing DNA and mRNA. *[Electromicrograph from O. L. Miller, B. R. Beatty, D. W. Fawcett/Visuals Unlimited.]*

✔ Concept Check 4

What evidence indicated that eukaryotic genes are not collinear with their proteins?

Introns

Many eukaryotic genes contain coding regions called **exons** and noncoding regions called intervening sequences or **introns**. For example, the ovalbumin gene has eight exons and seven introns; the gene for cytochrome *b* has five exons and four introns (**Figure 10.14**). The average human gene contains from 8 to 9 introns. All the introns and the exons are initially transcribed into RNA but, after transcription, the introns are removed by splicing and the exons are joined to yield the mature RNA.

Introns are common in eukaryotic genes but are rare in bacterial genes. All classes of eukaryotic genes—those that encode rRNA, tRNA, and proteins—may contain introns. The number and size of introns vary widely: some eukaryotic genes have no introns, whereas others may have more than 60; intron length varies from fewer than 200 nucleotides to more than 50,000. Introns tend to be longer than exons, and most eukaryotic genes contain more noncoding nucleotides than coding nucleotides. Finally, most introns do not encode proteins (an intron of one gene is not usually an exon for another), although geneticists are finding a growing number of exceptions.

Concepts

Many eukaryotic genes contain exons and introns, both of which are transcribed into RNA, but introns are later removed by RNA processing. The number and size of introns vary from gene to gene; they are common in many eukaryotic genes but uncommon in bacterial genes.

The Concept of the Gene Revisited

How does the presence of introns affect our concept of a gene? To define a gene as a sequence of nucleotides that encodes amino acids in a protein no longer seems appropriate, because this definition excludes introns, which do not specify amino acids. This definition also excludes nucleotides that encode the 5′ and 3′ ends of an mRNA molecule, which are required for translation but do not encode amino acids. And defining a gene in these terms also excludes sequences that encode rRNA, tRNA, and other RNAs that do not encode proteins. In view of our current understanding of DNA structure and function, we need a more satisfactory definition of gene.

Many geneticists have broadened the concept of a gene to include all sequences in the DNA that are transcribed into a single RNA molecule. Defined in this way, a gene includes all exons, introns, and those sequences at the beginning and end of the RNA that are not translated into a protein. This definition also includes DNA sequences that encode rRNAs, tRNAs, and other types of nonmessenger RNA. Some geneticists have expanded the definition of a gene even further, to include the entire transcription unit—the promoter, the RNA coding sequence, and the terminator.

Concepts

The discovery of introns forced a reevaluation of the definition of the gene. Today, a gene is often defined as a DNA sequence that encodes an RNA molecule or the entire DNA sequence required to transcribe and encode an RNA molecule.

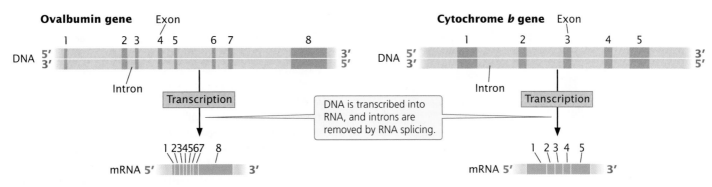

10.14 The coding sequences of many eukaryotic genes are disrupted by noncoding introns.

10.4 Many RNA Molecules Are Modified after Transcription in Eukaryotes

Many eukaryotic RNAs undergo extensive processing after transcription. In this section, we will examine different classes of RNA and the changes that they undergo after having been transcribed from the DNA template.

Messenger RNA Processing

Messenger RNA functions as the template for protein synthesis; it carries genetic information from DNA to a ribosome and helps to assemble amino acids in their correct order. In bacteria, mRNA is transcribed directly from DNA but, in eukaryotes, a pre-mRNA (the primary transcript) is first transcribed from DNA and then processed to yield the mature mRNA. We will reserve the term mRNA for RNA molecules that have been completely processed and are ready to undergo translation.

The Structure of mRNA In the mature mRNA, each amino acid in a protein is specified by a set of three nucleotides called a **codon**. Both prokaryotic and eukaryotic mRNAs contain three primary regions (**Figure 10.15**). The **5′ untranslated region** (5′ UTR; sometimes called the leader), a sequence of nucleotides at the 5′ end of the mRNA, does not encode any of the amino acids of a protein. In bacterial

mRNA, this region contains a consensus sequence called the **Shine–Dalgarno** sequence, which serves as the ribosome-binding site during translation; it is found approximately seven nucleotides upstream of the first codon translated into an amino acid (called the start codon). In eukaryotic cells, ribosomes bind to a modified 5′ end of mRNA, as discussed later in the chapter.

The next section of mRNA is the **protein-coding region**, which comprises the codons that specify the amino acid sequence of the protein. The protein-coding region begins with a start codon and ends with a stop codon. The last region of mRNA is the 3′ untranslated region (3′ UTR; sometimes called a trailer), a sequence of nucleotides that is at the 3′ end of the mRNA and not translated into protein. The **3′ untranslated region** affects the stability of mRNA and the translation of the mRNA protein-coding sequence.

Concepts

Messenger RNA molecules contain three main regions: a 5′ untranslated region, a protein-coding region, and a 3′ untranslated region. The 5′ and 3′ untranslated regions do not encode any amino acids of a protein.

✔ Concept Check 5

Which region of mRNA contains the Shine–Dalgarno sequence?

a. 5′ untranslated region   c. 3′ untranslated region

b. Protein-coding region   d. All three regions

In bacterial cells, transcription and translation take place simultaneously; while the 3′ end of an mRNA is undergoing transcription, ribosomes attach to the Shine–Dalgarno sequence near the 5′ end and begin translation. Because transcription and translation are coupled, there is little opportunity for the bacterial mRNA to be modified before protein synthesis. In contrast, transcription and translation are separated in both time and space in eukaryotic cells. Transcription takes place in the nucleus, whereas most translation takes place in the cytoplasm; this separation provides an opportunity for eukaryotic RNA to be modified before it

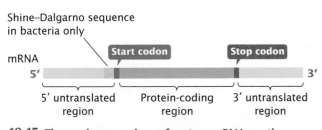

10.15 Three primary regions of mature mRNA are the 5′ untranslated region, the protein-coding region, and the 3′ untranslated region.

10.16 Most eukaryotic mRNAs have a 5′ cap. The cap consists of a nucleotide with 7-methylguanine attached to the pre-mRNA by a unique 5′–5′ bond, as well as methyl groups added to the 2′ position of the sugar in the second and third nucleotides and sometimes a methyl group added to the base (N) on the initial nucleotide.

is translated. Indeed, eukaryotic mRNA is extensively altered after transcription. Changes are made to the 5′ end, the 3′ end, and the protein-coding section of the RNA molecule. Recent research suggests that some translation in eukaryotes may take place in the nucleus, although these findings are controversial. If, indeed, translation takes place within the nucleus, then eukaryotic transcription and translation may be coupled as in prokaryotes. The significance of this coupling for RNA processing is not yet clear.

The Addition of the 5′ Cap   One type of modification of eukaryotic pre-mRNA is the addition at its 5′ end of a

structure called a **5′ cap**. This capping consists of the addition of an extra nucleotide at the 5′ end of the mRNA and methylation by the addition of a methyl group (CH_3) to the base in the newly added nucleotide and to the 2′-OH group of the sugar of one or more nucleotides at the 5′ end (**Figure 10.16**). Capping takes place rapidly after the initiation of transcription and, as will be discussed in more depth in Chapter 11, the 5′ cap functions in the initiation of translation. Cap-binding proteins recognize the cap and attach to it; a ribosome then binds to these proteins and moves downstream along the mRNA until the start codon is reached and translation begins. The presence of a 5′ cap also increases the stability of mRNA and influences the removal of introns.

The Addition of the Poly(A) Tail   A second type of modification to eukaryotic mRNA is the addition of 50 to 250 adenine nucleotides at the 3′ end, forming a **poly(A) tail**. These nucleotides are not encoded in the DNA but are added after transcription (**Figure 10.17**) in a process termed polyadenylation. Many eukaryotic genes are transcribed well beyond the end of the coding sequence; the extra material at the 3′ end is then cleaved and the poly(A) tail is added. For some pre-mRNA molecules, more than 1000 nucleotides may be removed from the 3′ end.

Processing of the 3′ end of pre-mRNA requires sequences both upstream and downstream of the cleavage site. The consensus sequence AAUAAA is usually from 11 to 30 nucleotides upstream of the cleavage site (see Figure 10.17) and determines the point at which cleavage will take place. A sequence rich in uracil nucleotides (or guanine and uracil nucleotides) is typically downstream of the cleavage site. A large number of proteins take part in finding the

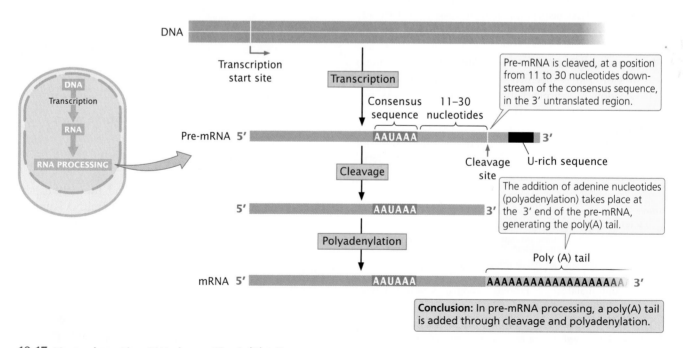

10.17 Most eukaryotic mRNAs have a 3′ poly(A) tail.

cleavage site and removing the 3′ end. After cleavage has been completed, adenine nucleotides are added to the new 3′ end, creating the poly(A) tail.

The poly(A) tail confers stability on many mRNAs, increasing the time during which the mRNA remains intact and available for translation before it is degraded by cellular enzymes. The stability conferred by the poly(A) tail depends on the proteins that attached to the tail. The poly(A) tail also facilitates attachment of the ribosome to the mRNA.

Concepts

Eukaryotic pre-mRNAs are processed at their 5′ and 3′ ends. A cap, consisting of a modified nucleotide and several methyl groups, is added to the 5′ end. The cap facilitates the binding of a ribosome, increases the stability of the mRNA, and may affect the removal of introns. Processing at the 3′ end includes cleavage downstream of an AAUAAA consensus sequence and the addition of a poly(A) tail.

RNA Splicing   The other major type of modification of eukaryotic pre-mRNA is the removal of introns by **RNA splicing**. This modification takes place in the nucleus, before the RNA moves to the cytoplasm. Splicing requires the presence of three sequences in the intron. One end of the intron is referred to as the **5′ splice site**, and the other end is the **3′ splice site** (**Figure 10.18**); these splice sites possess short consensus sequences. Most introns in pre-mRNAs begin with GU and end with AG, indicating that these sequences play a crucial role in splicing. Indeed, changing a single nucleotide at either of these sites prevents splicing.

The third sequence important for splicing is at the **branch point**, which is an adenine nucleotide that lies from 18 to 40 nucleotides upstream of the 3′ splice site (see Figure 10.18). The sequence surrounding the branch point does not have a strong consensus. The deletion or mutation of the adenine nucleotide at the branch point prevents splicing.

Splicing takes place within a large structure called the **spliceosome**, which is one of the largest and most complex of all molecular complexes. The spliceosome consists of five RNA molecules and almost 300 proteins. The RNA components are small nuclear RNAs; these snRNAs associate with proteins to form small ribonucleoprotein particles (snRNPs). Each snRNP contains a single snRNA molecule and multiple proteins. The spliceosome is composed of five snRNPs, named for the snRNAs that they contain (U1, U2, U4, U5, and U6), and some proteins not associated with an snRNA.

Concepts

Introns in nuclear genes contain three consensus sequences critical to splicing: a 5′ splice site, a 3′ splice site, and a branch point. Splicing of pre-mRNA takes place within a large complex called the spliceosome, which consists of snRNAs and proteins.

✔ Concept Check 6

If a splice site were mutated so that splicing did not take place, what would the effect be on the protein encoded by the mRNA?

a. It would be shorter than normal.

b. It would be longer than normal.

c. It would be the same length but would have different amino acids.

Before splicing takes place, an intron lies between an upstream exon (exon 1) and a downstream exon (exon 2), as shown in **Figure 10.19**. Pre-mRNA is spliced in two distinct steps. In the first step of splicing, the pre-mRNA is cut at the 5′ splice site. This cut frees exon 1 from the intron, and the 5′ end of the intron attaches to the branch point; that is, the intron folds back on itself, forming a structure called a **lariat**. In this reaction, the guanine nucleotide in the consensus sequence at the 5′ splice site bonds with the adenine nucleotide at the branch point through a transesterification reaction. The result is that the 5′ phosphate group of the guanine nucleotide is now attached to the 2′-OH group of the adenine nucleotide at the branch point (see Figure 10.19).

In the second step of RNA splicing, a cut is made at the 3′ splice site and, simultaneously, the 3′ end of exon 1 becomes covalently attached (spliced) to the 5′ end of exon 2. The intron is released as a lariat. The intron becomes linear when the bond breaks at the branch point and is then rapidly degraded by nuclear enzymes. The mature mRNA consisting of the exons spliced together is exported to the cytoplasm, where it is translated. These splicing reactions take place within the spliceosome, which carries out the splicing reactions.

Many eukaryotic mRNAs undergo **alternative processing**, in which a single pre-mRNA is processed in different ways to produce alternative types of mRNA, resulting in the production of different proteins from the same DNA sequence. One type of alternative processing is **alternative splicing**, in which the same pre-mRNA can be spliced in more than one way to yield multiple mRNAs that are

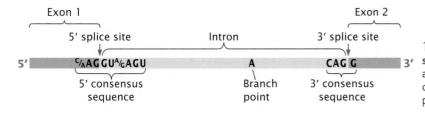

10.18 Splicing of pre-mRNA requires consensus sequences. Critical consensus sequences are present at the 5′ splice site and the 3′ splice site. A weak consensus sequence (not shown) exists at the branch point.

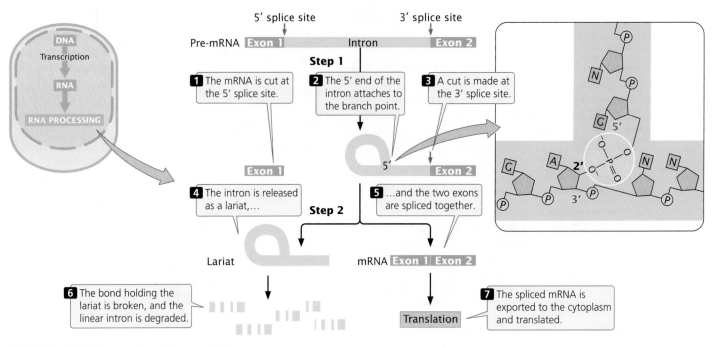

10.19 **The splicing of nuclear introns requires a two-step process.**

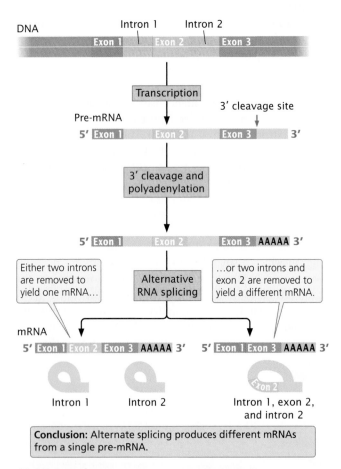

10.20 **Eukaryotic cells have alternative pathways for splicing pre-mRNA.**

translated into different amino acid sequences and thus different proteins (**Figure 10.20**). Alternative processing is an important source of protein diversity in vertebrates; an estimated 60% of all human genes are alternatively spliced.

Concepts

Intron splicing of nuclear genes is a two-step process: (1) the 5′ end of the intron is cleaved and attached to the branch point to form a lariat and (2) the 3′ end of the intron is cleaved and the two ends of the exon are spliced together. These reactions take place within the spliceosome. Alternative splicing enables exons to be spliced together in different combinations to yield mRNAs that encode different proteins.

Connecting Concepts

Eukaryotic Gene Structure and Pre-mRNA Processing

This chapter introduced a number of different components of genes and RNA molecules, including promoters, 5′ untranslated regions, coding sequences, introns, 3′ untranslated regions, poly(A) tails, and caps. Let's see how some of these components are combined to create a typical eukaryotic gene and how a mature mRNA is produced from them.

The promoter, which typically encompasses about 100 nucleotides upstream of the transcription start site, is necessary for transcription to take place but is itself not usually transcribed when protein-encoding genes are transcribed by RNA polymerase II (**Figure 10.21a**). Farther upstream or downstream of the start site, there may be enhancers that also regulate transcription.

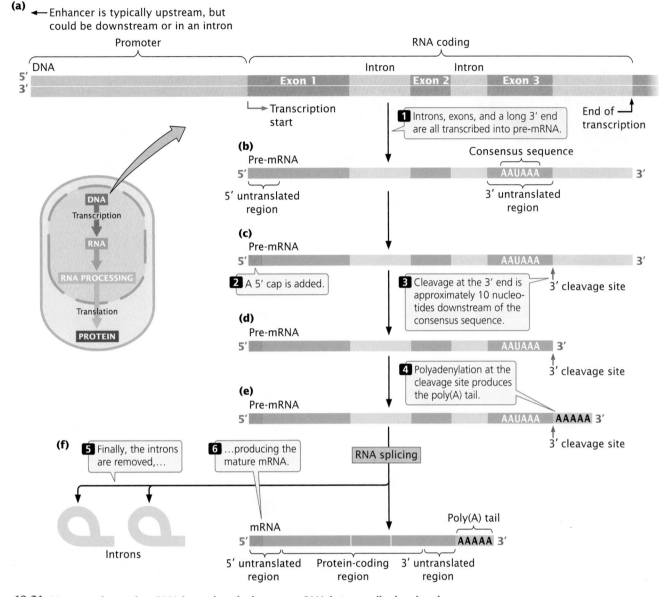

10.21 Mature eukaryotic mRNA is produced when pre-mRNA is transcribed and undergoes several types of processing.

In transcription, all the nucleotides between the transcription start site and the stop site are transcribed into pre-mRNA, including exons, introns, and a long 3′ end that is later cleaved from the transcript (**Figure 10.21b**). Notice that the 5′ end of the first exon contains the sequence that encodes the 5′ untranslated region and that the 3′ end of the last exon contains the sequence that encodes the 3′ untranslated region.

The pre-mRNA is then processed to yield a mature mRNA. The first step in this processing is the addition of a cap to the 5′ end of the pre-mRNA (**Figure 10.21c**). Next, the 3′ end is cleaved at a site downstream of the AAUAAA consensus sequence in the last exon (**Figure 10.21d**). Immediately after cleavage, a poly(A) tail is added to the 3′ end (**Figure 10.21e**). Finally, the introns are removed to yield the mature mRNA (**Figure 10.21f**). The mRNA now contains 5′ and 3′ untranslated regions, which are not translated into amino acids, and the nucleotides that carry the protein-coding sequences.

The Structure and Processing of Transfer RNAs

Transfer RNA serves as a link between the genetic code in mRNA and the amino acids that make up a protein. Each tRNA attaches to a particular amino acid and carries it to the ribosome, where the tRNA adds its amino acid to the growing polypeptide chain at the position specified by the genetic instructions in the mRNA.

The Structure of Transfer RNA   Each tRNA is capable of attaching to only one type of amino acid. The complex of tRNA plus its amino acid can be written in abbreviated form by adding a three-letter superscript representing the amino acid to the term tRNA. For example, a tRNA that attaches to the amino acid alanine is written as $tRNA^{Ala}$.

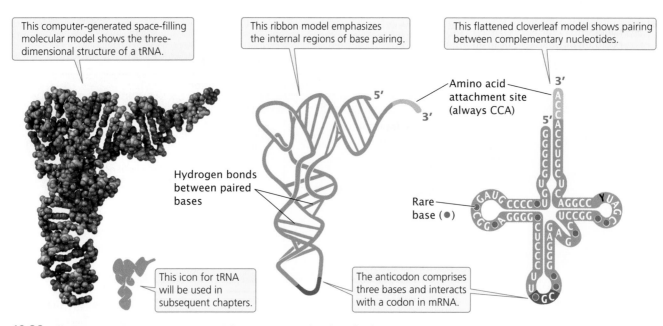

10.22 All tRNAs possess a common secondary structure, the cloverleaf structure. The base sequence in the flattened model is for tRNA^{Ala}.

A unique feature of tRNA is the presence of rare **modified bases**. All RNAs have the four standard bases (adenine, cytosine, guanine, and uracil) specified by DNA, but tRNAs have additional bases, including ribothymine, pseudouridine (which is also occasionally present in snRNAs and rRNA), and dozens of others.

The structures of all tRNAs are similar, a feature critical to tRNA function. Some of the nucleotides in a tRNA are complementary to each other and form intramolecular hydrogen bonds. As a result, each tRNA has a **cloverleaf structure** (Figure 10.22). All tRNAs have the same sequence (CCA) at the 3′ end, where the amino acid attaches to the tRNA. At the other end of the tRNA is a set of three nucleotides that make up the **anticodon**, which pairs with the corresponding codon on the mRNA during protein synthesis to ensure that the amino acids link in the correct order.

Although each tRNA molecule folds into a cloverleaf owing to the complementary pairing of bases, the cloverleaf is not the three-dimensional (tertiary) structure of tRNAs found in the cell. The results of X-ray crystallographic studies have shown that the cloverleaf folds on itself to form an L-shaped structure, as illustrated by the space-filling and ribbon models in Figure 10.22.

Transfer RNA Processing   Both bacterial and eukaryotic tRNAs are extensively modified after transcription. In *E. coli*, several tRNAs are usually transcribed together as one large precursor tRNA, which is then cut up into pieces, each containing a single tRNA. Additional nucleotides may then be removed one at a time from the 5′ and 3′ ends of the tRNA in a process known as trimming. Base-modifying enzymes may then change some of the standard bases into modified

bases. The CCA sequence found at the 3′ of all tRNAs is often synthesized by a special enzyme that adds these nucleotides without the use of any template. Eukaryotic tRNAs are processed in a manner similar to that for bacterial tRNAs: most are transcribed as larger precursors that are then cleaved, trimmed, and modified to produce mature tRNAs.

Concepts

All tRNAs are similar in size and have a common secondary structure known as the cloverleaf. Transfer RNAs contain modified bases and are extensively processed after transcription in both bacterial and eukaryotic cells.

The Structure and Processing of Ribosomal RNA

Within ribosomes, the genetic instructions contained in mRNA are translated into the amino acid sequences of polypeptides. Ribosomes are complex organelles, each consisting of more than 50 different proteins and RNA molecules (Table 10.4). A functional ribosome consists of two subunits, a **large ribosomal subunit** and a **small ribosomal subunit**, each of which consists of one or more pieces of RNA and a number of proteins. The sizes of the ribosomes and their RNA components are given in Svedberg (S) units (a measure of how rapidly an object sediments in a centrifugal field).

Ribosomal RNA is processed in both bacterial and eukaryotic cells. A precursor RNA molecule is methylated in several places and then cleaved and trimmed to produce the

	Table 10.4	Composition of ribosomes in bacterial and eukaryotic cells			
Cell Type	**Ribosome Size**	**Subunit**	**rRNA Component**		**Proteins**
Bacterial	70S	Large (50S)	23S (2900 nucleotides), 5S (120 nucleotides)		31
		Small (30S)	16S (1500 nucleotides)		21
Eukaryotic	80S	Large (60S)	28S (4700 nucleotides), 5.8S (160 nucleotides), 5S (120 nucleotides)		49
		Small (40S)	18S (1900 nucleotides)		33

Note: The letter "S" stands for Svedberg unit.

mature rRNAs that make up the ribosome. In eukaryotes, small nucleolar RNAs (snoRNAs) help to cleave and modify rRNA and assemble the processed rRNA into a mature ribosome.

> ## Concepts
>
> A ribosome is a complex organelle consisting of several rRNA molecules and many proteins. Each functional ribosome consists of a large and a small subunit. Ribosomal RNAs in both bacterial and eukaryotic cells are modified after transcription.

Small Interfering RNAs and MicroRNAs

In 1998, Andrew Fire, Craig Mello, and their colleagues observed what appeared to be a strange phenomenon. They were inhibiting the expression of genes in the nematode *Caenorhabditis elegans* by putting into the animals single-stranded RNA molecules that were complementary to a gene's DNA sequence. Called antisense RNA, such molecules are known to inhibit gene expression by binding to the mRNA sequences and inhibiting translation. Fire, Mello, and their colleagues found that even more potent gene silencing was triggered when double-stranded RNA was injected into the animals. This finding was puzzling, because no mechanism by which double-stranded RNA could inhibit translation was known. Several other, previously described types of gene silencing also were found to be triggered by double-stranded RNA. These studies led to the discovery of an abundant class of very small RNAs, called small interfering RNAs (siRNAs) or microRNAs (miRNAs), depending on their origin and mode of function. As mentioned earlier, we now know that siRNAs and miRNAs are found in many eukaryotes and are responsible for regulating gene expression through a process called RNA interference; microRNAs have also been implicated in a variety of other phenomena. For their discovery of RNA interference, Andrew Fire and Craig

Mello were awarded the Nobel Prize in physiology or medicine in 2006.

RNA interference (RNAi) is a powerful and precise mechanism used by eukaryotic cells to limit the invasion of foreign genes (from viruses and transposons) and to censor the expression of their own genes. RNA interference is triggered by double-stranded RNA molecules, which may arise in several ways (**Figure 10.23**): by the transcription of inverted repeats into an RNA molecule that then base pairs with itself to form double-stranded RNA; by the simultaneous transcription of two different RNA molecules that are complementary to one another and that pair, forming double-stranded RNA; or by infection by viruses that make double-stranded RNA. These double-stranded RNA molecules are chopped up by an enzyme appropriately called Dicer, resulting in tiny RNA molecules that are unwound to produce siRNAs and miRNAs (see Figure 10.23).

RNA interference is responsible for regulating a number of key genetic and developmental processes, including changes in chromatin structure, translation, cell fate and proliferation, and cell death. Geneticists also use the RNAi machinery as an effective tool for blocking the expression of specific genes (see Chapter 14).

Types of Small RNAs Small interfering RNAs and microRNAs constitute the two most abundant classes of RNA molecules in RNA interference. Although these two types of RNA differ in how they originate (**Table 10.5**; see Figure 10.23), they have a number of features in common and their functions overlap considerably.

Both siRNA and miRNA molecules combine with proteins to form an **RNA-induced silencing complex** (RISC; see Figure 10.23). The RISC pairs with an mRNA molecule that possesses a sequence complementary to RISC's siRNA or miRNA component and either cleaves the mRNA, leading to degradation of the mRNA, or represses translation of the mRNA. Some siRNAs also serve as guides for the methylation of complementary sequences in DNA, whereas others alter chromatin structure, both of which affect transcription.

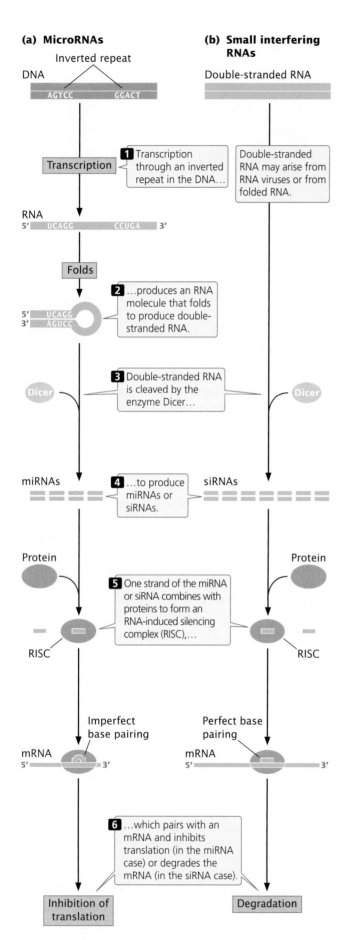

10.23 Small interfering RNAs and microRNAs are produced from double-stranded RNAs.

The Processing and Function of MicroRNAs   MicroRNAs have been found in all eukaryotic organisms examined to date, as well as viruses; they control the expression of genes taking part in many biological processes, including growth, development, and metabolism.

The genes that encode miRNAs are transcribed into longer precursors, called primary microRNA (pri-miRNA), that range from several hundred to several thousand nucleotides in length (**Figure 10.24**). The pri-miRNA is then cleaved into one or more smaller RNA molecules with a **hairpin**, which is a secondary structure that forms when sequences on the same strand are complementary and pair with one another. Dicer binds to this hairpin structure and removes the terminal loop. One of the miRNA strands is incorporated into the RISC; the other strand is released and degraded.

Concepts

Small interfering RNAs and microRNAs are tiny RNAs produced when larger double-stranded RNA molecules are cleaved by the enzyme Dicer. Small interfering RNAs and microRNAs participate in a variety of processes, including mRNA degradation, the inhibition of translation, the methylation of DNA, and chromatin remodeling.

Table 10.5	Differences between siRNAs and miRNAs	
Feature	**siRNA**	**miRNA**
Origin	mRNA, transposon, or virus	RNA transcribed from distinct gene
Cleavage of	RNA duplex or single-stranded RNA that forms long hairpins from folded RNA	Single-stranded RNA that forms short hairpins
Action	Some trigger degradation of mRNA, others inhibit transcription	Some trigger degradation of mRNA, others inhibit translation
Target	Genes from which they were transcribed	Genes other than those from which they were transcribed

Genes for miRNA are transcribed...

DNA

Transcription

...to produce a primary miRNA (pri-miRNA).

Pri-miRNA

Cleavage

The pri-miRNA is cleaved to produce a short RNA with a hairpin.

Dicer

Dicer removes the terminal loop of the hairpin.

Protein

One of the RNA strands of the miRNA combines with proteins to form the RISC.

RISC

The other strand of RNA is degraded.

10.24 MicroRNAs are cleaved from larger precursors (pri-miRNAs).

Model Genetic Organism
The Nematode Worm *Caenorhabditis elegans*

As we have seen, RNA interference was first demonstrated in the nematode *Caenorhabditis elegans* when geneticists discovered that they could silence specific genes in this species by injecting the animals with double-stranded DNA that was complementary to the genes. Geneticists were studying gene expression in *C. elegans* because this species had proved to be an excellent model genetic organism, particularly for studies of how genes influence development. For reasons that are not completely understood, RNAi is particularly effective in this species.

You may be asking, What is a nematode and why is it a model genetic organism? Although rarely seen, nematodes are

one of the most abundant organisms on Earth, inhabiting soils throughout the world. Most are free living and cause no harm, but a few are important parasites of plants and animals, including humans. Although *C. elegans* has no economic importance, it has become widely used in genetic studies because of its simple body plan, ease of culture, and high reproductive capacity (**Figure 10.25**). First introduced to the study of genetics by Sydney Brenner, who formulated plans in 1962 to use *C. elegans* for the genetic dissection of behavior, this species has made important contributions to the study of development, cell death, aging, and behavior.

Advantages of Caenorhabditis elegans as a model genetic organism   An ideal genetic organism, *C. elegans* is small, easy to culture, and produces large numbers of offspring. The adult *C. elegans* is about 1 mm in length. Most investigators grow *C. elegans* on agar-filled petri plates that are covered with a lawn of bacteria, which the nematodes devour. Thousands of worms can be easily cultured in a single laboratory. Compared with most multicellular animals, they have a very short generation time, about 3 days at room temperature. And they are prolific reproducers, with a single female producing from 250 to 1000 fertilized eggs in 3 to 4 days.

Another advantage of *C. elegans,* particularly for developmental studies, is that the worm is transparent, allowing easy observation of internal development at all stages. It has a simple body structure, with a small, invariant number of somatic cells: 959 cells in a mature hermaphroditic female and 1031 cells in a mature male.

Life cycle   Most mature adults are hermaphrodites, with the ability to produce both eggs and sperm and undergo self-fertilization. A few are male, which produce only sperm and mate with hermaphrodites. The hermaphrodites have two sex chromosomes (XX); the males possess a single sex chromosome (XO). Thus, hermaphrodites that self-fertilize produce only females (with the exception of a few males that result from nondisjunction of the X chromosomes). When hermaphrodites mate with males, half of the progeny are XX hermaphrodites and half are XO males.

Eggs are fertilized internally, either from sperm produced by the hermaphrodite or from sperm contributed by a male (see Figure 10.25). The eggs are then laid, and development is completed externally. Approximately 14 hours after fertilization, a larva hatches from the egg and goes through four larval stages—termed L1, L2, L3, and L4—that are separated by molts. The L4 larva undergoes a final molt to produce the adult worm. Under normal laboratory conditions, worms will live for 2 to 3 weeks.

Genetic techniques   Geneticists began developing plans in 1989 to sequence the genome of *C. elegans,* and the complete genome sequence was obtained in 1998. Compared with the genomes of most multicellular animals, that of *C. elegans,* at 103 million base pairs of DNA, is small, which

The Nematode Worm
Caenorhabditis elegans

ADVANTAGES

- Small size
- Short generation time of 3 days
- Each female can produce 200–1000 eggs
- Easy to culture in laboratory
- Simple body plan
- Transparent
- Capable of self-fertilization or crossing

STATS

Taxonomy:	Nematode
Size:	1 mm
Anatomy:	Unsegmented, elongated body
Habitat:	Lives and reproduces in soil

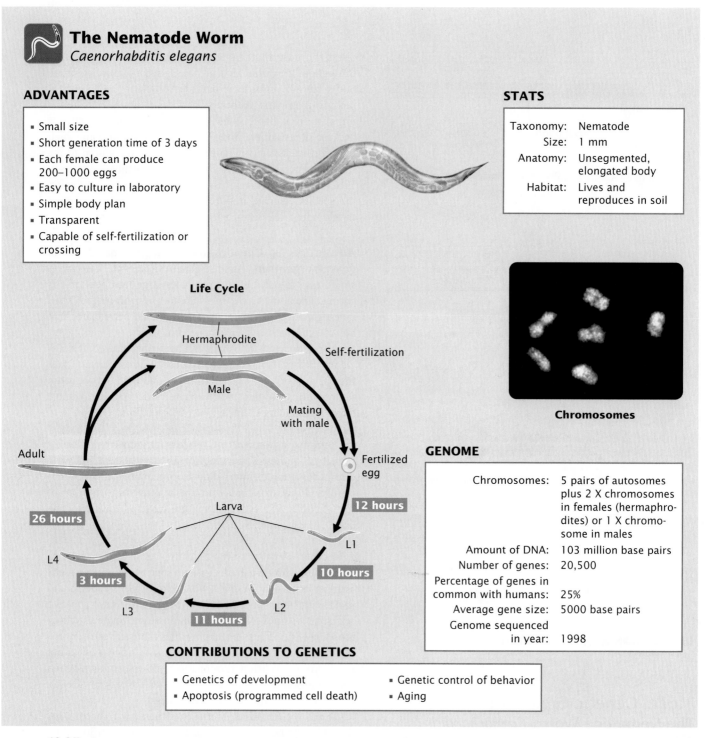

Life Cycle

Hermaphrodite

Self-fertilization

Male

Mating with male

Adult

Fertilized egg

12 hours

Larva

26 hours

L1

10 hours

L4

3 hours

L3

11 hours

L2

Chromosomes

GENOME

Chromosomes:	5 pairs of autosomes plus 2 X chromosomes in females (hermaphrodites) or 1 X chromosome in males
Amount of DNA:	103 million base pairs
Number of genes:	20,500
Percentage of genes in common with humans:	25%
Average gene size:	5000 base pairs
Genome sequenced in year:	1998

CONTRIBUTIONS TO GENETICS

- Genetics of development
- Apoptosis (programmed cell death)
- Genetic control of behavior
- Aging

10.25 The worm *Caenorhabditis elegans* is a model genetic organism. *[Micrograph: courtesy of William Goodyer and Monique Zetka.]*

facilitates genomic analysis. The availability of the complete genome sequence provides a great deal of information about gene structure, function, and organization in this species.

Chemical mutagens are routinely used to generate mutations in *C. elegans,* which are easy to identify and isolate. The ability of hermaphrodites to self-fertilize means that progeny homozygous for recessive mutations can be obtained in a single generation; the existence of males means that genetic crosses can be carried out.

Developmental studies are facilitated by the transparent body of the worms. As mentioned earlier, *C. elegans* has a small and exact number of somatic cells. Researchers studying the development of *C. elegans* have meticulously mapped the entire cell lineage of the species, and so the

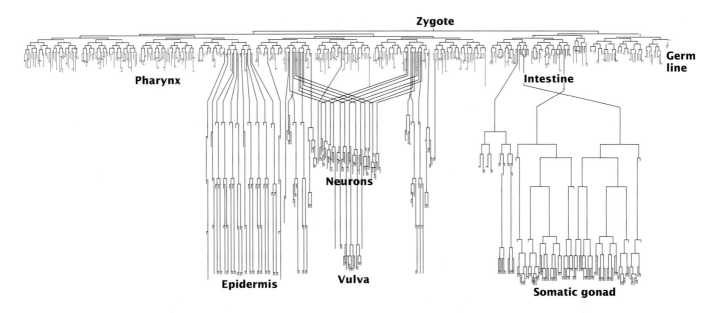

10.26 The developmental history of every cell in adult *C. elegans* has been determined. Shown here are the divisions that lead to adult cells in *C. elegans*.

developmental fate of every cell in the adult body can be traced back to the original single-celled fertilized egg (**Figure 10.26**). Developmental biologists often use lasers to destroy (ablate) specific cells in a developing worm and then study the effects on physiology, development, and behavior.

RNA interference has proved to be an effective tool for turning off genes in *C. elegans*. Geneticists inject double-stranded copies of RNA that is complementary to specific genes; the double-stranded RNA then silences the expression of these genes through the RNAi process. The worms can even be fed bacteria that have been genetically engineered to express the double-stranded RNA, thus avoiding the difficulties of microinjection.

Transgenic worms can be produced by injecting DNA into the ovary, where the DNA becomes incorporated into the oocytes. Geneticists have created a special reporter gene that produces the jellyfish green fluorescent protein (GFP). When this reporter gene is injected into the ovary and becomes inserted into the worm genome, its expression produces GFP, which fluoresces green, allowing the expression of the gene to be easily observed (**Figure 10.27**). ■

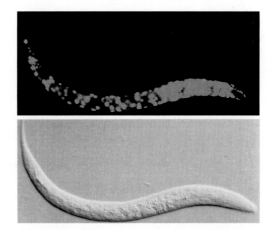

10.27 A sequence for the green fluorescent protein (GFP) has been used to visually determine the expression of genes inserted into *C. elegans* (lower photograph). The gene for GFP is injected into the ovary of a worm and becomes incorporated into the worm genome. The expression of this transgene produces GFP, which fluoresces green (upper photograph). *[Huaqi Jiang, Rong Guo, and Jo Anne Powell-Coffman, The Caenorhabditis elegans hif-1 gene encodes a bHLH-PAS protein that is required for adaptation to hypoxia. PNAS 98:7916–7921, 2001. © 2001 National Academy of Sciences, U.S.A.]*

Concepts Summary

- RNA is a polymer, consisting of nucleotides joined together by phosphodiester bonds. Each RNA nucleotide consists of a ribose sugar, a phosphate, and a base. RNA contains the base uracil and is usually single stranded.

- Cells possess a number of different classes of RNA. Ribosomal RNA is a component of the ribosome, messenger RNA carries

coding instructions for proteins, and transfer RNA helps incorporate the amino acids into a polypeptide chain.

- The template for RNA synthesis is single-stranded DNA. In transcription, RNA synthesis is complementary and antiparallel to the DNA template strand. A transcription unit consists of a promoter, an RNA-coding region, and a terminator.

- The substrates for RNA synthesis are ribonucleoside triphosphates.
- RNA polymerase in bacterial cells consists of a core enzyme, which catalyzes the addition of nucleotides to an RNA molecule, and other subunits. The sigma factor controls the binding of the core enzyme to the promoter. Eukaryotic cells contain three different RNA polymerases.
- Transcription begins at the start site, which is determined by consensus sequences. RNA is synthesized from a single strand of DNA as a template. RNA synthesis ceases after a terminator sequence has been transcribed.
- Introns—noncoding sequences that interrupt the coding sequences (exons) of genes—are common in eukaryotic cells but rare in bacterial cells.
- An mRNA molecule has three primary parts: a 5′ untranslated region, a protein-coding sequence, and a 3′ untranslated region.
- The pre-mRNA of a eukaryotic protein-encoding gene is extensively processed: a modified nucleotide and methyl groups, collectively termed the cap, are added to the 5′ end of pre-mRNA; the 3′ end is cleaved and a poly(A) tail is added; and introns are removed.
- Transfer RNAs, which attach to amino acids, are short molecules that assume a common secondary structure and contain modified bases. Ribosomes, the sites of protein synthesis, are composed of several ribosomal RNA molecules and numerous proteins.
- Small interfering RNAs and microRNAs are produced by cleavage of double-stranded RNA and play important roles in gene silencing and in a number of other phenonmena.
- *Caenorhabditis elegans* is a nematode that is widely used as a model genetic organism.

Important Terms

ribozyme (p. 243)
ribosomal RNA (rRNA) (p. 245)
messenger RNA (mRNA) (p. 245)
pre-messenger RNA (pre-mRNA) (p. 245)
transfer RNA (tRNA) (p. 245)
small nuclear RNA (snRNA) (p. 245)
small nuclear ribonucleoprotein (snRNP) (p. 245)
small nucleolar RNA (snoRNA) (p. 245)
microRNA (miRNA) (p. 245)
small interfering RNA (siRNA) (p. 245)
template strand (p. 247)
nontemplate strand (p. 247)
transcription unit (p. 247)
promoter (p. 247)
RNA-coding region (p. 247)
terminator (p. 247)
ribonucleoside triphosphate (rNTP) (p. 248)
RNA polymerase (p. 248)
core enzyme (p. 249)

sigma (σ) factor (p. 249)
holoenzyme (p. 249)
RNA polymerase I (p. 249)
RNA polymerase II (p. 249)
RNA polymerase III (p. 249)
RNA polymerase IV (p. 249)
consensus sequence (p. 250)
−10 consensus sequence (Pribnow box) (p. 250)
−35 consensus sequence (p. 250)
rho-dependent terminator (p. 252)
rho factor (p. 252)
rho-independent terminator (p. 252)
polycistronic RNA (p. 252)
colinearity (p. 253)
exon (p. 254)
intron (p. 254)
codon (p. 255)
5′ untranslated region (p. 255)
Shine–Dalgarno sequence (p. 255)
protein-coding region (p. 255)

3′ untranslated region (p. 255)
5′ cap (p. 256)
poly(A) tail (p. 256)
RNA splicing (p. 257)
5′ splice site (p. 257)
3′ splice site (p. 257)
branch point (p. 257)
spliceosome (p. 257)
lariat (p. 257)
alternative processing (p. 257)
alternative splicing (p. 257)
modified base (p. 260)
cloverleaf structure (p. 260)
anticodon (p. 260)
large ribosomal subunit (p. 260)
small ribosomal subunit (p. 260)
RNA interference (RNAi) (p. 261)
RNA-induced silencing complex (RISC) (p. 261)
hairpin (p. 262)

Answers to Concept Checks

1. b

2. The template strand is the DNA strand that is transcribed into an RNA molecule, whereas the nontemplate strand is not transcribed.

3. The sigma factor recognizes the promoter and controls the binding of RNA polymerase to the promoter.

4. When DNA was hybridized to the mRNA transcribed from it, regions of DNA that did not correspond to RNA looped out.

5. a

6. b

Worked Problems

1. DNA from a eukaryotic gene was isolated, denatured, and hybridized to the mRNA transcribed from the gene; the hybridized structure was then observed with the use of an electron microscope. The following structure was observed.

a. How many introns and exons are there in this gene? Explain your answer.

b. Identify the exons and introns in this hybridized structure.

• Solution

a. Each of the loops represents a region in which sequences in the DNA do not have corresponding sequences in the RNA; these regions are introns. There are five loops in the hybridized structure; so there must be five introns in the DNA and six exons.

b.

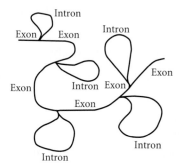

2. Draw a typical bacterial mRNA and the gene from which it was transcribed. Identify the 5′ and 3′ ends of the RNA and DNA molecules, as well as the following regions or sequences:

a. Promoter

b. 5′ untranslated region

c. 3′ untranslated region

d. Protein-coding sequence

e. Transcription start site

f. Terminator

g. Shine–Dalgarno sequence

h. Start and stop codons

• Solution

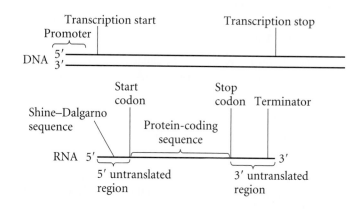

3. A test-tube splicing system contains all the components (snRNAs, proteins, splicing factors) necessary for the splicing of nuclear genes. When a piece of RNA containing an intron and two exons is added to the system, the intron is removed as a lariat and the exons are spliced together. If the RNA molecule added to the system has the following mutations, what intermediate products of the splicing reactions will accumulate? Explain your answer.

a. GU at the 5′ splice site is deleted.

b. A at the branch point is deleted.

c. AG at the 3′ splice site is deleted.

• Solution

a. The GU sequence at the 5′ splice site is required for the attachment of the spliceosome and the first cleavage reaction. If this sequence is mutated, cleavage will not take place. Thus, the original pre-mRNA with the intron will accumulate.

b. After cleavage at the 5′ splice site, the 5′ end of the intron attaches to the A at the branch point in a transesterification reaction. If the A at the branch point is deleted, no lariat structure will form. The separated first exon and the intron attached to the second exon will accumulate as intermediate products.

c. The AG sequence at the 3′ splice site is required for cleavage at the 3′ splice site. If this sequence is mutated, accumulated intermediate products will be: (1) the separated first exon and (2) the intron attached to the second exon, with the 5′ end of the intron attached to the branch point to form a lariat structure.

Comprehension Questions

Section 10.1

*1. Draw an RNA nucleotide and a DNA nucleotide, highlighting the differences. How is the structure of RNA similar to that of DNA? How is it different?

2. What are the major classes of cellular RNA? Where would you expect to find each class of RNA within eukaryotic cells?

Section 10.2

*3. What parts of DNA make up a transcription unit? Draw a typical bacterial transcription unit and identify its parts.

4. What is the substrate for RNA synthesis? How is this substrate modified and joined together to produce an RNA molecule?

5. What are the three basic stages of transcription? Describe what happens at each stage.

*6. Compare and contrast transcription and replication. How are these processes similar and how are they different?

Section 10.3

*7. What are the three principal elements in mRNA sequences in bacterial cells?

Section 10.4

8. What is the function of the Shine–Dalgarno consensus sequence?

*9. What is the 5′ cap?

*10. What is the function of the spliceosome?

11. What is alternative splicing? How does it lead to the production of multiple proteins from a single gene?

*12. Summarize the different types of processing that can take place in pre-mRNA.

13. Briefly describe the structure of tRNAs.

14. What is the origin of small interfering RNAs and microRNAs? What do these RNA molecules do in the cell?

Application Questions and Problems

Section 10.1

15. An RNA molecule has the following percentages of bases: A = 23%, U = 42%, C = 21%, and G = 14%.

   a. Is this RNA single stranded or double stranded? How can you tell?

   b. What would be the percentages of bases in the template strand of the DNA that contains the gene for this RNA?

Section 10.2

16. The following sequence of nucleotides is found in a single-stranded DNA template:

   ATTGCCAGATCATCCCAATAGAT

   Assume that RNA polymerase proceeds along this template from left to right.

   a. Which end of the DNA template is 5′ and which end is 3′?

   b. Give the sequence and identify the 5′ and 3′ ends of the RNA copied from this template.

17. Write the consensus sequence for the following set of nucleotide sequences.

   A G G A G T T
   A G C T A T T
   T G C A A T A
   A C G A A A A
   T C C T A A T
   T G C A A T T

*18. List at least five properties that DNA polymerases and RNA polymerases have in common. List at least three differences.

19. Most RNA molecules have *three* phosphate groups at the 5′ end, but DNA molecules never do. Explain this difference.

*20. What would be the most likely effect of a mutation at the following locations in an *E. coli* gene?

   a. –8   c. –20
   b. –35   d. Start site

21. A strain of bacteria possesses a temperature-sensitive mutation in the gene that encodes the sigma factor. At elevated temperatures, the mutant bacteria produce a sigma factor that is unable to bind to RNA polymerase. What effect will this mutation have on the process of transcription when the bacteria are raised at elevated temperatures?

*22. The following diagram represents a transcription unit on a DNA molecule.

Transcription start site

Template strand

Assume that this DNA molecule is from a bacterial cell. Draw the approximate location of the promoter and terminator for this transcription unit.

23. The following diagram represents the Christmas-tree-like structure of active transcription observed by Miller, Hamkalo, and Thomas (see Figure 10.3).

On the diagram, identify parts *a* through *i*:

a. DNA molecule

b. 5′ and 3′ ends of the template strand of DNA

c. At least one RNA molecule

d. 5′ and 3′ ends of at least one RNA molecule

e. Direction of movement of the transcription apparatus on the DNA molecule

f. Approximate location of the promoter

g. Possible location of a terminator

h. Upstream and downstream directions

i. Molecules of RNA polymerase (use dots to represent these molecules)

Section 10.3

*24. Duchenne muscular dystrophy is caused by a mutation in a gene that encompasses more than 2 million nucleotides and specifies a protein called dystrophin. However, less than 1% of the gene actually encodes the amino acids in the dystrophin protein. On the basis of what you now know about gene structure and RNA processing in eukaryotic cells, provide a possible explanation for the large size of the dystrophin gene.

*25. Draw a typical eukaryotic gene and the pre-mRNA and mRNA derived from it. Assume that the gene contains three exons. Identify the following items and, for each item, give a brief description of its function:

a. 5′ untranslated region

b. Promoter

c. AAUAAA consensus sequence

d. Transcription start site

e. 3′ untranslated region

f. Introns

g. Exons

h. Poly(A) tail

i. 5′ cap

26. How would the deletion of the Shine–Dalgarno sequence affect a bacterial mRNA?

27. Suppose that a mutation occurs in an intron of a gene encoding a protein. What will the most likely effect of the mutation be on the amino acid sequence of that protein? Explain your answer.

28. A geneticist isolates a gene that contains five exons. He then isolates the mature mRNA produced by this gene. After making the DNA single stranded, he mixes the single-stranded DNA and RNA. Some of the single-stranded DNA hybridizes (pairs) with the complementary mRNA. Draw a picture of what the DNA–RNA hybrids will look like under the electron microscope.

29. A geneticist discovers that two different proteins are encoded by the same gene. One protein has 56 amino acids, and the other has 82 amino acids. Provide a possible explanation for how the same gene can encode both of these proteins.

Section 10.4

30. In the early 1990s, Carolyn Napoli and her colleagues were working on petunias, attempting to genetically engineer a variety with dark purple petals by introducing numerous copies of a gene that codes for purple petals (C. Napoli, C. Lemieux, and R. Jorgensen. 1990. *Plant Cell* 2:279–289). Their thinking was that extra copies of the gene would cause more purple pigment to be produced and would result in a petunia with an even darker hue of purple. However, much to their surprise, many of the plants carrying extra copies of the purple gene were completely white or had only patches of color. Molecular analysis revealed that the level of the mRNA produced by the purple gene was reduced 50-fold in the engineered plants compared with levels of mRNA in wild-type plants. Somehow, the introduction of extra copies of the purple gene silenced both the introduced copies and the plant's own purple genes. Explain why the introduction of numerous copies of the purple gene silenced all copies of the purple gene.

Challenge Questions

Section 10.2

31. Many genes in both bacteria and eukaryotes contain numerous sequences that potentially cause pauses or premature terminations of transcription. Nevertheless, the transcription of these genes within a cell normally produces multiple RNA molecules thousands of nucleotides long without pausing or terminating prematurely. However, when a single round of transcription takes place on such templates in a test tube, RNA synthesis is frequently interrupted by pauses and premature terminations, which

reduce the rate at which transcription takes place and frequently shorten the length of the mRNA molecules produced. Most pauses and premature terminations occur when RNA polymerase temporarily backtracks (i.e., backs up) for one or two nucleotides along the DNA. Experimental findings have demonstrated that most transcriptional delays and premature terminations disappear if several RNA polymerases are simultaneously transcribing the DNA molecule. Propose an explanation for faster transcription and longer mRNA when the template DNA is being transcribed by multiple RNA polymerases.

Section 10.3

32. Alternative splicing occurs in approximately 60% of the human genes that code for proteins. Researchers have found that how a pre-mRNA is spliced is affected by the pre-mRNA's promoter sequence (D. Auboeuf et al. 2002. *Science* 298:416–419). In addition, factors that affect the rate of elongation of the RNA polymerase during transcription affect the type of splicing that takes place. These findings suggest that the process of transcription affects splicing. Propose one or more mechanisms that would explain how transcription might affect alternative splicing.

33. Duchenne muscular dystrophy (DMD) is an X-linked recessive genetic disease caused by mutations in the gene that encodes dystrophin, a large protein that plays an important role in the development of normal muscle fibers. The gene that encodes dystrophin is immense, spanning 2.5 million base pairs, and includes 79 exons and 78 introns. Many of the mutations that cause DMD produce premature stop codons, which bring protein synthesis to a halt, resulting in a greatly shortened and nonfunctional form of dystrophin. Some geneticists have proposed treating DMD patients by introducing small RNA molecules that cause the spliceosome to skip the exon containing the stop codon. The introduction of the small RNAs will produce a protein that is somewhat shortened (because an exon is skipped and some amino acids are missing) but may still result in a protein that has some function (A. Goyenvalle et al. 2004. *Science* 306:1796–1799). The small RNAs used for exon skipping are complementary to bases in the pre-mRNA. If you were designing small RNAs to bring about exon skipping for the treatment of DMD, what sequences should the small RNAs contain?

11

From DNA to Proteins: Translation

The bacterium *Corynebacterium diphtheriae* causes diphtheria. The diphtheria toxin that produces the symptoms of the disease is actually encoded by a gene carried by a bacteriophage that infects some strains of *C. diphtheriae*. [*Gary Gaugler/Visuals Unlimited.*]

The Deadly Diphtheria Toxin

Diphtheria—first described by Hippocrates in the fifth century B.C.—has an insidious onset, with initial symptoms that resemble the common cold followed by sore throat, loss of appetite, and a low-grade fever. A local infection in the nose, tonsils, or throat produces the diphtheria toxin, which accumulates and spreads through the circulatory system. Within 6 to 10 days, major organs are affected, leading to severe prostration, rapid pulse, paralysis, stupor, coma, and—in some cases—death.

Until recent times, diphtheria was a leading killer of children in temperate climates. In the seventeenth century, severe epidemics ravaged Europe; diphtheria became known in Spain as "El garatillo," the strangler. Diphtheria spread to the American colonies in the eighteenth century, often wiping out entire families. Even in modern times, diphtheria has been a leading killer of children. In the United States alone, between 100,000 and 200,000 cases were reported each year in the 1920s; from 13,000 to 15,000 of these patients—most of them children—died. The epidemics, which occurred primarily during winter and spring months, incited panic among parents of small children.

Today, the control of diphtheria is one of the success stories of modern medicine. Through the widespread application of immunization and antibiotic treatment, the disease is no longer a major public-health threat. Indeed, in the United States, only about one case per year is currently reported for the entire country. Although there are still epidemics and deaths resulting from the disease in some areas of the world, diphtheria is quickly disappearing as a serious infectious disease.

Diphtheria has an interesting pathogenesis. Its immediate cause is infection by the bacterium *Corynebacterium diphtheriae*, but the bacterium itself does not produce the deadly toxin that causes the disease. The toxin is instead encoded by a gene carried by a bacteriophage that infects the bacterium. The toxin, synthesized according to the genetic instructions in the phage DNA, is absorbed into an infected person's bloodstream and distributed throughout the body, where it causes the symptoms of diphtheria.

Why is the diphtheria toxin so deadly? The answer lies in its unique mode of action: the diphtheria toxin targets and inhibits a protein known as elongation factor 2 (EF-2), an essential component of the protein-synthesizing machinery in eukaryotic cells. In the course of translation, a transfer RNA carrying an amino acid specified by a codon on the mRNA enters the ribosome and donates its amino acid to the growing polypeptide chain. To read the next codon, the ribosome must physically move down the mRNA, and this process requires EF-2. Without a functional EF-2, the ribosome cannot travel down the mRNA and no polypeptide is assembled. Protein synthesis ceases, illness results, and—if the patient is not treated—death may ensue.

Diphtheria, caused by the inhibitory effect of its toxin on EF-2, illustrates the central importance of protein synthesis to normal cell function. Interrupting this process for just a few days can be deadly. Ironically, we often fight diphtheria and other infectious diseases by using the same strategy employed by the diphtheria toxin—that is, by inhibiting the protein-synthesizing machinery of the infectious agent, in this case with the use of antibiotics.

In this chapter, we will examine this process of translation, the mechanism by which the nucleotide sequence in mRNA specifies the amino acid sequence of a protein. We will begin by examining the genetic code—the instructions that specify the amino acid sequence of a protein—and then examine the mechanism of protein synthesis. Our primary focus will be on protein synthesis in bacterial cells, but we will examine some of the differences in eukaryotic cells. Finally, we will look at some additional aspects of protein synthesis.

11.1 The Genetic Code Determines How the Nucleotide Sequence Specifies the Amino Acid Sequence of a Protein

Many genes specify traits by encoding proteins. The first person to suggest the existence of a relation between genotype and proteins was Archibald Garrod. In 1908, Garrod correctly proposed that genes encode enzymes, but, unfortunately, his proposal made little impression on his contemporaries. Later, George Beadle and Edward Tatum examined the genetic basis of biochemical pathways in *Neurospora* and developed the **one-gene, one-enzyme hypothesis**, which suggested that genes function by encoding enzymes and that each gene encodes a separate enzyme. Later research findings showed that some proteins are composed of more than one polypeptide chain and that different polypeptide chains are encoded by separate genes; so this model was modified to become the **one-gene, one-polypeptide hypothesis**.

> ### Concepts
>
> Many genes specifiy traits by encoding proteins. The one-gene, one-enzyme hypothesis proposed that each gene encodes a separate enzyme. This hypothesis was later modified to the one-gene, one-polypeptide hypothesis.

The Structure and Function of Proteins

Proteins are central to all living processes (**Figure 11.1**). Many proteins are enzymes, the biological catalysts that drive the chemical reactions of the cell; others are structural components, providing scaffolding and support for membranes, filaments, bone, and hair. Some proteins help transport substances; others have a regulatory, communication, or defense function.

All proteins are composed of **amino acids**, linked end to end. Twenty common amino acids are found in proteins. The 20 common amino acids are similar in structure: each consists of a central carbon atom bonded to an amino group, a hydrogen atom, a carboxyl group, and an R (radical) group that differs for each amino acid (**Figure 11.2a**). The amino acids in proteins are joined together by **peptide bonds** (**Figure 11.2b**) to form polypeptide chains, and a protein consists of one or more **polypeptide** chains. Like nucleic acids, polypeptides have polarity, with one end having a free amino group (NH_3^+) and the other end possessing a free carboxyl group (COO^-). Some proteins consist of only a few amino acids, whereas others may have thousands.

Like that of nucleic acids, the molecular structure of proteins has several levels of organization. The primary structure of a protein is its sequence of amino acids (**Figure 11.3a** on p. 274). Through interactions between neighboring amino acids, a polypeptide chain folds and twists into a secondary structure (**Figure 11.3b**); two common secondary structures found in proteins are the beta (β) pleated sheet and the alpha (α) helix. Secondary structures interact and fold further to form a tertiary structure (**Figure 11.3c**), which is the overall, three-dimensional shape of the protein. The secondary and tertiary structures of a protein are

11.1 Proteins serve a number of biological functions. (a) The light produced by fireflies is the result of a light-producing reaction between luciferin and ATP catalyzed by the enzyme luciferase. (b) The protein fibroin is the major structural component of spider webs. (c) Castor beans contain a highly toxic protein called ricin. [*Part a: Gregory K. Scott/Photo Researchers. Part b: Rosemary Calvert/Imagestate. Part c: Gerald & Buff Corsi/Visuals Unlimited.*]

ultimately determined by the primary structure—the amino acid sequence—of the protein. Finally, some proteins consist of two or more polypeptide chains that associate to produce a quaternary structure (**Figure 11.3d**).

11.2 The common amino acids have similar structures. (a) Each amino acid consists of a central (α) carbon atom attached to: (1) an amino group (NH_3^+); (2) a carboxyl group (COO^-); (3) a hydrogen atom (H); and (4) a radical group, designated R. (b) Amino acids are joined together by peptide bonds. In a peptide bond (red), the carboxyl group of one amino acid is covalently attached to the amino group of another amino acid.

Concepts

The products of many genes are proteins whose actions produce the traits encoded by these genes. Proteins are polymers consisting of amino acids linked by peptide bonds. The amino acid sequence of a protein is its primary structure. This structure folds to create the secondary and tertiary structures; two or more polypeptide chains may associate to create a quaternary structure.

✔ Concept Check 1

What determines the secondary and tertiary structures of a protein?

Breaking the Genetic Code

In 1953, James Watson and Francis Crick solved the structure of DNA and identified the base sequence as the carrier of genetic information. However, the way in which the base sequence of DNA specifies the amino acid sequences of proteins (the genetic code) was not immediately obvious and remained elusive for another 10 years.

One of the first questions about the genetic code to be addressed was: *How many nucleotides are necessary to specify a single amino acid?* This basic unit of the genetic code—the set of bases that encode a single amino acid—is a codon (see p. 255 in Chapter 10). Many early investigators recognized that codons must contain a minimum of three nucleotides. Each nucleotide position in mRNA can be occupied by one of four bases: A, G, C, or U. If a codon consisted of a single nucleotide, only four different codons (A, G, C, and U) would be possible, which is not enough to encode the 20 different amino acids commonly found in proteins. If codons were made up of two nucleotides each (i.e., GU, AC, etc.), there would be $4 \times 4 = 16$ possible codons—still not enough to encode all 20 amino acids. With three nucleotides per codon, there are $4 \times 4 \times 4 = 64$ possible codons, which is more than enough to specify 20 different amino acids. Therefore, a

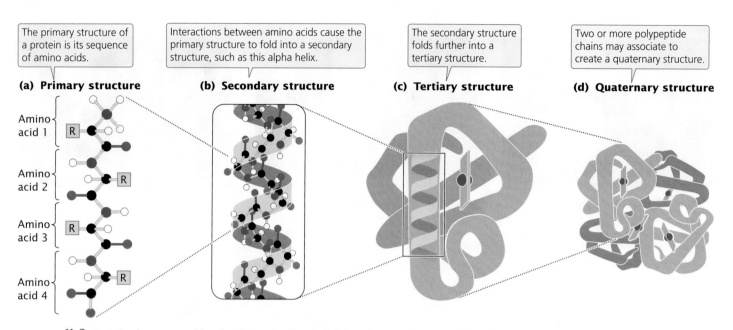

The primary structure of a protein is its sequence of amino acids.

(a) Primary structure

Interactions between amino acids cause the primary structure to fold into a secondary structure, such as this alpha helix.

(b) Secondary structure

The secondary structure folds further into a tertiary structure.

(c) Tertiary structure

Two or more polypeptide chains may associate to create a quaternary structure.

(d) Quaternary structure

Amino acid 1

Amino acid 2

Amino acid 3

Amino acid 4

11.3 Proteins have several levels of structural organization. Atoms are represented in color as follows: blue, nitrogen; white, hydrogen; black, carbon; and red, oxygen.

triplet code requiring three nucleotides per codon is the most efficient way to encode all 20 amino acids. Using mutations in bacteriophage, Francis Crick and his colleagues confirmed in 1961 that the genetic code is indeed a triplet code.

Concepts

The genetic code is a triplet code, in which three nucleotides encode each amino acid in a protein.

✔ Concept Check 2

A codon is

a. one of three nucleotides that encode an amino acid.

b. three nucleotides that encode an amino acid.

c. three amino acids that encode a nucleotide.

d. one of four bases in DNA.

When it had been firmly established that the genetic code consists of codons that are three nucleotides in length, the next step was to determine which groups of three nucleotides specify which amino acids. Logically, the easiest way to break the code would have been to determine the base sequence of a piece of RNA, add it to a test tube containing all the components necessary for translation, and allow it to direct the synthesis of a protein. The amino acid sequence of the newly synthesized protein could then be determined, and its sequence could be compared with that of the RNA. Unfortunately, there was no way at that time to determine

the nucleotide sequence of a piece of RNA; so indirect methods were necessary to break the code.

The first clues to the genetic code came in 1961, from the work of Marshall Nirenberg and Johann Heinrich Matthaei. These investigators created synthetic RNAs by using an enzyme called polynucleotide phosphorylase. Unlike RNA polymerase, polynucleotide phosphorylase does not require a template; it randomly links together any RNA nucleotides that happen to be available. The first synthetic mRNAs used by Nirenberg and Matthaei were homopolymers, RNA molecules consisting of a single type of nucleotide. For example, by adding polynucleotide phosphorylase to a solution of uracil nucleotides, they generated RNA molecules that consisted entirely of uracil nucleotides and thus contained only UUU codons (**Figure 11.4**). These poly(U) RNAs were then added to 20 tubes, each containing the components necessary for translation and all 20 amino acids. A different amino acid was radioactively labeled in each of the 20 tubes. Radioactive protein appeared in only one of the tubes—the one containing labeled phenylalanine (see Figure 11.4). This result showed that the codon UUU specifies the amino acid phenylalanine. The results of similar experiments using poly(C) and poly(A) RNA demonstrated that CCC encodes proline and AAA encodes lysine; for technical reasons, the results from poly(G) were uninterpretable.

Other experiments provided additional information about the genetic code, and the code was fully understood by 1968. The genetic code is so important to modern biology that Francis Crick compared its place to that of the periodic table of the elements in chemistry.

Characteristics of the Genetic Code

We will now examine a number of features of the genetic code.

The degeneracy of the code One amino acid is encoded by three consecutive nucleotides in mRNA, and each nucleotide can have one of four possible bases (A, G, C, and U) at each nucleotide position, thus permitting $4^3 = 64$ possible codons (**Figure 11.5**). Three of these codons are stop codons, specifying the end of translation. Thus, 61 codons, called **sense codons**, encode amino acids. Because there are 61 sense codons and only 20 different amino acids commonly found in proteins, the code contains more information than is needed to specify the amino acids and is said to be a **degenerate code**. This expression does not mean that the genetic code is depraved; *degenerate* is a term that Francis Crick borrowed from quantum physics, where it describes multiple physical states that have equivalent meaning. The degeneracy of the genetic code means that amino acids may be specified by more than one codon. Only tryptophan and methionine are encoded by a single codon (see Figure 11.5). Other amino acids are specified by two codons, and some, such as leucine, are specified by six different codons. Codons that specify the same amino acid are said to be **synonymous**, just as synonymous words are different words that have the same meaning.

As we learned in Chapter 10, tRNAs serve as adapter molecules, binding particular amino acids and delivering them to a ribosome, where the amino acids are then assembled into polypeptide chains. Each type of tRNA attaches to a single type of amino acid. The cells of most organisms possess from about 30 to 50 different tRNAs, and yet there are only 20 different amino acids in proteins. Thus, some amino acids are carried by more than one tRNA. Different tRNAs that accept the same amino acid but have different anticodons are called **isoaccepting tRNAs**. Some synonymous codons encode different isoacceptors.

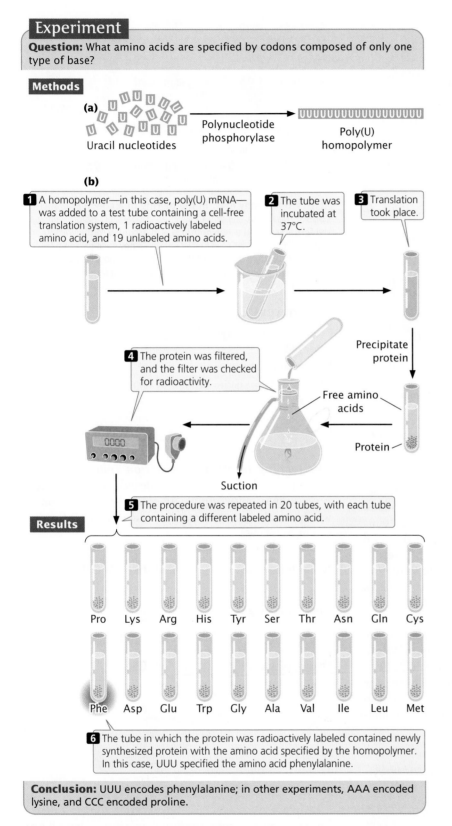

Experiment

Question: What amino acids are specified by codons composed of only one type of base?

Methods

(a) Uracil nucleotides → Polynucleotide phosphorylase → Poly(U) homopolymer

(b)

1. A homopolymer—in this case, poly(U) mRNA—was added to a test tube containing a cell-free translation system, 1 radioactively labeled amino acid, and 19 unlabeled amino acids.

2. The tube was incubated at 37°C.

3. Translation took place.

4. The protein was filtered, and the filter was checked for radioactivity.

Precipitate protein

Free amino acids

Protein

Suction

5. The procedure was repeated in 20 tubes, with each tube containing a different labeled amino acid.

Results

Pro Lys Arg His Tyr Ser Thr Asn Gln Cys

Phe Asp Glu Trp Gly Ala Val Ile Leu Met

6. The tube in which the protein was radioactively labeled contained newly synthesized protein with the amino acid specified by the homopolymer. In this case, UUU specified the amino acid phenylalanine.

Conclusion: UUU encodes phenylalanine; in other experiments, AAA encoded lysine, and CCC encoded proline.

11.4 **Nirenberg and Matthaei developed a method for identifying the amino acid specified by a homopolymer.**

Second base

	U	C	A	G	
U	UUU Phe UUC UUA Leu UUG	UCU UCC Ser UCA UCG	UAU Tyr UAC UAA Stop UAG Stop	UGU Cys UGC UGA Stop UGG Trp	U C A G
C	CUU CUC Leu CUA CUG	CCU CCC Pro CCA CCG	CAU His CAC CAA Gln CAG	CGU CGC Arg CGA CGG	U C A G
A	AUU AUC Ile AUA AUG Met	ACU ACC Thr ACA ACG	AAU Asn AAC AAA Lys AAG	AGU Ser AGC AGA Arg AGG	U C A G
G	GUU GUC Val GUA GUG	GCU GCC Ala GCA GCG	GAU Asp GAC GAA Glu GAG	GGU GGC Gly GGA GGG	U C A G

First base / *Third base*

11.5 The genetic code consists of 64 codons. The amino acids specified by each codon are given in their three-letter abbreviations. The codons are written 5′ → 3′, as they appear in the mRNA. AUG is an initiation codon; UAA, UAG, and UGA are termination (stop) codons.

Even though some amino acids have multiple (isoaccepting) tRNAs, there are still more codons than anticodons, because different codons can sometimes pair with the same anticodon through flexibility in base pairing at the third position of the codon. Examination of Figure 11.5 reveals that many synonymous codons differ only in the third position. For example, alanine is encoded by the codons GCU, GCC, GCA, and GCG, all of which begin with GC. When the codon on the mRNA and the anticodon of the tRNA join (**Figure 11.6**), the first (5′) base of the codon pairs with the third (3′) base of the anticodon, strictly according to Watson and Crick rules: A with U; C with G. Next, the middle bases of codon and anticodon pair, also strictly following the Watson and Crick rules. After these pairs have hydrogen bonded, the third bases pair weakly—there may be flexibility, or **wobble**, in their pairing.

Concepts

The genetic code consists of 61 sense codons that specify the 20 common amino acids; the code is degenerate, meaning that some amino acids are encoded by more than one codon. Isoaccepting tRNAs are different tRNAs with different anticodons that specify the same amino acid. Wobble exists when more than one codon can pair with the same anticodon.

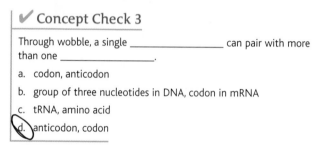

11.6 Wobble may exist in the pairing of a codon and anticodon. The mRNA and tRNA pair in an antiparallel fashion. Pairing at the first and second codon positions is in accord with the Watson-and-Crick pairing rules (A with U, G with C); however, pairing rules are relaxed at the third position of the codon, and G on the anticodon can pair with either U or C on the codon in this example.

✔ Concept Check 3

Through wobble, a single _____ can pair with more than one _____.

a. codon, anticodon
b. group of three nucleotides in DNA, codon in mRNA
c. tRNA, amino acid
d. anticodon, codon

The reading frame and initiation codons   Findings from early studies of the genetic code indicated that the code is generally **nonoverlapping**. An overlapping code is one in which a single nucleotide may be included in more than one codon, as follows:

Nucleotide sequence   A U A C G A G U C

Nonoverlapping code   A U A C G A G U C
                        Ile   Arg   Val

Overlapping code   A U A C G A G U C
                        Ile
                          U A C
                          Tyr
                            A C G
                            Thr

Usually, however, each nucleotide is part of a single codon. A few overlapping genes are found in viruses, but codons within the same gene do not overlap, and the genetic code is generally considered to be nonoverlapping.

For any sequence of nucleotides, there are three potential sets of codons—three ways in which the sequence can be read in groups of three. Each different way of reading the sequence is called a **reading frame**, and any sequence of nucleotides has three potential reading frames. The three reading frames have completely different sets of codons and will therefore specify proteins with entirely different amino acid sequences. Thus, it is essential for the translational machinery to use the correct reading frame. How is the correct reading frame established? The reading frame is set by the **initiation codon**, which is the first codon of the mRNA to specify an amino acid. After the initiation codon, the other codons are read as successive groups of three nucleotides. No bases are skipped between the codons; so there are no punctuation marks to separate the codons.

The initiation codon is usually AUG, although GUG and UUG are used on rare occasions. The initiation codon is not just a sequence that marks the beginning of translation; it specifies an amino acid. In bacterial cells, AUG encodes a modified type of methionine, *N*-formylmethionine; all proteins in bacteria initially begin with this amino acid, but the formyl group (or, in some cases, the entire amino acid) may be removed after the protein has been synthesized. When the codon AUG is at an internal position in a gene, it encodes unformylated methionine. In archaeal and eukaryotic cells, AUG specifies unformylated methionine both at the initiation position and at internal positions.

Termination codons   Three codons—UAA, UAG, and UGA—do not encode amino acids. These codons signal the end of the protein in both bacterial and eukaryotic cells and are called **stop codons**, **termination codons**, or **nonsense codons**. No tRNA molecules have anticodons that pair with termination codons.

The universality of the code   For many years, the genetic code was assumed to be **universal**, meaning that each codon specifies the same amino acid in all organisms. We now know that the genetic code is almost, but not completely, universal; a few exceptions have been found. Most of these exceptions are termination codons, but there are a few cases in which one sense codon substitutes for another. Most exceptions are found in mitochondrial genes; a few nonuniversal codons have also been detected in the nuclear genes of protozoans and in bacterial DNA.

Concepts

Each sequence of nucleotides possesses three potential reading frames. The correct reading frame is set by the initiation codon. The end of a protein-encoding sequence is marked by a termination codon. With a few exceptions, all organisms use the same genetic code.

Connecting Concepts

Characteristics of the Genetic Code

We have now considered a number of characteristics of the genetic code. Let's pause for a moment and review these characteristics.

1. The genetic code consists of a sequence of nucleotides in DNA or RNA. There are four letters in the code, corresponding to the four bases—A, G, C, and U (T in DNA).

2. The genetic code is a triplet code. Each amino acid is encoded by a sequence of three consecutive nucleotides, called a codon.

3. The genetic code is degenerate; that is, of 64 codons, 61 codons encode only 20 amino acids in proteins (3 codons are termination codons). Some codons are synonymous, specifying the same amino acid.

4. Isoaccepting tRNAs are tRNAs with different anticodons that accept the same amino acid; wobble allows the anticodon on one type of tRNA to pair with more than one type of codon on mRNA.

5. The code is generally nonoverlapping; each nucleotide in an mRNA sequence belongs to a single reading frame.

6. The reading frame is set by an initiation codon, which is usually AUG.

7. When a reading frame has been set, codons are read as successive groups of three nucleotides.

8. Any one of three termination codons (UAA, UAG, and UGA) can signal the end of a protein; no amino acids are encoded by the termination codons.

9. The code is almost universal.

11.2 Amino Acids Are Assembled into a Protein Through the Mechanism of Translation

Now that we are familiar with the genetic code, we can study the mechanism by which amino acids are assembled into proteins. Because more is known about translation in bacteria, we will focus primarily on bacterial translation. In most respects, eukaryotic translation is similar, although some significant differences will be noted as we proceed through the stages of translation.

Translation takes place on ribosomes; indeed, ribosomes can be thought of as moving protein-synthesizing machines. Through a variety of techniques, a detailed view of the structure of the ribosome has been produced in recent years, which has greatly improved our understanding of the translational process. A ribosome attaches near the 5′ end of an mRNA strand and moves toward the 3′ end, translating the codons as it goes (**Figure 11.7**). Synthesis begins at the amino end of the protein, and the protein is elongated by the addition of new amino acids to the carboxyl end.

Protein synthesis can be conveniently divided into four stages: (1) tRNA charging, which entails the binding of amino

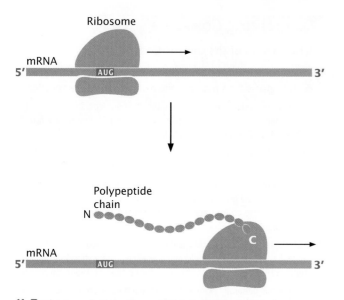

11.7 The translation of an mRNA molecule takes place on a ribosome. The letter N represents the amino end of the protein; C represents the carboxyl end.

acids to the tRNAs; (2) initiation, in which the components necessary for translation are assembled at the ribosome; (3) elongation, in which amino acids are joined, one at a time, to the growing polypeptide chain; and (4) termination, in which protein synthesis halts at the termination codon and the translation components are released from the ribosome.

The Binding of Amino Acids to Transfer RNAs

The first stage of translation is the binding of tRNA molecules to their appropriate amino acids, called tRNA charging. Although there may be several different tRNAs for a particular amino acid, each tRNA is specific for only one amino acid. The key to specificity between an amino acid and its tRNA is a set of enzymes called **aminoacyl-tRNA**

synthetases. A cell has 20 different aminoacyl-tRNA synthetases, one for each of the 20 amino acids. Each synthetase recognizes a particular amino acid, as well as all the tRNAs that accept that amino acid.

The attachment of a tRNA to its appropriate amino acid, termed **tRNA charging**, requires energy, which is supplied by adenosine triphosphate (ATP):

amino acid + tRNA + ATP $\rightarrow$

aminoacyl-tRNA + AMP + PP$_i$

The carboxyl group (COO$^-$) of the amino acid is attached to the adenine nucleotide at the 3′ end of the tRNA (**Figure 11.8**). To identify the resulting aminoacylated tRNA, we write the three-letter abbreviation for the amino acid in front of the tRNA; for example, the amino acid alanine (Ala) attaches to its tRNA (tRNA^{Ala}), giving rise to its aminoacyl-tRNA (Ala-tRNA^{Ala}).

Concepts

Amino acids are attached to specific tRNAs by aminoacyl-tRNA synthetases in a reaction that requires ATP.

✔ Concept Check 4

Amino acids bind to which part of the tRNA?

a. anticodon   c. 3′ end

b. codon   d. 5′ end

The Initiation of Translation

The second stage in the process of protein synthesis is initiation. At this stage, all the components necessary for protein synthesis assemble: (1) mRNA; (2) the small and large subunits of the ribosome; (3) a set of three proteins called initiation factors; (4) initiator tRNA with N-formylmethionine attached (fMet-tRNA^{fMet}); and (5) guanosine triphosphate (GTP).

11.8 An amino acid attaches to the 3′ end of a tRNA. The carboxyl group (COO$^-$) of the amino acid attaches to the oxygen of the 2′- or 3′-carbon atom of the final nucleotide at the 3′ end of the tRNA, in which the base is always adenine.

Initiation comprises three major steps. First, mRNA binds to the small subunit of the ribosome. Second, initiator tRNA binds to the mRNA through base pairing between the codon and the anticodon. Third, the large ribosome joins the initiation complex. Let's look at each of these steps more closely.

A functional ribosome exists as two subunits, the small 30S subunit and the large 50S subunit (in bacterial cells). When not actively translating, the two subunits are joined (**Figure 11.9a**). An mRNA molecule can bind to the small ribosome subunit only when the subunits are separate. **Initiation factor 3** (IF-3) binds to the small subunit of the ribosome and prevents the large subunit from binding during initiation (**Figure 11.9b**).

Next, the initiator tRNA, f Met-tRNA^{fMet}, attaches to the initiation codon (**Figure 11.9c**). This step requires **initiation factor 2** (IF-2), which forms a complex with GTP.

At this point, the initiation complex consists of (1) the small subunit of the ribosome; (2) the mRNA; (3) the initiator tRNA with its amino acid (f Met-tRNA^{fMet}); (4) one molecule of GTP; and (5) several initiation factors. These components are collectively known as the **30S initiation complex** (see Figure 11.9c). In the final step of initiation, initiation factors disassociate from the small subunit, allowing the large subunit of the ribosome to join the initiation complex (**Figure 11.9d**). When the large subunit has joined the initiation complex, the complex is called the **70S initiation complex**.

Similar events take place in the initiation of translation in eukaryotic cells, but there are some important differences. In bacterial cells, sequences in 16S rRNA of the small subunit of the ribosome bind to the Shine–Dalgarno sequence in mRNA. No analogous consensus sequence exists in eukaryotic mRNA. Instead, the cap at the 5′ end of eukaryotic mRNA plays a critical role in the initiation of translation. The small subunit of the eukaryotic ribosome, with the help of initiation factors, recognizes the cap and binds there; the small subunit then moves along (scans) the mRNA until it locates the first AUG codon. The identification of the start codon is facilitated by the presence of a consensus sequence (called the Kozak sequence) that surrounds the start codon:

Kozak sequence

5′–ACCAUGG–3′

Start codon

The poly(A) tail at the 3′ end of eukaryotic mRNA also plays a role in the initiation of translation. Proteins that attach to the poly(A) tail interact with proteins that bind to the 5′ cap, enhancing the binding of the small subunit of the ribosome to the 5′ end of the mRNA. This interaction between the 5′ cap and the 3′ tail suggests that the mRNA bends backward during the initiation of translation, forming a circular structure (**Figure 11.10**).

(a)

Ribosome

Large subunit (50S)

Small subunit (30S)

1 The ribosome consists of two subunits, which are normally joined.

IF-3

2 Initiation factor 3 binds to the small subunit, preventing the large subunit from binding…

(b)

Shine–Dalgarno sequence   Initiation codon

mRNA   AUGUGC

IF-3

3 …and thus allowing the small subunit to attach to mRNA.

GTP

tRNA

fMet   IF-2

IF-1

Anticodon   UAC

4 A tRNA charged with N-formylmethionine forms a complex with IF-2 and GTP…

(c)

30S initiation complex   fMet

IF-2 – GTP

UAC

mRNA   AUGUGC

IF-3   IF-1

5 …and binds to the initiation codon while IF-1 joins the small subunit.

IF-3

IF-1

IF-2 + GDP + Pᵢ

6 All initiation factors dissociate from the complex, GTP is hydrolyzed to GDP,…

(d)

70S initiation complex   fMet

UAC

mRNA   AUGUGC

Next codon

7 …and the large subunit joins to create a 70S initiation complex.

Conclusion: At the end of initiation, the ribosome is assembled on the mRNA and the first tRNA is attached to the initiation codon.

11.9 The initiation of translation requires several initiation factors and GTP.

11.10 The poly(A) tail of eukaryotic mRNA plays a role in the initiation of translation.

Concepts

In the initiation of translation in bacterial cells, the small ribosomal subunit attaches to mRNA, and initiator tRNA attaches to the initiation codon. This process requires several initiation factors (IF-1, IF-2, and IF-3) and GTP. In the final step, the large ribosomal subunit joins the initiation complex.

✔ Concept Check 5

During the initiation of translation, the small ribosome binds to which consensus sequence in bacteria?

Elongation

The next stage in protein synthesis is elongation, in which amino acids are joined to create a polypeptide chain. Elongation requires (1) the 70S complex just described; (2) tRNAs charged with their amino acids; (3) several elongation factors; and (4) GTP.

A ribosome has three sites that can be occupied by tRNAs; the **aminoacyl** (A) **site**, the **peptidyl** (P) **site**, and the **exit** (E) **site** (**Figure 11.11a**). The initiator tRNA immediately occupies the P site (the only site to which the fMet-tRNA^fMet^ is capable of binding), but all other tRNAs first enter the A site. After initiation, the ribosome is attached to the mRNA, and fMet-tRNA^fMet^ is positioned over the AUG start codon in the P site; the adjacent A site is unoccupied (see Figure 11.11a).

Elongation takes place in three steps. In the first step (**Figure 11.11b**), a charged tRNA binds to the A site. This binding takes place when **elongation factor Tu** (EF-Tu) joins with GTP and then with a charged tRNA to form a three-part complex. This complex enters the A site of the ribosome, where the anticodon on the tRNA pairs with the codon on the mRNA. After the charged tRNA is in the A site, GTP is cleaved to GDP, and the EF-Tu–GDP complex is released (**Figure 11.11c**). **Elongation factor Ts** (EF-Ts) regenerates EF-Tu–GDP to EF-Tu–GTP. In eukaryotic cells, a similar set of reactions delivers the charged tRNA to the A site.

The second step of elongation is the formation of a peptide bond between the amino acids that are attached to tRNAs in the P and A sites (**Figure 11.11d**). The formation of this peptide bond releases the amino acid in the P site from its tRNA. For many years, peptide-bond formation was thought to be catalyzed by one of the proteins in the large subunit of the ribosome. Evidence, however, now indicates that the catalytic activity is a property of the ribosomal RNA

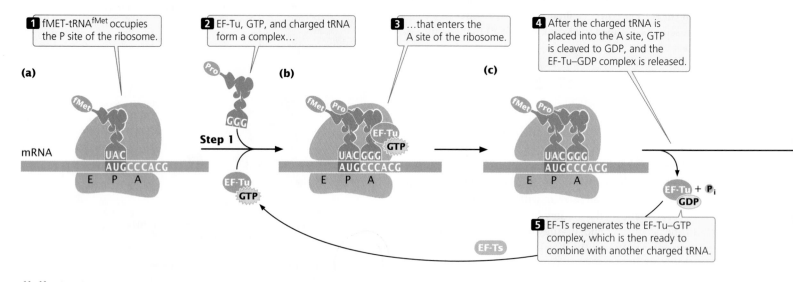

11.11 The elongation of translation comprises three steps.

in the large subunit of the ribosome; this rRNA acts as a ribozyme (see p. 243 in Chapter 10).

The third step in elongation is **translocation** (**Figure 11.11e**), the movement of the ribosome down the mRNA in the $5' \rightarrow 3'$ direction. This step positions the ribosome over the next codon and requires **elongation factor G** (EF-G) and the hydrolysis of GTP to GDP. Because the tRNAs in the P and A sites are still attached to the mRNA through codon–anticodon pairing, they do not move with the ribosome as it translocates. Consequently, the ribosome shifts so that the tRNA that previously occupied the P site now occupies the E site, from which it moves into the cytoplasm where it may be recharged with another amino acid. Translocation also causes the tRNA that occupied the A site (which is attached to the growing polypeptide chain) to be in the P site, leaving the A site open. Thus, the progress of each tRNA through the ribosome in the course of elongation can be summarized as follows: cytoplasm → A site → P site → E site → cytoplasm. As discussed earlier, the initiator tRNA is an exception: it attaches directly to the P site and never occupies the A site.

After translocation, the A site of the ribosome is empty and ready to receive the tRNA specified by the next codon. The elongation cycle (see Figure 11.11b through e) repeats itself: a charged tRNA and its amino acid occupy the A site, a peptide bond is formed between the amino acids in the A and P sites, and the ribosome translocates to the next codon. Throughout the cycle, the polypeptide chain remains attached to the tRNA in the P site. Elongation in eukaryotic cells takes place in a similar manner.

Concepts

Elongation consists of three steps: (1) a charged tRNA enters the A site, (2) a peptide bond is created between amino acids in the A and P sites, and (3) the ribosome translocates to the next codon. Elongation requires several elongation factors and GTP.

Termination

Protein synthesis terminates when the ribosome translocates to a termination codon. Because there are no tRNAs with anticodons complementary to the termination codons, no tRNA enters the A site of the ribosome when a termination codon is encountered (**Figure 11.12a**). Instead, proteins called **release factors** bind to the ribosome (**Figure 11.12b**). *E. coli* has three release factors—RF_1, RF_2, and RF_3. Release factor 1 binds to the termination codons UAA and UAG, and RF_2 binds to UGA and UAA. Release factor 3 forms a complex with GTP and binds to the ribosome. The release factors promote the cleavage of the tRNA in the P site from the polypeptide chain; in the process, GTP is hydrolyzed to GDP. The tRNA is released from the P site, mRNA is released from the ribosome, and the ribosome disassociates (**Figure 11.12c**).

Concepts

Termination takes place when the ribosome reaches a termination codon. Release factors bind to the termination codon, causing the release of the polypeptide from the last tRNA, of the tRNA from the ribosome, and of the mRNA from the ribosome.

6 A peptide bond forms between the amino acids in the P and A sites, and the tRNA in the P site releases its amino acid.

(d)

7 The ribosome moves down the mRNA to the next codon (translocation) which requires EF-G and GTP.

(e)

8 The tRNA that was in the P site is now in the E site from which it moves into the cytoplasm.

9 The tRNA that occupied the A site is now in the P site. The A site is now open and ready to receive another tRNA.

Dipeptide

Step 2

Step 3

EF-G

GTP   GDP + P_i

Conclusion: At the end of each cycle of elongation, the amino acid that was in the A site is added to the polypeptide chain and the A site is free to accept another tRNA.

11.12 Translation ends when a stop codon is encountered.

The overall process of protein synthesis, including tRNA charging, initiation, elongation, and termination, is summarized in **Figure 11.13**. The components taking part in this process are listed in **Table 11.1**.

11.13 Translation consists of tRNA charging, initiation, elongation, and termination. In this process, amino acids are linked together in the order specified by mRNA to create a polypeptide chain. A number of initiation, elongation, and release factors take part in the process, and energy is supplied by ATP and GTP.

Table 11.1   Components required for protein synthesis in bacterial cells

Stage	Component	Function
Binding of amino acid to tRNA	Amino acids	Building blocks of proteins
	tRNAs	Deliver amino acids to ribosomes
	Aminoacyl-tRNA synthetases	Attach amino acids to tRNAs
	ATP	Provides energy for binding amino acid to tRNA
Initiation	mRNA	Carries coding instructions
	fMet-tRNA^{fMet}	Provides first amino acid in peptide
	30S ribosomal subunit	Attaches to mRNA
	50S ribosomal subunit	Stabilizes tRNAs and amino acids
	Initiation factor 1	Enhances dissociation of large and small subunits of ribosome
	Initiation factor 2	Binds GTP; delivers fMet-tRNA^{fMet} to initiation codon
	Initiation factor 3	Binds to 30S subunit and prevents association with 50S subunit
Elongation	70S initiation complex	Functional ribosome with A, P, and E sites and peptidyl transferase activity where protein synthesis takes place
	Charged tRNAs	Bring amino acids to ribosome and help assemble them in order specified by mRNA
	Elongation factor Tu	Binds GTP and charged tRNA; delivers charged tRNA to A site
	Elongation factor Ts	Generates active elongation factor Tu
	Elongation factor G	Stimulates movement of ribosome to next codon
	GTP	Provides energy
	50S ribosomal subunit	Creates peptide bond between amino acids in A site and P site
Termination	Release factors 1, 2, and 3	Bind to ribosome when stop codon is reached and terminate translation

Connecting Concepts

A Comparison of Bacterial and Eukaryotic Translation

We have now considered the process of translation in bacterial cells and noted some distinctive differences that exist in eukaryotic cells. Let's take a few minutes to reflect on some of the important similarities and differences of protein synthesis in bacterial and eukaryotic cells.

First, we should emphasize that the genetic code of bacterial and eukaryotic cells is virtually identical; the only difference is in the amino acid specified by the initiation codon. In bacterial cells, AUG encodes a modified type of methionine, N-formylmethionine, whereas, in eukaryotic cells, AUG encodes unformylated methionine. One consequence of the fact that bacteria and eukaryotes use the same code is that eukaryotic genes can be translated in bacterial systems, and vice versa; this feature makes genetic engineering possible, as we will see in Chapter 14.

Another difference is that transcription and translation take place simultaneously in bacterial cells, but the nuclear envelope may separate these processes in eukaryotic cells. The physical separation of transcription and translation has important implications for the control of gene expression, which we will consider in Chapter 12, and it allows for extensive modification of eukaryotic mRNAs, as discussed in Chapter 10.

Yet another difference is that mRNA in bacterial cells is short-lived, typically lasting only a few minutes, but the longevity of mRNA in eukaryotic cells is highly variable and is frequently hours or days.

In both bacterial and eukaryotic cells, aminoacyl-tRNA synthetases attach amino acids to their appropriate tRNAs and the chemical process is the same. There are significant differences in the sizes and compositions of bacterial and eukaryotic ribosomal subunits. For example, the large subunit of the eukaryotic ribosome contains three rRNAs, whereas the bacterial ribosome contains only two. These differences allow antibiotics and other substances to inhibit bacterial translation while having no effect on the translation of eukaryotic nuclear genes, as will be discussed later in this chapter.

Other fundamental differences lie in the process of initiation. In bacterial cells, the small subunit of the ribosome attaches directly to the region surrounding the start codon through hydrogen bonding between the Shine–Dalgarno consensus sequence in the 5′ untranslated region of the mRNA and a sequence at the 3′ end of the 16S rRNA. In contrast, the small subunit of a eukaryotic ribosome first binds to proteins attached to the 5′ cap on mRNA and then migrates down the mRNA, scanning the sequence until it encounters the first AUG initiation codon. Additionally, a larger number of initiation factors take part in eukaryotic initiation than in bacterial initiation.

Elongation and termination are similar in bacterial and eukaryotic cells, although different elongation and termination factors are used. In both types of organisms, mRNAs are translated multiple times and are simultaneously attached to several ribosomes, forming polyribosomes, as discussed next.

11.3 Additional Properties of Translation and Proteins

Now that we have considered in some detail the process of translation, we will examine some additional aspects of protein synthesis.

Polyribosomes

In both prokaryotic and eukaryotic cells, mRNA molecules are translated simultaneously by multiple ribosomes (**Figure 11.14**). The resulting structure—an mRNA with several ribosomes attached—is called a **polyribosome**. Each ribosome successively attaches to the ribosome-binding site at the 5′ end of the mRNA and moves toward the 3′ end; the polypeptide associated with each ribosome becomes progressively longer as the ribosome moves along the mRNA.

In prokaryotic cells, transcription and translation are simultaneous; so multiple ribosomes may be attached to the 5′ end of the mRNA while transcription is still taking place at the 3′ end, as shown in Figure 11.14. Until recently, transcription and translation were thought not to be simultaneous in eukaryotes, because transcription takes place in the nucleus and all translation was assumed to take place in the cytoplasm. As mentioned in Chapter 10, recent research suggests that that translation of mRNAs of some genes takes place within the eukaryotic nucleus. If some translation does take place within the the nucleus, transcription and translation in eukaryotes may be simultaneous, much as in prokaryotes.

(a)

Direction of transcription

DNA

RNA

mRNA 5′

Incoming ribosomal subunits

Growing polypeptide chain

Direction of translation

(b)

One gene

mRNA

DNA

Directon of transcription

Ribosomes

11.14 An mRNA molecule may be transcribed simultaneously by several ribosomes. (a) Four ribosomes are translating an mRNA molecule; the ribosomes are depicted as moving from the 5′ end to the 3′ end of the mRNA. (b) In this electron micrograph of a polyribosome, the dark-staining spheres are ribosomes, and the thin filaments connecting the ribosomes are mRNA. *[Part b: O. L. Miller, Jr., and Barbara A. Hamaklo.]*

> ✔ **Concept Check 7**
>
> In a polyribosome, the polypeptides associated with which ribosomes will be the longest?
>
> a. Those at the 5′ end of mRNA
> b. Those at the 3′ end of mRNA
> c. Those in the middle of mRNA
> d. All polypeptides will be the same length.

The Posttranslational Modifications of Proteins

After translation, proteins in both prokaryotic and eukaryotic cells may undergo alterations termed posttranslational modifications. A number of different types of modifications are possible. Some proteins are synthesized as larger precursor proteins and must be cleaved and trimmed by enzymes before the proteins can become functional. For others, the attachment of carbohydrates may be required for activation. The functions of many proteins critically depend on the proper folding of the polypeptide chain; some proteins spontaneously fold into their correct shapes, but, for others, correct folding may initially require the participation of other molecules called **molecular chaperones**.

> **Concepts**
>
> In both prokaryotic and eukaryotic cells, multiple ribosomes may be attached to a single mRNA, generating a structure called a polyribosome.

Concepts

Many proteins undergo posttranslational modifications after their synthesis.

Translation and Antibiotics

Antibiotics are drugs that kill microorganisms. To make an effective antibiotic—not just any poison will do—the trick is to kill the microbe without harming the patient. Antibiotics must be carefully chosen so that they destroy bacterial cells but not the eukaryotic cells of their host.

Translation is frequently the target of antibiotics because translation is essential to all living organisms and differs significantly between bacterial and eukaryotic cells. For example, as already mentioned, bacterial and eukaryotic ribosomes differ in size and composition. A number of antibiotics bind selectively to bacterial ribosomes and inhibit various steps in translation, but they do not affect eukaryotic ribosomes. Tetracyclines, for instance, are a class of antibiotics that bind to the A site of a bacterial ribosome and block the entry of charged tRNAs, yet they have no effect on eukaryotic ribosomes. Chloramphenicol binds to the large subunit of the ribosome and blocks peptide-bond formation. Streptomycin binds to the small subunit of the ribosome and inhibits initiation, and erythromycin blocks translocation. Although chloramphenicol and streptomycin are potent inhibitors of translation in bacteria, they do not inhibit translation in archaebacteria.

Many antibiotics act by blocking specific steps in translation, and different antibiotics affect different steps in protein synthesis. Because of this specificity, antibiotics are frequently used to study the process of protein synthesis.

Concepts Summary

- Amino acids in a protein are linked together by peptide bonds. Chains of amino acids fold and associate to produce the secondary, tertiary, and quaternary structures of proteins.

- The genetic code is a triplet code: three nucleotides specify a single amino acid. It is also degenerate (meaning that more than one codon may specify an amino acid), nonoverlapping, and universal (almost).

- The reading frame is set by the initiation codon. The end of the protein-coding section of an mRNA is marked by one of three termination codons.

- Protein synthesis comprises four steps: (1) the binding of amino acids to the appropriate tRNAs, (2) initiation, (3) elongation, and (4) termination.

- The binding of an amino acid to a tRNA requires the presence of a specific aminoacyl-tRNA synthetase and ATP.

- In bacterial translation initiation, the small subunit of the ribosome attaches to the mRNA and is positioned over the initiation codon. It is joined by the first tRNA and its associated amino acid (N-formylmethionine in bacterial cells) and, later, by the large subunit of the ribosome. Initiation requires several initiation factors and GTP.

- In elongation, a charged tRNA enters the A site of a ribosome, a peptide bond is formed between amino acids in the A and P sites, and the ribosome moves (translocates) along the mRNA to the next codon. Elongation requires several elongation factors and GTP.

- Translation is terminated when the ribosome encounters one of the three termination codons. Release factors and GTP are required to bring about termination.

- Each mRNA may be simultaneously translated by several ribosomes, producing a structure called a polyribosome.

- Many proteins undergo posttranslational modification.

Important Terms

one-gene, one-enzyme hypothesis (p. 272)
one gene, one polypeptide hypothesis (p. 272)
amino acid (p. 272)
peptide bond (p. 272)
polypeptide (p. 272)
sense codon (p. 275)
degenerate genetic code (p. 275)
synonymous codons (p. 275)
isoaccepting tRNAs (p. 275)
wobble (p. 276)

nonoverlapping genetic code (p. 276)
reading frame (p. 277)
initiation codon (p. 277)
stop (termination or nonsense) codon (p. 277)
universal genetic code (p. 277)
aminoacyl-tRNA synthetase (p. 278)
tRNA charging (p. 278)
initiation factor (IF-1, IF-2, IF-3) (p. 279)
30S initiation complex (p. 279)

70S initiation complex (p. 279)
aminoacyl (A) site (p. 280)
peptidyl (P) site (p. 280)
exit (E) site (p. 280)
elongation factor Tu (EF-Tu) (p. 280)
elongation factor Ts (EF-Ts) (p. 280)
translocation (p. 281)
elongation factor G (EF-G) (p. 281)
release factor (p. 281)
polyribosome (p. 284)
molecular chaperone (p. 284)

Answers to Concept Checks

1. The amino acid sequence (primary structure) of the protein

2. b

3. d

4. c

5. The Shine–Dalgarno sequence

6. a

7. b

Worked Problems

1. If there were five different types of bases in mRNA instead of four, what would be the minimum codon size (number of nucleotides) required to specify the following numbers of different amino acid types: (a) 4, (b) 20, (c) 30?

• Solution

To answer this question, we must determine the number of combinations (codons) possible when there are different numbers of bases and different codon lengths. In general, the number of different codons possible will be equal to:

$$b^{lg} = \text{number of codons}$$

where b equals the number of different types of bases and lg equals the number of nucleotides in each codon (codon length). If there are five different types of bases, then:

$$5^1 = \quad 5 \text{ possible codons}$$

$$5^2 = \quad 25 \text{ possible codons}$$

$$5^3 = 125 \text{ possible codons}$$

The number of possible codons must be greater than or equal to the number of amino acids specified. Therefore, a codon length of one nucleotide could specify 4 different amino acids, a codon length of two nucleotides could specify 20 different amino acids, and a codon length of three nucleotides could specify 30 different amino acids: (a) one, (b) two, (c) three.

2. A template strand in bacterial DNA has the following base sequence:

$$5'\text{–AGGTTTAACGTGCAT–}3'$$

What amino acids are encoded by this sequence?

• Solution

To answer this question, we must first work out the mRNA sequence that will be transcribed from this DNA sequence. The mRNA must be antiparallel and complementary to the DNA template strand:

DNA template strand: $5'$–AGGTTTAACGTGCAT–$3'$

mRNA copied from DNA: $3'$–UCCAAAUUGCACGUA–$5'$

An mRNA is translated $5' \to 3'$; so it will be helpful if we turn the RNA molecule around with the $5'$ end on the left:

mRNA copied from DNA: $5'$–AUGCACGUUAAACCU–$3'$

The codons consist of groups of three nucleotides that are read successively after the first AUG codon; using Figure 11.5, we can determine that the amino acids are:

$$5'\text{–AUG—CAC—GUU—AAA—CCU–}3'$$
$$\downarrow \quad\quad \downarrow \quad\quad \downarrow \quad\quad \downarrow \quad\quad \downarrow$$
$$\text{f Met——His——Val——Lys——Pro}$$

Comprehension Questions

Section 11.1

1. What is the one-gene, one-enzyme hypothesis?

2. What are isoaccepting tRNAs?

*3. What is the significance of the fact that many synonymous codons differ only in the third nucleotide position?

*4. Define the following terms as they apply to the genetic code:
   a. Reading frame
   b. Overlapping code
   c. Nonoverlapping code
   d. Initiation codon
   e. Termination codon
   f. Sense codon
   g. Nonsense codon
   h. Universal code
   i. Nonuniversal codons

5. How is the reading frame of a nucleotide sequence set?

Section 11.2

*6. How are tRNAs linked to their corresponding amino acids?

7. What role do the initiation factors play in protein synthesis?

8. What events bring about the termination of translation?

9. Compare and contrast the process of protein synthesis in bacterial and eukaryotic cells, giving similarities and differences in the process of translation in these two types of cells.

Section 11.3

10. What are some types of posttranslational modification of proteins?

*11. Explain how some antibiotics work by affecting the process of protein synthesis.

Application Questions and Problems

Section 11.1

*12. Assume that the number of different types of bases in RNA is four. What would be the minimum codon size (number of nucleotides) required if the number of different types of amino acids in proteins were: (a) 2, (b) 8, (c) 17, (d) 45, (e) 75?

13. How many codons would be possible in a triplet code if only three bases (A, C, and U) were used?

*14. Using the genetic code presented in Figure 11.5, give the amino acids specified by the following bacterial mRNA sequences, and indicate the amino and carboxyl ends of the polypeptide produced.
   a. 5′–AUGUUUAAAUUUAAAUUUUGA–3′
   b. 5′–AUGUAUAUAUAUAUAUGA–3′
   c. 5′–AUGGAUGAAAGAUUUCUCGCUUGA–3′
   d. 5′–AUGGGUUAGGGGACAUCAUUUUGA–3′

15. A nontemplate strand on bacterial DNA has the following base sequence. What amino acid sequence will be encoded by this sequence?

   5′–ATGATACTAAGGCCC–3′

*16. The following amino acid sequence is found in a tripeptide: Met-Trp-His. Give all possible nucleotide sequences on the mRNA, on the template strand of DNA, and on the nontemplate strand of DNA that can encode this tripeptide.

17. How many different mRNA sequences can encode a polypeptide chain with the amino acid sequence Met-Leu-Arg? (Be sure to include the stop codon.)

18. Which of the following amino acid changes could result from a mutation that changed a single base? For each change that could result from the alteration of a single base, determine which position of the codon (first, second, or third nucleotide) in the mRNA must be altered for the change to result.
   a. Leu → Gln   **d.** Pro → Ala
   b. Phe → Ser   **e.** Asn → Lys
   c. Phe → Ile   **f.** Ile → Asn

Section 11.2

19. Arrange the following components of translation in the approximate order in which they would appear or be used in protein synthesis:

   70S initiation complex   30S initiation complex
   release factor 1   elongation factor Tu
   peptidyl transferase   initiation factor 3
   elongation factor G   f Met-Trna^{fMet}

20. The following diagram illustrates a step in the process of translation.

Sketch the diagram and identify the following elements on it.
   a. 5′ and 3′ ends of the mRNA
   b. A, P, and E sites
   c. Start codon
   d. Stop codon
   e. Amino and carboxyl ends of the newly synthesized polypeptide chain
   f. Approximate location of the next peptide bond that will be formed
   g. Place on the ribosome where release factor 1 will bind

21. Refer to the diagram in Problem 20 to answer the following questions.
   a. What will be the anticodon of the next tRNA added to the A site of the ribosome?
   b. What will be the next amino acid added to the growing polypeptide chain?

*22. A synthetic mRNA added to a cell-free protein-synthesizing system produces a peptide with the following amino acid sequence: Met-Pro-Ile-Ser-Ala. What would be the effect on translation if the following components were omitted from the cell-free protein-synthesizing system? What, if any, type of protein would be produced? Explain your reasoning.
   a. Initiation factor 1   **e.** Release factors RF$_1$, RF$_2$,
   b. Initiation factor 2   and RF$_3$
   c. Elongation factor Tu   **f.** ATP
   d. Elongation factor G   **g.** GTP

23. For each of the sequences in the table on page 288, place a check mark in the appropriate space to indicate the process *most immediately* affected by deleting the sequence. Choose only one process for each sequence (i.e., one check mark per sequence).

Sequence deleted	Process most immediately affected by deletion			
	Replication	Transcription	RNA processing	Translation
a. *ori* site	____	____	____	____
b. 3' splice-site consensus	____	____	____	____
c. poly(A) tail	____	____	____	____
d. terminator	____	____	____	____
e. start codon	____	____	____	____
f. −10 consensus	____	____	____	____
g. Shine–Dalgarno	____	____	____	____

Challenge Questions

Section 11.1

24. The redundancy of the genetic code means that some amino acids are specified by more than one codon. For example, the amino acid leucine is encoded by six different codons. Within a genome, synonymous codons are not present in equal numbers; some synonymous codons appear much more frequently than others, and the preferred codons differ among different species. For example, in one species, the codon UUA might be used most often to encode leucine, whereas, in another species, the codon CUU might be used most often. Speculate on a reason for this bias in codon usage and why the preferred codons are not the same in all organisms.

Section 11.2

*25. Several experiments were conducted to obtain information about how the eukaryotic ribosome recognizes the AUG start codon. In one experiment, the gene that encodes methionine initiator tRNA (tRNA$_i^{Met}$) was located and changed. The nucleotides that specify the anticodon on tRNA$_i^{Met}$ were mutated so that the anticodon in the tRNA was 5'-CCA-3' instead of 5'-CAU-3'. When this mutated gene was placed into a eukaryotic cell, protein synthesis took place but the proteins produced were abnormal. Some of the proteins produced contained extra amino acids, and others contained fewer amino acids than normal.

a. What do these results indicate about how the ribosome recognizes the starting point for translation in eukaryotic cells? Explain your reasoning.

b. If the same experiment had been conducted on bacterial cells, what results would you expect?

12 Control of Gene Expression

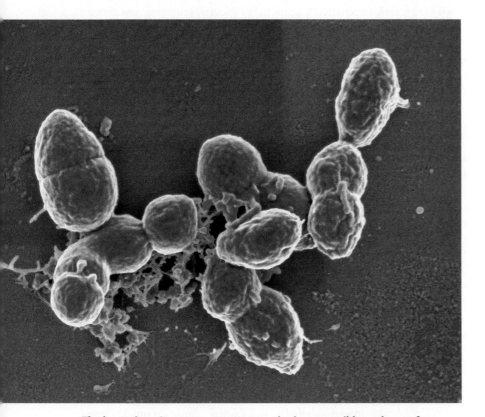

The bacterium _Streptococcus pneumoniae_ is responsible each year for millions of cases of pneumonia, inner-ear infection, sinus infection, and meningitis. Gene regulation is required for transformation, a process by which many strains of _S. pneumoniae_ have become resistant to antibiotics. [_Dr. Gary Gaugler/Photo Researchers._]

Stress, Sex, and Gene Regulation in Bacteria

Each year, _Streptococcus pneumoniae_, commonly known as pneumococcus, is responsible for millions of cases of pneumonia, inner-ear infection, sinus infection, and meningitis. Before the widespread use of antibiotics, pneumonia resulting from _S. pneumoniae_ infection was a leading cause of death, particularly among the young and the elderly. The development of penicillin and other antibiotics, beginning in the 1940s, provided an effective tool against _Streptococcus_ infections, and deaths from pneumonia and other life-threatening bacterial infections plummeted. In recent years, however, many strains of _S. pneumoniae_ and other pathogenic bacteria have developed resistance to penicillin and other antibiotics. Some bacterial strains have rapidly acquired resistance to multiple antibiotics, making treatment of _S. pneumoniae_ difficult. How has antibiotic resistance spread so rapidly among bacteria?

Although bacteria are simple organisms, they engage in complex and ingenious mechanisms to exchange genetic information and rapidly adapt to changing conditions. One such mechanism is transformation, first demonstrated by Fred Griffith in 1928, in which bacteria take up new DNA released by dead cells and integrate the foreign DNA into their own genome (see pp. 196–197 in Chapter 8). Transformation is considered a type of sexuality in bacteria, because it results in the exchange of genetic material between individual cells. It is widespread among bacteria and has been a major means by which antibiotic resistance has spread.

In _S. pneumoniae_, the ability to carry out transformation—called competence—requires from 105 to 124 genes, collectively termed the _com_ (for competence) regulon. Genes within the _com_ regulon facilitate the passage of DNA through the cell membrane and the DNA's recombination with homologous genes in the bacterial chromosome. For competence to take place, genes in the _com_ regulon must be precisely turned on and off in a carefully regulated sequence.

The _com_ regulon is activated in response to a protein called competence-stimulating peptide (CSP), which is produced by bacteria and is exported into the surrounding medium. When enough CSP accumulates, it attaches to a receptor on the bacterial cell membrane, which then activates a regulator protein that stimulates the transcription of genes within the *com* regulon and sets in motion a complex cascade of reactions that ultimately results in the development of competence.

Early steps in the development of competence are poorly understood, but most experts formerly assumed that CSP is continually produced and gradually accumulates in the medium; the assumption was that, when CSP reaches a critical threshold, transformation is induced. However, recent research has altered the view that induction of transformation is a passive process. This work has demonstrated that the production of CSP is itself regulated. When *S. pneumoniae* is subjected to stress, it produces more CSP, which causes transformation and allows the bacteria to genetically adapt to the stress by acquiring new genetic material from the environment. In fact, the presence of antibiotics is one type of stress that stimulates the production of CSP and induces transformation. This reaction to stress is probably a primary reason that bacteria such as *S. pneumoniae* have so quickly evolved resistance to antibiotics.

Not only is the regulation of gene expression important for the development of transformation; it is critical for the control of numerous life processes in all organisms. This chapter is about **gene regulation**, the mechanisms and systems that control the expression of genes. We begin by considering the levels at which gene expression is controlled and the difference between genes and regulatory elements. We then examine gene regulation in bacteria, including the structure and function of operons—groups of genes that are regulated together in bacteria. Finally, we look at gene regulation in eukaryotes and consider some of the similarities and differences in gene regulation of bacteria and eukaryotes.

12.1 The Regulation of Gene Expression Is Critical for All Organisms

A major theme of molecular genetics is the central dogma, which states that genetic information flows from DNA to RNA to proteins (see Figure 8.14). Although the central dogma provided a molecular basis for the connection between genotype and phenotype, it failed to address a critical question: How is the flow of information along the molecular pathway *regulated*?

Consider *Escherichia coli*, a bacterium that resides in your large intestine. Your eating habits completely determine the nutrients available to this bacterium: it can neither seek out nourishment when nutrients are scarce nor move away when confronted with an unfavorable environment. *Escherichia coli* makes up for its inability to alter the external environment by being internally flexible. For example, if glucose is present, *E. coli* uses it to generate ATP; if there's no glucose, it utilizes lactose, arabinose, maltose, xylose, or any of a number of other sugars. When amino acids are available,

E. coli uses them to synthesize proteins; if a particular amino acid is absent, *E. coli* produces the enzymes needed to synthesize that amino acid. Thus, *E. coli* responds to environmental changes by rapidly altering its biochemistry. This biochemical flexibility, however, has a high price. Producing all the enzymes necessary for every environmental condition would be energetically expensive. So how does *E. coli* maintain biochemical flexibility while optimizing energy efficiency?

The answer is through gene regulation. Bacteria carry the genetic information for synthesizing many proteins, but only a subset of this genetic information is expressed at any time. When the environment changes, new genes are expressed, and proteins appropriate for the new environment are synthesized. For example, if a carbon source appears in the environment, genes encoding enzymes that take up and metabolize this carbon source are quickly transcribed and translated. When this carbon source disappears, the genes that encode them are shut off.

Multicellular eukaryotic organisms face an additional dilemma. Individual cells in a multicellular organism are specialized for particular tasks. The proteins produced by a nerve cell, for example, are quite different from those produced by a white blood cell. The problem that a eukaryotic cell faces is how to specialize. Although they are quite different in shape and function, a nerve cell and a blood cell still carry the same genetic instructions.

A multicellular organism's challenge is to bring about the specialization of cells that have a common set of genetic instructions (the process of development). This challenge is met through gene regulation: all of an organism's cells carry the same genetic information, but only a subset of genes are expressed in each cell type. Genes needed for other cell types are not expressed. Gene regulation is therefore the key to both unicellular flexibility and multicellular specialization, and it is critical to the success of all living organisms.

Concepts

In bacteria, gene regulation maintains internal flexibility, turning genes on and off in response to environmental changes. In multicellular eukaryotic organisms, gene regulation also brings about cellular differentiation.

12.2 Many Aspects of Gene Regulation Are Similar in Bacteria and Eukaryotes

The mechanisms of gene regulation were first investigated in bacterial cells, in which the availability of mutants and the ease of laboratory manipulation made it possible to unravel the mechanisms. When the study of these mechanisms in eukaryotic cells began, bacterial gene regulation clearly seemed to differ from eukaryotic gene regulation. As more and more information has accumulated about gene regulation, however, a number of common themes have emerged, and, today, many aspects of gene regulation in bacterial and eukaryotic cells are recognized to be similar. Before examining specific elements of bacterial gene regulation and eukaryotic gene regulation, we will briefly consider some themes of gene regulation common to all organisms.

Genes and Regulatory Elements

In considering gene regulation in both bacteria and eukaryotes, we must distinguish between the DNA sequences that are transcribed and the DNA sequences that regulate the expression of other sequences. As defined in Chapter 10, a *gene* is any DNA sequence that is transcribed into an RNA molecule. Genes include DNA sequences that encode proteins, as well as sequences that encode rRNA, tRNA, snRNA, and other types of RNA. **Structural genes** encode proteins that are used in metabolism or biosynthesis or that play a structural role in the cell. **Regulatory genes** are genes whose products, either RNA or proteins, interact with other sequences and affect the transcription or translation of those sequences. In many cases, the products of regulatory genes are DNA-binding proteins. Bacteria and eukaryotes use regulatory genes to control the expression of many of their structural proteins. However, a few genes, particularly those that encode essential cellular functions, are expressed continually and are said to be **constitutive**.

We will also encounter DNA sequences that are not transcribed at all but still play a role in regulating other nucleotide sequences. These **regulatory elements** affect the expression of sequences to which they are physically linked. Regulatory elements are common in both bacterial and eukaryotic cells, and much of gene regulation in both types of organisms takes place through the action of proteins produced by regulatory genes that recognize and bind to regulatory elements.

The regulation of gene expression can be through processes that stimulate gene expression, termed *positive control*, or through processes that inhibit gene expression, termed *negative control*. Bacteria and eukaryotes use both positive and negative control mechanisms to regulate their genes.

Concepts

Genes are DNA sequences that are transcribed into RNA. Regulatory elements are DNA sequences that are not transcribed but affect the expression of genes. Positive control includes mechanisms that stimulate gene expression, whereas negative control inhibits gene expression.

✔ **Concept Check 1**

What is a constitutive gene?

Levels of Gene Regulation

In both bacteria and eukaryotes, genes can be regulated at a number of points along the pathway of information flow from genotype to phenotype (**Figure 12.1**). First, regulation may be through the alteration of gene structure; this type of gene regulation is primarily in eukaryotes. Modifications to DNA or its packaging may help to determine which sequences are available for transcription or the rate at which sequences are transcribed. DNA methylation and changes in chromatin are two processes that play a pivotal role in gene regulation.

A second point at which a gene can be regulated is at the level of transcription. For the sake of cellular economy, limiting the production of a protein early in the process makes sense, and transcription is an important point of gene regulation in both bacterial and eukaryotic cells. A third potential point of gene regulation is mRNA processing. Eukaryotic mRNA is extensively modified before it is translated: a 5′ cap is added, the 3′ end is cleaved and polyadenylated, and introns are removed (see Chapter 10). These modifications determine the stability of the mRNA, whether mRNA can be translated, the rate of translation, and the amino acid sequence of the protein produced. There is growing evidence that a number of regulatory mechanisms in eukaryotic cells operate at the level of mRNA processing.

A fourth point for the control of gene expression is the regulation of RNA stability. The amount of protein produced depends not only on the amount of mRNA synthesized, but also on the rate at which the mRNA is degraded. A fifth point of gene regulation is at the level of translation, a complex process requiring a large number of enzymes, protein factors, and RNA molecules (see Chapter 11). All of these factors, as well as the availability of amino acids, affect the rate at which proteins are produced and therefore provide points at which gene expression can be controlled. Translation can also be affected by sequences in mRNA.

Levels of gene control

Compact DNA

↓ Alteration of structure

Relaxed DNA

↓ Transcription

Pre-mRNA

↓ mRNA processing

Processed mRNA   AAAAA

RNA stability
Translation

Protein
(inactive)

↓ Posttranslational modification

Modified protein
(active)

12.1   Gene expression can be controlled at multiple levels.

Finally, many proteins are modified after translation (see Chapter 11), and these modifications affect whether the proteins become active; so genes can be regulated through processes that affect posttranslational modification. Gene expression can be affected by regulatory activities at any or all of these points.

Concepts

Gene expression can be controlled at any of a number of points along the molecular pathway from DNA to protein, including gene structure, transcription, mRNA processing, RNA stability, translation, and posttranslational modification.

✔ Concept Check 2

Why is transcription a particularly important level of gene regulation in both bacteria and eukaryotes?

12.3 Gene Regulation in Bacterial Cells

A significant difference between bacterial and eukaryotic gene control lies in the organization of functionally related genes. Many bacterial genes that have related functions are clustered and are under the control of a single promoter. These genes are often transcribed together into a single mRNA. A group of bacterial structural genes that are transcribed together (along with their promoter and additional sequences that control transcription) is called an **operon**. The operon regulates the expression of the structural genes by controlling transcription, which, in bacteria, is usually the most important level of gene regulation.

Operon Structure

The organization of a typical operon is illustrated in **Figure 12.2.** At one end of the operon is a set of structural genes, shown in Figure 12.2 as gene *a*, gene *b*, and gene *c*. These structural genes are transcribed into a single mRNA, which is translated to produce enzymes A, B, and C. These enzymes carry out a series of biochemical reactions that convert precursor molecule X into product Y. The transcription of structural genes *a*, *b*, and *c* is under the control of a promoter, which lies upstream of the first structural gene. RNA polymerase binds to the promoter and then moves downstream, transcribing the structural genes.

A **regulator gene** helps to control the transcription of the structural genes of the operon. Although it affects operon function, the regulator gene is not considered part of the operon. The regulator gene has its own promoter and is transcribed into a short mRNA, which is translated into a small protein. This **regulator protein** can bind to a region of DNA called the **operator** and affect whether transcription can take place. The operator usually overlaps the 3′ end of the promoter and sometimes the 5′ end of the first structural gene (see Figure 12.2).

Concepts

Functionally related genes in bacterial cells are frequently clustered together as a single transcriptional unit termed an operon. A typical operon includes several structural genes, a promoter for the structural genes, and an operator site where the product of a regulator gene binds.

✔ Concept Check 3

What is the difference between a structural gene and a regulator gene?

a. Structural genes are transcribed into mRNA, but regulator genes aren't.

b. Structural genes have complex structures; regulator genes have simple structures.

1 An operon is a group of structural genes plus sequences that control transcription.

Operon

Structural genes

RNA polymerase

Promoter | Regulator

Transcription

2 A separate regulator gene—with its own promoter—encodes a regulator protein...

mRNA

Translation

Regulator protein

Promoter | Operator | Gene *a* | Gene *b* | Gene *c*

3 ...that may bind to the operator site to regulate the transcription of mRNA.

Transcription

mRNA

Translation

Proteins (enzymes)   A   B   C

4 The products of mRNA catalyze reactions in a biochemical pathway.

Biochemical pathway

Precursor X → Intermediate products → Product Y

12.2 An operon is a single transcriptional unit that includes a series of structural genes, a promoter, and an operator.

c. Structural genes encode proteins that function in the structure of the cell; regulator genes carry out metabolic reactions.

d. Structural genes encode proteins; regulator genes control the transcription of structural genes.

Negative and Positive Control: Inducible and Repressible Operons

There are two types of transcriptional control: **negative control**, in which a regulatory protein is a repressor, binding to DNA and inhibiting transcription; and **positive control**, in which a regulatory protein is an activator, stimulating transcription. Operons can also be either inducible or repressible. **Inducible operons** are those in which transcription is normally off (not taking place); something must happen to induce transcription, or turn it on. **Repressible operons** are those in which transcription is normally on (taking place); something must happen to repress transcription, or turn it off. In the next sections, we will consider several varieties of these basic control mechanisms.

Negative inducible operons   In an operon with negative control at the operator site, the regulatory protein is a repressor: the binding of the regulator protein to the operator inhibits transcription. In a negative inducible operon, transcription and translation of the regulator gene produce an active *repressor* that readily binds to the operator (**Figure 12.3a**). Because the operator site overlaps the promoter site, the binding of this protein to the operator physically blocks the binding of RNA polymerase to the promoter and prevents transcription. For transcription to take place,

something must happen to prevent the binding of the repressor at the operator site. This type of system is said to be *inducible,* because transcription is normally off (inhibited) and must be turned on (induced).

Transcription is turned on when a small molecule, an **inducer**, binds to the repressor (**Figure 12.3b**). Regulatory proteins frequently have two binding sites: one that binds to DNA and another that binds to a small molecule such as an inducer. The binding of the inducer (precursor V in Figure 12.3b) alters the shape of the repressor, preventing it from binding to DNA. Proteins of this type, which change shape on binding to another molecule, are called **allosteric proteins**.

When the inducer is absent, the repressor binds to the operator, the structural genes are not transcribed, and enzymes D, E, and F (which metabolize precursor V) are not synthesized (see Figure 12.3a). This mechanism is an adaptive one: because no precursor V is available, synthesis of the enzymes would be wasteful when they have no substrate to metabolize. As soon as precursor V becomes available, some of it binds to the repressor, rendering the repressor inactive and unable to bind to the operator site. RNA polymerase can now bind to the promoter and transcribe the structural genes. The resulting mRNA is then translated into enzymes D, E, and F, which convert substrate V into product W (see Figure 12.3b). So, an operon with negative inducible control regulates the synthesis of the enzymes economically: the enzymes are synthesized only when their substrate (V) is available.

Inducible operons usually control proteins that carry out degradative processes—proteins that break down molecules. For these types of proteins, inducible control makes sense because the proteins are not needed unless the substrate (which is broken down by the proteins) is present.

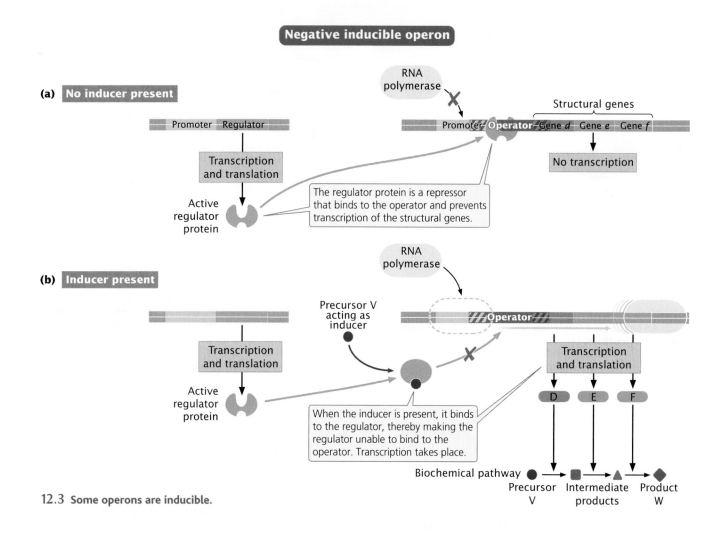

Negative inducible operon

(a) **No inducer present**

RNA polymerase

Promoter Regulator

Structural genes

Promoter Operator Gene *d* Gene *e* Gene *f*

Transcription and translation

No transcription

Active regulator protein

The regulator protein is a repressor that binds to the operator and prevents transcription of the structural genes.

(b) **Inducer present**

RNA polymerase

Precursor V acting as inducer

Operator

Transcription and translation

Transcription and translation

Active regulator protein

D   E   F

When the inducer is present, it binds to the regulator, thereby making the regulator unable to bind to the operator. Transcription takes place.

Biochemical pathway ● → ■ → ▲ → ◆

Precursor V   Intermediate products   Product W

12.3 Some operons are inducible.

Negative repressible operons   Some operons with negative control are *repressible*, meaning that transcription *normally* takes place and must be turned off, or repressed. The regulator protein in this type of operon also is a repressor but is synthesized in an *inactive* form that cannot by itself bind to the operator. Because no repressor is bound to the operator, RNA polymerase readily binds to the promoter and transcription of the structural genes takes place (**Figure 12.4a**).

To turn transcription off, something must happen to make the repressor active. A small molecule called a **corepressor** binds to the repressor and makes it capable of binding to the operator. In the example illustrated (see Figure 12.4a), the product (U) of the metabolic reaction is the corepressor. As long as the level of product U is high, it is available to bind to the repressor and activate it, preventing transcription (**Figure 12.4b**). With the operon repressed, enzymes G, H, and I are not synthesized, and no more U is produced from precursor T. However, when all of product U is used up, the repressor is no longer activated by U and cannot bind to the operator. The inactivation of the repressor allows the transcription of the structural genes and the

synthesis of enzymes G, H, and I, resulting in the conversion of precursor T into product U. Like inducible operons, repressible operons are economical: the enzymes are synthesized only as needed.

Repressible operons usually control proteins that carry out the biosynthesis of molecules, such as amino acids, needed in the cell. For these types of operons, repressible control makes sense because the product produced by the proteins is always needed by the cell. Thus, these operons are normally on and are turned off when there are adequate amounts of the product already present.

Note that both the inducible and the repressible systems that we have considered are forms of negative control, in which the regulatory protein is a repressor. We will now consider positive control, in which a regulator protein stimulates transcription.

Positive control   With positive control, a regulatory protein is an activator: it binds to DNA (usually at a site other than the operator) and stimulates transcription. Positive control can be inducible or repressible.

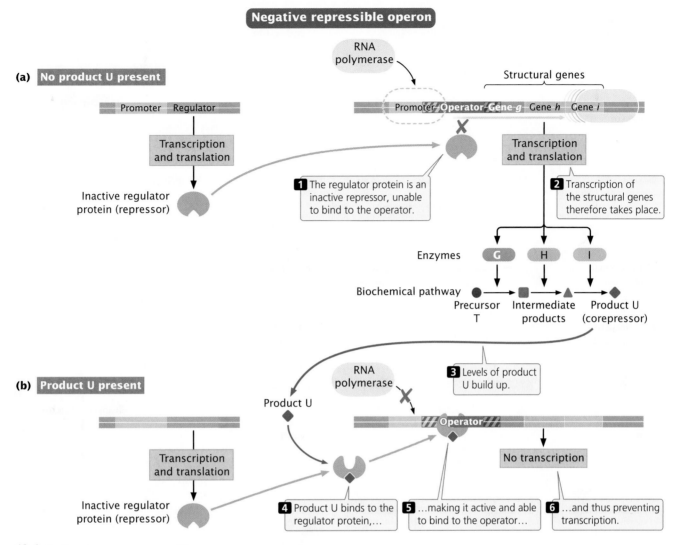

12.4 Some operons are repressible.

In a positive *inducible* operon, transcription is normally turned off because the regulator protein (an activator) is produced in an inactive form. Transcription takes place when an inducer became attached to the regulatory protein, rendering the regulator active. Logically, the inducer should be the precursor of the reaction controlled by the operon so that the necessary enzymes would be synthesized only when the substrate for their reaction was present.

A positive operon can also be repressible; transcription normally takes place and has to be repressed. In this case, the regulator protein is produced in a form that readily binds to DNA and stimulates transcription. Transcription is inhibited when a substance becomes attached to the activator and renders it unable to bind to the DNA so that transcription is no longer stimulated. Here, the product (P) of the reaction controlled by the operon would logically be the repressing substance, because it would be economical for the cell to prevent the transcription of genes that allow the synthesis of P when plenty of P was already available.

The characteristics of positive and negative control in inducible and repressible operons are summarized in **Figure 12.5.**

Concepts

There are two basic types of transcriptional control: negative and positive. In negative control, when a regulatory protein (repressor) binds to DNA, transcription is inhibited; in positive control, when a regulatory protein (activator) binds to DNA, transcription is stimulated. Some operons are inducible; transcription is normally off and must be turned on. Other operons are repressible; transcription is normally on and must be turned off.

✔ Concept Check 4

In a negative repressible operon, the regulator protein is synthesized as

a. an active activator.
b. an inactive activator.
c. an active repressor.
d. an inactive repressor.

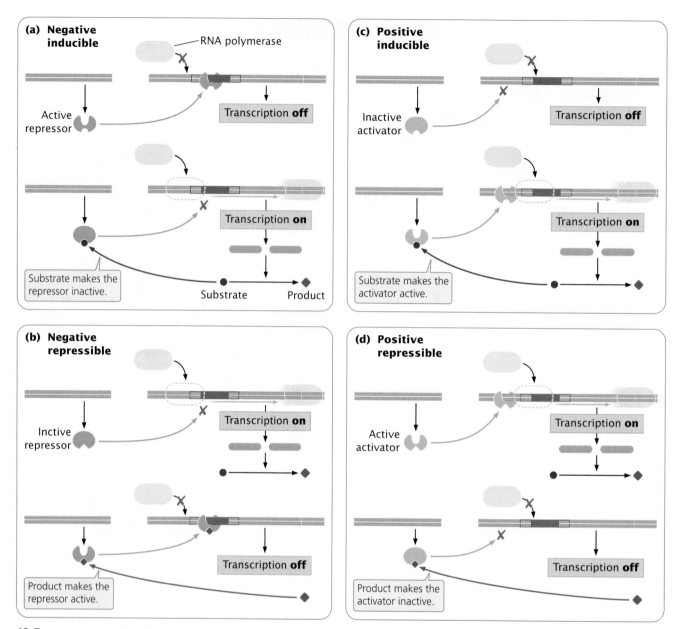

12.5 A summary of the characteristics of positive and negative control in inducible and repressible operons.

The *lac* Operon of *Escherichia coli*

In 1961, François Jacob and Jacques Monod described the "operon model" for the genetic control of lactose metabolism in *E. coli*. This work and subsequent research on the genetics of lactose metabolism established the operon as the basic unit of transcriptional control in bacteria. Despite the fact that, at the time, no methods were available for determining nucleotide sequences, Jacob and Monod deduced the structure of the operon *genetically* by analyzing the interactions of mutations that interfered with the normal regulation of lactose metabolism. We will examine the effects of some of these mutations after seeing how the *lac* operon regulates lactose metabolism.

Lactose is one of the major carbohydrates found in milk; it can be metabolized by *E. coli* bacteria that reside in the mammalian gut. Lactose does not easily diffuse across the *E. coli* cell membrane and must be actively transported into the cell by the protein permease (**Figure 12.6**). To utilize lactose as an energy source, *E. coli* must first break it into glucose and galactose, a reaction catalyzed by the enzyme β-galactosidase. This enzyme can also convert lactose into allolactose, a compound that plays an important role in regulating lactose metabolism. A third enzyme, thiogalactoside transacetylase, also is produced by the *lac* operon, but its function in lactose metabolism is not yet known.

The *lac* operon is an example of a negative inducible operon. The proteins β-galactosidase, permease, and

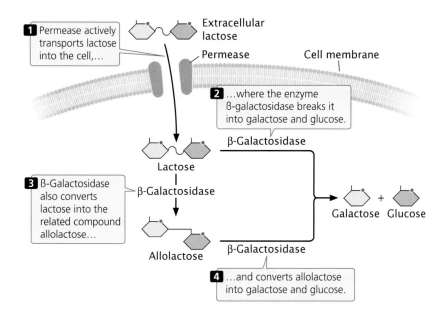

1 Permease actively transports lactose into the cell,...

Extracellular lactose

Permease

Cell membrane

2 ...where the enzyme β-galactosidase breaks it into galactose and glucose.

β-Galactosidase

Lactose

3 β-Galactosidase also converts lactose into the related compound allolactose...

β-Galactosidase

Galactose   Glucose

Allolactose

β-Galactosidase

4 ...and converts allolactose into galactose and glucose.

12.6 Lactose, a major carbohydrate found in milk, consists of 2 six-carbon sugars linked together.

transacetylase are encoded by adjacent structural genes in the *lac* operon of *E. coli* (**Figure 12.7a**) and have a common promoter (*lacP* in Figure 12.7a). β-Galactosidase is encoded by the *lacZ* gene, permease by the *lacY* gene, and transacetylase by the *lacA* gene. When lactose is absent from the medium in which *E. coli* grows, few molecules of each enzyme are produced. If lactose is added to the medium and glucose is absent, the rate of synthesis of all three enzymes simultaneously increases about a thousandfold within 2 to 3 minutes. This boost in enzyme synthesis results from the transcription of *lacZ*, *lacY*, and *lacA* and exemplifies **coordinate induction**, the simultaneous synthesis of several enzymes, stimulated by a specific molecule, the inducer (**Figure 12.7b**).

Although lactose appears to be the inducer here, allolactose is actually responsible for induction. Upstream of *lacP* is a regulator gene, *lacI*, which has its own promoter (P_I). The *lacI* gene is transcribed into a short mRNA that is translated into a repressor. Each repressor consists of four identical polypeptides and has two types of binding sites: one site binds to allolactose and the other binds to DNA. In the absence of lactose (and, therefore, allolactose), the repressor binds to the *lac* operator site *lacO* (see Figure 12.7a). The location of the operator site relative to the promoter and *lacZ* gene is shown in **Figure 12.8**.

RNA polymerase binds to the promoter and moves down the DNA molecule, transcribing the structural genes. When the repressor is bound to the operator, the binding of RNA polymerase is blocked, and transcription is prevented. When lactose is present, some of it is converted into allolactose, which binds to the repressor and causes the repressor to be released from the DNA. In the presence of lactose, then, the repressor is inactivated, the binding of RNA polymerase is no longer blocked, the transcription of *lacZ*, *lacY*, and *lacA* takes place, and the *lac* proteins are produced.

Have you spotted the flaw in the explanation just given for the induction of the *lac* proteins? You might recall that permease is required to transport lactose into the cell. If the

lac operon is repressed and no permease is being produced, how does lactose get into the cell to inactivate the repressor and turn on transcription? Furthermore, the inducer is actually allolactose, which must be produced from lactose by β-galactosidase. If β-galactosidase production is repressed, how can lactose metabolism be induced?

The answer is that repression never *completely* shuts down transcription of the *lac* operon. Even with active repressor bound to the operator, there is a low level of transcription and a few molecules of β-galactosidase, permease, and transacetylase are synthesized. When lactose appears in the medium, the permease that is present transports a small amount of lactose into the cell. There, the few molecules of β-galactosidase that are present convert some of the lactose into allolactose, which then induces transcription.

Concepts

The *lac* operon of *E. coli* controls the transcription of three genes needed in lactose metabolism: the *lacZ* gene, which encodes β-galactosidase; the *lacY* gene, which encodes permease; and the *lacA* gene, which encodes thiogalactoside transacetylase. The *lac* operon is negative inducible: a regulator gene produces a repressor that binds to the operator site and prevents the transcription of the structural genes. The presence of allolactose inactivates the repressor and allows the transcription of the *lac* operon.

✔ Concept Check 5

In the presence of allolactose, the *lac* repressor

a. binds to the operator.   c. cannot bind to the operator.

b. binds to the promoter.   d. binds to the regulator gene.

Mutations in *lac*

Jacob and Monod worked out the structure and function of the *lac* operon by analyzing mutations that affected lactose

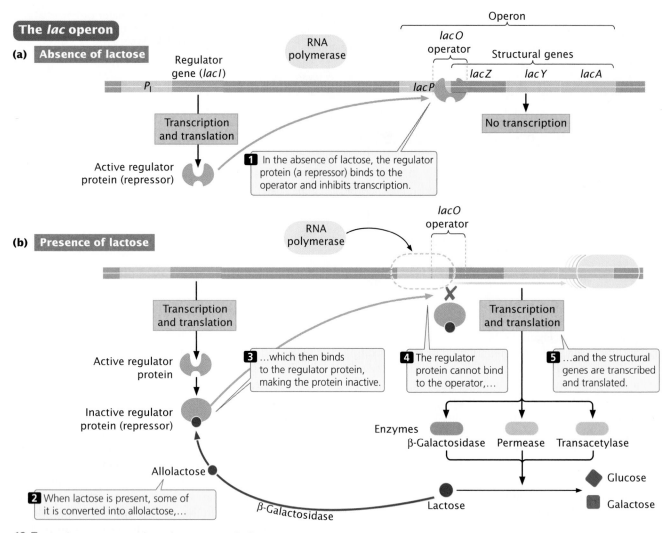

12.7 The *lac* operon regulates lactose metabolism.

metabolism. To help define the roles of the different components of the operon, they used **partial diploid** strains of *E. coli*. The cells of these strains possessed two different DNA molecules: the full bacterial chromosome and an extra piece of DNA. Jacob and Monod created these strains by allowing conjugation to take place between two bacteria (see Chapter 6). In conjugation, a small circular piece of DNA (the F plasmid, see Chapter 6) is transferred from one bacterium to another. The F plasmid used by Jacob and Monod contained the *lac* operon; so the recipient bacterium became partly diploid, possessing two copies of the *lac* operon. By using different combinations of mutations on the bacterial

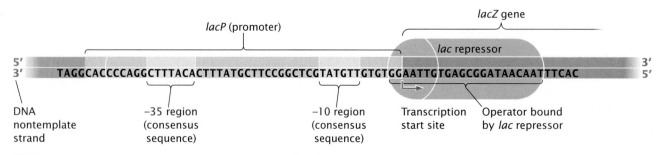

12.8 In the *lac* operon, the operator overlaps the promoter and the 5′ end of the first structural gene.

(a) **Absence of lactose**

RNA polymerase

P_I $lacI^+$

Active repressor

Mutant repressor

P_I

lacP$^+$ lacO$^+$ lacZ$^-$

1 The $lacI^+$ gene is trans dominant: the repressor produced by $lacI^+$ can bind to both operators and repress transcription in the absence of lactose.

RNA polymerase

Transcription inhibited

lacP$^+$ lacO$^+$ lacZ$^+$

(b) **Presence of lactose**

2 When lactose is present, it inactivates the repressor, and functional β-galactosidase is produced from the $lacZ^+$ gene.

P_I $lacI^+$

Active repressor

Mutant repressor

P_I $lacI^-$

Lactose

Inactive repressor

lacP$^+$ lacO$^+$ lacZ$^-$

Transcription and translation

Nonfunctional β-galactosidase

lacP$^+$ lacO$^+$ lacZ$^+$

Transcription and translation

β-Galactosidase

12.9 The partial diploid $lacI^+$ $lacZ^-$ / $lacI^-$ $lacZ^+$ produces β-galactosidase only in the presence of lactose because the $lacI$ gene is trans dominant.

and plasmid DNA, Jacob and Monod determined that some parts of the *lac* operon are cis acting (able to control the expression of genes only when on the same piece of DNA only), whereas other parts are trans acting (able to control the expression of genes on other DNA molecules).

Structural-gene mutations Jacob and Monod first discovered some mutant strains that had lost the ability to synthesize either β-galactosidase or permease. (They did not study in detail the effects of mutations on the transacetylase enzyme, and so transacetylase will not be considered here.) The mutations in the mutant strains mapped to the *lacZ* or *lacY* structural genes and altered the amino acid sequences of the enzymes encoded by the genes. These mutations clearly affected the *structure* of the enzymes and not the regulation of their synthesis.

Through the use of partial diploids, Jacob and Monod were able to establish that mutations at the *lacZ* and *lacY* genes were independent and usually affected only the product of the gene in which they occurred. Partial diploids with *lacZ$^+$ lacY$^-$* on the bacterial chromosome and *lacZ$^-$ lacY$^+$* on the plasmid functioned normally, producing β-galactosidase and permease in the presence of lactose. (The genotype of a partial diploid is written by separating the genes on each DNA molecule with a slash: *lacZ$^+$ lacY$^-$/lacZ$^-$ lacY$^+$*.) In

this partial diploid, a single functional β-galactosidase gene (*lacZ$^+$*) is sufficient to produce β-galactosidase; whether the functional β-galactosidase gene is coupled to a functional (*lacY$^+$*) or a defective (*lacY$^-$*) permease gene makes no difference. The same is true of the *lacY$^+$* gene.

Regulator-gene mutations Jacob and Monod also isolated mutations that affected the *regulation* of enzyme production. Mutations in the *lacI* gene affect the production of both β-galactosidase and permease, because genes for both proteins are in the same operon and are regulated coordinately.

Some of these mutations were **constitutive**, causing the *lac* proteins to be produced all the time, whether lactose was present or not. Such mutations in the regulator gene were designated *lacI$^-$*. The construction of partial diploids demonstrated that a *lacI$^+$* gene is dominant over a *lacI$^-$* gene; a single copy of *lacI$^+$* (genotype *lacI$^+$/lacI$^-$*) was sufficient to bring about normal regulation of enzyme production. Furthermore, *lacI$^+$* restored normal control to an operon even if the operon was located on a different DNA molecule, showing that *lacI$^+$* can be trans acting. A partial diploid with genotype *lacI$^+$ lacZ$^-$/lacI$^-$ lacZ$^+$* functioned normally, synthesizing β-galactosidase only when lactose was present (**Figure 12.9**). In this strain, the *lacI$^+$* gene on the bacterial chromosome was functional, but the *lacZ$^-$* gene was

(a) Partial diploid *lacI⁺ lacO⁺ lacZ⁻/lacI⁺ lacOᶜ lacZ⁺*

(b) Partial diploid *lacI⁺ lacO⁺ lacZ⁺/lacI⁺ lacOᶜ lacZ⁻*

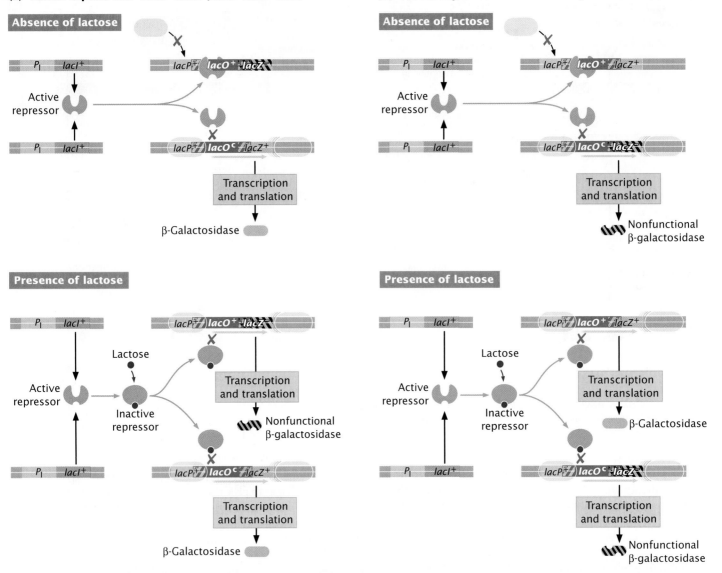

12.10 Mutations in *lacO* are constitutive and cis acting. (a) The partial diploid *lacI⁺ lacO⁺ lacZ⁻/lacI⁺ lacOᶜ lacZ⁺* is constitutive, producing β-galactosidase in the presence and absence of lactose. (b) The partial diploid *lacI⁺ lacO⁺ lacZ⁺/lacI⁺ lacOᶜ lacZ⁻* is inducible (produces β-galactosidase only when lactose is present), demonstrating that the *lacO* gene is cis acting.

defective; on the plasmid, the *lacI⁻* gene was defective, but the *lacZ⁺* gene was functional. The fact that a *lacI⁺* gene could regulate a *lacZ⁺* gene located on a different DNA molecule indicated to Jacob and Monod that the *lacI⁺* gene product was able to operate on either the plasmid or the chromosome.

Operator mutations   Jacob and Monod mapped a second set of constitutive mutants to a site adjacent to *lacZ*. These mutations occurred at the operator site and were referred to as *lacOᶜ* (O stands for operator and "c" for constitutive). The *lacOᶜ* mutations altered the sequence of DNA at the operator so that the repressor protein was no longer able to bind. A

partial diploid with genotype *lacI⁺ lacOᶜ lacZ⁺/lacI⁺ lacO⁺ lacZ⁺* exhibited constitutive synthesis of β-galactosidase, indicating that *lacOᶜ* is dominant over *lacO⁺*.

Analyses of other partial diploids showed that the *lacO* gene is cis acting, affecting only genes on the same DNA molecule. For example, a partial diploid with genotype *lacI⁺ lacO⁺ lacZ⁻/lacI⁺ lacOᶜ lacZ⁺* was constitutive, producing β-galactosidase in the presence or absence of lactose (**Figure 12.10a**), but a partial diploid with genotype *lacI⁺ lacO⁺ lacZ⁺/lacI⁺ lacOᶜ lacZ⁻* produced β-galactosidase only in the presence of lactose (**Figure 12.10b**). In the constitutive partial diploid (*lacI⁺ lacO⁺ lacZ⁻/lacI⁺ lacOᶜ lacZ⁺*; see Figure

12.10a), the *lacO*^{c} mutation and the functional *lacZ*$^+$ gene are present on the same DNA molecule; but, in *lacI*$^+$ *lacO*$^+$ *lacZ*$^+$/*lacI*$^+$ *lacO*^{c} *lacZ*$^-$ (see Figure 12.10b), the *lacO*^{c} mutation and the functional *lacZ*$^+$ gene are on different molecules. The *lacO* mutation affects only genes to which it is physically connected, as is true of all operator mutations. They prevent the binding of a repressor protein to the operator and thereby allow RNA polymerase to transcribe genes on the same DNA molecule. However, they cannot prevent a repressor from binding to normal operators on other DNA molecules.

Promoter mutations   Mutations affecting lactose metabolism have also been isolated at the promoter site; these mutations are designated *lacP*$^-$, and they interfere with the binding of RNA polymerase to the promoter. Because this binding is essential for the transcription of the structural genes, *E. coli* strains with *lacP*$^-$ mutations don't produce *lac* proteins either in the presence or in the absence of lactose. Like operator mutations, *lacP*$^-$ mutations are cis acting and thus affect only genes on the same DNA molecule. The partial diploid *lacI*$^+$ *lacP*$^+$ *lacZ*$^+$/*lacI*$^+$ *lacP*$^-$ *lacZ*$^+$ exhibits normal synthesis of β-galactosidase, whereas *lacI*$^+$ *lacP*$^-$ *lacZ*$^+$/*lacI*$^+$ *lacP*$^+$ *lacZ*$^-$ fails to produce β-galactosidase whether or not lactose is present.

Worked Problem

For *E. coli* strains with the following *lac* genotypes, use a plus sign (+) to indicate the synthesis of β-galactosidase and permease and a minus sign (−) to indicate no synthesis of the enzymes when lactose is absent and when it is present.

Genotype of strain

a. *lacI*$^+$ *lacP*$^+$ *lacO*$^+$ *lacZ*$^+$ *lacY*$^+$

b. *lacI*$^+$ *lacP*$^+$ *lacO*^{c} *lacZ*$^-$ *lacY*$^+$

c. *lacI*$^+$ *lacP*$^-$ *lacO*$^+$ *lacZ*$^+$ *lacY*$^-$

d. *lacI*$^+$ *lacP*$^+$ *lacO*$^+$ *lacZ*$^-$ *lacY*$^-$/*lacI*$^-$ *lacP*$^+$ *lacO*$^+$ *lacZ*$^+$ *lacY*$^+$

• Solution

Genotype of strain	Lactose absent		Lactose present	
	β-Galactosidase	Permease	β-Galactosidase	Permease
a. *lacI*$^+$ *lacP*$^+$ *lacO*$^+$ *lacZ*$^+$ *lacY*$^+$	−	−	+	+
b. *lacI*$^+$ *lacP*$^+$ *lacO*^{c} *lacZ*$^-$ *lacY*$^+$	−	+	−	+
c. *lacI*$^+$ *lacP*$^-$ *lacO*$^+$ *lacZ*$^+$ *lacY*$^-$	−	−	−	−
d. *lacI*$^+$ *lacP*$^+$ *lacO*$^+$ *lacZ*$^-$ *lacY*$^-$/ *lacI*$^-$ *lacP*$^+$ *lacO*$^+$ *lacZ*$^+$ *lacY*$^+$	−	−	+	+

a. All the genes possess normal sequences, and so the *lac* operon functions normally: when lactose is absent, the regulator protein binds to the operator and inhibits the transcription of the structural genes, and so β-galactosidase and permease are not produced. When lactose is present, some of it is converted into allolactose, which binds to the repressor and makes it inactive; the repressor does not bind to the operator, and so the structural genes are transcribed, and β-galactosidase and permease are produced.

b. The structural *lacZ* gene is mutated; so β-galactosidase will not be produced under any conditions. The *lacO* gene has a constitutive mutation, which means that the repressor is unable to bind to *lacO*, and so transcription takes place at all times. Therefore, permease will be produced in both the presence and the absence of lactose.

c. In this strain, the promoter is mutated, and so RNA polymerase is unable to bind and transcription does not take place. Therefore, β-galactosidase and permease are not produced under any conditions.

d. This strain is a partial diploid, which consists of two copies of the *lac* operon—one on the bacterial chromosome and the other on a plasmid. The *lac* operon represented in the upper part of the genotype has mutations in both the *lacZ* and the *lacY* genes, and so it is not capable of encoding β-galactosidase or permease under any conditions. The *lac* operon in the lower part of the genotype has a defective regulator gene, but the normal regulator gene in the upper operon produces a diffusible repressor (trans acting) that binds to the lower operon in the absence of lactose and inhibits transcription. Therefore, no β-galactosidase or permease is produced when lactose is absent. In the presence of lactose, the repressor cannot bind to the operator, and so the lower operon is transcribed and β-galactosidase and permease are produced.

> **?** Now try your own hand at predicting the outcome of different *lac* mutations by working Problem 20 at the end of the chapter.

Positive Control and Catabolite Repression

E. coli and many other bacteria metabolize glucose preferentially in the presence of lactose and other sugars. They do so because glucose enters glycolysis without further modification and therefore requires less energy to metabolize than do other sugars. When glucose is available, genes that participate in the metabolism of other sugars are repressed, in a phenomenon known as **catabolite repression**. For example, the efficient transcription of the *lac* operon takes place only if lactose is present and glucose is absent. But how is the expression of the *lac* operon influenced by glucose? What brings about catabolite repression?

Catabolite repression results from positive control in response to glucose. (This regulation is in addition to the negative control brought about by the repressor binding at the operator site of the *lac* operon when lactose is absent.) Positive control is accomplished through the binding of a dimeric protein called the **catabolite activator protein (CAP)** to a site that is about 22 nucleotides long and is located within or slightly upstream of the promoter of the *lac* genes (**Figure 12.11**). RNA polymerase does not bind efficiently to some promoters unless CAP is first bound to the DNA. Before CAP can bind to DNA, it must form a complex with a modified nucleotide called **adenosine-3′,5′-cyclic monophosphate** (cyclic AMP, or cAMP), which is important in cellular signaling processes in both bacterial and eukaryotic cells. In *E. coli*, the concentration of cAMP is regulated so that its concentration is inversely proportional to the level of available glucose. A high concentration of glucose within the cell lowers the amount of cAMP, and so little cAMP–CAP complex is available to bind to the DNA. Subsequently, RNA polymerase has poor affinity for the *lac*

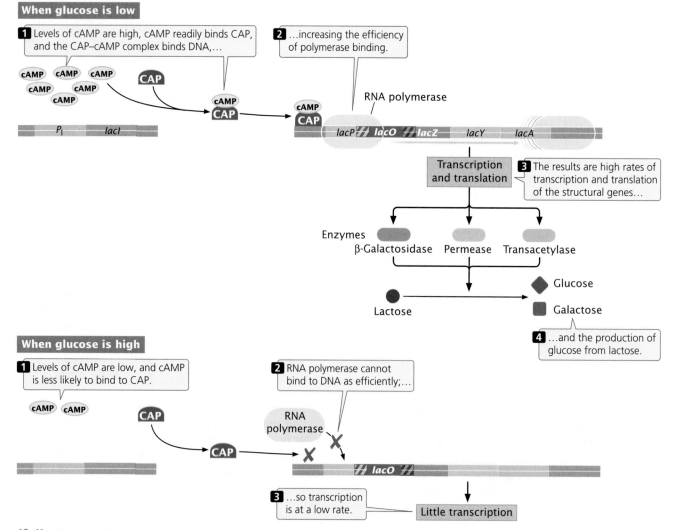

12.11 The catabolite activator protein (CAP) binds to the promoter of the *lac* operon and stimulates transcription. CAP must complex with adenosine-3′,5′-cyclic monophosphate (cAMP) before binding to the promoter of the *lac* operon. The binding of cAMP–CAP to the promoter activates transcription by facilitating the binding of RNA polymerase. Levels of cAMP are inversely related to glucose: low glucose stimulates high cAMP; high glucose stimulates low cAMP.

promoter, and little transcription of the *lac* operon takes place. Low concentrations of glucose stimulate high levels of cAMP, resulting in increased cAMP–CAP binding to DNA. This increase enhances the binding of RNA polymerase to the promoter and increases transcription of the *lac* genes by some 50-fold.

Concepts

In spite of its name, catabolite repression is a type of positive control in the *lac* operon. The catabolite activator protein (CAP), complexed with cAMP, binds to a site near the promoter and stimulates the binding of RNA polymerase. Cellular levels of cAMP in the cell are controlled by glucose; a low glucose level increases the abundance of cAMP and enhances the transcription of the *lac* structural genes.

The *trp* Operon of *Escherichia coli*

The *lac* operon just discussed is an inducible operon, one in which transcription does not normally take place and must be turned on. Other operons are repressible; transcription in these operons is normally turned on and must be repressed. The tryptophan (*trp*) operon in *E. coli*, which controls the biosynthesis of the amino acid tryptophan, is an example of a negative repressible operon.

The *trp* operon contains five structural genes (*trpE*, *trpD*, *trpC*, *trpB*, and *trpA*) that produce the components of three enzymes (two of the enzymes consist of two polypeptide chains). These enzymes convert chorismate into tryptophan (**Figure 12.12**). Upstream of the structural genes is the *trp* promoter. When tryptophan levels are low, RNA polymerase binds to the promoter and transcribes the five structural genes into a single mRNA, which is then translated into enzymes that convert chorismate into tryptophan.

Some distance from the *trp* operon is a regulator gene, *trpR*, which encodes a repressor that alone cannot bind DNA (see Figure 12.12). Like the *lac* repressor, the tryptophan repressor has two binding sites, one that binds to DNA at the operator site and another that binds to tryptophan (the activator). Binding with tryptophan causes a conformational change in the repressor that makes it capable of binding to DNA at the operator site, which overlaps the promoter. When the operator is occupied by the tryptophan repressor, RNA polymerase cannot bind to the promoter and the structural genes cannot be transcribed. Thus, when cellular levels of tryptophan are low, transcription of the *trp* operon takes place and more tryptophan is synthesized; when cellular levels of tryptophan are high, transcription of the *trp* operon is inhibited and the synthesis of more tryptophan does not take place.

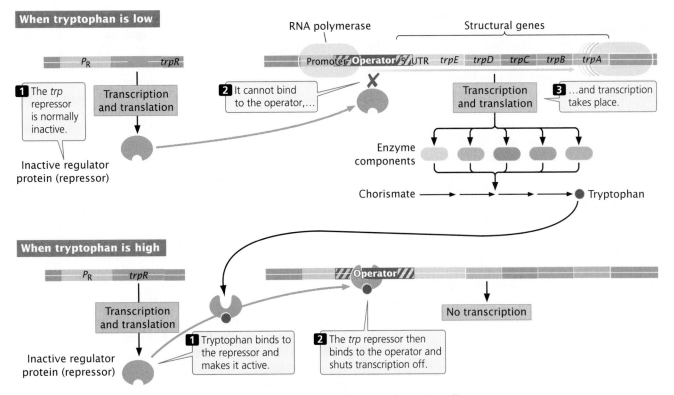

12.12 **The *trp* operon controls the biosynthesis of the amino acid tryptophan in *E. coli*.**

Concepts

The *trp* operon is a negative repressible operon that controls the biosynthesis of tryptophan. In a repressible operon, transcription is normally turned on and must be repressed. Repression is accomplished through the binding of tryptophan to the repressor, which renders the repressor active. The active repressor binds to the operator and prevents RNA polymerase from transcribing the structural genes.

✔ **Concept Check 6**

In the *trp* operon, what happens to the *trp* repressor in the absence of tryptophan?

a. It binds to the operator and represses transcription.

b. It cannot bind to the operator and transcription takes place.

c. It binds to the regulator gene and represses transcription.

d. It cannot bind to the regulator gene and transcription takes place.

12.4 Gene Regulation in Eukaryotic Cells Takes Place at Multiple Levels

Many features of gene regulation are common to both bacterial and eukaryotic cells. For example, in both types of cells, DNA-binding proteins influence the ability of RNA polymerase to initiate transcription. However, there are also some differences. First, most eukaryotic genes are not organized into operons and are rarely transcribed together into a single mRNA molecule; instead, each structural gene typically has its own promoter and is transcribed separately. Second, chromatin structure affects gene expression in eukaryotic cells; DNA must partly unwind from the histone proteins before transcription can take place. Finally, the presence of the nuclear membrane in eukaryotic cells separates transcription and translation in time and space. Therefore, the regulation of gene expression in eukaryotic cells is characterized by a greater diversity of mechanisms that act at different points in the transfer of information from DNA to protein.

Eukaryotic gene regulation is less well understood than bacterial regulation, partly owing to the larger genomes in eukaryotes, their greater sequence complexity, and the difficulty of isolating and manipulating mutations that can be used in the study of gene regulation. Nevertheless, great advances in our understanding of the regulation of eukaryotic genes have been made in recent years, and eukaryotic regulation continues to be a cutting-edge area of research in genetics.

Changes in Chromatin Structure

One type of gene control in eukaryotic cells is accomplished through the modification of gene structure. In the nucleus, histone proteins associate to form octamers, around which helical DNA tightly coils to create chromatin (see Figure 8.19). In a general sense, this chromatin structure represses gene expression. For a gene to be transcribed, transcription factors, activators, and RNA polymerase must bind to the DNA. How can these events take place with DNA wrapped tightly around histone proteins? The answer is that, before transcription, chromatin structure changes and the DNA becomes more accessible to the transcriptional machinery.

Histone modification   Histones in the octamer core of the nucleosome have two domains: (1) a globular domain that associates with other histones and the DNA and (2) a positively charged tail domain that probably interacts with the negatively charged phosphate groups on the backbone of DNA. The tails of histone proteins are often modified by the addition or removal of phosphate groups, methyl groups, or acetyl groups. These modifications have sometimes been called the **histone code**, because they encode information that affects how genes are expressed.

One type of histone modification is the addition of methyl groups to the tails of histone proteins. These modifications can bring about either the activation or the repression of transcription, depending on which particular amino acids in the histone tail are methylated. A common modification is the addition of three methyl groups to lysine 4 in the tail of the H3 histone protein, abbreviated H3K4me3 (K is the abbreviation for lysine). The H3K4me3 modification is frequently found near the transcription start site of genes in eukaryotes. Recent studies have identified proteins that recognize and bind to H3K4me3, including a protein called nucleosome-remodeling factor (NURF). NURF and other proteins that recognize H3K4me3 have a common protein-binding domain that binds to the H3 histone tail and then alters chromatin packing, allowing transcription to take place.

Another type of histone modification that affects chromatin structure is acetylation, the addition of acetyl groups (CH_3CO) to histone proteins (**Figure 12.13**). Acetylation of histones usually stimuates transcription. For example, the addition of a single acetyl group to lysine 16 in the tail of the H4 histone prevents the formation of the 30-nm chromatin fiber (see Figure 8.19), causing the chromatin to be in an open configuration and available for transcription. In general, acetyl groups destabilize the chromatin structure, allowing transcription to take place. Acetyl groups are added to histone proteins by acetyltransferase enzymes; other enzymes called deacetylases strip acetyl groups from histones and restore chromatin structure, which represses transcription. Certain proteins that regulate transcription either have acetyltransferase activity or attract acteyltransferases to the DNA.

The importance of histone acetylation in gene regulation is demonstrated by the control of flowering in

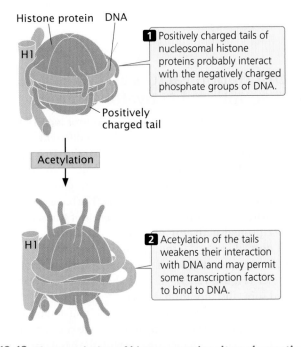

12.13 The acetylation of histone proteins alters chromatin structure and permits some transcription factors to bind to DNA.

Arabidopsis, a plant with a number of characteristics that make it an excellent genetic model for plant systems (see the section at the end of this chapter on the model genetic organism *Arabidopsis thaliana*). The time at which flowering takes place is critical to the life of a plant; if flowering is initiated at the wrong time of year, pollinators may not be available to fertilize the flowers or environmental conditions may be unsuitable for survival and germination of the seeds. Consequently, flowering time in most plants is carefully regulated in response to multiple internal and external cues, such as plant size, photoperiod, and temperature.

Among the many genes that control flowering in *Arabidopsis* is *flowering locus C* (*FLC*), which plays an important role in suppressing flowering until after an extended period of coldness (a process called vernalization). The *FLC* gene encodes a transcriptional activator protein, which acts on other genes that affect flowering (**Figure 12.14**).

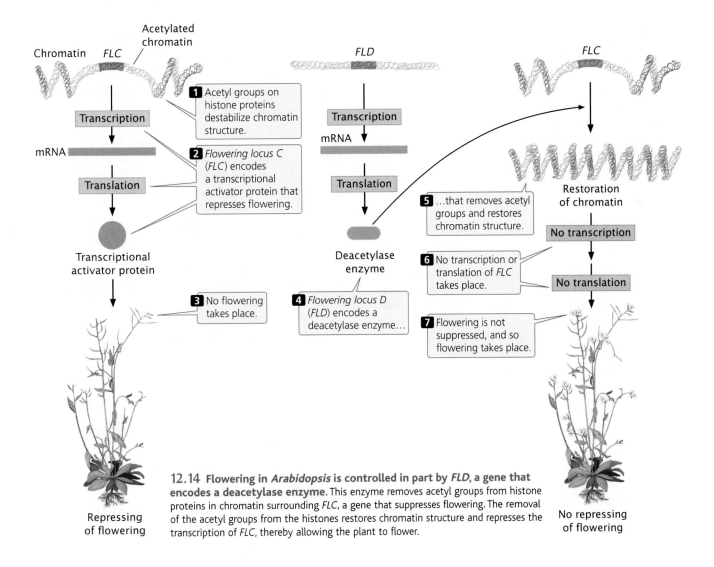

12.14 Flowering in *Arabidopsis* is controlled in part by *FLD*, a gene that encodes a deacetylase enzyme. This enzyme removes acetyl groups from histone proteins in chromatin surrounding *FLC*, a gene that suppresses flowering. The removal of the acetyl groups from the histones restores chromatin structure and represses the transcription of *FLC*, thereby allowing the plant to flower.

As long as *FLC* is active, flowering remains suppressed. The activity of *FLC* is controlled by another locus called *flowering locus D* (*FLD*), the key role of which is to stimulate flowering by repressing the action of *FLC*. How does *FLD* repress *FLC*?

FLD encodes a deacetylase enzyme, which removes acetyl groups from histone proteins in the chromatin surrounding *FLC* (see Figure 12.14). The removal of acetyl groups from histones restores chromatin structure and inhibits transcription. The inhibition of transcription prevents *FLC* from being transcribed and removes its repression on flowering. In short, *FLD* stimulates flowering in *Arabidopsis* by deacetylating the chromatin that surrounds *FLC*, thereby removing its inhibitory effect on flowering.

Chromatin remodeling   The changes to chromatin structure discussed so far have been through alteration of the structure of the histone proteins. Some transcription factors and other regulatory proteins alter chromatin structure without altering the chemical structure of the histones directly. These proteins are called **chromatin-remodeling complexes**. They bind directly to particular sites on DNA and reposition the nucleosomes, allowing transcription factors to bind to promoters and initiate transcription.

One of the best-studied examples of a chromatin-remodeling complex is SWI–SNF, which is found in yeast, humans, *Drosophila*, and other organisms. This complex utilizes energy derived from the hydrolysis of ATP to reposition nucleosomes, exposing promoters in the DNA to the action of other regulatory proteins and RNA polymerase. Research has shown that SWI–SNF is able to change the position of nucleosomes along a DNA molecule.

DNA methylation   Another change in chromatin structure associated with transcription is the methylation of cytosine bases, which yields 5-methylcytosine. The methylation of cytosine in DNA is distinct from the methylation of histone proteins mentioned earlier. Heavily methylated DNA is associated with the repression of transcription in vertebrates and plants, whereas transcriptionally active DNA is usually unmethylated in these organisms. Abnormal patterns of methylation are also associated with some types of cancer.

Concepts

Chromatin structure can be altered by modifications of histone proteins, by chromatin-remodeling complexes that reposition nucleosomes, and by the methylation of DNA.

Transcription Factors and Transcriptional Activator Proteins

Transcription is an important level of control in eukaryotic cells, and this control requires a number of different types of proteins and regulatory elements. **General transcription factors** and RNA polymerase assemble into a basal transcription apparatus, which binds to a core promoter located immediately upstream of a gene. The basal transcription apparatus is capable of minimal levels of transcription; **transcriptional activator proteins** are required to bring about normal levels of transcription. These proteins bind to a regulatory promoter, which is located upstream of the core promoter (**Figure 12.15**), and to **enhancers**, which may be located some distance from the gene.

Transcriptional activator proteins stimulate and stabilize the basal transcription apparatus at the core promoter. The activators may interact directly with the basal transcription apparatus or indirectly through protein **coactivators**. Some activators and coactivators, as well as the general transcription factors, also have acetyltransferase activity and so further stimulate transcription by altering chromatin structure (see earlier section on histone modification).

Within the regulatory promoter are typically several different consensus sequences to which different transcriptional activators can bind. Among different promoters, activator-binding sites are mixed and matched in different combinations (**Figure 12.16**), and so each promoter is regulated by a unique combination of transcriptional activator proteins.

Some regulatory proteins in eukaryotic cells act as repressors, inhibiting transcription. These repressors bind to sequences in the regulatory promoter or to distant sequences called **silencers**, which, like enhancers, are position and orientation independent. Unlike repressors in bacteria, most eukaryotic repressors do not directly block RNA polymerase. These repressors may compete with activators for DNA binding sites: when a site is occupied by an activator, transcription is activated, but, if a repressor occupies that site, there is no activation. Alternatively, a repressor may bind to sites near an activator site and prevent the activator from contacting the basal transcription apparatus. A third possible mechanism of repressor action is direct interference with the assembly of the basal transcription apparatus, thereby blocking the initiation of transcription.

Concepts

Transcriptional regulatory proteins in eukaryotic cells can influence the initiation of transcription by affecting the stability or assembly of the basal transcription apparatus. Some regulatory proteins are activators and stimulate transcription; others are repressors and inhibit transcription.

12.15 Transcriptional activator proteins bind to sites on DNA and stimulate transcription. Most act by stimulating or stabilizing the assembly of the basal transcription apparatus.

✔ Concept Check 7

Most transcriptional activator proteins affect transcription by interacting with

a. introns.

b. the basal transcription apparatus.

c. DNA polymerase.

d. nucleosomes.

Enhancers and insulators Enhancers are capable of affecting transcription at distant promoters. For example, an enhancer that regulates the gene encoding the alpha chain of the T-cell (T-lymphocyte) receptor is located 69,000 bp downstream of the gene's promoter. Furthermore, the exact position and orientation of an enhancer relative to the promoter can vary. How can an enhancer affect the initiation of transcription taking place at a promoter that is tens of thousands of base pairs away? In many cases, activator proteins bind to the enhancer and cause the DNA between the enhancer and the promoter to loop out, bringing the promoter and enhancer close to each other, and so the transcriptional activator proteins are able to directly interact with the basal transcription apparatus at the core promoter (see Figure 12.15).

Most enhancers are capable of stimulating any promoter in their vicinities. Their effects are limited, however, by **insulators** (also called boundary elements), which are DNA sequences that block or insulate the effect of enhancers in a position-dependent manner. If the insulator lies between the enhancer and the promoter, it blocks the action of the enhancer; but, if the insulator lies outside the region between the two, it has no effect (**Figure 12.17**). Specific proteins bind to insulators and play a role in their blocking activity. Some insulators also limit the spread of changes in chromatin structure that affect transcription.

12.16 The consensus sequences in the promoters of three eukaryotic genes illustrate the principle that different sequences can be mixed and matched in different combinations. A different transcriptional activator protein binds to each consensus sequence, and so each promoter responds to a unique combination of activator proteins.

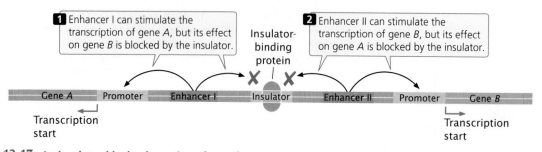

12.17 An insulator blocks the action of an enhancer on a promoter when the insulator lies between the enhancer and the promoter.

Concepts

Some activator proteins bind to enhancers, which are regulatory elements that are distant from the gene for which they stimulate transcription. Insulators are DNA sequences that block the action of enhancers.

✔ **Concept Check 8**

How does the binding of transcriptional activator proteins to enhancers affect transcription at genes that are thousands of base pairs away?

Coordinated gene regulation   Although most eukaryotic cells do not possess operons, several eukaryotic genes may be activated by the same stimulus. For example, many eukaryotic cells respond to extreme heat and other stresses by producing **heat-shock proteins** that help to prevent damage from such stressing agents. Heat-shock proteins are produced by a large number of different genes. During times of environmental stress, the transcription of all the heat-shock genes is greatly elevated. Groups of bacterial genes are often coordinately expressed (turned on and off together) because they are physically clustered as an operon and have the same promoter, but coordinately expressed genes in eukaryotic cells are not clustered. How, then, is the transcription of eukaryotic genes coordinately controlled if they are not organized into an operon?

Genes that are coordinately expressed in eukaryotic cells are able to respond to the same stimulus because they have regulatory sequences in common in their promoters or enhancers. For example, different eukaryotic heat-shock genes possess a common regulatory element upstream of their start sites. Such DNA regulatory sequences are called **response elements**; they typically contain short consensus sequences at varying distances from the gene being regulated, which provide binding sites for transcriptional activators. During times of stress, a transcriptional activator protein binds to this regulatory element and elevates transcription.

Gene Regulation by RNA Processing and Degradation

In bacteria, most gene regulation is at the level of transcription, because transcription and translation take place simultaneously, leaving little opportunity to control gene expression after transcription. In eukaryotes, transcription takes place in the nucleus and the pre-mRNAs are then processed before moving to the cytoplasm for translation, and so there are more opportunities for gene control after transcription. Consequently, posttranscriptional gene regulation assumes a more important role in eukaryotic cells. One place in eukaryotes where gene expression is frequently controlled is in RNA processing and degradation.

Gene regulation through RNA splicing   Alternative splicing allows a pre-mRNA to be spliced in multiple ways, generating different proteins in different tissues or at different times in development (see Chapter 10). Many eukaryotic genes undergo alternative splicing, and the regulation of splicing is probably an important means of controlling gene expression in eukaryotic cells.

An example of alternative mRNA splicing that regulates gene expression is the control of whether a fruit fly develops as male or female. Sex differentiation in *Drosophila* arises from a cascade of gene regulation. When the ratio of X chromosomes to the number of haploid sets of autosomes (the X : A ratio; see Chapter 4) is 1, a female-specific promoter is activated early in development and stimulates the transcription of the *sex-lethal* (*Sxl*) gene (**Figure 12.18**). The protein encoded by *Sxl* regulates the splicing of the pre-mRNA transcribed from another gene called *transformer* (*tra*). The splicing of *tra* pre-mRNA results in the production of Tra protein (see Figure 12.18). Together with another protein (Tra-2), Tra stimulates the female-specific splicing of pre-mRNA from yet another gene called *doublesex* (*dsx*). This event produces a female-specific Dsx protein, which causes the embryo to develop female characteristics.

In male embryos, which have an X : A ratio of 0.5 (see Figure 12.18), the promoter that transcribes the *Sxl* gene in females is inactive; so no Sxl protein is produced. In the

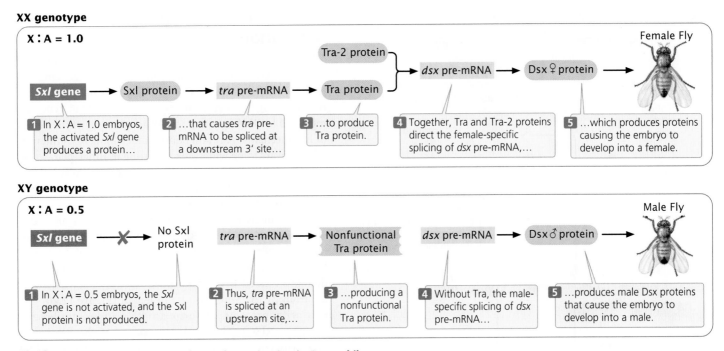

XX genotype

X∶A = 1.0

Sxl gene → Sxl protein → tra pre-mRNA → Tra protein

Tra-2 protein

Tra-2 protein + Tra protein → dsx pre-mRNA → Dsx ♀ protein → Female Fly

1. In X∶A = 1.0 embryos, the activated *Sxl* gene produces a protein…

2. …that causes *tra* pre-mRNA to be spliced at a downstream 3' site…

3. …to produce Tra protein.

4. Together, Tra and Tra-2 proteins direct the female-specific splicing of *dsx* pre-mRNA,…

5. …which produces proteins causing the embryo to develop into a female.

XY genotype

X∶A = 0.5

Sxl gene → ✗ No Sxl protein   *tra* pre-mRNA → Nonfunctional Tra protein   dsx pre-mRNA → Dsx ♂ protein → Male Fly

1. In X∶A = 0.5 embryos, the *Sxl* gene is not activated, and the Sxl protein is not produced.

2. Thus, *tra* pre-mRNA is spliced at an upstream site,…

3. …producing a nonfunctional Tra protein.

4. Without Tra, the male-specific splicing of *dsx* pre-mRNA…

5. …produces male Dsx proteins that cause the embryo to develop into a male.

12.18 Alternative splicing controls sex determination in *Drosophila*.

absence of Sxl protein, *tra* pre-mRNA is spliced at a different 3' splice site to produce a nonfunctional form of Tra protein (**Figure 12.19**). In turn, the presence of this nonfunctional Tra in males causes *dsx* pre-mRNAs to be spliced differently from that in females, and a male-specific Dsx protein is produced (see Figure 12.18). This event causes the development of male-specific traits.

In summary, the Tra, Tra-2, and Sxl proteins regulate alternative splicing that produces male and female phenotypes in *Drosophila*. Exactly how these proteins regulate alternative splicing is not yet known, but the Sxl protein (produced only in females) possibly blocks the upstream splice site on the *tra* pre-mRNA. This blockage would force the spliceosome to use the downstream 3' splice site, which causes the production of Tra protein and eventually results in female traits (see Figure 12.19).

Concepts

Eukaryotic genes can be regulated through the control of mRNA processing. The selection of alternative splice sites leads to the production of different proteins.

The degradation of RNA   The amount of a protein that is synthesized depends on the amount of corresponding mRNA available for translation. The amount of available mRNA, in turn, depends on both the rate of mRNA synthesis and the rate of mRNA degradation. Eukaryotic mRNAs are generally more stable than bacterial mRNAs, which typically last only a few minutes before being degraded. Nonetheless, there is great variability in the stability of eukaryotic mRNA: some mRNAs persist for only a few minutes; others last for hours, days, or even months. These variations can result in large differences in the amount of protein that is synthesized.

Various factors, including the 5' cap and the poly(A) tail, affect the stability of eukaryotic mRNA. Poly(A)-binding proteins (PABPs) normally bind to the poly(A) tail and contribute to its stability-enhancing effect. The presence of these proteins at the 3' end of the mRNA protects the 5' cap. When the poly(A) tail has been shortened below a critical limit, the 5' cap is removed, and the mRNA is degraded by removal of nucleotides from the 5' end. These observations suggest that the 5' cap and the 3' poly(A) tail of eukaryotic mRNA physically interact with each other, most likely by the poly(A) tail bending around so that the PABPs make contact with the 5' cap.

Much of RNA degradation takes place in specialized complexes called P bodies. P bodies help control the expression of genes by regulating which RNA molecules are degraded and which are sequestered for later release. RNA degradation facilitated by small interfering RNAs (siRNAs) also may take place within P bodies (see next section).

Other parts of eukaryotic mRNA, including sequences in the 5' untranslated region (5' UTR), the coding region, and the 3' UTR, also affect mRNA stability. Some short-lived eukaryotic mRNAs have one or more copies of a consensus

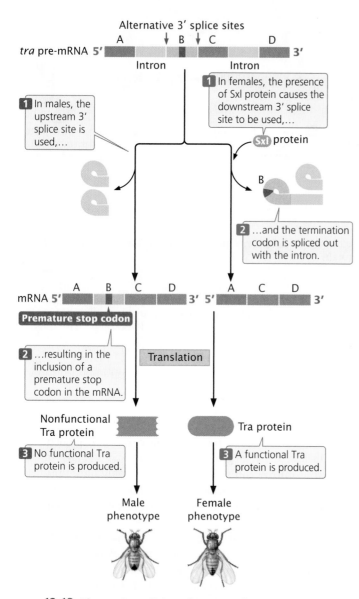

Alternative 3' splice sites

1 In males, the upstream 3' splice site is used,...

1 In females, the presence of Sxl protein causes the downstream 3' splice site to be used,...

2 ...and the termination codon is spliced out with the intron.

2 ...resulting in the inclusion of a premature stop codon in the mRNA.

Premature stop codon

Translation

Nonfunctional Tra protein

Tra protein

3 No functional Tra protein is produced.

3 A functional Tra protein is produced.

Male phenotype

Female phenotype

12.19 Alternative splicing of *tra* pre-mRNA. Two alternative 3' splice sites are present.

sequence consisting of 5'-AUUUAUAA-3', referred to as the AU-rich element, in the 3' UTR. The mRNAs containing AU-rich elements are degraded by a mechanism in which microRNAs take part (see next section).

Concepts

The stability of mRNA influences gene expression by affecting the amount of mRNA available to be translated. The stability of mRNA is affected by the 5' cap, the poly(A) tail, the 5' UTR, the coding section, and sequences in 3' UTR.

✔ Concept Check 9

How does the poly(A) tail affect stability?

RNA Interference and Gene Regulation

The expression of a number of eukaryotic genes is controlled through RNA interference, also known as RNA silencing and posttranscriptional gene silencing (see Chapter 10). Recent research suggests that as much as 30% of human genes are regulated by RNA interference. Although many of the details of this mechanism are still poorly understood, RNA interference appears to be widespread, existing in fungi, plants, and animals. This technique is also widely used as a powerful tool for artificially regulating gene expression in genetically engineered organisms (see Chapter 14).

RNA interference is triggered by small RNA molecules know as microRNAs (miRNAs) and small interfering RNAs (siRNAs), depending on their origin and mode of action (see Chapter 10). An enzyme called Dicer cleaves and processes double-stranded RNA to produce siRNAs or miRNAs that are from 21 to 25 nucleotides in length (**Figure 12.20**) and pair with proteins to form an RNA-induced silencing complex (RISC). The RNA component of the RISC then pairs with complementary base sequences of specific mRNA molecules, most often with sequences in the 3' UTR of the mRNA. Small interfering RNAs and microRNAs regulate gene expression through at least three distinct mechanisms: (1) cleavage of mRNA, (2) inhibition of translation, or (3) transcriptional silencing.

RNA cleavage RISCs that contain an siRNA (and some that contain an miRNA) pair with mRNA molecules and cleave the mRNA near the middle of the bound siRNA (see Figure 12.20a). This cleavage is sometimes referred to as "Slicer activity." After cleavage, the mRNA is further degraded. Thus, the presence of siRNAs and miRNAs increases the rate at which mRNAs are broken down and decreases the amount of protein produced.

Inhibition of translation Some miRNAs regulate genes by inhibiting the translation of their complementary mRNAs (see Figure 12.20b). For example, an important gene in flower development in *Arabidopsis thaliana* is *APETALA2*. The expression of this gene is regulated by an miRNA that base pairs with nucleotides in the coding region of *APETALA2* mRNA and inhibits its translation.

Transcriptional silencing Other siRNAs silence transcription by altering chromatin structure. These siRNAs combine with proteins to form a complex called RITS (for RNA transcriptional silencing), which is analogous to RISC (see Figure 12.20c). The siRNA component of a RITS then binds to its complementary sequence in DNA or an RNA molecule in the process of bring transcribed and represses transcription by attracting enzymes that methylate the tails of histone proteins. The addition of methyl groups to the histones causes them to bind DNA more tightly,

12.20 RNA silencing leads to the degradation of mRNA or to the inhibition of translation or transcription. (a) Small interfering RNAs (siRNAs) degrade mRNA by cleavage. (b) MicroRNAs (miRNAs) lead to the inhibition of translation. (c) Some siRNAs methylate histone proteins or DNA, inhibiting transcription.

restricting the access of proteins and enzymes necessary to carry out transcription (see earlier section on histone modification).

Concepts

RNA silencing is initiated by double-stranded RNA molecules that are cleaved and processed. The resulting siRNAs or miRNAs combine with proteins to form complexes that bind to complementary sequences in mRNA or DNA. The siRNAs and miRNAs affect gene expression by cleaving mRNA, inhibiting translation, or altering chromatin structure.

✔ Concept Check 10

In RNA silencing, siRNAs and miRNAs usually bind to which part of the mRNA molecules that they control?

a. 5′ UTR   c. 3′ poly(A) tail
b. 5′ cap   d. 3′ UTR

Gene Regulation in the Course of Translation and Afterward

Ribosomes, aminoacyl tRNAs, initiation factors, and elongation factors are all required for the translation of mRNA molecules. The availability of these components affects the rate of translation and therefore influences gene expression.

Mechanisms also exist for the regulation of translation of specific mRNAs. The initiation of translation in some mRNAs is regulated by proteins that bind to an mRNA's 5′ UTR and inhibit the binding of ribosomes, in a fashion similar to the way in which repressor proteins bind to operators and prevent the transcription of structural genes. The translation of some mRNAs is affected by the binding of proteins to sequences in the 3′ UTR.

Many eukaryotic proteins are extensively modified after translation by the selective cleavage and trimming of amino acids from the ends, by acetylation, or by the addition of phosphate groups, carboxyl groups, methyl groups, or carbohydrates to the protein. These modifications affect the transport, function, and activity of the proteins and have the capacity to affect gene expression.

Concepts

The initiation of translation may be affected by proteins that bind to specific sequences near the 5′ end of mRNA. The availability of ribosomes, tRNAs, initiation and elongation factors, and other components of the translational apparatus may affect the rate of translation.

Connecting Concepts

A Comparison of Bacterial and Eukaryotic Gene Control

Now that we have considered the major types of gene regulation in bacteria and eukaryotes, let's pause to consider some of the similarities and differences in bacterial and eukaryotic gene control.

1. Most gene regulation in bacterial cells is at the level of transcription (although it does exist at other levels). Gene regulation in eukaryotic cells most often takes place at multiple levels, including chromatin structure, transcription, mRNA processing, RNA stability, RNA interference, translation, and posttranslational control.

2. Complex biochemical and developmental events in bacterial and eukaryotic cells may require a cascade of gene regulation, in which the activation of one set of genes stimulates the activation of another set.

3. Much of gene regulation in both bacterial and eukaryotic cells is accomplished through proteins that bind to specific sequences in DNA.

4. Chromatin structure plays a role in eukaryotic (but not bacterial) gene regulation. In general, condensed chromatin represses gene expression; chromatin structure must be altered before transcription can take place. Chromatin structure is altered by the modification of histones, chromatin-remodeling proteins, and DNA methylation.

5. In bacterial cells, genes are often clustered in operons and are coordinately expressed by transcription into a single mRNA molecule. In contrast, each eukaryotic gene typically has its own promoter and is transcribed independently. Coordinate regulation in eukaryotic cells takes place through common response elements, present in the promoters and enhancers of the genes. Different genes that have the same response element in common are influenced by the same regulatory protein.

6. Regulatory proteins that affect transcription exhibit two basic types of control: *repressors* inhibit transcription (negative control); *activators* stimulate transcription (positive control). Both negative control and positive control are found in bacterial and eukaryotic cells.

7. The initiation of transcription is a relatively simple process in bacterial cells, and regulatory proteins function by blocking or stimulating the binding of RNA polymerase to DNA. In contrast, eukaryotic transcription requires complex machinery that includes RNA polymerase, general transcription factors, and transcriptional activators, which allows transcription to be influenced by multiple factors.

8. Some eukaryotic transcriptional activator proteins function at a distance from the gene by binding to enhancers, causing the formation of a loop in the DNA, which brings the promoter and enhancer into close proximity. Some distant-acting sequences analogous to enhancers have been described in bacterial cells, but they appear to be less common.

9. The greater time lag between transcription and translation in eukaryotic cells than in bacterial cells allows mRNA stability and mRNA processing to play larger roles in eukaryotic gene regulation.

10. Regulation by siRNAs and miRNAs, which is extensive in eukaryotes, is absent from bacterial cells.

Model Genetic Organism
The Plant *Arabidopsis thaliana*

 Much of the early work in genetics was carried out on plants, including Mendel's seminal discoveries in pea plants as well as in important aspects of heredity, gene mapping, chromosome genetics, and quantitative inheritance in corn, wheat, beans, and other plants.

However, by the mid-twentieth century, many geneticists had turned to bacteria, viruses, yeast, *Drosophila*, and mouse genetic models. Because a good genetic plant model did not exist, plants were relatively neglected, particularly for the study of molecular genetic processes.

This neglect of plants changed in the last part of the twentieth century with the widespread introduction of a new genetic model organism, the plant *Arabidopsis thaliana* (**Figure 12.21**). *Arabidopsis thaliana* was identified in the sixteenth century, and the first mutant was reported in 1873; but this species was not commonly studied until the first detailed genetic maps appeared in the early 1980s. Today, *Arabidopsis* figures prominently in the study of genome structure, gene regulation, development, and evolution in plants, and it provides important basic information about plant genetics that is applied to other economically important plant species.

Advantages of Arabidopsis as a model genetic organism

The mustard *Arabidopsis thaliana* is a member of the Brassicaceae family and grows as a weed in many parts of the world. Except in its role as a model genetic organism, *Arabidopsis* has no economic importance, but it has a number of characteristics that make it well suited to the study of genetics. As an angiosperm, it has features in common with other flowering plants, some of which play critical roles in the ecosystem or are important sources of food, fiber, building materials, and pharmaceuticals. *Arabidopsis*'s chief advantages are its small size (maximum height of 10 to 20 cm), prolific reproduction, and small genome (see Figure 12.21).

Arabidopsis thaliana completes development—from seed germination to seed production—in about 6 weeks. Its small size and ability to grow under low illumination make it ideal for laboratory culture. Each plant is capable of producing from 10,000 to 40,000 seeds, and the seeds typically have a high rate of germination; so large numbers of progeny can be obtained from single genetic crosses.

Another key advantage for molecular studies is *Arabidopsis*'s small genome, which consists of only 125 million base pairs of DNA on five pairs of chromosomes, compared with 2.5 billion base pairs of DNA in the maize genome and 16 billion base pairs in the wheat genome. The genome of *A. thaliana* was completely sequenced in 2000, providing detailed information about gene structure and organization in this species. A number of variants of *A. thaliana*—called ecotypes—that vary in shape, size, physiological characteristics, and DNA sequence are available for study.

Life cycle of Arabidopsis
The *Arabidopsis* life cycle is fairly typical of most flowering plants (see Figures 2.17 and 12.21). The main, vegetative part of the plant is diploid; haploid gametes are produced in the pollen and ovaries. When a pollen grain lands on the stigma of a flower, a pollen

12.21 *Arabidopsis thaliana* is a model genetic organism that serves as an important subject for research on genetic processes in plants. *[Photograph courtesy of Dr. Paul Franz.]*

tube grows into the pistil and ovary. Two haploid sperm nuclei contained in each pollen grain travel down the pollen tube and enter the embryo sac. There, one of the haploid sperm cells fertilizes the haploid egg cell to produce a diploid zygote. The other haploid sperm cell fuses with two haploid nuclei to form the $3n$ endosperm, which provides tissue that will nourish the growing embryonic plant. The zygotes develop within the seeds, which are produced in a long pod.

Under appropriate conditions, the embryo germinates and begins to grow into a plant. The shoot grows upward and the roots downward, a compact rosette of leaves is produced and, under the right conditions, the shoot enlarges and differentiates into flower structures. At maturity, *A. thaliana* is a low-growing plant with roots, a main shoot with branches that bear mature leaves, and small white flowers at the tips of the branches.

Genetic techniques with Arabidopsis A number of traditional and modern molecular techniques are commonly used with *Arabidopsis* and provide it with special advantages for genetic studies. *Arabidopsis* can self-fertilize, which means that any recessive mutation appearing in the germ line can be recovered in the immediate progeny.

Cross-fertilization also is possible by removing the anther from one plant and dusting pollen on the stigma of another plant—essentially the same technique used by Gregor Mendel with pea plants (see Figure 3.3).

As already mentioned, many naturally occurring variants of *Arabidopsis* are available for study, and new mutations can be produced by exposing its seeds to chemical mutagens, radiation, or transposable elements that randomly insert into genes. The large number of offspring produced by *Arabidopsis* facilitates screening for rare mutations.

Genes from other organisms can be transferred to *Arabidopsis* by means of the Ti plasmid from the bacterium *Agrobacterium tumefaciens*, which naturally infects plants and transfers the Ti plasmid to plant cells. Subsequent to the transfer, the Ti plasmid randomly inserts into the DNA of the plant that it infects, thereby generating mutations in the plant DNA in a process called *insertional mutagenesis*. Geneticists have modified the Ti plasmid to carry a *GUS* gene, which has no promoter of its own. The *GUS* gene encodes an enzyme that converts a colorless compound (X-Glu) into a blue dye. Because the *GUS* gene has no promoter, it is expressed only when inserted into the coding sequence of a plant gene. When that happens, the enzyme encoded by *GUS* is synthesized and converts X-Glu into a blue dye that stains the cell. This dye provides a means to visually determine the expression pattern of a gene that has been interrupted by Ti DNA, producing information about the expression of genes that are mutated by insertional mutagenesis. ∎

Concepts Summary

- Gene expression can be controlled at different levels, including the alteration of gene structure, transcription, mRNA processing, RNA stability, translation, and posttranslational modification. Much of gene regulation is through the action of regulatory proteins binding to specific sequences in DNA.

- Genes in bacterial cells are typically clustered into operons—groups of functionally related structural genes and the sequences that control their transcription. Structural genes in an operon are transcribed together as a single mRNA molecule.

- In negative control, a repressor protein binds to DNA and inhibits transcription. In positive control, an activator protein binds to DNA and stimulates transcription. In inducible operons, transcription is normally off and must be turned on; in repressible operons, transcription is normally on and must be turned off.

- The *lac* operon of *E. coli* is a negative inducible operon. In the absence of lactose, a repressor binds to the operator and prevents the transcription of genes that encode β-galactosidase, permease, and transacetylase. When lactose is present, some of it is converted into allolactose, which binds to the repressor and makes it inactive, allowing the structural genes to be transcribed.

- Positive control in the *lac* and other operons is through catabolite repression. When complexed with cAMP, the catabolite activator protein (CAP) binds to a site in or near the promoter and stimulates the transcription of the structural genes. Levels of cAMP are inversely correlated with glucose; so low levels of glucose stimulate transcription and high levels inhibit transcription.

- The *trp* operon of *E. coli* is a negative repressible operon that controls the biosynthesis of tryptophan.

- Eukaryotic cells differ from bacteria in several ways that affect gene regulation, including, in eukaryotes, the absence of operons, the presence of chromatin and a nuclear membrane, and the more common use of activators.

- In eukaryotic cells, chromatin structure represses gene expression. In transcription, chromatin structure may be altered by the modification of histone proteins, including acetylation, phosphorylation, and methylation. The repositioning of nucleosomes and the methylation of DNA also affect transcription.

- The initiation of eukaryotic transcription is controlled by general transcription factors that assemble into the basal transcription apparatus and by transcriptional activator proteins that stimulate normal levels of transcription by binding to regulatory promoters and enhancers.

- Enhancers affect the transcription of distant genes. Transcriptional activators bind to enhancers and interact with the basal transcription apparatus by causing the intervening DNA to loop out. Insulators limit the action of enhancers by blocking their action in a position-dependent manner.

- Coordinately controlled genes in eukaryotic cells respond to the same factors because they have common response elements that are stimulated by the same transcriptional activator.

- Gene expression in eukaryotic cells can be influenced by RNA processing and by changes in RNA stability. The 5′ cap, the coding sequence, the 3′ UTR, and the poly(A) tail are important in controlling the stability of eukaryotic mRNAs.

- RNA silencing plays an important role in eukaryotic gene regulation. Small RNA molecules (siRNAs and miRNAs) combine with proteins and bind to sequences on mRNA or DNA. These complexes cleave RNA, inhibit translation, affect RNA degradation, and silence transcription.

- Control of the posttranslational modification of proteins may play a role in gene expression.

- *Arabidopsis thaliana* possesses a number of characteristics that make it an ideal model genetic organism.

Important Terms

gene regulation (p. 290)
structural gene (p. 291)
regulatory gene (p. 291)
constitutive gene (p. 291)
regulatory element (p. 291)
operon (p. 292)
regulator gene (p. 292)
regulator protein (p. 292)
operator (p. 292)
negative control (p. 293)
positive control (p. 293)
inducible operon (p. 293)

repressible operon (p. 293)
inducer (p. 293)
allosteric protein (p. 293)
corepressor (p. 294)
coordinate induction (p. 297)
partial diploid (p. 298)
constitutive mutation (p. 299)
catabolite repression (p. 302)
catabolite activator protein (CAP) (p. 302)
adenosine-3′,5′-cyclic monophosphate
  (cAMP) (p. 302)

histone code (p. 304)
chromatin-remodeling complex (p. 306)
general transcription factor (p. 306)
transcriptional activator protein (p. 306)
enhancer (p. 306)
coactivator (p. 306)
silencer (p. 306)
insulator (p. 307)
heat-shock protein (p. 308)
response element (p. 308)

Answers to Concept Checks

1. A constitutive gene is not regulated and is expressed continually.

2. Because it is the first step in the process of information transfer from DNA to protein. For cellular efficiency, gene expression is often regulated early in the process of protein production.

3. d

4. d

5. c

6. b

7. b

8. The DNA between the enhancer and the promoter loops out, and so transcription activators bound to the enhancer are able to interact directly with the transcription apparatus.

9. The poly(A) tail stabilizes the 5′ cap, which must be removed before the mRNA molecule can be degraded from the 5′ end.

10. d

Worked Problems

1. A regulator gene produces a repressor in an inducible operon. A geneticist isolates several constitutive mutations affecting this operon. Where might these constitutive mutations occur? How would the mutations cause the operon to be constitutive?

• Solution
An inducible operon is normally not being transcribed, meaning that the repressor is active and binds to the operator, inhibiting transcription. Transcription takes place when the inducer binds to the repressor, making it unable to bind to the operator. Constitutive mutations cause transcription to take place at all times, whether the inducer is present or not. Constitutive mutations might occur in the regulator gene, altering the repressor so that it is never able to bind to the operator. Alternatively, constitutive mutations might occur in the operator, altering the binding site for the repressor so that the repressor is unable to bind under any conditions.

2. The *fox* operon, which has sequences A, B, C, and D (which may represent either structural genes or regulatory sequences), encodes enzymes 1 and 2. Mutations in sequences A, B, C, and D have the following effects, where a plus sign (+) indicates that the enzyme is synthesized and a minus sign (−) indicates that the enzyme is not synthesized.

	Fox absent		Fox present	
Mutation in sequence	**Enzyme 1**	**Enzyme 2**	**Enzyme 1**	**Enzyme 2**
No mutation	−	−	+	+
A	−	−	−	+
B	−	−	−	−
C	−	−	+	−
D	+	+	+	+

a. Is the *fox* operon inducible or repressible?

b. Indicate which sequence (A, B, C, or D) is part of the following components of the operon:

Regulator gene	_____
Promoter	_____
Structural gene for enzyme 1	_____
Structural gene for enzyme 2	_____

• Solution
Because the structural genes in an operon are coordinately expressed, mutations that affect only one enzyme are likely to

occur in the structural genes; mutations that affect both enzymes must occur in the promoter or regulator.

a. When no mutations are present, enzymes 1 and 2 are produced in the presence of Fox but not in its absence, indicating that the operon is inducible and Fox is the inducer.

b. The mutation in A allows the production of enzyme 2 in the presence of Fox, but enzyme 1 is not produced in the presence or absence of Fox, and so A must have a mutation in the structural gene for enzyme 1. With the mutation in B, neither enzyme is produced under any conditions, and so this mutation most likely occurs in the promoter and prevents RNA polymerase from binding. The mutation in C affects only enzyme 2, which is not produced in the presence or absence of Fox; enzyme 1 is produced normally (only in the presence of Fox), and so the mutation in C most likely occurs in the structural gene for enzyme 2. The mutation in D is constitutive, allowing the production of enzymes 1 and 2 whether or not Fox is present. This mutation most likely occurs in the regulator gene, producing a defective repressor that is unable to bind to the operator under any conditions.

Regulator gene	*D*
Promoter	*B*
Structural gene for enzyme 1	*A*
Structural gene for enzyme 2	*C*

3. What would be the effect of a mutation that caused poly(A)-binding proteins to be nonfunctional?

• Solution

Degradation of mRNA from the 5′ end requires the removal of the 5′ cap and is usually preceded by the shortening of the poly(A) tail. Poly(A)-binding proteins bind to the poly(A) tail and prevent it from being shortened. Thus, the presence of these proteins on the poly(A) tail protects the 5′ cap, which prevents RNA degradation. If the gene for poly(A)-binding proteins were mutated in such a way that nonfunctional poly(A) proteins were produced, the proteins would not bind to the poly(A) tail. The tail would be shortened prematurely, the 5′ cap removed, and mRNA degraded more easily. The end result would be less mRNA and thus less protein synthesis.

Comprehension Questions

Section 12.1

1. Why is gene regulation important for bacterial cells?

Section 12.2

*2. Name six different levels at which gene expression might be controlled.

Section 12.3

*3. Draw a picture illustrating the general structure of an operon and identify its parts.

4. What is the difference between positive and negative control? What is the difference between inducible and repressible operons?

*5. Briefly describe the *lac* operon and how it controls the metabolism of lactose.

6. What is catabolite repression? How does it allow a bacterial cell to use glucose in preference to other sugars?

Section 12.4

*7. What changes take place in chromatin structure and what role do these changes play in eukaryotic gene regulation?

8. What is the histone code?

9. Briefly explain how transcriptional activator proteins and repressors affect the level of transcription of eukaryotic genes.

*10. What is an enhancer? How does it affect the transcription of distant genes?

*11. What role does RNA stability play in gene regulation? What controls RNA stability in eukaryotic cells?

*12. Briefly list some of the ways in which siRNAs and miRNAs regulate genes.

*13. How does bacterial gene regulation differ from eukaryotic gene regulation? How are they similar?

Application Questions and Problems

Section 12.3

*14. For each of the following types of transcriptional control, indicate whether the protein produced by the regulator gene will be synthesized initially as an active repressor, inactive repressor, active activator, or inactive activator.

a. Negative control in a repressible operon

b. Positive control in a repressible operon

c. Negative control in an inducible operon

d. Positive control in an inducible operon

*15. A mutation at the operator site prevents the regulator protein from binding. What effect will this mutation have in the following types of operons?

    a. Regulator protein is a repressor in a repressible operon.

    b. Regulator protein is a repressor in an inducible operon.

16. The *blob* operon produces enzymes that convert compound A into compound B. The operon is controlled by a regulatory gene *S*. Normally, the enzymes are synthesized only in the absence of compound B. If gene *S* is mutated, the enzymes are synthesized in the presence *and* in the absence of compound B. Does gene *S* produce a repressor or an activator? Is this operon inducible or repressible?

*17. A mutation prevents the catabolite activator protein (CAP) from binding to the promoter in the *lac* operon. What will the effect of this mutation be on the transcription of the operon?

18. Under which of the following conditions would a *lac* operon produce the greatest amount of β-glactosidase? The least? Explain your reasoning.

	Lactose present	Glucose present
Condition 1	Yes	No
Condition 2	No	Yes
Condition 3	Yes	Yes
Condition 4	No	No

19. A mutant strain of *E. coli* produces β-galactosidase in the presence *and* in the absence of lactose. Where in the operon might the mutation in this strain occur?

*20. For *E. coli* strains with the *lac* genotypes shown below, use a plus sign (+) to indicate the synthesis of β-galactosidase and permease and a minus sign (−) to indicate no synthesis of the enzymes.

	Lactose absent		Lactose present	
Genotype of strain	β-Galactosidase	Permease	β-Galactosidase	Permease
$lacI^+\ lacP^+\ lacO^+\ lacZ^+\ lacY^+$	+	+	+	+
$lacI^-\ lacP^+\ lacO^+\ lacZ^+\ lacY^+$				
$lacI^+\ lacP^+\ lacO^c\ lacZ^+\ lacY^+$	+	+	+	+
$lacI^-\ lacP^+\ lacO^+\ lacZ^+\ lacY^-$	+	−	+	−
$lacI^-\ lacP^-\ lacO^+\ lacZ^+\ lacY^+$				
$lacI^+\ lacP^+\ lacO^+\ lacZ^-\ lacY^+$ / $lacI^-\ lacP^+\ lacO^+\ lacZ^+\ lacY^-$				
$lacI^-\ lacP^+\ lacO^c\ lacZ^+\ lacY^+$ / $lacI^+\ lacP^+\ lacO^+\ lacZ^-\ lacY^-$				
$lacI^-\ lacP^+\ lacO^+\ lacZ^+\ lacY^-$ / $lacI^+\ lacP^-\ lacO^+\ lacZ^-\ lacY^+$				
$lacI^+\ lacP^-\ lacO^c\ lacZ^-\ lacY^+$ / $lacI^-\ lacP^+\ lacO^+\ lacZ^+\ lacY^-$				
$lacI^+\ lacP^+\ lacO^+\ lacZ^+\ lacY^+$ / $lacI^+\ lacP^+\ lacO^+\ lacZ^+\ lacY^+$				

[Handwritten annotations in left margin: "constitutive", "Inhibitor", "promoter", "operator", "bgal", "permease", "Still getting express"]

21. Give all possible genotypes of a *lac* operon that produces β-galactosidase and permease under the following conditions. Do not give partial diploid genotypes.

	Lactose absent		Lactose present	
	β-Galactosidase	Permease	β-Galactosidase	Permease
a.	−	−	+	+
b.	−	−	−	+
c.	−	−	+	−
d.	+	+	+	+
e.	−	−	−	−
f.	+	−	+	−
g.	−	+	−	+

*22. Explain why mutations in the *lacI* gene are trans in their effects, but mutations in the *lacO* gene are cis in their effects.

*23. The *mmm* operon, which has sequences A, B, C, and D (which may be structural genes or regulatory sequences), encodes enzymes 1 and 2. Mutations in sequences A, B, C, and D have the following effects, where a plus sign (+) indicates that the enzyme is synthesized and a minus sign (−) indicates that the enzyme is not synthesized.

Mutation in sequence	mmm absent		mmm present	
	Enzyme 1	Enzyme 2	Enzyme 1	Enzyme 2
No mutation	+	+	−	−
A	−	+	−	−
B	+	+	+	+
C	+	−	−	−
D	−	−	−	−

a. Is the *mmm* operon inducible or repressible?

b. Indicate which sequence (*A*, *B*, *C*, or *D*) is part of the following components of the operon:

Regulator gene   _____

Promoter   _____

Structural gene for enzyme 1   _____

Structural gene for enzyme 2   _____

24. Ellis Engelsberg and his coworkers examined the regulation of genes taking part in the metabolism of arabinose, a sugar (E. Engelsberg et al. 1965. *Journal of Bacteriology* 90:946–957). Four structural genes encode enzymes that help metabolize arabinose (genes *A*, *B*, *D*, and *E*). An additional gene *C* is linked to genes *A*, *B*, and *D*. These genes are in the order *D-A-B-C*. Gene *E* is distant from the other genes. Engelsberg and his colleagues isolated mutations at the *C* gene that affected the expression of structural genes *A*, *B*, *D*, and *E*. In one set of experiments, they created various genotypes at the *A* and *C* loci and determined whether arabinose isomerase (the enzyme encoded by gene *A*) was produced in the presence or absence of arabinose (the substrate of arabinose isomerase). Results from this experiment are shown in the following table, where a plus sign (+) indicates that the arabinose isomerase was synthesized and a minus sign (−) indicates that the enzyme was not synthesized.

Genotype	Arabinose absent	Arabinose present
1. $C^+ A^+$	−	+
2. $C^- A^+$	−	−
3. $C^- A^+/C^+ A^-$	−	+
4. $C^c A^-/C^- A^+$	+	+

a. On the basis of the results of these experiments, is the *C* gene an operator or a regulator gene? Explain your reasoning.

b. Do these experiments suggest that the arabinose operon is negatively or positively controlled? Explain your reasoning.

c. What type of mutation is C^c?

25. In *E. coli*, three structural genes (*A*, *D*, and *E*) encode enzymes A, D and E respectively. Gene *O* is an operator. The genes are in the order *O-A-D-E* on the chromosome. These enzymes catalyze the biosynthesis of valine. Mutations were isolated at the *A*, *D*, *E*, and *O* genes to study the production of enzymes A, D, and E when cellular levels of valine were low (T. Ramakrishnan and E. A. Adelberg. 1965. *Journal of Bacteriology* 89:654–660). Levels of the enzymes produced by partial-diploid *E. coli* with various combinations of mutations are shown in the following table.

Genotype	Amount of enzyme produced		
	E	**D**	**A**
1. $E^+ D^+ A^+ O^+/$ $E^+ D^+ A^+ O^+$	2.40	2.00	3.50
2. $E^+ D^+ A^+ O^-/$ $E^+ D^+ A^+ O^+$	35.80	38.60	46.80
3. $E^+ D^- A^+ O^-/$ $E^+ D^+ A^- O^+$	1.80	1.00	47.00
4. $E^+ D^+ A^- O^-/$ $E^+ D^- A^+ O^+$	35.30	38.00	1.70
5. $E^- D^+ A^+ O^-/$ $E^+ D^- A^+ O^+$	2.38	38.00	46.70

a. Is the regulator protein that binds to the operator of this operon a repressor (negative control) or an activator (positive control)? Explain your reasoning.

b. Are genes *A*, *D*, and *E* all under the control of operator *O*? Explain your reasoning.

c. Propose an explanation for the low level of enzyme E produced in genotype 3.

Section 12.4

26. A geneticist is trying to determine how many genes are found in a 300,000-bp region of DNA. Analysis shows that four H3K4me3 modifications are found in this piece of DNA. What might their presence suggest about the number of genes located there?

27. In a line of human cells grown in culture, a geneticist isolates a temperature-sensitive mutation at a locus that encodes an acetyltransferase enzyme; at temperatures above 38°C, the mutant cells produce a nonfunctional form of the enzyme. What would be the most likely effect of this mutation when the cells are raised at 40°C?

28. X31b is an experimental compound that is taken up by rapidly dividing cells. Research has shown that X31b stimulates the methylation of DNA. Some cancer researchers are interested in testing X31b as a possible drug for treating prostate cancer. Offer a possible explanation for why X31b might be an effective anticancer drug.

29. An enhancer is surrounded by four genes (*A*, *B*, *C*, and *D*), as shown in the below diagram. An insulator lies between gene *C* and gene *D*. On the basis of the positions of the genes, the enhancer, and the insulator, the transcription of which genes is most likely to be stimulated by the enhancer? Explain your reasoning.

| Gene A | Gene B | Enhancer | Gene C | Insulator | Gene D |

30. Some eukaryotic mRNAs have an AU-rich element in the 3′ untranslated region. What would be the effect on gene expression if this element were mutated or deleted?

31. A strain of *Arabidopsis thaliana* possesses a mutation in the *APETALA2* gene, in which much of the 3′ untranslated region of mRNA transcribed from the gene is deleted. What is the most likely effect of this mutation on the expression of the *APETALA2* gene?

Challenge Questions

Section 12.4

32. A yeast gene termed *SER3*, which has a role in serine biosynthesis, is repressed during growth in nutrient-rich medium, and so little transcription takes place and little SER3 enzyme is produced. In an investigation of the nature of the repression of the *SER3* gene, a region of DNA upstream of the *SER3* gene was found to be heavily transcribed when the *SER3* gene is repressed (J. A. Martens, L. Laprade, and F. Winston. 2004. *Nature* 429:571–574). Within this upstream region is a promoter that stimulates the transcription of an RNA molecule called *SRG1* (for *SER3* regulatory gene). This RNA molecule has none of the sequences necessary for translation. Mutations in the promoter for *SRG1* result in the disappearance of *SRG1* RNA, and these mutations remove the repression of *SER3*. When RNA polymerase binds to the *SRG1* promoter, the polymerase was found to travel downstream, transcribing the *SGR1* RNA, and to pass through and transcribe the promoter for *SER3*. This activity leads to the repression of *SER3*. Propose a possible explanation for how the transcription of *SGR1* might repress the transcription of *SER3*. (Hint: Remember that the *SGR1* RNA does not encode a protein.)

33. A common feature of many eukaryotic mRNAs is the presence of a rather long 3′ UTR, which often contains consensus sequences. Creatine kinase B (CK-B) is an enzyme important in cellular metabolism. Certain cells—termed U937D cells—have lots of CK-B mRNA, but no CK-B enzyme is present. In these cells, the 5′ end of the CK-B mRNA is bound to ribosomes, but the mRNA is apparently not translated. Something inhibits the translation of the CK-B mRNA in these cells.

    Researchers introduced numerous short segments of RNA containing only 3′ UTR sequences into U937D cells. As a result, the U937D cells began to synthesize the CK-B enzyme, but the total amount of CK-B mRNA did not increase. The introduction of short segments of other RNA sequences did not stimulate the synthesis of CK-B; only the 3′ UTR sequences turned on the translation of the enzyme.

    On the basis of these results, propose a mechanism for how CK-B translation is inhibited in the U937D cells. Explain how the introduction of short segments of RNA containing the 3′ UTR sequences might remove the inhibition.

Gene Mutations, Transposable Elements, and DNA Repair

A Fly Without a Heart

The heart of a fruit fly is a simple organ, an open-ended tube that rhythmically contracts, pumping fluid—rather inefficiently—around the body of the fly. Although simple and inelegant, the fruit fly's heart is nevertheless essential, at least for a fruit fly. Remarkably, a few rare mutant fruit flies never develop a heart and die (not surprisingly) at an early embryonic stage. Geneticist Rolf Bodmer analyzed these mutants in the 1980s and made an important discovery—a gene that specifies the development of a heart. He named the gene *tinman*, after the character in *The Wizard of Oz* who also lacked a heart. Bodmer's research revealed that *tinman* encodes a transcription factor, which binds to DNA and turns on other genes that are essential for the normal development of a heart. In the mutant flies studied by Bodmer, this gene was lacking, the transcription factor was never produced, and the heart never developed. Findings from subsequent research revealed the existence of a human gene (called *Nkx2.5*) with a sequence similar to that of *tinman*, but the function of the human gene was unknown.

Then, in the 1990s, physicians Jonathan and Christine Seidman began studying people born with abnormal hearts, such as those with a hole in the septum that separates the chambers on the left and right sides of the heart. Such heart defects can cause abnormal blood flow through the heart, causing the heart to work harder than normal and the mixing of oxygenated and deoxyganted blood. Congenital heart defects are not uncommon; they're found in about 1 of every 125 babies. Some of the defects heal on their own, but others require corrective surgery. Although surgery is often successful in reversing congenital heart problems, many of these patients begin to have irregular heartbeats in their 20s and 30s. The Seidmans and their colleagues found several

Many people who have a mutation in the *tinman* gene, named after Tin Man in *The Wizard of Oz*, have congenital heart defects. [*Mary Evans Picture Library/Alamy.*]

families in which congenital heart defects and irregular heartbeats were inherited together in an autosomal dominant fashion. Detailed molecular analysis of one of these families revealed that the gene responsible for the heart problems was located on chromosome 5, at a spot where the human *tinman* gene (*Nkx2.5*) had been previously mapped. All members of this family who inherited the heart defects also inherited a mutation in the *tinman* gene. Subsequent studies with additional patients found that many people with congenital heart defects have a mutation in the *tinman* gene. The human version of this gene, like its counterpart in flies, encodes a transcription factor that controls heart development. Despite tremendous differences in size, anatomy, and physiology, humans and flies use the same gene to make a heart.

The story of *tinman* illustrates the central importance of studying mutations: the analysis of mutants is often a source of key insights into important biological processes. This chapter focuses on gene mutations. We begin with a brief examination of the different types of mutations, including their phenotypic effects, how they can be suppressed, and mutation rates. The next section explores how mutations spontaneously arise, as well as how chemicals and radiation induce mutations. We then consider the analysis of mutations, followed by a look at transposable elements, which often generate mutations. Finally, we look at DNA repair.

13.1 Mutations Are Inherited Alterations in the DNA Sequence

DNA is a highly stable molecule that replicates with amazing accuracy (see Chapters 8 and 9), but changes in DNA structure and errors of replication do occur. These changes may alter the genetic information of the DNA. A **mutation** is defined as an inherited change in genetic information; the descendants may be cells or organisms.

The Importance of Mutations

Mutations are both the sustainer of life and the cause of great suffering. On the one hand, mutation is the source of all genetic variation, the raw material of evolution. On the other hand, many mutations have detrimental effects, and mutation is the source of many human diseases and disorders.

Much of the study of genetics focuses on how variants produced by mutation are inherited; genetic crosses are meaningless if all individual members of a species are identically homozygous for the same alleles. Much of Gregor Mendel's success in unraveling the principles of inheritance can be traced to his use of carefully selected variants of the garden pea; similarly, Thomas Hunt Morgan and his students discovered many basic principles of genetics by analyzing mutant fruit flies.

Mutations are also useful for probing fundamental biological processes. Finding or creating mutations that affect different components of a biological system and studying their effects can often lead to an understanding of the system. This method, referred to as genetic dissection, is analogous to figuring out how an automobile works by breaking different parts of a car and observing the effects; for example, smash the radiator and the engine overheats, revealing that the radiator cools the engine. Using mutations to disrupt function can likewise can be a source of insight into biological processes. For example, geneticists have begun to unravel the molecular details of development by studying mutations, such as *tinman*, that interrupt various embryonic stages in *Drosophila*. Although this method of breaking "parts" to determine their function might seem like a crude approach to understanding a system, it is actually very powerful and has been used extensively in biochemistry, developmental biology, physiology, and behavioral science (but this method is *not* recommended for learning how your car works).

Concepts

Mutations are heritable changes in DNA. They are essential to the study of genetics and are useful in many other biological fields.

✔ Concept Check 1

How are mutations used to help in understanding basic biological processes?

Categories of Mutations

In multicellular organisms, we can distinguish between two broad categories of mutations: somatic mutations and germ-line mutations. **Somatic mutations** arise in somatic tissues, which do not produce gametes (**Figure 13.1**). When a somatic cell with a mutation divides (mitosis), the mutation is passed on to the daughter cells, leading to a population of genetically identical cells (a clone). The earlier in development that a somatic mutation occurs, the larger the clone of cells in that individual organism containing the mutation.

Because of the huge number of cells present in a typical eukaryotic organism, somatic mutations are numerous. For example, there are about 10^{14} cells in the human body. Typically, a mutation arises once in every million cell divisions, and so hundreds of millions of somatic mutations must arise in each person. Many somatic mutations have no obvious effect on the phenotype of the organism, because the function of the mutant cell (even the cell itself) is replaced by that of normal cells. However, cells with a somatic mutation that stimulates cell division can increase in number and spread; this type of mutation can give rise to cells with a selective advantage and is the basis for all cancers (see Chapter 15).

Germ-line mutations arise in cells that ultimately produce gametes. A germ-line mutation can be passed to future generations, producing individual organisms that carry the

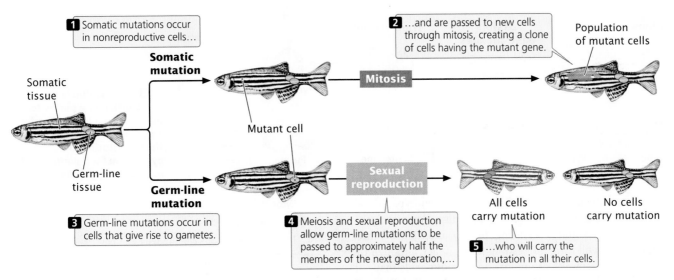

1 Somatic mutations occur in nonreproductive cells...

Somatic mutation

Somatic tissue

Mutant cell

2 ...and are passed to new cells through mitosis, creating a clone of cells having the mutant gene.

Mitosis

Population of mutant cells

Germ-line tissue

Germ-line mutation

3 Germ-line mutations occur in cells that give rise to gametes.

4 Meiosis and sexual reproduction allow germ-line mutations to be passed to approximately half the members of the next generation,...

Sexual reproduction

All cells carry mutation

No cells carry mutation

5 ...who will carry the mutation in all their cells.

13.1 **The two basic classes of mutations are somatic mutations and germ-line mutations.**

mutation in all their somatic and germ-line cells (see Figure 13.1). When we speak of mutations in multicellular organisms, we're usually talking about germ-line mutations.

Historically, mutations have been partitioned into those that affect a single gene, called *gene mutations*, and those that affect the number or structure of chromosomes, called *chromosome mutations*. This distinction arose because chromosome mutations could be observed directly, by looking at chromosomes with a microscope, whereas gene mutations could be detected only by observing their phenotypic effects. Now, with the development of DNA sequencing, gene mutations and chromosome mutations are distinguished somewhat arbitrarily on the basis of the size of the DNA lesion. Nevertheless, it is useful to use the term *chromosome mutation* for a large-scale genetic alteration that affects chromosome structure or the number of chromosomes and to use the term **gene mutation** for a relatively small DNA lesion that affects a single gene. This chapter focuses on gene mutations; chromosome mutations were discussed in Chapter 7.

Types of Gene Mutations

There are a number of ways to classify gene mutations. Some classification schemes are based on the nature of the phenotypic effect, others are based on the causative agent of the mutation, and still others focus on the molecular nature of the defect. Here, we will categorize mutations primarily on the basis of their molecular nature, but we will also encounter some terms that relate the causes and the phenotypic effects of mutations.

Base substitutions   The simplest type of gene mutation is a **base substitution**, the alteration of a single nucleotide in the DNA (**Figure 13.2a**). Base substitutions are of two types. In a **transition**, a purine is replaced by a different purine or, alternatively, a pyrimidine is replaced by a different pyrimidine (**Figure 13.3**). In a **transversion**, a purine is replaced by

a pyrimidine or a pyrimidine is replaced by a purine. The number of possible transversions (see Figure 13.3) is twice the number of possible transitions, but transitions arise more frequently.

Insertions and deletions   The second major class of gene mutations contains **insertions** and **deletions**—the addition or the removal, respectively, of one or more nucleotide pairs (**Figure 13.2b and c**). Although base substitutions are often assumed to be the most common type of mutation, molecular analysis has revealed that insertions and deletions are more frequent. Insertions and deletions within sequences that encode proteins may lead to **frameshift mutations**, changes in the reading frame (see p. 276–277 in Chapter 11) of the gene. Frameshift mutations usually alter all amino acids encoded by nucleotides following the mutation, and so

Original DNA sequence

GGG AGT GTA GAT CGT

A base substitution alters a single codon.

(a) Base substitution

GGG AGT GCA GAT CGT

One codon changed

(b) Base insertion

T

GGG AGT GTT AGA TCG T

An insertion or a deletion alters the reading frame and may change many codons.

(c) Base deletion

T

GGG AGT GAG ATC GT

13.2 **Three basic types of gene mutations are base substitutions, insertions, and deletions.**

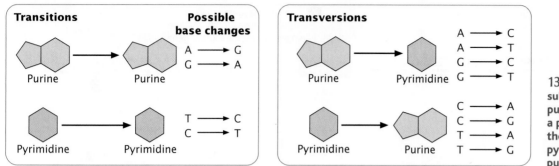

13.3 **A transition is the substitution of a purine for a purine or of a pyrimidine for a pyrimidine; a transversion is the substitution of a pyrimidine for a purine or of a purine for a pyrimidine.**

they generally have drastic effects on the phenotype. Not all insertions and deletions lead to frameshifts, however; insertions and deletions consisting of any multiple of three nucleotides will leave the reading frame intact, although the addition or removal of one or more amino acids may still affect the phenotype. These mutations are called **in-frame insertions** and **deletions**, respectively.

Concepts

Gene mutations consist of changes in a single gene and can be base substitutions (a single pair of nucleotides is altered) or insertions or deletions (nucleotides are added or removed). A base substitution can be a transition (substitution of like bases) or a transversion (substitution of unlike bases). Insertions and deletions often lead to a change in the reading frame of a gene.

✔ Concept Check 2

Which of the following changes is a transition base substitution?

a. Adenine is replaced by thymine.

b. Cytosine is replaced by adenine.

c. Guanine is replaced by adenine.

d. Three nucleotide pairs are inserted into DNA.

Expanding trinucleotide repeats   Mutations in which the number of copies of a trinucleotide (a set of three nucleotides) increase in number are called **expanding trinucleotide repeats**. This type of mutation was first observed in 1991 in a gene called *FMR-1*, which causes fragile-X syndrome, the most common hereditary cause of mental retardation. The disorder is so named because, in specially treated cells from persons having the condition, the tip of each long arm of the X chromosome is attached by only a slender thread (**Figure 13.4**). The normal *FMR-1* allele (not containing the mutation) has 60 or fewer copies of the trinucleotide CGG but, in persons with fragile-X syndrome, the allele may harbor hundreds or even thousands of copies.

Expanding trinucleotide repeats have been found in other human genetic diseases (**Table 13.1**). The number of

Table 13.1   Examples of genetic diseases caused by expanding trinucleotide repeats

Disease	Repeated Sequence	Number of Copies of Repeat	
		Normal Range	Disease Range
Spinal and bulbar muscular atrophy	CAG	11–33	40–62
Fragile-X syndrome	CGG	6–54	50–1500
Jacobsen syndrome	CGG	11	100–1000
Spinocerebellar ataxia (several types)	CAG	4–44	21–130
Autosomal dominant cerebellar ataxia	CAG	7–19	37–220
Myotonic dystrophy	CTG	5–37	44–3000
Huntington disease	CAG	9–37	37–121
Friedreich ataxia	GAA	6–29	200–900
Dentatorubral-pallidoluysian atrophy	CAG	7–25	49–75
Myoclonus epilepsy of the Unverricht–Lundborg type*	CCCCGCCCCGCG	2–3	12–13

*Technically not a trinucleotide repeat but does entail a multiple of three nucleotides that expands and contracts in similar fashion to trinucleotide repeats.

13.4 The fragile-X chromosome is associated with a characteristic constriction (fragile site) on the long arm. [*Visuals Unlimited.*]

copies of the trinucleotide repeat often correlates with the severity or age of onset of the disease. The number of copies of the repeat also corresponds to the instability of trinucleotide repeats: when more repeats are present, the probability of expansion to even more repeats increases.

How an increase in the number of trinucleotides produces disease symptoms is not yet understood. In several of the diseases (e.g., Huntington disease), the trinucleotide expands within the coding part of a gene, producing a toxic protein that has extra glutamine residues (the amino acid encoded by CAG). In other diseases (e.g., fragile-X syndrome and myotonic dystrophy), the repeat is outside the coding region of the gene and therefore must have some other mode of action.

The mechanism that leads to the expansion of trinucleotide repeats also is not completely understood. A possible source of expansion is the formation of hairpins and other special DNA structures, which can cause nucleotides in the template strand to be replicated twice, thus increasing the number of repeats on the newly synthesized strand (**Figure 13.5**).

Concepts

Expanding trinucleotide repeats are regions of DNA that consist of repeated copies of three nucleotides. Increased numbers of trinucleotide repeats are associated with several genetic diseases.

Phenotypic Effects of Mutations

Mutations have a variety of phenotypic effects. The phenotypic effect of a mutation is realized when the mutant is compared with the wild-type phenotype. A mutation that alters the wild-type allele is called a **forward mutation**, whereas a **reverse mutation** (a *reversion*) changes a mutant allele back into the wild-type allele.

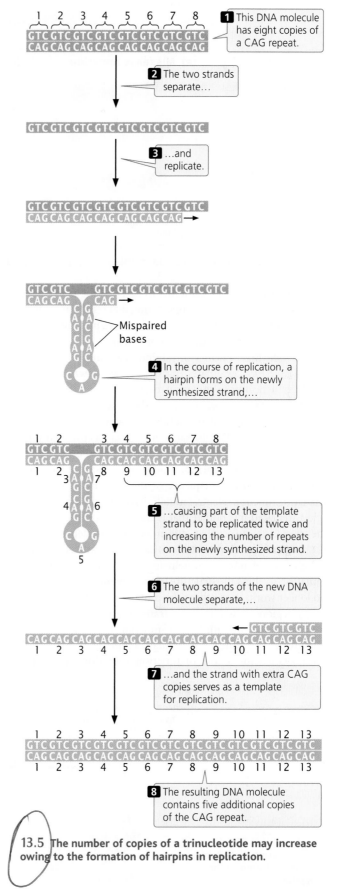

13.5 The number of copies of a trinucleotide may increase owing to the formation of hairpins in replication.

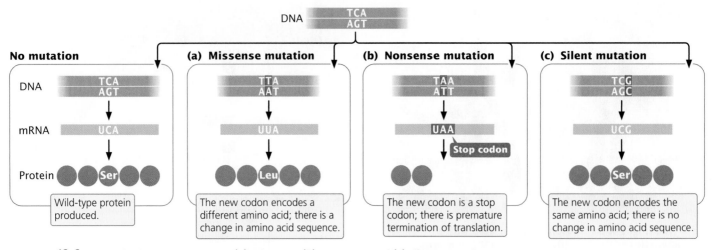

13.6 **Base substitutions can cause (a) missense, (b) nonsense, and (c) silent mutations.**

Geneticists use special terms to describe the phenotypic effects of mutations. A base substitution that results in a different amino acid in the protein is referred to as a **missense mutation** (**Figure 13.6a**). A **nonsense mutation** changes a sense codon (one that specifies an amino acid) into a nonsense codon (one that terminates translation), as shown in **Figure 13.6b**. If a nonsense mutation occurs early in the mRNA sequence, the protein will be greatly shortened and will usually be nonfunctional. A **silent mutation** creates a different DNA sequence that specifies the same amino acid as the wild-type sequence does, thanks to the redundancy of the genetic code (**Figure 13.6c**). A **neutral mutation** is a missense mutation that alters the amino acid sequence of the protein but does not change its function. Neutral mutations occur when one amino acid is replaced by another that is chemically similar or when the affected amino acid has little influence on protein function.

Loss-of-function mutations cause the complete or partial absence of normal protein function. A loss-of-function mutation so alters the structure of the protein that the protein no longer works correctly or the mutation can occur in regulatory regions that affect the transcription, translation, or splicing of the protein. Loss-of-function mutations are frequently recessive, and an individual diploid organism must be homozygous for a loss-of-function mutation before the effects of the loss of the functional protein can be exhibited. In contrast, a **gain-of-function mutation** produces an entirely new trait or it causes a trait to appear in an inappropriate tissue or at an inappropriate time in development. For example, a mutation might occur in a gene that encodes a receptor for a growth factor so that the mutated receptor stimulates growth all the time, even in the absence of the growth factor. Gain-of-function mutations are frequently dominant in their expression. Still other types of mutations are **conditional mutations**, which are expressed only under certain conditions, and **lethal mutations**, which cause premature death.

Suppressor Mutations

A **suppressor mutation** is a genetic change that hides or suppresses the effect of another mutation. This type of mutation is distinct from a reverse mutation, in which the mutated site changes back into the original wild-type sequence (**Figure 13.7**). A suppressor mutation occurs at a site that is distinct from the site of the original mutation; thus, an individual with a suppressor mutation is a double mutant, possessing both the original mutation and the suppressor mutation but exhibiting the phenotype of an unmutated wild type. Like other mutations, suppressors arise randomly.

13.7 **Relation of forward, reverse, and suppressor mutations.**

13.8 An intragenic suppressor mutation occurs in the gene containing the mutation being suppressed.

Geneticists distinguish between two classes of suppressor mutations: intragenic and intergenic. An **intragenic suppressor mutation** is in the same gene as that containing the mutation being suppressed and may work in several ways. The suppressor may change a second nucleotide in the same codon altered by the original mutation, producing a codon that specifies the same amino acid as that specified by the original, unmutated codon (**Figure 13.8**). Intragenic suppressors may also work by suppressing a frameshift mutation. If the original mutation is a one-base deletion, then the addition of a single base elsewhere in the gene will restore the former reading frame. Consider the following nucleotide sequence in DNA and the amino acids that it encodes:

DNA	AAA	TCA	CTT	GGC	GTA	CAA
Amino acids	Phe	Ser	Glu	Pro	His	Val

Suppose a one-base deletion occurs in the first nucleotide of the second codon. This deletion shifts the reading frame by one nucleotide and alters all the amino acids that follow the mutation.

One-nucleotide deletion

AAA	TCAC	TTG	GCG	TAC	AA
Phe	Val	Asn	Arg	Met	

If a single nucleotide is added to the third codon (the suppressor mutation), the reading frame is restored, although two of the amino acids differ from those specified by the original sequence.

One-nucleotide insertion

AAA	CAC	TTT	GGC	GTA	CAA
Phe	Val	Lys	Pro	His	Val

Similarly, a mutation due to an insertion may be suppressed by a subsequent deletion in the same gene.

A third way in which an intragenic suppressor may work is by making compensatory changes in the protein. A first missense mutation may alter the folding of a polypeptide chain by changing the way in which amino acids in the protein interact with one another. A second missense mutation at a different site (the suppressor) may recreate the original folding pattern by restoring interactions between the amino acids.

An **intergenic suppressor mutation**, in contrast, occurs in a gene other than the one bearing the original mutation. These suppressors sometimes work by changing the way that the mRNA is translated. In the example illustrated in **Figure 13.9a**, the original DNA sequence is AAC (UUG in the mRNA) and specifies leucine. This sequence mutates to A̲TC (U̲AG in mRNA), a termination codon (**Figure 13.9b**). The ATC nonsense mutation could be suppressed by a second mutation in a different gene that encodes a tRNA; this second mutation would result in a codon capable of pairing with the UAG termination codon (**Figure 13.9c**). For example, the gene that encodes the tRNA for tyrosine (tRNA^{Tyr}), which has the anticodon AUA, might be mutated to have the anticodon AU̲C, which will then pair with the UAG stop codon. Instead of translation terminating at the UAG codon, tyrosine would be inserted into the protein and a full-length protein would be produced, although tyrosine would now substitute for leucine. The effect of this change would depend on the role of this amino acid in the overall structure of the protein, but the effect of the suppressor mutation is likely to be less detrimental than the effect of the nonsense mutation, which would halt translation prematurely.

Because cells in many organisms have multiple copies of tRNA genes, other unmutated copies of tRNA^{Tyr} would remain available to recognize tyrosine codons in the transcripts of the mutant gene in question and other genes being expressed concurrently. However, we might expect that the tRNAs that have undergone a suppressor mutation would also suppress the normal termination codons at the ends of coding sequences, resulting in the production of longer-than-normal proteins, but this event does not usually take place.

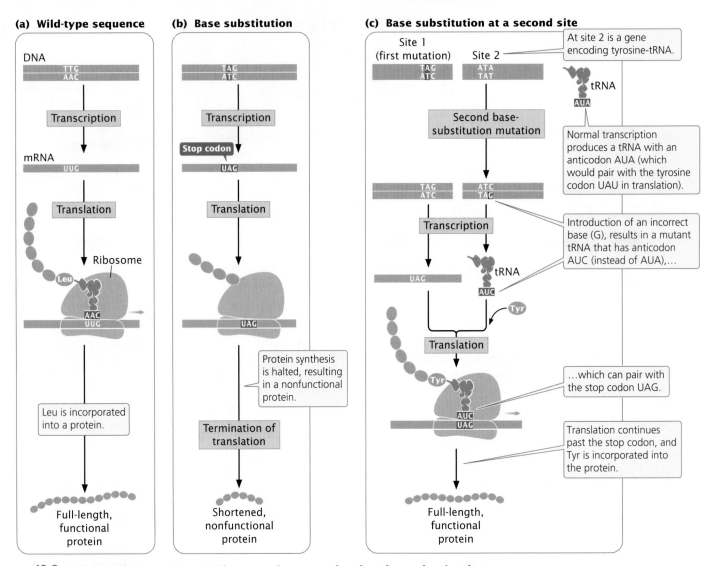

13.9 An intergenic suppressor mutation occurs in a gene other than the one bearing the original mutation. (a) The wild-type sequence produces a full-length, functional protein. (b) A base substitution at a site in the same gene produces a premature stop codon, resulting in a shortened, nonfunctional protein. (c) A base substitution at a site in another gene, which in this case encodes tRNA, alters the anticodon of tRNA^Tyr so that tRNA^Tyr can pair with the stop codon produced by the original mutation, allowing tyrosine to be incorporated into the protein and translation to continue.

Characteristics of some of the different types of mutations are summarized in **Table 13.2**.

Concepts

A suppressor mutation overrides the effect of an earlier mutation at a different site. An intragenic suppressor mutation occurs within the *same* gene as that containing the original mutation, whereas an intergenic suppressor mutation occurs in a *different* gene.

✔ Concept Check 3

How does a suppressor mutation differ from a reverse mutation?

Mutation Rates

The frequency with which a wild-type allele at a locus changes into a mutant allele is referred to as the **mutation rate** and is generally expressed as the number of mutations per biological unit, which may be mutations per cell division, per gamete, or per round of replication. For example, achondroplasia is a type of hereditary dwarfism in humans that results from a dominant mutation. On average, about four achondroplasia mutations arise in every 100,000 gametes, and so the mutation rate is $^4/_{100,000}$, or 0.00004 mutations per gamete. The mutation rate provides information about how often a mutation arises.

Mutation rates vary among genes and species (**Table 13.3** on page 330), but we can draw several general conclusions about mutation rates. First, spontaneous mutation rates are low for all organisms studied. Typical mutation rates for bacterial genes range from about 1 to 100 mutations per 10 billion cells (from 1×10^{-8} to 1×10^{-10}). The mutation rates for most eukaryotic genes are a bit higher, from about 1 to 10 mutations per million gametes (from 1×10^{-5} to 1×10^{-6}).

Table 13.2 Characteristics of different types of mutations

Type of Mutation	Definition
Base substitution	Changes the base of a single DNA nucleotide
Transition	Base substitution in which a purine replaces a purine or a pyrimidine replaces a pyrimidine
Transversion	Base substitution in which a purine replaces a pyrimidine or a pyrimidine replaces a purine
Insertion	Addition of one or more nucleotides
Deletion	Deletion of one or more nucleotides
Frameshift mutation	Insertion or deletion that alters the reading frame of a gene
In-frame deletion or insertion	Deletion or insertion of a multiple of three nucleotides that does not alter the reading frame
Expanding trinucleotide repeats	Repeated sequence of three nucleotides (trinucleotide) in which the number of copies of the trinucleotide increases
Forward mutation	Changes the wild-type phenotype to a mutant phenotype
Reverse mutation	Changes a mutant phenotype back to the wild-type phenotype
Missense mutation	Changes a sense codon into a different sense codon, resulting in the incorporation of a different amino acid in the protein
Nonsense mutation	Changes a sense codon into a nonsense codon, causing premature termination of translation
Silent mutation	Changes a sense codon into a synonymous codon, leaving the amino acid sequence of the protein unchanged
Neutral mutation	Changes the amino acid sequence of a protein without altering its ability to function
Loss-of-function mutation	Causes a complete or partial loss of function
Gain-of-function mutation	Causes the appearance of a new trait or function or causes the appearance of a trait in inappropriate tissue or at an inappropriate time
Lethal mutation	Causes premature death
Suppressor mutation	Suppresses the effect of an earlier mutation at a different site
Intragenic suppressor mutation	Suppresses the effect of an earlier mutation within the same gene
Intergenic suppressor mutation	Suppresses the effect of an earlier mutation in another gene

These higher values in eukaryotes may be due to the fact that the rates are calculated per gamete, and several cell divisions are required to produce a gamete, whereas mutation rates in prokaryotic cells are calculated *per cell division.*

The differences in mutation rates among species may be due to differing abilities to repair mutations, unequal exposures to mutagens, or biological differences in rates of spontaneously arising mutations. Even within a single species, spontaneous rates of mutation vary among genes. The reason for this variation is not entirely understood, but some regions of DNA are known to be more susceptible to mutation than others.

Concepts

Mutation rate is the frequency with which a specific mutation arises. Rates of mutations are generally low and are affected by environmental and genetic factors.

13.2 Mutations Are Potentially Caused by a Number of Different Natural and Unnatural Factors

Mutations result from both internal and external factors. Those that are a result of natural changes in DNA structure are termed **spontaneous mutations**, whereas those that result from changes caused by environmental chemicals or radiation are **induced mutations**.

Spontaneous Replication Errors

Replication is amazingly accurate: less than one error in a billion nucleotides arises in the course of DNA synthesis (see Chapter 9). However, spontaneous replication errors do occasionally occur.

Organism	Mutation	Rate	Unit
Bacteriophage T2	Lysis inhibition	1×10^{-8}	Per replication
	Host range	3×10^{-9}	
Escherichia coli	Lactose fermentation	2×10^{-7}	Per cell division
	Histidine requirement	2×10^{-8}	
Neurospora crassa	Inositol requirement	8×10^{-8}	Per asexual spore
	Adenine requirement	4×10^{-8}	
Corn	Kernel color	2.2×10^{-6}	Per gamete
Drosophila	Eye color	4×10^{-5}	Per gamete
	Allozymes	5.14×10^{-6}	
Mouse	Albino coat color	4.5×10^{-5}	Per gamete
	Dilution coat color	3×10^{-5}	
Human	Huntington disease	1×10^{-6}	Per gamete
	Achondroplasia	1×10^{-5}	
	Neurofibromatosis (Michigan)	1×10^{-4}	
	Hemophilia A (Finland)	3.2×10^{-5}	
	Duchenne muscular dystrophy (Wisconsin)	9.2×10^{-5}	

Table 13.3 Mutation rates of different genes in different organisms

The primary cause of spontaneous replication errors was formerly thought to be tautomeric shifts, in which the positions of protons in the DNA bases change. Purine and pyrimidine bases exist in different chemical forms called tautomers (**Figure 13.10a**). The two tautomeric forms of each base are in dynamic equilibrium, although one form is more common than the other. The standard Watson-and-Crick base pairings—adenine with thymine, and cytosine with guanine—are between the common forms of the bases, but, if the bases are in their rare tautomeric forms, other base pairings are possible (**Figure 13.10b**).

Watson and Crick proposed that tautomeric shifts might produce mutations, and, for many years, their proposal was the accepted model for spontaneous replication errors. However, there has never been convincing evidence that the rare tautomers are the cause of spontaneous mutations. Furthermore, research now shows little evidence of tautomers in DNA.

Mispairing can also occur through wobble, in which normal, protonated, and other forms of the bases are able to pair because of flexibility in the DNA helical structure (**Figure 13.11**). These structures have been detected in DNA molecules and are now thought to be responsible for many of the mispairings in replication.

When a mismatched base has been incorporated into a newly synthesized nucleotide chain, an **incorporated error** is said to have occurred. Suppose that, in replication, thymine (which normally pairs with adenine) mispairs with guanine through wobble (**Figure 13.12** on page 332). In the next round of replication, the two mismatched bases separate, and each serves as template for the synthesis of a new nucleotide strand. This time, thymine pairs with adenine, producing another copy of the original DNA sequence. On the other strand, however, the incorrectly incorporated guanine serves as the template and pairs with cytosine, producing a new DNA molecule that has an error—a C · G pair in place of the original T · A pair (a T · A → C · G base substitution). The original incorporated error leads to a **replicated error**, which creates a permanent mutation, because all the base pairings are correct and there is no mechanism for repair systems to detect the error.

Mutations due to small insertions and deletions also can arise spontaneously in replication and crossing over. **Strand slippage** can occur when one nucleotide strand forms a small loop (**Figure 13.13** on page 332). If the looped-out nucleotides are on the newly synthesized strand, an insertion results. At the next round of replication, the insertion will be replicated so that both strands contain the insertion. If the looped-out nucleotides are on the template strand, then the newly replicated strand has a deletion, and this deletion will be perpetuated in subsequent rounds of replication.

Another process that produces insertions and deletions is unequal crossing over. During normal crossing over, the homologous sequences of the two DNA molecules align, and crossing over produces no net change in the number of

(a)

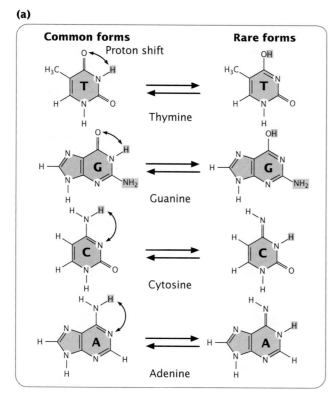

13.10 Purine and pyrimidine bases exist in different forms called tautomers. (a) A tautomeric shift takes place when a proton changes its position, resulting in a rare tautomeric form. (b) Standard and anomalous base-pairing arrangements that arise if bases are in the rare tautomeric forms. Base mispairings due to tautomeric shifts were originally thought to be a major source of errors in replication, but such structures have not been detected in DNA, and most evidence now suggests that other types of anomalous pairings (see also Figure 13.11) are responsible for replication errors.

nucleotides in either molecule. Misaligned pairing can cause **unequal crossing over**, which results in one DNA molecule with an insertion and the other with a deletion (**Figure 13.14**). Some DNA sequences are more likely than others to undergo strand slippage or unequal crossing over. Stretches of repeated sequences, such as trinucleotide repeats or homopolymeric repeats (more than five repeats of the same base in a row), are prone to strand slippage. Duplicated or repetitive sequences may misalign during pairing, leading to unequal crossing over. Both strand slippage and unequal crossing over produce duplicated copies of sequences, which in turn promote further strand slippage and unequal crossing over.

(b)

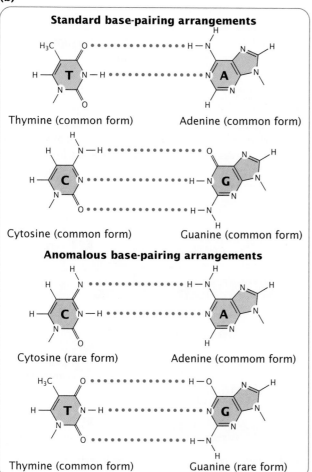

Concepts

Spontaneous replication errors arise from altered base structures and from wobble base pairing. Small insertions and deletions can occur through strand slippage in replication and through unequal crossing over.

✔ Concept Check 4

How does an incorporated error differ from a replicated error?

13.11 Nonstandard base pairings can occur as a result of the flexibility in DNA structure. Thymine and guanine can pair through wobble between normal bases. Cytosine and adenine can pair through wobble when adenine is protonated (has an extra hydrogen).

13.12 Wobble base pairing leads to a replicated error.

Spontaneous Chemical Changes

In addition to spontaneous mutations that arise in replication, mutations also result from spontaneous chemical changes in DNA. One such change is **depurination**, the loss of a purine base from a nucleotide. Depurination results when the covalent bond connecting the purine to the 1′-carbon atom of the deoxyribose sugar breaks (**Figure 13.15a**), producing an apurinic site, a nucleotide that lacks its purine base. An apurinic site cannot act as a template for a complementary base in replication. In the absence of base-pairing constraints, an incorrect nucleotide (most often adenine) is incorporated into the newly synthesized DNA strand opposite the apurinic site (**Figure 13.15b**), frequently leading to an incorporated error. The incorporated error is then transformed into a replication error at the next round of replication. Depurination is a common cause of spontaneous mutation; a mammalian cell in culture loses approximately 10,000 purines every day.

Another spontaneously occurring chemical change that takes place in DNA is **deamination**, the loss of an amino group (NH_2) from a base. Deamination can be spontaneous or be induced by mutagenic chemicals.

Deamination can alter the pairing properties of a base: the deamination of cytosine, for example, produces uracil

(**Figure 13.16a**), which pairs with adenine in replication. After another round of replication, the adenine will pair with thymine, creating a T·A pair in place of the original C·G pair (C·G → U·A → T·A); this chemical change is a transition mutation. This type of mutation is usually repaired by enzymes that remove uracil whenever it is found in DNA. The ability to recognize the product of cytosine deamination may explain why thymine, not uracil, is found in DNA. Some cytosine bases in DNA are naturally methylated and exist in the form of 5-methylcytosine, which, when deaminated, becomes thymine (**Figure 13.16b**). Because thymine pairs with adenine in replication, the deamination of 5-methylcytosine changes an original C·G pair to T·A (C·G → 5mC·G → T·G → T·A). This change cannot be detected by DNA repair systems, because it produces a normal base. Consequently, C·G → T·A transitions are frequent in eukaryotic cells, and 5mC sites are mutation hotspots in humans.

Concepts

Some mutations arise from spontaneous alterations to DNA structure, such as depurination and deamination, which may alter the pairing properties of the bases and cause errors in subsequent rounds of replication.

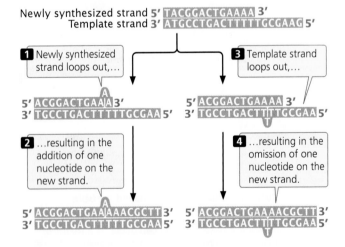

13.13 Insertions and deletions may result from strand slippage.

13.14 Unequal crossing over produces insertions and deletions.

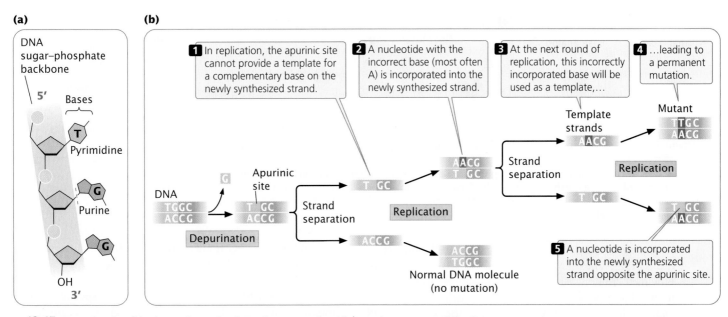

(a)

DNA sugar–phosphate backbone

5′

Bases

T — Pyrimidine

G — Purine

G

OH

3′

(b)

1 In replication, the apurinic site cannot provide a template for a complementary base on the newly synthesized strand.

2 A nucleotide with the incorrect base (most often A) is incorporated into the newly synthesized strand.

3 At the next round of replication, this incorrectly incorporated base will be used as a template,…

4 …leading to a permanent mutation.

Template strands

Mutant

DNA

Depurination

Apurinic site

Strand separation

Replication

Strand separation

Replication

Normal DNA molecule (no mutation)

5 A nucleotide is incorporated into the newly synthesized strand opposite the apurinic site.

13.15 Depurination (the loss of a purine base from a nucleotide) produces an apurinic site.

Chemically Induced Mutations

Although many mutations arise spontaneously, a number of environmental agents are capable of damaging DNA, including certain chemicals and radiation. Any environmental agent that significantly increases the rate of mutation above the spontaneous rate is called a **mutagen**.

Base analogs   One class of chemical mutagens consists of **base analogs**, chemicals with structures similar to that of any of the four standard bases of DNA. DNA polymerases cannot distinguish these analogs from the standard bases; so, if base analogs are present during replication, they may be incorporated into newly synthesized DNA molecules. For example, 5-bromouracil (5BU) is an analog of thymine; it has the same structure as that of thymine except that it has a bromine (Br) atom on the 5-carbon atom instead of a methyl group (**Figure 13.17a**). Normally, 5-bromouracil pairs with adenine just as thymine does, but it occasionally mispairs with guanine (**Figure 13.17b**), leading to a transition (T·A → 5BU·A → 5BU·G → C·G), as shown in **Figure 13.18**. Through mispairing, 5-bromouracil can also be

incorporated into a newly synthesized DNA strand opposite guanine. In the next round of replication 5-bromouracil pairs with adenine, leading to another transition (G·C → G·5BU → A·5BU → A·T).

Another mutagenic chemical is 2-aminopurine (2AP), which is a base analog of adenine. Normally, 2-aminopurine base pairs with thymine, but it may mispair with cytosine, causing a transition mutation (T·A → T·2AP → C·2AP → C·G). Alternatively, 2-aminopurine may be incorporated through mispairing into the newly synthesized DNA opposite cytosine and then later pair with thymine, leading to a C·G → C·2AP → T·2AP → T·A transition.

Thus, both 5-bromouracil and 2-aminopurine can produce transition mutations. In the laboratory, mutations caused by base analogs can be reversed by treatment with the same analog or by treatment with a different analog.

Alkylating agents   Alkylating agents are chemicals that donate alkyl groups, such as methyl (CH_3) and ethyl ($CH_3–CH_2$) groups, to nucleotide bases. For example, ethylmethylsulfonate (EMS) adds an ethyl group to guanine,

(a)

NH_2

C — Cytosine

Deamination

NH_2

U — Uracil

(b)

NH_2

H_3C

C — 5-Methylcytosine (5mC)

Deamination

NH_2

H_3C

T — Thymine

13.16 Deamination alters DNA bases.

(a) Normal base — Thymine; Base analog — 5-Bromouracil

(b) Normal pairing — 5-Bromouracil / Adenine; Mispairing — 5-Bromouracil (ionized) / Guanine

13.17 5-Bromouracil (a base analog) resembles thymine, except that it has a bromine atom in place of a methyl group on the 5-carbon atom. Because of the similarity in their structures, 5-bromouracil may be incorporated into DNA in place of thymine. Like thymine, 5-bromouracil normally pairs with adenine but, when ionized, it may pair with guanine through wobble.

producing O^6-ethylguanine, which pairs with thymine (**Figure 13.19a**). Thus, EMS produces C·G → T·A transitions. Ethylmethylsulfonate is also capable of adding an ethyl group to thymine, producing 4-ethylthymine, which then pairs with guanine, leading to a T·A → C·G transition. Because EMS produces both C·G → T·A and T·A → C·G transitions, mutations produced by EMS can be reversed by additional treatment with EMS. Mustard gas is another alkylating agent.

Deamination   In addition to its spontaneous occurrence (see Figure 13.16), deamination can be induced by some chemicals. For instance, nitrous acid deaminates cytosine, creating uracil, which in the next round of replication pairs with adenine (**Figure 13.19b**), producing a C·G → T·A transition mutation. Nitrous acid changes adenine into hypoxanthine, which pairs with cytosine, leading to a T·A → C·G transition. Nitrous acid also deaminates guanine, producing xanthine, which pairs with cytosine just as guanine does; however, xanthine can also pair with thymine, leading to a C·G → T·A transition. Nitrous acid produces exclusively transition mutations and, because both C·G →

T·A and T·A → C·G transitions are produced, these mutations can be reversed with nitrous acid.

Hydroxylamine   Hydroxylamine is a very specific base-modifying mutagen that adds a hydroxyl group to cytosine, converting it into hydroxylaminocytosine (**Figure 13.19c**). This conversion increases the frequency of a rare tautomer that pairs with adenine instead of guanine and leads to C·G → T·A transitions. Because hydroxylamine acts only on cytosine, it will not generate T·A → C·G transitions; thus, hydroxylamine will not reverse the mutations that it produces.

Intercalating agents   Proflavin, acridine orange, ethidium bromide, and dioxin are **intercalating agents** (**Figure 13.20a**), which produce mutations by sandwiching themselves (intercalating) between adjacent bases in DNA, distorting the three-dimensional structure of the helix and causing single-nucleotide insertions and deletions in replication (**Figure 13.20b**). These insertions and deletions frequently produce frameshift mutations, and so the mutagenic effects of intercalating agents are often severe. Because

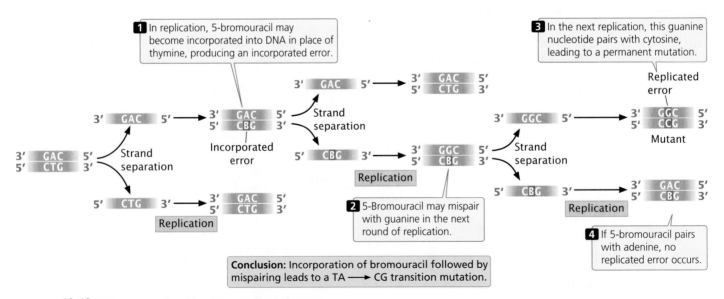

1 In replication, 5-bromouracil may become incorporated into DNA in place of thymine, producing an incorporated error.

2 5-Bromouracil may mispair with guanine in the next round of replication.

3 In the next replication, this guanine nucleotide pairs with cytosine, leading to a permanent mutation.

4 If 5-bromouracil pairs with adenine, no replicated error occurs.

Conclusion: Incorporation of bromouracil followed by mispairing leads to a TA ⟶ CG transition mutation.

13.18 5-Bromouracil can lead to a replicated error.

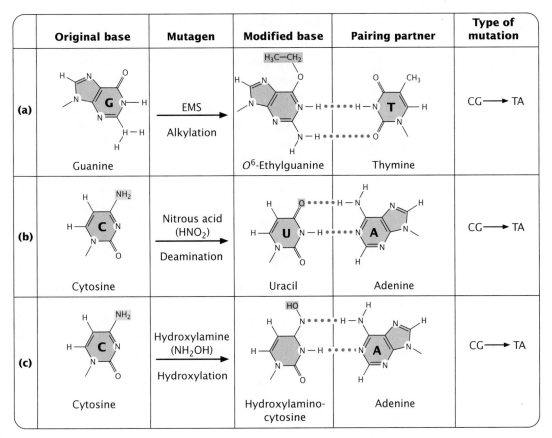

	Original base	Mutagen	Modified base	Pairing partner	Type of mutation
(a)	Guanine	EMS Alkylation	O^6-Ethylguanine	Thymine	CG → TA
(b)	Cytosine	Nitrous acid (HNO_2) Deamination	Uracil	Adenine	CG → TA
(c)	Cytosine	Hydroxylamine (NH_2OH) Hydroxylation	Hydroxylamino-cytosine	Adenine	CG → TA

13.19 Chemicals can alter DNA bases.

intercalating agents generate both additions and deletions, they can reverse the effects of their own mutations.

Concepts

Chemicals can produce mutations by a number of mechanisms. Base analogs are inserted into DNA and frequently pair with the wrong base. Alkylating agents, deaminating chemicals, hydroxylamine, and other chemicals change the structure of DNA bases, thereby altering their pairing properties. Intercalating agents wedge between the bases and cause single-base insertions and deletions in replication.

✔ Concept Check 5

Base analogs are mutagenic because of which characteristic?

a. They produce changes in DNA polymerase that cause it to malfunction.
b. They distort the structure of DNA.
c. They are similar in structure to the normal bases.
d. They chemically modify the normal bases.

Radiation

In 1927, Hermann Muller demonstrated that mutations in fruit flies could be induced by X-rays. The results of subsequent studies showed that X-rays greatly increase mutation rates in all organisms. Because of their high energies, X-rays, gamma rays, and cosmic rays are all capable of penetrating tissues and damaging DNA. These forms of radiation, called ionizing radiation, dislodge electrons from the atoms that they encounter, changing stable molecules into free radicals and reactive ions, which then alter the structures of bases and break phosphodiester bonds in DNA. Ionizing radiation also frequently results in double-strand breaks in DNA. Attempts to repair these breaks can produce chromosome mutations (discussed in Chapter 7).

13.20 Intercalating agents such as proflavin and acridine orange insert themselves between adjacent bases in DNA, distorting the three-dimensional structure of the helix and causing single-nucleotide insertions and deletions in replication.

(a)

(b)

13.21 Pyrimidine dimers result from ultraviolet light.
(a) Formation of thymine dimer. (b) Distorted DNA.

Ultraviolet (UV) light has less energy than that of ionizing radiation and does not eject electrons and cause ionization but is nevertheless highly mutagenic. Purine and pyrimidine bases readily absorb UV light, resulting in the formation of chemical bonds between adjacent pyrimidine molecules on the same strand of DNA and in the creation of **pyrimidine dimers** (**Figure 13.21a**). Pyrimidine dimers consisting of two thymine bases (called thymine dimers) are most frequent, but cytosine dimers and thymine–cytosine dimers also can form. Dimers distort the configuration of DNA (**Figure 13.21b**) and often block replication. Most pyrimidine dimers are immediately repaired by mechanisms discussed later in this chapter, but some escape repair and inhibit replication and transcription.

When pyrimidine dimers block replication, cell division is inhibited and the cell usually dies; for this reason, UV light kills bacteria and is an effective sterilizing agent. For a mutation to occur, the replication block must be overcome. Bacteria can sometimes circumvent replication blocks produced by pyrimidine dimers and other types of DNA damage by means of the **SOS system**. This system allows replication blocks to be overcome but, in the process, makes numerous mistakes and greatly increases the rate of mutation. Indeed, the very reason that replication can proceed in the presence of a block is that the enzymes in the SOS system do not strictly adhere to the base-pairing rules. The trade-off is that replication may continue and the cell survives, but only by sacrificing the normal accuracy of DNA synthesis.

Concepts

Ionizing radiation such as X-rays and gamma rays damage DNA by dislodging electrons from atoms; these electrons then break phosphodiester bonds and alter the structure of bases. Ultraviolet light causes mutations primarily by producing pyrimidine dimers that disrupt replication and transcription. The SOS system enables bacteria to overcome replication blocks but introduces mistakes in replication.

Detecting Mutations with the Ames Test

People in industrial societies are surrounded by a multitude of artificially produced chemicals: more than 50,000 different chemicals are in commercial and industrial use today, and from 500 to 1000 new chemicals are introduced each year. Some of these chemicals are potential carcinogens and may cause harm to humans. One method for testing the cancer-causing potential of chemicals is to administer them to laboratory animals (rats or mice) and compare the incidence of cancer in the treated animals with that of control animals. These tests are unfortunately time consuming and expensive. Furthermore, the ability of a substance to cause cancer in rodents is not always indicative of its effect on humans. After all, we aren't rats!

In 1974, Bruce Ames developed a simple test for evaluating the potential of chemicals to cause cancer. The **Ames test** is based on the principle that both cancer and mutations result from damage to DNA, and the results of experiments have demonstrated that 90% of known carcinogens are also mutagens. Ames proposed that mutagenesis in bacteria could serve as an indicator of carcinogenesis in humans.

The Ames test uses four auxotrophic strains of the bacterium *Salmonella typhimurium* that have defects in the lipopolysaccharide coat, which normally protects the bacteria from chemicals in the environment. Furthermore, the DNA-repair system in these strains has been inactivated, enhancing their susceptibility to mutagens.

One of the four auxotrophic strains used in the Ames test detects base-pair substitutions; the other three detect different types of frameshift mutations. Each strain carries a *his⁻* mutation, which renders it unable to synthesize the amino acid histidine, and the bacteria are plated onto medium that lacks histidine (**Figure 13.22**). Only bacteria that have undergone a reverse mutation of the histidine gene (*his⁻ → his⁺*) are able to synthesize histidine and grow on the medium. Different dilutions of a chemical to be tested are added to plates inoculated with the bacteria, and the number of mutant bacterial colonies that appear on each plate is compared with the number that appear on control plates with no chemical (i.e., that arose through spontaneous mutation). Any chemical that significantly increases the number of colonies appearing on a treated plate is mutagenic and is probably also carcinogenic.

Some compounds are not active carcinogens but can be converted into cancer-causing compounds in the body. To make the Ames test sensitive for such *potential* carcinogens, a compound to be tested is first incubated in mammalian liver extract that contains metabolic enzymes.

The Ames test has been applied to thousands of chemicals and commercial products. An early demonstration of its usefulness was the discovery, in 1975, that many hair dyes sold in the United States contained compounds that were mutagenic to bacteria. These compounds were then removed from most hair dyes.

Question: How can chemicals be quickly screened for their ability to cause cancer?

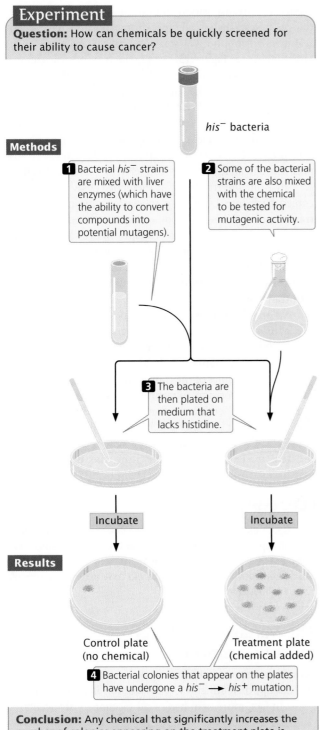

his⁻ bacteria

Methods

1 Bacterial *his⁻* strains are mixed with liver enzymes (which have the ability to convert compounds into potential mutagens).

2 Some of the bacterial strains are also mixed with the chemical to be tested for mutagenic activity.

3 The bacteria are then plated on medium that lacks histidine.

Incubate   Incubate

Results

Control plate (no chemical)   Treatment plate (chemical added)

4 Bacterial colonies that appear on the plates have undergone a *his⁻* → *his⁺* mutation.

Conclusion: Any chemical that significantly increases the number of colonies appearing on the treatment plate is mutagenic and therefore probably also carcinogenic.

13.22 The Ames test is used to identify chemical mutagens.

Concepts

The Ames test uses *his⁻* strains of bacteria to test chemicals for their ability to produce *his⁻* → *his⁺* mutations. Because mutagenic activity and carcinogenic potential are closely correlated, the Ames test is widely used to screen chemicals for their cancer-causing potential.

13.3 Transposable Elements Are Mobile DNA Sequences Capable of Inducing Mutations

Transposable elements are DNA sequences capable of moving and are found in the genomes of all organisms. In many genomes, they are quite abundant: for example, they make up at least 45% of human DNA. Most transposable elements are able to insert at many different locations, relying on mechanisms that are distinct from homologous recombination. They often cause mutations, either by inserting into another gene and disrupting it or by promoting DNA rearrangements such as deletions, duplications, and inversions (see Chapter 7).

General Characteristics of Transposable Elements

There are many different types of transposable elements: some have simple structures, encompassing only those sequences necessary for their own transposition (movement), whereas others have complex structures and encode a number of functions not directly related to transposition. Despite this variation, many transposable elements have certain features in common.

Short **flanking direct repeats** from 3 to 12 bp long are present on both sides of most transposable elements. They are not a part of a transposable element and do not travel with it. Rather, they are generated in the process of transposition, at the point of insertion. The sequences of these repeats vary, but the length is constant for each type of transposable element. The presence of flanking direct repeats indicates that staggered cuts are made in the target DNA when a transposable element inserts itself, as shown in **Figure 13.23**. The staggered cuts leave short, single-stranded pieces of DNA on either side of the transposable element. Replication of the single-stranded DNA then creates the flanking direct repeats.

At the ends of many, but not all, transposable elements are **terminal inverted repeats**, which are sequences from 9 to 40 bp in length that are inverted complements of one another. For example, the following sequences are inverted repeats:

$$5'-\text{ACAGTTCAG} \ldots \text{CTGAACTGT}-3'$$
$$3'-\text{TGTCAAGTC} \ldots \text{GACTTGACA}-5'$$

On the same strand, the two sequences are not simple inversions, as their name might imply; rather, they are both inverted and complementary. (Notice that the sequence from left to right in the top strand is the same as the sequence from right to left in the bottom strand.) Terminal inverted repeats are recognized by enzymes that catalyze transposition and are required for transposition to take place. **Figure 13.24** summarizes the general characteristics of transposable elements.

1 Staggered cuts are made in the target DNA.

CGTCGATAG
GCAGCTATC

CGTCGAT   AG
GC   AGCTATC

Transposable element

2 A transposable element inserts itself into the DNA.

CGTCGAT   AG
GC   AGCTATC

Gaps filled in by DNA polymerase

3 The staggered cuts leave short, single-stranded pieces of DNA.

CGTCGAT   TCGATAG
GCAGCTA   AGCTATC

Flanking direct repeats

4 Replication of this single-stranded DNA creates the flanking direct repeats.

13.23 Flanking direct repeats are generated when a transposable element inserts into DNA.

Concepts

Transposable elements are mobile DNA sequences that often cause mutations. There are many different types of transposable elements; most generate short flanking direct repeats at the target sites as they insert. Many transposable elements also possess short terminal inverted repeats.

✔ **Concept Check 6**

How are flanking direct repeats created in transposition?

Transposition

Transposition is the movement of a transposable element from one location to another. Several different mechanisms

are used for transposition in both prokaryotic and eukaryotic cells. Nevertheless, all types of transposition have several features in common: (1) staggered breaks are made in the target DNA (see Figure 13.23); (2) the transposable element is joined to single-stranded ends of the target DNA; and (3) DNA is replicated at the single-strand gaps.

Some transposable elements transpose as DNA (instead of being first copied into RNA, as retrotransposons are) and are referred to as **DNA transposons** (also called Class I transposable elements). Other transposable elements transpose through an RNA intermediate. In this case, RNA is transcribed from the transposable element (DNA) and is then copied back into DNA by a special enzyme called reverse transcriptase. Elements that transpose through an RNA intermediate are called **retrotransposons** (also called Class II transposons). Most transposable elements found in bacteria are DNA transposons. Both DNA transposons and retrotransposons are found in eukaryotes, although retrotransposons are more common.

Among DNA transposons, transposition may be replicative or nonreplicative. In **replicative transposition**, a new copy of the transposable element is introduced at a new site while the old copy remains behind at the original site, and so the number of copies of the transposable element increases as a result of transposition. In **nonreplicative transposition**, the transposable element excises from the old site and inserts at a new site without any increase in the number of its copies. Nonreplicative transposition requires the replication of only the few nucleotides that constitute the direct repeats. Retrotransposons use replicative transposition only.

Concepts

Transposition may take place through DNA or an RNA intermediate. In replicative transposition, a new copy of the transposable element inserts in a new location and the old copy stays behind; in nonreplicative transposition, the old copy excises from the old site and moves to a new site.

The Mutagenic Effects of Transposition

Because transposable elements may insert into other genes and disrupt their function, transposition is generally mutagenic. In fact, more than half of all spontaneously occurring

(a)

Transposable element

TGCAAATCGCA   TGCGATTGCAA
ACGTTTAGCGT   ACGCTAACGTT

Terminal inverted repeat

Flanking direct repeat

(b)

Transposable element

Terminal inverted repeat

Flanking direct repeat

13.24 Many transposable elements have common characteristics. (a) Most transposable elements generate flanking direct repeats on each side of the point of insertion into target DNA. Many transposable elements also possess terminal inverted repeats. (b) Representations of direct and indirect repeats.

mutations in *Drosophila* result from the insertion of a transposable element in or near a functional gene.

A number of cases of human genetic disease have been traced to the insertion of a transposable genetic element into a vital gene. Although most mutations resulting from transposition are detrimental, transposition may occasionally activate a gene or change the phenotype of the cell in a beneficial way. For instance, bacterial transposable elements sometimes carry genes that encode antibiotic resistance, and several transposable elements have created mutations that confer insecticide resistance in insects.

A dramatic example of the mutagenic effect of transposable elements is seen in the color of grapes, which come in black, red, and white varieties (**Figure 13.25**). Black and red grapes result from the production of red pigments—called anthocyanins—in the skin and are lacking in white grapes. A mutation in black grapes that turned off the production of anthocyanin pigments produced white grapes. This mutation consisted of the insertion of a 10,422-bp retrotransposon called *Gret1* near a gene that promotes the production of anthocyanins (see Figure 13.25). The *Gret1* retrotransposon apparently disrupted sequences that regulate the gene, effectively shutting down pigment production and producing a white grape with no anthocyanins. Interestingly, red grapes resulted from a second mutation taking place in the white grapes (see Figure 13.25). This mutation (probably due to faulty recombination) removed most but not all of the retrotransposon, switching pigment production back on, though not as intensely as in the original black grapes.

Because transposition entails the exchange of DNA sequences and recombination, it often leads to DNA rearrangements. Homologous recombination between multiple copies of transposons also leads to duplications, deletions, and inversions, as shown in **Figure 13.26**. The *Bar* mutation in *Drosophila* (see Figures 7.6 and 7.7) is a tandem duplication thought to have arisen through homologous recombination between two copies of a transposable element present in different locations on the X chromosome.

Transposable elements in humans   About 45% of the human genome consists of sequences derived from transposable elements, although most of these elements are now inactive and no longer capable of transposing. A comparison of human and chimpanzee genomes suggests that almost 11,000 transposition events have taken place since these two species diverged approximately 6 million years ago.

One of the most common transposable elements in the human genome is *Alu*. Every human cell contains more than 1 million related, but not identical, copies of *Alu* in its chromosomes. *Alu* sequences are similar to the gene that encodes the 7S RNA molecule, which transports newly synthesized proteins across the endoplasmic reticulum. *Alu* sequences create short flanking direct repeats when they insert into DNA and have characteristics that suggest that they have transposed through an RNA intermediate.

The Evolutionary Significance of Transposable Elements

Transposable elements have clearly played an important role in shaping the genomes of many organisms. The large size of many eukaryotic genomes is due primarily to the abundance of transposable elements, particularly retrotransposons. About 50% of all spontaneously occurring mutations in *Drosophila* are due to transposition. Homologous recombination between copies of transposable elements has been an important force in producing gene duplications and other chromosome rearrangements. Furthermore, some transposable elements may carry extra DNA with them when they transpose to a new site, providing the potential to move DNA sequences that regulate genes to new sites, where they may alter the expression of genes.

1 In black grapes, the *VvmybA1* gene regulates the synthesis of anthocyanin pigments.

VvmybA1

Mutation

Gret1 retrotransposon

2 In white grapes, a retrotransposon has inserted near the *VvmybA1* gene, disrupting the synthesis of anthocyanins.

VvmybA1

Mutation

3 In red grapes, a second mutation has removed most of the retrotransposon, but a piece is left behind. Anthocyanin production is partly restored.

VvmybA1

13.25 Red and white color in grapes resulted from the insertion and deletion of a retrotransposon.

> ## Concepts
>
> Increases in copy number of transposable elements have contributed to the large size of many eukaryotic genomes.

13.4 A Number of Pathways Repair Changes in DNA

The integrity of DNA is under constant assault from radiation, chemical mutagens, and spontaneously arising changes. In spite of this onslaught of damaging agents, the rate of mutation

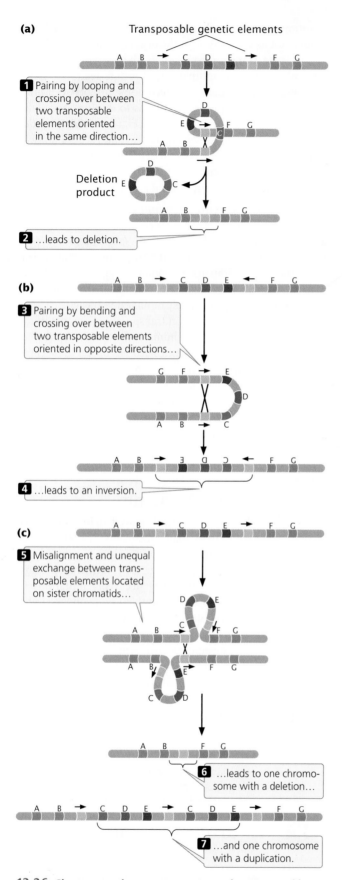

(a) Transposable genetic elements

1 Pairing by looping and crossing over between two transposable elements oriented in the same direction...

Deletion product

2 ...leads to deletion.

(b)

3 Pairing by bending and crossing over between two transposable elements oriented in opposite directions...

4 ...leads to an inversion.

(c)

5 Misalignment and unequal exchange between transposable elements located on sister chromatids...

6 ...leads to one chromosome with a deletion...

7 ...and one chromosome with a duplication.

13.26 Chromosomal rearrangements can be generated by transposition.

remains remarkably low, thanks to the efficiency with which DNA is repaired. Less than one in a thousand DNA lesions is estimated to become a mutation; all the others are corrected.

There are a number of complex pathways for repairing DNA, but several general statements can be made about DNA repair. First, most DNA-repair mechanisms require two nucleotide strands of DNA because most replace whole nucleotides, and a template strand is needed to specify the base sequence.

A second general feature of DNA repair is redundancy, meaning that many types of DNA damage can be corrected by more than one pathway of repair. This redundancy testifies to the extreme importance of DNA repair to the survival of the cell: it ensures that almost all mistakes are corrected. If a mistake escapes one repair system, it's likely to be repaired by another system.

One type of DNA repair is **mismatch repair**, which corrects incorrectly inserted nucleotides that arise in the course of replication (**Figure 13.27**). Incorrectly paired bases distort the three-dimensional structure of DNA, and mismatch-repair enzymes detect these distortions. A complex of mismatch-repair enzymes cuts out the distorted section of the newly synthesized strand and fills the gap with new nucleotides, by using the original DNA strand as a template. The template strand is recognized by the presence of methyl groups on special sequences of the old strand.

Another type of DNA-repair mechanism is **direct repair**, which does not replace altered nucleotides but, instead, changes them back into their original (correct) structures. For example, direct repair corrects O^6-methylguanine, an alkylation product of guanine that pairs with adenine, producing G·C $\rightarrow$ T·A transversions. An enzyme called O^6-methylguanine-DNA methyltransferase removes the methyl group from O^6-methyl-guanine, restoring the base to guanine (**Figure 13.28**).

In **base-excision repair**, a modified base is first excised and then the entire nucleotide is replaced. The excision of modified bases is catalyzed by a set of enzymes called DNA glycosylases, each of which recognizes and removes a specific type of modified base. Uracil glycosylase, for example, recognizes and removes uracil produced by the deamination of cytosine. Other glycosylases recognize hypoxanthine, 3-methyladenine, 7-methylguanine, and other modified bases.

A final repair pathway that we'll consider is **nucleotide-excision repair**, which removes bulky DNA lesions (such as pyrimidine dimers) that distort the double helix. Nucleotide-excision repair is quite versatile and can repair many different types of DNA damage. It is found in cells of all organisms from bacteria to humans and is among the most important of all repair mechanisms.

Concepts

A number of pathways exist for the repair of DNA. Most require two nucleotide strands because a template strand is needed to specify the correct base sequence.

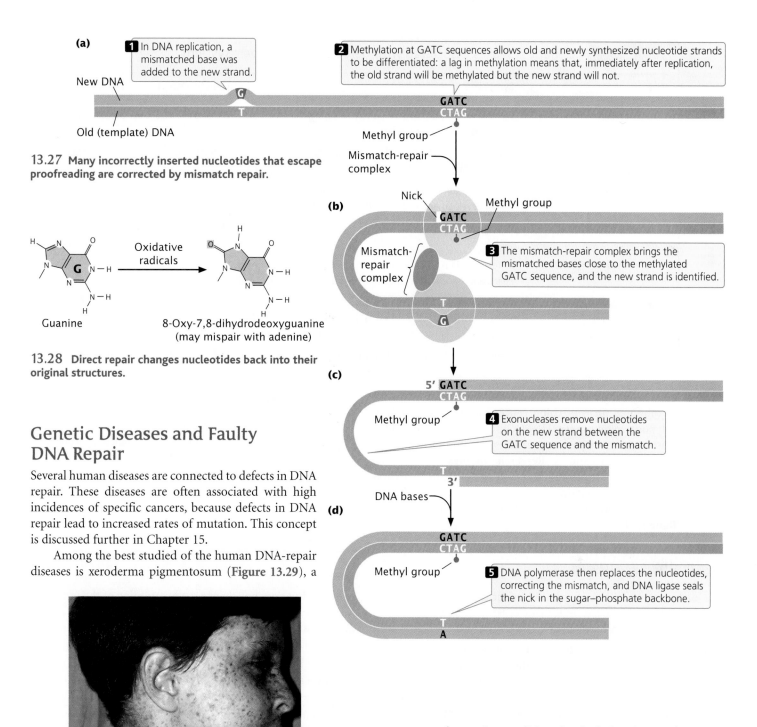

(a)

1 In DNA replication, a mismatched base was added to the new strand.

New DNA

Old (template) DNA

2 Methylation at GATC sequences allows old and newly synthesized nucleotide strands to be differentiated: a lag in methylation means that, immediately after replication, the old strand will be methylated but the new strand will not.

13.27 Many incorrectly inserted nucleotides that escape proofreading are corrected by mismatch repair.

Guanine

8-Oxy-7,8-dihydrodeoxyguanine (may mispair with adenine)

Oxidative radicals

13.28 Direct repair changes nucleotides back into their original structures.

Methyl group

Mismatch-repair complex

(b)

Nick   Methyl group

Mismatch-repair complex

3 The mismatch-repair complex brings the mismatched bases close to the methylated GATC sequence, and the new strand is identified.

(c)

Methyl group

4 Exonucleases remove nucleotides on the new strand between the GATC sequence and the mismatch.

DNA bases

(d)

Methyl group

5 DNA polymerase then replaces the nucleotides, correcting the mismatch, and DNA ligase seals the nick in the sugar–phosphate backbone.

Genetic Diseases and Faulty DNA Repair

Several human diseases are connected to defects in DNA repair. These diseases are often associated with high incidences of specific cancers, because defects in DNA repair lead to increased rates of mutation. This concept is discussed further in Chapter 15.

Among the best studied of the human DNA-repair diseases is xeroderma pigmentosum (**Figure 13.29**), a

13.29 Xeroderma pigmentosum results from defects in DNA repair. The disease is characterized by frecklelike spots on the skin (shown here) and predisposition to skin cancer. *[Ken Greer/Visuals Unlimited.]*

rare autosomal recessive condition that includes abnormal skin pigmentation and acute sensitivity to sunlight. Persons who have this disease also have a strong predisposition to skin cancer, with an incidence ranging from 1000 to 2000 times that found in unaffected people.

Sunlight includes a strong UV component; so exposure to sunlight produces pyrimidine dimers in the DNA of skin cells. Most pyrimidine dimers in humans can be corrected by nucleotide-excision repair. However, the cells of most people with xeroderma pigmentosum are defective in nucleotide-excision repair, and many of their pyrimidine dimers remain uncorrected and may lead to cancer.

Table 13.4	Genetic diseases associated with defects in DNA-repair systems	
Disease	**Symptoms**	**Genetic Defect**
Xeroderma pigmentosum	Frecklelike spots on skin, sensitivity to sunlight, predisposition to skin cancer	Defects in nucleotide-excision repair
Cockayne syndrome	Dwarfism, sensitivity to sunlight, premature aging, deafness, mental retardation	Defects in nucleotide-excision repair
Trichothiodystrophy	Brittle hair, skin abnormalities, short stature, immature sexual development, characteristic facial features	Defects in nucleotide-excision repair
Hereditary nonpolyposis colon cancer	Predisposition to colon cancer	Defects in mismatch repair
Fanconi anemia	Increased skin pigmentation, abnormalities of skeleton, heart, and kidneys, predisposition to leukemia	Possibly defects in the repair of interstrand cross-links
Ataxia telangiectasia	Defective muscle coordination, dilation of blood vessels in skin and eyes, immune deficiencies, sensitivity to ionizing radiation, predisposition to cancer	Defects in DNA-damage detection and response
Li–Fraumeni syndrome	Predisposition to cancer in many different tissues	Defects in DNA-damage response

Another genetic disease caused by faulty DNA repair is an inherited form of colon cancer called hereditary non-polyposis colon cancer (HNPCC). This cancer is one of the most common hereditary cancers, accounting for about 15% of colon cancers. Research findings indicate that HNPCC arises from mutations in the proteins that carry out mismatch repair. Some additional genetic diseases associated with defective DNA repair are summarized in **Table 13.4**.

Concepts

Defects in DNA repair are the underlying cause of several genetic diseases. Many of these diseases are characterized by a predisposition to cancer.

✔ **Concept Check 7**

Why are defects in DNA repair often associated with increases in cancer?

Concepts Summary

- Mutations are heritable changes in genetic information. Somatic mutations occur in somatic cells; germ-line mutations occur in cells that give rise to gametes.

- The simplest type of mutation is a base substitution, a change in a single base pair of DNA. Transitions are base substitutions in which purines are replaced by purines or pyrimidines are replaced by pyrimidines. Transversions are base substitutions in which a purine replaces a pyrimidine or a pyrimidine replaces a purine.

- Insertions are the addition of nucleotides, and deletions are the removal of nucleotides; these mutations often change the reading frame of the gene.

- Expanding trinucleotide repeats are mutations in which the number of copies of a trinucleotide increases through time; they are responsible for several human genetic diseases.

- A missense mutation alters the coding sequence so that one amino acid substitutes for another. A nonsense mutation changes a codon that specifies an amino acid into a termination codon. A silent mutation produces a synonymous codon that specifies the same amino acid as does the original sequence, whereas a neutral mutation alters the amino acid

sequence but does not change the functioning of the protein. A suppressor mutation reverses the effect of a mutation at a different site and may be intragenic (within the same gene as the original mutation) or intergenic (within a different gene).

- Mutation rate is the frequency with which a particular mutation arises in a population. Mutation rates are influenced by both genetic and environmental factors.

- Some mutations occur spontaneously. These mutations include the mispairing of bases in replication and spontaneous depurination and deamination.

- Insertions and deletions can arise from strand slippage in replication or from unequal crossing over.

- Base analogs can become incorporated into DNA in the course of replication and pair with the wrong base in subsequent replication events. Alkylating agents and hydroxylamine modify the chemical structure of bases and lead to mutations. Intercalating agents insert into the DNA molecule and cause single-nucleotide additions and deletions.

- Ionizing radiation is mutagenic, altering base structures and breaking phosphodiester bonds. Ultraviolet light produces pyrimidine dimers, which block replication.

- The Ames tests uses bacteria to assess the mutagenic potential of chemical substances.

- Transposable elements are mobile DNA sequences that insert into many locations within a genome and often cause mutations and DNA rearrangements.

- Most transposable elements have two common characteristics: terminal inverted repeats and the generation of short direct repeats in DNA at the point of insertion.

- Transposition may take place through a DNA molecule or through the production of an RNA molecule that is then reverse transcribed into DNA. Transposition may be replicative, in which the transposable element is copied and the copy moves to a new site, or nonreplicative, in which the transposable element excises from the old site and moves to a new site.

- Transposons are mutagenic and have played an important role in genome evolution.

- Damage to DNA is often corrected by DNA-repair mechanisms. Most repair pathways require two strands of DNA and exhibit some overlap in the types of damage repaired.

- Defects in DNA repair are the underlying cause of several genetic diseases.

Important Terms

mutation (p. 322)
somatic mutation (p. 322)
germ-line mutation (p. 322)
gene mutation (p. 323)
base substitution (p. 323)
transition (p. 323)
transversion (p. 323)
insertion (p. 323)
deletion (p. 323)
frameshift mutation (p. 323)
in-frame insertion (p. 324)
in-frame deletion (p. 324)
expanding trinucleotide repeat (p. 324)
forward mutation (p. 325)
reverse mutation (reversion) (p. 325)
missense mutation (p. 326)
nonsense mutation (p. 326)
silent mutation (p. 326)

neutral mutation (p. 326)
loss-of-function mutation (p. 326)
gain-of-function mutation (p. 326)
conditional mutation (p. 326)
lethal mutation (p. 326)
suppressor mutation (p. 326)
intragenic suppressor mutation (p. 327)
intergenic suppressor mutation (p. 327)
mutation rate (p. 328)
spontaneous mutation (p. 329)
induced mutation (p. 329)
incorporated error (p. 330)
replicated error (p. 330)
strand slippage (p. 330)
unequal crossing over (p. 331)
depurination (p. 332)
deamination (p. 332)

mutagen (p. 333)
base analog (p. 333)
intercalating agent (p. 334)
pyrimidine dimer (p. 336)
SOS system (p. 336)
Ames test (p. 336)
flanking direct repeat (p. 337)
terminal inverted repeat (p. 337)
transposition (p. 338)
DNA transposon (p. 338)
retrotransposon (p. 338)
replicative transposition (p. 338)
nonreplicative transposition (p. 338)
mismatch repair (p. 340)
direct repair (p. 340)
base-excision repair (p. 340)
nucleotide-excision repair (p. 340)

Answers to Concept Checks

1. Studying mutations that disrupt normal processes often leads to the identification of genes that normally play a role in the process and can help in understanding the molecular details of a process.

2. c

3. A reverse mutation restores the original phenotype by changing the DNA sequence back to the wild type. A suppressor mutation restores the phenotype by causing an additional change in the DNA at a site that is different from that of the original mutation.

4. An incorporated error is due to a change that takes place in DNA. This change may be corrected by a DNA-repair pathway.

However, if the error has been replicated, it is permanent and cannot be detected by repair pathways.

5. c

6. In transposition, staggered cuts are made in DNA and the transposable element inserts into the cut. Later, replication of the single-stranded pieces of DNA creates short flanking repeats on either side of the inserted transposable element.

7. Changes in DNA structure do not undergo repair in people with defects in DNA-repair mechanisms. Consequently, increased numbers of mutations occur at all genes, including those that predispose to cancer. This observation indicates that cancer arises from mutations in DNA.

Worked Problem

1. A codon that specifies the amino acid Asp undergoes a single-base substitution that yields a codon that specifies Ala. Refer to the genetic code in Figure 11.5 and give all possible DNA sequences for the original and the mutated codon. Is the mutation a transition or a transversion?

• Solution

There are two possible RNA codons for Asp: GAU and GAC. The DNA sequences that encode these codons will be complementary to the RNA codons: CTA and CTG. There are four possible RNA codons for Ala: GCU, GCC, GCA, and GCG, which correspond to DNA sequences CGA, CGG, CGT, and CGC. If we organize the original and the mutated sequences as shown in the following table, the types of mutations that may have occurred can be easily seen:

Possible original sequence for Asp	Possible mutated sequence for Ala
CTA	CGA
CTG	CGG
	CGT
	CGC

If the mutation is confined to a single-base substitution, then the only mutations possible are that CTA mutated to CGA or that CTG mutated to CGG. In both, there is a T → G transversion in the middle nucleotide of the codon.

Comprehension Questions

Section 13.1

*1. What is the difference between a transition and a transversion? Which type of base substitution is usually more common?

*2. Briefly describe expanding trinucleotide repeats.

3. What is the difference between a missense mutation and a nonsense mutation? A silent mutation and a neutral mutation?

4. Briefly describe two different ways that intragenic suppressors may reverse the effects of mutations.

Section 13.2

*5. What causes errors in DNA replication?

6. How do insertions and deletions arise?

*7. How do base analogs lead to mutations?

8. What types of mutations are produced by ionizing and UV radiation?

9. What is the purpose of the Ames test? How are *his⁻* bacteria used in this test?

Section 13.3

*10. What general characteristics are found in many transposable elements? Describe the differences between replicative and nonreplicative transposition.

*11. What is a retrotransposon and how does it move?

Section 13.4

*12. List at least three different types of DNA repair and briefly explain how each is carried out.

Application Questions and Problems

Section 13.1

*13. A codon that specifies the amino acid Gly undergoes a single-base substitution to become a nonsense mutation. In accord with the genetic code given in Figure 11.5, is this mutation a transition or a transversion? At which position of the codon does the mutation occur?

*14. a. If a single transition occurs in a codon that specifies Phe, what amino acids can be specified by the mutated sequence?

b. If a single transversion occurs in a codon that specifies Phe, what amino acids can be specified by the mutated sequence?

c. If a single transition occurs in a codon that specifies Leu, what amino acids can be specified by the mutated sequence?

d. If a single transversion occurs in a codon that specifies Leu, what amino acids can be specified by the mutated sequence?

15. Hemoglobin is a complex protein that contains four polypeptide chains. The normal hemoglobin found in adults—called adult hemoglobin—consists of two alpha and two beta polypeptide chains, which are encoded by different loci. Sickle-cell hemoglobin, which causes sickle-cell anemia, arises from a mutation in the beta chain of adult hemoglobin. Adult hemoglobin and sickle-cell hemoglobin differ in a single amino acid: the sixth amino acid from one end in adult hemoglobin is glutamic acid, whereas sickle-cell hemoglobin has valine at this position. After consulting the genetic code provided in Figure 11.5, indicate the type and location of the mutation that gave rise to sickle-cell anemia.

*16. The following nucleotide sequence is found on the template strand of DNA. First, determine the amino acids of the protein encoded by this sequence by using the genetic code provided in Figure 11.5. Then, give the altered amino acid sequence of the protein that will be found in each of the following mutations:

Sequence of
DNA template

 3′–TAC TGG CCG TTA GTT GAT ATA ACT–5′

 1   24

Nucleotide
number

a. Mutant 1:   A transition at nucleotide 11

b. Mutant 2:   A transition at nucleotide 13

c. Mutant 3:   A one-nucleotide deletion at nucleotide 7

d. Mutant 4:   A T → A transversion at nucleotide 15

e. Mutant 5:   An addition of TGG after nucleotide 6

f. Mutant 6:   A transition at nucleotide 9

17. A polypeptide has the following amino acid sequence:

Met-Ser-Pro-Arg-Leu-Glu-Gly

The amino acid sequence of this polypeptide was determined in a series of mutants listed in parts *a* through *e*. For each mutant, indicate the type of mutation that occurred in the DNA (single-base substitution, insertion, deletion) and the phenotypic effect of the mutation (nonsense mutation, missense mutation, frameshift, etc.).

a. Mutant 1:   Met-Ser-Ser-Arg-Leu-Glu-Gly

b. Mutant 2:   Met-Ser-Pro

c. Mutant 3:   Met-Ser-Pro-Asp-Trp-Arg-Asp-Lys

d. Mutant 4:   Met-Ser-Pro-Glu-Gly

e. Mutant 5:   Met-Ser-Pro-Arg-Leu-Leu-Glu-Gly

*18. A gene encodes a protein with the following amino acid sequence:

Met-Trp-His-Arg-Ala-Ser-Phe

A mutation occurs in the gene. The mutant protein has the following amino acid sequence:

Met-Trp-His-Ser-Ala-Ser-Phe

An intragenic suppressor restores the amino acid sequence to that of the original protein:

Met-Trp-His-Arg-Ala-Ser-Phe

Give at least one example of base changes that could produce the original mutation and the intragenic suppressor. (Consult the genetic code in Figure 11.5.)

19. XG syndrome is a rare genetic disease that is due to an autosomal dominant gene. A complete census of a small European country reveals that 77,536 babies were born in 2004, of whom 3 had XG syndrome. In the same year, this country had a population of 5,964,321 people, and there were 35 living persons with XG syndrome. What is the mutation rate of XG syndrome in this country?

Section 13.2

*20. The following nucleotide sequence is found in a short stretch of DNA:

5′–ATGT–3′
3′–TACA–5′

If this sequence is treated with hydroxylamine, what sequences will result after replication?

21. The following nucleotide sequence is found in a short stretch of DNA:

5′–AG–3′
3′–TC–5′

a. Give all the mutant sequences that may result from spontaneous depurination in this stretch of DNA.

b. Give all the mutant sequences that may result from spontaneous deamination in this stretch of DNA.

22. Mary Alexander studied the effects of radiation on mutation rates in the sperm of *Drosophila melanogaster*. She irradiated

Drosophila larvae with either 3000 roentgens (r) or 3975 r, collected the adult males that developed from irradiated larvae, mated them with unirradiated females, and then counted the number of mutant F_1 flies produced by each male. All mutant flies that appeared were used in subsequent crosses to determine if their mutant phenotypes were genetic. She obtained the following results (M. L. Alexander. 1954. *Genetics* 39:409–428):

Group	Number of offspring	Offspring with a genetic mutation
Control (0 r)	45,504	0
Irradiated (3000 r)	49,512	71
Irradiated (3975 r)	50,159	70

a. Calculate the mutation rates of the control group and the two groups of irradiated flies.

b. On the basis of these data, do you think radiation has any effect on mutation? Explain your answer.

Section 13.3

*23. A particular transposable element generates flanking direct repeats that are 4 bp long. Give the sequence that will be found on both sides of the transposable element if this transposable element inserts at the position indicated on each of the following sequences.

a.   Transposable element

5′–ATTCGAACTGACCGATCA–3′

b.   Transposable element

5′–ATTCGAACTGACCGATCA–3′

*24. What factor do you think determines the length of the flanking direct repeats that are produced in transposition?

25. Zidovudine (AZT) is a drug used to treat patients with AIDS. AZT works by blocking the reverse-transcriptase enzyme used by the human immunodeficiency virus (HIV), the causative agent of AIDS. Do you expect that AZT would have any effect on transposable elements? If so, what type of transposable elements would be affected and what would be the most likely effect?

26. A transposable element is found to encode a reverse-transcriptase enzyme. On the basis of this information, what conclusions can you make about the likely method of transposition of this element?

Section 13.4

*27. A plant breeder wants to isolate mutants in tomatoes that are defective in DNA repair. However, this breeder does not have the expertise or equipment to study enzymes in DNA-repair systems. How can the breeder identify tomato plants that are deficient in DNA repair? What are the traits to look for?

Challenge Questions

Section 13.1

28. Robert Bost and Richard Cribbs studied a strain of *E. coli* (*araB14*) that possessed a nonsense mutation in the structural gene that encodes L-ribulokinase, an enzyme that allows the bacteria to metabolize the sugar arabinose (R. Bost and R. Cribbs. 1969. *Genetics* 62:1–8). From the *araB14* strain, they isolated some bacteria that possessed mutations that caused them to revert back to the wild type. Genetic analysis of these revertants showed that they possessed two different suppressor mutations. One suppressor mutation (*R1*) was linked to the original mutation in the L-ribulokinase and probably occurred at the same locus. By itself, this mutation allowed the production of L-ribulokinase, but the enzyme was not as effective in metabolizing arabinose as the enzyme encoded by the wild-type allele. The second suppressor mutation (Su^B) was not linked to the original mutation. In conjunction with the *R1* mutation, Su^B allowed the production of L-ribulokinase, but Su^B by itself was not able to suppress the original mutation.

a. On the basis of this information, are the *R1* and Su^B mutations intragenic suppressors or intergenic suppressors? Explain your reasoning.

b. Propose an explanation for how *R1* and Su^B restore the ability of *araB14* to metabolize arabinose and why Su^B is able to more fully restore the ability to metabolize arabinose.

29. Achondroplasia is an autosomal dominant disorder characterized by disproportionate short stature—the legs and arms are short compared with the head and trunk. The disorder is due to a base substitution in the gene, located on the short arm of chromosome 4, for fibroblast-growth-factor receptor 3 (FGFR3).

Although achondroplasia is clearly inherited as an autosomal dominant trait, more than 80% of the people who have achondroplasia are born to parents with normal stature. This high percentage indicates that most cases are caused by newly arising mutations; these cases (not inherited from an affected parent) are referred to as sporadic. Findings from molecular studies have demonstrated that sporadic cases of achondroplasia are almost always caused by mutations inherited from the father (paternal mutations). In addition, the occurrence of achondroplasia is higher among older fathers; indeed, approximately 50% of children with achondroplasia are born to fathers older than 35 years of age. There is no association with maternal age. The mutation rate for achondroplasia (about 4×10^{-5} mutations per gamete) is high compared with those for other genetic disorders. Explain why most spontaneous mutations for achondroplasia are paternal in origin and why the occurrence of achondroplasia is higher among older fathers.

A family of three who have achondroplasia. [*Gail Burton/AP.*]

30. Mutations *ochre* and *amber* are two types of nonsense mutations. Before the genetic code was worked out, Sydney Brenner, Anthony O. Stretton, and Samuel Kaplan applied different types of mutagens to bacteriophages in an attempt to determine the bases present in the codons responsible for *amber* and *ochre* mutations. They knew that *ochre* and *amber* mutants were suppressed by different types of mutations, demonstrating that each is a different termination codon. They obtained the following results:

1. A single-base substitution could convert an *ochre* mutation into an *amber* mutation.

2. Hydroxylamine induced both *ochre* and *amber* mutations in wild-type phages.

3. 2-Aminopurine caused *ochre* to mutate to *amber*.

4. Hydroxylamine did not cause *ochre* to mutate to *amber*.

These data do not allow the complete nucleotide sequence of the *amber* and *ochre* codons to be worked out, but they do provide some information about the bases found in the nonsense mutations.

a. What conclusions about the bases found in the codons of *amber* and *ochre* mutations can be made from these observations?

b. Of the three nonsense codons (UAA, UAG, UGA), which represents the *ochre* mutation?

Section 13.2

31. To determine whether radiation associated with the atomic bombings of Hiroshima and Nagasaki produced recessive germ-line mutations, scientists examined the sex ratio of the children of the survivors of the blasts. Can you explain why an increase in germ-line mutations might be expected to alter the sex ratio?

14 Molecular Genetic Analysis, Biotechnology, and Genomics

Genetic engineering is being used to modify rice and other crops to grow in environments that are currently unable to support agriculture. [Friedrich Stark/Peter Arnold.]

Feeding the Future Population of the World

In the year 2000, the world's population reached 6 billion. Because the human population has exhibited exponential growth, the pace of increase in the number of people is ever quickening: more than 100,000 years were required for humans to reach 1 billion in number (in 1830); only another 100 years were required for the population to double to 2 billion (in 1930); and only 45 years were required for it to double again (in 1975) to 4 billion. The United Nations projects that the world population will reach somewhere between 7.3 billion and 10.7 billion by the year 2050 and, because the tendency is for people to have smaller families, will eventually level off or even drop in the last part of the twenty-first century.

How will we feed the additional billions of people that will populate the planet Earth 40 years from now? So far, we have been able to sustain the tremendous increase in human numbers because advances in agriculture have greatly increased worldwide food production. Much of this increase was between 1950 and 1980 through the Green Revolution, which utilized traditional techniques of plant breeding and genetics to develop new varieties of corn, wheat, and rice. For example, worldwide grain production increased 260% between 1950 and 1990; worldwide cereal production increased from 275 kg/person in the 1950s to 370 kg/person in the 1980s, during a time in which human population almost doubled. Thus, even though human numbers have increased tremendously in the past 60 years, the world's farmers today produce more food per person than they did in 1950.

What about the next 40 years, when there will be between 1 billion and 5 billion more people to feed? Most of the world's cultivatable land is already in use, and increases in crop yield achievable through traditional breeding and genetics have leveled off. Many experts propose that feeding the future population of the world can be achieved only through the application of genetic engineering to bring about a "second" Green Revolution. Already, genetic engineering has been used to produce crops that are resistant to pests, disease, and herbicides. Genetically engineered (often called genetically modified) crops are today cultivated on more than 125 million hectares (1 hectare = 2.471 acres) of land worldwide; in 2008, 80% of corn, 92% of soybeans, and 86% of cotton grown in the United States was genetically engineered.

The potential of genetic engineering to help feed the future world population must be weighed against concerns about the widespread use of genetically modified crops. Although recent scientific reviews contain little evidence of risk to human health from eating genetically modified foods, many consumers remain wary of eating them. Findings from recent studies in the United Kingdom demonstrated that genetically modified beets and oilseed rape reduce the biodiversity of native plants and insects in agricultural fields, and there are concerns that genetically modified plants may hybridize with native plants and cause ecological disruption.

This chapter introduces some of the techniques being used to create genetically engineered crops and other organisms. We begin by considering molecular genetic technology and some of its effects. We examine a number of methods used to isolate, study, alter, and recombine DNA sequences and place them back into cells. We then explore some of the applications of molecular genetic analysis. The last part of the chapter deals with genomics, the study of whole genomes. We consider genetic and physical maps, methods for sequencing entire genomes, and functional genomics—how genes are identified in genomic sequences and how their functions are defined. Finally, we compare the genomes of different organisms and consider methods for studying the proteins of a cell (proteomics).

14.1 Molecular Techniques Are Used to Isolate, Recombine, and Amplify Genes

Recombinant DNA technology is a set of molecular techniques for locating, isolating, altering, and studying DNA segments. The term *recombinant* is used because, frequently, the goal is to combine DNA from two distinct sources. Genes from two different bacteria might be joined, for example, or a human gene might be inserted into a viral chromosome. Commonly called **genetic engineering**, recombinant DNA technology now encompasses many molecular techniques that can be used to analyze, alter, and recombine virtually any DNA sequences from any number of sources.

The Molecular Genetics Revolution

The techniques of recombinant DNA technology are just a part of a vast array of molecular methods that are now available for the study of genetics. These molecular techniques have drastically altered the way that genes are studied. Previously, information about the structure and organization of genes was gained by examining their phenotypic effects, but molecular genetic analysis allows the nucleotide sequences themselves to be read. Methods in molecular genetics have provided new information about the structure and function of genes

and has altered many fundamental concepts of genetics. Our detailed understanding of genetic processes such as replication, transcription, translation, RNA processing, and gene regulation has been learned through the use of molecular genetic techniques. These techniques are used in many other fields as well, including biochemistry, microbiology, developmental biology, neurobiology, evolution, and ecology.

Recombinant DNA technology and other molecular techniques are also being used to create a number of commercial products, including drugs, hormones, enzymes, and crops (**Figure 14.1**). A complete industry—**biotechnology**—has grown up around the use of these techniques to develop new products. In medicine, molecular genetics is being used to probe the nature of cancer, diagnose genetic and infectious diseases, produce drugs, and treat hereditary disorders.

Concepts

Molecular genetics and recombinant DNA technology are used to locate, analyze, alter, study, and recombine DNA sequences. These techniques are used to probe the structure and function of genes, address questions in many areas of biology, create commercial products, and diagnose and treat diseases.

Working at the Molecular Level

The manipulation of genes at the molecular level presents a serious challenge, often requiring strategies that may not, at first, seem obvious. The basic problem is that genes are minute and every cell contains thousands of them. Individual nucleotides cannot be seen, and no physical features mark the beginning or the end of a gene.

Let's consider a typical situation faced by a molecular geneticist. Suppose we wanted to use bacteria to produce large quantities of a human protein. The first and most formidable problem is to find the gene that encodes the desired protein. A haploid human genome consists of 3.2 billion base pairs of DNA. Let's assume that the gene that we want to isolate is 3000 bp long. Our target gene occupies only one-millionth of the genome; so searching for our gene in the huge expanse of genomic DNA is more difficult than looking for a needle in the proverbial haystack. But, even if we are able to locate the gene, how are we to separate it from the rest of the DNA?

14.1 Recombinant DNA technology has been used to create genetically modified crops. Genetically engineered corn, which produces a toxin that kills insect pests, now constitutes 57% of all corn grown in the United States. [Chris Knapton/Photo Researchers.]

If we did succeed in locating and isolating the desired gene, we would next need to insert it into a bacterial cell. Linear fragments of DNA are quickly degraded by bacteria; so the gene must be inserted in a stable form. It must also be able to successfully replicate or it will not be passed on when the cell divides. If we succeed in transferring our gene to bacteria in a stable form, we must still ensure that the gene is properly transcribed and translated.

Finally, the methods used to isolate and transfer genes are inefficient and, of a million cells that are subjected to these procedures, only *one* cell might successfully take up and express the human gene. So we must search through many bacterial cells to find the one containing the recombinant DNA. We are back to the problem of the needle in the haystack.

Although these problems might seem insurmountable, molecular techniques have been developed to overcome all of them, and human genes are routinely transferred to bacterial cells, where the genes are expressed.

Concepts

Molecular genetic analyses require special methods because individual genes make up a tiny fraction of the cellular DNA and they cannot be seen.

✔ Concept Check 1

Briefly outline the steps required to genetically engineer bacteria that will produce a protein encoded by a human gene.

Cutting and Joining DNA Fragments

A first step in the molecular analysis of a DNA segment or gene is to isolate it from the remainder of the DNA. A key discovery in the development of molecular genetic methods was the discovery in the late 1960s of **restriction enzymes** (also called **restriction endonucleases**) that recognize and make double-stranded cuts in DNA at specific nucleotide sequences. These enzymes are produced naturally by bacteria, where they are used in defense against viruses. A bacterium protects its own DNA from a restriction enzyme by modifying the recognition sequence, usually by adding methyl groups to its DNA.

More than 800 different restriction enzymes that recognize and cut DNA at more than 100 different sequences have been isolated from bacteria. Many of these enzymes are commercially available; examples of some commonly used restriction enzymes are given in **Table 14.1**. The name of each restriction enzyme begins with an abbreviation that signifies its bacterial origin.

The sequences recognized by restriction enzymes are usually from 4 to 8 bp long; most enzymes recognize a sequence of 4 or 6 bp. Most recognition sequences are palindromic—sequences that read the same forward and backward.

Some of the enzymes make staggered cuts in the DNA. For example, *Hin*dIII recognizes the following sequence:

$$\downarrow$$
$$5'–AAGCTT–3'$$
$$3'–TTCGAA–5'$$
$$\uparrow$$

*Hin*dIII cuts the sugar–phosphate backbone of each strand at the point indicated by the arrow, generating fragments with short, single-stranded overhanging ends:

$$5'–A \qquad AGCTT–3'$$
$$3'–TTCGA \qquad A–5'$$

Such ends are called **cohesive ends** or sticky ends, because they are complementary to each other and can spontaneously pair to connect the fragments. Thus, DNA fragments can be "glued" together: any two fragments cleaved by the same enzyme will have complementary ends and will pair (**Figure 14.2** on p. 351). When their cohesive ends have paired, two DNA fragments can be joined together permanently by DNA ligase, which seals nicks between the sugar–phosphate groups of the fragments.

Not all restriction enzymes produce staggered cuts and sticky ends. *Pvu*II cuts in the middle of its recognition site, producing blunt-ended fragments:

$$\downarrow$$
$$5'–CAGCTG–3'$$
$$3'–GTCGAC–5'$$
$$\downarrow$$

$$5'–CAG \qquad CTG–3'$$
$$3'–GTC \qquad GAC–5'$$

Fragments that have blunt ends must be joined together in other ways. One option is to use the enzyme DNA ligase,

Table 14.1   Characteristics of some common restriction enzymes used in recombinant DNA technology

Enzyme	Microorganism from Which Enzyme Is Produced	Recognition Sequence	Type of Fragment End Produced
BamHI	Bacillus amyloliquefaciens	5'–G↓GATCC–3' 3'–CCTAG↑G–5'	Cohesive
CofI	Clostridium formicoaceticum	5'–GCG↓C–3' 3'–C↑GCG–5'	Cohesive
EcoRI	Escherichia coli	5'–G↓AATTC–3' 3'–CTTAA↑G–5'	Cohesive
EcoRII	Escherichia coli	5'–↓CCAGG–3' 3'–GGTCC↑–5'	Cohesive
HaeIII	Haemophilus aegyptius	5'–GG↓CC–3' 3'–CC↑GG–5'	Blunt
HindIII	Haemophilus influenzae	5'–A↓AGCTT–3' 3'–TTCGA↑A–5'	Cohesive
PvuII	Proteus vulgaris	5'–CAG↓CTG–3' 3'–GTC↑GAC–5'	Blunt

Note: The first three letters of the abbreviation for each restriction enzyme refer to the bacterial species from which the enzyme was isolated (e.g., *Eco* refers to *E. coli*). A fourth letter may refer to the strain of bacteria from which the enzyme was isolated (the "R" in *Eco*RI indicates that this enzyme was isolated from the RY13 strain of *E. coli*). Roman numerals that follow the letters allow different enzymes from the same species to be identified.

which can join together any two blunt-ended pieces of DNA. However, because DNA ligase connects *any* blunt-ended DNA fragments, it is nonspecific in its joining and may produce undesired products.

The sequences recognized by a restriction enzyme are located randomly within the genome. Consequently, there is a relation between the length of the recognition sequence and the number of times that it is present in a genome: there will be fewer longer recognition sequences than shorter recognition sequences, because the probability of the occurrence of a particular sequence consisting of, say, six specific bases is less than the probability of the occurrence of a particular sequence of four specific bases. Consequently, restriction enzymes that recognize longer sequences will cut a given piece of DNA into fewer and longer fragments than will restriction enzymes that recognize shorter sequences.

Restriction enzymes are used whenever DNA fragments must be cut or joined. In a typical restriction reaction, a concentrated solution of purified DNA is placed in a small tube with a buffer solution and a small amount of restriction enzyme. Within a few hours, the enzyme cuts all the appropriate restriction sites in the DNA, producing a mixture of DNA fragments.

Concepts

Restriction enzymes cut DNA at specific base sequences that are palindromic. Some restriction enzymes make staggered cuts, producing DNA fragments with cohesive ends; others cut both strands straight across, producing blunt-ended fragments. There are fewer long recognition sequences in DNA than short sequences.

✔ Concept Check 2

Where do restriction enzymes come from?

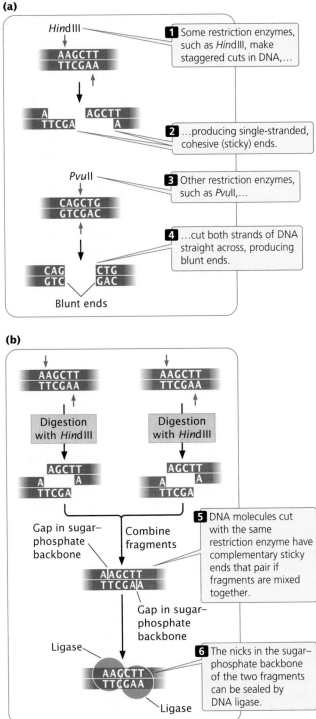

(a)

*Hind*III

AAGCTT
TTCGAA

1 Some restriction enzymes, such as *Hind*III, make staggered cuts in DNA,…

A | AGCTT
TTCGA | A

2 …producing single-stranded, cohesive (sticky) ends.

*Pvu*II

CAGCTG
GTCGAC

3 Other restriction enzymes, such as *Pvu*II,…

4 …cut both strands of DNA straight across, producing blunt ends.

CAG | CTG
GTC | GAC

Blunt ends

(b)

AAGCTT
TTCGAA

AAGCTT
TTCGAA

Digestion with *Hind*III

Digestion with *Hind*III

AGCTT
A   A
TTCGA

AGCTT
A   A
TTCGA

Gap in sugar–phosphate backbone

Combine fragments

5 DNA molecules cut with the same restriction enzyme have complementary sticky ends that pair if fragments are mixed together.

A AGCTT
TTCGA A

Gap in sugar–phosphate backbone

Ligase

6 The nicks in the sugar–phosphate backbone of the two fragments can be sealed by DNA ligase.

AAGCTT
TTCGAA

Ligase

14.2 Restriction enzymes make double-stranded cuts in DNA, producing cohesive, or sticky, ends.

Viewing DNA Fragments

After the completion of a restriction reaction, a number of questions arise. Did the restriction enzyme cut the DNA? Into how many fragments was the DNA cut? What are the

sizes of the resulting fragments? Gel electrophoresis provides us with a means of answering these questions.

Electrophoresis is a standard biochemical technique for separating molecules on the basis of their size and electrical charge. There are a number of different types of electrophoresis; to separate DNA molecules, **gel electrophoresis** is used. A porous gel is often made from agarose (a polysaccharide isolated from seaweed), which is melted in a buffer solution and poured into a plastic mold. As it cools, the agarose solidifies, making a gel that looks something like stiff gelatin.

Small wells are made at one end of the gel to hold solutions of DNA fragments (**Figure 14.3a**), and an electrical current is passed through the gel. Because the phosphate group of each DNA nucleotide carries a negative charge, the DNA fragments migrate toward the positive end of the gel (**Figure 14.3b**). In this migration, the porous gel acts as a sieve, separating the DNA fragments by size. Small DNA fragments migrate more rapidly than do large ones and, with the passage of time, the fragments separate on the basis of their size. Typically, DNA fragments of known length (a marker sample) are placed in another well. By comparing the migration distance of the unknown fragments with the distance traveled by the marker fragments, one can determine the approximate size of the unknown fragments (**Figure 14.3c**).

The DNA fragments are still too small to see; so the problem of visualizing the DNA needs to be addressed. Visualization can be accomplished in several ways. The simplest procedure is to stain the gel with a dye specific for nucleic acids, such as ethidium bromide, which wedges itself tightly (intercalates) between the bases of DNA and fluoresces orange when exposed to UV light, producing brilliant orange bands on the gel (**Figure 14.3d**).

Alternatively, DNA fragments can be visualized by adding a radioactive or chemical label to the DNA before it is placed in the gel. Nucleotides with radioactively labeled phosphate (^{32}P) can be used as the substrate for DNA synthesis and will be incorporated into the newly synthesized DNA strand. Radioactively labeled DNA can be detected with a technique called **autoradiography** in which a piece of X-ray film is placed on top of the gel. Radiation from the labeled DNA exposes the film, just as light exposes photographic film in a camera. The developed autoradiograph gives a picture of the fragments in the gel, with each DNA fragment appearing as a dark band on the film. Chemical labels can be detected by adding antibodies or other substances that carry a dye and will attach to the relevant DNA, which can be visualized directly.

Concepts

DNA fragments can be separated, and their sizes can be determined with the use of gel electrophoresis. The fragments can be viewed by using a dye that is specific for nucleic acids or by labeling the fragments with a radioactive or chemical tag.

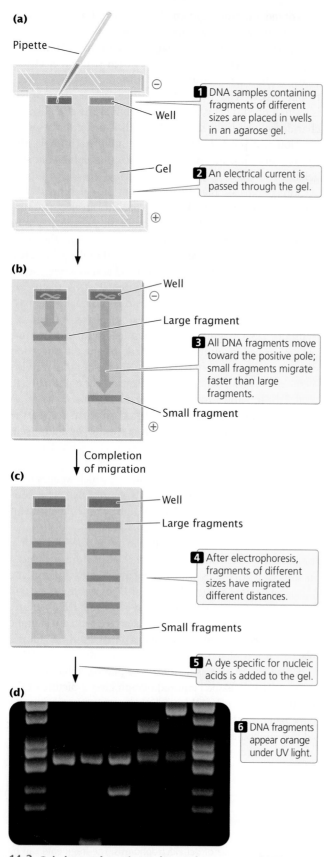

(a)

Pipette

Well

1 DNA samples containing fragments of different sizes are placed in wells in an agarose gel.

Gel

2 An electrical current is passed through the gel.

(b)

Well

Large fragment

3 All DNA fragments move toward the positive pole; small fragments migrate faster than large fragments.

Small fragment

Completion of migration

(c)

Well

Large fragments

4 After electrophoresis, fragments of different sizes have migrated different distances.

Small fragments

5 A dye specific for nucleic acids is added to the gel.

(d)

6 DNA fragments appear orange under UV light.

14.3 Gel electrophoresis can be used to separate DNA molecules on the basis of their size and electrical charge.
[Photograph courtesy of Carol Eng.]

✔ **Concept Check 3**

DNA fragments that are 500 bp, 1000 bp, and 2000 bp in length are separated by gel electrophoresis. Which fragment will migrate farthest in the gel?

a. 2000-bp fragment   c. 500-bp fragment

b. 1000-bp fragment   d. All will migrate equal distances.

If a small piece of DNA, such as a plasmid, is cut by a restriction enzyme, the few fragments produced can be seen as distinct bands on an electrophoretic gel. In contrast, if genomic DNA from a cell is cut by a restriction enzyme, a large number of fragments of different sizes are produced. Usually, one is interested in only a few of these fragments, perhaps those carrying a specific gene. How does one locate the desired fragments in such a large pool of DNA?

One approach is to use a **probe**, which is a DNA or RNA molecule with a base sequence complementary to a sequence in the gene of interest. The bases on a probe will pair only with the bases on a complementary sequence and, if suitably labeled, the probe can be used to locate a specific gene or other DNA sequence.

Cloning Genes

Many recombinant DNA methods require numerous copies of a specific DNA fragment. One way to amplify a specific piece of DNA is to place the fragment in a bacterial cell and allow the cell to replicate the DNA. This procedure is termed **gene cloning**, because identical copies (clones) of the original piece of DNA are produced.

A **cloning vector** is a stable, replicating DNA molecule to which a foreign DNA fragment can be attached for introduction into a cell. An effective cloning vector has three important characteristics (**Figure 14.4**): (1) an origin of replication, which ensures that the vector is replicated within the cell; (2) selectable markers, which enable any cells containing the vector to be selected or identified; and (3) one or more unique restriction sites into which a DNA fragment can be inserted. The restriction sites used for cloning must be unique; if a vector is cut at multiple recognition sites, generating several pieces of DNA, there will be no way to get the pieces back together in the correct order.

Plasmid vectors   Plasmids, circular DNA molecules that exist naturally in bacteria (see Chapter 6), are commonly used vectors for cloning DNA fragments in bacteria. They contain origins of replication and are therefore able to replicate independently of the bacterial chromosome. The plasmids typically used in cloning have been constructed from the larger, naturally occurring bacterial plasmids and have multiple restriction sites, an origin of replication site, and selectable markers.

The easiest method for inserting a gene into a plasmid vector is to cut the foreign DNA (containing the gene) and

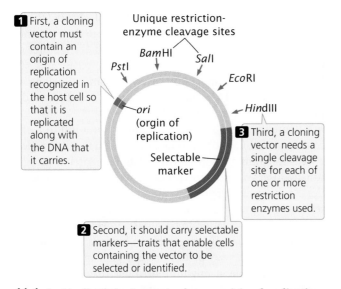

1 First, a cloning vector must contain an origin of replication recognized in the host cell so that it is replicated along with the DNA that it carries.

Unique restriction-enzyme cleavage sites

*Bam*HI   *Sal*I

*Pst*I

*Eco*RI

ori (orgin of replication)

*Hind*III

Selectable marker

3 Third, a cloning vector needs a single cleavage site for each of one or more restriction enzymes used.

2 Second, it should carry selectable markers—traits that enable cells containing the vector to be selected or identified.

14.4 An idealized cloning vector has an origin of replication, one or more selectable markers, and one or more unique restriction sites.

the plasmid with the same restriction enzyme (**Figure 14.5**). If the restriction enzyme makes staggered cuts in the DNA, complementary sticky ends are produced on the foreign DNA and the plasmid. The DNA and plasmid are then mixed together; some of the foreign DNA fragments will pair with the cut ends of the plasmid. DNA ligase is used to seal the nicks in the sugar–phosphate backbone, creating a recombinant plasmid that contains the foreign DNA fragment.

Transformation When a gene has been placed inside a plasmid, the plasmid must be introduced into bacterial cells. This task is usually accomplished by *transformation*, which is the capacity of bacterial cells to take up DNA from the external environment (see Chapter 6). Some types of cells undergo transformation naturally; others must be treated chemically or physically before they will undergo transformation. Inside the cell, the plasmids replicate and multiply.

The use of selectable markers Cells bearing recombinant plasmids can be detected by using the selectable markers on the plasmid. One type of selectable marker commonly used with plasmids is a copy of the *lacZ* gene (**Figure 14.6**). The *lacZ* gene contains a series of unique restriction sites into which can be inserted a fragment of DNA to be cloned. In the absence of an inserted fragment, the *lacZ* gene is active and produces β-galactosidase. When foreign DNA is inserted into the restriction site, it disrupts the *lacZ* gene, and β-galactosidase is not produced. The plasmid also usually contains a second selectable marker, which may be a gene that confers resistance to an antibiotic such as ampicillin.

Bacteria that are *lacZ⁻* are transformed by the plasmids and plated on medium that contains ampicillin. Only cells

that have been successfully transformed and contain a plasmid with the ampicillin-resistance gene will survive and grow. Some of these cells will contain an intact plasmid, whereas others possess a recombinant plasmid. The medium also contains the chemical X-gal, which produces a blue substance when cleaved. Bacterial cells with an intact original plasmid—without an inserted fragment—have a functional *lacZ* gene and can synthesize β-galactosidase, which cleaves X-gal and turns the bacteria blue. Bacterial cells with a recombinant plasmid, however, have a β-galactosidase gene that is disrupted by the inserted DNA; they do not synthesize β-galactosidase and remain white. (In these experiments, the bacterium's own β-galactosidase gene has been inactivated, and so only bacteria with the plasmid turn blue.) Thus, the color of the colony allows quick determination of whether a recombinant or intact plasmid is present in the cell.

Plasmids make ideal cloning vectors but can hold only DNA less than about 15 kb in size. When large DNA

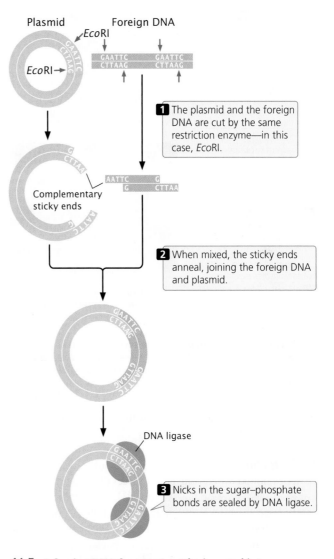

Plasmid   Foreign DNA

*Eco*RI

*Eco*RI

GAATTC   GAATTC
CTTAAG   CTTAAG

1 The plasmid and the foreign DNA are cut by the same restriction enzyme—in this case, *Eco*RI.

G
CTTAA

AATTC   G
G   CTTAA

Complementary sticky ends

2 When mixed, the sticky ends anneal, joining the foreign DNA and plasmid.

DNA ligase

3 Nicks in the sugar–phosphate bonds are sealed by DNA ligase.

14.5 A foreign DNA fragment can be inserted into a plasmid with the use of restriction enzymes.

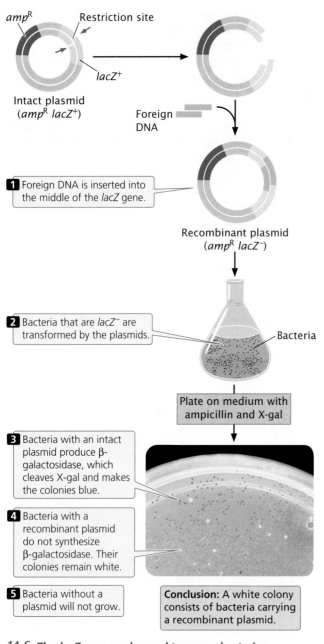

1 Foreign DNA is inserted into the middle of the *lacZ* gene.

2 Bacteria that are *lacZ⁻* are transformed by the plasmids.

3 Bacteria with an intact plasmid produce β-galactosidase, which cleaves X-gal and makes the colonies blue.

4 Bacteria with a recombinant plasmid do not synthesize β-galactosidase. Their colonies remain white.

5 Bacteria without a plasmid will not grow.

Conclusion: A white colony consists of bacteria carrying a recombinant plasmid.

14.6 The *lacZ* gene can be used to screen bacteria containing recombinant plasmids. A special plasmid carries a copy of the *lacZ* gene and an ampicillin-resistance gene. *[Photograph: Cytographics/Visuals Unlimited.]*

fragments are inserted into a plasmid vector, the plasmid becomes unstable.

Bacteriophage vectors   A number of other vectors have been developed for cloning larger pieces of DNA in bacteria. For example, bacteriophage λ (lambda), which infects *E. coli*, can be used to clone as much as about 23,000 bp of foreign DNA; it transfers DNA into bacteria cells with high efficiency. **Cosmids** are plasmids that are packaged into empty viral protein coats and transferred to bacteria by viral infection. They can carry more than twice as much foreign DNA

as can a phage vector. **Bacterial artificial chromosomes** (BACs) are vectors originally constructed from the F plasmid (a special plasmid that controls mating and the transfer of genetic material in some bacteria; see Chapter 6) and can hold very large fragments of DNA that can be as long as 300,000 bp. **Table 14.2** compares the properties of plasmids, phage λ vectors, cosmids, and BACs.

Sometimes the goal in gene cloning is not just to replicate the gene, but also to produce the protein that it encodes. To ensure transcription and translation, a foreign gene is usually inserted into an **expression vector**, which, in addition to the usual origin of replication, restriction sites, and selectable markers, contains sequences required for transcription and translation in bacterial cells (**Figure 14.7**).

Although manipulating genes in bacteria is simple and efficient, the goal may be to transfer a gene into eukaryotic cells. For example, it might be desirable to transfer a gene conferring herbicide resistance into a crop plant or to transfer a gene for clotting factor into a person suffering from hemophilia. Many eukaryotic proteins are modified after translation (e.g., sugar groups may be added). Such modifications are essential for proper function, but bacteria do not have the capacity to carry out the modification; thus a functional protein can be produced only in a eukaryotic cell. A number of cloning vectors have been developed that allow the insertion of genes into eukaryotic cells. For example, the Ti plasmid, a large plasmid from the soil bacterium *Agrobacterium tumefaciens*, has been genetically engineered to transfer genes to plants, including genes that confer economically significant attributes such as resistances to herbicides, plant viruses, and insect pests.

Concepts

DNA fragments can be inserted into cloning vectors, stable pieces of DNA that will replicate within a cell. A cloning vector must have an origin of replication, one or more unique restriction sites, and selectable markers. An expression vector contains sequences that allow a cloned gene to be transcribed and translated. Special cloning vectors have been developed for introducing genes into eukaryotic cells.

✔ **Concept Check 4**

How is a gene inserted into a plasmid cloning vector?

Amplifying DNA Fragments by Using the Polymerase Chain Reaction

The manipulation and analysis of genes require multiple copies of the DNA sequences used. A major problem in working at the molecular level is that each gene is a tiny fraction of the total cellular DNA. Because each gene is rare, it must be isolated and amplified before it can be studied. As already stated, one way to amplify a DNA fragment is to

Cloning Vector	Size of DNA That Can Be Cloned	Method of Propagation	Introduction into Bacteria
Plasmid	As large as 15 kb	Plasmid replication	Transformation
Phage λ	As large as 23 kb	Phage reproduction	Phage infection
Cosmid	As large as 44 kb	Plasmid reproduction	Phage infection
Bacterial artificial chromosome	As large as 300 kb	Plasmid reproduction	Electroporation

Table 14.2 Comparison of plasmids, phage λ vectors, cosmids, and bacterial artificial chromosomes

Note: 1 kb = 1000 bp. Electroporation consists of electrical pulses that increase the permeability of a membrane.

clone it in bacterial cells. Indeed, for many years, gene cloning was the only way to quickly amplify a DNA fragment, and cloning was a prerequisite for many other molecular methods. Cloning is labor intensive and requires at least several days to grow the bacteria. Today, the polymerase chain reaction makes the amplification of short DNA fragments possible without cloning, but cloning is still widely used for amplifying large DNA fragments and for other manipulations of DNA sequences.

The **polymerase chain reaction** (PCR), first developed in 1983, allows DNA fragments to be amplified a billionfold within just a few hours. It can be used with extremely small amounts of original DNA, even a single molecule. The polymerase chain reaction has revolutionized molecular biology and is now one of the most widely used of all molecular techniques. The basis of PCR is replication catalyzed by a DNA polymerase. Replication in this case has two essential requirements: (1) a single-stranded DNA template from which a new DNA strand can be copied and (2) a primer with a 3′-OH group to which new nucleotides can be added.

Because a DNA molecule consists of two nucleotide strands, each of which can serve as a template to produce a new molecule of DNA, the amount of DNA doubles with each replication event. The primers used in PCR to replicate the templates are short fragments of DNA, typically from 17 to 25 nucleotides long, that are complementary to known sequences on the template.

To carry out PCR, we begin with a solution that includes the target DNA, DNA polymerase, all four deoxyribonucleoside triphosphates (dNTPs—the substrates for DNA polymerase), primers, and magnesium ions and other salts that are necessary for the reaction to proceed. A typical polymerase chain reaction includes three steps (**Figure 14.8**). In the first step, the DNA is heated to high temperature (typically, from 90° to 100°C), which breaks the hydrogen bonds between the strands and produces single-stranded templates. In the second step, the DNA solution is cooled quickly to between 30° and 65°C, which allows the primers to attach to the template strands. In the third step, the solution is heated to between 60° and 70°C, the temperature at which DNA polymerase can synthesize new DNA strands. At the end of the cycle, two new double-stranded DNA molecules are produced for each original molecule of target DNA.

The whole cycle is then repeated. With each cycle, the amount of target DNA doubles. One molecule of DNA increases to more than 1000 molecules in 10 PCR cycles, to more than 1 million molecules in 20 cycles, and to more than 1 billion molecules in 30 cycles. Each cycle is completed within a few minutes; so a large amplification of DNA can be achieved within a few hours.

A key innovation that facilitated the use of PCR in the laboratory was the discovery of a DNA

1 Expression vectors contain operon sequences that allow inserted DNA to be transcribed and translated.

Transcription-initiation sequences

Restriction sites

Transcription-termination sequence

Operator (O)

Bacterial promoter (P) sequences

Ribosome-binding site

2 They also include sequences that regulate—turn on or turn off—the desired gene.

Gene-encoding repressor that binds O and regulates P

ori

Selectable genetic marker (e.g., antibiotic resistance)

14.7 To ensure transcription and translation, a foreign gene may be inserted into an expression vector—in this example, an E. coli expression vector.

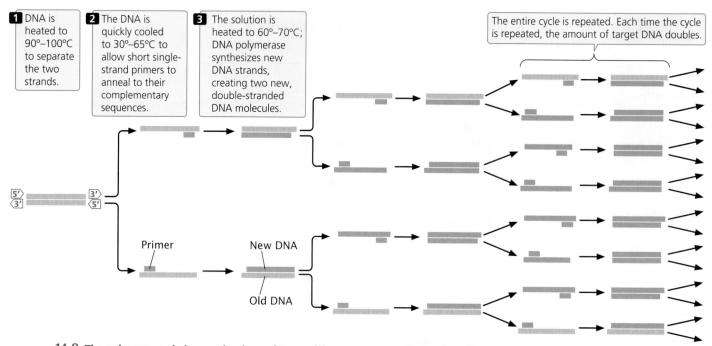

1 DNA is heated to 90°–100°C to separate the two strands.

2 The DNA is quickly cooled to 30°–65°C to allow short single-strand primers to anneal to their complementary sequences.

3 The solution is heated to 60°–70°C; DNA polymerase synthesizes new DNA strands, creating two new, double-stranded DNA molecules.

The entire cycle is repeated. Each time the cycle is repeated, the amount of target DNA doubles.

Primer   New DNA

Old DNA

14.8 The polymerase chain reaction is used to amplify even very small samples of DNA.

polymerase that is stable at the high temperatures used in the first step of PCR. The DNA polymerase from *E. coli* that was originally used in PCR denatures at 90°C, and fresh enzyme had to be added to the reaction mixture in *each* cycle. This obstacle was overcome when DNA polymerase was isolated from the bacterium *Thermus aquaticus*, which lives in the boiling springs of Yellowstone National Park. This enzyme, dubbed **Taq polymerase**, is remarkably stable at high temperatures and is not denatured in the strand-separation step of PCR; so it can be added to the reaction mixture at the beginning of the PCR process and will continue to function through many cycles.

Concepts

The polymerase chain reaction is an enzymatic in vitro (in a test tube) method for rapidly amplifying DNA. In this process, DNA is heated to separate the two strands, short primers attach to the target DNA, and DNA polymerase synthesizes new DNA strands from the primers. Each cycle of PCR doubles the amount of DNA.

14.2 Molecular Techniques Can Be Used to Find Genes of Interest

To analyze a gene or to transfer it to another organism, the gene must first be located and isolated. For instance, if we want to transfer a human gene for growth hormone to bacteria, we must first find the human gene that encodes growth

hormone and separate it from the 3.2 billion base pairs of human DNA. In our consideration of gene cloning, we've glossed over the problem of finding the DNA sequence to be cloned. A discussion of how to solve this problem has been purposely delayed until now because, paradoxically, we must often clone a gene to find it.

This approach—to clone first and search later—is called "shotgun cloning," because it is like hunting with a shotgun: we spray shots widely in the general direction of the quarry, knowing that there is a good chance that one or more of the pellets will hit the intended target. In shotgun cloning, we first clone a large number of DNA fragments, knowing that one or more contains the DNA of interest, and then search for the fragment of interest among the clones.

Gene Libraries

A collection of clones containing all the DNA fragments from one source is called a **DNA library**. For example, we might isolate genomic DNA from human cells, break it into fragments, and clone all of them in bacterial cells or phages. The set of bacterial colonies or phages containing these fragments is a human **genomic library**, containing all the DNA sequences found in the human genome. In contrast, a **cDNA library** contains only DNA sequences that are transcribed into mRNA; a cDNA library is created from mRNA that is first converted into DNA and then cloned into bacteria.

Creating a genomic library   To create a genomic library, cells are collected and disrupted, which causes them to release their DNA and other cellular contents into an aqueous solution, and the DNA is extracted from the solution. After the DNA has been isolated, it is cut into fragments by a restriction enzyme for a limited amount of time only (a partial digestion) so that only *some* of the restriction sites in each DNA molecule are cut. Because which sites are cut is random, different DNA molecules will be cut in different places, and a set of overlapping fragments will be produced (**Figure 14.9**). The fragments are then joined to vectors, which can be transferred to bacteria. A few of the clones contain the entire gene of interest, a few contain parts of the gene, but most contain fragments that have no part of the gene of interest.

A genomic library must contain a large number of clones to ensure that all DNA sequences in the genome are represented in the library. A library of the human genome formed by using cosmids, each carrying a random DNA fragment from 35,000 to 44,000 bp long, would require about 350,000 cosmid clones to provide a 99% chance that every sequence is included in the library.

Concepts

One method of finding a gene is to create and screen a DNA library. A genomic library is created by cutting genomic DNA into overlapping fragments and cloning each fragment in a separate bacterial cell. A cDNA library is created from mRNA that is converted into cDNA and cloned in bacteria.

Screening DNA libraries   Creating a genomic or cDNA library is relatively easy compared with screening the library to find clones that contain the gene of interest. The screening procedure used depends on what is known about the gene.

The first step in screening is to plate the clones of the library. If a plasmid or cosmid vector was used to construct the library, the cells are diluted and plated so that each bacterium grows into a distinct colony. If a phage vector was used, the phages are allowed to infect a lawn of bacteria on a petri plate. Each plaque or bacterial colony contains a single, cloned DNA fragment that must be screened for the gene of interest.

A common way to screen libraries is with probes. To use a probe, replicas of the plated colonies or plaques in the library must first be made. **Figure 14.10** illustrates this procedure for a cosmid library.

How is a probe obtained when the gene has not yet been isolated? One option is to use a similar gene from another organism as the probe. For example, if we wanted to screen a human genomic library for the growth-hormone gene and the gene had already been isolated from rats, we could use a purified rat-gene sequence as the probe to find the human gene for growth hormone. Successful hybridization does not require perfect complementarity between the probe and

1 Multiple copies of genomic DNA are digested by a restriction enzyme for a limited time so that only some of the restriction sites in each molecule are cut.

2 Different DNA molecules are cut in different places, providing a set of overlapping fragments.

3 Each fragment is then joined to a cloning vector…

4 …and transferred to a bacterial cell,…

5 …producing a set of clones containing overlapping genomic fragments, some of which may include segments of the gene of interest.

Conclusion: Some clones contain the entire gene of interest, others include part of the gene, and most contain none of the gene of interest.

14.9 **A genomic library contains all of the DNA sequences found in an organism's genome.**

14.10 Genomic and cDNA libraries can be screened with a probe to find the gene of interest.

the target sequence; so a related sequence can often be used as a probe.

Another method of screening a library is to look for the protein product of a gene. The clones can be tested for the presence of the protein by using an antibody that recognizes the protein or by using a chemical test for the protein product. This method depends on the existence of a test for the protein produced by the gene.

Almost any method used to screen a library will identify several clones, some of which will be false positives that do not contain the gene of interest; several screening methods may be needed to determine which clones actually contain the gene.

> **Concepts**
>
> A DNA library can be screened for a specific gene by using complementary probes that hybridize to the gene. Alternatively, clones in a gene library can be examined for the protein product of the gene.

Positional Cloning

For many genes with important functions, no associated protein product is yet known. The biochemical bases of many human genetic diseases, for example, are still unknown. How can these genes be isolated? One approach is to first determine the general location of the gene on the chromosome by using recombination frequencies derived from crosses or pedigrees (see Chapter 5). After the chromosomal region where the gene is found has been identified, any genes in this region can be cloned and identified. Then other techniques can be used to identify which of the "candidate" genes might be the one that causes the disease. This approach—to isolate genes on the basis of their position on a gene map—is called **positional cloning**. Through positional cloning, we can identify genes that encode a phenotype without a detailed understanding of the underlying biochemical nature of the phenotype. A number of important human diseases have been identified through positional cloning.

In Silico Gene Discovery

The complete genomes of more and more species are sequenced each year (see sections that follow), and partial sequences of many other organisms are continually being added to DNA databases. With this growth in sequence information, the task of finding genes is now often carried out by using high-speed computers that analyze and search DNA databases rather than by using gene cloning and DNA libraries. Sometimes called "in silico," these methods of locating and characterizing genes rely on the identification of characteristic sequences associated with genes and on comparisons with sequences of known genes in DNA databases. These methods will be discussed in more detail later in the chapter.

14.3 DNA Sequences Can Be Determined and Analyzed

In addition to cloning and amplifying DNA, molecular techniques are used to analyze DNA molecules through the study and determination of their sequences.

Restriction Fragment Length Polymorphisms

A significant contribution of molecular genetics has been to provide numerous genetic markers that can be used in gene mapping. One group of such markers comprises **restriction fragment length polymorphisms** (RFLPs), which are variations (polymorphisms) in the patterns of fragments produced when DNA molecules are cut with the same restriction enzyme (**Figure 14.11**). These differences are inherited and can be used in mapping, similar to the way in which allelic differences are used to map conventional genes.

To illustrate mapping with RFLPs, consider Huntington disease, an autosomal dominant disorder. In the family

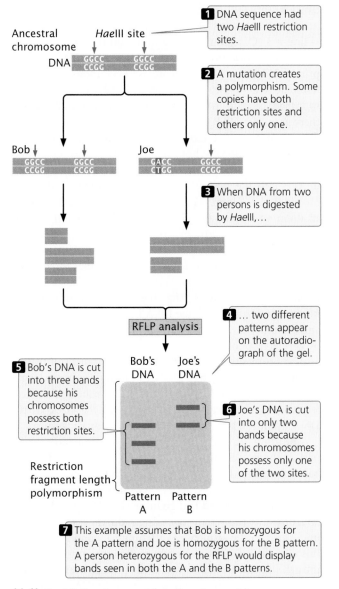

1. DNA sequence had two *Hae*III restriction sites.

2. A mutation creates a polymorphism. Some copies have both restriction sites and others only one.

3. When DNA from two persons is digested by *Hae*III,…

4. … two different patterns appear on the autoradiograph of the gel.

5. Bob's DNA is cut into three bands because his chromosomes possess both restriction sites.

6. Joe's DNA is cut into only two bands because his chromosomes possess only one of the two sites.

7. This example assumes that Bob is homozygous for the A pattern and Joe is homozygous for the B pattern. A person heterozygous for the RFLP would display bands seen in both the A and the B patterns.

14.11 Restriction fragment length polymorphisms are genetic markers that can be used in mapping.

genes encoding the RFLP and Huntington disease are assorting independently and are not linked.

Concepts

Restriction fragment length polymorphisms are variations in the pattern of fragments produced by restriction enzymes, which reveal variations in DNA sequences. They are used extensively in gene mapping.

DNA Sequencing

A powerful molecular method for analyzing DNA is a technique known as **DNA sequencing**, which quickly determines the sequence of bases in DNA. Sequencing allows the genetic information in DNA to be read, providing an enormous amount of information about gene structure and function. In the mid-1970s, Frederick Sanger and his colleagues created the dideoxy-sequencing method based on the elongation of DNA, and it quickly became the standard procedure for sequencing any purified fragment of DNA.

The Sanger, or dideoxy, method of DNA sequencing is based on replication. The fragment to be sequenced is used as a template to make a series of new DNA molecules. In the process, replication is sometimes (but not always) terminated when a specific base is encountered, producing DNA strands of different lengths, each of which ends in the same base.

The method relies on the use of a special substrate for DNA synthesis. Normally, DNA is synthesized from deoxyribonucleoside triphosphates (dNTPs), which have an OH group on the 3′-carbon atom (**Figure 14.13a**). In the Sanger method, a special nucleotide, called a **dideoxyribonucleoside triphosphate** (ddNTP; **Figure 14.13b**), is used as one of the substrates. The ddNTPs are identical with dNTPs, except that they lack a 3′-OH group. In the course of DNA synthesis, ddNTPs are incorporated into a growing DNA strand. However, after a ddNTP has been incorporated into the DNA

shown in **Figure 14.12**, the father is heterozygous both for Huntington disease (*Hh*) and for a restriction pattern (*AC*). From the father, each child inherits either a Huntington-disease allele (*H*) or a normal allele (*h*); any child inheriting the Huntington-disease allele develops the disease, because it is an autosomal dominant disorder. The child also inherits one of the two RFLP alleles from the father, either *A* or *C*, which produces the corresponding RFLP pattern. In Figure 14.12, every child who inherits the *C* pattern from the father also inherits Huntington disease (and therefore the *H* allele), because the locus for the RFLP is closely linked to the locus for the disease-causing gene. If we had observed no correspondence between the inheritance of the RFLP pattern and the inheritance of the disease, it would indicate that the

Every child who inherits the *C* allele has the disease. Thus, the RFLP is closely linked to the disease gene.

14.12 Restriction fragment length polymorphisms can be used to detect linkage. There is a close correspondence between the inheritance of the RFLP alleles and the presence of Huntington disease, indicating that the genes that encode the RFLP and Huntington disease are closely linked.

(a)   **(b)**

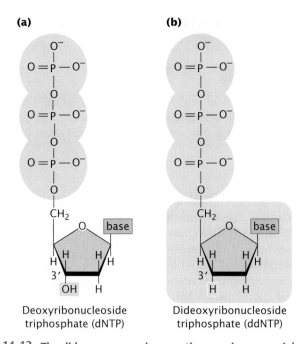

Deoxyribonucleoside   Dideoxyribonucleoside
triphosphate (dNTP)   triphosphate (ddNTP)

14.13 The dideoxy sequencing reaction requires a special substrate for DNA synthesis. (a) Structure of deoxyribonucleoside triphosphate, the normal substrate for DNA synthesis. (b) Structure of dideoxyribonucleoside triphosphate, which lacks an OH group on the 3′-carbon atom.

strand, no more nucleotides can be added, because there is no 3′-OH group to form a phosphodiester bond with an incoming nucleotide. Thus, ddNTPs terminate DNA synthesis.

Although the sequencing of a single DNA molecule is technically possible, most sequencing procedures in use today require a considerable amount of DNA; so any DNA fragment to be sequenced must first be amplified by PCR or by cloning in bacteria. Copies of the target DNA are isolated and split into four parts (**Figure 14.14**). Each part is placed in a different tube, to which are added:

1. many copies of a primer that is complementary to one end of the target DNA strand;

2. all four types of deoxyribonucleoside triphosphates, the normal precursors of DNA synthesis;

3. a small amount of one of the four types of dideoxyribonucleoside triphosphates, which will terminate DNA synthesis as soon as it is incorporated into any growing chain (each of the four tubes received a different ddNTP); and

4. DNA polymerase.

Either the primer or one of the dNTPs is radioactively or chemically labeled so that newly produced DNA can be detected.

Within each of the four tubes, the DNA polymerase enzyme synthesizes DNA. Let's consider the reaction in one of the four tubes; the one that received ddATP. Within this tube, each of the single strands of target DNA serves as a template for DNA synthesis. The primer pairs to its complementary sequence at one end of each template strand, providing a 3′-OH group for the initiation of DNA synthesis. DNA polymerase elongates a new strand of DNA from this primer. Wherever DNA polymerase encounters a T on the template strand, it uses at random either a dATP or a ddATP to introduce an A in the newly synthesized strand. Because there is more dATP than ddATP in the reaction mixture, dATP is incorporated most often, allowing DNA synthesis to continue. Occasionally, however, ddATP is incorporated into the strand and synthesis terminates. The incorporation of ddATP into the new strand occurs randomly at different positions in different copies, producing a set of DNA chains of different lengths (12, 7, and 2 nucleotides long in the example illustrated in Figure 14.14), each ending in a nucleotide that contains adenine.

Equivalent reactions take place in the other three tubes, except that synthesis is terminated at nucleotides with a different base in each tube. After the completion of the polymerization reactions, all the DNA in the tubes is denatured, and the single-strand products of each reaction are separated by gel electrophoresis.

The contents of the four tubes are separated side by side on an acrylamide gel so that DNA strands differing in length by only a single nucleotide can be distinguished. After electrophoresis, the locations, and therefore the sizes, of the DNA strands in the gel are revealed by autoradiography.

Reading the DNA sequence is the simplest and shortest part of the procedure. In Figure 14.14, you can see that the band closest to the bottom of the gel is from the tube that contained the ddGTP reaction, which means that the first nucleotide synthesized had guanine (G). The next band up is from the tube that contained ddATP; so the next nucleotide in the sequence is adenine (A), and so forth. In this way, the sequence is read from the bottom to the top of the gel, with the nucleotides near the bottom corresponding to the 5′ end of the newly synthesized DNA strand and those near the top corresponding to the 3′ end. Keep in mind that the sequence obtained is not that of the target DNA but that of its *complement*.

For many years, DNA sequencing was done largely by hand and was laborious and expensive. Today, sequencing is often carried out by automated machines that use fluorescent dyes and laser scanners to sequence thousands of base pairs in a few hours (**Figure 14.15** on page 362). The dideoxy reaction is also used here, but the ddNTPs used in the reaction are labeled with a fluorescent dye, and a different colored dye is used for each type of dideoxynucleotide. In this case, the four sequencing reactions can take place in the same test tube and can be placed in the same well during electrophoresis. Sequencing machines carry out electrophoresis in gel-containing capillary tubes. The different-size fragments produced by the sequencing reaction separate within a tube and migrate past a laser beam and detector. As the fragments pass the laser, their fluorescent dyes are activated

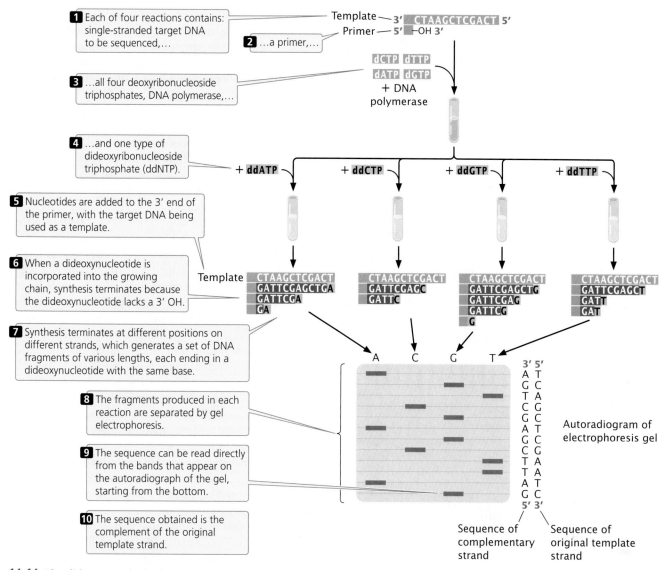

1 Each of four reactions contains: single-stranded target DNA to be sequenced,…

Template — 3' ⬛CTAAGCTCGACT⬛ 5'
Primer — 5' ⬛—OH 3'

2 …a primer,…

dCTP dTTP
dATP dGTP
+ DNA polymerase

3 …all four deoxyribonucleoside triphosphates, DNA polymerase,…

4 …and one type of dideoxyribonucleoside triphosphate (ddNTP).

+ **ddATP**   + **ddCTP**   + **ddGTP**   + **ddTTP**

5 Nucleotides are added to the 3' end of the primer, with the target DNA being used as a template.

6 When a dideoxynucleotide is incorporated into the growing chain, synthesis terminates because the dideoxynucleotide lacks a 3' OH.

Template CTAAGCTCGACT   CTAAGCTCGACT   CTAAGCTCGACT   CTAAGCTCGACT
          GATTCGAGCTGA   GATTCGAGC   GATTCGAGCTG   GATTCGAGCT
          GATTCGA   GATTC   GATTCGAG   GATT
          GA   GATTCG   GAT
                                              G

7 Synthesis terminates at different positions on different strands, which generates a set of DNA fragments of various lengths, each ending in a dideoxynucleotide with the same base.

A   C   G   T

8 The fragments produced in each reaction are separated by gel electrophoresis.

9 The sequence can be read directly from the bands that appear on the autoradiograph of the gel, starting from the bottom.

10 The sequence obtained is the complement of the original template strand.

3' 5'
A T
G C
T A
C G
G C
A T
G C
C G
T A
T A
A T
G C
5' 3'

Autoradiogram of electrophoresis gel

Sequence of complementary strand

Sequence of original template strand

14.14 **The dideoxy method of DNA sequencing is based on the termination of DNA synthesis.**

and the resulting fluorescence is detected by an optical scanner. Each colored dye emits fluorescence of a characteristic wavelength, which is read by the optical scanner. The information is fed into a computer for interpretation, and the results are printed out as a set of peaks on a graph (see Figure 14.15). Automated sequencing machines may contain 96 or more capillary tubes, allowing from 50,000 to 60,000 bp of sequence to be read in a few hours.

Concepts

DNA can be rapidly sequenced by the dideoxy method, in which ddNTPs are used to terminate DNA synthesis at specific bases. Automated sequencing methods allow tens of thousands of base pairs to be read in just a few hours.

✔ **Concept Check 5**

In the dideoxy sequencing reaction, what terminates DNA synthesis at a particular base?

a. The absence of a base on the ddNTP halts the DNA polymerase.

b. The ddNTP causes a break in the sugar–phosphate backbone.

c. DNA polymerase will not incorporate a ddNTP into the growing DNA strand.

d. The absence of a 3'-OH group on the ddNTP prevents the addition of another nucleotide.

DNA Fingerprinting

The use of DNA sequences to identify individual persons is called **DNA fingerprinting** or DNA profiling. Because some parts of the genome are highly variable, each person's DNA

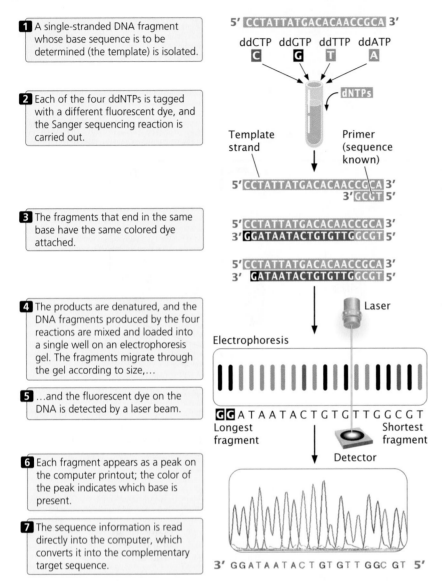

1 A single-stranded DNA fragment whose base sequence is to be determined (the template) is isolated.

2 Each of the four ddNTPs is tagged with a different fluorescent dye, and the Sanger sequencing reaction is carried out.

3 The fragments that end in the same base have the same colored dye attached.

4 The products are denatured, and the DNA fragments produced by the four reactions are mixed and loaded into a single well on an electrophoresis gel. The fragments migrate through the gel according to size,…

5 …and the fluorescent dye on the DNA is detected by a laser beam.

6 Each fragment appears as a peak on the computer printout; the color of the peak indicates which base is present.

7 The sequence information is read directly into the computer, which converts it into the complementary target sequence.

14.15 The dideoxy sequencing method can be automated.

microsatellite repeats, so that a DNA fragment containing the repeated sequences is amplified. The length of the amplified segment depends on the number of repeats; DNA from a person with more repeats will produce a longer amplified segment than will DNA from a person with fewer repeats. After PCR has been completed, the amplified fragments are separated with gel electrophoresis and stained, producing a series of bands on a gel. When several different microsatellite loci are examined, the probability that two people have the same set of patterns becomes vanishingly small, unless they are identical twins (**Figure 14.16**).

In a typical application, DNA fingerprinting might be used to confirm that a suspect was present at the scene of a crime (**Figure 14.17**). A sample of DNA from blood, semen, hair, or other body tissue is collected from the crime scene. If the sample is very small, PCR can be used to amplify it so that enough DNA is available for testing. Additional DNA samples are collected from one or more suspects. The pattern of bands produced by DNA fingerprinting from the sample collected at the crime scene is then compared with the patterns produced by DNA fingerprinting of the DNA from the suspects. A match between the sample from the crime scene and one from a suspect can provide evidence that the suspect was present at the scene of the crime.

Since its introduction in the 1980s, DNA fingerprinting has helped convict a number of suspects in murder and rape cases. Suspects in other cases have been proved innocent when their DNA failed to match that from the crime scenes. Initially, calculating the odds of a match (the probability that two people could have the same pattern) was controversial,

sequence is unique and, like a traditional fingerprint, provides a distinctive characteristic that allows identification.

Today, most DNA fingerprinting utilizes **microsatellites**, or **short tandem repeats** (STRs), which are very short DNA sequences repeated in tandem and are found widely in the human genome. People vary in the number of copies of repeat sequences that they possess. Microsatellites are typically detected with the use of PCR, with primers flanking the

14.16 Variation in banding patterns reveals inherited differences in microsatellite sequences. Microsatellite variation within a family. All bands found in the children are present in the parents. [*From A. Griffiths, S. Wessler, R. Lewontin, W. Gelbart, D. Suzuki, and J. Miller,* Introduction to Genetic Analysis, *8th ed. © 2005 by W. H. Freeman and Company.*]

Experiment

Question: How can the identity of DNA from blood, hair, or semen be determined?

Methods

Sample collected at a scene of crime   Suspect 1   Suspect 2

DNA samples are collected…

DNA   DNA   DNA

Microsatellite sequence   8 repeats of CA   2 repeats   8 repeats

CACACACACACACACA
GTGTGTGTGTGTGTGT

CACA
GTGT

CACACACACACACACA
GTGTGTGTGTGTGTGT

…and subjected to PCR.

CACACACACACACACA
GTGTGTGTGTGTGTGT

Template DNA

CACA
GTGT

CACACACACACACACA
GTGTGTGTGTGTGTGT

Primer

CACACACACACACACA
GTGTGTGTGTGTGTGT

Primer

CACA
GTGT

CACACACACACACACA
GTGTGTGTGTGTGTGT

Template DNA

The length of the DNA fragment produced by PCR depends on the number of copies of the microsatellite sequence.

CACACACACACACACA
GTGTGTGTGTGTGTGT

CACA
GTGT

CACACACACACACACA
GTGTGTGTGTGTGTGT

Results

The fragments are separated by gel electrophoresis. Different-size fragments appear as different bands.

The DNA of the sample collected at the scene of the crime matches DNA from suspect 2.

Results of one STR locus

λ 1kb TS   D   jeans   4µg 8µg
shirt   V   λ 1kb

Multiple microsatellite loci produce multiple bands on the gel.

Conclusion: The patterns of bands produced by different samples are compared. The bloodstain specimen matches DNA from suspect 2.

14.17 DNA fingerprinting can be used to identify a person. [Gel courtesy of Orchard Cellmark, Germantown, Maryland.]

and there were concerns about quality control (such as the accidental contamination of samples and the reproducibility of results) in laboratories where DNA analysis is done. Today, DNA fingerprinting has become an important tool in forensic investigations. In addition to its application in the analysis of crimes, DNA fingerprinting is used to assess paternity, study genetic relationships among individual organisms in natural populations, identify specific strains of pathogenic bacteria, and identify human remains. For example, DNA fingerprinting was used to determine that several samples of anthrax mailed to different people in 2001 were all from the same source.

Concepts

DNA fingerprinting detects genetic differences among people by using probes for highly variable regions of chromosomes.

14.4 Molecular Techniques Are Increasingly Used to Analyze Gene Function

In the preceding sections, we have seen how powerful molecular techniques are available for isolating, recombining, and analyzing DNA sequences. Although these methods provide a great deal of information about the organization and nature of gene sequences, the ultimate goal of many molecular studies is to better understand the function of the sequences. In this section, we will explore some advanced molecular techniques that are frequently used to determine gene function and to better understand the genetic processes that these sequences undergo.

Forward and Reverse Genetics

The traditional approach to the study of gene function begins with the isolation of mutant organisms. For example, suppose a geneticist is interested in genes that affect cardiac function in mammals. A first step would be to find individuals—perhaps mice—that have hereditary defects in heart function. The mutations causing the cardiac problems in the mice could then be mapped, and the implicated genes could be isolated, cloned, and sequenced. The proteins produced by the genes could then be predicted from the gene sequences and isolated. Finally, the biochemistry of the proteins could be studied and their role in heart function discerned. This approach, which begins with a phenotype (a mutant individual) and proceeds to a gene that encodes the phenotype, is called **forward genetics**.

An alternative approach, made possible by advances in molecular genetics, is to begin with a genotype—a DNA sequence—and proceed to the phenotype by altering the sequence or inhibiting its expression. A geneticist might begin with a gene of unknown function, induce mutations in it, and then look to see what effect these mutations have on the phenotype of the organism. This approach is called **reverse genetics**. Today, both forward and reverse genetic approaches are widely used in analysis of gene function. This section introduces some of the molecular techniques that are used in forward and reverse genetics.

Concepts

Forward genetics begins with a phenotype and detects and analyzes the genotype that causes the phenotype. Reverse genetics begins with a gene sequence and through analysis determines the phenotype that it encodes.

✔ Concept Check 6

A geneticist interested in immune function induces random mutations in a number of genes in mice and then determines which of the resulting mutant mice have impaired immune function. This is an example of

a. forward genetics.   c. both forward and reverse genetics.

b. reverse genetics.   d. neither forward nor reverse genetics.

Transgenic Animals

Another way that gene function can be analyzed is by adding DNA sequences of interest to the genome of an organism that normally lacks such sequences and then seeing what effect the introduced sequence has on the organism's phenotype. This method is a form of reverse genetics. An organism that has been permanently altered by the addition of a DNA sequence to its genome is said to be *transgenic*, and the foreign DNA that it carries is called a **transgene**. Here, we consider techniques for the creation of transgenic mice, which are often used in the study of the function of human genes because they can be genetically manipulated in ways that are impossible with humans and, as mammals, they are more similar to humans than are fruit flies, fish, and other model genetic organisms.

The oocytes of mice and other mammals are large enough that DNA can be injected into them directly. Immediately after penetration by a sperm, a fertilized mouse egg contains two pronuclei, one from the sperm and one from the egg; these pronuclei later fuse to form the nucleus of the embryo. Mechanical devices can manipulate extremely fine, hollow glass needles to inject DNA directly into one of the pronuclei of a fertilized egg (**Figure 14.18**). Typically, a few hundred copies of cloned, linear DNA are injected into a pronucleus, and, in a few of the injected eggs, copies of the cloned DNA integrate randomly into one of the chromosomes through a process called nonhomologous recombination. After injection, the embryos are implanted in a pseudopregnant female—a surrogate mother that has been physiologically prepared for pregnancy by mating with a vasectomized male.

Only about 10% to 30% of the eggs survive and, of those that do survive, only a few have a copy of the cloned DNA stably integrated into a chromosome. Nevertheless, if several hundred embryos are injected and implanted, there is a good chance that one or more mice whose chromosomes contain the foreign DNA will be born. Moreover, because the DNA was injected at the one-cell stage of the embryo, these mice usually carry the cloned DNA in every cell of their bodies, including their reproductive cells, and will therefore pass the foreign DNA on to their progeny. Through interbreeding, a strain of mice that is homozygous for the foreign gene can be created.

Transgenic mice have proved useful in the study of gene function. For example, proof that the *SRY* gene (see Chapter 4) is the male-determining gene in mice was obtained by injecting a copy of the *SRY* gene into XX embryos and observing that these mice developed as males.

14.18 Transgenic animals have genomes that have been permanently altered through recombinant DNA technology. In the photograph, a mouse embryo is being injected with DNA. [*Photograph: Chad Davis/PhotoDisc.*]

1 Mice are mated and fertilized eggs are removed from the female mouse.

Fertilized egg

Pronuclei

2 Foreign DNA is injected into one of the pronuclei.

Injecting needle

Gene of interest

Suction pipette

3 Embryos are implanted in a pseudopregnant female.

Offspring

4 Offspring are tested for the presence of the introduced transgene.

5 Mice carrying the gene are bred to produce a strain of mice homozygous for the foreign gene.

In addition, researchers have created a number of transgenic mouse strains that serve as experimental models for human genetic diseases.

Knockout Mice

A useful variant of the transgenic approach is to produce mice in which a normal gene has been not just mutated, but fully disabled. These animals, called **knockout mice**, are particularly helpful in determining the function of a gene: the phenotype of the knockout mouse often gives a good indication of the function of the missing gene.

A variant of the knockout procedure is to insert in mice a particular DNA sequence into a known chromosome location. For example, researchers might insert the sequence of a human disease-causing allele into the same locus in mice, creating a precise mouse model of the human disease. Mice that carry inserted sequences at specific locations are called **knock-in mice**.

Concepts

A transgenic mouse is produced by the injection of cloned DNA into the pronucleus of a fertilized egg, followed by implantation of the egg into a female mouse. In knockout mice, the injected DNA contains a mutation that disables a gene.

Model Genetic Organism
The Mouse *Mus musculus*

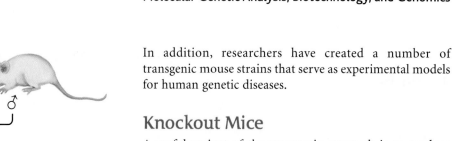 The ability to create transgenic, knockout, and knock-in mice has greatly facilitated the study of human genetics, and these techniques illustrate the power of the mouse as a model genetic organism. The common house mouse, *Mus musculus*, is among the oldest and most valuable subjects for genetic study (**Figure 14.19**). It's an excellent genetic organism—small, prolific, and easy to keep with a short generation time.

Advantages of the mouse as a model genetic organism
Foremost among many advantages that *Mus musculus* has as a model genetic organism is its close evolutionary relationship to humans. Being a mammal, the mouse is genetically, behaviorally, and physiologically more similar to humans than are other organisms used in genetics studies, making the mouse the model of choice for many studies of human and medical genetics. Other advantages include a short generation time compared with that of most other mammals. *Mus musculus* is well adapted to life in the laboratory and can be easily raised and bred in cages that require little space; thus several thousand mice can be raised within the confines of a small laboratory room. Mice have large litters (8–10 pups), and are docile and easy to handle. Finally, a large number of mutations have been isolated and studied in captive-bred mice, providing an important source of variation for genetic analysis.

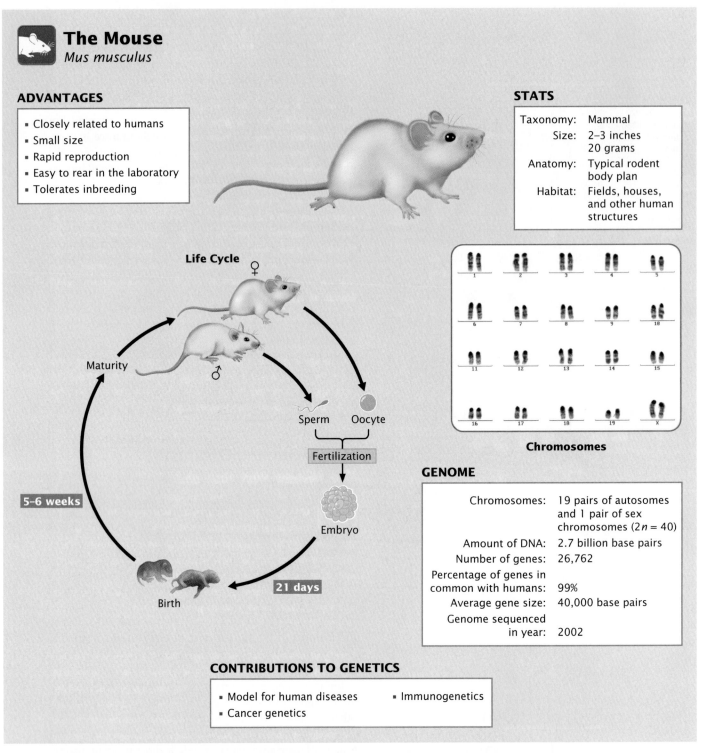

The Mouse
Mus musculus

ADVANTAGES

- Closely related to humans
- Small size
- Rapid reproduction
- Easy to rear in the laboratory
- Tolerates inbreeding

STATS

Taxonomy:	Mammal
Size:	2–3 inches 20 grams
Anatomy:	Typical rodent body plan
Habitat:	Fields, houses, and other human structures

Life Cycle

Maturity

Sperm   Oocyte

Fertilization

Embryo

5–6 weeks

21 days

Birth

Chromosomes

GENOME

Chromosomes:	19 pairs of autosomes and 1 pair of sex chromosomes ($2n = 40$)
Amount of DNA:	2.7 billion base pairs
Number of genes:	26,762
Percentage of genes in common with humans:	99%
Average gene size:	40,000 base pairs
Genome sequenced in year:	2002

CONTRIBUTIONS TO GENETICS

- Model for human diseases
- Cancer genetics
- Immunogenetics

14.19 The mouse *Mus musculus* is a model genetic organism. [*Chromosome photograph courtesy of Ellen C. Akeson and Muriel T. Davisson, The Jackson Laboratory, Bar Harbor, Maine.*]

Life cycle of the mouse The production of gametes and reproduction in the mouse are very similar to those in humans (see Figure 14.19). Diploid germ cells in the gonads undergo meiosis to produce sperm and oocytes, as outlined in Chapter 2. Male mice begin producing sperm at puberty and continue sperm production throughout the remainder of their lives. Starting at puberty, female mice go through an estrus cycle about every 4 days. If mating takes place during estrus, sperm are deposited into the vagina and swim into the oviduct, where one penetrates the outer layer of the ovum and the nuclei of sperm and ovum fuse. After fertilization, the diploid embryo implants into the uterus. Gestation typically takes about 21 days. Mice reach puberty in about 5 to 6 weeks and will live for about 2 years. A complete generation can be completed in about 8 weeks.

Genetic techniques with the mouse The mouse genome contains about 2.6 billion base pairs of DNA, which is similar in size to the human genome. For most human genes, there are homologous genes in the mouse. An important tool for determining the function of an unknown gene in humans is to search for a homologous gene whose function has already been determined in the mouse. Furthermore, the linkage relations of many mouse genes are similar to those in humans, and the linkage relations of genes in mice often provide important clues to linkage relations among genes in humans. The mouse genome is distributed across 19 pairs of autosomes and one pair of sex chromosomes (see Figure 14.19).

We have already considered three powerful techniques that have been developed for use in the mouse: (1) the creation of transgenic mice by the injection of DNA into a mouse embryo, (2) the ability to disrupt specific genes by the creation of knockout mice, and (3) the ability to insert specific sequences into specific loci. These techniques are made possible by the ability to manipulate the mouse reproductive cycle, including the ability to hormonally induce ovulation, isolate unfertilized oocytes from the ovary, and implant fertilized embryos back into the uterus of a surrogate mother.

A large number of mouse models of specific human diseases have been created—in some cases, by isolating and inbreeding mice with naturally occurring mutations and, in other cases, by using knockout and knock-in techniques to disable and modify specific genes. Mice tolerate inbreeding well, and inbred strains of mice are easily created by brother–sister mating. ■

Silencing Genes by Using RNA interference

In the preceding sections, we considered the analysis of gene function by introducing mutations or new DNA sequences into the genome and analyzing the resulting phenotype to provide information about the function of the altered or introduced DNA. We could also analyze gene function by temporarily turning a gene off and seeing what effect the absence of the gene product has on the phenotype. Until recently, there was no method for selectively affecting gene expression. However, the recent discoveries of siRNAs (small interfering RNAs) and miRNAs (microRNAs; see Chapters 10 and 12) have provided powerful tools for controlling the expression of individual genes.

Recall that siRNAs and miRNAs are small RNA molecules that combine with proteins to form the RNA-induced silencing complex (RISC). The RISC pairs with complementary sequences on mRNA and either cleaves the mRNA or prevents the mRNA from being translated. Molecular geneticists have exploited this natural machinery for turning off the expression of specific genes. Studying the effect of silencing a gene with the use of siRNA can often be a source of insight into the gene's function.

14.5 Biotechnology Harnesses the Power of Molecular Genetics

In addition to providing valuable new information about the nature and function of genes, techniques of molecular genetics have many practical applications. These applications include the production of pharmaceuticals and other chemicals, specialized bacteria, agriculturally important plants, and genetically engineered farm animals. The technology is also used extensively in medical testing and, in a few cases, is even being used to correct human genetic defects. Hundreds of firms now specialize in developing products through genetic engineering, and many large multinational corporations have invested enormous sums of money in molecular genetics research. As discussed earlier, the analysis of DNA is also used in criminal investigations and for the identification of human remains.

Pharmaceuticals

The first commercial products to be developed with the use of genetic engineering were pharmaceuticals used in the treatment of human diseases and disorders. In 1979, the Eli Lilly corporation began selling human insulin produced with the use of recombinant DNA technology. The gene for human insulin was inserted into plasmids and transferred to bacteria that then produced human insulin. Pharmaceuticals produced through recombinant DNA technology include human growth hormone (for children with growth deficiencies), clotting factors (for hemophiliacs), and tissue plasminogen activator (used to dissolve blood clots in heart-attack patients).

Specialized Bacteria

Bacteria play an important role in many industrial processes, including the production of ethanol from plant material, the leaching of minerals from ore, and the treatment of sewage and other wastes. The bacteria used in these processes are being modified by genetic engineering so that they work more efficiently. New strains of technologically useful bacteria are being developed that will break down toxic chemicals and pollutants, enhance oil recovery, increase nitrogen uptake by plants, and inhibit the growth of pathogenic bacteria and fungi.

Agricultural Products

Recombinant DNA technology has had a major effect on agriculture, where it is now used to create crop plants and domestic animals with valuable traits. For many years, plant pathologists had recognized that plants infected with mild strains of viruses are resistant to infection by virulent strains. Using this knowledge, geneticists have created viral resistance in plants by transferring genes for viral proteins to the plant cells. A genetically engineered squash, called Freedom II, carries genes from the watermelon mosaic virus 2 and the zucchini yellow mosaic virus that protect the squash against viral infections.

Another objective has been to genetically engineer pest resistance into plants to reduce dependence on chemical pesticides. A gene from the bacterium *Bacillus thuringiensis* that produces an insecticidal toxin has been transferred into corn, tomato, potato, and cotton plants. These BT crops are now grown worldwide. Other genes that confer resistance to viruses and herbicides have been introduced into a number of crop plants. In 2008, 65.8 million hectares of genetically engineered soybeans and 37.3 million hectares of genetically engineered corn were grown throughout the world.

The genetic engineering of agricultural products is controversial. One area of concern focuses on the potential effects of releasing novel organisms produced by genetic engineering into the environment. There is also concern that transgenic organisms may hybridize with native organisms and transfer their genetically engineered traits. Other concerns focus on health-safety matters associated with the presence of engineered products in natural foods; some critics have advocated required labeling of all genetically engineered foods that contain transgenic DNA or protein. Such labeling is required in countries of the European Union but not in the United States.

On the other hand, the use of genetically engineered crops and domestic animals has potential benefits. Genetically engineered crops that are pest resistant have the potential to reduce the use of environmentally harmful chemicals, and research findings indicate that lower amounts of pesticides are used in the United States as a result of the adoption of transgenic plants. Transgenic crops also increase yields, providing more food per acre, which reduces the amount of land that must be used for agriculture. As discussed in the introduction to the chapter, genetically engineered plants offer the potential for greater yields that may be necessary to feed the world's future population.

Concepts

Recombinant DNA technology is used to create a wide range of commercial products, including pharmaceuticals, specialized bacteria, genetically engineered crops, and transgenic domestic animals.

Genetic Testing

The identification and cloning of many important disease-causing human genes have allowed the development of probes for detecting disease-causing mutations. Prenatal testing is already available for several hundred genetic disorders. Additionally, presymptomatic genetic tests for adults and children are available for an increasing number of disorders.

Gene Therapy

Perhaps the ultimate application of recombinant DNA technology is **gene therapy**, the direct transfer of genes into humans to treat disease. In 1990, gene therapy became reality. W. French Anderson and his colleagues at the U.S. National Institutes of Health (NIH) transferred a functional gene for adenosine deaminase to a young girl with severe combined immunodeficiency disease, an autosomal recessive condition that produces impaired immune function. Today, thousands of patients have received gene therapy, and many clinical trials are underway. Gene therapy is being used to treat genetic diseases, cancer, heart disease, and even some infectious diseases such as AIDS.

In spite of the growing number of clinical trials for gene therapy, significant problems remain in transferring foreign genes into human cells, getting them expressed, and limiting immune responses to the gene products and the vectors used to transfer the genes to the cells. There are also heightened concerns about safety. In 1999, a patient participating in a gene-therapy trial had a fatal immune reaction after he was injected with a viral vector carrying a gene to treat his metabolic disorder. And in 2002, two children who had undergone gene therapy for severe combined immunodeficiency disease developed leukemia that appeared to be directly related to the insertion of the retroviral gene vectors into cancer-causing genes. Despite these setbacks, gene-therapy research has moved on. Unequivocal results demonstrating positive benefits from gene therapy for a severe combined immunodeficiency disease and for head and neck cancer were announced in 2000.

Gene therapy conducted to date has targeted only nonreproductive, somatic cells. Correcting a genetic defect in these cells (termed *somatic gene therapy*) may provide positive benefits to patients but will not affect the genes of future generations. Gene therapy that alters reproductive, or germ-line, cells (termed germ-line gene therapy) is technically possible but raises a number of significant ethical issues, because it has the capacity to alter the gene pool of future generations.

Concepts

Gene therapy is the direct transfer of genes into humans to treat disease. Gene therapy was first successfully implemented in 1990 and is now being used to treat genetic diseases, cancer, and infectious diseases.

14.6 Genomics Determines and Analyzes the DNA Sequences of Entire Genomes

Genomics is the field of genetics that attempts to understand the content, organization, function, and evolution of genetic information contained in whole genomes. The field of genomics is at the cutting edge of modern biology; information resulting from research in this field has made significant contributions to human health, agriculture, and numerous other areas. It has provided gene sequences necessary for producing medically important proteins through recombinant DNA technology, and comparisons of genome sequences from different organisms are leading to a better understanding of evolution and the history of life.

Structural genomics concerns the organization and sequence of genetic information contained within a genome. Often, one of the early steps in characterizing a genome is to prepare genetic and physical maps of its chromosomes. These maps provide information about the relative locations of genes, molecular markers, and chromosome segments, which are often essential for positioning chromosome segments and aligning stretches of sequenced DNA into a whole-genome sequence.

Genetic Maps

Everyone has used a map at one time or another. Maps are indispensable for finding a new friend's house, the way to an unfamiliar city in your state, or the location of a country. Each of these examples requires a map with a different scale. To find a friend's house, you would probably use a city street map; to find your way to an unknown city, you might pick up a state highway map; to find a country such as Kazakhstan, you would need a world atlas. Similarly, navigating a genome requires maps of different types and scales.

Genetic maps (also called linkage maps) provide a rough approximation of the locations of genes relative to the locations of other known genes (**Figure 14.20**). These maps are based on the genetic function of recombination (hence the name genetic map). The basic principles of constructing genetic maps are discussed in detail in Chapter 5. In short, individual organisms of known genotype are crossed, and the frequency of recombination between loci is determined by examining the progeny. For linked genes, the rate of recombination is proportional to the physical distance between the loci. Distances on genetic maps are measured in percent recombination (centimorgans, cM), or map units.

Physical Maps

Physical maps are based on the direct analysis of DNA, and they place genes in relation to distances measured in

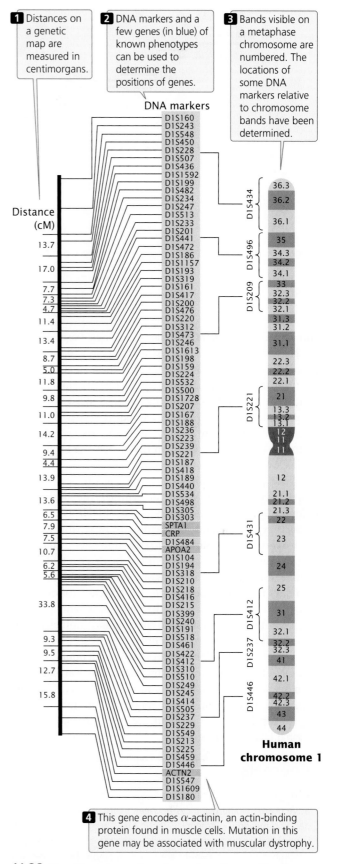

1 Distances on a genetic map are measured in centimorgans.

2 DNA markers and a few genes (in blue) of known phenotypes can be used to determine the positions of genes.

3 Bands visible on a metaphase chromosome are numbered. The locations of some DNA markers relative to chromosome bands have been determined.

4 This gene encodes α-actinin, an actin-binding protein found in muscle cells. Mutation in this gene may be associated with muscular dystrophy.

14.20 Genetic maps are based on rates of recombination. Shown here is a genetic map of human chromosome 1.

14.21 Physical maps are often used to order cloned DNA fragments. A part of a physical map of a set of overlapping YAC (yeast artificial chromosome) clones from one end of the human Y chromosome.

number of base pairs, kilobases, or megabases (**Figure 14.21**). A common type of physical map is one that connects isolated pieces of genomic DNA that have been cloned in bacteria or yeast. Physical maps generally have higher resolution and are more accurate than genetic maps. A physical map is analogous to a neighborhood map that shows the location of every house along a street, whereas a genetic map is analogous to a highway map that shows the locations of major towns and cities.

One of the techniques for creating physical maps is restriction mapping, which determines the position of restriction sites on DNA. When a piece of DNA is cut with a restriction enzyme and the fragments are separated by gel electrophoresis, the number of restriction sites in the DNA and the distances between them can be determined by the number and positions of bands on the gel, but this information does not tell us the order or the precise location of the restriction sites. To map restriction sites, a sample of the DNA is cut with one restriction enzyme, and another sample is cut with a different restriction enzyme. A third sample is cut with both restriction enzymes together (a double digest). The DNA fragments produced by these restriction digests are then separated by gel electrophoresis, and their sizes are compared. Overlap in size of fragments produced by the digests can be used to position the restriction sites on the original DNA molecule.

Concepts

Both genetic and physical maps provide information about the relative positions and distances between genes, molecular markers, and chromosome segments. Genetic maps are based on rates of recombination and are measured in percent recombination, or centimorgans. Physical maps are based on physical distances and are measured in base pairs.

Sequencing an Entire Genome

The ultimate goal of structural genomics is to determine the ordered nucleotide sequences of entire genomes of organisms. Earlier in this chapter, we considered the use of the dideoxy sequencing method to sequence small fragments of DNA. The main obstacle to sequencing a whole genome is the immense size of most genomes. Bacterial genomes are usually at least several million base pairs long; many eukaryotic genomes are billions of base pairs long and are distributed among dozens of chromosomes. Furthermore, for technical reasons, sequencing cannot begin at one end of a chromosome and continue straight through to the other end; only small fragments of DNA—usually from 500 to 700 nucleotides—can be sequenced at one time. Therefore, determining the sequence for an entire genome requires that the DNA be broken into thousands or millions of smaller fragments that can then be sequenced. The difficulty lies in putting these short sequences back together in the correct order. Two different approaches have been used to assemble the short sequenced fragments into a complete genome: map-based sequencing and whole-genome shotgun sequencing. We will consider these two approaches in the context of the Human Genome Project.

The Human Genome Project

By 1980, methods for mapping and sequencing DNA fragments had been sufficiently developed that geneticists began seriously proposing that the entire human genome could be sequenced. An international collaboration was planned to undertake the Human Genome Project (**Figure 14.22**); initial estimates suggested that 15 years and $3 billion would be required to accomplish the task.

The Human Genome Project officially got underway in October 1990. Initial efforts focused on developing new and

The initial effort to sequence the genome was a public project consisting of the international collaboration of 20 research groups and hundreds of individual researchers who formed the International Human Genome Sequencing Consortium. This group used a map-based strategy for sequencing the human genome.

Map-based sequencing   In **map-based sequencing**, short sequenced fragments are assembled into a whole-genome sequence by first creating detailed genetic and physical maps of the genome, which provide known locations of genetic markers (restriction sites, other genes, or known DNA sequences) at regularly spaced intervals along each chromosome. These markers are later used to help align the short sequenced fragments into their correct order.

After the genetic and physical maps are available, chromosomes or large pieces of chromosomes are separated by pulsed-field gel electrophoresis (PFGE) or by flow cytometry in which chromosomes are sorted optically by size. Each chromosome (or sometimes the entire genome) is then cut up by partial digestion with restriction enzymes (**Figure 14.24**). Partial digestion means that the restriction enzymes are allowed to act for only a limited time so that not all restriction sites in every DNA molecule are cut. Thus, partial digestion produces a set of large overlapping DNA fragments, which are then cloned with the use of cosmids, yeast artificial chromosomes (YACs), or bacterial artificial chromosomes (BACs).

Next, these large-insert clones are put together in their correct order on the chromosome (see Figure 14.24). This assembly can be done in several ways. One method relies on the presence of a high-density map of genetic markers. A complementary DNA probe is made for each genetic marker, and a library of the large-insert clones is screened with the probe, which will hybridize to any colony containing a clone with the marker. The library is then screened for neighboring markers. Because the clones are much larger than the markers used as probes, some clones will have more than one marker. For example, clone A might have markers M1 and M2, clone B markers M2, M3, and M4, and clone C markers M4 and M5. Such a result would indicate that these clones contain areas of overlap, as shown here:

A set of two or more overlapping DNA fragments that form a contiguous stretch of DNA is called a **contig**. This approach was used in 1993 to create a contig consisting of

14.22 A rough draft of the complete sequence of the human genome was completed in 2000. Craig Venter (left), President of Celera Genomics, and Francis Collins (right), Director of the National Human Genome Research Institute, NIH, announced this landmark achievement at a press conference in Washington on June 26, 2000. [*Alex Wong/Newsmakers/Getty Images.*]

automated methods for cloning and sequencing DNA and on generating detailed physical and genetic maps of the human genome. The methods described earlier for mapping, sequencing, and assembling DNA fragments were pivotal in these early stages of the project. By 1993, large-scale physical maps were completed for all 23 pairs of human chromosomes. At the same time, automated sequencing techniques (**Figure 14.23**) had been developed that made large-scale sequencing feasible.

14.23 Automated sequencers and powerful computers allowed the human genome sequence to be completed in just 13 years. [*Sam Ogden/Photo Researchers.*]

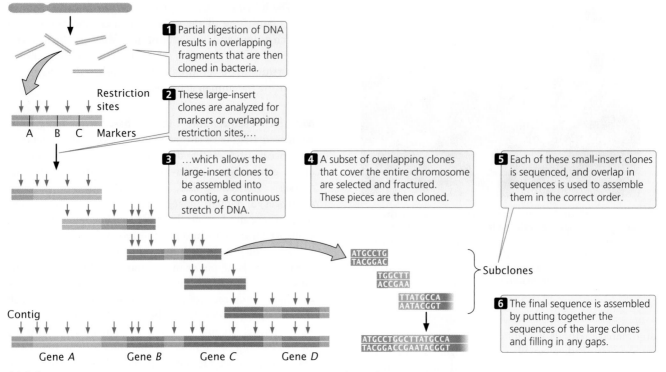

14.24 Map-based approaches to whole-genome sequencing rely on detailed genetic and physical maps to align sequenced fragments.

196 overlapping YAC clones (see Figure 14.21) of the human Y chromosome.

The order of clones can also be determined without the use of preexisting genetic maps. For example, each clone can be cut with a series of restriction enzymes and the resulting fragments then separated by gel electrophoresis. This method generates a unique set of restriction fragments, called a fingerprint, for each clone. The restriction patterns for the clones are stored in a database. A computer program is then used to examine the restriction patterns of all the clones and look for areas of overlap. The overlap is then used to arrange the clones in order, as shown here:

Other genetic markers can be used to help position contigs along the chromosome.

When the large-insert clones have been assembled into the correct order on the chromosome, a subset of overlapping clones that efficiently cover the entire chromosome can

be chosen for sequencing. Each of the selected large-insert clones is fractured into smaller overlapping fragments, which are themselves cloned (see Figure 14.24). These smaller clones (called small-insert clones) are then sequenced. The sequences of the small-insert clones are examined for overlap, which allows them to be correctly assembled to give the sequence of the larger insert clones. Enough overlapping small-insert clones are usually sequenced to ensure that the entire genome is sequenced several times. Finally, the whole genome is assembled by putting together the sequences of all overlapping contigs (see Figure 14.24). Often, gaps in the genome map still exist and must be filled in by using other methods.

The Human Genome Sequencing Consortium used a map-based approach to sequencing the human genome; many copies of the human genome were cut up into fragments of about 150,000 bp each, which were inserted into bacterial artificial chromosomes. Restriction fingerprints were used to assemble the BAC clones into contigs, which were positioned on the chromosomes with the use of genetic markers and probes. The individual BAC clones were sheared into smaller overlapping fragments and sequenced, and the whole genome was assembled by putting together the sequence of the BAC clones.

In 1998, Craig Venter announced that he would lead a company called Celera Genomics in a private effort to sequence the human genome. He proposed using a shotgun sequencing approach, which he suggested would be quicker

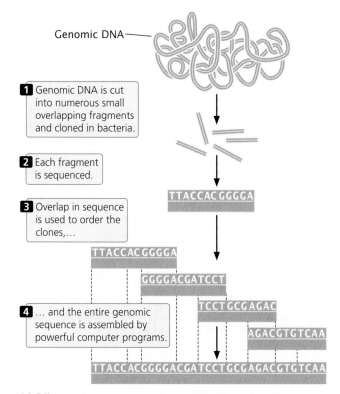

Genomic DNA

1 Genomic DNA is cut into numerous small overlapping fragments and cloned in bacteria.

2 Each fragment is sequenced.

TTACCACGGGGA

3 Overlap in sequence is used to order the clones,…

TTACCACGGGGA

GGGGACGATCCT

TCCTGCGAGAC

4 … and the entire genomic sequence is assembled by powerful computer programs.

AGACGTGTCAA

TTACCACGGGGACGATCCTGCGAGACGTGTCAA

14.25 Whole-genome shotgun sequencing utilizes sequence overlap to align sequenced fragments.

than the map-based approach employed by the Human Genome Sequencing Consortium.

Whole-genome shotgun sequencing In **whole-genome shotgun sequencing** (**Figure 14.25**), small-insert clones are prepared directly from genomic DNA and sequenced. Powerful computer programs then assemble the entire genome by examining overlap among the small-insert clones. One advantage of shotgun sequencing is that the small-insert clones can be placed into plasmids, which are simple and easy to manipulate. The requirement for overlap means that most of the genome will be sequenced multiple (often from 10 to 15) times. Shotgun sequencing can be carried out in a highly automated way, with few decisions to be made by the researcher, because the computer assembles the final draft of the sequence.

Concepts

Sequencing a genome requires breaking it up into small overlapping fragments whose DNA sequences can be determined in a sequencing reaction. In map-based sequencing, sequenced fragments are ordered into the final genome sequence with the use of genetic and physical maps. In whole-genome shotgun sequencing, the genome is assembled by comparing overlap in the sequences of small fragments.

✔ Concept Check 7

A contig is

a. a set of molecular markers used in genetic mapping.

b. a set of overlapping fragments that form a continuous stretch of DNA.

c. a set of fragments generated by a restriction enzyme.

d. a small DNA fragment used in sequencing.

Shotgun sequencing was initially used for assembling small genomes such as those of bacteria. When Venter proposed the use of this approach for sequencing the human genome, it was not at all clear that the approach could successfully assemble a complex genome consisting of billions of base pairs such as the human genome.

For several years, the public effort by the Human Genome Sequencing Consortium, using a map-based approach, and the private Celera effort, using shotgun sequencing, moved forward simultaneously. In the summer of 2000, both public and private sequencing projects announced the completion of a rough draft that included most of the sequence of the human genome, 5 years ahead of schedule. Analysis of this sequence was published 6 months later. The human genome sequence was declared completed in the spring of 2003, although some gaps still remain. For most chromosomes, the finished sequence is 99.999% accurate, with less than one base-pair error per 100,000 bp, which is 10 times as accurate as the initial goal.

The availability of the complete sequence of the human genome is proving to be of enormous benefit. It has made it easier to identify and isolate genes that contribute to many human diseases and to create probes that can be used in genetic testing, diagnosis, and drug development. The sequence is also providing important information about many basic cellular processes. Comparisons of the human genome with those of other organisms are adding to our understanding of evolution and the history of life.

Concepts

The Human Genome Project was an effort to sequence the entire human genome. Begun in 1990, a rough draft of the sequence was completed by two competing teams, an international consortium of publicly supported investigators and a private company, both of which finished a rough draft of the genome sequence in 2000. The entire sequence was completed in 2003.

✔ Concept Check 8

The Human Genome Sequencing Consortium used which approach in sequencing the human genome?

a. Whole-genome shotgun sequencing

b. Map-based sequencing

c. A combination of whole-genome shotgun sequencing and map-based sequencing

Single-Nucleotide Polymorphisms

Since the completion of the sequencing of the human genome, much of the effort of sequencers has focused on mapping differences among people in their genomic sequences.

Imagine that you are riding the elevator with a random stranger. How much of your genome do you have in common with this person? Studies of variation in the human genome indicate that you and the stranger will be identical at about 99.9% of your DNA sequences. This difference is very small in *relative* terms but, because the human genome is so large (3.2 billion base pairs), you and the stranger will actually be different at more than 3 million base pairs of your genomic DNA. These differences are what makes each of us unique, and they greatly affect our physical features, our health, and possibly even our intelligence and personality.

A site in the genome where individual members of a species differ in a single base pair is called a **single-nucleotide polymorphism** (SNP, pronounced "snip"). Arising through mutation, SNPs are inherited as allelic variants (just as are alleles that produce phenotypic differences, such as blood types), although SNPs do not usually produce a phenotypic difference. Single-nucleotide polymorphisms are numerous and are present throughout genomes. In a comparison of the same chromosome from two different people, a SNP can be found approximately every 1000 bp.

Most SNPs present within a population arose once from a single mutation that occurred on a particular chromosome and subsequently spread through the population. Thus, each SNP is initially associated with other SNPs (as well as other types of genetic variants or alleles) that were present on the particular chromosome on which the mutation arose. The specific set of SNPs and other genetic variants observed on a single chromosome or part of a chromosome is called a **haplotype** (**Figure 14.26**). SNPs within a haplotype are physically linked and therefore tend to be inherited together. New haplotypes can arise through mutation or crossing over, which breaks up the particular set of SNPs in a haplotype.

Because of their variability and widespread occurrence throughout the genome, SNPs are valuable as markers in linkage studies. When a SNP is physically close to a disease-causing locus, it will tend to be inherited along with the disease-causing allele. People with the disease will tend to have different SNPs from those of healthy people. A comparison of SNP haplotypes in people with a disease and in healthy people can reveal the presence of genes that affect the disease; because the disease gene and the SNP are closely linked, the location of the disease-causing gene can be determined from the location of associated SNPs. This approach is the same as that used in gene mapping with RFLPs, but there are many more SNPs than RFLPs, providing a dense set of

14.26 A haplotype is a specific set of single-nucleotide polymorphisms (SNPs) and other genetic variants observed on a single chromosome or part of a chromosome. Chromosomes 1a, 1b, 1c, and 1d represent different copies of a chromosome that might be found in a population.

variable markers covering the entire genome that can be used more effectively in mapping.

As expected, the availability of SNPs has greatly facilitated the search for genes that cause human diseases. In one of the most successful applications of SNPs for finding disease associations, a consortium of 50 researchers genotyped each of 17,000 people in the United Kingdom for 500,000 SNPs in 2007. They detected strong associations between 24 genes and chromosome segments and the incidence of seven common diseases, including coronary artery disease, Crohn disease, rheumatoid arthritis, bipolar disorder, hypertension, and two types of diabetes. The importance of this study is its demonstration that genomewide association studies utilizing SNPs can successfully locate genes that contribute to complex diseases caused by multiple genetic and environmental factors.

Bioinformatics

Complete genome sequences have now been determined for more than 1000 organisms, with many additional projects underway. These studies are producing tremendous quantities of sequence data. Cataloging, storing, retrieving, and analyzing this huge data set are major challenges of modern genetics. **Bioinformatics** is an emerging field consisting of molecular biology and computer science that centers on developing databases, computer-search algorithms, gene-prediction software, and other analytical tools that are used to make sense of DNA-, RNA-, and protein-sequence data. Bioinformatics develops and applies these tools to "mine the data," extracting the useful information from sequencing projects. The development and use of algorithms and computer software for analyzing DNA-, RNA-, and protein-sequence data have helped to make molecular biology a more quantitative field. Sequence data in publicly available databases, freely searchable with an Internet connection, enable scientists and students throughout the world to access this tremendous resource.

> ### Concepts
>
> Genomic projects are collecting databases of nucleotides that vary among individual organisms (single-nucleotide polymorphisms, SNPs). Bioinformatics is a interdisciplinary field that combines molecular biology and computer science. It develops databases of DNA, RNA, and protein sequences and tools for analyzing those sequences.

14.7 Functional Genomics Determines the Function of Genes by Using Genome-Based Approaches

A genomic sequence is, by itself, of limited use. Merely knowing the sequence would be like having a huge set of encyclopedias without being able to read: you could recognize the different letters but the text would be meaningless. **Functional genomics** characterizes what the sequences do—their function. The goals of functional genomics include the identification of all the RNA molecules transcribed from a genome, called the **transcriptome** of that genome, and all the proteins encoded by the genome, called the **proteome**. Functional genomics exploits both bioinformatics and laboratory-based experimental approaches in its search to define the function of DNA sequences.

Predicting Function from Sequence

The nucleotide sequence of a gene can be used to predict the amino acid sequence of the protein that it encodes. The protein can then be synthesized or isolated and its properties studied to determine its function. However, this biochemical approach to understanding gene function is both time consuming and expensive. A major goal of functional genomics has been to develop computational methods that allow gene function to be identified from DNA sequence alone, bypassing the laborious process of isolating and characterizing individual proteins.

One computational method (often the first employed) for determining gene function is to conduct a homology search, which relies on comparisons of DNA and protein sequences from the same organism and from different organisms. Genes that are evolutionarily related are said to be **homologous**. Databases containing genes and proteins found in a wide array of organisms are available for homology searches. Powerful computer programs, such as BLAST, have been developed for scanning these databases to look for particular sequences. Suppose a geneticist sequences a genome and locates a gene that encodes a protein of unknown function. A homology search conducted on databases containing the DNA or protein sequences of other organisms may identify one or more homologous sequences. If a function is known for a protein encoded by one of these sequences, that function may provide information about the function of the newly discovered protein.

> ### Concepts
>
> The function of an unknown gene can sometimes be determined by finding genes with similar sequence whose function is known.

Gene Expression and Microarrays

Many important clues about gene function come from knowing when and where the genes are expressed. The development of microarrays has allowed the expression of thousands of genes to be monitored simultaneously.

Microarrays rely on nucleic acid hybridization, in which a known DNA fragment is used as a probe to find complementary sequences (**Figure 14.27**). Numerous known DNA fragments are fixed to a solid support in an orderly pattern or array, usually as a series of dots. These DNA fragments (the probes) usually correspond to known genes.

After the microarray has been constructed, mRNA, DNA, or cDNA isolated from experimental cells is labeled with fluorescent nucleotides and applied to the array. Any of the DNA or RNA molecules that are complementary to probes on the array will hybridize with them and emit fluorescence, which can be detected by an automated scanner. An array containing tens of thousands of probes can be applied to a glass slide or silicon wafer just a few square centimeters in size.

Used with cDNA, microarrays can provide information about the expression of thousands of genes, enabling scientists to study which genes are active in particular tissues. They can also be used to investigate how gene expression changes in the course of biological processes such as development or disease progression. In one study, researchers used microarrays to examine the expression patterns of 25,000 genes from primary tumors of 78 young women who

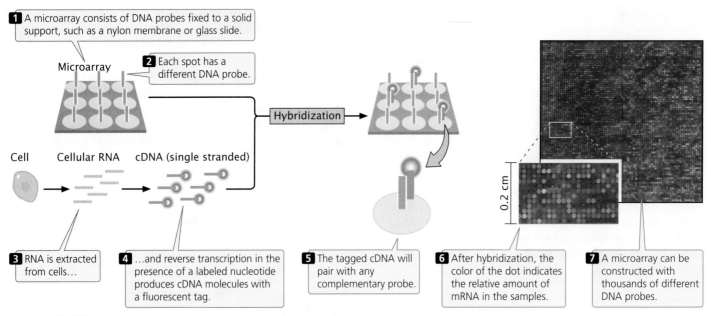

14.27 Microarrays are used to simultaneously detect the expression of many genes. [*After D. Lockhart and E. Winzeler,* Nature *405:827, 2000.*]

had breast cancer (**Figure 14.28**). Messenger mRNA from cancer cells and noncancer cells was converted into cDNA and labeled with red fluorescent nucleotides and with green fluorescent nucleotides, respectively. The labeled cDNAs were mixed and hybridized to a DNA chip, which contains DNA probes from different genes. Hybridization of the red (cancer) and green (noncancer) cDNAs is proportional to the relative amounts of mRNA in the samples. The fluorescence of each spot is assessed with microscopic scanning and appears as a single color. Red indicates the overexpression of a gene in the cancer cells relative to that in the noncancer cells (more red-labeled cDNA hybridizes), whereas green indicates the underexpression of a gene in the cancer cells relative to that in the noncancer cells (more green-labeled cDNA hybridizes). Yellow indicates equal expression in both types of cells (equal hybridization of red- and green-labeled cDNAs), and no color indicates no expression in either type of cell.

In 34 of the 78 patients, the cancer later spread to other sites; the other 44 patients remained free of breast cancer for 5 years after their initial diagnoses. The researchers identified a subset of 70 genes whose expression patterns in the initial tumors accurately predicted whether the cancer would later spread (see Figure 14.28). This degree of prediction was much higher than that of traditional predictive measures, which are based on the size and histology of the tumor. These results, though preliminary and confined to a small sample of cancer patients, suggest that gene-expression data obtained from microarrays can be a powerful tool in determining the nature of cancer treatment.

Concepts

Microarrays, consisting of DNA probes attached to a solid support, can be used to determine which RNA and DNA sequences are present in a mixture of nucleic acids. They are capable of determining which RNA molecules are being synthesized and can thus be used to examine changes in gene expression.

14.8 Comparative Genomics Studies How Genomes Evolve

Genome-sequencing projects provide detailed information about gene content and organization in different species and even in different members of the same species, allowing inferences about how genes function and genomes evolve. They also provide important information about evolutionary relationships among organisms and about factors that influence the speed and direction of evolution. **Comparative genomics** is the field of genomics that compares similarities and differences in gene content, function, and organization among genomes of different organisms.

Prokaryotic Genomes

Hundreds of bacterial genomes have now been sequenced. Most prokaryotic genomes consist of a single circular chromosome, but there are exceptions, such as *Vibrio cholerae,*

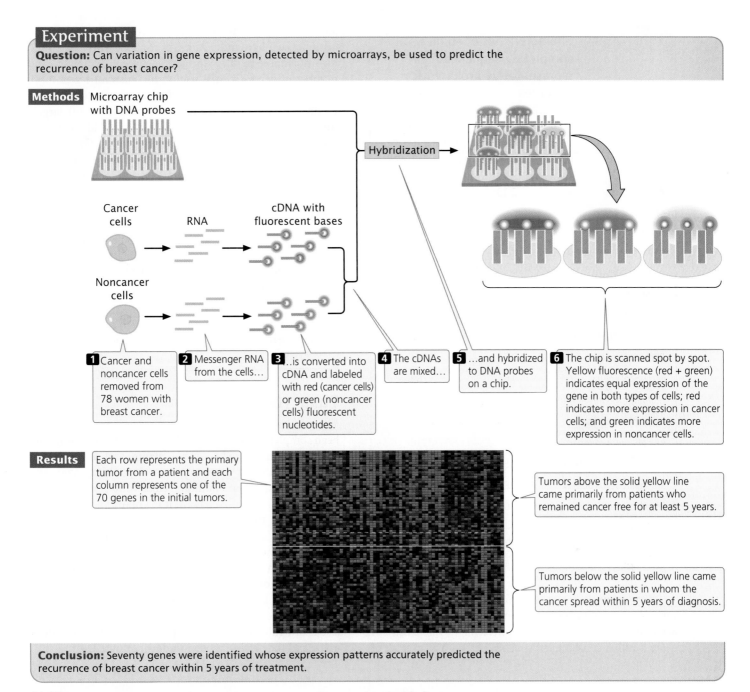

Experiment

Question: Can variation in gene expression, detected by microarrays, be used to predict the recurrence of breast cancer?

Methods Microarray chip with DNA probes

Hybridization

Cancer cells

RNA

cDNA with fluorescent bases

Noncancer cells

1 Cancer and noncancer cells removed from 78 women with breast cancer.

2 Messenger RNA from the cells…

3 …is converted into cDNA and labeled with red (cancer cells) or green (noncancer cells) fluorescent nucleotides.

4 The cDNAs are mixed…

5 …and hybridized to DNA probes on a chip.

6 The chip is scanned spot by spot. Yellow fluorescence (red + green) indicates equal expression of the gene in both types of cells; red indicates more expression in cancer cells; and green indicates more expression in noncancer cells.

Results Each row represents the primary tumor from a patient and each column represents one of the 70 genes in the initial tumors.

Tumors above the solid yellow line came primarily from patients who remained cancer free for at least 5 years.

Tumors below the solid yellow line came primarily from patients in whom the cancer spread within 5 years of diagnosis.

Conclusion: Seventy genes were identified whose expression patterns accurately predicted the recurrence of breast cancer within 5 years of treatment.

14.28 Microarrays can be used to examine gene expression associated with disease progression. [After L. J. van't Veer, Nature 405:532, 2002.]

the bacterium that causes cholera, which has two circular chromosomes, and *Borrelia burgdorferi*, which has one large linear chromosome and 21 smaller chromosomes.

Genome size and number of genes The total amount of DNA in prokaryotic genomes ranges from 490,885 bp in *Nanoarchaeum equitans*, an archaeon that lives entirely within another archaeon, to 9,105,828 bp in *Bradyrhizobium japonicum*, a soil bacterium (**Table 14.3**). Although this range in genome size might seem extensive, it is much less than the enormous range of genome sizes seen in eukaryotes, which can vary from a few million base pairs to hundreds of billions of base pairs. *Escherichia coli*, the most widely used bacterium for genetic studies, has a fairly typical genome size at 4.6 million base pairs. Archaea and bacteria are similar in their ranges of genome size. Surprisingly, genome size shows extensive variation within some species; for example, different strains of *E. coli* vary in genome size by more than 1 million base pairs.

Table 14.3	Characteristics of some completely sequenced representative prokaryotic genomes	
Species	Size (millions of base pairs)	Number of Predicted Genes
Archaea		
Archaeoglobus fulgidus	2.18	2407
Methanobacterium thermoautotrophicum	1.75	1869
Methanococcus jannaschii	1.66	1715
Nanoarchaeum equitans	0.490	536
Eubacteria		
Bacillus subtilis	4.21	4100
Bradyrhizobium japonicum	9.11	8317
Buchnera species	0.64	564
Escherichia coli	4.64	4289
Haemophilus influenzae	1.83	1709
Mesorhizobium loti	7.04	6752
Mycobacterium tuberculosis	4.41	3918
Mycoplasma genitalium	0.58	480
Staphylococcus aureus	2.88	2697
Vibrio cholerae	4.03	3828

Source: Data from the Genome Atlas of the Center for Biological Sequence Analysis, http://www.cbs.dtu.dk/services/GenomeAtlas/.

Among prokaryotes, the number of genes typically varies from 1000 to 2000, but some species have as many as 6700 and others as few as 480. Interestingly, the density of genes is rather constant across all species, with an average of about 1 gene per 1000 bp. Thus, prokaryotes with larger genomes will have more genes, in contrast with eukaryotes, for which there is little association between genome size and number of genes (see the section on eukaryotic genomes). The evolutionary factors that determine the size of genomes in prokaryotes (as well as in eukaryotes) is still largely unknown. Only about half of the genes identified in prokaryotic genomes can be assigned a function (**Figure 14.29**). Almost a quarter of the genes have no significant sequence similarity to any known genes in other bacteria.

Concepts

Comparative genomics compares the content and organization of whole genomic sequences from different organisms. Prokaryotic genomes are small, usually ranging from 1 million to 3 million base pairs of DNA, with several thousand genes.

✔ Concept Check 9

What is the relation between genome size and gene number in prokaryotes?

Eukaryotic Genomes

The genomes of more than 100 eukaryotic organisms have been completely sequenced, including a number of fungi and protists, several insects, several plants, and a number of vertebrates such as the mouse, rat, dog, chimpanzee, and human. Hundreds of additional eukaryotic genomes are in the process of being sequenced. It is important to note that, even though the genomes of these organisms have been "completely sequenced," many of the final assembled sequences contain gaps, and regions of heterochromatin may not have been sequenced at all. Thus, the sizes of eukaryotic genomes are often estimates, and the number of base pairs given for the genome size of a particular species may vary. Predicting the number of genes that are present in a genome also is difficult and may vary, depending on the assumptions made and the particular gene-finding software used.

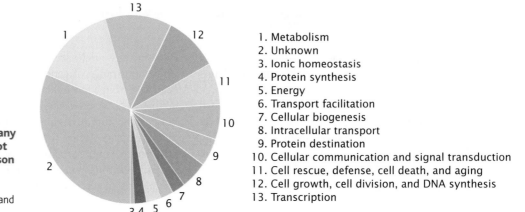

14.29 The functions of many genes in prokaryotes cannot be determined by comparison with genes in other prokaryotes. Percentages of genes affecting various known and unknown functions in *E. coli*.

1. Metabolism
2. Unknown
3. Ionic homeostasis
4. Protein synthesis
5. Energy
6. Transport facilitation
7. Cellular biogenesis
8. Intracellular transport
9. Protein destination
10. Cellular communication and signal transduction
11. Cell rescue, defense, cell death, and aging
12. Cell growth, cell division, and DNA synthesis
13. Transcription

Table 14.4	Characteristics of some eukaryotic genomes that have been completely sequenced	
Species	Genome Size (millions of base pairs)	Number of Predicted Genes
Saccharomyces cerevisiae (yeast)	12	6,144
Arabidopsis thaliana (plant)	125	25,706
Caenorhabditis elegans (nematode worm)	103	20,598
Drosophila melanogaster (fruit fly)	170	13,525
Anopheles gambiae (mosquito)	278	14,707
Danio rerio (zebrafish)	1,465	22,409
Takifugu rubripes (tiger pufferfish)	329	22,089
Mus musculus (mouse)	2,627	26,762
Ratus novegicus (Norway rat)	2,571	23,761
Pan troglodytes (chimpanzee)	2,733	22,524
Homo sapiens (human)	3,223	~24,000

Source: Ensembl Web site: http//www.ensembl.org.

Genome size and number of genes The genomes of eukaryotic organisms (Table 14.4) are larger than those of prokaryotes, and, in general, multicellular eukaryotes have more DNA than do simple, single-celled eukaryotes such as yeast (see p. 212 in Chapter 8). However, there is no close relation between genome size and complexity among the multicellular eukaryotes. For example, the nematode worm *Caenorhabditis elegans* is structurally more complex than the plant *Arabidopsis thaliana* but has considerably less DNA.

In general, eukaryotic genomes also contain more genes than do prokaryotes (but there are some large bacteria that have more genes than single-celled yeasts do), and the genomes of multicellular eukaryotes have more genes than do the genomes of single-celled eukaryotes. In contrast with bacteria, there is no correlation between genome size and number of genes in eukaryotes. The *number* of genes among multicellular eukaryotes also is not obviously related to phenotypic complexity: humans have more genes than do invertebrates but only twice as many as fruit flies and only slightly

more than the plant *A. thaliana.* The nematode *C. elegans* has more genes than does *D. melanogaster* but is less complex. Additionally, the pufferfish has only about one-tenth the amount of DNA present in humans and mice but has almost as many genes. Eukaryotic genomes contain multiple copies of many genes, indicating that gene duplication has been an important process in genome evolution.

Gene deserts The density of genes in a typical eukaryotic genome varies greatly, with some chromosomes having a high density of genes and others being relatively gene poor. In some areas of the genome, long stretches of DNA, often consisting of hundreds of thousands to millions of base pairs are completely devoid of any known genes or other functional sequences; these regions are known as **gene deserts**. Gene deserts are surprisingly common in eukaryotes. The human genome contains about 500 gene deserts or more, making up approximately 25% of the total euchromatin in the human genome. Gene deserts are particularly common on human chromosomes 4, 5, and 13, where they cover as much as 40% of the entire chromosome. What is the purpose of a gene desert? Why does it exist? A possible answer is that it contains DNA sequences that have a functional role—perhaps in regulating genes or in the overall architecture of the chromosome—but we are unable to recognize the function of the sequences that it contains.

Transposable elements A substantial part of the genomes of most multicellular organisms consists of moderately and highly repetitive sequences (see Chapter 8), and the percentage of repetitive sequences is usually higher in those species with larger genomes (**Table 14.5**). Most of these repetitive sequences appear to have arisen through transposition and are particularly evident in the human genome: 45% of the DNA in the human genome is derived from transposable elements, many of which are defective and no longer able to move. Most of the DNA in multicellular

Table 14.5	Percentage of genome consisting of interspersed repeats derived from transposable elements
Organism	Percentage of Genome
Plant (Arabidopsis thaliana)	10.5
Nematode worm (Caenorhabditis elegans)	6.5
Fly (Drosophila melanogaster)	3.1
Tiger pufferfish (Takifugu rubripes)	2.7
Human (Homo sapiens)	44.4

organisms is noncoding, and many genes are interrupted by introns. In the more complex eukaryotes, both the number and the length of the introns are greater.

Protein diversity   In spite of only a modest increase in gene number, vertebrates have considerably more protein diversity than do invertebrates. One way to measure the amount of protein diversity is by counting the number of protein domains, which are characteristic parts of proteins that are often associated with a function. Vertebrate genomes do not encode more protein domains than do invertebrate genomes; for example, there are 1262 domains in humans compared with 1035 in fruit flies. However, the existing domains in humans are assembled into more combinations, leading to many more types of proteins. For example, the human genome contains almost twice as many arrangements of protein domains as do worms or flies and almost six times as many as does yeast.

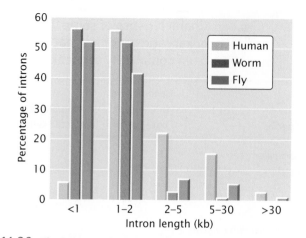

14.30 The introns of genes in humans are generally longer than the introns of genes in worms and flies.

Concepts

Genome size varies greatly among eukaryotic species. For multicellular eukaryotic organisms, there is no clear relation between organismal complexity and amount of DNA or gene number. A substantial part of the genome in eukaryotic organisms consists of repetitive DNA, much of which is derived from transposable elements. Vast regions of DNA may contain no genes or other functional sequences.

The Human Genome

The human genome, which is fairly typical of mammalian genomes, has been extensively studied and analyzed because of its importance to human health and evolution. It is

3.2 billion base pairs in length, but only about 25% of the DNA is transcribed into RNA, and less than 2% encodes proteins. Active genes are often separated by vast regions of noncoding DNA, much of which consists of repeated sequences derived from transposable elements.

The average gene in the human genome is approximately 27,000 bp in length, with about 9 exons. (One exceptional gene has 234 exons.) The introns of human genes are much longer, and there are more of them than in other genomes (**Figure 14.30**). The human genome does not encode substantially more protein domains, but the domains are combined in more ways to produce a relatively diverse proteome. Gene functions encoded by the human genome are presented in **Figure 14.31**. Similarly to the situation in bacteria, the function of many genes in the human genome is still unknown. A single gene often encodes multiple proteins through alternative splicing; each gene encodes, on the

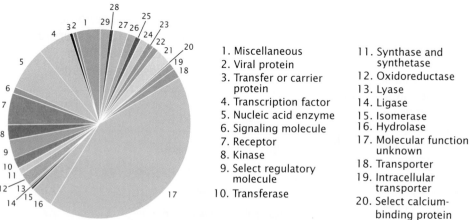

1. Miscellaneous
2. Viral protein
3. Transfer or carrier protein
4. Transcription factor
5. Nucleic acid enzyme
6. Signaling molecule
7. Receptor
8. Kinase
9. Select regulatory molecule
10. Transferase
11. Synthase and synthetase
12. Oxidoreductase
13. Lyase
14. Ligase
15. Isomerase
16. Hydrolase
17. Molecular function unknown
18. Transporter
19. Intracellular transporter
20. Select calcium-binding protein
21. Proto-oncogene
22. Structural protein of muscle
23. Motor
24. Ion channel
25. Immunoglobulin
26. Extracellular matrix
27. Cytoskeletal structural protein
28. Chaperone
29. Cell adhesion

14.31 Functions for many human genes have yet to be determined. Percentages of genes affecting various known and unknown functions.

average, two or three different mRNAs, meaning that the human genome, with approximately 24,000 genes, might encode 72,000 mRNAs or more.

Proteomics

DNA sequence data are tremendous sources of insight into the biology of an organism, but they are not the whole story. In recent years, molecular biologists have turned their attention to analysis of the protein content of cells. The ultimate goal is to determine the proteome, the complete set of proteins found in a given cell, and the study of the proteome is termed **proteomics**.

The traditional method for identifying a protein is to remove its amino acids one at a time and determine the identity of each amino acid removed. This method is far too slow and labor intensive for analyzing the thousands of proteins present in a typical cell. Today, researchers use **mass spectrometry**, which is a method for precisely determining the molecular mass of a molecule. In mass spectrometry, a molecule is ionized and its migration rate in an electrical field is determined. Because small molecules migrate more rapidly than larger molecules, the migration rate can accurately determine the mass of the molecule.

To analyze proteins with mass spectrometry, a protein is broken up into small peptide fragments and mass spectrometry is then used to separate the peptides on the basis of their mass-to-charge (m/z) ratio (**Figure 14.32**). A computer program then searches through a database of proteins to find a match between the profile generated and the profile expected with a known protein. Using bioinformatics, the computer creates "virtual digests" and predicts the profiles of all proteins found in a genome, given the DNA sequences of the protein-encoding genes.

Protein–protein interactions can be analyzed with **protein microarrays**, which are similar to the microarrays used for examining gene expression. With this technique, a large number of different proteins are applied to a glass slide as a series of spots, with each spot containing a different protein. In one application, each spot is an antibody for a different protein and is labeled with a tag that fluoresces when bound. An extract of tissue is applied to the protein microarray. A spot of fluorescence appears when a protein in the extract binds to antibody, indicating the presence of that particular protein in the tissue.

Concepts

The proteome is the complete set of proteins found in a cell. Techniques of protein separation and mass spectrometry are used to identify the proteins present within a cell. Microarrays are used to determine sets of interacting proteins. Structural proteomics attempts to determine the structure of all proteins.

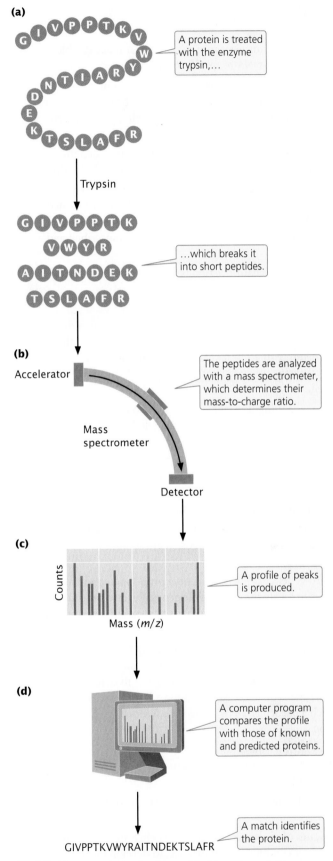

(a) A protein is treated with the enzyme trypsin,…

Trypsin

…which breaks it into short peptides.

(b)

Accelerator

The peptides are analyzed with a mass spectrometer, which determines their mass-to-charge ratio.

Mass spectrometer

Detector

(c)

Counts

A profile of peaks is produced.

Mass (m/z)

(d)

A computer program compares the profile with those of known and predicted proteins.

GIVPPTKVWYRAITNDEKTSLAFR

A match identifies the protein.

14.32 Mass spectrometry is used to identify proteins.

Concepts Summary

- Restriction endonucleases are enzymes that make double-stranded cuts in DNA at specific base sequences.

- DNA fragments can be separated with the use of gel electrophoresis and visualized by staining the gel with a dye that is specific for nucleic acids or by labeling the fragments with a radioactive or chemical tag.

- In gene cloning, a gene or a DNA fragment is placed into a bacterial cell, where it will be multiplied as the cell divides.

- Plasmids, small circular pieces of DNA, are often used as vectors to ensure that a cloned gene is stable and replicated within the recipient cells. Expression vectors contain sequences necessary for foreign DNA to be transcribed and translated.

- The polymerase chain reaction is a method for amplifying DNA enzymatically without cloning.

- Genes can be isolated by creating a DNA library—a set of bacterial colonies or viral plaques that each contain a different cloned fragment of DNA. A genomic library contains the entire genome of an organism; a cDNA library contains DNA fragments complementary to all the different mRNAs in a cell.

- Positional cloning uses linkage relations to determine the location of genes without any knowledge of their products.

- The Sanger (dideoxy) method of DNA sequencing uses special substrates for DNA synthesis (dideoxynucleoside triphosphates, ddNTPs) that terminate synthesis after they are incorporated into the newly made DNA.

- Short tandem repeats (STRs) and microsatellites are used to identify people by their DNA sequences (DNA fingerprinting).

- Forward genetics begins with a phenotype and conducts analyses to locate the responsible genes. Reverse genetics starts with a DNA sequence and conducts analyses to determine its phenotypic effect.

- Transgenic animals, produced by injecting DNA into fertilized eggs, contain foreign DNA that is integrated into a chromosome. Knockout mice are transgenic mice in which a normal gene is disabled. Knock-in mice are transgenic mice in which a particular DNA sequence is inserted into a known location.

- The mouse *Mus musculus* is an excellent model genetic organism because of its similarity to humans, small size, and short generation time.

- RNA interference is used to silence the expression of specific genes.

- Techniques of molecular genetics are being used to create products of commercial importance, to develop diagnostic tests, and to treat diseases.

- Genomics is the field of genetics that attempts to understand the content, organization, and function of genetic information contained in whole genomes.

- Genetic maps position genes relative to other genes by determining rates of recombination and are measured in percent recombination. Physical maps are based on the physical distances between genes and are measured in base pairs.

- The Human Genome Project is an effort to determine the entire sequence of the human genome.

- Sequencing a whole genome requires breaking the genome into small overlapping fragments whose DNA sequences can be determined in sequencing reactions. The individual sequences can be ordered into a whole-genome sequence with the use of a map-based approach, in which fragments are assembled in order by using previously created genetic and physical maps, or with the use of a whole-genome shotgun approach, in which overlap between fragments is used to assemble them into a whole-genome sequence.

- Single-nucleotide polymorphisms are single-base differences in DNA between individual organisms and are valuable as markers in linkage studies.

- Bioinformatics is a synthesis of molecular biology and computer science that develops tools to store, retrieve, and analyze DNA-, cDNA-, and protein-sequence data.

- Homologous genes are evolutionarily related. Gene function may be determined by looking for homologous sequences whose function has been previously determined.

- A microarray consists of DNA fragments fixed in an orderly pattern to a solid support, such as a nylon filter or glass slide. Microarrays can be used to monitor the expression of thousands of genes simultaneously.

- Most prokaryotic species have between 1 million and 3 million base pairs of DNA and from 1000 to 2000 genes. Compared with that of eukaryotic genomes, the density of genes in prokaryotic genomes is relatively uniform, with about one gene per 1000 bp.

- Eukaryotic genomes are larger and more variable in size than prokaryotic genomes. There is no clear relation between organismal complexity and the amount of DNA or number of genes among multicellular organisms.

- Proteomics determines the protein content of a cell and the functions of those proteins. Proteins within a cell can be separated and identified with the use of mass spectrometry.

Important Terms

recombinant DNA technology (p. 348)
genetic engineering (p. 348)
biotechnology (p. 348)
restriction enzyme (p. 349)
restriction endonuclease (p. 349)
cohesive end (p. 349)
gel electrophoresis (p. 351)
autoradiography (p. 351)
probe (p. 352)
gene cloning (p. 352)
cloning vector (p. 352)
cosmid (p. 354)
bacterial artificial chromosome (BAC) (p. 354)
expression vector (p. 354)
polymerase chain reaction (PCR) (p. 355)
Taq polymerase (p. 356)
DNA library (p. 356)
genomic library (p. 356)

cDNA library (p. 356)
positional cloning (p. 358)
restriction fragment length polymorphism (RFLP) (p. 358)
DNA sequencing (p. 359)
dideoxyribonucleoside triphosphate (ddNTP) (p. 359)
DNA fingerprinting (p. 361)
microsatellite (p. 362)
short tandem repeat (STR) (p. 362)
forward genetics (p. 364)
reverse genetics (p. 364)
transgene (p. 364)
knockout mice (p. 365)
knock-in mice (p. 365)
gene therapy (p. 368)
genomics (p. 369)
structural genomics (p. 369)
genetic map (p. 369)

physical map (p. 369)
map-based sequencing (p. 371)
contig (p. 371)
whole-genome shotgun sequencing (p. 373)
single-nucleotide polymorphism (SNP) (p. 374)
haplotype (p. 374)
bioinformatics (p. 374)
functional genomics (p. 375)
transcriptome (p. 375)
proteome (p. 375)
homologous genes (p. 375)
microarray (p. 375)
comparative genomics (p. 376)
gene desert (p. 379)
proteomics (p. 381)
mass spectrometry (p. 381)
protein microarray (p. 381)

Answers to Concept Checks

1. First, the gene must be located and isolated from the rest of the genomic DNA. Then, the gene must be inserted into bacteria in a form that is stable and will be replicated. The gene must be placed in the bacteria in a way that ensures that it will be transcribed and translated. Finally, those bacteria that have taken up an active form of the gene must be separated from other bacteria.

2. Restriction enzymes exist naturally in bacteria, which use them to prevent the entry of viral DNA.

3. c

4. The gene and plasmid are cut with the same restriction enzyme and mixed together. DNA ligase is used to seal nicks in the sugar–phosphate bonds.

5. d

6. b

7. b

8. b

9. Species with larger genomes generally have more genes than do species with smaller genomes, and so gene density is relatively constant.

Worked Problems

1. A molecule of double-stranded DNA that is 5 million base pairs long has a base composition that is 62% G + C. How many times, on average, are the following restriction sites likely to be present in this DNA molecule?

a. *Bam*HI (recognition sequence is GGATCC)
b. *Hind*III (recognition sequence is AAGCTT)
c. *Hpa*II (recognition sequence is CCGG)

• **Solution**

The percentages of G and C are equal in double-stranded DNA; so, if G + C = 62%, then %G = %C = 62%/2 = 31%. The percentage of A + T = (100% − G + C) = 38%, and %A= %T = 38%/2 = 19%. To determine the probability of finding a particular base sequence, we use the multiplication rule, multiplying together the probably of finding each base at a particular site.

a. The probability of finding the sequence GGATCC = 0.31 × 0.31 × 0.19 × 0.19 × 0.31 × 0.31 = 0.0003333. To determine the average number of recognition sequences in a 5-million-base-pair piece of DNA, we multiply 5,000,000 bp × 0.00033 = 1666.5 recognition sequences.

b. The number of AAGCTT recognition sequences is 0.19 × 0.19 × 0.31 × 0.31 × 0.19 × 0.19 × 5,000,000 = 626 recognition sequences.

c. The number of CCGG recognition sequences is 0.31 × 0.31 × 0.31 × 0.31 × 5,000,000 = 46,176 recognition sequences.

2. You are given the following DNA fragment to sequence: 5′-GCTTAGCATC-3′. You first clone the fragment in bacterial cells to produce sufficient DNA for sequencing. You isolate the DNA from the bacterial cells and carry out the dideoxy-sequencing method. You then separate the products of the polymerization reactions by gel electrophoresis. Draw the bands that should appear on the gel from the four sequencing reactions.

• Solution

In the dideoxy-sequencing reaction, the original fragment is used as a template for the synthesis of a new DNA strand; the sequence of the new strand is the sequence that is actually determined. The first task, therefore, is to write out the sequence of the newly synthesized fragment, which will be complementary and antiparallel to the original fragment. The sequence of the newly synthesized strand, written 5′ → 3′ is: 5′-GATGCTAAGC-3′. Bands representing this sequence will appear on the gel, with the bands representing nucleotides near the 5′ end of the molecule at the bottom of the gel.

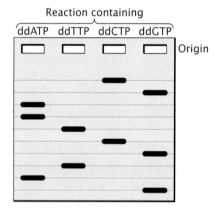

3. A linear piece of DNA that is 30 kb long is first cut with *Bam*HI, then with *Hpa*II, and, finally, with both *Bam*HI and *Hpa*II together. Fragments of the following sizes were obtained from this reaction:

　*Bam*HI: 20-kb, 6-kb, and 4-kb fragments

　*Hpa*II: 21-kb and 9-kb fragments

　*Bam*HI and HpaII: 20-kb, 5-kb, 4-kb, and 1-kb fragments

Draw a restriction map of the 30-kb piece of DNA, indicating the locations of the *Bam*HI and *Hpa*II restriction sites.

• Solution

This problem can be solved correctly through a variety of approaches; this solution applies one possible approach.

　When cut by *Bam*HI alone, the linear piece of DNA is cleaved into three fragments; so there must be two *Bam*HI restriction sites. When cut with *Hpa*II alone, a clone of the same piece of DNA is cleaved into only two fragments; so there is a single *Hpa*II site.

Let's begin to determine the location of these sites by examining the *Hpa*II fragments. Notice that the 21-kb fragment produced when the DNA is cut by *Hpa*II is not present in the fragments produced when the DNA is cut by *Bam*HI and *Hpa*II together (the double digest); this result indicates that the 21-kb *Hpa*II fragment has within it a *Bam*HI site. If we examine the fragments produced by the double digest, we see that the 20-kb and 1-kb fragments sum to 21 kb; so a *Bam*HI site must be 20 kb from one end of the fragment and 1 kb from the other end.

20 kb   1 kb

Similarly, we see that the 9-kb *Hpa*II fragment does not appear in the double digest and that the 5-kb and 4-kb fragments in the double digest add up to 9 kb; so another *Bam*HI site must be 5 kb from one end of this fragment and 4 kb from the other end.

5 kb   4 kb

Now, let's examine the fragments produced when the DNA is cut by *Bam*HI alone. The 20-kb and 4-kb fragments are also present in the double digest; so neither of these fragments contains an *Hpa*II site. The 6-kb fragment, however, is not present in the double digest, and the 5-kb and 1-kb fragments in the double digest sum to 6 kb; so this fragment contains an *Hpa*II site that is 5 kb from one end and 1 kb from the other end.

5 kb   1 kb

We have accounted for all the restriction sites, but we must still determine the order of the sites on the original 30-kb fragment. Notice that the 5-kb fragment must be adjacent to both the 1-kb and the 4-kb fragments; so it must be in between these two fragments.

Hpall site   BamHI site

1 kb   5 kb   4 kb

We have also established that the 1-kb and 20-kb fragments are adjacent; because the 5-kb fragment is on one side, the 20-kb fragment must be on the other, completing the restriction map:

20 kb   1 kb   5 kb   4 kb

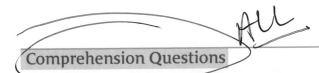

Comprehension Questions

Section 14.1

1. What role do restriction enzymes play in bacteria? How do bacteria protect their own DNA from the action of restriction enzymes?

*2. Explain how gel electrophoresis is used to separate DNA fragments of different lengths.

*3 Give three important characteristics of cloning vectors.

*4. Briefly explain how an antibiotic-resistance gene and the *lacZ* gene can be used as markers to determine which cells contain a particular plasmid.

*5. Briefly explain how the polymerase chain reaction is used to amplify a specific DNA sequence.

Section 14.2

*6. How does a genomic library differ from a cDNA library? How is a genomic library created?

7. Briefly explain how a gene can be isolated through positional cloning.

Section 14.3

8. What is the purpose of the dideoxynucleoside triphosphate in the dideoxy sequencing reaction?

*9. What is DNA fingerprinting? What types of sequences are examined in DNA fingerprinting?

Section 14.4

10. How does a reverse genetics approach differ from a forward genetics approach?

*11. What are knockout mice and for what are they used?

12. How is RNA interference used in the analysis of gene function?

Section 14.5

13. What is gene therapy?

Section 14.6

*14. What is the difference between a genetic map and a physical map? Which generally has higher resolution and accuracy and why?

*15. What is the difference between a map-based approach to sequencing a whole genome and a whole-genome shotgun approach?

16. What is a single-nucleotide polymorphism (SNP)? How are SNPs used in genomic studies?

17. What is a haplotype?

Section 14.7

18. What is a microarray? How can it be used to obtain information about gene function?

Section 14.8

19. What is the relation between genome size and gene number in prokaryotes?

20. DNA content varies considerably among different multicellular organisms. Is this variation closely related to the number of genes and the complexity of the organism? If not, what accounts for the differences?

21. What is a gene desert?

22. How does proteomics differ from genomics?

23. How is mass spectrometry used to identify proteins in a cell?

Application Questions and Problems

Section 14.1

24. How often, on average, would you expect a restriction endonuclease to cut a DNA molecule if the recognition sequence for the enzyme had 5 bp? (Assume that the four types of bases are equally likely to be found in the DNA and that the bases in a recognition sequence are independent.) How often would the endonuclease cut the DNA if the recognition sequence had 8 bp?

*25. A microbiologist discovers a new restriction endonuclease. When DNA is digested by this enzyme, fragments that average 1,048,500 bp in length are produced. What is the most likely number of base pairs in the recognition sequence of this enzyme?

*26. Restriction mapping of a linear piece of DNA reveals the following *Eco*RI restriction sites.

EcoRI site 1   EcoRI site 2

2 kb ↓ 4 kb ↓ 5 kb

a. This piece of DNA is cut by *Eco*RI, the resulting fragments are separated by gel electrophoresis, and the gel is stained with ethidium bromide. Draw a picture of the bands that will appear on the gel.

b. If a mutation that alters *Eco*RI site 1 occurs in this piece of DNA, how will the banding pattern on the gel differ from the one that you drew in part *a*?

c. If mutations that alter *Eco*RI sites 1 and 2 occur in this piece of DNA, how will the banding pattern on the gel differ from the one that you drew in part *a?*

d. If 1000 bp of DNA were inserted between the two restriction sites, how would the banding pattern on the gel differ from the one that you drew in part *a?*

e. If 500 bp of DNA between the two restriction sites were deleted, how would the banding pattern on the gel differ from the one that you drew in part *a?*

*27. Which vectors (plasmid, phage λ, cosmid, bacterial artificial chromosome) can be used to clone a continuous fragment of DNA with the following lengths?

a. 4 kb   **c.** 35 kb

b. 20 kb   **d.** 100 kb

28. A geneticist uses a plasmid for cloning that has the *lacZ* gene and a gene that confers resistance to penicillin. The geneticist inserts a piece of foreign DNA into a restriction site that is located within the *lacZ* gene and uses the plasmid to transform bacteria. Explain how the geneticist can identify bacteria that contain a copy of a plasmid with the foreign DNA.

Section 14.2

*29. Suppose that you have just graduated from college and have started working at a biotechnology firm. Your first job assignment is to clone the pig gene for the hormone prolactin. Assume that the pig gene for prolactin has not yet been isolated, sequenced, or mapped; however, the mouse gene for prolactin has been cloned and the amino acid sequence of mouse prolactin is known. Briefly explain two different strategies that you might use to find and clone the pig gene for prolactin.

Section 14.3

30. Suppose that you want to sequence the following DNA fragment:

5′–TCCCGGGAAA-primer site–3′

You first clone the fragment in bacterial cells to produce sufficient DNA for sequencing. You isolate the DNA from the bacterial cells and apply the dideoxy sequencing method. You then separate the products of the polymerization reactions by gel electrophoresis. Draw the bands that should appear on the gel from the four sequencing reactions.

Reaction containing

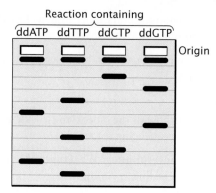

*31. Suppose that you are given a short fragment of DNA to sequence. You clone the fragment, isolate the cloned DNA fragment, and set up a series of four dideoxy reactions. You then separate the products of the reactions by gel electrophoresis and obtain the following banding pattern:

Reaction containing

Write out the base sequence of the original fragment that you were given.

Original sequence: 5′– _____ –3′

32. The adjoining autoradiograph is from the original study that first sequenced the cystic fibrosis gene (J. R. Riordan et al. 1989. *Science* 245:1066–1073). From the autoradiograph, determine the sequence of the normal copy of the gene and the sequence of the mutated copy of the gene. Identify the location of the mutation that causes cystic fibrosis (CF).

CTAG   CTAG

DNA from a healthy person   DNA from a person with CF

Section 14.4

*33. You have discovered a gene in mice that is similar to a gene in yeast. How might you determine whether this gene is essential for development in mice?

Section 14.6

*34. A piece of DNA that is 14 kb long is cut first by *Eco*RI alone, then by *Sma*I alone, and, finally, by both *Eco*RI and *Sma*I together. The following results are obtained:

Digestion by *Eco*RI alone	Digestion by *Sma*I alone	Digestion by both *Eco*RI and *Sma*I
3-kb fragment	7-kb fragment	2-kb fragment
5-kb fragment	7-kb fragment	3-kb fragment
6-kb fragment		4-kb fragment
		5-kb fragment

Draw a map of the *Eco*RI and *Sma*I restriction sites on this 14-kb piece of DNA, indicating the relative positions of the restriction sites and the distances between them.

Section 14.7

35. Microarrays can be used to determine the levels of gene expression. In one type of microarray, hybridization of the red (experimental) and green (control) cDNAs is proportional to the relative amounts of mRNA in the samples. Red indicates the overexpression of a gene and green indicates the underexpression of a gene in the experimental cells relative to the control cells, yellow indicates equal expression in experimental and control cells, and no color indicates no expression in either experimental or control cells.

In one experiment, mRNA from a strain of antibiotic-resistant bacteria (experimental cells) is converted into cDNA and labeled with red fluorescent nucleotides; mRNA from a nonresistant strain of the same bacteria (control cells) is converted into cDNA and labeled with green fluorescent nucleotides. The cDNAs from the resistant and nonresistant cells are mixed and hybridized to a chip containing spots of DNA from genes 1 through 25. The results are shown in the adjoining illustration. What conclusions can you make about which genes might be implicated in antibiotic resistance in these bacteria? How might this information be used to design new antibiotics that are less vulnerable to resistance?

Section 14.8

36. *Dictyostelium discoideum* is a soil-dwelling, social amoeba: much of the time, the organism consists of single, solitary cells, but, during times of starvation, individual amoebae come together to form aggregates that have many characteristics of multicellular organisms. Biologists have long debated whether *D. discoideum* is a unicellular or multicellular organism. In 2005, the genome of *D. discoideum* was completely sequenced. The table at the bottom of the page lists some genomic characteristics of *D. discoideum* and other eukaryotes (L. Eichinger et al. 2005. *Nature* 435:43–57).

a. On the basis of the organisms listed in the table other than *D. discoideum*, what are some differences in genome characteristics between unicellular and multicellular organisms?

b. On the basis of these data, do you think the genome of *D. discoideum* is more like those of other unicellular eukaryotes or more like those of multicellular eukaryotes? Explain your answer.

37. What are some of the major differences in the ways in which genetic information is organized in the genomes of prokaryotes versus eukaryotes?

Table for Problem 36

Feature	*Dictyostelium discoideum*	*Plasmodium falciparium*	*Saccharhomyces cerevisiae*	*Arabidopsis thaliana*	*Drosophila melanogaster*	*Caenorhabditis elegans*	*Homo sapiens*
Organism	amoeba	malaria parasite	yeast	plant	fruit fly	worm	human
Cellularity	?	uni	uni	multi	multi	multi	multi
Genome size (millions bp)	34	23	13	125	180	103	2,851
Number of genes	12,500	5,268	5,538	25,498	13,676	19,893	22,287
Average gene length (bp)	1,756	2,534	1,428	2,036	1,997	2,991	27,000
Genes with introns (%)	69	54	5	79	38	5	85
Mean number of introns*	1.9	2.6	1.0	5.4	4.0	5.0	8.1
Mean intron size (bp)	146	179	nd*	170	nd*	270	3,365

*nd = not determined

Challenge Questions

Section 14.4

38. Suppose that you are hired by a biotechnology firm to produce a giant strain of fruit flies by using recombinant DNA technology so that genetics students will not be forced to strain their eyes when looking at tiny flies. You go to the library and learn that growth in fruit flies is normally inhibited by a hormone called shorty substance P (SSP). You decide that you can produce giant fruit flies if you can somehow turn off the production of SSP. Shorty substance P is synthesized from a compound called XSP in a single-step reaction catalyzed by the enzyme runtase:

$$XSP \xrightarrow[\text{Runtase}]{} SSP$$

A researcher has already isolated cDNA for runtase and has sequenced it, but the location of the runtase gene in the *Drosophila* genome is unknown.

In attempting to devise a strategy for turning off the production of SSP and producing giant flies by using standard recombinant DNA techniques, you discover that deleting, inactivating, or otherwise mutating this DNA sequence in *Drosophila* turns out to be extremely difficult. Therefore you must restrict your genetic engineering to gene augmentation (adding new genes to cells). Describe the methods that you will use to turn off SSP and produce giant flies by using recombinant DNA technology.

Section 14.5

39. Much of the controversy over genetically engineered foods has centered on whether special labeling should be required on all products made from genetically modified crops. Some people have advocated labeling that identifies the product as having been made from genetically modified plants. Others have argued that labeling should be required only to identify the ingredients, not the process by which they were produced. Take one side in this issue and justify your stand.

Section 14.8

*40. Some researchers have proposed creating an entirely new, free-living organism with a minimal genome—the smallest set of genes that allows for replication of the organism in a particular environment. This genome could be used to design and create, from "scratch," novel organisms that might perform specific tasks such as the breakdown of toxic materials in the environment.

a. How might the minimal genome required for life be determined?

b. What, if any, social and ethical concerns might be associated with the construction of an entirely new organism with a minimal genome?

15 Cancer Genetics

Palladin and the Spread of Cancer

Villa designed by Renaissance architect Andrea Palladio, for whom the *palladin* gene is named. The *palladin* gene encodes an essential component of the a cell's cytoskeleton; when mutated, *palladin* contributes to the spread of pancreatic cancer. [© "La Rotunda a Vicenzo" by Giovanni Giaconi.]

Pancreatic cancer is among the most serious of all cancers. Although only the ninth most common form of cancer, with about 42,000 new cases each year in the United States, pancreatic cancer is the fourth leading cause of death due to cancer, killing more than 35,000 people each year. Most people with pancreatic cancer survive less than 6 months after the cancer is diagnosed; only 5% survive more than 5 years. A primary reason for pancreatic cancer's lethality is its propensity to spread rapidly to the lymph nodes and other organs. Most symptoms don't appear until the disease is advanced and the cancer has invaded other organs. So what makes pancreatic cancer so likely to spread?

In 2006, researchers identified a key gene that contributes to the development of pancreatic cancer—an important source of insight into pancreatic cancer's aggressive nature. Geneticists at the University of Washington in Seattle had found a unique family in which nine members over three generations were diagnosed with pancreatic cancer (**Figure 15.1**). Nine additional family members had precancerous growths that were likely to develop into pancreatic cancer. In this family, pancreatic cancer was inherited as an autosomal dominant trait.

Using gene-mapping techniques, geneticists determined that the gene causing pancreatic cancer in the family was located within a region on the long arm of chromosome 4. Unfortunately, this region encompasses 16 million base pairs and included 250 genes.

To determine which of the 250 genes in the delineated region might be responsible for cancer in the family, researchers designed a unique microarray (see Chapter 14) that contained sequences from the region. They used this microarray to examine gene expression in pancreatic tumors and precancerous growths in family members, as well as in sporadic pancreatic tumors in other people and in normal pancreatic tissue from unaffected people. The researchers reasoned that the cancer gene might be overexpressed or underexpressed in the tumors relative to normal tissue. Data from the microarray revealed that the most overexpressed gene in the pancreatic tumors and precancerous growths was a gene encoding a critical component of the cytoskeleton, referred to as the *palladin* gene. Sequencing demonstrated that all members of the family with pancreatic cancer had an identical mutation in exon 2 of the *palladin* gene.

The *palladin* gene is named for Renaissance architect Andrea Palladio, because *palladin* plays a central role in the architecture of the cell. Palladin protein functions as a scaffold for the binding of the other cytoskeleton proteins that are necessary for maintaining cell shape, movement, and differentiation. The ability of a cancer cell to spread is directly related to its

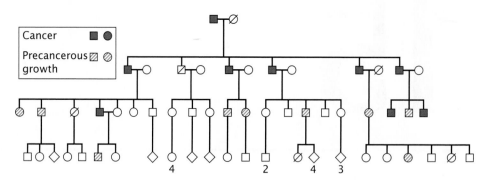

15.1 Pancreatic cancer is inherited as an autosomal dominant trait in a family that possesses a mutant *palladin* gene. [After K. L. Pogue et al., Plos Medicine 3:2216–2228, 2006.]

cytoskeleton; cells that spread typically have poor cytoskeleton architecture, enabling them to detach easily from a primary tumor mass and migrate through other tissues. To determine if mutations in the *palladin* gene affected cell mobility, researchers genetically engineered cells with a mutant copy of the *palladin* gene and tested the ability of these cells to migrate. The cells with mutated *palladin* were 33% more efficient at migrating than cells with normal *palladin*, demonstrating that the *palladin* gene contributes to the ability of pancreatic cancer cells to spread.

The discovery of *palladin*'s link to pancreatic cancer illustrates the power of modern molecular genetics for unraveling the biological nature of cancer. In this chapter, we examine the genetic nature of cancer, a peculiar disease that is fundamentally genetic but is often not inherited. We begin by considering the nature of cancer and how multiple genetic alterations are required to transform a normal cell into a cancerous one. We then consider some of the types of genes that contribute to cancer, including oncogenes and tumor-suppressor genes, genes that control the cell cycle, genes encoding DNA-repair systems and telomerase, and genes that, like *palladin*, contribute to the spread of cancer. Next, we take a look at chromosome mutations associated with cancer and genomic instability. We examine the role that viruses play in some cancers. Finally, we take a detailed look at how specific genes contribute to the progression of colon cancer.

15.1 Cancer Is a Group of Diseases Characterized by Cell Proliferation

Every year, cancer kills one of every five people in the United States, and cancer treatments cost billions of dollars per year. Cancer is not a single disease; rather, it is a heterogeneous group of disorders characterized by the presence of cells that do not respond to the normal controls on division. Cancer cells divide rapidly and continuously, creating tumors that crowd out normal cells and eventually rob healthy tissues of nutrients (**Figure 15.2**). The cells of an advanced tumor can

(a)

(b)

15.2 Abnormal proliferation of cancer cells produces a tumor that crowds out normal cells. (a) Metastatic lung-tumor masses (white protrusions) growing on a human liver. (b) A light micrograph of the section in part *a* showing areas of small, dark tumor cells invading a region of larger, lighter normal liver cells. [Courtesy of J. Braun.]

Table 15.1	Estimated incidences of various cancers and cancer mortality in the United States in 2009	
Type of Cancer	New Cases per Year	Deaths per Year
Lung and bronchus	219,440	159,390
Breast	194,280	40,610
Prostate	192,280	27,360
Colon and rectum	146,970	49,920
Lymphoma	74,490	20,790
Bladder	70,980	14,330
Melanoma	68,720	8,650
Leukemias	44,790	21,870
Pancreas	42,470	35,240
Uterus	42,160	7,780
Oral cavity and pharynx	35,720	7,600
Liver	22,620	18,160
Brain and nervous system	22,070	12,920
Ovary	21,550	14,600
Stomach	21,130	10,620
Uterine cervix	11,270	4,070
Cancers of soft tissues including heart	10,660	3,820
All cancers	1,479,350	562,340

Source: American Cancer Society, *Cancer Facts and Figures, 2009* (Atlanta: American Cancer Society, 2009), p. 4.

separate from the tumor and travel to distant sites in the body, where they may take up residence and develop into new tumors. The most common cancers in the United States are those of the breast, prostate, lung, colon and rectum, and blood (**Table 15.1**).

Tumor Formation

Normal cells grow, divide, mature, and die in response to a complex set of internal and external signals. A normal cell receives both stimulatory and inhibitory signals, and its growth and division are regulated by a delicate balance between these opposing forces. In a cancer cell, one or more of the signals has been disrupted, which causes the cell to proliferate at an abnormally high rate. As they lose their response to the normal controls, cancer cells gradually lose their regular shape and boundaries, eventually forming a distinct mass of abnormal cells—a tumor. If the cells of the tumor remain localized, the tumor is said to be

benign; if the cells invade other tissues, the tumor is said to be **malignant**. Cells that travel to other sites in the body, where they establish secondary tumors, have undergone **metastasis**.

Cancer As a Genetic Disease

Cancer arises as a result of fundamental defects in the regulation of cell division, and its study therefore has significance not only for public health, but also for our basic understanding of cell biology. Through the years, a large number of theories have been put forth to explain cancer, but we now recognize that most, if not all, cancers arise from defects in DNA.

Early observations suggested that cancer might result from genetic damage. First, many agents, such as ionizing radiation and chemicals that cause mutations also cause cancer (are carcinogens). Second, some cancers are consistently associated with particular chromosome abnormalities. About 90% of people with chronic myeloid leukemia, for example, have a reciprocal translocation between chromosome 22 and chromosome 9. Third, some specific types of cancers tend to run in families. Retinoblastoma, a rare childhood cancer of the retina, appears with high frequency in a few families and is inherited as an autosomal dominant trait, suggesting that a single gene is responsible for these cases of the disease.

Although these observations hinted that genes play some role in cancer, the theory of cancer as a genetic disease had several significant problems. If cancer is inherited, every cell in the body should receive the cancer-causing gene, and therefore every cell should become cancerous. In those types of cancer that run in families, however, tumors typically appear only in certain tissues and often only when the person reaches an advanced age. Finally, many cancers do not run in families at all and, even in regard to those cancers that generally do, isolated cases crop up in families with no history of the disease.

Knudson's multistep model of cancer   In 1971, Alfred Knudson proposed a model to explain the genetic basis of cancer. Knudson was studying retinoblastoma, a cancer that usually develops in only one eye but occasionally appears in both. Knudson found that, when retinoblastoma appears in both eyes, it presents itself at an early age and many affected children have close relatives who also are affected.

Knudson proposed that retinoblastoma results from two separate genetic defects, both of which are necessary for cancer to develop (**Figure 15.3**). He suggested that, in the cases in which the disease affects just one eye, a single cell in one eye undergoes two successive mutations. Because the chance of these two mutations occurring in a single cell is remote, retinoblastoma is rare and typically develops in only one eye. For the bilateral case, Knudson proposed that the

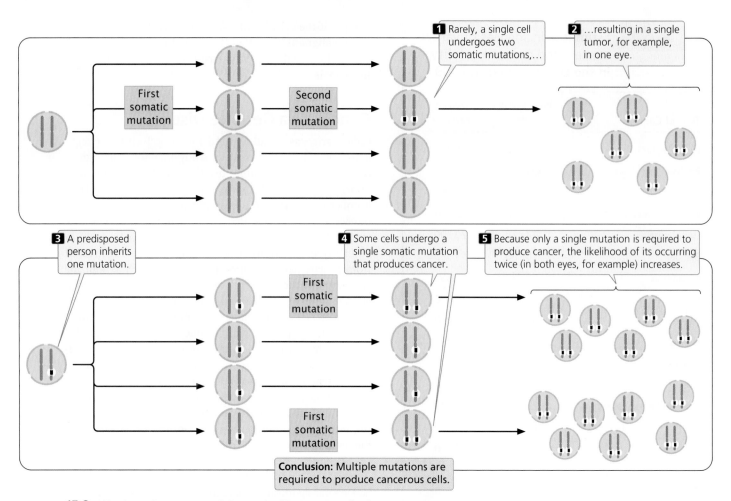

1 Rarely, a single cell undergoes two somatic mutations,…

2 …resulting in a single tumor, for example, in one eye.

First somatic mutation

Second somatic mutation

3 A predisposed person inherits one mutation.

4 Some cells undergo a single somatic mutation that produces cancer.

5 Because only a single mutation is required to produce cancer, the likelihood of its occurring twice (in both eyes, for example) increases.

First somatic mutation

First somatic mutation

Conclusion: Multiple mutations are required to produce cancerous cells.

15.3 Alfred Knudson proposed that retinoblastoma results from two separate genetic defects, both of which are necessary for cancer to develop.

child inherited one of the two mutations required for the cancer, and so every cell contains this initial mutation. In these cases, all that is required for cancer to develop is for one eye cell to undergo the second mutation. Because each eye possesses millions of cells, the probability that the second mutation will occur in at least one cell of each eye is high, producing tumors in both eyes at an early age.

Knudson's proposal suggests that cancer is the result of a multistep process that requires several mutations. If one or more of the required mutations is inherited, fewer additional mutations are required to produce cancer, and the cancer will tend to run in families. The idea that cancer results from multiple mutations turns out to be correct for most cancers.

Knudson's genetic theory for cancer has been confirmed by the identification of genes that, when mutated, cause cancer. Today, we recognize that cancer is fundamentally a genetic disease, although few cancers are actually inherited. Most tumors arise from somatic mutations that accumulate

in a person's life span, either through spontaneous mutation or in response to environmental mutagens.

The clonal evolution of tumors   Cancer begins when a single cell undergoes a mutation that causes the cell to divide at an abnormally rapid rate. The cell proliferates, giving rise to a clone of cells, each of which carries the same mutation. Because the cells of the clone divide more rapidly than normal, they soon outgrow other cells. An additional mutation that arises in some of the clone's cells may further enhance the ability of those cells to proliferate, and cells carrying both mutations soon become dominant in the clone. Eventually, they may be overtaken by cells that contain yet more mutations that enhance proliferation. In this process, called **clonal evolution**, the tumor cells acquire more mutations that allow them to become increasingly more aggressive in their proliferative properties (**Figure 15.4**).

The rate of clonal evolution depends on the frequency with which new mutations arise. Any genetic defect that

Mutations in genes that affect chromosome segregation also may contribute to the clonal evolution of tumors. Many cancer cells are aneuploid, and it is clear that chromosome mutations contribute to cancer progression by duplicating some genes (those on extra chromosomes) and eliminating others (those on deleted chromosomes). Cellular defects that interfere with chromosome separation increase aneuploidy and may therefore accelerate cancer progression.

Concepts

Cancer is fundamentally a genetic disease. Mutations in several genes are usually required to produce cancer. If one of these mutations is inherited, fewer somatic mutations are necessary for cancer to develop, and the person may have a predisposition to cancer. Clonal evolution is the accumulation of mutations in a clone of cells.

✔ Concept Check 1

How does the multistep model of cancer explain the observation that sporadic cases of retinoblastoma usually appear in only one eye, whereas inherited forms of the cancer appear in both eyes?

The Role of Environmental Factors in Cancer

Although cancer is fundamentally a genetic disease, most cancers are not inherited, and there is little doubt that many cancers are influenced by environmental factors. The role of environmental factors in cancer is suggested by differences in the incidence of specific cancers throughout the world (**Table 15.2**). The results of studies show that migrant populations typically take on the cancer incidence of their host country. For example, the overall rates of cancer are considerably lower in Japan than in Hawaii. However, within a single generation after migration to Hawaii, Japanese people develop cancer at rates similar to those of native Hawaiians.

Smoking is a good example of an environmental factor that is strongly associated with cancer. Other environmental factors that induce cancer are certain types of chemicals, such as benzene (used as an industrial solvent), benzo[a]pyrene (found in cigarette smoke), and polychlorinated biphenyls (PCBs; used in industrial transformers and capacitors). Ultraviolet light, ionizing radiation, and viruses are other known carcinogens and are associated with variation in the incidence of many cancers. Most environmental factors associated with cancer cause somatic mutations that stimulate cell division or otherwise affect the process of cancer progression.

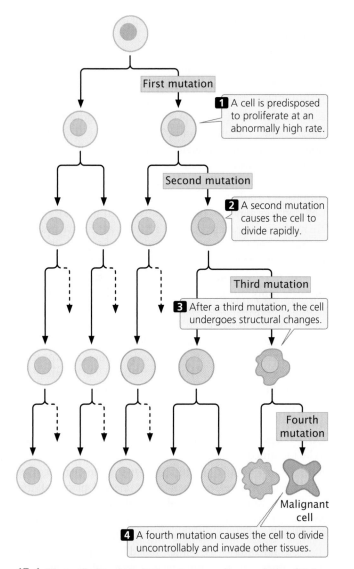

15.4 **Through clonal evolution, tumor cells acquire multiple mutations that allow them to become increasingly aggressive and proliferative.** To conserve space, a dashed arrow is used to represent a second cell of the same type in each case.

First mutation

1 A cell is predisposed to proliferate at an abnormally high rate.

Second mutation

2 A second mutation causes the cell to divide rapidly.

Third mutation

3 After a third mutation, the cell undergoes structural changes.

Fourth mutation

Malignant cell

4 A fourth mutation causes the cell to divide uncontrollably and invade other tissues.

allows more mutations to arise will accelerate cancer progression. Genes that regulate DNA repair are often found to have been mutated in the cells of advanced cancers, and inherited disorders of DNA repair are usually characterized by increased incidences of cancer. Because DNA-repair mechanisms normally eliminate many of the mutations that arise, cells with defective DNA-repair systems are more likely to retain mutations than are normal cells, including mutations in genes that regulate cell division. Xeroderma pigmentosum, for example, is a rare disorder caused by a defect in DNA repair (see the introduction to Chapter 9 and pp. 341–342 in Chapter 13). People with this condition have elevated rates of skin cancer when exposed to sunlight (which induces mutation).

Table 15.2	Examples of geographic variation in the incidence of cancer	
Type of Cancer	Location	Incidence Rate*
Lip	Canada (Newfoundland)	15.1
	Brazil (Fortaleza)	1.2
Nasopharynx	Hong Kong	30.0
	United States (Utah)	0.5
Colon	United States (Iowa)	30.1
	India (Bombay)	3.4
Lung	United States (New Orleans, African Americans)	110.0
	Costa Rica	17.8
Prostate	United States (Utah)	70.2
	China (Shanghai)	1.8
Bladder	United States (Connecticut, Whites)	25.2
	Philippines (Rizal)	2.8
All cancer	Switzerland (Basel)	383.3
	Kuwait	76.3

Source: C. Muir et al., *Cancer Incidence in Five Continents*, vol. 5 (Lyon: International Agency for Research on Cancer, 1987), Table 12-2.

*The incidence rate is the age-standardized rate in males per 100,000 population.

15.2 Mutations in a Number of Different Types of Genes Contribute to Cancer

As we have seen, cancer is a disease caused by alterations in the DNA. There are, however, many different types of genetic alterations that may contribute to cancer. In the next several sections, we will outline some of the different types of genes that frequently have roles in cancer.

Oncogenes and Tumor-Suppressor Genes

The signals that regulate cell division fall into two basic types: molecules that stimulate cell division and those that inhibit it. These control mechanisms are similar to the accelerator and brake of an automobile. In normal cells (but, one would hope, not your car), both accelerators and brakes are applied at the same time, causing cell division to proceed at the proper speed.

Because cell division is affected by both accelerators and brakes, cancer can arise from mutations in either type of signal, and there are several fundamentally different routes to cancer (**Figure 15.5**). A stimulatory gene can be made hyperactive or active at inappropriate times, analogous to having the accelerator of an automobile stuck in the floored position. Mutations in stimulatory genes are usually dominant because a mutation in a single copy of the gene is usually sufficient to produce a stimulatory effect. Dominant-acting stimulatory genes that cause cancer are termed **oncogenes**. Cell division may also be stimulated when inhibitory genes are made *inactive*, analogously to having a defective brake in an automobile. Mutated inhibitory genes generally have recessive effects because both copies must be mutated to remove all inhibition. Inhibitory genes in cancer are termed **tumor-suppressor genes**. Many cancer cells have mutations in both oncogenes and tumor-suppressor genes.

Although oncogenes or mutated tumor-suppressor genes or both are required to produce cancer, mutations in DNA-repair genes can increase the likelihood of acquiring mutations in these genes. Having mutated DNA-repair genes is analogous to having a lousy car mechanic who does not make the necessary repairs on a broken accelerator or brake.

Oncogenes   Oncogenes were the first cancer-causing genes to be identified. In 1909, a farmer brought physician Peyton Rous a hen with a large connective-tissue tumor (sarcoma) growing on its breast. When Rous injected pieces of this tumor into other hens, they also developed sarcomas. Rous conducted experiments that demonstrated that the tumors were being transmitted by a virus, which became known as the Rous sarcoma virus, as mentioned in Chapter 6. A number of other cancer-causing viruses were subsequently isolated from various animal tissues. These viruses were generally assumed to carry a cancer-causing gene that was transferred to the host cell. The first oncogene, called *src*, was isolated from the Rous sarcoma virus in 1970.

In 1975, Michael Bishop, Harold Varmus, and their colleagues began to use probes for viral oncogenes to search for related sequences in normal cells. They discovered that the genomes of all normal cells carry DNA sequences that are closely related to viral oncogenes. These cellular genes are called **proto-oncogenes**. They are responsible for basic cellular functions in normal cells but, when mutated, they become oncogenes that contribute to the development of cancer. When a virus infects a cell, a proto-oncogene may become incorporated into the viral genome through recombination. Within the viral genome, the proto-oncogene may mutate to an oncogene that, when inserted back into a cell, causes rapid cell division and cancer. Because the proto-oncogenes are more likely to undergo mutation or recombination within a virus, viral infection is often associated with the cancer.

15.5 Both oncogenes (a) and tumor-suppressor genes (b) contribute to cancer but differ in their modes of action and dominance.

Proto-oncogenes can be converted into oncogenes in viruses by several different ways. The sequence of the proto-oncogene may be altered or truncated as it is being incorporated into the viral genome. This mutated copy of the gene may then produce an altered protein that causes uncontrolled cell proliferation. Alternatively, through recombination, a proto-oncogene may end up next to a viral promoter or enhancer, which then causes the gene to be overexpressed. Finally, sometimes the function of a proto-oncogene in the host cell may be altered when a virus inserts its own DNA into the gene, disrupting its normal function.

Many oncogenes have been identified by experiments in which selected fragments of DNA are added to cells in culture. Some of the cells take up the DNA and, if these cells become cancerous, then the DNA fragment that was added to the culture must contain an oncogene. The fragments can then be sequenced, and the oncogene can be identified. A large number of oncogenes have now been discovered (**Table 15.3**).

Tumor-suppressor genes Tumor-suppressor genes are more difficult than oncogenes to identify because they *inhibit* cancer and are recessive; both alleles must be mutated before the inhibition of cell division is removed. Because it is the *failure* of their function that promotes cell proliferation, tumor-suppressor genes cannot be identified by adding them to cells and looking for cancer. Defects in both copies of a tumor-suppressor gene are usually required to cause cancer; an organism can inherit one defective copy of the tumor-suppressor gene (is heterozygous for the cancer-causing mutation) and not have cancer, because the remaining normal allele produces the tumor-suppressing product. However, these heterozygotes are often predisposed to cancer, because inactivation or loss of the one remaining allele is all that is required to completely eliminate the tumor-suppressor product and is referred to as **loss of heterozygosity**. A common mechanism for loss of heterozygosity is a deletion on the chromosome that carried the normal copy of the tumor-suppressor gene (**Figure 15.6**).

One of the first tumor-suppressor genes to be identified was the retinoblastoma gene. In 1985, Raymond White and

| Table 15.3 | Some oncogenes and functions of their corresponding proto-oncogenes |

Oncogene	Cellular Location of Product	Function of Proto-oncogene
sis	Secreted	Growth factor
erbB	Cell membrane	Part of growth-factor receptor
erbA	Cytoplasm	Thyroid-hormone receptor
src	Cell membrane	Protein tyrosine kinase
ras	Cell membrane	GTP binding and GTPase
myc	Nucleus	Transcription factor
fos	Nucleus	Transcription factor
jun	Nucleus	Transcription factor
bcl-1	Nucleus	Cell cycle

This person is heterozygous (*Aa*) for a tumor-suppressor gene.

Loss of the wild-type allele, in this case through a chromosome deletion, causes loss of the tumor-suppressor activity.

Chromosome deletion

Conclusion: People heterozygous for a tumor-suppressor gene are predisposed to cancer.

15.6 Loss of heterozygosity often leads to cancer in a person heterozygous for a tumor-suppressor gene.

Webster Cavenne showed that large segments of chromosome 13 were missing in cells of retinoblastoma tumors, and, later, the tumor-suppressor gene was isolated from these segments. A number of tumor-suppressor genes have now been discovered (**Table 15.4**).

Sometimes the mutation or loss of a single allele of a recessive tumor-suppressor gene is sufficient to cause cancer. This effect—the appearance of the trait in an individual cell or organism that is heterozygous for a normally recessive trait—is called **haploinsufficiency**. This phenomenon is thought to be due to dosage effects: the heterozygote produces only half as much of the product encoded by the tumor-suppressing gene. Normally, this amount is sufficient for the cellular processes that prevent tumor formation, but it is less than the optimal amount, and other factors may

Table 15.4	Some tumor-suppressor genes and their functions	
Gene	Cellular Location of Product	Function
NF1	Cytoplasm	GTPase activator
p53	Nucleus	Transcription factor, regulates apoptosis
RB	Nucleus	Transcription factor
WT-1	Nucleus	Transcription factor

Source: J. Marx, Learning how to suppress cancer, *Science* 261:1385, 1993.

sometimes combine with the lowered tumor-suppressor product to cause cancer.

Haploinsufficiency is seen in some inherited predispositions to cancer. Bloom syndrome is an autosomal recessive disease characterized by short stature, male infertility, and a predisposition to cancers of many types. The disease results from a defect in the *BLM* locus, which encodes a DNA helicase enzyme that plays a key role in the repair of double-strand breaks. Persons homozygous for mutations at the *BLM* locus have a greatly elevated risk of cancer. Persons heterozygous for mutations at the *BLM* locus were thought to be unaffected. However, a recent survey of Ashkenazi Jews (who have a high frequency of Bloom syndrome) showed that heterozygous carriers of a *BLM* mutation were at increased risk of colorectal cancer. Similarly, mice with one mutated copy of the *BLM* gene are more than twice as likely to develop intestinal tumors as are mice with no *BLM* mutations.

Concepts

Proto-oncogenes are genes that control normal cellular functions. When mutated, proto-oncogenes become oncogenes that stimulate cell proliferation. They tend to be dominant in their action. Tumor-suppressor genes normally inhibit cell proliferation; when mutated, they allow cells to proliferate. Tumor-suppressor genes tend to be recessive in their action. Individual organisms that are heterozygous for tumor-suppressor genes are often predisposed to cancer.

✔ **Concept Check 2**

Why are oncogenes usually dominant in their action, whereas tumor-suppressor genes are recessive?

Genes That Control the Cycle of Cell Division

The cell cycle is the normal process by which cells undergo growth and division. Normally, progression through the cell cycle is tightly regulated so that cells divide only when additional cells are needed, when all the components necessary for division are present, and when the DNA has been replicated without damage. Sometimes, however, errors arise in one or more of the components that regulate the cell cycle. These errors often cause cells to divide at inappropriate times or rates, leading to cancer. Indeed, many proto-oncogenes and tumor-suppressor genes function normally by helping to control the cell cycle. Before considering how errors in this system contribute to cancer, we must first understand how the cell cycle is usually regulated.

Control of the cell cycle As discussed in Chapter 2, the cell cycle consists of the period from one cell division to the next. Cells that are actively dividing pass through the G_1, S, and G_2 phases of interphase and then move directly into the M phase, when cell division takes place. Nondividing cells exit from G_1 into the G_0 stage, in which they are functional but not actively growing or dividing. Progression from one stage of the cell cycle to another is influenced by a number of internal and external signals and is regulated at key points in the cycle called checkpoints.

For many years, the biochemical events that control the progression of cells through the cell cycle were completely unknown, but research findings have now revealed many of the details of this process. Key events of the cell cycle are controlled by **cyclin-dependent kinases** (CDKs), which are enzymes that add phosphate groups to other proteins. Sometimes, phosphorylation activates the other protein and, other times, it inactivates the protein. As their name implies, CDKs are functional only when they associate with another protein called a **cyclin**. The level of cyclin oscillates in the course of the cell cycle; when bound to a CDK, cyclin specifies which proteins the CDK will phosphorylate. Each cyclin appears at a specific point in the cell cycle, usually because its synthesis and destruction are regulated by another cyclin. Cyclins and CDKs are called by different names in different organisms; here, we will use the terms applied to these molecules in mammals.

Let's look at the G_1-to-S transition. As stated in Chapter 2, progression through the cell cycle is regulated at several checkpoints, which ensure that all cellular components are present and in good working order before the cell proceeds to the next stage. The G_1/S checkpoint is in G_1, just before the cell enters into the S phase and replicates its DNA. The cell is prevented from passing through the G_1/S checkpoint by a molecule called the retinoblastoma (RB) protein (**Figure 15.7**), which binds to another molecule called E2F and keeps it inactive. In G_1, cyclin D and cyclin E continuously increase in concentration and combine with their associated CDKs. Cyclin-D–CDK and cyclin-E–CDK both phosphorylate molecules of RB. Late in G_1, phosphorylation of RB is completed, which inactivates RB. Without the inhibitory effects of RB, the E2F protein is released. E2F is a transcription factor that stimulates the transcription of genes that produce enzymes necessary for replication of the DNA, and the cell moves into the S stage of the cell cycle.

Other checkpoints control the G_2-to-M transition, the assembly of the spindle apparatus, and the cell's exit from mitosis.

Mutations in cell-cycle control and cancer   Many cancers are caused by defects in the cell cycle's regulatory machinery. For example, mutations in the gene that encodes the RB protein—which normally holds the cell in G_1 until the DNA is ready to be replicated—are associated with many cancers, including retinoblastoma (from which the RB protein gets it name). When the *RB* gene is mutated, cells pass through the G_1/S checkpoint without the normal controls that prevent cell proliferation. Overexpression of the gene

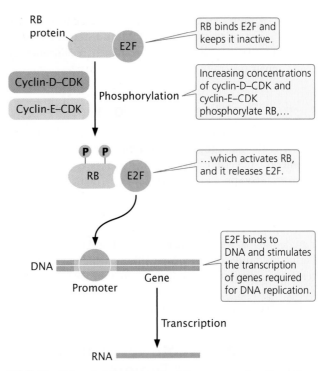

15.7 The RB protein helps control the progression through the G_1/S checkpoint by binding transcription factor E2F.

that encodes cyclin D (thus stimulating the passage of cells through the G_1/S checkpoint) takes place in about 50% of all breast cancers, as well as some cases of esophageal and skin cancer. Likewise, the tumor-suppressor gene *p53*, which is mutated in about 75% of all colon cancers, regulates a potent inhibitor of CDK activity.

Concepts

Progression through the cell cycle is controlled at checkpoints, which are regulated by interactions between cyclins and cyclin-dependent kinases. Genes that control the cell cycle are frequently mutated in cancer cells.

DNA-Repair Genes

Cancer arises from the accumulation of multiple mutations in a single cell. The rate at which mutations occur is affected not only by the rate at which they arise, but also by the efficiency with which errors are corrected by DNA-repair systems (see pp. 339–341 in Chapter 13). Defects in genes that encode components of these repair systems have been consistently associated with a number of cancers. People with xeroderma pigmentosum, for example, are defective in nucleotide-excision repair, an important cellular repair system that normally corrects DNA damage caused by a number of mutagens, including ultraviolet light. Likewise, about 13% of colorectal, endometrial, and stomach cancers have cells that are defective in mismatch repair, another major repair system in the cell.

Genes That Regulate Telomerase

Another factor that may contribute to the progression of cancer is the inappropriate activation of an enzyme called telomerase. In germ cells, telomerase replicates the chromosome ends (see pp. 234–235 in Chapter 9), thereby maintaining the telomeres, but this enzyme is not normally expressed in somatic cells. In many tumor cells, however, sequences that regulate the expression of the telomerase gene are mutated so that the enzyme is expressed, and the cell is capable of unlimited cell division. Although the expression of telomerase appears to contribute to the development of many cancers, its precise role in tumor progression is unknown and is under investigation.

Genes That Promote Vascularization and the Spread of Tumors

A final set of factors that contribute to the progression of cancer includes genes that affect the growth and spread of tumors. Oxygen and nutrients, which are essential to the survival and growth of tumors, are supplied by blood vessels, and the growth of new blood vessels (angiogenesis) is important to tumor progression. Angiogenesis is stimulated by growth factors and other proteins encoded by genes whose expression is carefully regulated in normal cells. In tumor cells, genes encoding these proteins are often overexpressed compared with normal cells, and inhibitors of angiogenesis-promoting factors may be inactivated or underexpressed. At least one inherited cancer syndrome—van Hippel–Lindau disease, in which people develop multiple types of tumors—is caused by the mutation of a gene that affects angiogenesis.

In the development of many cancers, the primary tumor gives rise to cells that spread to distant sites, producing secondary tumors. This process of metastasis is the cause of death in 90% of human cancer cases; it is influenced by cellular changes induced by somatic mutation. As discussed in the introduction to this chapter, the *palladin* gene, when mutated, contributes to the metastasis of pancreatic tumors. By using microarrays to measure levels of gene expression, researchers have identified other genes that are transcribed at a significantly higher rate in metastatic cells compared with nonmetastatic cells. For example, one study detected a set of 95 genes that were overexpressed or underexpressed in a population of metastatic breast cancer cells that were strongly metastatic to the lung, compared with a population of cells that were only weakly metastatic to the lung. Genes that contribute to metastasis often encode components of the extracellular matrix and the cytoskeleton. Others encode adhesion proteins, which help hold cells together.

Concepts

Mutations in genes that encode components of DNA-repair systems are often associated with cancer; these mutations increase the rate at which mutations are retained and result in an increased number of mutations in proto-oncogenes, tumor-suppressor genes, and other genes that contribute to cell proliferation. Mutations that allow telomerase to be expressed in somatic cells and those that affect vascularization and metastasis also may contribute to cancer progression.

✔ Concept Check 3

Which type of mutation in telomerase could be associated with cancer cells?

a. Mutations that produce an inactive form of telomerase

b. Mutations that decrease the expression of telomerase

c. Mutations that increase the expression of telomerase

d. All of the above

15.3 Changes in Chromosome Number and Structure Are Often Associated with Cancer

Most tumors contain cells with chromosome mutations. For many years, geneticists argued about whether these chromosome mutations were the cause or the result of cancer. Some types of tumors are consistently associated with *specific* chromosome mutations; for example, most cases of chronic myelogenous leukemia are associated with a reciprocal translocation between chromosomes 22 and 9. These types of associations suggest that chromosome mutations contribute to the cause of the cancer. Yet many cancers are not associated with specific types of chromosome abnormalities, and individual *gene* mutations are now known to contribute to many types of cancer. Nevertheless, chromosome instability is a general feature of cancer cells, causing them to accumulate chromosome mutations, which then affect individual genes that may contribute to the cancer process. Thus, chromosome mutations appear to be both a *cause* and a *result* of cancer.

At least three types of chromosome rearrangements— deletions, inversions, and translocations—are associated with certain types of cancer. Deletions may result in the loss of one or more tumor-suppressor genes. Inversions and translocations contribute to cancer in several ways. First, the chromosomal breakpoints that accompany these mutations may lie within tumor-suppressor genes, disrupting their function and leading to cell proliferation. Second, translocations and inversions may bring together sequences from two different genes, generating a fused protein that stimulates some aspect of the cancer process.

15.8 A reciprocal translocation between chromosomes 9 and 22 causes chronic myelogenous leukemia.

sequences that normally activate the production of immunoglobulins, and c-*MYC* is expressed in B cells. The c-MYC protein stimulates the division of the B cells and leads to Burkitt lymphoma.

Concepts

Many tumors contain a variety of types of chromosome mutations. Some tumors are associated with specific deletions, inversions, and translocations. Deletions can eliminate or inactivate genes that control the cell cycle; inversions and translocations can cause breaks in genes that suppress tumors, fuse genes to produce cancer-causing proteins, or move genes to new locations, where they are under the influence of different regulatory sequences.

✔ **Concept Check 4**

Chronic myelogenous leukemia is usually associated with which type of chromosome rearrangement?

a. Duplication   c. Inversion

b. Deletion   d. Translocation

Most advanced tumors contain cells that exhibit a dramatic variety of chromosome anomalies, including extra chromosomes, missing chromosomes, and chromosome rearrangements (**Figure 15.10**). Some cancer researchers believe that cancer is initiated when genetic changes take

Fusion proteins are seen in most cases of chronic myelogenous leukemia, a form of leukemia affecting bone-marrow cells. As mentioned earlier, most patients with chronic myelogenous leukemia have a reciprocal translocation between the long arm of chromosome 22 and the tip of the long arm of chromosome 9 (**Figure 15.8**). This translocation produces a shortened chromosome 22, called the Philadelphia chromosome, because it was first discovered in Philadelphia. At the end of a normal chromosome 9 is a potential cancer-causing gene called c-*ABL*. As a result of the translocation, part of the c-*ABL* gene is fused with the *BCR* gene from chromosome 22. The protein produced by this *BCR*–c-*ABL* fusion gene is much more active than the protein produced by the normal c-*ABL* gene; the fusion protein stimulates increased, unregulated cell division and eventually leads to leukemia.

A third mechanism by which chromosome rearrangements may produce cancer is by the transfer of a potential cancer-causing gene to a new location, where it is activated by different regulatory sequences. Burkitt lymphoma is a cancer of the B cells, the lymphocytes that produce antibodies. Many people with Burkitt lymphoma possess a reciprocal translocation between chromosome 8 and chromosome 2, 14, or 22 (**Figure 15.9**). This translocation relocates a gene called c-*MYC* from the tip of chromosome 8 to a position in chromosome 2, 14, or 22 that is next to a gene that encodes an immunoglobulin protein. At this new location, c-*MYC*, a cancer-causing gene, comes under the control of regulatory

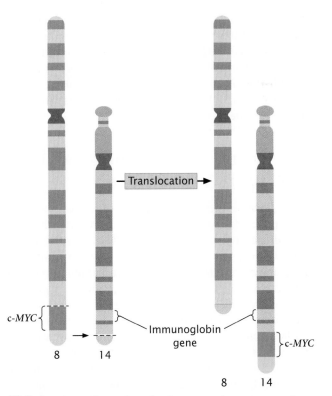

15.9 A reciprocal translocation between chromosomes 8 and 14 causes Burkitt lymphoma.

15.10 Cancer cells often possess chromosome abnormalities, including extra chromosomes, missing chromosomes, and chromosome rearrangements. Shown here are chromosomes from a colon-cancer cell, which has numerous chromosome abnormalities. *[Courtesy of Dr. Peter Duesberg, University of California at Berkeley.]*

place that cause the genome to become unstable, generating numerous chromosome abnormalities that then alter the expression of oncogenes and tumor-suppressor genes.

A number of genes that contribute to genomic instability and lead to missing or extra chromosomes (aneuploidy) have now been identified. For example, mutations in genes that encode parts of the spindle apparatus may contribute to abnormal segregation and lead to chromosome abnormalities. *APC* is a tumor-suppressor gene that is often mutated in colon-cancer cells. *APC* has several functions, one of which is to interact with the ends of the microtubules that associate with the kinetochore. Dividing mouse cells that have defective copies of the *APC* gene give rise to cells with many chromosome defects.

15.4 Viruses Are Associated with Some Cancers

As mentioned earlier in the chapter, viruses are responsible for a number of cancers in animals, and there is evidence that some viruses contribute to at least a few cancers in humans (**Table 15.5**). For example, about 95% of all women with cervical cancer are infected with **human papilloma viruses** (HPVs). Similarly, infection with the virus that causes hepatitis B increases the risk of liver cancer in some people. The Epstein–Barr virus, which is responsible for mononucleosis, has been linked to several types of cancer that are prevalent in parts of Africa, including Burkitt's lymphoma.

Many of the viruses that cause cancer in animals are retroviruses; earlier, we learned how studies of the Rous sarcoma retrovirus in chickens led to the identification of oncogenes in humans. Retroviruses sometimes cause cancer by

mutating and rearranging host genes, converting proto-oncogenes into oncogenes (**Figure 15.11a**) Another way in which viruses may contribute to cancer is by altering the expression of host genes (**Figure 15.11b**). Retroviruses often contain strong promoters to ensure that their own genetic material is transcribed by the host cell. If the provirus inserts near a proto-oncogene, viral promoters may stimulate high levels of expression of the proto-oncogene, leading to cell proliferation. Although viruses may contribute to cancer in all of these ways, only a few viruses have been clearly shown to cause cancer in humans.

Table 15.5	Some human cancers associated with viruses
Virus	**Cancer**
Human papilloma viruses (HPVs)	Cervical, penile, and vulvar cancers
Hepatitis B virus	Liver cancer
Human T-cell leukemia virus 1 (HTLV-1)	Adult T-cell leukemia
Human T-cell leukemia virus 2 (HTLV-2)	Hairy-cell leukemia
Epstein–Barr virus	Burkitt's lymphoma, nasopharyngeal cancer, Hodgkins lymphoma
Human herpes virus	Kaposi sarcoma

Note: Some of these associations between cancer and viruses exist only in certain populations and geographic areas.

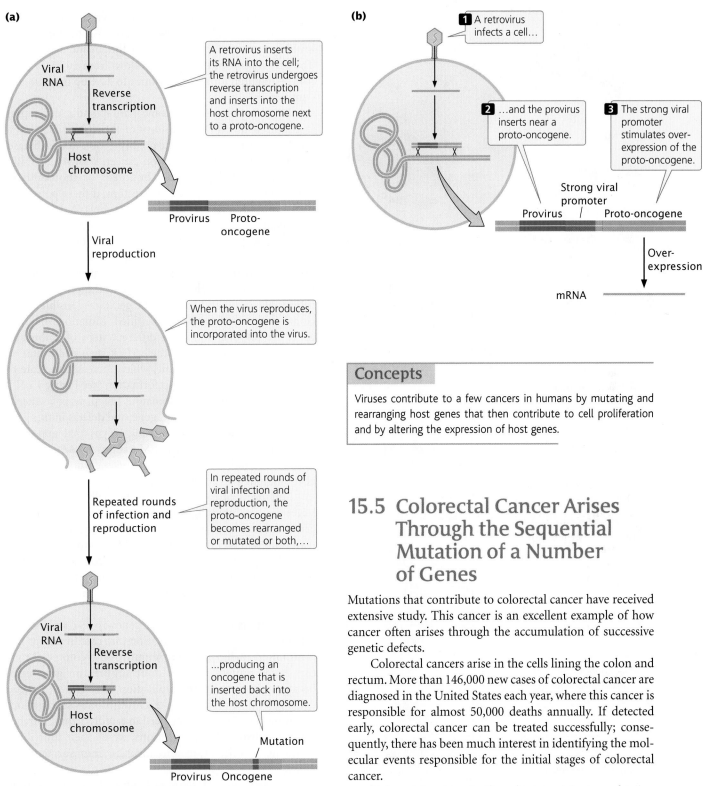

(a)

Viral RNA

Reverse transcription

> A retrovirus inserts its RNA into the cell; the retrovirus undergoes reverse transcription and inserts into the host chromosome next to a proto-oncogene.

Host chromosome

Provirus   Proto-oncogene

Viral reproduction

> When the virus reproduces, the proto-oncogene is incorporated into the virus.

Repeated rounds of infection and reproduction

> In repeated rounds of viral infection and reproduction, the proto-oncogene becomes rearranged or mutated or both,...

Viral RNA

Reverse transcription

Host chromosome

> ...producing an oncogene that is inserted back into the host chromosome.

Mutation

Provirus   Oncogene

15.11 Retroviruses cause cancer by (a) mutating and rearranging proto-oncogenes and (b) inserting strong promoters near proto-oncogenes.

(b)

1 A retrovirus infects a cell...

2 ...and the provirus inserts near a proto-oncogene.

3 The strong viral promoter stimulates over-expression of the proto-oncogene.

Strong viral promoter
Provirus   Proto-oncogene

Over-expression

mRNA

Concepts

Viruses contribute to a few cancers in humans by mutating and rearranging host genes that then contribute to cell proliferation and by altering the expression of host genes.

15.5 Colorectal Cancer Arises Through the Sequential Mutation of a Number of Genes

Mutations that contribute to colorectal cancer have received extensive study. This cancer is an excellent example of how cancer often arises through the accumulation of successive genetic defects.

Colorectal cancers arise in the cells lining the colon and rectum. More than 146,000 new cases of colorectal cancer are diagnosed in the United States each year, where this cancer is responsible for almost 50,000 deaths annually. If detected early, colorectal cancer can be treated successfully; consequently, there has been much interest in identifying the molecular events responsible for the initial stages of colorectal cancer.

Colorectal cancer is thought to originate as benign tumors called adenomatous polyps (**Figure 15.12**). Initially, these polyps are microscopic but, in time, they enlarge and

15.12 Mutations in multiple genes contribute to the progression of colorectal cancer.

Section through normal colon

Normal cells

Loss of normal tumor-suppressor gene *APC*

1 A polyp (small growth) forms on the colon wall.

2 A benign, precancerous tumor grows.

Blood vessel

Activation of oncogene *ras*

3 An adenoma (benign tumor) grows.

Loss of tumor-suppressor gene *p53*

4 A carcinoma (malignant tumor) develops.

Other changes; loss of antimetastasis gene

5 The cancer metastasizes (spreads to other tissue through the bloodstream).

the cells of the polyp acquire the abnormal characteristics of cancer cells. In the later stages of the disease, the tumor may invade the muscle layer surrounding the gut and metastasize. The progression of the disease is slow; from 10 to 35 years may be required for a benign tumor to develop into a malignant tumor.

Most cases of colorectal cancer are sporadic, developing in people with no family history of the disease, but a few families display a clear genetic predisposition to this disease. In one form of hereditary colon cancer, known as familial adenomatous polyposis coli, hundreds or thousands of polyps develop in the colon and rectum; if these polyps are not removed, one or more almost invariably become malignant.

Because polyps and tumors of the colon and rectum can be easily observed and removed with a colonoscope (a fiber-optic instrument used to view the interior of the rectum and colon), much is known about the progression of colorectal cancer, and some of the genes responsible for its clonal evolution have been identified. Mutations in these genes are responsible for the different steps of colorectal-cancer progression. One of the earliest steps is a mutation that inactivates the *APC* gene, which increases the rate of cell division, leading to polyp formation (see Figure 15.12). A person with familial adenomatous polyposis coli inherits one defective copy of the *APC* gene, and defects in this gene are associated with the numerous polyps that appear in those who have this disorder. Mutations in *APC* are also found in the polyps that develop in people who do not have adenomatous polyposis coli.

Mutations of the *ras* oncogene usually occur later, in larger polyps consisting of cells that have acquired some genetic mutations. The normal *ras* proto-oncogene is a key player in a pathway that relays signals from growth factors to the nucleus, where the signal stimulates cell divsion. When *ras* is mutated, the protein that it encodes continually relays a stimulatory signal for cell division, even when growth factor is absent.

Mutations in *p53* and other genes appear still later in tumor progression; these mutations are rare in polyps but common in malignant cells. About 75% of colorectal cancers have mutations in tumor-suppressor gene *p53*. Because *p53* prevents the replication of cells with genetic damage and controls proper chromosome segregation, mutations in *p53* may allow a cell to rapidly acquire further gene and chromosome mutations, which then contribute to further proliferation and invasion into surrounding tissues.

The sequence of steps just outlined is not the only route to colorectal cancer, and the mutations need not occur in the order presented here, but this sequence is a common pathway by which colon and rectal cells become cancerous.

Concepts Summary

- Cancer is fundamentally a genetic disorder, arising from somatic mutations in multiple genes that affect cell division and proliferation. If one or more mutations are inherited, then fewer additional mutations are required for cancer to develop.

- A mutation that allows a cell to divide rapidly provides the cell with a growth advantage; this cell gives rise to a clone of cells having the same mutation. Within this clone, other mutations occur that provide additional growth advantages, and cells with these additional mutations become dominant in the clone. In this way, the clone evolves.

- Environmental factors play an important role in the development of many cancers by increasing the rate of somatic mutations.

- Oncogenes are dominant mutated copies of normal genes (proto-oncogenes) that normally stimulate cell division. Tumor-suppressor genes normally inhibit cell division; recessive mutations in these genes may contribute to cancer. Sometimes, the mutation of a single allele of a tumor-suppressor gene is sufficient to cause cancer, a phenomenon known as haploinsufficiency.

- The cell cycle is controlled by cyclins and cyclin-dependent kinases. Mutations in genes that control the cell cycle are often associated with cancer.

- Defects in DNA-repair genes often increase the overall mutation rate of other genes, leading to defects in proto-oncogenes and tumor-suppressor genes that may contribute to cancer progression.

- Mutations in sequences that regulate telomerase allows cells to divide indefinitely, contributing to cancer progression. Tumor progression is also affected by mutations in genes that promote vascularization and the spread of tumors.

- Some cancers are associated with specific chromosome mutations, including chromosome deletions, inversions, and translocations. Mutations in some genes cause or allow the missegregation of chromosomes, leading to aneuploidy that may contribute to cancer.

- Viruses are associated with some cancers; they contribute to cell proliferation by mutating and rearranging host genes and by altering the expression of host genes.

- Colorectal cancer offers a model system for understanding tumor progression in humans. Initial mutations stimulate cell division, leading to a small benign polyp. Additional mutations allow the polyp to enlarge, invade the muscle layer of the gut, and eventually spread to other sites. Mutations in particular genes affect different stages of this progression.

Important Terms

malignant tumor (p. 391)
metastasis (p. 391)
clonal evolution (p. 392)
oncogene (p. 394)

tumor-suppressor gene (p. 394)
proto-oncogene (p. 394)
loss of heterozygosity (p. 395)
haploinsufficiency (p. 396)

cyclin-dependent kinase (CDK) (p. 397)
cyclin (p. 397)
human papilloma virus (HPV) (p. 400)

Answers to Concept Checks

1. Retinoblastoma results from at least two separate genetic defects, both of which are necessary for cancer to develop. In sporadic cases, two successive mutations must occur in a single cell, which is unlikely and therefore typically occurs in only one eye. In people who have inherited one of the two required mutations, every cell contains this mutation so that a single additional mutation is all that is required for cancer to develop. Given the millions of cells in each eye, there is a high probability that the second mutation will occur in at least one cell of each eye, producing tumors in both eyes and the inheritance of this type of retinoblastoma.

2. Oncogenes have a stimulatory effect on cell proliferation. Mutations in oncogenes are usually dominant because a mutation in a single copy of the gene is usually sufficient to produce a stimulatory effect. Tumor-suppressor genes inhibit cell proliferation. Mutations in tumor-suppressor genes are generally recessive, because both copies must be mutated to remove all inhibition.

3. c

4. d

Worked Problem

1. In some cancer cells, a specific gene has become duplicated many times. Is this gene likely to be an oncogene or a tumor-suppressor gene? Explain your reasoning.

• Solution

The gene is likely to be an oncogene. Oncogenes stimulate cell proliferation and act in a dominant manner. Therefore, extra copies of an oncogene will result in cell proliferation and cancer. Tumor-suppressor genes, on the other hand, suppress cell proliferation and act in a recessive manner; a single copy of a tumor-suppressor gene is sufficient to prevent cell proliferation. Therefore, extra copies of the tumor-suppressor gene will not lead to cancer.

Comprehension Questions

Section 15.1

*1. What types of evidence indicate that cancer arises from genetic changes?

2. How is it true that many types of cancer are genetic and yet not inherited?

Section 15.2

*3. Outline Knudson's multistage theory of cancer and describe how it helps to explain unilateral and bilateral cases of retinoblastoma.

4. Briefly explain how cancer arises through clonal evolution.

*5. What is the difference between an oncogene and a tumor-suppressor gene? Give some examples of the functions of proto-oncogenes and tumor suppressers in normal cells.

*6. How do cyclins and CDKs differ? How do they interact in controlling the cell cycle?

7. Briefly outline the events that control the progression of cells through the G_1/S checkpoint in the cell cycle.

8. Why do mutations in genes that encode DNA-repair enzymes often produce a predisposition to cancer?

*9. What role do telomeres and telomerase play in cancer progression?

Section 15.3

*10. Explain how chromosome deletions, inversions, and translocations may cause cancer.

11. Briefly outline how the Philadelphia chromosome leads to chronic myelogenous cancer.

12. What is genomic instability? Give some ways in which genomic instability may arise.

Section 15.4

*13. How do viruses contribute to cancer?

Section 15.5

14. Briefly outline some of the genetic changes that are commonly associated with the progression of colorectal cancer.

Application Questions and Problems

Section 15.1

*15. The *palladin* gene, which plays a role in pancreatic cancer (see the introduction to this chapter), is said to be an oncogene. Which of its characteristics suggest that it is an oncogene rather than a tumor-suppressor gene?

16. If cancer is fundamentally a genetic disease, how might an environmental factor such as smoking cause cancer?

17. Both genes and environmental factors contribute to cancer. Table 15.2 shows that prostate cancer is 39 times as common among Caucasians in Utah as among Chinese in Shanghai. Briefly outline how you might go about determining if these differences in the incidence of prostate cancer are due to differences in the genetic makeup of two populations or to differences in their environments.

Section 15.2

*18. A couple has one child with bilateral retinoblastoma. The mother is free from cancer, but the father has unilateral retinoblastoma and he has a brother who has bilateral retinoblastoma.

a. If the couple has another child, what is the probability that this next child will have retinoblastoma?

b. If the next child has retinoblastoma, is it likely to be bilateral or unilateral?

c. Explain why the father's case of retinoblastoma is unilateral, whereas his son's and brother's cases are bilateral.

19. Mutations in the *RB* gene are often associated with cancer. Explain how a mutation that results in a nonfunctional RB protein contributes to cancer.

20. Cells in a tumor contain mutated copies of a particular gene that promotes tumor growth. Gene therapy can be used to introduce a normal copy of this gene into the tumor cells. Would you expect this therapy to be effective if the mutated gene were an oncogene? A tumor-suppressor gene? Explain your reasoning.

21. Radiation is known to cause cancer, yet radiation is often used as treatment for some types of cancer. How can radiation be a contributor to both the cause and the treatment of cancer?

Section 15.3

22. Some cancers are consistently associated with the deletion of a particular part of a chromosome. Does the deleted region contain an oncogene or a tumor-suppressor gene? Explain.

Challenge Questions

Section 15.2

23. Many cancer cells are immortal (will divide indefinitely) because they have mutations that allow telomerase to be expressed. How might this knowledge be used to design anticancer drugs?

24. Bloom syndrome is an autosomal recessive disease that exhibits haploinsufficiency. As described on page 396, a recent survey showed that people heterozygous for mutations at the *BLM* locus are at increased risk of colon cancer. Suppose you are a genetic counselor. A young woman is referred to you whose mother has Bloom syndrome; the young woman's father has no family history of Bloom syndrome. The young woman asks whether she is likely to experience any other health problems associated with her family history of Bloom syndrome. What advice would you give her?

25. Imagine that you discover a large family in which bladder cancer is inherited as an autosomal dominant trait. Briefly outline a series of studies that you might conduct to identify the gene that causes bladder cancer in this family.

16

Quantitative Genetics

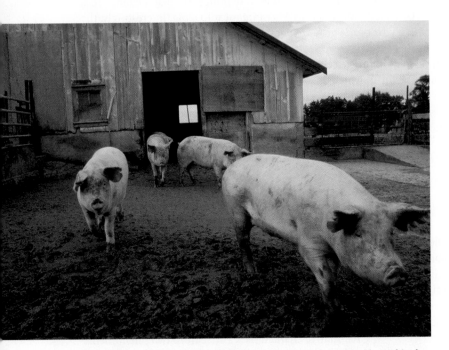

Methods of quantitative genetics are being used to identify and isolate genes that are important in determining muscle mass in pigs. [USDA.]

Porkier Pigs Through Quantitative Genetics

What makes a bigger and tastier pig? The answer to this question is worth billions of dollars to the pork industry. Weight in pigs entails muscle growth and fat deposition and is influenced by a combination of genes and environmental factors. Consumers today expect pork with less fat, and pork producers have responded: today's pigs have 50% less fat than the typical pig of the 1950s. Identifying the genes that promote muscle mass and growth is critical to producing larger, leaner pigs and has long been the goal of agricultural geneticists.

Muscle mass in pigs is not, however, a simple genetic characteristic such as seed shape in peas. Numerous genes and environmental factors, such as diet, rearing practices, and health contribute to the muscle mass of a pig. The inheritance of muscle mass in pigs is more complex than that of any of the characteristics that we have studied so far. Can the inheritance of a complex characteristic such as the muscle mass of pigs be studied? Is it possible to predict the muscle mass of a pig on the basis of its pedigree? The answers are yes—at least in part—but these questions cannot be addressed with the methods that we used for simple genetic characteristics. Instead, we must use statistical procedures that have been developed for analyzing complex characteristics. The genetic analysis of complex characteristics such as muscle mass in pigs is known as **quantitative genetics**.

Although the mathematical methods for analyzing complex characteristics may seem imposing at first, most people can intuitively grasp the underlying logic of quantitative genetics. We all recognize family resemblance: we talk about inheriting our father's height or our mother's intelligence. Family resemblance lies at the heart of the statistical methods used in quantitative genetics. When genes influence variation in a characteristic, related individuals resemble one another more than unrelated individuals. Closely related individuals (such as siblings) should resemble one another more than distantly related individuals (such as cousins). Comparing individuals with different degrees of relatedness, then, provides information about the extent to which genes influence a characteristic.

In 2003, geneticists used a combination of quantitative genetics and molecular techniques to identify and isolate chromosomal regions that play an important role in determining increased muscle mass in pigs. Chromosome regions containing genes that influence a quantitative trait are termed **quantitative trait loci** (QTLs). To locate QTLs affecting muscle mass, the geneticists started with crosses between European wild boars and

Large White domestic pigs. The alleles from some of the domestic pigs in these crosses markedly increased muscle mass and back-fat thickness in the offspring, indicating that the domestic pigs possess genes that stimulated muscle growth.

The geneticists then used molecular markers to map the position of the QTLs that influence muscle mass. They were able to narrow their search for the location of one important QTL to a 250,000-bp interval on pig chromosome 2. This region is known to contain several genes, including one for insulin-like growth factor 2 (IGF2). Because IGF2 is known to stimulate muscle mass in mammals, the gene immediately attracted their attention. By sequencing the *IGF2* gene of the more-muscled pigs and comparing their sequences with those from less-muscled pigs, the geneticists were able to demonstrate that a change in a single nucleotide, from a G to an A, added 3% to 5% more meat to a pig. Interestingly, the nucleotide change is not in a part of the gene that encodes the protein, but instead is in an intron. Findings from further research revealed that this substitution increases the expression of *IGF2* mRNA threefold in muscle cells. The increased levels of *IGF2* mRNA result in more insulin-like growth factor 2, which stimulates muscle growth and results in more-muscled, leaner pigs. This study demonstrates the power of quantitative genetics coupled with modern molecular techniques to identify and exploit genetic variation that influences economically important characteristics such as muscle mass in pigs.

This chapter is about the genetic analysis of complex characteristics such as muscle mass. We begin by considering the differences between quantitative and qualitative characteristics and why the expression of some characteristics varies continuously. We'll see how quantitative characteristics are often influenced by many genes, each of which has a small effect on the phenotype. Next, we will examine statistical procedures for describing and analyzing quantitative characteristics. We will consider the question of how much of phenotypic variation can be attributed to genetic and environmental influences and will conclude by looking at the effects of selection on quantitative characteristics. Of importance, however, is that we recognize that the methods of quantitative genetics are not designed to identify individual genes and genotypes. Rather, the focus is on statistical predictions based on groups of individuals.

16.1 Quantitative Characteristics Vary Continuously and Many Are Influenced by Alleles at Multiple Loci

Qualitative, or discontinuous, characteristics possess only a few distinct phenotypes (**Figure 16.1a**); these characteristics are the types studied by Mendel and have been the focus of our attention thus far. However, many characteristics vary continuously along a scale of measurement with many overlapping phenotypes (**Figure 16.1b**). They are referred to as

continuous characteristics; they are also called *quantitative characteristics* because any individual's phenotype must be described with a quantitative measurement. Quantitative characteristics might include height, weight, and blood pressure in humans, growth rate in mice, seed weight in plants, and milk production in cattle.

Quantitative characteristics arise from two phenomena. First, many are polygenic: they are influenced by genes at many loci. If many loci take part, many genotypes are possible, each producing a slightly different phenotype. Second, quantitative characteristics often arise when environmental factors affect the phenotype, because environmental differences result in a single genotype producing a range of phenotypes. Most continuously varying characteristics are *both* polygenic *and* influenced by environmental factors, and these characteristics are said to be multifactorial.

The Relation Between Genotype and Phenotype

For many discontinuous characteristics, the relation between genotype and phenotype is straightforward. Each genotype produces a single phenotype, and most phenotypes are encoded by a single genotype. Dominance and epistasis may allow two or three genotypes to produce the same phenotype, but the relation remains simple. This simple relation between genotype and phenotype allowed Mendel to decipher the basic rules of inheritance from his crosses with pea plants; it also permits us both to predict the outcome of genetic crosses and to assign genotypes to individuals.

(a) Discontinuous characteristic

1 A discontinuous (qualitative) characteristic exhibits only a few, easily distinguished phenotypes.

2 The plants are either dwarf or tall.

Dwarf

Tall

Number of individuals →

Phenotype (height)

(b) Continuous characteristic

3 A continuous (quantitative) characteristic exhibits a continuous range of phenotypes.

4 The plants exhibit a wide range of heights.

Number of individuals →

Dwarf   Tall

Phenotype (height)

16.1 Discontinuous and continuous characteristics differ in the number of phenotypes exhibited.

For quantitative characteristics, the relation between genotype and phenotype is often more complex. If the characteristic is polygenic, many different genotypes are possible, several of which may produce the same phenotype. For instance, consider a plant whose height is determined by three loci (A, B, and C), each of which has two alleles. Assume that one allele at each locus (A^+, B^+, and C^+) encodes a plant hormone that causes the plant to grow 1 cm above its baseline height of 10 cm. The other allele at each locus (A^-, B^-, and C^-) does not encode a plant hormone and thus does not contribute to additional height. If we consider only the two alleles at a single locus, 3 genotypes are possible (A^+A^+, A^+A^-, and A^-A^-). If all three loci are taken into account, there are a total of $3^3 = 27$ possible multilocus genotypes (A^+A^+ B^+B^+ C^+C^+, A^+A^- B^+B^+ C^+C^+, etc.). Although there are 27 genotypes, they produce only seven phenotypes (10 cm, 11 cm, 12 cm, 13 cm, 14 cm, 15 cm, and 16 cm in height). Some of the genotypes produce the same phenotype (**Table 16.1**); for example, genotypes A^+A^- B^-B^- C^-C^-, A^-A^- B^+B^- C^-C^-, and A^-A^- B^-B^- C^+C^-

all have one gene that encodes a plant hormone. These genotypes produce one dose of the hormone and a plant that is 11 cm tall. Even in this simple example of only three loci, the relation between genotype and phenotype is quite complex. The more loci encoding a characteristic, the greater the complexity.

The influence of environment on a characteristic also can complicate the relation between genotype and

Table 16.1	Hypothetical example of plant height determined by pairs of alleles at each of three loci	
Plant Genotype	**Doses of Hormone**	**Height (cm)**
A^-A^- B^-B^- C^-C^-	0	10
A^+A^- B^-B^- C^-C^-	1	11
A^-A^- B^+B^- C^-C^-		
A^-A^- B^-B^- C^-C^+		
A^+A^+ B^-B^- C^-C^-	2	12
A^-A^- B^+B^+ C^-C^-		
A^-A^- B^-B^- C^+C^+		
A^+A^- B^+B^- C^-C^-		
A^+A^- B^-B^- C^+C^-		
A^-A^- B^+B^- C^+C^-		
A^+A^+ B^+B^- C^-C^-	3	13
A^+A^+ B^-B^- C^+C^-		
A^+A^- B^+B^+ C^-C^-		
A^-A^- B^+B^+ C^+C^-		
A^+A^- B^-B^- C^+C^+		
A^-A^- B^+B^- C^+C^+		
A^+A^- B^+B^- C^+C^-		
A^+A^+ B^+B^+ C^-C^-	4	14
A^+A^+ B^+B^- C^+C^-		
A^+A^- B^+B^+ C^+C^-		
A^-A^- B^+B^+ C^+C^+		
A^+A^+ B^-B^- C^+C^+		
A^+A^- B^+B^- C^+C^+		
A^+A^+ B^+B^+ C^+C^-	5	15
A^+A^- B^+B^+ C^+C^+		
A^+A^+ B^+B^- C^+C^+		
A^+A^+ B^+B^+ C^+C^+	6	16

Note: Each + allele contributes 1 cm in height above a baseline of 10 cm.

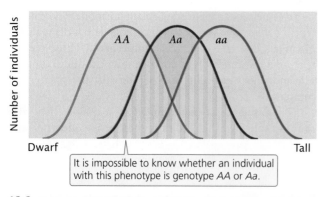

It is impossible to know whether an individual with this phenotype is genotype *AA* or *Aa*.

16.2 For a quantitative characteristic, each genotype may produce a range of possible phenotypes. In this hypothetical example, the phenotypes produced by genotypes *AA*, *Aa*, and *aa* overlap.

phenotype. Because of environmental effects, the same genotype may produce a range of potential phenotypes (the norm of reaction; see p. 96 in Chapter 4). The phenotypic ranges of different genotypes may overlap, making it difficult to know whether individuals differ in phenotype because of genetic or environmental differences (**Figure 16.2**).

In summary, the simple relation between genotype and phenotype that exists for many qualitative (discontinuous) characteristics is absent in quantitative characteristics, and it is impossible to assign a genotype to an individual on the basis of its phenotype alone. The methods used for analyzing qualitative characteristics (examining the phenotypic ratios of progeny from a genetic cross) will not work with quantitative characteristics. Our goal remains the same: we wish to make predictions about the phenotypes of offspring produced in a genetic cross. We may also want to know how much of the variation in a characteristic results from genetic differences and how much results from environmental differences. To answer these questions, we must turn to statistical methods that allow us to make predictions about the inheritance of phenotypes in the absence of information about the underlying genotypes.

Types of Quantitative Characteristics

Before we look more closely at polygenic characteristics and relevant statistical methods, we need to more clearly define what is meant by a quantitative characteristic. Thus far, we have considered only quantitative characteristics that vary continuously in a population. A *continuous characteristic* can theoretically assume any value between two extremes; the number of phenotypes is limited only by our ability to precisely measure the phenotype. Human height is a continuous characteristic because, within certain limits, people can theoretically have any height. Although the number of phenotypes possible with a continuous characteristic is infinite, we often group similar phenotypes together for convenience; we

may say that two people are both 5 feet 11 inches tall, but careful measurement may show that one is slightly taller than the other.

Some characteristics are not continuous but are nevertheless considered quantitative because they are determined by multiple genetic and environmental factors. **Meristic characteristics**, for instance, are measured in whole numbers. An example is litter size: a female mouse may have 4, 5, or 6 pups but not 4.13 pups. A meristic characteristic has a limited number of distinct phenotypes, but the underlying determination of the characteristic may still be quantitative. These characteristics must therefore be analyzed with the same techniques that we use to study continuous quantitative characteristics.

Another type of quantitative characteristic is a **threshold characteristic**, which is simply present or absent. For example, the presence of some diseases can be considered a threshold characteristic. Although threshold characteristics exhibit only two phenotypes, they are considered quantitative because they, too, are determined by multiple genetic and environmental factors. The expression of the characteristic depends on an underlying susceptibility (usually referred to as liability or risk) that varies continuously. When the susceptibility is larger than a threshold value, a specific trait is expressed (**Figure 16.3**). Diseases are often threshold characteristics because many factors, both genetic and environmental, contribute to disease susceptibility. If enough of the susceptibility factors are present, the disease develops; otherwise, it is absent. Although we focus on the genetics of continuous characteristics in this chapter, the same principles apply to many meristic and threshold characteristics.

It is important to point out that just because a characteristic can be measured on a continuous scale does not mean that it exhibits quantitative variation. One of the characteristics studied by Mendel was height of the pea plant, which can be described by measuring the length of the plant's stem. However, Mendel's particular plants exhibited only two distinct phenotypes (some were tall and others

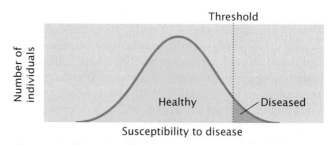

16.3 Threshold characteristics display only two possible phenotypes—the trait is either present or absent—but they are quantitative because the underlying susceptibility to the characteristic varies continuously. When the susceptibility exceeds a threshold value, the characteristic is expressed.

short), and these differences were determined by alleles at a single locus. The differences that Mendel studied were therefore discontinuous in nature.

Concepts

Characteristics whose phenotypes vary continuously are called quantitative characteristics. For most quantitative characteristics, the relation between genotype and phenotype is complex. Some characteristics whose phenotypes do not vary continuously also are considered quantitative because they are influenced by multiple genes and environmental factors.

Polygenic Inheritance

After the rediscovery of Mendel's work in 1900, questions soon arose about the inheritance of continuously varying characteristics. These characteristics had already been the focus of a group of biologists and statisticians, led by Francis Galton, who used statistical procedures to examine the inheritance of quantitative characteristics such as human height and intelligence. The results of these studies showed that quantitative characteristics are inherited, although the mechanism of inheritance was not yet known. Some biometricians argued that the inheritance of quantitative characteristics could not be explained by Mendelian principles, whereas others felt that Mendel's principles acting on numerous genes (polygenes) could adequately account for the inheritance of quantitative characteristics.

This conflict began to be resolved through independent work by Wilhelm Johannsen, George Udny Yule, and Herman Nilsson-Ehle, who each studied continuous variation in plants. The argument was finally laid to rest in 1918, when Ronald Fisher demonstrated that the inheritance of quantitative characteristics could indeed be explained by the cumulative effects of many genes, each following Mendel's rules.

Kernel Color in Wheat

To illustrate how multiple genes acting on a characteristic can produce a continuous range of phenotypes, let us examine one of the first demonstrations of polygenic inheritance. Nilsson-Ehle studied kernel color in wheat and found that the intensity of red pigmentation was determined by three unlinked loci, each of which had two alleles.

Nilsson-Ehle obtained several homozygous varieties of wheat that differed in color. Like Mendel, he performed crosses between these homozygous varieties and studied the ratios of phenotypes in the progeny. In one experiment, he crossed a variety of wheat that possessed white kernels with a variety that possessed purple (very dark red) kernels and obtained the following results:

P   Plants with white kernels $\times$ Plants with purple kernels

$\downarrow$

F_1 Plants with red kernels

$\downarrow$

F_2

$\frac{1}{16}$ plants with purple kernels

$\frac{4}{16}$ plants with dark-red kernels

$\frac{6}{16}$ plants with red kernels

$\frac{4}{16}$ plants with light-red kernels

$\frac{1}{16}$ plants with white kernels

Nilsson-Ehle interpreted this phenotypic ratio as the result of the segregation of alleles at two loci. (Although he found alleles at three loci that affect kernel color, the two varieties used in this cross differed at only two of the loci.) He proposed that there were two alleles at each locus: one that produced red pigment and another that produced no pigment. We'll designate the alleles that encoded pigment A^+ and B^+ and the alleles that encoded no pigment A^- and B^-. Nilsson-Ehle recognized that the effects of the genes were additive. Each gene seemed to contribute equally to color; so the overall phenotype could be determined by adding the effects of all the genes, as shown in the following table.

Genotype	Doses of pigment	Phenotype
$A^+A^+\ B^+B^+$	4	purple
$A^+A^+B^+B^-$ $A^+A^-B^+B^+$	3	dark red
$A^+A^+B^-B^-$ $A^-A^-B^+B^+$ $A^+A^-B^+B^-$	2	red
$A^+A^-B^-B^-$ $A^-A^-B^+B^-$	1	light red
$A^-A^-\ B^-B^-$	0	white

Notice that the purple and white phenotypes are each encoded by a single genotype, but other phenotypes may result from several different genotypes.

From these results, we see that five phenotypes are possible when alleles at two loci influence the phenotype and the effects of the genes are additive. When alleles at more than two loci influence the phenotype, more phenotypes are possible, and the color would appear to vary continuously between white and purple. If environmental factors had influenced the characteristic, individuals of the same genotype would vary somewhat in color, making it even more difficult to distinguish between discrete phenotypic classes. Luckily, environment played little role in determining kernel color in Nilsson-Ehle's crosses, and only a few loci encoded color; so Nilsson-Ehle was able to distinguish among the different phenotypic classes. This ability allowed him to see the Mendelian nature of the characteristic.

Let's now see how Mendel's principles explain the ratio obtained by Nilsson-Ehle in his F_2 progeny. Remember that Nilsson-Ehle crossed the homozygous purple variety ($A^+A^+ B^+B^+$) with the homozygous white variety ($A^-A^- B^-B^-$), producing F_1 progeny that were heterozygous at both loci ($A^+A^- B^+B^-$). This is a dihybrid cross, like those that we worked in Chapter 3, except that both loci encode the same trait. All the F_1 plants possessed two pigment-producing alleles that allowed two doses of color to make red kernels. The types and proportions of progeny expected in the F_2 can be found by applying Mendel's principles of segregation and independent assortment.

Let's first examine the effects of each locus separately. At the first locus, two heterozygous F_1s are crossed ($A^+A^- \times A^+A^-$). As we learned in Chapter 3, when two heterozygotes are crossed, we expect progeny in the proportions $\frac{1}{4} A^+A^+$, $\frac{1}{2} A^+A^-$, and $\frac{1}{4} A^-A^-$. At the second locus, two heterozygotes also are crossed, and, again, we expect progeny in the proportions $\frac{1}{4} B^+B^+$, $\frac{1}{2} B^+B^-$, and $\frac{1}{4} B^-B^-$.

To obtain the probability of combinations of genes at both loci, we must use the multiplication rule of probability (see Chapter 3), which is based on Mendel's principle of independent assortment. The expected proportion of F_2 progeny with genotype $A^+A^+ B^+B^+$ is the product of the probability of obtaining genotype A^+A^+ ($\frac{1}{4}$) and the probability of obtaining genotype B^+B^+ ($\frac{1}{4}$), or $\frac{1}{4} \times \frac{1}{4} = \frac{1}{16}$ (**Figure 16.4**). The probabilities of each of the phenotypes can then be obtained by adding the probabilities of all the genotypes that produce that phenotype. For example, the red phenotype is produced by three genotypes:

Genotype	Probability
$A^+A^+ B^-B^-$	$\frac{1}{16}$
$A^-A^- B^+B^+$	$\frac{1}{16}$
$A^+A^- B^+B^-$	$\frac{1}{4}$

Thus, the overall probability of obtaining red kernels in the F_2 progeny is $\frac{1}{16} + \frac{1}{16} + \frac{1}{4} = \frac{6}{16}$. Figure 16.4 shows that the phenotypic ratio expected in the F_2 is $\frac{1}{16}$ purple, $\frac{4}{16}$ dark red, $\frac{6}{16}$ red, $\frac{4}{16}$ light red, and $\frac{1}{16}$ white. This phenotypic ratio is precisely what Nilsson-Ehle observed in his F_2 progeny, demonstrating that the inheritance of a

16.4 Nilsson-Ehle demonstrated that kernel color in wheat is inherited according to Mendelian principles. He crossed two varieties of wheat that differed in pairs of alleles at two loci affecting kernel color. A purple strain ($A^+A^+ B^+B^+$) was crossed with a white strain ($A^-A^- B^-B^-$), and the F_1 was intercrossed to produce F_2 progeny. The ratio of phenotypes in the F_2 can be determined by breaking the dihybrid cross into two simple single-locus crosses and combining the results by using the multiplication rule.

Experiment

Question: How is a continous trait, such as kernel color in wheat, inherited?

Methods Cross wheat having white kernels and wheat having purple kernels. Intercross the F_1 to produce F_2.

P generation
$A^+A^+B^+B^+$ × $A^-A^-B^-B^-$
Purple   White

Results

F_1 generation
$A^+A^- B^+B^-$ Red

Break into simple crosses

$A^+A^- \times A^+A^-$ $B^+B^- \times B^+B^-$

$\frac{1}{4}A^+A^+$ $\frac{1}{2}A^+A^-$ $\frac{1}{4}A^-A^-$ $\frac{1}{4}B^+B^+$ $\frac{1}{2}B^+B^-$ $\frac{1}{4}B^-B^-$

Combine results

F_2 generation — **Number of pigment genes** — **Phenotype**

$\frac{1}{4}A^+A^+$
- $\frac{1}{4}B^+B^+ \rightarrow \frac{1}{4} \times \frac{1}{4} = \frac{1}{16}$ $A^+A^+B^+B^+$ 4 Purple
- $\frac{1}{2}B^+B^- \rightarrow \frac{1}{4} \times \frac{1}{2} = \frac{2}{16}$ $A^+A^+B^+B^-$ 3 Dark red
- $\frac{1}{4}B^-B^- \rightarrow \frac{1}{4} \times \frac{1}{4} = \frac{1}{16}$ $A^+A^+B^-B^-$ 2 Red

$\frac{1}{2}A^+A^-$
- $\frac{1}{4}B^+B^+ \rightarrow \frac{1}{2} \times \frac{1}{4} = \frac{2}{16}$ $A^+A^-B^+B^+$ 3 Dark red
- $\frac{1}{2}B^+B^- \rightarrow \frac{1}{2} \times \frac{1}{2} = \frac{4}{16}$ $A^+A^-B^+B^-$ 2 Red
- $\frac{1}{4}B^-B^- \rightarrow \frac{1}{2} \times \frac{1}{4} = \frac{2}{16}$ $A^+A^-B^-B^-$ 1 Light red

$\frac{1}{4}A^-A^-$
- $\frac{1}{4}B^+B^+ \rightarrow \frac{1}{4} \times \frac{1}{4} = \frac{1}{16}$ $A^-A^-B^+B^+$ 2 Red
- $\frac{1}{2}B^+B^- \rightarrow \frac{1}{4} \times \frac{1}{2} = \frac{2}{16}$ $A^-A^-B^+B^-$ 1 Light red
- $\frac{1}{4}B^-B^- \rightarrow \frac{1}{4} \times \frac{1}{4} = \frac{1}{16}$ $A^-A^-B^-B^-$ 0 White

Combine common phenotypes

F_2 ratio

Frequency	Number of pigment genes	Phenotype
$\frac{1}{16}$	4	Purple
$\frac{4}{16}$	3	Dark red
$\frac{6}{16}$	2	Red
$\frac{4}{16}$	1	Light red
$\frac{1}{16}$	0	White

Conclusion: Kernel color in wheat is inherited according to Mendel's principles acting on alleles at two loci.

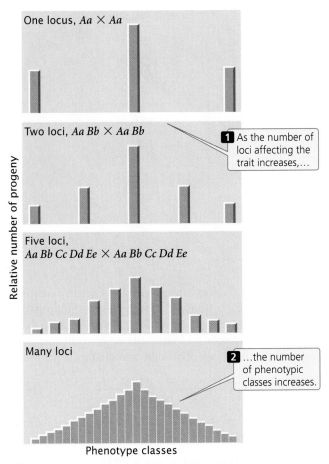

16.5 The results of crossing individuals heterozygous for different numbers of loci affecting a characteristic.

phenotypic classes. Second, the genes affecting kernel color had strictly additive effects, making the relation between genotype and phenotype simple. Third, environment played almost no role in the phenotype; had environmental factors modified the phenotypes, distinguishing between the five phenotypic classes would have been difficult. Finally, the loci that Nilsson-Ehle studied were not linked; so the genes assorted independently. Nilsson-Ehle was fortunate: for many polygenic characteristics, these simplifying conditions are not present and Mendelian inheritance of these characteristics is not obvious.

Concepts

The principles that determine the inheritance of quantitative characteristics are the same as the principles that determine the inheritance of discontinuous characteristics, but more genes take part in the determination of quantitative characteristics.

16.2 Analyzing Quantitative Characteristics

Because quantitative characteristics are described by a measurement and are influenced by multiple factors, their inheritance must be analyzed statistically.

Distributions

Understanding the genetic basis of any characteristic begins with a description of the numbers and kinds of phenotypes present in a group of individuals. Phenotypic variation in a group, such as the progeny of a cross, can be conveniently represented by a **frequency distribution**, which is a graph of the frequencies (numbers or proportions) of the different phenotypes (**Figure 16.6**). In a typical frequency distribution, the phenotypic classes are plotted on the horizontal (x)

continuously varying characteristic such as kernel color is indeed according to Mendel's basic principles.

Nilsson-Ehle's crosses demonstrated that the difference between the inheritance of genes influencing quantitative characteristics and the inheritance of genes influencing discontinuous characteristics is in the *number* of loci that determine the characteristic. When multiple loci affect a character, more genotypes are possible; so the relation between the genotype and the phenotype is less obvious. For example, in a cross of F_1 individuals heterozygous for alleles at a single locus with additive effects, 3 phenotypes appear among the progeny (**Figure 16.5**). When parents of the cross are heterozygous at two loci, there are 5 phenotypes in the progeny, and, when the parents are heterozygous at five loci, there are 11 phenotypes in the progeny. As the number of loci affecting a character increases, the number of phenotypic classes in the F_2 increases.

Several conditions of Nilsson-Ehle's crosses greatly simplified the polygenic inheritance of kernel color and made it possible for him to recognize the Mendelian nature of the characteristic. First, genes affecting color segregated at only two or three loci. If genes at many loci had been segregating, he would have had difficulty in distinguishing the

16.6 A frequency distribution is a graph that displays the number or proportion of different phenotypes. Phenotypic values are plotted on the horizontal axis, and the numbers (or proportions) of individuals in each class are plotted on the vertical axis.

(a) Sugar beet percentage of sucrose

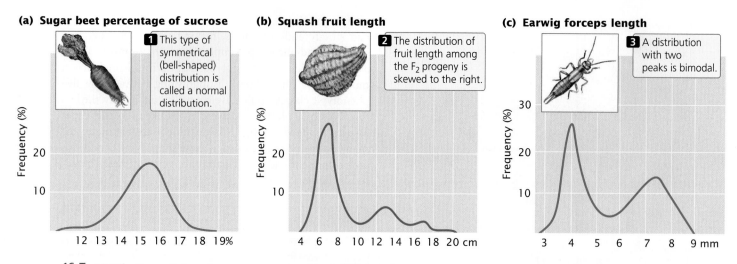

1 This type of symmetrical (bell-shaped) distribution is called a normal distribution.

Frequency (%)

12 13 14 15 16 17 18 19%

(b) Squash fruit length

2 The distribution of fruit length among the F$_2$ progeny is skewed to the right.

Frequency (%)

4 6 8 10 12 14 16 18 20 cm

(c) Earwig forceps length

3 A distribution with two peaks is bimodal.

Frequency (%)

3   4   5   6   7   8   9 mm

16.7 Distributions of phenotypes can assume several different shapes.

axis, and the numbers (or proportions) of individuals in each class are plotted on the vertical (y) axis. A frequency distribution is a concise method of summarizing all phenotypes of a quantitative characteristic.

Connecting the points of a frequency distribution with a line creates a curve that is characteristic of the distribution. Many quantitative characteristics exhibit a symmetrical (bell-shaped) curve called a **normal distribution** (**Figure 16.7a**). Normal distributions arise when a large number of independent factors contribute to a measurement, as is often the case in quantitative characteristics. Two other common types of distributions (skewed and bimodal) are illustrated in **Figure 16.7b** and **c**.

The Mean

The **mean**, also called the average, provides information about the center of the distribution. If we measured the heights of 10-year-old and 18-year-old boys and plotted a frequency distribution for each group, we would find that both distributions are normal, but the two distributions would be centered at different heights, and this difference would be indicated in their different means (**Figure 16.8**).

Suppose we have five measurements of height in centimeters: 160, 161, 167, 164, and 165. If we represent a group of measurements as x_1, x_2, x_3, and so forth, then the mean ($\bar{x}$) is calculated by adding all the individual measurements and dividing by the total number of measurements in the sample (n):

$$\bar{x} = \frac{x_1 + x_2 + x_3 + \cdots + x_n}{n} \qquad (16.1)$$

In our example, $x_1 = 160$, $x_2 = 161$, $x_3 = 167$, and so forth. The mean height ($\bar{x}$) equals:

$$\bar{x} = \frac{160 + 161 + 167 + 164 + 165}{5} = \frac{817}{5} = 163.4$$

A shorthand way to represent this formula is

$$\bar{x} = \frac{\sum x_i}{n} \qquad (16.2)$$

or

$$\bar{x} = \frac{1}{n}\sum x_i \qquad (16.3)$$

where the symbol Σ means "the summation of" and x_i represents individual x values.

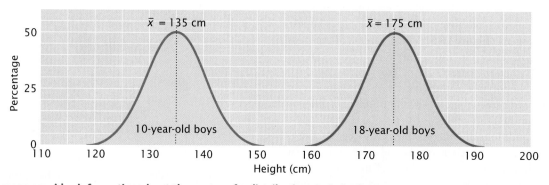

Percentage

$\bar{x} = 135$ cm

$\bar{x} = 175$ cm

50

25

0

110   120   130   140   150   160   170   180   190   200

10-year-old boys

18-year-old boys

Height (cm)

16.8 The mean provides information about the center of a distribution. Both distributions of heights of 10-year-old and 18-year-old boys are normal, but they have different locations along a continuum of height, which makes their means different.

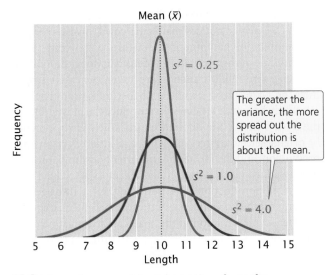

16.9 The variance provides information about the variability of a group of phenotypes. Shown here are three distributions with the same mean but different variances.

The Variance

A statistic that provides key information about a distribution is the **variance**, which indicates the variability of a group of measurements, or how spread out the distribution is. Distributions may have the same mean but different variances (**Figure 16.9**). The larger the variance, the greater the spread of measurements in a distribution about its mean.

The variance (s^2) is defined as the average squared deviation from the mean:

$$s^2 = \frac{\sum (x_i - \overline{x})^2}{n - 1} \qquad (16.4)$$

To calculate the variance, we (1) subtract the mean from each measurement and square the value obtained, (2) add all the squared deviations, and (3) divide this sum by the number of original measurements minus 1. For example, suppose we wanted to calculate the variance for the five heights mentioned earlier (160, 161, 167, 164, and 165 cm). As already shown, the mean of these heights is 163.4 cm. The variance for the heights is:

$$
\begin{aligned}
s^2 &= \frac{\begin{array}{c}(160 - 163.4)^2 + (161 - 163.4)^2 + (167 - 163.4)^2 \\ + (164 - 163.4)^2 + (165 - 163.4)^2\end{array}}{5 - 1} \\
&= \frac{(-3.4)^2 + (-2.4)^2 + (3.6)^2 + (0.6)^2 + (1.6)^2}{4} \\
&= \frac{11.56 + 5.76 + 12.96 + 0.36 + 2.56}{4} \\
&= 8.3
\end{aligned}
$$

Concepts

The mean and variance describe a distribution of measurements: the mean provides information about the location of the center of a distribution, and the variance provides information about its variability.

Applying Statistics to the Study of a Polygenic Characteristic

Edward East carried out one early statistical study of polygenic inheritance on the length of flowers in tobacco (*Nicotiana longiflora*). He obtained two varieties of tobacco that differed in flower length: one variety had a mean flower length of 40.5 mm, and the other had a mean flower length of 93.3 mm (**Figure 16.10**). These two varieties had been inbred for many generations and were homozygous at all loci contributing to flower length. Thus, there was no genetic variation in the original parental strains; the small differences in flower length within each strain were due to environmental effects on flower length.

When East crossed the two strains, he found that flower length in the F_1 was about halfway between that in the two parents (see Figure 16.10), as would be expected if the genes determining the differences in the two strains were additive in their effects. The variance of flower length in the F_1 was similar to that seen in the parents because all the F_1 had the same genotype, as did each parental strain (the F_1 were all heterozygous at the genes that differed between the two parental varieties).

East then interbred the F_1 to produce F_2 progeny. The mean flower length of the F_2 was similar to that of the F_1, but the variance of the F_2 was much greater (see Figure 16.10). This greater variability indicates that not all of the F_2 progeny had the same genotype.

East selected some F_2 plants and interbred them to produce F_3 progeny. He found that flower length of the F_3 depended on flower length in the plants selected as their parents. This finding demonstrated that flower-length differences in the F_2 were partly genetic and were therefore passed to the next generation.

16.3 Heritability Is Used to Estimate the Proportion of Variation in a Trait That Is Genetic

In addition to being polygenic, quantitative characteristics are frequently influenced by environmental factors. It is often useful to know how much of the variation in a quantitative characteristic is due to genetic differences and how much is due to environmental differences. The proportion of the total phenotypic variation that is due to genetic differences is known as the **heritability**.

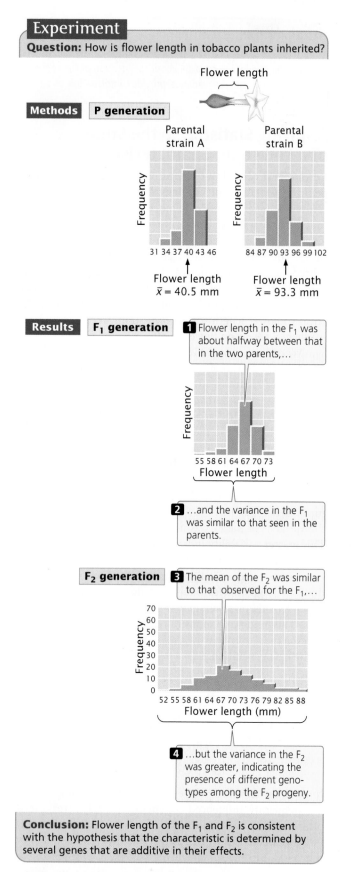

Experiment

Question: How is flower length in tobacco plants inherited?

Flower length

Methods   **P generation**

Parental strain A   Parental strain B

31 34 37 40 43 46   84 87 90 93 96 99 102

Flower length
$\bar{x} = 40.5$ mm

Flower length
$\bar{x} = 93.3$ mm

Results   **F₁ generation**

1 Flower length in the F₁ was about halfway between that in the two parents,...

55 58 61 64 67 70 73
Flower length

2 ...and the variance in the F₁ was similar to that seen in the parents.

F₂ generation   **3** The mean of the F₂ was similar to that observed for the F₁,...

70
60
50
40
30
20
10
0

52 55 58 61 64 67 70 73 76 79 82 85 88
Flower length (mm)

4 ...but the variance in the F₂ was greater, indicating the presence of different genotypes among the F₂ progeny.

Conclusion: Flower length of the F₁ and F₂ is consistent with the hypothesis that the characteristic is determined by several genes that are additive in their effects.

16.10 Edward East conducted an early statistical study of the inheritance of flower length in tobacco.

Consider a dairy farmer who owns several hundred milk cows. The farmer notices that some cows consistently produce more milk than others. The nature of these differences is important to the profitability of his dairy operation. If the differences in milk production are largely genetic in origin, then the farmer may be able to boost milk production by selectively breeding the cows that produce the most milk. On the other hand, if the differences are largely environmental in origin, selective breeding will have little effect on milk production, and the farmer might better boost milk production by adjusting the environmental factors associated with higher milk production. To determine the extent of genetic and environmental influences on variation in a characteristic, phenotypic variation in the characteristic must be partitioned into components attributable to different factors.

Phenotypic Variance

To determine how much of phenotypic differences in a population is due to genetic and environmental factors, we must first have some quantitative measure of the phenotype under consideration. Consider a population of wild plants that differ in size. We could collect a representative sample of plants from the population, weigh each plant in the sample, and calculate the mean and variance of plant weight. This **phenotypic variance** is represented by V_P.

Components of phenotypic variance   First, some of the phenotypic variance may be due to differences in genotypes among individual members of the population. These differences are termed the **genetic variance** and are represented by V_G.

Second, some of the differences in phenotype may be due to environmental differences among the plants; these differences are termed the **environmental variance**, V_E. Environmental variance includes differences in environmental factors such as the amount of light or water that the plant receives; it also includes random differences in development that cannot be attributed to any specific factor. Any variation in phenotype that is not inherited is, by definition, a part of the environmental variance.

Third, **genetic–environmental interaction variance** (V_{GE}) arises when the effect of a gene depends on the specific environment in which it is found. An example is shown in **Figure 16.11**. In a dry environment, genotype AA produces a plant that averages 12 g in weight, and genotype aa produces a smaller plant that averages 10 g. In a wet environment, genotype aa produces the larger plant, averaging 24 g in weight, whereas genotype AA produces a plant that averages 20 g. In this example, there are clearly differences in the two environments: both genotypes produce heavier plants in the wet environment. There are also differences in the weights of the two genotypes, but the relative performances of the genotypes depend on whether the plants are grown in a wet or a dry environment. In this case, the influences on

16.11 Genetic–environmental interaction variance is obtained when the effect of a gene depends on the specific environment in which it is found. In this example, the genotype affects plant weight, but the environmental conditions determine which genotype produces the heavier plant.

phenotype cannot be neatly allocated into genetic and environmental components, because the expression of the genotype depends on the environment in which the plant grows. The phenotypic variance must therefore include a component that accounts for the way in which genetic and environmental factors interact.

In summary, the total phenotypic variance can be apportioned into three components:

$$V_P = V_G + V_E + V_{GE} \qquad (16.5)$$

Components of genetic variance Genetic variance can be further subdivided into components consisting of different types of genetic effects. First, **additive genetic variance** (V_A) comprises the additive effects of genes on the phenotype, which can be summed to determine the overall effect on the phenotype. For example, suppose that, in a plant, allele A^1 contributes 2 g in weight and allele A^2 contributes 4 g. If the alleles are strictly additive, then the genotypes would have the following weights:

$$A^1A^1 = 2 + 2 = 4 \text{ g}$$

$$A^1A^2 = 2 + 4 = 6 \text{ g}$$

$$A^2A^2 = 4 + 4 = 8 \text{ g}$$

The genes that Nilsson-Ehle studied, which affect kernel color in wheat, are additive in this way. The additive genetic

variance primarily determines the resemblance between parents and offspring. For example, if all of the phenotypic variance is due to additive genetic variance, then the phenotypes of the offspring will be exactly intermediate between those of the parents, but, if some genes have dominance, then offspring may be phenotypically different from both parents (i.e., $Aa \times Aa \rightarrow aa$ offspring).

Second, there is **dominance genetic variance** (V_D) when some genes have a dominance component. In this case, the alleles at a locus are not additive; rather, the effect of an allele depends on the identity of the other allele at that locus. For example, with a dominant allele (T), genotypes TT and Tt have the same phenotype. Here, we cannot simply add the effects of the alleles together, because the effect of the small t allele is masked by the presence of the large T allele. Instead, we must add a component (V_D) to the genetic variance to account for the way in which alleles interact.

Third, genes at different loci may interact in the same way that alleles at the same locus interact. When this genic interaction takes place, the effects of the genes are not additive. For example, Chapter 4 described how coat color in Labrador retrievers exhibits genic interaction; genotypes BB ee and bb ee both produce yellow dogs because the effect of alleles at the B locus are masked when ee alleles are present at the E locus. With genic interaction, we must include a third component, called **genic interaction variance** (V_I), to the genetic variance:

$$V_G = V_A + V_D + V_I \qquad (16.6)$$

Summary equation We can now integrate these components into one equation to represent all the potential contributions to the phenotypic variance:

$$V_P = V_A + V_D + V_I + V_E + V_{GE} \qquad (16.7)$$

Equation 16.7 provides us with a model that describes the potential causes of differences that we observe among individual phenotypes. Importantly, note that this model deals strictly with the observable *differences* (variance) in phenotypes among individual members of a population; it says nothing about the absolute value of the characteristic or about the underlying genotypes that produce these differences.

Types of Heritability

The model of phenotypic variance that we've just developed can be used to address the question of how much of the phenotypic variance in a characteristic is due to genetic differences. **Broad-sense heritability** (H^2) represents the proportion of phenotypic variance that is due to genetic variance and is calculated by dividing the genetic variance by the phenotypic variance:

$$\text{broad-sense heritability} = H^2 = \frac{V_G}{V_P} \qquad (16.8)$$

The symbol H^2 represents broad-sense heritability because it is a measure of variance, which is in units squared.

Broad-sense heritability can potentially range from 0 to 1. A value of 0 indicates that none of the phenotypic variance results from differences in genotype and all of the differences in phenotype result from environmental variation. A value of 1 indicates that all of the phenotypic variance results from differences in genotypes. A heritability value between 0 and 1 indicates that both genetic and environmental factors influence the phenotypic variance.

Often, we are more interested in the proportion of the phenotypic variance that results from the additive genetic variance because, as mentioned earlier, the additive genetic variance primarily determines the resemblance between parents and offspring. **Narrow-sense heritability** (h^2) is equal to the additive genetic variance divided by the phenotypic variance:

$$\text{narrow-sense heritability} = h^2 = \frac{V_A}{V_P} \qquad (16.9)$$

Calculating Heritability

Having considered the components that contribute to phenotypic variance and having developed a general concept of heritability, we can ask, How do we go about estimating these different components and calculating heritability? There are several ways to measure the heritability of a characteristic. The mathematical theory that underlies these calculations of heritability is complex and beyond the scope of this book. Nevertheless, we can develop a general understanding of how heritability is measured.

Most methods for calculating heritability compare the degree of resemblance between related and unrelated individuals or between individuals with different degrees of relatedness. For example, one method compares the phenotypes of parents and offspring with that of unrelated individuals. Another method compares the similarity of identical twins (which have 100% of their genes in common) with the similarity of nonidentical twins (which have 50% of their genes in common). The general idea is that, if genes influence genetic differences among individuals, individuals that are more closely related will be more similar in phenotype. Statistical methods called regression and correlation are used to determine the degree to which more closely related individuals are more similar in phenotype.

An example of calculating heritability by comparing the phenotypes of parents and offspring is illustrated in **Figure 16.12**. Here, the mean phenotype of the parents is plotted against the mean phenotype of the offspring. Each data point on the graph represents one family. The line represents the best fit of the all of the points of the graph (deviations of the points from the line are minimized). A

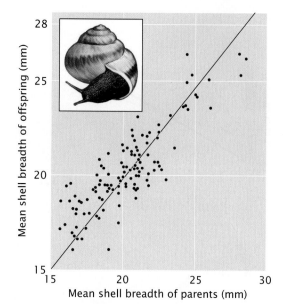

16.12 The heritability of shell breadth in snails can be determined by plotting the mean phenotype of offspring against the mean phenotype of the parents. [From L. M. Cook, Evolution 19:86–94, 1965.]

complex mathematical proof (which we will not go into here) demonstrates that the slope of the line equals the narrow-sense heritability.

All estimates of heritability depend on the assumption that the environments of related individuals are not more similar than those of unrelated individuals. This assumption is difficult to meet in human studies because related people are usually reared together. Heritability estimates for humans should therefore always be viewed with caution.

Concepts

Broad-sense heritability is the proportion of phenotypic variance that is due to genetic variance. Narrow-sense heritability is the proportion of phenotypic variance that is due to additive genetic variance. Heritability can be measured by comparing the degree of resemblance between related and unrelated individuals or between individuals having different degrees of relatedness.

✔ Concept Check 1

If the environmental variance (V_E) increases and all other variance components remain the same, what will the effect be?

a. Broad-sense heritability will decrease.

b. Broad-sense heritability will increase.

c. Narrow-sense heritability will increase.

d. Broad-sense heritability will increase, but narrow-sense heritability will decrease.

The Limitations of Heritability

Knowledge of heritability has great practical value because it allows us to statistically predict the phenotypes of offspring on the basis of their parent's phenotype. It also provides useful information about how characteristics will respond to selection (see Section 16.4). In spite of its importance, heritability is frequently misunderstood. Heritability does not provide information about an individual's genes or the environmental factors that control the development of a characteristic, and it says nothing about the nature of differences between groups. This section outlines some limitations and common misconceptions concerning broad- and narrow-sense heritability.

Heritability does not indicate the degree to which a characteristic is genetically determined Heritability is the proportion of the phenotypic variance that is due to genetic variance; it says nothing about the degree to which genes determine a characteristic. Heritability indicates only the degree to which genes determine *variation* in a characteristic. The determination of a characteristic and the determination of variation in a characteristic are two very different things.

Consider polydactyly (the presence of extra digits) in rabbits, which can be caused either by environmental factors or by a dominant gene. Suppose we have a group of rabbits all homozygous for a gene that produces the usual numbers of digits. None of the rabbits in this group carries a gene for polydactyly, but a few of the rabbits are polydactylous because of environmental factors. Broad-sense heritability for polydactyly in this group is zero because there is no genetic variation for polydactyly; all of the variation is due to environmental factors. However, it would be incorrect for us to conclude that genes play no role in determining the number of digits in rabbits. Indeed, we know that there are specific genes that can produce extra digits. Heritability indicates nothing about whether genes control the development of a characteristic; it provides information only about causes of the variation in a characteristic within a defined group.

An individual does not have heritability Broad- and narrow-sense heritabilities are statistical values based on the genetic and phenotypic variances found in a *group* of individuals. Heritability cannot be calculated for an individual, and heritability has no meaning for a specific individual. Suppose we calculate the narrow-sense heritability of adult body weight for the students in a biology class and obtain a value of 0.6. We could conclude that 60% of the variation in adult body weight among the students in this class is determined by additive genetic variation. We should not, however, conclude that 60% of any particular student's body weight is due to additive genes.

There is no universal heritability for a characteristic The value of heritability for a characteristic is specific for a given population in a given environment. Recall that broad-sense heritability is genetic variance divided by phenotypic variance. Genetic variance depends on which genes are present, which often differs between populations. In the example of polydactyly in rabbits, there were no genes for polydactyly in the group; so the heritability of the characteristic was zero. A different group of rabbits might contain many genes for polydactyly, and the heritability of the characteristic might then be high.

Environmental differences may affect heritability because V_P is composed of both genetic and environmental variance. When the environmental differences that affect a characteristic differ between two groups, the heritabilities for the two groups also will often differ. Because heritability is specific to a defined population in a given environment, it is important not to extrapolate heritabilities from one population to another.

Even when heritability is high, environmental factors may influence a characteristic High heritability does not mean that environmental factors cannot influence the expression of a characteristic. High heritability indicates only that the environmental variation to which the population is *currently* exposed is not responsible for variation in the characteristic. Let's look at human height. In most developed countries, heritability of human height is high, indicating that genetic differences are responsible for most of the variation in height. It would be wrong for us to conclude that human height cannot be changed by alteration of the environment. Indeed, in several European cities during World War II, height decreased owing to hunger and disease, and height can be increased dramatically by the administration of growth hormone to children. The absence of environmental variation in a characteristic does not mean that the characteristic will not respond to environmental change.

Heritabilities indicate nothing about the nature of population differences in a characteristic A common misconception about heritability is that it provides information about population differences in a characteristic. Heritability is specific for a given population in a given environment, and so it cannot be used to draw conclusions about why populations differ in a characteristic.

Suppose we measured heritability for human height in two groups. One group is from a small town in a developed country, where everyone consumes a high-protein diet. Because there is little variation in the environmental factors that affect human height and there is some genetic variation, the heritability of height in this group is high. The second group comprises the inhabitants of a single village in a developing country. The consumption of protein by these people is only 25% of that consumed by those in the first group; so their average adult height is several centimeters less than that in the developed country. Again, there is little variation in the environmental factors that determine height in this group because everyone in the village eats the same types of food and is exposed to the same diseases. Because there is little environmental variation and there is some genetic variation, the heritability of height in this group also is high.

Thus, the heritability of height in both groups is high, and the average height in the two groups is considerably different. We might be tempted to conclude that the difference in height between the two groups is genetically based—that the people in the developed country are genetically taller than the people in the developing country. This conclusion is obviously wrong, however, because the differences in height are due largely to diet—an environmental factor. Heritability provides no information about the causes of differences between populations.

These limitations of heritability have often been ignored, particularly in arguments about possible social implications of genetic differences between humans. For example, the results of a number of modern studies indicate that human intelligence as measured by IQ (intelligence quotient) and other intelligence tests has a moderately high heritability (usually from 0.4 to 0.8). On the basis of this observation, some people have argued that intelligence is innate and that enhanced educational opportunities cannot boost intelligence. This argument is based on the misconception that, when heritability is high, changing the environment will not alter the characteristic. In addition, because heritabilities of intelligence range from 0.4 to 0.8, a considerable amount of the variance in intelligence originates from environmental differences.

Another argument based on a misconception about heritability is that ethnic differences in measures of intelligence are genetically based. Because the results of some genetic studies show that IQ has moderately high heritability and because other studies find differences in the average IQ of ethnic groups, some people have suggested that ethnic differences in IQ are genetically based. As in the example of the effects of diet on human height, heritability provides no information about causes of differences among groups; it indicates only the degree to which phenotypic variance within a single group is genetically based. High heritability for a characteristic does not mean that phenotypic differences between ethnic groups are genetic. We should also remember that separating genetic and environmental effects in humans is very difficult; so heritability estimates themselves may be unreliable.

Concepts

Heritability provides information only about the degree to which *variation* in a characteristic is genetically determined. There is no universal heritability for a characteristic; heritability is specific for a given population in a specific environment. Environmental factors can potentially affect characteristics with high heritability, and heritability says nothing about the nature of population differences in a characteristic.

✔ Concept Check 2

Suppose that you just learned that the narrow-sense heritability of blood pressure measured among a group of African Americans in Detroit, Michigan, is 0.40. What does this heritability tell us about genetic and environmental contributions to blood pressure?

Locating Genes That Affect Quantitative Characteristics

The statistical methods described for use in analyzing quantitative characteristics can be used both to make predictions about the average phenotype expected in offspring and to estimate the overall contribution of genes to variation in the characteristic. These methods do not, however, allow us to identify and determine the influence of individual genes that affect quantitative characteristics. As stated in the introduction to this chapter, chromosome regions with genes that control polygenic characteristics are referred to as quantitative trait loci. Although quantitative genetics has made important contributions to basic biology and to plant and animal breeding, the past inability to identify QTLs and measure their individual effects severely limited the application of quantitative genetic methods.

Mapping QTLs   In recent years, numerous genetic markers have been identified and mapped with the use of molecular techniques, making it possible to identify QTLs by linkage analysis. The underlying idea is simple: if the inheritance of a genetic marker is associated consistently with the inheritance of a particular characteristic (such as increased height), then that marker must be linked to a QTL that affects height. The key is to have enough genetic markers so that QTLs can be detected throughout the genome. With the introduction of restriction fragment length polymorphisms, microsatellite

16.13 The availability of molecular markers makes the mapping of QTLs possible in many organisms. QTL mapping is used to identify genes that affect yield in corn and other agriculturally important plants. *[Brand X Pictures.]*

variations, and single-nucleotide polymorphisms (see Chapter 14), variable markers are now available for mapping QTLs in a number of different organisms (**Figure 16.13**; see also the story at the beginning of this chapter).

A common procedure for mapping QTLs is to cross two homozygous strains that differ in alleles at many loci. The resulting F_1 progeny are then intercrossed or backcrossed to allow the genes to recombine through independent assortment and crossing over. Genes on different chromosomes and genes that are far apart on the same chromosome will recombine freely; genes that are closely linked will be inherited together. The offspring are measured for one or more quantitative characteristics; at the same time, they are genotyped for numerous genetic markers that span the genome. Any correlation between the inheritance of a particular marker allele and a quantitative phenotype indicates that a QTL is linked to that marker. If enough markers are used, the detection of all the QTLs affecting a characteristic is theoretically possible. It is important to recognize that a QTL is not a gene; rather, a QTL is a map location for a chromosome region that is associated with that trait. After a QTL has been identified, it can be studied for the presence of specific genes that influence the quantitative trait. This approach has been used to detect genes affecting various characteristics in several plant and animal species (**Table 16.2**).

Applications of QTL mapping The number of genes affecting a quantitative characteristic can be estimated by locating QTLs with genetic markers and adding up the number of QTLs detected. This method will always be an underestimation because QTLs that are located close together on the same chromosome will be counted together and those with small effects are likely to be missed.

QTL mapping also provides information about the magnitude of the effects that genes have on a quantitative characteristic. The polygenic model assumes that many genes affect a quantitative characteristic, that the effect of each gene is small, and that the effects of the genes are equal and additive. The results of studies of QTLs in a number of organisms now show that these assumptions are not always valid. Polygenes appear to vary widely in their effects. In many of the characteristics that have been studied, a few QTLs account for much of the phenotypic variation. In some instances, individual QTLs have been mapped that account for more than 20% of the variance in the characteristic.

Table 16.2	Quantitative characteristics for which QTLs have been detected	
Organism	**Quantitative Characteristic**	**Number of QTLs Detected**
Tomato	Soluble solids	7
	Fruit mass	13
	Fruit pH	9
	Growth	5
	Leaflet shape	9
	Height	9
Corn	Height	11
	Leaf length	7
	Tiller number	1
	Glume hardness	5
	Grain yield	18
	Number of ears	9
	Thermotolerance	6
Common bean	Number of nodules	4
Mung bean	Seed weight	4
Cow pea	Seed weight	2
Wheat	Preharvest sprout	4
Pig	Growth	2
	Length of small intestine	1
	Average back fat	1
	Abdominal fat	1
Mouse	Epilepsy	2
Rat	Hypertension	2

Source: After S. D. Tanksley, Mapping polygenes, *Annual Review of Genetics* 27:218, 1993.

Concepts

The availability of numerous genetic markers revealed by molecular methods makes it possible to map chromosome segments containing genes that contribute to polygenic characteristics.

16.4 Genetically Variable Traits Change in Response to Selection

Evolution is genetic change among members of a population. Several different forces are potentially capable of producing evolution, and we will explore these forces and the process of evolution more fully in Chapter 17. Here, we consider how one of these forces—natural selection—may bring about genetic change in a quantitative characteristic.

Charles Darwin proposed the idea of natural selection in his book *On the Origin of Species* in 1859. **Natural selection** arises through the differential reproduction of individuals with different genotypes, allowing individuals with certain genotypes to produce more offspring than others. Natural selection is one of the most important of the forces that brings about evolutionary change and can be summarized as follows:

Observation 1 Many more individuals are produced each generation than are capable of surviving long enough to reproduce.

Observation 2 There is much phenotypic variation within natural populations.

Observation 3 Some phenotypic variation is heritable. In the terminology of quantitative genetics, some of the phenotypic variation in these characteristics is due to genetic variation, and these characteristics have heritability.

Logical consequence Individuals with certain characters (called adaptive traits) survive and reproduce better than others. Because the adaptive traits are heritable, offspring will tend to resemble their parents with regard to these traits, and there will be more individuals with these adaptive traits in the next generation. Thus, adaptive traits will tend to increase in the population through time.

In this way, organisms become genetically suited to their environments; as environments change, groups of organisms change in ways that make them better able to survive and reproduce.

For thousands of years, humans have practiced a form of selection by promoting the reproduction of organisms with traits perceived as desirable. This form of selection is **artificial selection**, and it has produced the domestic plants and animals that make modern agriculture possible.

Predicting the Response to Selection

When a quantitative characteristic is subjected to natural or artificial selection, it will frequently change with the passage of time, provided that there is genetic variation for that characteristic in the population. Suppose that a dairy farmer breeds only those cows in his herd that have the highest milk production. If there is genetic variation in milk production, the mean milk production in the offspring of the selected cows should be higher than the mean milk production of the original herd. This increased production is due to the fact that the selected cows possess more genes for high milk production than does the average cow, and these genes are passed on to the offspring. The offspring of the selected cows possess a higher proportion of genes for greater milk yield and therefore produce more milk than the average cow in the initial herd.

The extent to which a characteristic subjected to selection changes in one generation is termed the **response to selection**. Suppose that the average cow in a dairy herd produces 80 liters of milk per week. A farmer selects for increased milk production by breeding the highest milk producers, and the progeny of these selected cows produce 100 liters of milk per week on average. The response to selection is calculated by subtracting the mean phenotype of the original population (80 liters) from the mean phenotype of the offspring (100 liters), obtaining a response to selection of $100 - 80 = 20$ liters per week.

The response to selection is determined primarily by two factors. First, it is affected by the narrow-sense heritability, which largely determines the degree of resemblance between parents and offspring. When the narrow-sense heritability is high, offspring will tend to resemble their parents; conversely, when the narrow-sense heritability is low, there will be little resemblance between parents and offspring.

The second factor that determines the response to selection is how much selection there is. If the farmer is very stringent in the choice of parents and breeds only the highest milk producers in the herd (say, the top 2 cows), then all the offspring will receive genes for high-quality milk production. If the farmer is less selective and breeds the top 20 milk producers in the herd, then the offspring will not carry as many superior genes for high milk production, and they will not, on average, produce as much milk as the offspring of the top 2 producers. The response to selection depends on the phenotypic difference of the individuals that are selected as parents; this phenotypic difference is measured by the **selection differential**, defined as the difference between the mean phenotype of the selected parents and the mean phenotype of the original population. If the average milk production of the original herd is 80 liters and the farmer breeds cows with an average milk production of 120 liters, then the selection differential is $120 - 80 = 40$ liters.

The response to selection (R) depends on the narrow-sense heritability (h^2) and the selection differential (S):

$$R = h^2 \times S \qquad (16.10)$$

This equation can be used to predict the magnitude of change in a characteristic when a given selection differential is applied. G. A. Clayton and his colleagues estimated the response to selection that would take place in abdominal bristle number of *Drosophila melanogaster*. Using several different methods, they first estimated the narrow-sense heritability of abdominal bristle number in one population of fruit flies to be 0.52. The mean number of bristles in the original population was 35.3. They selected individual flies with a mean bristle number of 40.6 and intercrossed them to produce the next generation. The selection differential was $40.6 - 35.3 = 5.3$; so they predicted a response to selection to be

$$R = 0.52 \times 5.3 = 2.8$$

The response to selection of 2.8 is the expected increase in the characteristic of the offspring above the mean of the original population. They therefore expected the average

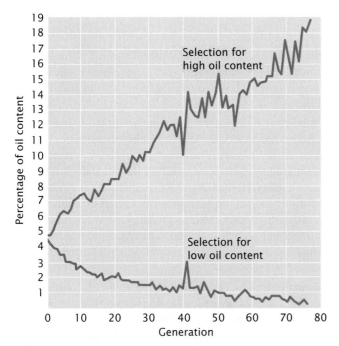

16.14 In a long-term response-to-selection experiment, selection for oil content in corn increased oil content in one line to about 20%, whereas oil content was almost eliminated in another line.

number of abdominal bristles in the offspring of their selected flies to be 35.3 + 2.8 = 38.1. Indeed, they found an average bristle number of 37.9 in these flies.

Rearranging Equation 16.10 provides another way to calculate the narrow-sense heritability:

$$h^2 = \frac{R}{S} \qquad (16.11)$$

In this way, h^2 can be calculated by conducting a response-to-selection experiment. First, the selection differential is obtained by subtracting the population mean from the mean of selected parents. The selected parents are then interbred, and the mean phenotype of their offspring is measured. The difference between the mean of the offspring and that of the initial population is the response to selection, which can be used with the selection differential to estimate the heritability. Heritability determined by a response-to-selection experiment is usually termed the **realized heritability**. If certain assumptions are met, the realized heritability is identical with the narrow-sense heritability.

One of the longest selection experiments is a study of oil and protein content in corn seeds (**Figure 16.14**). This experiment began at the University of Illinois on 163 ears of corn with an oil content ranging from 4% to 6%. Corn plants having high oil content and those having low oil content were selected and interbred. Response to selection for increased oil content (the upper line in Figure 16.14) reached about 20%, whereas response to selection for decreased oil content reached a lower limit near zero. Genetic analyses of the

high- and low-oil-content strains revealed that at least 20 loci take part in determining oil content.

Concepts

The response to selection is influenced by narrow-sense heritability and the selection differential.

✔ Concept Check 3

The narrow-sense heritability for a trait is 0.4 and the selection differential is 0.5. What is the predicted response to selection?

Limits to Selection Response

When a characteristic has been selected for many generations, the response eventually levels off and the characteristic no longer responds to selection (**Figure 16.15**). A potential reason for this leveling off is that the genetic variation in the population may be exhausted; at some point, all individuals in the population have become homozygous for alleles that encode the selected trait. When there is no more additive genetic variation, heritability equals zero and there can be no further response to selection.

The response to selection may level off even while some genetic variation remains in the population, however, because natural selection opposes further change in the characteristic. Response to selection for small body size in mice, for example, eventually levels off because the smallest animals are sterile and cannot pass on their genes for small body size. In this case, artificial selection for small size is opposed by natural selection for fertility, and the population can no longer respond to the artificial selection.

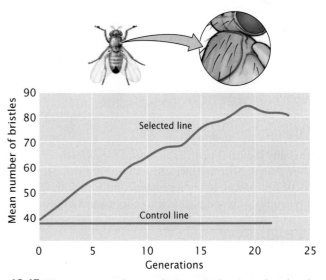

16.15 The response of a population to selection often levels off at some point in time. In a response-to-selection experiment for increased number of bristles on the abdomen of female fruit flies, the number of bristles increased steadily for about 20 generations and then leveled off.

Concepts Summary

- Quantitative genetics focuses on the inheritance of complex characteristics whose phenotypes vary continuously. For many quantitative characteristics, the relation between genotype and phenotype is complex because many genes and environmental factors influence a characteristic.

- The individual genes that influence a polygenic characteristic follow the same Mendelian principles that govern discontinuous characteristics, but, because many genes participate, the expected ratios of phenotypes are obscured.

- A frequency distribution, in which the phenotypes are represented on one axis and the number of individuals possessing each phenotype is represented on the other, is a convenient means of summarizing phenotypes found in a group of individuals.

- The mean and variance provide key information about a distribution: the mean gives the central location of the distribution, and the variance provides information about how the phenotype varies within a group.

- Phenotypic variance in a characteristic can be divided into components that are due to additive genetic variance, dominance genetic variance, genic interaction variance, environmental variance, and genetic–environmental interaction variance.

- Broad-sense heritability is the proportion of the phenotypic variance due to genetic variance; narrow-sense heritability is the proportion of the phenotypic variance due to additive genetic variance.

- Heritability provides information only about the degree to which variation in a characteristic results from genetic differences. Heritability is based on the variances present within a group of individuals, and an individual does not have heritability. Heritability of a characteristic varies among populations and among environments. Even if heritability for a characteristic is high, the characteristic may still be altered by changes in the environment. Heritabilities provide no information about the nature of population differences in a characteristic.

- Quantitative trait loci are chromosome segments containing genes that control polygenic characteristics. QTLs can be mapped by examining the association between the inheritance of a quantitative characteristic and the inheritance of genetic markers.

- The amount that a quantitative characteristic changes in a single generation when subjected to selection (the response to selection) is directly related to the selection differential and narrow-sense heritability.

Important Terms

quantitative genetics (p. 407)
quantitative trait locus (QTL) (p. 407)
meristic characteristic (p. 410)
threshold characteristic (p. 410)
frequency distribution (p. 413)
normal distribution (p. 414)
mean (p. 414)
variance (p. 415)

heritability (p. 415)
phenotypic variance (p. 416)
genetic variance (p. 416)
environmental variance (p. 416)
genetic–environmental interaction
   variance (p. 416)
additive genetic variance (p. 417)
dominance genetic variance (p. 417)

genic interaction variance (p. 417)
broad-sense heritability (p. 417)
narrow-sense heritability (p. 418)
natural selection (p. 421)
artificial selection (p. 422)
response to selection (p. 422)
selection differential (p. 422)
realized heritability (p. 423)

Answers to Concept Checks

1. a

2. The heritability indicates that about 40% of the differences in blood pressure among African Americans in Detroit are due to additive genetic differences. It neither provides information about the heritability of blood pressure in other groups of people

nor indicates anything about the nature of differences in blood pressure between African Americans in Detroit and people in other groups.

3. 0.2

Worked Problems

1. Seed weight in a particular plant species is determined by pairs of alleles at two loci (a^+a^- and b^+b^-) that are additive and equal in their effects. Plants with genotype $a^-a^-\ b^-b^-$ have seeds that average 1 g in weight, whereas plants with genotype $a^+a^+\ b^+b^+$ have seeds that average 3.4 g in weight. A plant with genotype $a^-a^-\ b^-b^-$ is crossed with a plant of genotype $a^+a^+\ b^+b^+$.

a. What is the predicted weight of seeds from the F_1 progeny of this cross?

b. If the F_1 plants are intercrossed, what are the expected seed weights and proportions of the F_2 plants?

P $\qquad$ $a^-a^-\,b^-b^-\;\times\;a^+a^+\,b^+b^+$

$\qquad\qquad$ 1 g $\qquad$ 3.4 g

F_1 $\qquad\qquad$ $a^+a^-\,b^+b^-$

$\qquad\qquad\qquad$ 2.2 g

Genotype	Probability		Number of contributing genes	Average seed weight
$a^-a^-b^-b^-$	$\tfrac{1}{4}\times\tfrac{1}{4}=\tfrac{1}{16}$		0	1 g + (0 × 0.6 g) = 1 g
$a^+a^-b^-b^-$	$\tfrac{1}{2}\times\tfrac{1}{4}=\tfrac{1}{8}$	$\left.\right\}\; \tfrac{2}{8}=\tfrac{4}{16}$	1	1 g + (1 × 0.6 g) = 1.6 g
$a^-a^-b^+b^-$	$\tfrac{1}{4}\times\tfrac{1}{2}=\tfrac{1}{8}$			
$a^+a^+b^-b^-$	$\tfrac{1}{4}\times\tfrac{1}{4}=\tfrac{1}{16}$			
$a^-a^-b^+b^+$	$\tfrac{1}{4}\times\tfrac{1}{4}=\tfrac{1}{16}$	$\left.\right\}\;\tfrac{2}{16}+\tfrac{1}{4}=\tfrac{6}{16}$	2	1 g + (2 × 0.6 g) = 2.2 g
$a^+a^-b^+b^-$	$\tfrac{1}{2}\times\tfrac{1}{2}=\tfrac{1}{4}$			
$a^+a^+b^+b^-$	$\tfrac{1}{4}\times\tfrac{1}{2}=\tfrac{1}{8}$	$\left.\right\}\;\tfrac{2}{8}=\tfrac{4}{16}$	3	1 g + (3 × 0.6 g) = 2.8 g
$a^+a^-b^+b^+$	$\tfrac{1}{2}\times\tfrac{1}{4}=\tfrac{1}{8}$			
$a^+a^+b^+b^+$	$\tfrac{1}{4}\times\tfrac{1}{4}=\tfrac{1}{16}$		4	1 g + (4 × 0.6 g) = 3.4 g

(F_2 label appears to the left of the $a^+a^+b^-b^-$ through $a^+a^-b^+b^-$ group.)

• Solution

The difference in average seed weight of the two parental genotypes is 3.4 g − 1 g = 2.4 g. These two genotypes differ in four genes; so, if the genes have equal and additive effects, each gene difference contributes an additional 2.4 g/4 = 0.6 g of weight to the 1-g weight of a plant ($a^-a^-\,b^-b^-$), which has none of the contributing genes.

The cross between the two homozygous genotypes produces the F_1 and F_2 progeny shown in the table at the top of this page.

a. The F_1 are heterozygous at both loci ($a^+a^-\,b^+b^-$) and possess two genes that contribute an additional 0.6 g each to the 1-g weight of a plant that has no contributing genes. Therefore, the seeds of the F_1 should average 1 g + 2(0.6 g) = 2.2 g.

b. The F_2 will have the following phenotypes and proportions: $\tfrac{1}{16}$ 1 g; $\tfrac{4}{16}$ 1.6 g; $\tfrac{6}{16}$ 2.2 g; $\tfrac{4}{16}$ 2.8 g; and $\tfrac{1}{16}$ 3.4 g.

2. Phenotypic variation is analyzed for milk production in a herd of dairy cattle and the following variance components are obtained:

Additive genetic variance (V_A)	= 0.4
Dominance genetic variance (V_D)	= 0.1
Genic interaction variance (V_I)	= 0.2
Environmental variance (V_E)	= 0.5
Genetic–environmental interaction variance (V_{GE})	= 0.0

a. What is the narrow-sense heritability of milk production?

b. What is the broad-sense heritability of milk production?

• Solution

To determine the heritabilities, we first need to calculate V_P and V_G:

$$V_P = V_A + V_D + V_I + V_E + V_{GE}$$
$$= 0.4 + 0.1 + 0.2 + 0.5 + 0$$
$$= 1.2$$

$$V_G = V_A + V_D + V_I$$
$$= 0.7$$

a. The narrow-sense heritability is: $h^2 = \dfrac{V_A}{V_P} = \dfrac{0.4}{1.2} = 0.33$

b. The broad-sense heritability is: $H^2 = \dfrac{V_G}{V_P} = \dfrac{0.7}{1.2} = 0.58$

3. A farmer is raising rabbits. The average body weight in his population of rabbits is 3 kg. The farmer selects the 10 largest rabbits in his population, whose average body weight is 4 kg, and interbreeds them. If the heritability of body weight in the rabbit population is 0.7, what is the expected body weight among offspring of the selected rabbits?

• Solution

The farmer has carried out a response-to-selection experiment, in which the response to selection will equal the selection differential times the narrow-sense heritability. The selection differential equals the difference in average weights of the selected rabbits and the entire population: 4 kg − 3 kg = 1 kg. The narrow-sense heritability is given as 0.7; so the expected response to selection is: $R = h^2 \times S = 0.7 \times 1$ kb $= 0.7$ kg. This value is the increase in weight that is expected in the offspring of the selected parents; so the average weight of the offspring is expected to be: 3 kg + 0.7 kg = 3.7 kg.

Comprehension Questions

Section 16.1

*1. How does a quantitative characteristic differ from a discontinuous characteristic?

2. Briefly explain why the relation between genotype and phenotype is frequently complex for quantitative characteristics.

*3. Why do polygenic characteristics have many phenotypes?

Section 16.2

4. What information do the mean and variance provide about a distribution?

Section 16.3

*5. List all the components that contribute to the phenotypic variance and define each component.

*6. How do the broad-sense and narrow-sense heritabilities differ?

7. Briefly describe common misunderstandings or misapplications of the concept of heritability.

8. Briefly explain how genes affecting a polygenic characteristic are located with the use of QTL mapping.

Section 16.4

*9. How is the response to selection related to the narrow-sense heritability and the selection differential? What information does the response to selection provide?

10. Why does the response to selection often level off after many generations of selection?

Application Questions and Problems

Section 16.1

*11. Indicate whether each of the following characteristics would be considered a discontinuous characteristic or a quantitative characteristic. Briefly justify your answer.

  a. Kernel color in a strain of wheat, in which two codominant alleles segregating at a single locus determine the color. Thus, there are three phenotypes present in this strain: white, light red, and medium red.

  b. Body weight in a family of Labrador retrievers. An autosomal recessive allele that causes dwarfism is present in this family. Two phenotypes are recognized: dwarf (less than 13 kg) and normal (greater than 13 kg).

  c. Presence or absence of leprosy. Susceptibility to leprosy is determined by multiple genes and numerous environmental factors.

  d. Number of toes in guinea pigs, which is influenced by genes at many loci.

  e. Number of fingers in humans. Extra (more than five) fingers are caused by the presence of an autosomal dominant allele.

*12. Assume that plant weight is determined by a pair of alleles at each of two independently assorting loci (A and a, B and b) that are additive in their effects. Further assume that each allele represented by an uppercase letter contributes 4 g to weight and that each allele represented by a lowercase letter contributes 1 g to weight.

  a. If a plant with genotype AA BB is crossed with a plant with genotype aa bb, what weights are expected in the F_1 progeny?

  b. What is the distribution of weight expected in the F_2 progeny?

*13. Assume that three loci, each with two alleles (A and a, B and b, C and c), determine the differences in height between two homozygous strains of a plant. These genes are additive and equal in their effects on plant height. One strain (aa bb cc) is 10 cm in height. The other strain (AA BB CC) is 22 cm in height. The two strains are crossed, and the resulting F_1 are interbred to produce F_2 progeny. Give the phenotypes and the expected proportions of the F_2 progeny.

14. Seed size in a plant is a polygenic characteristic. A grower crosses two pure-breeding varieties of the plant and measures seed size in the F_1 progeny. She then backcrosses the F_1 plants to one of the parental varieties and measures seed size in the backcross progeny. The grower finds that

seed size in the backcross progeny has a higher variance than does seed size in the F_1 progeny. Explain why the backcross progeny are more variable.

Section 16.3

*15. Phenotypic variation in tail length of mice has the following components:

Additive genetic variance (V_A)	= 0.5
Dominance genetic variance (V_D)	= 0.3
Genic interaction variance (V_I)	= 0.1
Environmental variance (V_E)	= 0.4
Genetic–environmental interaction variance (V_{GE}) = 0.0	

    a. What is the narrow-sense heritability of tail length?

    b. What is the broad-sense heritability of tail length?

16. The narrow-sense heritability of ear length in Reno rabbits is 0.4. The phenotypic variance (V_P) is 0.8, and the environmental variance (V_E) is 0.2. What is the additive genetic variance (V_A) for ear length in these rabbits?

17. A characteristic has a narrow-sense heritability of 0.6.

    a. If the dominance variance (V_D) increases and all other variance components remain the same, what will happen to the narrow-sense heritability? Will it increase, decrease, or remain the same? Explain.

    b. What will happen to the broad-sense heritability? Explain.

    c. If the environmental variance (V_E) increases and all other variance components remain the same, what will happen to the narrow-sense heritability? Explain.

    d. What will happen to the broad-sense heritability? Explain.

18. Many researchers have estimated the heritability of human traits by comparing the correlation coefficients of monozygotic and dizygotic twins. An assumption in using this method is that two monozygotic twins experience environments that are no more similar to each other than those experienced by two dizygotic twins. How might this assumption be violated? Give some specific examples of ways in which the environments of two monozygotic twins might be more similar than the environments of two dizyotic twins.

19. A genetics researcher determines that the broad-sense heritability of height among Southwestern University undergraduate students is 0.90. Which of the following conclusions would be reasonable? Explain your answer.

    a. Because Sally is a Southwestern University undergraduate student, 10% of her height is determined by nongenetic factors.

    b. Ninety percent of variation in height among all undergraduate students in the United States is due to genetic differences.

    c. Ninety percent of the height of Southwestern University undergraduate students is determined by genes.

    d. Ten percent of the variation in height among Southwestern University undergraduate students is determined by variation in nongenetic factors.

    e. Because the heritability of height among Southwestern Unversity students is so high, any change in the students' environment will have minimal effect on their height.

20. *Drosophila buzzati* is a fruit fly that feeds on the rotting fruits of cacti in Australia. Timothy Prout and Stuart Barker calculated the heritabilities of body size, as measured by thorax length, for a natural population of *D. buzzati* raised in the wild and for a populatios of *D. buzzati* collected in the wild but raised in the laboratory (T. Prout and J. S. F. Barker. 1989. *Genetics* 123:803–813). They found the following heritabilities.

Population	Heritability of body size ($\pm$ standard error)
Wild population	0.0595 $\pm$ 0.0123
Laboratory-reared population	0.3770 $\pm$ 0.0203

Why do you think the heritability measured in the laboratory-reared population is higher than that measured in the natural population raised in the wild?

*21. Mr. Jones is a pig farmer. For many years, he has fed his pigs the food left over from the local university cafeteria, which is known to be low in protein, deficient in vitamins, and downright untasty. However, the food is free, and his pigs don't complain. One day a salesman from a feed company visits Mr. Jones. The salesman claims that his company sells a new, high-protein, vitamin-enriched feed that enhances weight gain in pigs. Although the food is expensive, the salesman claims that the increased weight gain of the pigs will more than pay for the cost of the feed, increasing Mr. Jones's profit. Mr. Jones responds that he took a genetics class when he went to the university and that he has conducted some genetic experiments on his pigs; specifically, he has calculated the narrow-sense heritability of weight gain for his pigs and found it to be 0.98. Mr. Jones says that this heritability value indicates that 98% of the variance in weight gain among his pigs is determined by genetic differences, and therefore the new pig feed can have little effect on the growth of his pigs. He concludes that the feed would be a waste of his money. The salesman doesn't dispute Mr. Jones' heritability estimate, but he still claims that the new feed can significantly increase weight gain in Mr. Jones' pigs. Who is correct and why?

Section 16.4

22. Joe is breeding cockroaches in his dorm room. He finds that the average wing length in his population of cockroaches is 4 cm. He chooses six cockroaches that have the largest wings; the average wing length among these selected cockroaches is 10 cm. Joe interbreeds these

selected cockroaches. From earlier studies, he knows that the narrow-sense heritability for wing length in his population of cockroaches is 0.6.

   a. Calculate the selection differential and expected response to selection for wing length in these cockroaches.

   b. What should be the average wing length of the progeny of the selected cockroaches?

23. Three characteristics in beef cattle—body weight, fat content, and tenderness—are measured, and the following variance components are estimated:

	Body weight	Fat content	Tenderness
V_A	22	45	12
V_D	10	25	5
V_I	3	8	2
V_E	42	64	8
V_{GE}	0	0	1

In this population, which characteristic would respond best to selection? Explain your reasoning.

*24. A rancher determines that the average amount of wool produced by a sheep in her flock is 22 kg per year. In an attempt to increase the wool production of her flock, the rancher picks five male and five female sheep with the greatest wool production; the average amount of wool produced per sheep by those selected is 30 kg. She interbreeds these selected sheep and finds that the average wool production among the progeny of the selected sheep is 28 kg. What is the narrow-sense heritability for wool production among the sheep in the rancher's flock?

25. A strawberry farmer determines that the average weight of individual strawberries produced by plants in his garden is 2 g. He selects the 10 plants that produce the largest strawberries; the average weight of strawberries among these selected plants is 6 g. He interbreeds these selected strawberry plants. The progeny of these selected plants produce strawberries that weigh 5 g. If the farmer were to select plants that produce an average strawberry weight of 4 g, what would be the predicted weight of strawberries produced by the progeny of these selected plants?

26. Pigs have been domesticated from wild boars. Would you expect to find higher heritability for weight among domestic pigs or among wild boars? Explain your answer.

Challenge Questions

Section 16.1

27. Bipolar illness is a psychiatric disorder that has a strong hereditary basis, but the exact mode of inheritance is not known. Research has shown that siblings of patients with bipolar illness are more likely to develop the disorder than are siblings of unaffected persons. Findings from a recent study demonstrated that the ratio of bipolar brothers to bipolar sisters is higher when the patient is male than when the patient is female. In other words, relatively more brothers of bipolar patients also have the disease when the patient is male than when the patient is female. What does this observation suggest about the inheritance of bipolar illness?

Section 16.3

28. We have explored some of the difficulties in separating the genetic and environmental components of human behavioral characteristics. Considering these difficulties and what you know about calculating heritability, propose an experimental design for accurately measuring the heritability of musical ability.

29. A student who has just learned about quantitative genetics says, "Heritability estimates are worthless! They don't tell you anything about the genes that affect a characteristic. They don't provide any information about the types of offspring to expect from a cross. Heritability estimates measured in one population can't be used for other populations; so they don't even give you any general information about how much of a characteristic is genetically determined. I can't see that heritabilities do anything other than make undergraduate students sweat during tests." How would you respond to this statement? Is the student correct? What good are heritabilities, and why do geneticists bother to calculate them?

Section 16.4

30. Eugene Eisen selected for increased 12-day litter weight (total weight of a litter of offspring 12 days after birth) in a population of mice (E. J. Eisen. 1972. *Genetics* 72:129–142). The 12-day litter weight of the population steadily increased, but then leveled off after about 17 generations. At generation 17, Eisen took one family of mice from the selected population and reversed the selection procedure: in this group, he selected for *decreased* 12-day litter size. This group immediately responded to decreased selection; the 12-day litter weight dropped 4.8 g within one generation and dropped 7.3 g after 5 generations. Based on the results of the reverse selection, what is the most likely explanation for the leveling off of 12-day litter weight in the original population?

17 Population and Evolutionary Genetics

Genetic Rescue of Bighorn Sheep

Rocky Mountain bighorn sheep (*Ovis canadensis*). A population of bighorn sheep at the National Bison Range suffered the loss of genetic variation owing to genetic drift; the introduction of sheep from other populations dramatically increased genetic variation and the fitness of the sheep. *[Tom J. Ulrich/Visuals Unlimited.]*

Rocky Mountain bighorn sheep (*Ovis canadensis*) are among North America's most spectacular animals, characterized by the male's magnificent horns that curve gracefully back over the ears, spiraling down and back up beside the face. Two hundred years ago, bighorn sheep were numerous throughout western North America, ranging from Mexico to southern Alberta and from Colorado to California. Meriwether Lewis and William Clark reported numerous sightings of these beautiful animals in their expedition across the western United States from 1804 to 1806. Before 1900, the number of bighorn sheep in North America was about 2 million.

Unfortunately, settlement of the west by Europeans was not kind to the bighorns. Beginning in the late 1800s, hunting, loss of habitat, competition from livestock, and diseases carried by domestic sheep decimated the bighorns. Today, fewer than 70,000 bighorn sheep remain, scattered across North America in fragmented and isolated populations.

In 1922, wildlife biologists established a population of bighorn sheep at the National Bison Range, an isolated tract of 18,000 acres nestled between the mountains of northwestern Montana. In that year, 12 bighorn sheep—4 males (rams) and 8 females (ewes)— were trapped at Banff National Park in Canada and transported to the National Bison Range. No additional animals were introduced to this population for the next 60 years.

At first, the population of bighorns at the National Bison Range flourished, protected there from hunting and livestock. Within 8 years, it had grown to 90 sheep but then began to slowly decrease in size. Population size waxed and waned through the years, but the number of sheep had dropped to about 50 animals by 1985, and the population was in trouble. The amount of genetic variation was low compared with other native populations of bighorn sheep. The reproductive rate of both male and female sheep had dropped, and the size and survival of the sheep were lower than in more-healthy populations. The population at the National Bison Range was suffering from genetic drift, an evolutionary force operating in small populations that causes random changes in the gene pool and the loss of genetic variation.

To counteract the negative effects of genetic drift, biologists added 5 new rams from other herds in Montana and Wyoming in 1985, mimicking the effects of natural migration among herds. Another 10 sheep were introduced between 1990 and 1994. This influx of new genes had a dramatic effect on the genetic health of the population. Genetic variation among individual sheep increased significantly. Outbred rams (those containing the new genes) were more dominant, were more likely to copulate, and were more likely to produce offspring. Outbred ewes had more than twice the annual reproductive success of inbred females. Adult survival increased after the introduction of new genes, and, slowly, the population grew in size, reaching 69 sheep by 2003.

The bighorn sheep at the National Bison Range illustrate an important principle of genetics: small populations lose genetic variation with the passage of time through genetic drift, often with catastrophic consequences for survival and reproduction. The introduction of new genetic variation into an inbred population, called **genetic rescue**, often dramatically improves the health of the population and can better ensure its long-term survival. These effects have important implications for wildlife management, as well as for how organisms evolve in the natural world.

This chapter introduces *population genetics*, the branch of genetics that studies the genetic makeup of groups of individuals and how a *group's* genetic composition changes with time. Population geneticists usually focus their attention on a **Mendelian population**, which is a group of interbreeding, sexually reproducing individuals that have a common set of genes, the **gene pool**. A population evolves through changes in its gene pool; therefore, population genetics is also the study of *evolution*. Population geneticists study the variation in alleles within and between groups and the evolutionary forces responsible for shaping the patterns of genetic variation found in nature. In this chapter, we will learn how the gene pool of a population is measured and what factors are responsible for shaping it. At the end of the chapter, we turn to the evolutionary changes that bring about the appearance of new species and patterns of evolutionary change at the molecular level.

(a)

(b)

17.1 All organisms exhibit genetic variation. (a) Extensive variation among humans. (b) Variation in the spotting patterns of Asian lady beetles. *[Part a: Paul Warner/AP.]*

17.1 Genotypic and Allelic Frequencies Are Used to Describe the Gene Pool of a Population

An obvious and pervasive feature of life is variability. Consider a group of students in a typical college class, the members of which vary in eye color, hair color, skin pigmentation, height, weight, facial features, blood type, and susceptibility to numerous diseases and disorders. No two students in the class are likely to be even remotely similar in appearance.

Humans are not unique in their extensive variability (**Figure 17.1a**); almost all organisms exhibit variation in phenotype. For instance, lady beetles are highly variable in their patterns of spots (**Figure 17.1b**), mice vary in body size, snails have different numbers of stripes on their shells, and plants vary in their susceptibility to pests. Much of this phenotypic variation is hereditary. Recognition of the extent of phenotypic variation led Charles Darwin to the idea of evolution through natural selection. Genetic variation is the basis of all evolution, and the extent of genetic variation within a population affects its potential to adapt to environmental change.

In fact, even more genetic variation exists in populations than is visible in the phenotype. Much variation exists at the molecular level owing, in part, to the redundancy of the genetic code, which allows different codons to specify the same amino acid. Thus, two members of a population can produce the same protein even if their DNA sequences are different. DNA sequences between the genes and introns within genes do not encode proteins; much of the variation in these sequences probably also has little effect on the phenotype.

An important, but frequently misunderstood, tool used in population genetics is the mathematical model. Let's take a moment to consider what a model is and how it can be used. A mathematical model usually describes a process as an equation. Factors that may influence the process are represented by variables in the equation; the equation defines the way in which the variables influence the process. Most models are simplified representations of a process because the simultaneous consideration of all of the influencing factors is impossible; some factors must be ignored in order to examine the effects of others. At first, a model might consider only one factor or a few factors, but, after their effects have been understood, the model can be improved by the addition of more details. Importantly, even a simple model can be a source of valuable insight into how a process is influenced by key variables.

Before we can explore the evolutionary processes that shape genetic variation, we must be able to describe the genetic structure of a population. The usual way of doing so is to enumerate the types and frequencies of genotypes and alleles in a population.

Calculating Genotypic Frequencies

A frequency is simply a proportion or a percentage, usually expressed as a decimal fraction. For example, if 20% of the alleles at a particular locus in a population are A, we would say that the frequency of the A allele in the population is 0.20. For large populations, where it is not practical to determine the genes of all individual members, a sample of the population is usually taken and the genotypic and allelic frequencies are calculated for this sample. The genotypic and allelic frequencies of the sample are then used to represent the gene pool of the population.

To calculate a **genotypic frequency**, we simply add up the number of individuals possessing the genotype and divide by the total number of individuals in the sample (N). For a locus with three genotypes AA, Aa, and aa, the frequency (f) of each genotype is

$$f(AA) = \frac{\text{number of } AA \text{ individuals}}{N}$$

$$f(Aa) = \frac{\text{number of } Aa \text{ individuals}}{N} \qquad (17.1)$$

$$f(aa) = \frac{\text{number of } aa \text{ individuals}}{N}$$

The sum of all the genotypic frequencies always equals 1.

Calculating Allelic Frequencies

The gene pool of a population can also be described in terms of the allelic frequencies. There are always fewer alleles than genotypes; so the gene pool of a population can be described in fewer terms when the allelic frequencies are used. In a sexually reproducing population, the genotypes are only temporary assemblages of the alleles: the genotypes break down each generation when individual alleles are passed to the next generation through the gametes, and so the types and numbers of alleles, rather than genotypes, have real continuity from one generation to the next and make up the gene pool of a population.

Allelic frequencies can be calculated from (1) the numbers or (2) the frequencies of the genotypes. To calculate the **allelic frequency** from the numbers of genotypes, we count the number of copies of a particular allele present in a sample and divide by the total number of all alleles in the sample:

$$\text{frequency of an allele} = \frac{\begin{array}{c}\text{number of copies}\\\text{of the allele}\end{array}}{\begin{array}{c}\text{number of copies of all}\\\text{alleles at the locus}\end{array}} \quad (17.2)$$

For a locus with only two alleles (A and a), the frequencies of the alleles are usually represented by the symbols p and q, and can be calculated as follows:

$$p = f(A) = \frac{2n_{AA} + n_{Aa}}{2N}$$

$$q = f(a) = \frac{2n_{aa} + n_{Aa}}{2N} \qquad (17.3)$$

where n_{AA}, n_{Aa}, and n_{aa} represent the numbers of AA, Aa, and aa individuals, and N represents the total number of individuals in the sample. We divide by $2N$ because each diploid individual has two alleles at a locus. The sum of the allelic frequencies always equals 1 ($p + q = 1$); so, after p has been obtained, q can be determined by subtraction: $q = 1 - p$.

Alternatively, allelic frequencies can be calculated from the genotypic frequencies. To do so, we add the frequency of the homozygote for each allele to half the frequency of the heterozygote (because half of the heterozygote's alleles are of each type):

$$p = f(A) = f(AA) + \frac{1}{2} f(Aa)$$
$$q = f(a) = f(aa) + \frac{1}{2} f(Aa) \qquad (17.4)$$

We obtain the same values of p and q whether we calculate the allelic frequencies from the numbers of genotypes (see Equation 17.3) or from the genotypic frequencies (see Equation 17.4). A sample calculation of allelic frequencies is provided in the next Worked Problem.

Loci with multiple alleles We can use the same principles to determine the frequencies of alleles for loci with more than two alleles. To calculate the allelic frequencies from the numbers of genotypes, we count up the number of copies of an allele by adding twice the number of homozygotes to the number of heterozygotes that possess the allele and divide this sum by twice the number of individuals in the sample. For a locus with three alleles (A^1, A^2, and A^3) and six genotypes ($A^1A^1, A^1A^2, A^2A^2, A^1A^3, A^2A^3$, and A^3A^3), the frequencies (p, q, and r) of the alleles are

$$p = f(A^1) = \frac{2n_{A^1A^1} + n_{A^1A^2} + n_{A^1A^3}}{2N}$$

$$q = f(A^2) = \frac{2n_{A^2A^2} + n_{A^1A^2} + n_{A^2A^3}}{2N} \qquad (17.5)$$

$$r = f(A^3) = \frac{2n_{A^3A^3} + n_{A^1A^3} + n_{A^2A^3}}{2N}$$

Alternatively, we can calculate the frequencies of multiple alleles from the genotypic frequencies by extending Equation 17.4. Once again, we add the frequency of the homozygote to half the frequency of each heterozygous genotype that possesses the allele:

$$p = f(A^1A^1) + \frac{1}{2}f(A^1A^2) + \frac{1}{2}f(A^1A^3)$$
$$q = f(A^2A^2) + \frac{1}{2}f(A^1A^2) + \frac{1}{2}f(A^2A^3) \qquad (17.6)$$
$$r = f(A^3A^3) + \frac{1}{2}f(A^1A^3) + \frac{1}{2}f(A^2A^3)$$

Concepts

Population genetics concerns the genetic composition of a population and how it changes with time. The gene pool of a population can be described by the frequencies of genotypes and alleles in the population.

Worked Problem

The human MN blood-type antigens are determined by two codominant alleles, L^M and L^N (see p. 83 in Chapter 4). The MN blood types and corresponding genotypes of 398 Finns in Karjala are tabulated here.

Phenotype	Genotype	Number
MM	$L^M L^M$	182
MN	$L^M L^N$	172
NN	$L^N L^N$	44

Source: W. C. Boyd, *Genetics and the Races of Man* (Boston: Little, Brown, 1950.)

Calculate the genotypic and allelic frequencies at the MN locus for the Karjala population.

• Solution

The genotypic frequencies for the population are calculated with the following formula:

$$\text{genotypic frequency} = \frac{\text{number of individuals with genotype}}{\text{total number of individuals in sample } (N)}$$

$$f(L^M L^M) = \frac{\text{number of } L^M L^M \text{individuals}}{N} = \frac{182}{398} = 0.457$$

$$f(L^M L^N) = \frac{\text{number of } L^M L^N \text{individuals}}{N} = \frac{172}{398} = 0.432$$

$$f(L^N L^N) = \frac{\text{number of } L^N L^N \text{individuals}}{N} = \frac{44}{398} = 0.111$$

The allelic frequencies can be calculated from either the numbers or the frequencies of the genotypes. To calculate allelic frequencies from numbers of genotypes, we add the number of copies of the allele and divide by the number of copies of all alleles at that locus.

$$\text{frequency of an allele} = \frac{\text{number of copies of the allele}}{\text{number of copies of all alleles}}$$

$$p = f(L^M) = \frac{(2n_{L^M L^M}) + (n_{L^M L^N})}{2N}$$
$$= \frac{2(182) + 172}{2(398)} = \frac{536}{796} = 0.673$$

$$q = f(L^N) = \frac{(2n_{L^N L^N}) + (n_{L^M L^N})}{2N}$$
$$= \frac{2(44) + 172}{2(398)} = \frac{260}{796} = 0.327$$

To calculate the allelic frequencies from genotypic frequencies, we add the frequency of the homozygote for that genotype to half the frequency of each heterozygote that contains that allele:

$$p = f(L^M) = f(L^M L^M) + \frac{1}{2} f(L^M L^N)$$
$$= 0.457 + \frac{1}{2} (0.432) = 0.673$$

$$q = f(L^N) = f(L^N L^N) + \frac{1}{2} f(L^M L^N)$$
$$= 0.111 + \frac{1}{2} (0.432) = 0.327$$

> **?** Now try your hand at calculating genotypic and allelic frequencies by working Problem 24 at the end of the chapter.

17.2 The Hardy–Weinberg Law Describes the Effect of Reproduction on Genotypic and Allelic Frequencies

The primary goal of population genetics is to understand the processes that shape a population's gene pool. First, we must ask what effects reproduction and Mendelian principles have on the genotypic and allelic frequencies: How do the segregation of alleles in gamete formation and the combining of alleles in fertilization influence the gene pool? The answer to this question lies in the **Hardy–Weinberg law**, among the most important principles of population genetics.

The Hardy–Weinberg law was formulated independently by both Godfrey H. Hardy and Wilhelm Weinberg in 1908. (Similar conclusions were reached by several other geneticists at about the same time.) The law is actually a mathematical model that evaluates the effect of reproduction on the genotypic and allelic frequencies of a population. It makes several simplifying assumptions about the population and provides two key predictions if these assumptions are met. For an autosomal locus with two alleles, the Hardy–Weinberg law can be stated as follows:

Assumptions If a population is large, randomly mating, and not affected by mutation, migration, or natural selection, then:

Prediction 1 the allelic frequencies of a population do not change; and

Prediction 2 the genotypic frequencies stabilize (will not change) after one generation in the proportions p^2 (the frequency of AA), $2pq$ (the frequency of Aa), and q^2 (the frequency of aa), where p equals the frequency of allele A and q equals the frequency of allele a.

The Hardy–Weinberg law indicates that, when the assumptions are met, reproduction alone does not alter allelic or genotypic frequencies and the allelic frequencies determine the frequencies of genotypes.

The statement that genotypic frequencies stabilize after one generation means that they may change in the first generation after random mating, because one generation of random mating is required to produce Hardy–Weinberg proportions of the genotypes. Afterward, the genotypic frequencies, like allelic frequencies, do not change as long as the population continues to meet the assumptions of the

Hardy–Weinberg law. When genotypes are in the expected proportions of p^2, $2pq$, and q^2, the population is said to be in **Hardy–Weinberg equilibrium**.

Concepts

The Hardy–Weinberg law describes how reproduction and Mendelian principles affect the allelic and genotypic frequencies of a population.

✔ Concept Check 1

Which statement is not an assumption of the Hardy–Weinberg law?

a. The allelic frequencies (p and q) are equal.

b. The population is randomly mating.

c. The population is large.

d. Natural selection has no effect.

Genotypic Frequencies at Hardy–Weinberg Equilibrium

How do the conditions of the Hardy–Weinberg law lead to genotypic proportions of p^2, $2pq$, and q^2? Mendel's principle of segregation says that each individual organism possesses two alleles at a locus and that each of the two alleles has an equal probability of passing into a gamete. Thus, the frequencies of alleles in gametes will be the same as the frequencies of alleles in the parents. Suppose we have a Mendelian population in which the frequencies of alleles A and a are p and q, respectively. These frequencies will also be those in the gametes. If mating is random (one of the assumptions of the Hardy–Weinberg law), the gametes will come together in random combinations, which can be represented by a Punnett square as shown in **Figure 17.2**.

Conclusion: Random mating will produce genotypes of the next generation in proportions $p^2(AA)$, $2pq(Aa)$, and $q^2(aa)$

17.2 Random mating produces genotypes in the proportions p^2, $2pq$, and q^2.

The multiplication rule of probability can be used to determine the probability of various gametes pairing. For example, the probability of a sperm containing allele A is p and the probability of an egg containing allele A is p. Applying the multiplicative rule, we find that the probability that these two gametes will combine to produce an AA homozygote is $p \times p = p^2$. Similarly, the probability of a sperm containing allele a combining with an egg containing allele a to produce an aa homozygote is $q \times q = q^2$. An Aa heterozygote can be produced in one of two ways: (1) a sperm containing allele A may combine with an egg containing allele a ($p \times q$) or (2) an egg containing allele A may combine with a sperm containing allele a ($p \times q$). Thus, the probability of alleles A and a combining to produce an Aa heterozygote is $2pq$. In summary, whenever the frequencies of alleles in a population are p and q, the frequencies of the genotypes in the next generation will be p^2, $2pq$, and q^2.

Closer Examination of the Assumptions of the Hardy–Weinberg Law

Before we consider the implications of the Hardy–Weinberg law, we need to take a closer look at the three assumptions that it makes about a population. First, it assumes that the population is large. How big is "large"? Theoretically, the Hardy–Weinberg law requires that a population be infinitely large in size, but this requirement is obviously unrealistic. In practice, many large populations are in the predicated Hardy–Weinberg proportions, and significant deviations arise only when population size is rather small. Later in the chapter, we will examine the effects of small population size on allelic frequencies.

The second assumption of the Hardy–Weinberg law is that members of the population mate randomly, which means that each genotype mates relative to its frequency. For example, suppose that three genotypes are present in a population in the following proportions: $f(AA) = 0.6$, $f(Aa) = 0.3$, and $f(aa) = 0.1$. With random mating, the frequency of mating between two AA homozygotes ($AA \times AA$) will be equal to the multiplication of their frequencies: $0.6 \times 0.6 = 0.36$, whereas the frequency of mating between two aa homozygotes ($aa \times aa$) will be only $0.1 \times 0.1 = 0.01$.

The third assumption of the Hardy–Weinberg law is that the allelic frequencies of the population are not affected by natural selection, migration, and mutation. Although mutation occurs in every population, its rate is so low that it has little short-term effect on the predictions of the Hardy–Weinberg law (although it may largely shape allelic frequencies over long periods of time when no other forces are acting). Although natural selection and migration are significant factors in real populations, we must remember that the purpose of the Hardy–Weinberg law is to examine only the effect of reproduction on the gene pool. When this effect is known, the effects of other factors (such as migration and natural selection) can be examined.

A final point is that the assumptions of the Hardy–Weinberg law apply to a single locus. No real population mates randomly for all traits; and a population is not completely free of natural selection for all traits. The Hardy–Weinberg law, however, does not require random mating and the absence of selection, migration, and mutation for all traits; it requires these conditions only for the locus under consideration. A population may be in Hardy–Weinberg equilibrium for one locus but not for others.

Implications of the Hardy–Weinberg Law

The Hardy–Weinberg law has several important implications for the genetic structure of a population. One implication is that a population cannot evolve if it meets the Hardy–Weinberg assumptions, because evolution consists of change in the allelic frequencies of a population. Therefore the Hardy–Weinberg law tells us that reproduction alone will not bring about evolution. Other processes such as natural selection, mutation, migration, or chance are required for populations to evolve.

A second important implication is that, when a population is in Hardy–Weinberg equilibrium, the genotypic frequencies are determined by the allelic frequencies. The heterozygote frequency never exceeds 0.5 when the population is in Hardy–Weinberg equilibrium. Furthermore, when the frequency of one allele is low, homozygotes for that allele will be rare and most of the copies of a rare allele will be present in heterozygotes.

A third implication of the Hardy–Weinberg law is that a single generation of random mating produces the equilibrium frequencies of p^2, $2pq$, and q^2. The fact that genotypes are in Hardy–Weinberg proportions does not prove that the population is free from natural selection, mutation, and migration. It means only that these forces have not acted since the last time random mating took place.

Testing for Hardy–Weinberg Proportions

To determine if a population's genotypes are in Hardy–Weinberg equilibrium, the genotypic proportions expected under the Hardy–Weinberg law must be compared with the observed genotypic frequencies. To do so, we first calculate the allelic frequencies, then find the expected genotypic frequencies by using the square of the allelic frequencies, and, finally, compare the observed and expected genotypic frequencies by using a chi-square test.

Worked Problem

Jeffrey Mitton and his colleagues found three genotypes (R^2R^2, R^2R^3, and R^3R^3) at a locus encoding the enzyme peroxidase in ponderosa pine trees growing at Glacier Lake, Colorado. The observed numbers of these genotypes were:

Genotypes	Number observed
R^2R^2	135
R^2R^3	44
R^3R^3	11

Are the ponderosa pine trees at Glacier Lake in Hardy–Weinberg equilibrium at the peroxidase locus?

• Solution

If the frequency of the R^2 allele equals p and the frequency of the R^3 allele equals q, the frequency of the R^2 allele is

$$p = f(R^2) = \frac{(2n_{R^2R^2}) + (n_{R^2R^3})}{2N} = \frac{2(135) + 44}{2(190)} = 0.826$$

The frequency of the R^3 allele is obtained by subtraction:

$$q = f(R^3) = 1 - p = 0.174$$

The frequencies of the genotypes expected under Hardy–Weinberg equilibrium are then calculated by using p^2, $2pq$, and q^2:

$$R^2R^2 = p^2 = (0.826)^2 = 0.683$$

$$R^2R^3 = 2pq = 2(0.826)(0.174) = 0.287$$

$$R^3R^3 = q^2 = (0.174)^2 = 0.03$$

Multiplying each of these expected genotypic frequencies by the total number of observed genotypes in the sample (190), we obtain the numbers expected for each genotype:

$$R^2R^2 = 0.683 \times 190 = 129.8$$

$$R^2R^3 = 0.287 \times 190 = 54.5$$

$$R^3R^3 = 0.03 \times 190 = 5.7$$

By comparing these expected numbers with the observed numbers of each genotype, we see that there are more R^2R^2 homozygotes and fewer R^2R^3 heterozygotes and R^3R^3 homozygotes in the population than we expect at equilibrium.

A goodness-of-fit chi-square test is used to determine whether the differences between the observed and the expected numbers of each genotype are due to chance:

$$X^2 = \sum \frac{(\text{observed} - \text{expected})^2}{\text{expected}}$$

$$= \frac{(135 - 129.8)^2}{129.8} + \frac{(44 - 54.5)^2}{54.5} + \frac{(11 - 5.7)^2}{5.7}$$

$$= 0.21 + 2.02 + 4.93 = 7.16$$

The calculated chi-square value is 7.16; to obtain the probability associated with this chi-square value, we determine the appropriate degrees of freedom.

So far, the chi-square test for assessing Hardy–Weinberg equilibrium has been identical with the chi-square tests that we used in Chapter 3 to assess progeny ratios in a genetic cross, where the degrees of freedom were $n - 1$ and n equaled the number of expected genotypes. For the Hardy–Weinberg test, however, we must subtract an additional degree of freedom because the expected numbers are based on the observed allelic frequencies; therefore, the observed numbers are not completely free to vary. In general, the degrees of freedom for a chi-square test of Hardy–Weinberg equilibrium equal the number of expected genotypic classes minus the number of associated alleles. For this particular Hardy–Weinberg test, the degree of freedom is $3 - 2 = 1$.

After we have calculated both the chi-square value and the degrees of freedom, the probability associated with this value can be sought in a chi-square table (see Table 3.4). With one degree of freedom, a chi-square value of 7.16 has a probability between 0.01 and 0.001. The peroxidase genotypes observed at Glacier Lake are not likely to be in Hardy–Weinberg proportions.

> **?** For additional practice, determine whether the genotypic frequencies in Problem 26 at the end of the chapter are in Hardy–Weinberg equilibrium.

Concepts

The observed number of genotypes in a population can be compared with the Hardy–Weinberg expected proportions by using a goodness-of-fit chi-square test.

Estimating Allelic Frequencies by Using the Hardy–Weinberg Law

A practical use of the Hardy–Weinberg law is that it allows us to calculate allelic frequencies when dominance is present. For example, cystic fibrosis is an autosomal recessive disorder characterized by respiratory infections, incomplete digestion, and abnormal sweating (see pp. 83–84 in Chapter 4). Among North American Caucasians, the incidence of the disease is approximately 1 person in 2000. The formula for calculating allelic frequency (see Equation 17.3) requires that we know the numbers of homozygotes and heterozygotes, but cystic fibrosis is a recessive disease and so we cannot easily distinguish between homozygous normal persons and heterozygous carriers. Although molecular tests are available for identifying heterozygous carriers of the cystic fibrosis gene, the low frequency of the disease makes widespread screening impractical. In such situations, the Hardy–Weinberg law can be used to estimate the allelic frequencies.

If we assume that a population is in Hardy–Weinberg equilibrium with regard to this locus, then the frequency of the recessive genotype (aa) will be q^2, and the allelic frequency is the square root of the genotypic frequency:

$$q = 2\,\overline{f(aa)} \qquad (17.7)$$

If the frequency of cystic fibrosis in North American Caucasians is approximately 1 in 2000, or 0.0005, then $q = \sqrt{0.0005} = 0.02$. Thus, about 2% of the alleles in the Caucasian population encode cystic fibrosis. We can calculate the frequency of the normal allele by subtracting: $p = 1 - q = 1 - 0.02 = 0.98$. After we have calculated p and q, we can use the Hardy–Weinberg law to determine the frequencies of homozygous normal people and heterozygous carriers of the gene:

$$f(AA) = p^2 = (0.98)^2 = 0.960$$

$$f(Aa) = 2pq = 2(0.02)(0.98) = 0.0392$$

Thus, about 4% (1 of 25) of Caucasians are heterozygous carriers of the allele that causes cystic fibrosis.

Concepts

Although allelic frequencies cannot be calculated directly for traits that exhibit dominance, the Hardy–Weinberg law can be used to estimate the allelic frequencies if the population is in Hardy–Weinberg equilibrium for that locus. The frequency of the recessive allele will be equal to the square root of the frequency of the recessive trait.

✔ Concept Check 2

In cats, all-white color is dominant over not all-white. In a population of 100 cats, 19 are all-white cats. Assuming that the population is in Hardy–Weinberg equilibrium, what is the frequency of the all-white allele in this population?

Nonrandom Mating

An assumption of the Hardy–Weinberg law is that mating is random with respect to genotype. Nonrandom mating affects the way in which alleles combine to form genotypes and alters the genotypic frequencies of a population.

One form of nonrandom mating is **inbreeding**, which is preferential mating between related individuals. Inbreeding causes a departure from the Hardy–Weinberg equilibrium frequencies of p^2, $2pq$, and q^2. More specifically, it leads to an increase in the proportion of homozygotes and a decrease in the proportion of heterozygotes in a population.

Concepts

Nonrandom mating alters the frequencies of the genotypes but not the frequencies of the alleles. Inbreeding is preferential mating between related individuals. With inbreeding, the frequency of homozygotes increases, whereas the frequency of heterozygotes decreases.

17.3 Several Evolutionary Forces Potentially Cause Changes in Allelic Frequencies

The Hardy–Weinberg law indicates that allelic frequencies do not change as a result of reproduction; thus, other processes must cause alleles to increase or decrease in frequency. Processes that bring about change in allelic frequency include mutation, migration, genetic drift (random effects due to small population size), and natural selection.

Mutation

Before evolution can take place, genetic variation must exist within a population; consequently, all evolution depends on processes that generate genetic variation. Although new *combinations* of existing genes may arise through recombination in meiosis, all genetic variants ultimately arise through mutation.

The effect of mutation on allelic frequencies Mutation can influence the rate at which one genetic variant increases at the expense of another. Consider a single locus in a population of 25 diploid individuals. Each individual possesses two alleles at the locus under consideration; so the gene pool of the population consists of 50 allelic copies. Let us assume that there are two different alleles, designated G^1 and G^2 with frequencies p and q, respectively. If there are 45 copies of G^1 and 5 copies of G^2 in the population, $p = 0.90$ and $q = 0.10$. Now suppose that a mutation changes a G^1 allele into a G^2 allele. After this mutation, there are 44 copies of G^1 and 6 copies of G^2, and the frequency of G^2 has increased from 0.10 to 0.12. Mutation has changed the allelic frequency.

If copies of G^1 continue to mutate to G^2, the frequency of G^2 will increase and the frequency of G^1 will decrease (**Figure 17.3**). The amount that G^2 will change as a result of mutation depends on: (1) the rate of G^1-to-G^2 mutation; and (2) p, the frequency of G^1 in the population. When p is large, many copies of G^1 are available to mutate to G^2 and the amount of change will be relatively large. As more mutations occur and p decreases, fewer copies of G^1 will be available to mutate to G^2.

So far, we have considered only the effects of $G^1 \rightarrow G^2$ forward mutations. Reverse $G^2 \rightarrow G^1$ mutations also occur but at a rate that will probably differ from the forward mutation rate. Whenever a reverse mutation occurs, the frequency of G^2 decreases and the frequency of G^1 increases (see Figure 17.3).

Reaching equilibrium of allelic frequencies Consider an allele that begins with a high frequency of G^1 and a low frequency of G^2 (see Figure 17.3a). In this population, many copies of G^1 are initially available to mutate to G^2 and the

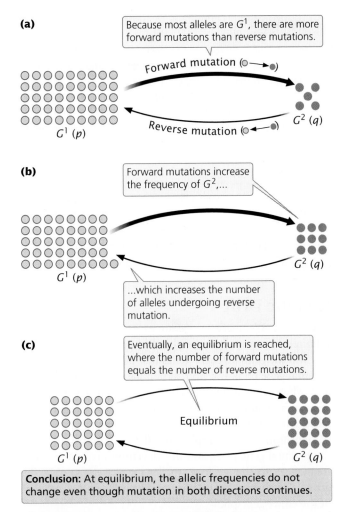

(a) Because most alleles are G^1, there are more forward mutations than reverse mutations.

Forward mutation (○ → ●)

Reverse mutation (○ ← ●)

G^1 (p)

G^2 (q)

(b) Forward mutations increase the frequency of G^2,...

...which increases the number of alleles undergoing reverse mutation.

G^1 (p)

G^2 (q)

(c) Eventually, an equilibrium is reached, where the number of forward mutations equals the number of reverse mutations.

Equilibrium

G^1 (p)

G^2 (q)

Conclusion: At equilibrium, the allelic frequencies do not change even though mutation in both directions continues.

17.3 Recurrent mutation changes allelic frequencies. Forward and reserve mutations eventually lead to a stable equilibrium.

increase in G^2 due to forward mutation will be relatively large. However, as the frequency of G^2 increases as a result of forward mutations, fewer copies of G^1 are available to mutate; so the number of forward mutations decreases. On the other hand, few copies of G^2 are initially available to undergo a reverse mutation to G^1 but, as the frequency of G^2 increases, the number of copies of G^2 available to undergo reverse mutation to G^1 increases; so the number of genes undergoing reverse mutation will increase (see Figure 17.3b). Eventually, the number of genes undergoing forward mutation will be counterbalanced by the number of genes undergoing reverse mutation (see Figure 17.3c). At this point, the increase in q due to forward mutation will be equal to the decrease in q due to reverse mutation and there will be no net change in allelic frequency, in spite of the fact that forward and reverse mutations continue to occur. The point at which there is no change in the allelic frequency of a population is referred to as **equilibrium** (see Figure 17.3c). At equilibrium, the frequency of G^2 is determined solely by the forward and reverse mutation rates.

Summary of effects When the only evolutionary force acting on a population is mutation, allelic frequencies change with the passage of time because some alleles mutate into others. Eventually, these allelic frequencies reach equilibrium and are determined only by the forward and reverse mutation rates.

The mutation rates for most genes are low; so change in allelic frequency due to mutation in one generation is very small, and long periods of time are required for a population to reach mutational equilibrium. Nevertheless, if mutation is the only force acting on a population for long periods of time, mutation rates will determine allelic frequencies.

Concepts

Recurrent mutation causes changes in the frequencies of alleles. At equilibrium, the allelic frequencies are determined by the forward and reverse mutation rates. Because mutation rates are low, the effect of mutation per generation is very small.

Migration

Another process that may bring about change in the allelic frequencies is the influx of genes from other populations, commonly called **migration** or **gene flow**. One of the assumptions of the Hardy–Weinberg law is that migration does not take place, but many natural populations do experience migration from other populations. The overall effect of migration is twofold: (1) it prevents populations from becoming genetically different from one another and (2) it increases genetic variation within populations.

The effect of migration on allelic frequencies Let us consider the effects of migration by looking at a simple, unidirectional model of migration between two populations that differ in the frequency of an allele *a* (**Figure 17.4**). In each generation, a representative sample of the individuals in population I migrates to population II and reproduces, adding its genes to population II's gene pool. Migration is only from population I to population II (is unidirectional), and all the conditions of the Hardy–Weinberg law apply except the absence of migration.

After migration, population II consists of two types of individuals: (1) migrants with genes from population I, and (2) the original residents with genes from population II. The allelic frequencies in population II after migration depend on the contributions of alleles from the migrants and from the original residents. The amount of change in the frequency of allele *a* in population II is directly proportional to the amount of migration; as the amount of migration increases, the change in allelic frequency increases. The magnitude of change is also affected by the differences in allelic

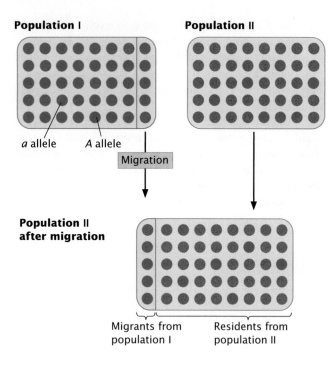

Population I   **Population II**

a allele   *A* allele

Migration

Population II after migration

Migrants from population I   Residents from population II

17.4 The amount of change in allelic frequency due to migration between populations depends on the difference in allelic frequency and the extent of migration. Shown here is a model of the effect of unidirectional migration on allelic frequencies.

frequencies of the two populations; when the difference is large, the change in allelic frequency will be large.

With each generation of migration, the frequencies of the two populations become more and more similar until, eventually, the allelic frequency of population II equals that of population I. When the allelic frequencies are equal, there will be no further change in the allelic frequency of population II, in spite of the fact that migration continues. If migration between two populations takes place for a number of generations with no other evolutionary forces present, an equilibrium is reached at which the allelic frequency of the recipient population equals that of the source population.

The simple model of unidirectional migration between two populations just outlined can be expanded to accommodate multidirectional migration between several populations.

The overall effect of migration   Migration has two major effects. First, it causes the gene pools of populations to become more similar. Later, we will see how genetic drift and natural selection lead to genetic differences between populations; migration counteracts this tendency and tends to keep populations homogeneous in their allelic frequencies. Second, migration adds genetic variation to populations. Different alleles may arise in different populations owing to rare mutational events, and these alleles can be spread to new populations by migration, increasing the genetic variation within the recipient population.

Concepts

Migration causes changes in the allelic frequency of a population by introducing alleles from other populations. The magnitude of change due to migration depends on both the extent of migration and the difference in allelic frequencies between the source and the recipient populations. Migration decreases genetic differences between populations and increases genetic variation within populations.

✔ Concept Check 3

Each generation, 10 random individuals migrate from population A to population B. What will happen to allelic frequency q as a result of migration when q is equal in populations A and B?

a. q in A will decrease.   c. q will not change in either A or B.

b. q in B will increase.   d. q in B will become q^2.

Genetic Drift

The Hardy–Weinberg law assumes random mating in an infinitely large population; only when population size is infinite will the gametes carry genes that perfectly represent the parental gene pool. But no real population is infinitely large and, when population size is limited, the gametes that unite to form individuals of the next generation carry a sample of alleles present in the parental gene pool. Just by chance, the composition of this sample will often deviate from that of the parental gene pool, and this deviation may cause allelic frequencies to change. The smaller the gametic sample, the greater the chance that its composition will deviate from that of the entire gene pool.

The role of chance in altering allelic frequencies is analogous to the flip of a coin. Each time we flip a coin, we have a 50% chance of getting a head and a 50% chance of getting a tail. If we flip a coin 1000 times, the observed ratio of heads to tails will be very close to the expected 50 : 50 ratio. If, however, we flip a coin only 10 times, there is a good chance that we will obtain not exactly 5 heads and 5 tails but perhaps 7 heads and 3 tails or 8 tails and 2 heads. This kind of deviation from an expected ratio due to limited sample size is referred to as **sampling error**.

Sampling error arises when gametes unite to produce progeny. Many organisms produce a large number of gametes but, when population size is small, a limited number of gametes unite to produce the individuals of the next generation. Chance influences which alleles are present in this limited sample and, in this way, sampling error may lead to **genetic drift**, or changes in allelic frequency. Because the deviations from the expected ratios are random, the direction of change is unpredictable. We can nevertheless predict the magnitude of the changes.

The magnitude of genetic drift   The amount of change resulting from genetic drift is determined largely by the population size (N): genetic drift will be higher when the population size is small. For ecological and demographic studies, population size is usually defined as the number of individuals in a group. The evolution of a gene pool depends, however, only on those individuals who contribute genes to the next generation. Population geneticists usually define population size as the equivalent number of breeding adults, the **effective population size** (N_e).

Concepts

Genetic drift is change in allelic frequency due to chance factors. The amount of change in allelic frequency due to genetic drift is inversely related to the effective population size (the equivalent number of breeding adults in a population).

✔ Concept Check 4

Which of the following statements is an example of genetic drift?

a. Allele *g* for fat production increases in a small population because birds with more body fat have higher survivorship in a harsh winter.

b. Random mutation increases the frequency of allele *A* in one population but not in another.

c. Allele *R* reaches a frequency of 1.0 because individuals with genotype *rr* are sterile.

d. Allele *m* is lost when a virus kills all but a few individuals and just by chance none of the survivors possess allele *m*.

Causes of genetic drift   All genetic drift arises from sampling error, but sampling error can arise in several different ways. First, a population can be reduced in size for a number of generations because of limitations in space, food, or some other critical resource. Genetic drift in a small population for multiple generations can significantly affect the composition of a population's gene pool.

A second way in which sampling error can arise is through the **founder effect**, which is due to the establishment of a population by a small number of individuals; the population of bighorn sheep at the National Bison Range, discussed in the introduction to this chapter, underwent a founder effect. Although a population can increase and become quite large, the genes carried by all its members are derived from the few genes originally present in the founders (assuming no migration or mutation). Chance events affecting which genes were present in the founders will have an important influence on the makeup of the entire population.

A third way in which genetic drift arises is through a **genetic bottleneck**, which develops when a population undergoes a drastic reduction in population size. A genetic bottleneck developed in northern elephant seals (**Figure 17.5**). Before 1800, thousands of elephant seals were found along the California coast, but the population was devastated by hunting between 1820 and 1880. By 1884, as few as 20 seals survived on a remote beach of Isla de Guadelupe west of Baja California. Restrictions on hunting enacted by the

17.5 Northern elephant seals underwent a severe genetic bottleneck between 1820 and 1880. Today, these seals have low levels of genetic variation. *[PhotoDisc.]*

United States and Mexico allowed the seals to recover, and there are now estimated to be almost 100,000 seals. All seals in the population today are genetically similar, because they have genes that were carried by the few survivors of the population bottleneck.

The effects of genetic drift   Genetic drift has several important effects on the genetic composition of a population. First, it produces change in allelic frequencies within a population. Because drift is random, allelic frequency is just as likely to increase as it is to decrease and will wander with the passage of time (hence the name genetic drift). **Figure 17.6** illustrates a computer simulation of genetic drift in five

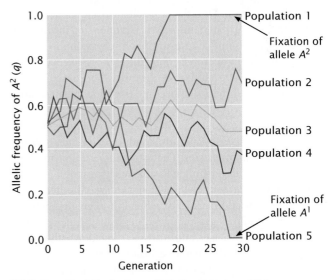

17.6 Genetic drift changes allelic frequencies within populations, leading to a reduction in genetic variation through fixation and genetic divergence among populations. Shown here is a computer simulation of changes in the frequency of allele A^2 (q) in five different populations due to random genetic drift. Each population consists of 10 males and 10 females and begins with $q = 0.5$.

populations over 30 generations, starting with $q = 0.5$ and maintaining a constant population size of 10 males and 10 females. These allelic frequencies change randomly from generation to generation.

A second effect of genetic drift is to reduce genetic variation within populations. Through random change, an allele may eventually reach a frequency of either 1 or 0, at which point all individuals in the population are homozygous for one allele. When an allele has reached a frequency of 1, we say that it has reached **fixation**. Other alleles are lost (reach a frequency of 0) and can be restored only by migration from another population or by mutation. Fixation, then, leads to a loss of genetic variation within a population. This loss can be seen in the northern elephant seals described earlier. Today, these seals have low levels of genetic variation; a study of 24 protein-encoding genes found no individual or population differences in these genes. A subsequent study of sequence variation in mitochondrial DNA also revealed low levels of genetic variation. In contrast, the southern elephant seal had much higher levels of mitochondrial DNA variation. The southern elephant seals also were hunted, but their population size never dropped below 1000; therefore, unlike the northern elephant seals, they did not experience a genetic bottleneck.

Given enough time, all small populations will become fixed for one allele or the other. Which allele becomes fixed is random and is determined by the initial frequency of the allele. If the population begins with two alleles, each with a frequency of 0.5, both alleles have an equal probability of fixation. However, if one allele is initially common, it is more likely to become fixed.

A third effect of genetic drift is that different populations diverge genetically with time. In Figure 17.6, all five populations begin with the same allelic frequency ($q = 0.5$) but, because drift is random, the frequencies in different populations do not change in the same way, and so populations gradually acquire genetic differences. Eventually, all the populations reach fixation; some will become fixed for one allele, and others will become fixed for the alternative allele.

The effect of genetic drift on variation among populations is illustrated by a study conducted by Luca Cavalli-Sforza and his colleagues. They studied variation in blood types among villagers in the Parma Valley of Italy, where the amount of migration between villages was limited. They found that variation in allelic frequency was greatest between small isolated villages in the upper valley but decreased between larger villages and towns farther down the valley. This result is exactly what we expect with genetic drift: there should be more genetic drift and thus more variation among villages when population size is small.

The three results of genetic drift (allelic frequency change, loss of variation within populations, and genetic divergence between populations) take place simultaneously, and all result from sampling error. The first two results take place *within* populations, whereas the third takes place *between* populations.

Concepts

Genetic drift results from continuous small population size, the founder effect (establishment of a population by a few founders), and the bottleneck effect (population reduction). Genetic drift causes a change in allelic frequencies within a population, a loss of genetic variation through the fixation of alleles, and genetic divergence between populations.

Natural Selection

A final process that brings about changes in allelic frequencies is *natural selection*, the differential reproduction of genotypes (see p. 421–422 in Chapter 16). Natural selection takes place when individuals with adaptive traits produce a greater number of offspring than that produced by others in the population. If the adaptive traits have a genetic basis, they are inherited by the offspring and appear with greater frequency in the next generation. A trait that provides a reproductive advantage thereby increases with the passage of time, enabling populations to become better suited to their environments—to become better adapted. Natural selection is unique among evolutionary forces in that it promotes adaptation (**Figure 17.7**).

Fitness and the selection coefficient   The effect of natural selection on the gene pool of a population depends on the fitness values of the genotypes in the population. **Fitness** is defined as the relative reproductive success of a genotype. Here, the term *relative* is critical: fitness is the reproductive

17.7 Natural selection produces adaptations, such as those seen in the polar bears that inhabit the extreme Arctic environment. These bears blend into the snowy background, which helps them in hunting seals. The hairs of their fur stay erect even when wet, and thick layers of blubber provide insulation, which protects against subzero temperatures. Their digestive tracts are adapted to a seal-based carnivorous diet. [*Digital Vision.*]

success of one genotype compared with the reproductive successes of other genotypes in the population.

Fitness (W) ranges from 0 to 1. Suppose the average number of viable offspring produced by three genotypes is

Genotypes: A^1A^1 A^1A^2 A^2A^2

Mean number of offspring produced: 10 5 2

To calculate fitness for each genotype, we take the mean number of offspring produced by a genotype and divide it by the mean number of offspring produced by the most prolific genotype:

$$A^1A^1 \qquad A^1A^2$$

$$\text{Fitness } (W): \quad W_{11} = \frac{10}{10} = 1.0 \quad W_{12} = \frac{5}{10} = 0.5$$

$$A^2A^2$$

$$W_{22} = \frac{2}{10} = 0.2 \tag{17.8}$$

The fitness of genotype A^1A^1 is designated W_{11}, that of A^1A^2 is W_{12}, and that of A^2A^2 is W_{22}. A related variable is the **selection coefficient** (s), which is the relative intensity of selection against a genotype. We usually speak of selection for a particular genotype, but keep in mind that, when selection is *for* one genotype, selection is automatically *against* at least one other genotype. The selection coefficient is equal to $1 - W$; so the selection coefficients for the preceding three genotypes are

$$A^1A^1 \qquad A^1A^2 \qquad A^2A^2$$

Selection coefficient $(1 - W)$: $s_{11} = 0$ $s_{12} = 0.5$ $s_{22} = 0.8$

Concepts

Natural selection is the differential reproduction of genotypes. It is measured as fitness, which is the reproductive success of a genotype compared with other genotypes in a population.

✔ Concept Check 5

The average numbers of offspring produced by three genotypes are: $GG = 6$; $Gg = 3$, $gg = 2$. What is the fitness of Gg?

a. 3   c. 0.3

b. 0.5   d. 0.27

The results of selection The results of selection depend on the relative fitnesses of the genotypes. If we have three genotypes (A^1A^1, A^1A^2, and A^2A^2) with fitnesses W_{11}, W_{12}, and W_{22}, we can identify six different types of natural selection (**Table 17.1**). In type 1 selection, a dominant allele A^1 confers a fitness advantage; in this case, the fitnesses of genotypes A^1A^1 and A^1A^2 are equal and higher than the fitness of A^2A^2 ($W_{11} = W_{12} > W_{22}$). Because both the heterozygote and the A^1A^1 homozygote have copies of the A^1 allele and produce more offspring than the A^2A^2 homozygote does, the frequency of the A^1 allele will increase with time, and the frequency of the A^2 allele will decrease. This form of selection, in which one allele or trait is favored over another, is termed **directional selection**.

Type 2 selection (see Table 17.1) is directional selection against a dominant allele A^1 ($W_{11} = W_{12} < W_{22}$). Type 3 and type 4 selection also are directional selection but, in these cases, there is incomplete dominance and the heterozygote has a fitness that is intermediate between the two homozygotes ($W_{11} > W_{12} > W_{22}$ for type 3; $W_{11} < W_{12} < W_{22}$ for type 4). Eventually, directional selection leads to fixation of the favored allele and elimination of the other allele, as long as no other evolutionary forces act on the population.

Two types of selection (types 5 and 6) are special situations that lead to equilibrium, where there is no further change in allelic frequency. Type 5 selection is referred to as **overdominance** or heterozygote advantage. Here, the heterozygote has higher fitness than the fitnesses of the two homozygotes ($W_{11} < W_{12} > W_{22}$). With overdominance,

Type	Fitness Relation	Form of Selection	Result
1	$W_{11} = W_{12} > W_{22}$	Directional selection against recessive allele A^2	A^1 increases, A^2 decreases
2	$W_{11} = W_{12} < W_{22}$	Directional selection against dominant allele A^1	A^2 increases, A^1 decreases
3	$W_{11} > W_{12} > W_{22}$	Directional selection against incompletely dominant allele A^2	A^1 increases, A^2 decreases
4	$W_{11} < W_{12} < W_{22}$	Directional selection against incompletely dominant allele A^1	A^2 increases, A^1 decreases
5	$W_{11} < W_{12} > W_{22}$	Overdominance	Stable equilibrium, both alleles maintained
6	$W_{11} > W_{12} < W_{22}$	Underdominance	Unstable equilibrium

Table 17.1 Types of natural selection

Note: W_{11}, W_{12}, and W_{22} represent the fitnesses of genotypes A^1A^1, A^1A^2, and A^2A^2, respectively.

both alleles are favored in the heterozygote, and neither allele is eliminated from the population. The allelic frequencies change with overdominant selection until a stable equilibrium is reached, at which point there is no further change. The allelic frequency at equilibrium ($\hat{q}$) depends on the relative fitnesses (usually expressed as selection coefficients) of the two homozygotes:

$$\hat{q} = f(A^2) = \frac{s_{11}}{s_{11} + s_{22}} \qquad (17.9)$$

where s_{11} represents the selection coefficient of the A^1A^1 homozygote and s_{22} represents the selection coefficient of the A^2A^2 homozygote.

The last type of selection (type 6) is **underdominance**, in which the heterozygote has lower fitness than both homozygotes ($W_{11} > W_{12} < W_{22}$). Underdominance leads to an unstable equilibrium; here, allelic frequencies will not change as long as they are at equilibrium but, if they are disturbed from the equilibrium point by some other evolutionary force, they will move away from equilibrium until one allele eventually becomes fixed.

Table 17.2	Effects of different evolutionary forces on allelic frequencies within populations	
Force	**Short-Term Effect**	**Long-Term Effect**
Mutation	Change in allelic frequency	Equilibrium reached between forward and reverse mutations
Migration	Change in allelic frequency	Equilibrium reached when allelic frequencies of source and recipient population are equal
Genetic drift	Change in allelic frequency	Fixation of one allele
Natural selection	Change in allelic frequency	Directional selection: fixation of one allele Overdominant selection: equilibrium reached

Concepts

Natural selection changes allelic frequencies; the direction and magnitude of change depend on the intensity of selection, the dominance relations of the alleles, and the allelic frequencies. Directional selection favors one allele over another and eventually leads to fixation of the favored allele. Overdominance leads to a stable equilibrium with maintenance of both alleles in the population. Underdominance produces an unstable equilibrium because the heterozygote has lower fitness than those of the two homozygotes.

Connecting Concepts

The General Effects of Forces That Change Allelic Frequencies

You now know that four processes bring about change in the allelic frequencies of a population: mutation, migration, genetic drift, and natural selection. Their short- and long-term effects on allelic frequencies are summarized in **Table 17.2**. In some cases, these changes continue until one allele is eliminated and the other becomes fixed in the population. Genetic drift and directional selection will eventually result in fixation, provided these forces are the only ones acting on a population. With the other evolutionary forces, allelic frequencies change until an equilibrium point is reached, and then there is no additional change in allelic frequency. Mutation, migration, and some forms of natural selection can lead to stable equilibria (see Table 17.2).

The different evolutionary forces affect both genetic variation within populations and genetic divergence between populations. Evolutionary forces that maintain or increase genetic variation within populations are listed in the upper-left quadrant

of **Figure 17.8**. These forces include some types of natural selection, such as overdominance, in which both alleles are favored. Mutation and migration also increase genetic variation within populations because they introduce new alleles to the population. Evolutionary forces that decrease genetic variation within populations are listed in the lower-left quadrant of Figure 17.8. These forces include genetic drift, which decreases variation through fixation of alleles, and some forms of natural selection such as directional selection.

The various evolutionary forces also affect the amount of genetic divergence between populations. Natural selection increases divergence between populations if different alleles are favored in the different populations, but it can also decrease divergence between populations by favoring the same allele in the dif-

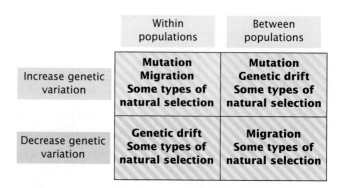

	Within populations	Between populations
Increase genetic variation	**Mutation Migration Some types of natural selection**	**Mutation Genetic drift Some types of natural selection**
Decrease genetic variation	**Genetic drift Some types of natural selection**	**Migration Some types of natural selection**

17.8 Mutation, migration, genetic drift, and natural selection have different effects on genetic variation within populations and on genetic divergence between populations.

ferent populations. Mutation almost always increases divergence between populations because different mutations arise in each population. Genetic drift also increases divergence between populations because changes in allelic frequencies due to drift are random and are likely to change in different directions in separate populations. Migration, on the other hand, decreases divergence between populations because it makes populations similar in their genetic composition.

Migration and genetic drift act in opposite directions: migration increases genetic variation within populations and decreases divergence between populations, whereas genetic drift decreases genetic variation within populations and increases divergence between populations. Mutation increases both variation within populations and divergence between populations. Natural selection can either increase or decrease variation within populations, and it can increase or decrease divergence between populations.

An important point to keep in mind is that real populations are simultaneously affected by many evolutionary forces. We have examined the effects of mutation, migration, genetic drift, and natural selection in isolation so that the effect of each process would be clear. However, in the real world, populations are commonly affected by several evolutionary forces at the same time, and evolution results from the complex interplay of numerous processes.

17.4 Organisms Evolve Through Genetic Change Taking Place Within Populations

Evolution is one of the foundational principles of all of biology. Theodosius Dobzhansky, an important early leader in the field of evolutionary genetics, once remarked "Nothing in biology makes sense except in the light of evolution." Indeed, evolution is an all-encompassing theory that helps to make sense of much of natural world, from the sequences of DNA found in our cells to the types of plants and animals that surround us. The evidence for evolution is overwhelming. Evolution has been directly observed numerous times; for example, hundreds of different insects evolved resistance to common pesticides introduced after World War II. Evolution is supported by the fossil record, comparative anatomy, embryology, the distribution of plants and animals (biogeography), and molecular genetics.

In spite of its vast importance to all fields of biology, evolution is often misunderstood and misinterpreted. In our society, the term *evolution* frequently refers to any type of change. However, biological **evolution** refers only to a specific type of change—genetic change taking place in a group of organisms. Two aspects of this definition should be emphasized. First, evolution includes genetic change only. Many nongenetic changes take place in living organisms, such as the development of a complex intelligent person from an original single-celled zygote. Although remarkable,

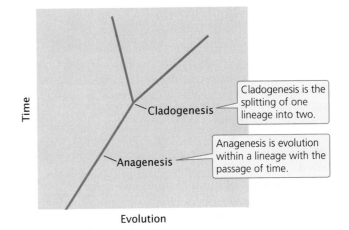

17.9 Anagenesis and cladogenesis are two different types of evolutionary change. Anagenesis is change within an evolutionary lineage; cladogenesis is the splitting of lineages (speciation).

this change isn't evolution, because it does not include changes in genes. The second aspect to emphasize is that evolution takes place in *groups* of organisms. An individual organism does not evolve; what evolves is the gene pool common to a group of organisms.

Evolution can be thought of as a two-step process. First, genetic variation arises. Genetic variation has its origin in the processes of mutation, which produces new alleles, and recombination, which shuffles alleles into new combinations. Both of these processes are random and produce genetic variation continually, regardless of evolution's need for it. The second step in the process of evolution is the increase and decrease in the frequencies of genetic variants. Various evolutionary forces discussed earlier in this chapter cause some alleles in the gene pool to increase in frequency and other alleles to decrease in frequency. This shift in the composition of the gene pool common to a group of organisms constitutes evolutionary change.

We can differentiate between two types of evolution that take place within a group of organisms connected by reproduction. **Anagenesis** refers to evolution taking place in a single group (a lineage) with the passage of time (**Figure 17.9**). Another type of evolution is **cladogenesis**, the splitting of one lineage into two. When a lineage splits, the two branches no longer have a common gene pool and evolve independently of one another. New species arise through cladogenesis.

Concepts

Biological evolution is genetic change that takes place within a group of organisms. Anagenesis is evolution that takes place within a single lineage; cladogenesis is the splitting of one lineage into two.

17.5 New Species Arise Through the Evolution of Reproductive Isolation

The term *species* literally means kind or appearance; **species** are different kinds or types of living organisms. In many cases, species differences are easy to recognize: a horse is clearly a different species from a chicken. Sometimes, however, species differences are not so clear cut. Some species of *Plethodon* salamanders are so similar in appearance that they can be distinguished only by looking at their proteins or genes.

The concept of a species has two primary uses in biology. First, a species is a name given to a particular type of organism. For effective communication, biologists must use a standard set of names for the organisms that they study, and species names serve that purpose. When a geneticist talks about conducting crosses with *Drosophila melanogaster*, other biologists immediately understand which organism was used. The second use of the term species is in an evolutionary context: a species is considered an evolutionarily independent group of organisms.

The Biological Species Concept

What kinds of differences are required to consider two organisms different species? A widely used definition of species is the **biological species concept**, first fully developed by evolutionary biologist Ernst Mayr in 1942. Mayr was primarily interested in the biological characteristics that are responsible for separating organisms into independently evolving units. He defined a species as a group of organisms whose members are capable of interbreeding with one another but are reproductively isolated from the members of other species. In other words, members of the same species have the biological potential to exchange genes, and members of different species cannot exchange genes. Because different species do not exchange genes, each species evolves independently.

Not all biologists adhere to the biological species concept, and several problems are associated with it. In practice, most species are distinguished on the basis of phenotypic (usually anatomical) differences. Biologists often assume that phenotypic differences represent underlying genetic differences; if the phenotypes of two organisms are quite different, then they probably cannot and do not interbreed in nature.

Reproductive Isolating Mechanisms

The key to species differences under the biological species concept is reproductive isolation—biological characteristics that prevent genes from being exchanged between different species. Any biological factor or mechanism that prevents gene exchange is termed a **reproductive isolating mechanism**.

Some species are separated by **prezygotic reproductive isolating mechanisms**, which prevent gametes from two different species from fusing and forming a hybrid zygote.

Table 17.3	Types of reproductive isolating mechanisms
Type	**Characteristic**
Prezygotic	**Mechanisms Before a Zygote Has Formed**
Ecological	Differences in habitat; individuals do not meet
Temporal	Reproduction takes place at different times
Mechanical	Anatomical differences prevent copulation
Behavioral	Differences in mating behavior prevent mating
Gametic	Gametes incompatible or not attracted to each other
Postzygotic	**Mechanisms After a Zygote Has Formed**
Hybrid inviability	Hybrid zygote does not survive to reproduction
Hybrid sterility	Hybrid is sterile
Hybrid breakdown	F_1 hybrids are viable and fertile, but F_2 are inviable or sterile

This type of reproductive isolation can arise in a number of different ways (**Table 17.3**). Other species are separated by **postzygotic reproductive isolating mechanisms**, in which gametes of two species fuse and form a zygote, but there is no gene flow between the two species, either because the resulting hybrids are inviable or sterile or because reproduction breaks down in subsequent generations (Table 17.3).

Concepts

The biological species concept defines a species as a group of potentially interbreeding organisms that are reproductively isolated from the members of other species. Under this concept, species are separated by reproductive isolating mechanisms, which may intervene before a zygote is formed (prezygotic reproductive isolating mechanisms) or after a zygote is formed (postzygotic reproductive isolating mechanisms).

✔ Concept Check 6

Which statement is an example of postzygotic reproductive isolation?

a. Sperm of species A dies in the oviduct of species B before fertilization can take place.

b. Hybrid zygotes between species A and B are spontaneously aborted early in development.

c. The mating seasons of species A and B do not overlap.

d. Males of species A are not attracted to the pheromones produced by the females of species B.

Modes of Speciation

Speciation is the process by which new species arise. In regard to the biological species concept, speciation comes about through the evolution of reproductive isolating mechanisms—mechanisms that prevent the exchange of genes between groups of organisms.

New species arise in two principle ways. **Allopatric speciation** arises when a geographic barrier first splits a population into two groups and blocks the exchange of genes between them. The interruption of gene flow then leads to the evolution of genetic differences that result in reproductive isolation. **Sympatric speciation** arises in the absence of any external barrier to gene flow; reproductive isolating mechanisms evolve within a single population. We will take a more detailed look at both of these mechanisms next.

Allopatric speciation   Allopatric speciation is initiated when a geographic barrier splits a population into two or more groups and prevents gene flow between the isolated groups (**Figure 17.10a**). Geographic barriers can take a number of forms. Uplifting of a mountain range may split a population of lowland plants into separate groups on each side of the mountains. Oceans serve as effective barriers for many types of terrestrial organisms, separating individuals on different islands from one another and from those on the mainland. Rivers often separate populations of fish located in separate drainages. The erosion of mountains may leave populations of alpine plants isolated on separate mountain peaks.

After two populations have been separated by a geographic barrier that prevents gene flow between them, they evolve independently (**Figure 17.10b**). The genetic isolation allows each population to accumulate genetic differences that are not found in the other population through natural selection, unique mutations, and genetic drift (if the populations are small). These genetic differences eventually lead to prezygotic and postzygotic isolation. It is important to note that prezygotic isolation and postzygotic isolation arise simply as a consequence of genetic divergence.

If the populations come into secondary contact (**Figure 17.10c**), several outcomes are possible. If limited genetic differentiation has taken place during the separation of the populations, reproductive isolating mechanisms may not have evolved or may be incomplete. In this case, the populations will remain a single species. A second possible outcome is that genetic differentiation during separation leads to prezygotic reproductive isolating mechanisms; in this case, the two populations are different species. A third possible outcome is that, during their time apart, some genetic differentiation took place between the populations, leading to postzygotic isolation. If postzygotic isolating mechanisms have evolved, any mating between individuals from the different populations will produce

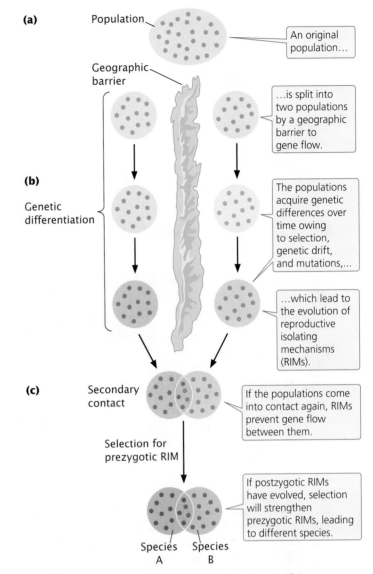

17.10 Allopatric speciation is initiated by a geographic barrier to gene flow between two populations.

hybrid offspring that are inviable or sterile. Individuals that mate only with members of the same population will have higher fitness than that of individuals that mate with members of the other population; so natural selection will increase the frequency of any trait that prevents interbreeding between members of the different populations. With the passage of time, prezygotic reproductive isolating mechanisms will evolve. In short, if some postzygotic reproductive isolation exists, natural selection will favor the evolution of prezygotic reproductive isolating mechanisms to prevent wasted reproduction by individuals mating with members of the other population.

An excellent example of allopatric speciation can be found in Darwin's finches, a group of birds on the Galápagos Islands; these finches were discovered by Charles Darwin in his voyage aboard the *Beagle*. The Galápagos are an archipel-

17.11 **The Galápagos Islands are relatively young geologically and are volcanic in origin.** The oldest islands are to the east. [*After* Philosophical Transactions at the Royal Society of London, Series B *351:756–772, 1996.*]

ago of islands located some 900 km off the coast of South America (**Figure 17.11**). Consisting of more than a dozen large islands and many smaller ones, the Galápagos formed from volcanoes that erupted over a geological hot spot that has remained stationary while the geological plate over it moved eastward in the past 3 million years. Thus, the islands to the east (San Cristóbal and Española) are older than those to the west (Isabela and Fernandina). With the passage of time, the number of islands in the archipelago increased as new volcanoes arose.

Darwin's finches consist of 14 species found on various islands in the Galápagos archipelago (**Figure 17.12**). The birds vary in the shape and sizes of their beaks, which are adapted for eating different types of food items. Genetic studies have demonstrated that all the birds are closely related and evolved from a single ancestral species that migrated to the islands from the coast of South America some 2 million to 3 million years ago. The evolutionary relationships among the 14 species, based on studies of microsatellite data, are depicted in the evolutionary tree shown in Figure 17.12. Most of the species are separated by a behavioral isolating mechanism (song in particular), but some of the species can and occasionally do hybridize in nature.

The first finches to arrive in the Galápagos probably colonized one of the larger eastern islands. A breeding population became established and increased with time. At some point, a few birds dispersed to other islands, where they were effectively isolated from the original population, and established a new population. This population underwent genetic differentiation owing to genetic drift and adaptation to the local conditions of the island. It eventually became reproductively isolated from the original population. Individual birds from the new population then dispersed to other islands and gave rise to additional species. This process was repeated many times. Occasionally, newly evolved birds dispersed to an island where another species was already present, giving rise to secondary contact between the species. Today, many of the islands have more than one resident finch.

The age of the 14 species has been estimated with data from mitochondrial DNA. **Figure 17.13** shows that there is a strong correspondence between the number of bird species present at various times in the past and the number of islands in the archipelago. This correspondence is one of the most compelling pieces of evidence for the theory that the different species of finches arose through allopatric speciation.

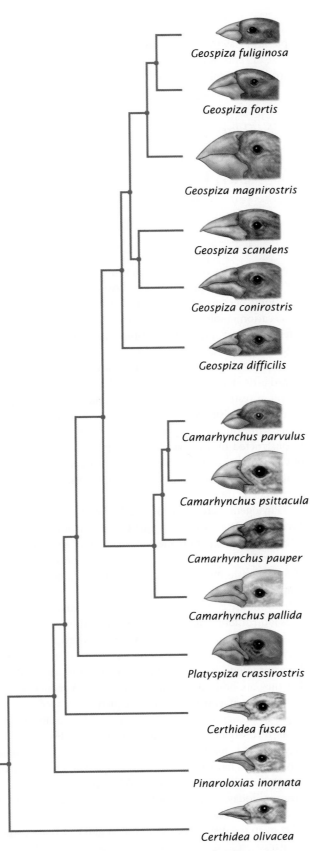

17.12 Darwin's finches consist of 14 species that evolved from a single ancestral species that migrated to the Galápagos Islands and underwent repeated allopatric speciation. [*After B. R. Grant and P. R. Grant.* Bioscience 53: 965–975, 2003.]

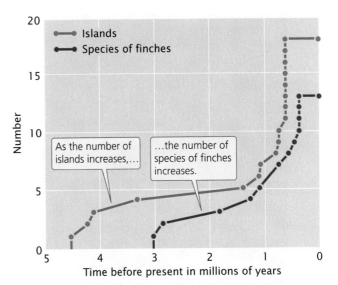

17.13 The number of species of Darwin's finches present at various times in the past corresponds to the number of islands in the Galápagos archipelago. [*Data from P. R. Grant, B. R. Grant, and J. C. Deutsch. Speciation and hybridization in island birds.* Philosophical Transactions of the Royal Society of London Series B 351:765–772, 1996.]

Concepts

Allopatric speciation is initiated by a geographic barrier to gene flow. A single population is split into two or more populations by a geographic barrier. With the passage of time, the populations evolve genetic differences that bring about reproductive isolation. After postzygotic reproductive isolating mechanisms have evolved, selection favors the evolution of prezygotic reproductive isolating mechanisms.

✔ Concept Check 7

What role does genetic drift play in allopatric speciation?

Sympatric speciation Sympatric speciation arises in the absence of any geographic barrier to gene flow; reproductive isolating mechanisms evolve within a single interbreeding population. Sympatric speciation has long been controversial within evolutionary biology. Ernst Mayr believed that sympatric speciation was impossible, and he demonstrated that many apparent cases of sympatric speciation could be explained by allopatric speciation. More recently, however, evidence that sympatric speciation can and has arisen under special circumstances has accumulated. The difficulty with sympatric speciation is that isolating mechanisms arise as a *consequence* of genetic differentiation, which takes place only if gene flow between groups is interrupted. But, without reproductive isolation (or some external barrier), how can gene flow be interrupted? How can genetic differentiation arise within a single group that is freely exchanging genes?

Most models of sympatric speciation assume that genetic differentiation is initiated by strong disruptive

17.14 Host races of the apple maggot fly, *Rhagoletis pomenella*, have evolved some reproductive isolation without any geographic barrier to gene flow. [Tom Murray.]

selection taking place within a single population. One example of how sympatric speciation might arise is seen in apple maggot flies, *Rhagoletis pomonella* (**Figure 17.14**), studied by Guy Bush. The flies of this species feed on the fruits of a specific host tree. Mating takes place near the fruits, and the flies lay their eggs on the ripened fruits, where their larvae grow and develop. *Rhagoletis pomnella* originally existed only on fruits of hawthorn trees, which are native to North America; 150 years ago, *R. pomnella* was first observed on cultivated apples, which are related to hawthorns but a different species. Infestations of apples by this new apple host race of *R. pomnella* quickly spread, and, today, many apple trees throughout North America are infested with the flies.

The apple host race of *R. pomnella* probably originated when a few flies acquired a mutation that allowed them to feed on apples instead of the hawthorn fruits. Because mating takes place on and near the fruits, flies that utilize apples are more likely to mate with other flies utilizing apples, leading to genetic isolation between flies using hawthorns and those utilizing applies. Indeed, Bush found that some genetic differentiation has already taken place between the two host races. Flies lay their eggs on ripening fruit, and there has been strong selection for the flies to synchronize their reproduction with the period when their host species has ripening fruit. Apples ripen several weeks earlier than hawthorns. Correspondingly, the peak mating period of the apple host races is 3 weeks earlier than that of the hawthorn race. These differences in the timing of reproduction between apple and hawthorn races have further reduced gene flow—to about 2%—between the two host races and have led to significant genetic differentiation between them. All of it has evolved in the past 150 years. Although genetic differentiation has taken place between

apple and hawthorn host races of *R. pomnella* and some degree of reproductive isolation has evolved between them, reproductive isolation is not yet complete and speciation has not fully taken place.

17.6 The Evolutionary History of a Group of Organisms Can Be Reconstructed by Studying Changes in Homologous Characteristics

The evolutionary relationships among a group of organisms are termed a **phylogeny**. Because most evolution takes place over long periods of time and is not amenable to direct observation, biologists must reconstruct phylogenies by inferring the evolutionary relationships among present-day organisms. The discovery of fossils of ancestral organisms can aid in the reconstruction of phylogenies, but the fossil record is often too poor to be of much help. Thus, biologists are often restricted to the analysis of characteristics in present-day organisms to determine their evolutionary relationships. In the past, phylogenetic relationships were reconstructed on the basis of phenotypic characteristics—often, anatomical traits. Today, molecular data, including protein and DNA sequences, are frequently used to construct phylogenetic trees.

Phylogenies are reconstructed by inferring changes that have taken place in homologous characteristics. Such characteristics evolved from the same character in a common ancestor. For example, the front leg of a mouse and the wing of a bat are homologous structures, because both evolved from the forelimb of an early mammal that was an ancestor to both mouse and bat. Although these two anatomical features look different and have different functions, close examination of their structure and development reveals that they are indeed homologous. And, because mouse and bat have these homologous features and others in common, we know that they are both mammals. Similarly, DNA sequences are homologous if two present-day sequences evolved from a single sequence found in an ancestor. For example, all eukaryotic organisms have a gene for cytochrome *c*, an enzyme that helps carry out oxidative respiration. This gene is assumed to have arisen in a single organism in the distant past and was then passed down to descendants of that early ancestor. Today, all copies of the gene for cytochrome *c* are homologous, because they all evolved from the same original copy in the distant ancestor of all organisms that possess this gene.

A graphical representation of a phylogeny is called a **phylogenetic tree**. As shown in **Figure 17.15**, a phylogenetic tree depicts the evolutionary relationships among different organisms, similarly to the way in which a pedigree repre-

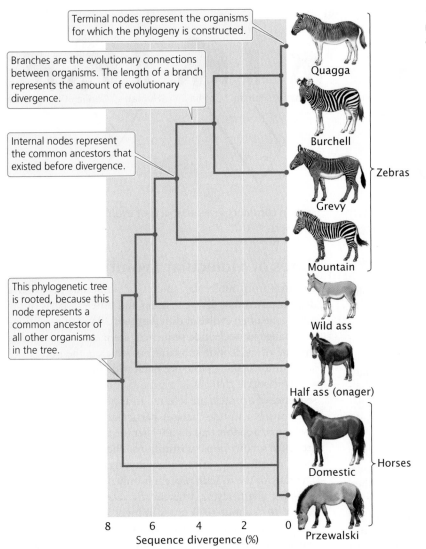

Terminal nodes represent the organisms for which the phylogeny is constructed.

Branches are the evolutionary connections between organisms. The length of a branch represents the amount of evolutionary divergence.

Internal nodes represent the common ancestors that existed before divergence.

This phylogenetic tree is rooted, because this node represents a common ancestor of all other organisms in the tree.

17.15 A phylogenetic tree is a graphical representation of the evolutionary relations among a group of organisms.

sents the genealogical relationships among family members. A phylogenetic tree consists of **nodes** that represent the different organisms being compared, which might be different individuals, populations, or species. Terminal nodes (those at the end of the outermost branches of the tree) represent organisms for which data have been obtained, usually present-day organisms. Internal nodes represent common ancestors that existed before divergence between organisms took place. In most cases, the internal nodes represent past ancestors that are inferred from the analysis. The nodes are connected by **branches**, which may represent the evolutionary connections between organisms. In many phylogenetic trees, the lengths of the branches represent the amount of evolutionary divergence that has taken place between organisms. When one internal node represents a common ancestor to all other nodes on the tree, the tree is said to be **rooted**. Trees are often rooted by including in the analysis an organism that is distantly related to all the others; this distantly related organism is referred to as an outgroup.

Concepts

A phylogeny represents the evolutionary relationships among a group of organisms and is often depicted graphically by a phylogenetic tree, which consists of nodes representing the organisms and branches representing their evolutionary connections.

The Construction of Phylogenetic Trees

Consider a simple phylogeny that depicts the evolutionary relationships among three organisms—humans, chimpanzees, and gorillas. Charles Darwin originally proposed that chimpanzees and gorillas were closely related to humans. However, subsequent study placed humans in the family Hominidae and the great apes (chimpanzees, gorilla, orangutan, and gibbon) in the family Pongidae. There are three possible phylogenetic trees for human, chimpanzees,

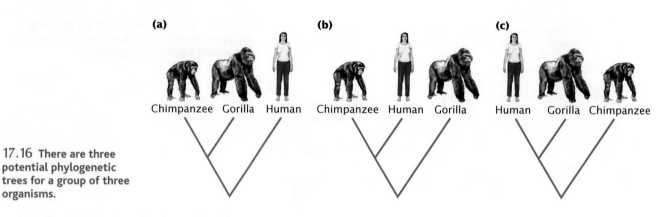

(a)

Chimpanzee   Gorilla   Human

(b)

Chimpanzee   Human   Gorilla

(c)

Human   Gorilla   Chimpanzee

17.16 There are three potential phylogenetic trees for a group of three organisms.

and gorillas (**Figure 17.16**). The goal of the evolutionary biologist is to determine which of the trees is correct. Molecular data have been applied to this question; they strongly suggest a close relationship between humans and chimpanzees.

There are two different approaches to inferring evolutionary relationships and constructing phylogenetic trees. In the first approach, termed the *distance approach*, evolutionary relationships are inferred on the basis of the overall degree of similarity between organisms. Typically, a number of different phenotypic characteristics or gene sequences are examined and the organisms are grouped on the basis of their overall similarity, considering all the examined characteristics and sequences. The second approach, called the *parsimony approach*, infers phylogenetic relationships on the basis of the minimum number of evolutionary changes that must have taken place since the organisms last had an ancestor in common. With both the distance and the parsimony approaches, a number of different numerical methods are available for the construction of phylogenetic trees. These methods are beyond the scope of this book. All include certain assumptions that help limit the number of different trees that must be considered; most rely on computer programs that compare phenotypic characteristics or sequence data to sequentially group organisms in the construction of the tree.

Concepts

Molecular data can be used to infer phylogenies (evolutionary histories) of groups of living organisms. Two approaches to reconstructing a phylogeny are the distance approach, which uses the overall degree of similarity in the organisms, and the parsimony approach, which uses the minimum number of evolutionary steps required to connect organisms.

17.7 Patterns of Evolution Are Revealed by Changes at the Molecular Level

In recent years, the ability to analyze genetic variation at the molecular level has revealed a number of evolutionary processes and features that were formerly unsuspected. This section considers several aspects of evolution at the molecular level.

Rates of Molecular Evolution

Findings from molecular studies of numerous genes have demonstrated that different genes and different parts of the same gene often evolve at different rates. Rates of evolutionary change in nucleotide sequences are usually measured as the rate of nucleotide substitution, which is the number of substitutions taking place per nucleotide site per year.

Nucleotide changes in a gene that alter the amino acid sequence of a protein are referred to as nonsynonymous substitutions. Nucleotide changes, particularly those at the third position of a codon, that do not alter the amino acid sequence are called synonymous substitutions. The rate of nonsynonymous substitution varies widely among mammalian genes. The rate for the α-actin protein is only 0.01×10^{-9} substitutions per site per year, whereas the rate for interferon γ is 2.79×10^{-9}, about 1000 times as high. The rate of synonymous substitution also varies among genes, but not to the extent of variation in the nonsynonymous rate. For most protein-encoding genes, the synonymous rate of change is considerably higher than the nonsynonymous rate because synonymous mutations are tolerated by natural selection (**Table 17.4**). Nonsynonymous mutations, on the other hand, alter the amino acid sequence of the protein and, in many cases, are detrimental to the fitness of the organism; so most of these mutations are eliminated by natural selection.

Different parts of a gene also evolve at different rates, with the highest rates of substitutions in regions of the gene that have the least effect on function, such as the third position of a codon, flanking regions, and introns (**Figure 17.17**). The 5′ and 3′ flanking regions of genes are not transcribed into RNA; therefore, substitutions in these regions do not alter the amino acid sequence of the protein, although they may affect gene expression (see Chapter 12). Rates of substitution in introns are nearly as high. Although these nucleotides do not encode amino acids, introns must be spliced out of the pre-mRNA for a functional protein to be produced, and particular sequences are required at the 5′ splice site, 3′ splice site, and branch point for correct splicing (see Chapter 10).

Substitution rates are somewhat lower in the 5′ and 3′ untranslated regions of a gene. As we know from Chapters 10 and 11, these regions are transcribed into RNA but do not encode amino acids. The 5′ untranslated region contains the ribosome-binding site, which is essential for translation, and the 3′ untranslated region contains sequences that may function in regulating mRNA stability and translation; so substitutions in these regions may have deleterious effects on organismal fitness and may not be tolerated.

The lowest rates of substitution are seen in nonsynonymous changes in the coding region, because these substitutions always alter the amino acid sequence of the protein and are often deleterious. The highest rates of substitution are in pseudogenes, which are duplicated nonfunctional copies of genes that have acquired mutations. Such genes no longer produce a functional product; so mutations in pseudogenes have no effect on the fitness of the organism.

Table 17.4	Rates of nonsynonymous and synonymous substitutions in mammalian genes based on human–rodent comparisons	
Gene	Nonsynonymous Rate (per site per 10^9 years)	Synonymous Rate (per site per 10^9 years)
α-Actin	0.01	3.68
β-Actin	0.03	3.13
Albumin	0.91	6.63
Aldolase A	0.07	3.59
Apoprotein E	0.98	4.04
Creatine kinase	0.15	3.08
Erythropoietin	0.72	4.34
α-Globin	0.55	5.14
β-Globin	0.80	3.05
Growth hormone	1.23	4.95
Histone 3	0.00	6.38
Immunoglobulin heavy chain (variable region)	1.07	5.66
Insulin	0.13	4.02
Interferon α 1	1.41	3.53
Interferon γ	2.79	8.59
Luteinizing hormone	1.02	3.29
Somatostatin-28	0.00	3.97

Source: After W. Li and D. Graur, *Fundamentals of Molecular Evolution* (Sunderland, Mass.: Sinauer, 1991), p. 69.

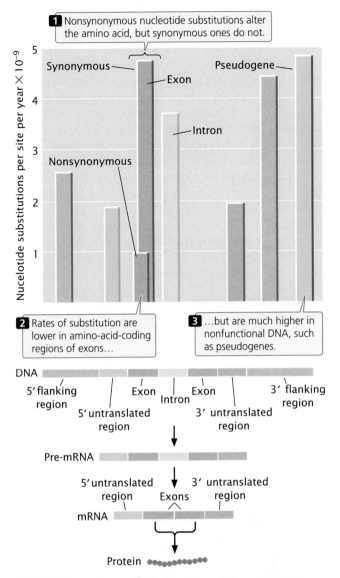

1 Nonsynonymous nucleotide substitutions alter the amino acid, but synonymous ones do not.

2 Rates of substitution are lower in amino-acid-coding regions of exons…

3 …but are much higher in nonfunctional DNA, such as pseudogenes.

17.17 Different parts of genes evolve at different rates. The highest rates of nucleotide substitution are in sequences that have the least effect on protein function.

In summary, there is a relation between the function of a sequence and its rate of evolution. Higher rates are found where they have the least effect on function.

The Molecular Clock

The neutral-mutation hypothesis is the idea that evolutionary change at the molecular level takes place primarily through the fixation of neutral mutations (mutations that have little or no effect on fitness) by genetic drift. The rate at which one neutral mutation replaces another depends only on the mutation rate, which should be fairly constant for any particular gene. If the rate at which a protein evolves is roughly constant with the passage of time, the amount of molecular change that a protein has undergone can be used as a **molecular clock** to date evolutionary events.

For example, we could examine the enzyme cytochrome *c* in two organisms known from fossil evidence to have had a common ancestor 400 million years ago. By determining the number of differences in the cytochrome *c* amino acid sequences in each organism, we could calculate the number of substitutions that have occurred per amino acid site. The occurrence of 20 amino acid substitutions since the two organisms diverged indicates an average rate of 5 substitutions per 100 million years. Knowing how fast the molecular clock ticks allows us to use molecular changes in cytochrome *c* to date other evolutionary events: if we found that cytochrome *c* in two organisms differed by 15 amino acid substitutions, our molecular clock would suggest that they diverged some 300 million years ago. If we assumed some error in our estimate of the rate of amino acid substitution, statistical analysis would show that the true divergence time might range from 160 million to 440 million years. The molecular clock is analogous to geological dating based on the radioactive decay of elements.

The molecular clock was proposed by Emile Zuckerandl and Linus Pauling in 1965 as a possible means of dating evolutionary events on the basis of molecules in present-day organisms. A number of studies have examined the rate of evolutionary change in proteins (**Figure 17.18**), and the molecular clock has been widely used to date evolutionary events when the fossil record is absent or ambiguous. However, the results of several studies have shown that the molecular clock does not always tick at a constant rate, particularly over shorter time periods, and this method remains controversial.

Concepts

Different genes and different parts of the same gene evolve at different rates. Those parts of genes that have the least effect on function tend to evolve at the highest rates. The idea of the molecular clock is that individual proteins and genes evolve at a constant rate and that the differences in the sequences of present-day organisms can be used to date past evolutionary events.

✔ Concept Check 8

In general, which types of sequences are expected to exhibit the slowest evolutionary change?

a. Synonymous changes in amino acid coding regions of exons

b. Nonsynonymous changes in amino acid coding regions of exons

c. Introns

d. Pseudogenes

Genome Evolution

The rapid growth of sequence data available in DNA databases has been a source of insight into evolutionary processes. Whole-genome sequences also are providing new information about how genomes evolve and the

(a)

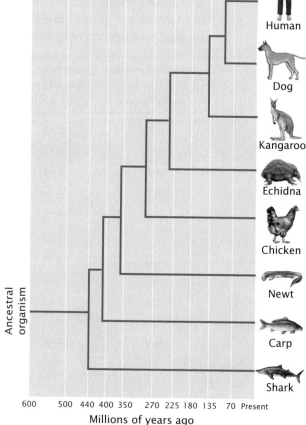

(b)

17.18 The molecular clock is based on the assumption of a constant rate of change in protein or DNA sequence.
(a) Relation between the rate of amino acid substitution and time since divergence, based in part on amino acid sequences of α-hemoglobin from the eight species shown in part *b*. The constant rate of evolution in protein and DNA sequences has been used as a molecular clock to date past evolutionary events. (b) Phylogeny of eight species and their approximate times of divergence, based on the fossil record.

17.19 Human globin genes constitute a multigene family that has evolved through successive gene duplications.

processes that shape the size, complexity, and organization of genomes.

Gene duplication   New genes have also evolved through the duplication of whole genes and their subsequent divergence. This process creates **multigene families**—sets of genes that are similar in sequence but encode different products. For example, humans possess 13 different genes found on chromosomes 11 and 16 that encode globinlike molecules, which take part in oxygen transport (**Figure 17.19**). All of these genes have a similar structure, with three exons separated by two introns, and are assumed to have evolved through repeated duplication and divergence from a single globin gene in a distant ancestor. This ancestral gene is thought to have been most similar to the present-day myoglobin gene and first duplicated to produce an α/β-globin precursor gene and the myoglobin gene. The α/β gene then underwent another duplication to give rise to a primordial α-globin gene and a primordial β-globin gene. Subsequent duplications led to multiple α-globin and β-globin genes. Similarly, vertebrates contain four clusters of *Hox* genes, each cluster comprising from 9 to 11 genes. *Hox* genes play an important role in development.

Some gene families include genes that are arrayed in tandem on the same chromosome; others are dispersed among different chromosomes. Gene duplication is a common occurrence in eukaryotic genomes; for example, about 5% of the human genome consists of duplicated segments.

Gene duplication provides a mechanism for the addition of new genes with novel functions; after a gene duplicates, there are two copies of the sequence, one of which is free to change and potentially take on a new function. The extra copy of the gene may, for example, become active at a different time in development or be expressed in a different tissue or even diverge and encode a protein having different amino acids. However, the most common fate of gene duplication is that one copy acquires a mutation that renders it nonfunc-

tional, giving rise to a pseudogene. Pseudogenes are common in genomes of complex eukaryotes; the human genome is estimated to contain as many as 20,000 pseudogenes.

Whole-genome duplication   In addition to the duplication of individual genes, whole genomes of some organisms have apparently duplicated in the past. For example, a comparison of the genome of the yeast *Saccharomyces cerevisiae* with the genomes of other fungi reveals that *S. cerevisiae* or one of its immediate ancestors underwent a whole-genome duplication, generating two copies of every gene. Many of the copies subsequently acquired new functions; others acquired mutations that destroyed the original function and then diverged into random DNA sequences. Whole-genome duplication can take place through polyploidy.

Horizontal gene transfer   Most organisms acquire their genomes through vertical transmission—transfer through the reproduction of genetic information from parents to offspring. Most phylogenetic trees assume vertical transmission of genetic information. Findings from DNA sequence studies reveal that DNA sequences are sometimes exchanged by horizontal gene transfer, in which DNA is transferred between different species. This process is especially common among bacteria, and there are a number of documented cases in which genes are transferred from bacteria to eukaryotes. The extent of horizontal gene transfer among eukaryotic organisms is controversial, with few well-documented cases. Horizontal gene transfer can obscure phylogenetic relationships and make the reconstruction of phylogenetic trees difficult.

Concepts

New genes may evolve through the duplication of genes and through the duplication of whole genomes. Genes can be passed among distantly related organisms through horizontal gene transfer.

Concepts Summary

- A Mendelian population is a group of interbreeding, sexually reproducing individuals, whose set of genes constitutes the population's gene pool. A population's genetic composition can be described by its genotypic and allelic frequencies.

- The Hardy–Weinberg law describes the effects of reproduction and Mendel's laws on the allelic and genotypic frequencies of a population. When a population is large, randomly mating, and free from the effects of mutation, migration, and natural selection, the allelic frequencies do not change and the genotypic frequencies stabilize after one generation in the Hardy–Weinberg equilibrium proportions p^2, $2pq$, and q^2, where p and q equal the frequencies of the alleles.

- Nonrandom mating affects the frequencies of genotypes but not those of alleles. Inbreeding increases the frequency of homozygotes while decreasing the frequency of heterozygotes.

- Recurrent mutation eventually leads to an equilibrium, with the allelic frequencies being determined by the relative rates of forward and reverse mutation.

- Migration, the movement of genes between populations, increases the amount of genetic variation within populations and decreases the number of differences between populations.

- Genetic drift is change in allelic frequencies due to chance factors. Genetic drift arises when a population consists of a small number of individuals, is established by a small number of founders, or undergoes a major reduction in size. Genetic drift changes allelic frequencies, reduces genetic variation within populations, and causes genetic divergence among populations.

- Natural selection is the differential reproduction of genotypes; it is measured by the relative reproductive successes (fitnesses) of genotypes.

- Evolution is genetic change taking place within a group of organisms. It is a two-step process: (1) genetic variation arises and (2) genetic variants change in frequency. Anagenesis refers to change within a single lineage; cladogenesis is the splitting of one lineage into two.

- A species can be defined as a group of organisms that are capable of interbreeding with one another and are reproductively isolated from the members of other species.

- Species are prevented from exchanging genes by reproductive isolating mechanisms, either before a zygotes has formed (prezygotic reproductive isolation) or after a zygote has formed (postzygotic reproductive isolation).

- Allopatric speciation arises when a geographic barrier prevents gene flow between two populations. Sympatric speciation arises when reproductive isolation exists in the absence of any geographic barrier.

- Evolutionary relationships (a phylogeny) can be represented by a phylogenetic tree, consisting of nodes that represent organisms and branches that represent their evolutionary connections. Two different approaches to constructing phylogenetic trees are the distance approach and the parsimony approach.

- Different parts of the genome show different amounts of genetic variation. In general, those parts that have the least effect on function evolve at the highest rates.

- The molecular-clock hypothesis proposes a constant rate of nucleotide substitution, providing a means of dating evolutionary events by looking at nucleotide differences between organisms.

- Genome evolution takes place through the duplication of genes to form gene families, whole-genome duplication, and the horizontal transfer of genes between organisms.

Important Terms

genetic rescue (p. 430)
Mendelian population (p. 430)
gene pool (p. 430)
genotypic frequency (p. 431)
allelic frequency (p. 431)
Hardy–Weinberg law (p. 433)
Hardy–Weinberg equilibrium (p. 433)
inbreeding (p. 436)
equilibrium (p. 437)
migration (gene flow) (p. 437)
sampling error (p. 438)
genetic drift (p. 438)
effective population size (p. 439)
founder effect (p. 439)

genetic bottleneck (p. 439)
fixation (p. 440)
fitness (p. 440)
selection coefficient (p. 441)
directional selection (p. 441)
overdominance (p. 441)
underdominance (p. 442)
evolution (p. 443)
anagenesis (p. 443)
cladogenesis (p. 443)
species (p. 444)
biological species concept (p. 444)
reproductive isolating mechanism
  (p. 444)

prezygotic reproductive isolating
  mechanism (p. 444)
postzygotic reproductive isolating
  mechanism (p. 444)
speciation (p. 445)
allopatric speciation (p. 445)
sympatric speciation (p. 445)
phylogeny (p. 448)
phylogenetic tree (p. 448)
node (p. 449)
branch (p. 449)
rooted tree (p. 449)
molecular clock (p. 451)
multigene family (p. 453)

Answers to Concept Checks

1. a

2. 0.10

3. c

4. d

5. b

6. b

7. Genetic drift can bring about changes in the allelic frequencies of populations and lead to genetic differences among populations. Genetic differentiation is the cause of postzygotic and prezygotic reproductive isolation between populations that leads to speciation.

8. b

Comprehension Questions

Section 17.1

1. What is a Mendelian population? How is the gene pool of a Mendelian population usually described?

Section 17.2

2. What are the predictions given by the Hardy–Weinberg law?

*3. What assumptions must be met for a population to be in Hardy–Weinberg equilibrium?

4. Define inbreeding and briefly describe its effects on a population.

Section 17.3

5. What determines the allelic frequencies at mutational equilibrium?

*6. What factors affect the magnitude of change in allelic frequencies due to migration?

7. Define genetic drift and give three ways in which it can arise. What effect does genetic drift have on a population?

*8. What is effective population size? How does it affect the amount of genetic drift?

9. Define natural selection and fitness.

10. Briefly describe the differences between directional selection, overdominance, and underdominance. Describe the effect of each type of selection on the allelic frequencies of a population.

*11. Compare and contrast the effects of mutation, migration, genetic drift, and natural selection on genetic variation within populations and on genetic divergence between populations.

Section 17.4

*12. What are the two steps in the process of evolution?

13. How does anagenesis differ from cladogenesis?

Section 17.5

*14. What is the biological species concept?

15. What is the difference between prezygotic and postzygotic reproductive isolating mechanisms?

16. What is the basic difference between allopatric and sympatric modes of speciation?

Section 17.6

*17. Briefly describe the difference between the distance approach and the parsimony approach to the reconstruction of phylogenetic trees.

Section 17.7

18. Outline the different rates of evolution that are typically seen in different parts of a protein-encoding gene. What might account for these differences?

*19. What is the molecular clock?

20. What is a multigene family? What processes produce multigene families?

21. Define horizontal gene transfer. What problems does it cause for evolutionary biologists?

Application Questions and Problems

Section 17.1

22. How would you respond to someone who said that models are useless in studying population genetics because they represent oversimplifications of the real world?

*23. Voles (*Microtus ochrogaster*) were trapped in old fields in southern Indiana and were genotyped for a transferrin locus. The following numbers of genotypes were recorded, where T^E and T^F represent different alleles.

$T^E T^E$	$T^E T^F$	$T^F T^F$
407	170	17

Calculate the genotypic and allelic frequencies of the transferrin locus for this population.

24. Jean Manning, Charles Kerfoot, and Edward Berger studied the allelic frequencies at the glucose phosphate isomerase (GPI) locus in the cladoceran *Bosmina longirostris*. At one location, they collected 176 animals from Union Bay in Seattle, Washington, and determined their GPI genotypes by using electrophoresis (J. Manning, W. C. Kerfoot, and E. M. Berger. 1978. *Evolution* 32:365–374).

Genotype	Number
S^1S^1	4
S^1S^2	38
S^2S^2	134

Determine the genotypic and allelic frequencies for this population.

Section 17.2

25. A total of 6129 North American Caucasians were blood typed for the MN locus, which is determined by two codominant alleles, L^M and L^N. The following data were obtained:

Blood type	Number
M	1787
MN	3039
N	1303

Carry out a chi-square test to determine whether this population is in Hardy–Weinberg equilibrium at the MN locus.

26. Most black bears (*Ursus americanus*) are black or brown in color. However, occasional white bears of this species appear in some populations along the coast of British Columbia. Kermit Ritland and his colleagues determined that white coat color in these bears results from a recessive mutation (*G*) caused by a single nucleotide replacement in which guanine substitutes for adenine at the melanocortin-1 receptor locus (*mcr1*), the same locus responsible for red hair in humans (K. Ritland, C. Newton, and H. D. Marshall. 2001. *Current Biology* 11:1468–1472). The wild-type allele at this locus (*A*) encodes black or brown color. Ritland and his colleagues collected samples from bears on three islands and determined their genotypes at the *mcr1* locus.

Genotype	Number
AA	42
AG	24
GG	21

a. What are the frequencies of the *A* and *G* alleles in these bears?

b. Give the genotypic frequencies expected if the population is in Hardy–Weinberg equilibrium.

c. Use a chi-square test to compare the number of observed genotypes with the number expected under Hardy–Weinberg equilibrium. Is this population in Hardy–Weinberg equilibrium? Explain your reasoning.

27. Genotypes of leopard frogs from a population in central Kansas were determined for a locus (*M*) that encodes the enzyme malate dehydrogenase. The following numbers of genotypes were observed:

Genotype	Number
M^1M^1	20
M^1M^2	45
M^2M^2	42
M^1M^3	4
M^2M^3	8
M^3M^3	6
Total	125

a. Calculate the genotypic and allelic frequencies for this population.

b. What would the expected numbers of genotypes be if the population were in Hardy–Weinberg equilibrium?

28. Full color (*D*) in domestic cats is dominant over dilute color (*d*). Of 325 cats observed, 194 have full color and 131 have dilute color.

a. If these cats are in Hardy–Weinberg equilibrium for the dilution locus, what is the frequency of the dilute allele?

b. How many of the 194 cats with full color are likely to be heterozygous?

29. Tay–Sachs disease is an autosomal recessive disorder. Among Ashkenazi Jews, the frequency of Tay–Sachs disease is 1 in 3600. If the Ashkenazi population is mating randomly for the Tay–Sachs gene, what proportion of the population consists of heterozygous carriers of the Tay–Sachs allele?

*30. The human MN blood type is determined by two codominant alleles, L^M and L^N. The frequency of L^M in Eskimos on a small Arctic island is 0.80. If the inbreeding coefficient for this population is 0.05, what are the expected frequencies of the M, MN, and N blood types on the island?

Section 17.3

31. Pikas are small mammals that live at high elevation in the talus slopes of mountains. Most populations located on mountain tops in Colorado and Montana in North America are isolated from one another, because the pikas don't occupy the low-elevation habitats that separate the mountain tops and don't venture far from the talus slopes. Thus, there is little gene flow between populations. Furthermore, each population is small in size and was founded by a small number of pikas. A group of population geneticists propose to study the amount of genetic variation in a series of pika populations and to compare the allelic frequencies in different populations. On the bases of the biology and the distribution of pikas, predict what the population geneticists will find concerning the within- and between-population genetic variation.

32. Two chromosomal inversions are commonly found in populations of *Drosophila pseudoobscura:* Standard (*ST*) and Arrowhead (*AR*). When treated with the insecticide DDT, the genotypes for these inversions exhibit overdominance, with the following fitnesses:

Genotype	Fitness
ST/ST	0.47
ST/AR	1
AR/AR	0.62

What will the frequencies of *ST* and *AR* be after equilibrium has been reached?

33. The fruit fly *Drosophila melanogaster* normally feeds on rotting fruit, which may ferment and contain high levels of alcohol. Douglas Cavener and Michael Clegg studied allelic frequencies at the locus for alcohol dehydrogenase (*Adh*) in experimental populations of *D. melanogaster* (D. R. Cavener and M. T. Clegg. 1981. *Evolution* 35:1–10). The experimental populations were established from wild-caught flies and were raised in cages in the laboratory. Two control populations (C1 and C2) were raised on a standard cornmeal–molasses–agar diet. Two ethanol populations (E1 and E2) were raised on a cornmeal–molasses–agar diet to which was added 10% ethanol. The four populations were periodically sampled to determine the allelic frequencies of two alleles at the alcohol dehydrogenase locus, Adh^S and Adh^F. The frequencies of these alleles in the experimental populations are shown in the graph.

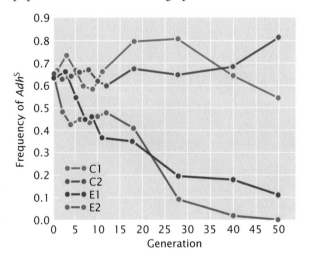

a. One the basis of these data, what conclusion might you draw about the evolutionary forces that are affecting the *Adh* alleles in these populations?

b. Cavener and Clegg measured the viability of the different *Adh* genotypes in the alcohol environment and obtained the following values:

Genotype	Relative viability
Adh^F/Adh^F	0.932
Adh^F/Adh^S	1.288
Adh^S/Adh^S	0.596

Using these relative viabilities, calculate relative fitnesses for the three genotypes.

Section 17.4

34. The following illustrations represent two different patterns of evolution. Briefly discuss the differences in these two patterns, particularly in regard to the role of cladogenesis in evolutionary change.

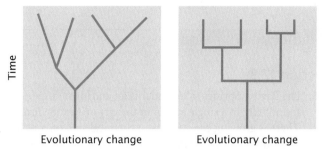

Section 17.6

35. Michael Bunce and his colleagues in England, Canada, and the United States extracted and sequenced mitochondrial DNA from fossils of Haast's eagle, a gigantic eagle that was driven to extinction 700 years ago when humans first arrived in New Zealand (M. Bunce et al. 2005. *Plos Biology* 3:44–46). Using mitochondrial DNA sequences from living eagles and those from Haast-eagle fossils, they created the following phylogenetic tree. On this phylogenetic tree, identify (a) all terminal nodes; (b) all internal nodes; (c) one example of a branch; and (d) the outgroup.

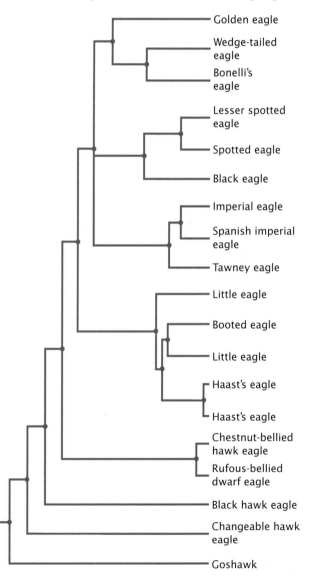

Challenge Questions

Section 17.3

36. The Barton Springs salamander is an endangered species found only in a single spring in the city of Austin, Texas. There is growing concern that a chemical spill on a nearby freeway could pollute the spring and wipe out the species. To provide a source of salamanders to repopulate the spring in the event of such a catastrophe, a proposal has been made to establish a captive breeding population of the salamander in a local zoo. You are asked to provide a plan for the establishment of this captive breeding population, with the goal of maintaining as much of the genetic variation of the species as possible in the captive population. What factors might cause loss of genetic variation in the establishment of the captive population? How could loss of such variation be prevented? With the assumption that only a limited number of salamanders can be maintained in captivity, what procedures should be instituted to ensure the long-term maintenance of as much of the variation as possible?

Section 17.5

37. Explain why natural selection may cause prezygotic reproductive isolating mechanisms to evolve if postzygotic reproductive isolating mechanisms are already present, but natural selection can never cause the evolution of postzygotic reproductive isolating mechanisms.

Glossary

acentric chromatid Lacks a centromere; produced when crossing over takes place within a paracentric inversion. The acentric chromatid does not attach to a spindle fiber and does not segregate in meiosis or mitosis; so it is usually lost after one or more rounds of cell division.

acrocentric chromosome Chromosome in which the centromere is near one end, producing a long arm at one end and a knob, or satellite, at the other end.

activator *See* **transcriptional activator protein**.

addition rule States that the probability of any of two or more mutually exclusive events occurring is calculated by adding the probabilities of the individual events.

additive genetic variance Component of the genetic variance that can be attributed to the additive effect of different genotypes.

adenine (A) Purine base in DNA and RNA.

adenosine-3′,5′-cyclic monophosphate (cAMP) Modified nucleotide that functions in catabolite repression. Low levels of glucose stimulate high levels of cAMP; cAMP then attaches to CAP, which binds to the promoter of certain operons and stimulates transcription.

A-DNA Right-handed helical structure of DNA that exists when little water is present.

allele One of two or more alternate forms of a gene.

allelic frequency Proportion of a particular allele in a population.

allopatric speciation Arises when a geographic barrier first splits a population into two groups and blocks the exchange of genes between them. Compare **sympatric speciation**.

allopolyploidy Condition in which the sets of chromosomes of a polyploid individual possessing more than two haploid sets are derived from two or more species.

allosteric protein Protein that changes its conformation on binding with another molecule.

alternative processing One of several pathways by which a single pre-mRNA can be processed in different ways to produce alternative types of mRNA.

alternative splicing Process by which a single pre-mRNA can be spliced in more than one way to produce different types of mRNA.

Ames test Test in which special strains of bacteria are used to evaluate the potential of chemicals to cause cancer.

amino acid Repeating unit of proteins; consists of an amino group, a carboxyl group, a hydrogen atom, and a variable R group.

aminoacyl (A) site One of three sites in a ribosome occupied by a tRNA in translation. All charged tRNAs (with the exception of the initiator tRNA) first enter the A site in translation.

aminoacyl-tRNA synthetase Enzyme that attaches an amino acid to a tRNA. Each aminoacyl-tRNA synthetase is specific for a particular amino acid.

amphidiploidy Type of allopolyploidy in which two different diploid genomes are combined, so that every chromosome has one and only one homologous partner and the genome is functionally diploid.

anagenesis Evolutionary change within a single lineage.

anaphase Stage of mitosis in which chromatids separate and move toward the spindle poles.

anaphase I Stage of meiosis I. In anaphase I, homologous chromosomes separate and move toward the spindle poles.

anaphase II Stage of meiosis II. In anaphase II, chromatids separate and move toward the spindle poles.

aneuploidy Change from the wild type in the number of chromosomes; most often an increase or decrease of one or two chromosomes.

anticodon Sequence of three nucleotides in tRNA that pairs with the corresponding codon in mRNA in translation.

antiparallel Refers to a characteristic of the DNA double helix in which the two polynucleotide strands run in opposite directions.

apoptosis Programmed cell death, in which a cell degrades its own DNA, the nucleus and cytoplasm shrink, and the cell undergoes phagocytosis by other cells without leakage of its contents.

archaea One of the three primary divisions of life. Archaea consist of unicellular organisms with prokaryotic cells.

artificial selection Selection practiced by humans.

autopolyploidy Condition in which all the sets of chromosomes of a polyploid individual possessing more than two haploid sets are derived from a single species.

autoradiography Method for visualizing DNA or RNA molecules labeled with radioactive substances. A piece of X-ray film is placed on top of a slide, gel, or other substance that contains DNA labeled with radioactive chemicals. Radiation from the labeled DNA exposes the film, providing a picture of the labeled molecules.

autosome Chromosome that is the same in males and females; nonsex chromosome.

auxotroph Bacterium or fungus that possesses a nutritional mutation that disrupts its ability to synthesize an essential biological molecule; cannot grow on minimal medium but can grow on minimal medium to which has been added the biological molecule that it cannot synthesize.

backcross Cross between an F_1 individual and one of the parental (P) genotypes.

bacterial artificial chromosome (BAC) Cloning vector used in bacteria that is capable of carrying DNA fragments as large as 500 kb.

bacterial colony Clump of genetically identical bacteria derived from a single bacterial cell that undergoes repeated rounds of division.

bacteriophage Virus that infects bacterial cells.

Barr body Condensed, darkly staining structure that is found in most cells of female placental mammals and is an inactivated X chromosome.

base *See* **nitrogenous base**.

base analog Chemical substance that has a structure similar to that of one of the four standard bases of DNA and may be incorporated into newly synthesized DNA molecules in replication.

base-excision repair DNA repair that first excises modified bases and then replaces the entire nucleotide.

base substitution Mutation in which a single pair of bases in DNA is altered.

B-DNA Right-handed helical structure of DNA that exists when water is abundant; the secondary structure described by Watson and Crick and probably the most common DNA structure in cells.

bidirectional replication Replication at both ends of a replication bubble.

bioinformatics Synthesis of molecular biology and computer science that develops databases and computational tools to store, retrieve, and analyze nucleic acid and protein sequence data.

biological species concept Defines a species as a group of organisms whose members are capable of interbreeding with one another but are reproductively isolated from the members of other species. Because different species do not exchange genes, each species evolves independently. Not all biologists adhere to this concept.

biotechnology Use of biological processes, particularly molecular genetics and recombinant DNA technology, to produce products of commercial value.

bivalent Refers to a synapsed pair of homologous chromosomes.

blending inheritance Early concept of heredity proposing that offspring possess a mixture of the traits from both parents.

branch Evolutionary connections between organisms in a phylogenetic tree.

branch point Adenine nucleotide in nuclear pre-mRNA introns that lies from 18 to 40 nucleotides upstream of the 3′ splice site.

broad-sense heritability Proportion of the phenotypic variance that can be attributed to genetic variance.

catabolite activator protein (CAP) Protein that functions in catabolite repression. When bound with cAMP, CAP binds to the promoter of certain operons and stimulates transcription.

catabolite repression System of gene control in some bacterial operons in which glucose is used preferentially and the metabolism of other sugars is repressed in the presence of glucose.

cDNA (complementary DNA) library Collection of bacterial colonies or phage colonies containing DNA fragments that have been produced by reverse transcription of cellular mRNA.

cell cycle Stages through which a cell passes from one cell division to the next.

cell theory States that all life is composed of cells, that cells arise only from other cells, and that the cell is the fundamental unit of structure and function in living organisms.

centiMorgan Another name for map unit.

central dogma Concept that genetic information passes from DNA to RNA to protein in a one-way information pathway.

centriole Cytoplasmic organelle consisting of microtubules; present at each pole of the spindle apparatus in animal cells.

centromere Constricted region on a chromosome that stains less strongly than the rest of the chromosome; region where spindle microtubules attach to a chromosome.

centromeric sequence DNA sequence found in functional centromeres.

Chargaff's rules Rules developed by Erwin Chargaff and his colleagues concerning the ratios of bases in DNA.

checkpoint A key transition point at which progression to the next stage in the cell cycle is regulated.

chiasma (pl., **chiasmata**) Point of attachment between homologous chromosomes at which crossing over took place.

chromatin Material found in the eukaryotic nucleus; consists of DNA and proteins.

chromatin-remodeling complex Complex of proteins that alters chromatin structure without acetylating histone proteins.

chromatosome Consists of a nucleosome and an H1 histone protein.

chromosomal scaffold protein Protein that plays a role in the folding and packing of the chromosome, revealed when chromatin is treated with a concentrated salt solution, which removes histones and some other chromosomal proteins.

chromosome deletion Loss of a chromosome segment.

chromosome duplication Mutation that doubles a segment of a chromosome.

chromosome inversion Rearrangement in which a segment of a chromosome has been inverted 180 degrees.

chromosome mutation Difference from the wild type in the number or structure of one or more chromosomes; often affects many genes and has large phenotypic effects.

chromosome rearrangement Change from the wild type in the structure of one or more chromosomes.

chromosome theory of heredity States that genes are located on chromosomes.

cis configuration Arrangement in which two or more wild-type genes are on one chromosome and their mutant alleles are on the homologous chromosome; also called coupling configuration.

cladogenesis Evolution in which one lineage is split into two.

clonal evolution Process by which mutations that enhance the ability of cells to proliferate predominate in a clone of cells, allowing the clone to become increasingly rapid in growth and increasingly aggressive in proliferation properties.

cloning vector Stable, replicating DNA molecule to which a foreign DNA fragment can be attached and transferred to a host cell.

cloverleaf structure Secondary structure common to all tRNAs.

coactivator Protein that cooperates with an activator of transcription. In eukaryotic transcriptional control, coactivators often physically interact with transcriptional activators and the basal transcription apparatus.

codominance Type of allelic interaction in which the heterozygote simultaneously expresses traits of both homozygotes.

codon Sequence of three nucleotides that encodes one amino acid in a protein.

coefficient of coincidence Ratio of observed double crossovers to expected double crossovers.

cohesive end Short, single-stranded overhanging end on a DNA molecule produced when the DNA is cut by certain restriction enzymes. Cohesive ends are complementary and can spontaneously pair to rejoin DNA fragments that have been cut with the same restriction enzyme.

colinearity Concept that there is a direct correspondence between the nucleotide sequence of a gene and the continuous sequence of amino acids in a protein.

colony *See* **bacterial colony**.

comparative genomics Comparative studies of the genomes of different organisms.

competent cell Capable of taking up DNA from its environment (capable of being transformed).

complementary DNA strands The relation between the two nucleotide strands of DNA in which each purine on one strand pairs with a specific pyrimidine on the opposite strand (A pairs with T, and G pairs with C).

complementation Two different mutations in the heterozygous condition are exhibited as the wild-type phenotype; indicates that the mutations are at different loci.

complementation test Test designed to determine whether two different mutations are at the same locus (are allelic) or at different loci (are nonallelic). Two individuals that are homozygous for two independently derived mutations are crossed, producing F_1 progeny that are heterozygous for the mutations. If the mutations are at the same locus, the F_1 will have a mutant phenotype. If the mutations are at different loci, the F_1 will have a wild-type phenotype.

complete linkage Linkage between genes that are located close together on the same chromosome with no crossing over between them.

complete medium Used to culture bacteria or some other microorganism; contains all the nutrients required for growth and synthesis, including those normally synthesized by the organism. Nutritional mutants can grow on complete medium.

concept of dominance Principle of heredity discovered by Mendel stating that, when two different alleles are present in a genotype, only one allele may be expressed in the phenotype. The dominant allele is the allele that is expressed, and the recessive allele is the allele that is not expressed.

conditional mutation Expressed only under certain conditions.

conjugation Mechanism by which genetic material may be exchanged between bacterial cells. During conjugation, two bacteria lie close together and a cytoplasmic connection forms between them. A plasmid or sometimes a part of the bacterial chromosome passes through this connection from one cell to the other.

consanguinity Mating between related individuals.

consensus sequence Comprises the most commonly encountered nucleotides found at a specific location in DNA or RNA.

−10 consensus sequence (Pribnow box) Consensus sequence (TATAAT) found in most bacterial promoters approximately 10 bp upstream of the transcription start site.

−35 consensus sequence Consensus sequence (TTGACA) found in many bacterial promoters approximately 35 bp upstream of the transcription start site.

constitutive gene A gene that is not regulated and is expressed continually.

contig Set of overlapping DNA fragments that have been assembled in the correct order to form a continuous stretch of DNA sequence.

continuous characteristic Displays a large number of possible phenotypes that are not easily distinguished, such as human height.

continuous replication Replication of the leading strand in the same direction as that of unwinding, allowing new nucleotides to be added continuously to the 3′ end of the new strand as the template is exposed.

coordinate induction Simultaneous synthesis of several enzymes that is stimulated by a single environmental factor.

core enzyme Part of bacterial RNA polymerase that, during transcription, catalyzes the elongation of the RNA molecule by the addition of RNA nucleotides; consists of four subunits: two copies of alpha (α), a single copy of beta (β), and a single copy of beta prime (β').

corepressor Substance that inhibits transcription in a repressible system of gene regulation; usually a small molecule that binds to a repressor protein and alters it so that the repressor is able to bind to DNA and inhibit transcription.

correlation Degree of association between two or more variables.

cosmid Cloning vector that combines the properties of plasmids and phage vectors and is used to clone large pieces of DNA in bacteria. Cosmids are small plasmids that carry λ *cos* sites, allowing the plasmid to be packaged into viral coats.

cotransduction Process in which two or more genes are transferred together from one bacterial cell to another. Only genes located close together on a bacterial chromosome will be cotransduced.

cotransformation Process in which two or more genes are transferred together during cell transformation.

coupling configuration *See* **cis configuration**.

crossing over Exchange of genetic material between homologous but nonsister chromatids.

C value Haploid amount of DNA found in a cell of an organism.

cyclin A key protein in the control of the cell cycle; combines with a cyclin-dependent kinase (CDK). The levels of cyclin rise and fall in the course of the cell cycle.

cyclin-dependent kinase (CDK) A key protein in the control of the cell cycle; combines with cyclin.

cytokinesis Process by which the cytoplasm of a cell divides.

cytoplasmic inheritance Inheritance of characteristics encoded by genes located in the cytoplasm. Because the cytoplasm is usually contributed entirely by only one parent, most cytoplasmically inherited characteristics are inherited from a single parent.

cytosine (C) Pyrimidine base in DNA and RNA.

deamination Loss of an amino group (NH_2) from a base.

degenerate genetic code Refers to the fact that the genetic code contains more information than is needed to specify all 20 common amino acids.

deletion Mutation in which one or more nucleotides are deleted from a DNA sequence.

deoxyribonucleotide Basic building block of DNA, consisting of deoxyribose, a phosphate, and a nitrogenous base.

deoxyribose Five-carbon sugar in DNA; lacks a hydroxyl group on the 2′-carbon atom.

depurination Break in the covalent bond connecting a purine base to the 1′-carbon atom of deoxyribose, resulting in the loss of the purine base. The resulting apurinic site cannot provide a template in replication, and a nucleotide with another base may be incorporated into the newly synthesized DNA strand opposite the apurinic site.

dicentric bridge Structure produced when the two centromeres of a dicentric chromatid are pulled toward opposite poles, stretching the dicentric chromosome across the center of the nucleus. Eventually, the dicentric bridge breaks as the two centromeres are pulled apart.

dicentric chromatid Chromatid that has two centromeres; produced when crossing over takes place within a paracentric inversion. The two centromeres of the dicentric chromatid are frequently pulled toward opposite poles in mitosis or meiosis, breaking the chromosome.

dideoxyribonucleoside triphosphate (ddNTP) Special substrate for DNA synthesis used in the Sanger dideoxy sequencing method; identical with dNTP (the usual substrate for DNA synthesis) except that it lacks a 3′-OH group. The incorporation of a ddNTP into DNA terminates DNA synthesis.

dihybrid cross A cross between two individuals that differ in two characteristics—more specifically, a cross between individuals that are homozygous for different alleles at the two loci (*AA BB* $\times$ *aa bb*); also refers to a cross between two individuals that are both heterozygous at two loci (*Aa Bb* $\times$ *Aa Bb*).

diploid Possessing two sets of chromosomes (two genomes).

directional selection Selection in which one trait or allele is favored over another.

direct repair DNA repair in which modified bases are changed back into their original structures.

discontinuous characteristic Exhibits only a few, easily distinguished phenotypes. An example is seed shape in which seeds are either round or wrinkled.

discontinuous replication Replication of the lagging strand in the direction opposite that of unwinding, which means that DNA must be synthesized in short stretches (Okazaki fragments).

displaced duplication Chromosome rearrangement in which the duplicated segment is some distance from the original segment, either on the same chromosome or on a different one.

dizygotic twins Nonidentical twins that arise when two different eggs are fertilized by two different sperm; also called fraternal twins.

DNA fingerprinting Technique used to identify individuals by examining their DNA sequences.

DNA gyrase *E. coli* topoisomerase enzyme that relieves the torsional strain that builds up ahead of the replication fork.

DNA helicase Enzyme that unwinds double-stranded DNA by breaking hydrogen bonds.

DNA library Collection of bacterial colonies containing all the DNA fragments from one source.

DNA ligase Enzyme that catalyzes the formation of a phosphodiester bond between adjacent 3′-OH and 5′-phosphate groups in a DNA molecule.

DNA methylation Modification of DNA by the addition of methyl groups to certain positions on the bases.

DNA polymerase Enzyme that synthesizes DNA.

DNA polymerase I Bacterial DNA polymerase that removes and replaces RNA primers with DNA nucleotides.

DNA polymerase III Bacterial DNA polymerase that synthesizes new nucleotide strands off the primers.

DNA polymerase α Eukaryotic DNA polymerase that initiates replication.

DNA polymerase β Eukaryotic DNA polymerase that participates in DNA repair.

DNA polymerase δ Eukaryotic DNA polymerase that replicates the lagging strand.

DNA polymerase ε Eukaryotic DNA polymerase that replicates the leading strand during DNA synthesis.

DNA polymerase γ Eukaryotic DNA polymerase that replicates mitochondrial DNA. A γ-like DNA polymerase replicates chloroplast DNA.

DNA sequencing Process of determining the sequence of bases along a DNA molecule.

DNA transposon *See* **transposable element**.

dominance genetic variance Component of the genetic variance that can be attributed to dominance (interaction between genes at the same locus).

dominant Refers to an allele or a phenotype that is expressed in homozygotes (*AA*) and in heterozygotes (*Aa*); only the dominant allele is expressed in a heterozygote phenotype.

dosage compensation Equalization in males and females of the amount of protein produced by X-linked genes. In placental mammals, dosage compensation is accomplished by the random inactivation of one X chromosome in the cells of females.

Down syndrome (trisomy 21) Characterized by variable degrees of mental retardation, characteristic facial features, some retardation of growth and development, and an increased incidence of heart defects, leukemia, and other abnormalities; caused by the duplication of all or part of chromosome 21.

Edward syndrome (trisomy 18) Characterized by severe retardation, low-set ears, a short neck, deformed feet, clenched fingers, heart problems, and other disabilities; results from the presence of three copies of chromosome 18.

effective population size Effective number of breeding adults in a population; influenced by the number of individuals contributing genes to the next generation, their sex ratio, variation between individuals in reproductive success, fluctuations in population size, the age structure of the population, and whether mating is random.

egg Female gamete.

elongation factor G (EF-G) Protein that combines with GTP and is required for movement of the ribosome along the mRNA during translation.

elongation factor Ts (EF-Ts) Protein that regenerates elongation factor Tu in the elongation stage of protein synthesis.

elongation factor Tu (EF-Tu) Protein taking part in the elongation stage of protein synthesis; forms a complex with GTP and a charged amino acid and then delivers the charged tRNA to the ribosome.

enhancer Sequence that stimulates maximal transcription of distant genes; affects only genes on the same DNA molecule (is cis acting), contains short consensus sequences, is not fixed in relation to the transcription start site, can stimulate almost any promoter in its vicinity, and may be upstream or downstream of the gene. The function of an enhancer is independent of sequence orientation.

environmental variance Component of the phenotypic variance that is due to environmental differences among individual members of a population.

epigenetics Phenomena due to alterations to DNA that do not include changes in the base sequence; often affect the way in which the DNA sequences are expressed. Such alterations are often stable and heritable in the sense that they are passed from one cell to another.

episome Plasmid capable of integrating into a bacterial chromosome.

epistasis Type of gene interaction in which a gene at one locus masks or suppresses the effects of a gene at a different locus.

epistatic gene Masks or suppresses the effect of a gene at a different locus.

equilibrium Situation in which no further change takes place; in population genetics, refers to a population in which allelic frequencies do not change.

equilibrium density gradient centrifugation Method used to separate molecules or organelles of different density by centrifugation.

eubacteria One of the three primary divisions of life. Eubacteria consist of unicellular organisms with prokaryotic cells and include most of the common bacteria.

euchromatin Chromatin that undergoes condensation and decondensation in the course of the cell cycle.

eukaryote Organism with a complex cell structure including a nuclear envelope and membrane-bounded organelles. One of the three primary divisions of life, eukaryotes include unicellular and multicellular forms.

evolution Genetic change taking place in a group of organisms.

exit (E) site One of three sites in a ribosome occupied by a tRNA. In the elongation stage of translation, the tRNA moves from the peptidyl (P) site to the E site from which it then exits the ribosome.

exon Coding region of a split gene (a gene that is interrupted by introns). After processing, the exons remain in messenger RNA.

expanding trinucleotide repeat Mutation in which the number of copies of a trinucleotide (or some multiple of three nucleotides) increases in succeeding generations.

expression vector Cloning vector containing DNA sequences such as a promoter, a ribosome-binding site, and transcription initiation and termination sites that allow DNA fragments inserted into the vector to be transcribed and translated.

expressivity Degree to which a trait is expressed.

familial Down syndrome Caused by a Robertsonian translocation in which the long arm of chromosome 21 is translocated to another chromosome; tends to run in families.

fertilization Fusion of gametes, or sex cells, to form a zygote.

F (fertility) factor Episome of *E. coli* that controls conjugation and gene exchange between *E. coli* cells. The F factor contains an origin of replication and genes that enable the bacterium to undergo conjugation.

F$_1$ (first filial) generation Offspring of the initial parents (P) in a genetic cross.

F$_2$ (second filial) generation Offspring of the F$_1$ generation in a genetic cross; the third generation of a genetic cross.

first polar body One of the products of meiosis I in oogenesis; contains half the chromosomes but little of the cytoplasm.

fitness Reproductive success of a genotype compared with that of other genotypes in a population.

5′ cap Modified 5′ end of eukaryotic mRNA, consisting of an extra nucleotide (methylated) and methylation of the 2′ position of the ribose sugar in one or more subsequent nucleotides; plays a role in the binding of the ribosome to mRNA and affects mRNA stability and the removal of introns.

5′ end End of the polynucleotide chain where a phosphate is attached to the 5′-carbon atom of the nucleotide.

5′ splice site The 5′ end of an intron where cleavage takes place in RNA splicing.

5′ untranslated (UTR) region Sequence of nucleotides at the 5′ end of mRNA; does not encode the amino acids of a protein.

fixation Point at which one allele reaches a frequency of 1. At this point, all members of the population are homozygous for the same allele.

flanking direct repeat Short, directly repeated sequence produced on either side of a transposable element when the element inserts into DNA.

forward genetics Traditional approach to the study of gene function that begins with a phenotype (a mutant organism) and proceeds to a gene that encodes the phenotype.

forward mutation Alters a wild-type phenotype.

founder effect Sampling error that arises when a population is established by a small number of individuals; leads to genetic drift.

fragile site Constriction or gap that appears at a particular location on a chromosome when cells are cultured under special conditions. One fragile site on the human X chromosome is associated with mental retardation (fragile-X syndrome) and results from an expanding trinucleotide repeat.

frameshift mutation Alters the reading frame of a gene.

fraternal twins Nonidentical twins that arise when two different eggs are fertilized by two different sperm; also called dizygotic twins.

frequency distribution Graphical way of representing values. In genetics, usually the phenotypes found in a group of individuals are displayed as a frequency distribution. Typically, the phenotypes are plotted on the horizontal (*x*) axis and the numbers (or proportions) of individuals with each phenotype are plotted on the vertical (*y*) axis.

functional genomics Area of genomics that studies the functions of genetic information contained within genomes.

G$_0$ (gap 0) Nondividing stage of the cell cycle.

G$_1$ (gap 1) Stage in interphase of the cell cycle in which the cell grows and develops.

G$_2$ (gap 2) Stage of interphase in the cell cycle that follows DNA replication. In G$_2$, the cell prepares for division.

gain-of-function mutation Produces a new trait or causes a trait to appear in inappropriate tissues or at inappropriate times in development.

gamete Male or female sex cell.

gametophyte Haploid phase of the life cycle in plants.

gel electrophoresis Technique for separating charged molecules (such as proteins or nucleic acids) on the basis of molecular size or charge or both.

gene Genetic factor that helps determine a trait; often defined at the molecular level as a DNA sequence that is transcribed into an RNA molecule.

gene cloning Insertion of DNA fragments into bacteria in such a way that the fragments will be stable and copied by the bacteria.

gene desert In reference to the density of genes in the genome, a region that is gene poor—that is, a long stretch of DNA possibly consisting of hundreds of thousands to millions of base pairs completely devoid of any known genes or other functional sequences.

gene family *See* **multigene family.**

gene flow Movement of genes from one population to another; also called migration.

gene interaction Interaction between genes at different loci that affect the same characteristic.

gene mutation Affects a single gene or locus.

gene pool Total of all genes in a population.

gene regulation Mechanisms and processes that control the phenotypic expression of genes.

gene therapy Use of recombinant DNA to treat a disease or disorder by altering the genetic makeup of the patient's cells.

generalized transduction Transduction in which any gene may be transferred from one bacterial cell to another by a virus.

general transcription factor Protein that binds to eukaryotic promoters near the start site and is a part of the basal transcription apparatus that initiates transcription.

genetic bottleneck Sampling error that arises when a population undergoes a drastic reduction in population size; leads to genetic drift.

genetic drift Change in allelic frequency due to sampling error.

genetic engineering Common term for recombinant DNA technology.

genetic–environmental interaction variance Component of the phenotypic variance that results from an interaction between genotype and environment. Genotypes are expressed differently in different environments.

genetic map Map of the relative distances between genetic loci, markers, or other chromosome regions determined by rates of recombination; measured in percent recombination or map units.

genetic marker Any gene or DNA sequence used to identify a location on a genetic or physical map.

genetic maternal effect Determines the phenotype of an offspring. With genetic maternal effect, an offspring inherits genes for the characteristics from both parents, but the offspring's phenotype is determined not by its own genotype but by the nuclear genotype of its mother.

genetic rescue Introduction of new genetic variation into an inbred population that often dramatically improves the health of the population in an effort to increase its chances of long-term survival.

genetic variance Component of the phenotypic variance that is due to genetic differences among individual members of a population.

genic balance system Sex-determining system in which sexual phenotype is controlled by a balance between genes on the X chromosome and genes on the autosomes.

genic interaction variance Component of the genetic variance that can be attributed to genic interaction (interaction between genes at different loci).

genic sex determination Sex determination in which the sexual phenotype is specified by genes at one or more loci, but there are no obvious differences in the chromosomes of males and females.

genome Complete set of genetic instructions for an organism.

genomic imprinting Differential expression of a gene that depends on the sex of the parent that transmitted the gene. If the gene is inherited from the father, its expression is different from that if it is inherited from the mother.

genomic library Collection of bacterial or phage colonies containing DNA fragments that consist of the entire genome of an organism.

genomics Study of the content, organization, and function of genetic information in whole genomes.

genotype The set of genes possessed by an individual organism.

genotypic frequency Proportion of a particular genotype.

germ-line mutation Mutation in a germ-line cell (one that gives rise to gametes).

germ-plasm theory States that cells in the reproductive organs carry a complete set of genetic information.

G_2/M (gap 2/mitotic) checkpoint Important point in the cell cycle near the end of G_2. After this checkpoint has been passed, the cell undergoes mitosis.

goodness-of-fit chi-square test Statistical test used to evaluate how well a set of observed values fit the expected values. The probability associated with a calculated chi-square value is the probability that the differences between the observed and the expected values may be due to chance.

G_1/S (gap 1/synthesis) checkpoint Important point in the cell cycle. After the G_1/S checkpoint has been passed, DNA replicates and the cell is committed to dividing.

guanine (G) Purine base in DNA and RNA.

gyrase See **DNA gyrase.**

hairpin Secondary structure formed when sequences of nucleotides on the same strand are complementary and pair with each other.

haploid Possessing a single set of chromosomes (one genome).

haploinsufficiency The appearance of a mutant phenotype in an individual cell or organism that is heterozygous for a normally recessive trait.

haploinsufficient gene Must be present in two copies for normal function. If one copy of the gene is missing, a mutant phenotype is produced.

haplotype A specific set of linked genetic variants or alleles on a single chromosome or on part of a chromosome.

Hardy–Weinberg equilibrium Frequencies of genotypes when the conditions of the Hardy–Weinberg law are met.

Hardy–Weinberg law Important principle of population genetics stating that, in a large, randomly mating population not affected by mutation, migration, or natural selection, allelic frequencies will not change and genotypic frequencies stabilize after one generation in the proportions p^2 (the frequency of AA), $2pq$ (the frequency of Aa), and q^2 (the frequency of aa), where p equals the frequency of allele A and q equals the frequency of allele a.

heat-shock protein Produced by many cells in response to extreme heat and other stresses; helps cells prevent damage from such stressing agents.

helicase See **DNA helicase.**

hemizygosity Possession of a single allele at a locus. Males of organisms with XX-XY sex determination are hemizygous for X-linked loci, because their cells possess a single X chromosome.

heritability Proportion of phenotypic variation due to genetic differences. See also **broad-sense heritability** and **narrow-sense heritability.**

hermaphroditism Condition in which an individual organism possesses both male and female reproductive structures. True hermaphrodites produce both male and female gametes.

heterochromatin Chromatin that remains in a highly condensed state throughout the cell cycle; found at the centromeres and telomeres of most chromosomes.

heteroduplex DNA DNA consisting of two strands, each of which is from a different chromosome.

heterogametic sex The sex (male or female) that produces two types of gametes with respect to sex chromosomes. For example, in the XX-XY sex-determining system, the male produces both X-bearing and Y-bearing gametes.

heterozygous Refers to an individual organism that possesses two different alleles at a locus.

highly repetitive DNA DNA that consists of short sequences that are present in hundreds of thousands to millions of copies; clustered in certain regions of chromosomes.

histone Low-molecular-weight protein found in eukaryotes that complexes with DNA to form chromosomes.

histone code Modification of histone proteins, such as the addition or removal of phosphate groups, methyl groups, or acetyl groups, that encode information affecting how genes are expressed.

Holliday junction Model of homologous recombination that is initiated by single-strand breaks in a DNA molecule.

holoenzyme Complex of an enzyme and other protein factors necessary for complete function.

homogametic sex The sex (male or female) that produces gametes that are all alike with regard to sex chromosomes. For example, in the XX-XY sex-determining system, the female produces only X-bearing gametes.

homologous genes Evolutionarily related genes, having descended from a gene in a common ancestor.

homologous pair of chromosomes Two chromosomes that are alike in structure and size and that carry genetic information for the same set of hereditary characteristics. One chromosome of a homologous pair is inherited from the male parent and the other is inherited from the female parent.

homologous recombination Exchange of genetic information between homologous DNA molecules.

homozygous Refers to an individual organism that possesses two identical alleles at a locus.

horizontal gene transfer Transfer of genes from one organism to another by a mechanism other than reproduction.

human papilloma virus (HPV) Virus associated with cervical cancer.

hypostatic gene Gene that is masked or suppressed by the action of a gene at a different locus.

identical twins Twins that arise when a single egg fertilized by a single sperm splits into two separate embryos; also called monozygotic twins.

inbreeding Mating between related individuals that takes place more frequently than expected on the basis of chance.

incomplete dominance Refers to the phenotype of a heterozygote that is intermediate between the phenotypes of the two homozygotes.

incomplete linkage Linkage between genes that exhibit some crossing over; intermediate in its effects between independent assortment and complete linkage.

incomplete penetrance Refers to a genotype that does not always express the expected phenotype. Some individuals possess the genotype for a trait but do not express the phenotype.

incorporated error Incorporation of a damaged nucleotide or mismatched base pair into a DNA molecule.

independent assortment Independent separation of chromosome pairs in anaphase I of meiosis; contributes to genetic variation.

induced mutation Results from an environmental agent, such as a chemical or radiation.

inducer Substance that stimulates transcription in an inducible system of gene regulation; usually a small molecule that binds to a repressor protein and alters that repressor so that it can no longer bind to DNA and inhibit transcription.

inducible operon Operon or other system of gene regulation in which transcription is normally off. Something must happen for transcription to be induced, or turned on.

in-frame deletion Deletion of some multiple of three nucleotides, which does not alter the reading frame of the gene.

in-frame insertion Insertion of some multiple of three nucleotides, which does not alter the reading frame of the gene.

inheritance of acquired characteristics Early notion of inheritance proposing that acquired traits are passed to descendants.

initiation codon The codon in mRNA that specifies the first amino acid (fMet in bacterial cells; Met in eukaryotic cells) of a protein; most commonly AUG.

initiation factor 1 (IF-1) Protein required for the initiation of translation in bacterial cells; enhances the dissociation of the large and small subunits of the ribosome.

initiation factor 2 (IF-2) Protein required for the initiation of translation in bacterial cells; forms a complex with GTP and the charged initiator tRNA and then delivers the charged tRNA to the initiation complex.

initiation factor 3 (IF-3) Protein required for the initiation of translation in bacterial cells; binds to the small subunit of the ribosome and prevents the large subunit from binding during initiation.

initiator protein Binds to an origin of replication and unwinds a short stretch of DNA, allowing helicase and other single-strand-binding proteins to bind and initiate replication.

insertion Mutation in which nucleotides are added to a DNA sequence.

insulator DNA sequence that blocks or insulates the effect of an enhancer; must be located between the enhancer and the promoter to have blocking activity; also may limit the spread of changes in chromatin structure.

integrase Enzyme that inserts prophage, or proviral, DNA into a chromosome.

intercalating agent Chemical substance that is about the same size as a nucleotide and may become sandwiched between adjacent bases in DNA, distorting the three-dimensional structure of the helix and causing single-nucleotide insertions and deletions in replication.

interchromosomal recombination Recombination among genes on different chromosomes.

interference Degree to which one crossover interferes with additional crossovers.

intergenic suppressor mutation Occurs in a gene (locus) that is different from the gene containing the original mutation.

interkinesis Period between meiosis I and meiosis II.

interphase Period in the cell cycle between the cell divisions. In interphase, the cell grows, develops, and prepares for cell division.

interspersed repeat sequence Repeated sequence at multiple locations throughout the genome.

intrachromosomal recombination Recombination among genes located on the same chromosome.

intragenic suppressor mutation Occurs in the same gene (locus) as the mutation that it suppresses.

intron Intervening sequence in a split gene; removed from the RNA after transcription.

inverted repeats Sequences on the same strand that are inverted and complementary.

isoaccepting tRNAs Different tRNAs with different anticodons that specify the same amino acid.

isotopes Different forms of an element that have the same number of protons and electrons but differ in the number of neutrons in the nucleus.

karyotype Picture of an individual organism's complete set of metaphase chromosomes.

kinetochore Set of proteins that assemble on the centromere, providing the point of attachment for spindle microtubules.

Klinefelter syndrome Human condition in which cells contain one or more Y chromosomes along with multiple X chromosomes (most commonly XXY but may also be XXXY, XXXXY, or XXYY). Persons with Klinefelter syndrome are male in appearance but frequently possess small testes, some breast enlargement, and reduced facial and pubic hair; often taller than normal and sterile, most have normal intelligence.

knock-in mouse Mouse that carries a foreign sequence inserted at a specific chromosome location.

knockout mouse Mouse in which a normal gene has been disabled ("knocked out").

labeling Method for adding a radioactive or chemical label to the ends of DNA molecules.

lagging strand DNA strand that is replicated discontinuously.

large ribosomal subunit The larger of the two subunits of a functional ribosome.

lariat Looplike structure created in the splicing of nuclear pre-mRNA in which the 5′ end of an intron is attached to a branch point in pre-mRNA.

leading strand DNA strand that is replicated continuously.

lethal allele Causes the death of an individual organism, often early in development, and so the organism does not appear in the progeny of a genetic cross. Recessive lethal alleles kill individual organisms that are homozygous for the allele; dominant lethals kill both heterozygotes and homozygotes.

lethal mutation Causes premature death.

LINE *See* **long interspersed element.**

linkage group Genes located together on the same chromosome.

linked genes Genes located on the same chromosome.

linker DNA Stretch of DNA separating two nucleosomes.

locus Position on a chromosome where a specific gene is located.

long interspersed element (LINE) Long DNA sequence repeated many times and interspersed throughout the genome.

loss-of-function mutation Causes the complete or partial absence of normal function.

loss of heterozygosity At a locus having a normal allele and a mutant allele, inactivation or loss of the normal allele.

Lyon hypothesis Proposed by Mary Lyon in 1961, this hypothesis proposes that one X chromosome in each female cell becomes inactivated (a Barr body) and suggests that which X becomes inactivated is random and varies from cell to cell.

lysogenic cycle Life cycle of a bacteriophage in which phage genes first integrate into the bacterial chromosome and are not immediately transcribed and translated.

lytic cycle Life cycle of a bacteriophage in which phage genes are transcribed and translated, new phage particles are produced, and the host cell is lysed.

malignant tumor Consists of cells that are capable of invading other tissues.

map-based sequencing Method of sequencing a genome in which sequenced fragments are ordered into contigs with the use of genetic or physical maps.

mapping function Relates recombination frequencies to actual physical distances between genes.

map unit (m.u.) Unit of measure for distances on a genetic map; 1 map unit equals 1% recombination.

mass spectrometry Method for precisely determining the molecular mass of a molecule by using the migration rate of an ionized molecule in an electrical field.

mean Statistic that describes the center of a distribution of measurements; calculated by dividing the sum of all measurements by the number of measurements; also called the average.

megaspore One of the four products of meiosis in plants.

megasporocyte In the ovary of a plant, a diploid reproductive cell that undergoes meiosis to produce haploid macrospores.

meiosis Process in which chromosomes of a eukaryotic cell divide to give rise to haploid reproductive cells. Consists of two divisions: meiosis I and meiosis II.

meiosis I First division of meiosis. In meiosis I, chromosome number is reduced by half.

meiosis II Second division of meiosis. Events in meiosis II are essentially the same as those in mitosis.

Mendelian population Group of interbreeding, sexually reproducing individuals.

meristic characteristic Characteristic whose phenotype varies in whole numbers, such as number of vertebrae.

merozygote Bacterial cell that has two copies of some genes—one copy on the bacterial chromosome and a second copy on an introduced F plasmid; also called partial diploid.

messenger RNA (mRNA) RNA molecule that carries genetic information for the amino acid sequence of a protein.

metacentric chromosome Chromosome in which the two chromosome arms are approximately the same length.

metaphase Stage of mitosis. In metaphase, chromosomes align in the center of the cell.

metaphase I Stage of meiosis I. In metaphase I, homologous pairs of chromosomes align in the center of the cell.

metaphase II Stage of meiosis II. In metaphase II, individual chromosomes align on the metaphase plate.

metaphase plate Plane in a cell between two spindle poles. In metaphase, chromosomes align on the metaphase plate.

metastasis Refers to cells that separate from malignant tumors and travel to other sites, where they establish secondary tumors.

microarray Ordered array of DNA fragments fixed to a solid support, which serve as probes to detect the presence of complementary sequences; often used to assess the expression of genes in various tissues and under different conditions.

microRNA (miRNA) Small RNAs, typically 21 or 22 bp in length, that are produced by cleavage of double-stranded RNA arising from small hairpins within RNA that is mostly single stranded. The miRNAs combine with proteins to form a complex that binds (imperfectly) to mRNA molecules and inhibits their translation.

microsatellite *See* **variable number of tandem repeats.**

microspore Haploid product of meiosis in plants.

microsporocyte Diploid reproductive cell in the stamen of a plant; undergoes meiosis to produce four haploid microspores.

microtubule Long fiber composed of the protein tubulin; plays an important role in the movement of chromosomes in mitosis and meiosis.

migration Movement of genes from one population to another; also called gene flow.

minimal medium Used to culture bacteria or some other microorganism; contains only the nutrients required by prototrophic (wild-type) cells—typically, a carbon source, essential elements such as nitrogen and phosphorus, certain vitamins, and other required ions and nutrients.

mismatch repair Process that corrects mismatched nucleotides in DNA after replication has been completed. Enzymes excise incorrectly paired nucleotides from the newly synthesized strand and use the original nucleotide strand as a template when replacing them.

missense mutation Alters a codon in the mRNA, resulting in a different amino acid in the protein encoded.

mitochondrial DNA (mtDNA) DNA in mitochondria; has some characteristics in common with eubacterial DNA and typically consists of a circular molecule that lacks histone proteins and encodes some of the rRNAs, tRNAs, and proteins found in mitochondria.

mitosis Process by which the nucleus of a eukaryotic cell divides.

mitotic spindle Array of microtubules that radiate from two poles; moves chromosomes in mitosis and meiosis.

model genetic organism An organism that is widely used in genetic studies because it has characteristics, such as short generation time and large numbers of progeny, that make it well suited to genetic analysis.

moderately repetitive DNA DNA consisting of sequences that are from 150 to 300 bp in length and are repeated thousands of times.

modified base Rare base found in some RNA molecules. Such bases are modified forms of the standard bases (adenine, guanine, cytosine, and uracil).

molecular chaperone Molecule that assists in the proper folding of another molecule.

molecular clock Refers to the use of molecular differences to estimate the time of divergence between organisms; assumes a roughly constant rate at which one neutral mutation replaces another.

molecular genetics Study of the chemical nature of genetic information and how it is encoded, replicated, and expressed.

monohybrid cross A cross between two individuals that differ in a single characteristic—more specifically, a cross between individuals that are homozygous for different alleles at the same locus ($AA \times aa$); also refers to a cross between two individuals that are both heterozygous for two alleles at a single locus ($Aa \times Aa$).

monosomy Absence of one of the chromosomes of a homologous pair.

monozygotic twins Identical twins that arise when a single egg fertilized by a single sperm splits into two separate embryos.

Morgan 100 map units.

M (mitotic) phase Period of active cell division; includes mitosis (nuclear division) and cytokinesis (cytoplasmic division).

multifactorial characteristic Determined by multiple genes and environmental factors.

multigene family Set of genes similar in sequence that arose through repeated duplication events; often encode different proteins.

multiple alleles Presence in a group of individuals of more than two alleles at a locus. Although, for the group, the locus has more than two alleles, each member of the group has only two of the possible alleles.

multiplication rule States that the probability of two or more independent events occurring together is calculated by multiplying the probabilities of each of the individual events.

mutagen Any environmental agent that significantly increases the rate of mutation above the spontaneous rate.

mutation Heritable change in genetic information.

mutation rate Frequency with which a gene changes from the wild type to a specific mutant; generally expressed as the number of mutations per biological unit (that is, mutations per cell division, per gamete, or per round of replication).

narrow-sense heritability Proportion of the phenotypic variance that can be attributed to additive genetic variance.

natural selection Differential reproduction of genotypes.

negative control Gene regulation in which the binding of a regulatory protein to DNA inhibits transcription (the regulatory protein is a repressor).

negative supercoiling *See* **supercoiling.**

neutral mutation Changes the amino acid sequence of a protein but does not alter the function of the protein.

nitrogenous base Nitrogen-containing base that is one of the three parts of a nucleotide.

node Point in a phylogenetic tree that represents an organism. Terminal nodes are those that are at the outermost branches of the tree and represent organisms for which data have been obtained. Internal nodes represent ancestors common to organisms on different branches of the tree.

nondisjunction Failure of homologous chromosomes or sister chromatids to separate in meiosis or mitosis.

nonhistone chromosomal protein One of a heterogeneous assortment of nonhistone proteins in chromatin.

nonidentical twins Twins that arise when two different eggs are fertilized by two different sperm; also called dizygotic twins or fraternal twins.

nonoverlapping genetic code Refers to the fact that, generally, each nucleotide is a part of only one codon and encodes only one amino acid in a protein.

nonreciprocal translocation Movement of a chromosome segment to a nonhomologous chromosome or region without any (or with unequal) reciprocal exchange of segments.

nonrecombinant (parental) gamete Contains only the original combinations of genes present in the parents.

nonrecombinant (parental) progeny Possesses the original combinations of traits possessed by the parents.

nonreplicative transposition Type of transposition in which a transposable element excises from an old site and moves to a new site, resulting in no net increase in the number of copies of the transposable element.

nonsense codon Codon in mRNA that signals the end of translation; also called a stop codon or termination codon. The three common nonsense codons are UAA, UAG, and UGA.

nonsense mutation Changes a sense codon (one that specifies an amino acid) into a stop codon.

nontemplate strand The DNA strand that is complementary to the template strand; not ordinarily used as a template during transcription.

normal distribution Common type of frequency distribution that exhibits a symmetrical, bell-shaped curve; usually arises when a large number of independent factors contribute to the measurement.

norm of reaction Range of phenotypes produced by a particular genotype in different environmental conditions.

nuclear envelope Membrane that surrounds the genetic material in eukaryotic cells to form a nucleus; segregates the DNA from other cellular contents.

nucleoid Bacterial DNA confined to a definite region of the cytoplasm.

nucleoside Ribose or deoxyribose bonded to a base.

nucleosome Basic repeating unit of chromatin, consisting of a core of eight histone proteins (two each of H2A, H2B, H3, and H4) and about 146 bp of DNA that wraps around the core about two times.

nucleotide Repeating unit of DNA and RNA made up of a sugar, a phosphate, and a base.

nucleotide-excision repair DNA repair that removes bulky DNA lesions and other types of DNA damage.

nucleus Space in eukaryotic cells that is enclosed by the nuclear envelope and contains the chromosomes.

nullisomy Absence of both chromosomes of a homologous pair ($2n - 2$).

Okazaki fragment Short stretch of newly synthesized DNA. Produced by discontinuous replication on the lagging strand, these fragments are eventually joined together.

oncogene Dominant-acting gene that stimulates cell division, leading to the formation of tumors and contributing to cancer; arises from mutated copies of a normal cellular gene (proto-oncogene).

one-gene, one-enzyme hypothesis Idea proposed by Beadle and Tatum that each gene encodes a separate enzyme.

one-gene, one-polypeptide hypothesis Modification of the one gene, one enzyme hypothesis; proposes that each gene encodes a separate polypeptide chain.

oogenesis Egg production in animals.

oogonium Diploid cell in the ovary; capable of undergoing meiosis to produce an egg cell.

operator DNA sequence in the operon of a bacterial cell. A regulator protein binds to the operator and affects the rate of transcription of structural genes.

operon Set of structural genes in a bacterial cell along with a common promoter and other sequences (such as an operator) that control the transcription of the structural genes.

origin of replication Site where DNA synthesis is initiated.

overdominance Selection in which the heterozygote has higher fitness than that of either homozygote; also called heterozygote advantage.

ovum Final product of oogenesis.

palindrome Sequence of nucleotides that reads the same on complementary strands; inverted repeats.

pangenesis Early concept of heredity proposing that particles carry genetic information from different parts of the body to the reproductive organs.

paracentric inversion Chromosome inversion that does not include the centromere in the inverted region.

parental gamete *See* **nonrecombinant (parental) gamete.**

parental progeny *See* **nonrecombinant (parental) progeny.**

partial diploid Bacterial cell that possesses two copies of genes, including one copy on the bacterial chromosome and the other on an extra piece of DNA (usually a plasmid); also called merozygote.

Patau syndrome (trisomy 13) Characterized by severe mental retardation, a small head, sloping forehead, small eyes, cleft lip and palate, extra fingers and toes, and other disabilities; results from the presence of three copies of chromosome 13.

pedigree Pictorial representation of a family history outlining the inheritance of one or more traits or diseases.

penetrance Percentage of individuals with a particular genotype that express the phenotype expected of that genotype.

peptide bond Chemical bond that connects amino acids in a protein.

peptidyl (P) site One of three sites in a ribosome occupied by a tRNA in translation. In the elongation stage of protein synthesis, tRNAs move from the aminoacyl (A) site into the P site.

pericentric inversion Chromosome inversion that includes the centromere in the inverted region.

P (parental) generation First set of parents in a genetic cross.

phage *See* **bacteriophage.**

phenocopy Phenotype that is produced by environmental effects and is the same as the phenotype produced by a genotype.

phenotype Appearance or manifestation of a characteristic.

phenotypic variance Measures the degree of phenotypic differences among a group of individuals; composed of genetic, environmental, and genetic–environmental interaction variances.

phenylketonuria (PKU) Genetic disease characterized by mental retardation, light skin, and eczema; caused by mutations in the gene that encodes phenylalanine hydroxylase (PAH), a liver enzyme that normally metabolizes the amino acid phenylalanine. When the enzyme is defective, phenylalanine is not metabolized and builds up to high levels in the body, eventually causing mental retardation and other characteristics of the disease. The disease is inherited as an autosomal recessive disorder and can be effectively treated by limiting phenylalanine in the diet.

phosphate group A phosphorus atom attached to four oxygen atoms; one of the three components of a nucleotide.

phosphodiester Molecule containing R–O–P–O–R, where R is a carbon-containing group, O is oxygen, and P is phosphorus.

phosphodiester linkage Phosphodiester bond connecting two nucleotides in a polynucleotide strand.

phylogenetic tree Graphical representation of the evolutionary connections between organisms or genes.

phylogeny Evolutionary relationships among a group of organisms or genes, usually depicted as a family tree or branching diagram.

physical map Map of physical distances between loci, genetic markers, or other chromosome segments; measured in base pairs.

pilus (pl., **pili**) Extension of the surface of some bacteria that allows conjugation to take place. When a pilus on one cell makes contact with a receptor on another cell, the pilus contracts and pulls the two cells together.

plaque Clear patch of lysed cells on a continuous layer of bacteria on the agar surface of a petri plate. Each plaque represents a single original phage that multiplied and lysed many cells.

plasmid Small, circular DNA molecule found in bacterial cells that is capable of replicating independently from the bacterial chromosome.

pleiotropy A single genotype influences multiple phenotypes.

poly(A)-binding protein (PABP) Binds to the poly(A) tail of eukaryotic mRNA and makes the mRNA more stable. There are several types of PABPs, one of which is PABII.

poly(A) tail String of adenine nucleotides added to the 3′ end of a eukaryotic mRNA after transcription.

polycistronic mRNA Single bacterial RNA molecule that encodes more than one polypeptide chain; uncommon in eukaryotes.

polygenic characteristic Encoded by genes at many loci.

polymerase chain reaction (PCR) Method of enzymatically amplifying DNA fragments.

polynucleotide strand Series of nucleotides linked together by phosphodiester bonds.

polypeptide Chain of amino acids linked by peptide bonds; also called a protein.

polyploidy Possession of more than two haploid sets of chromosomes.

polyribosome Messenger RNA molecule with several ribosomes attached to it.

population genetics Study of the genetic composition of populations (groups of members of the same species) and how a population's collective group of genes changes with the passage of time.

positional cloning Method that allows for the isolation and identification of a gene by examining the cosegregation of a phenotype with previously mapped genetic markers.

position effect Dependence of the expression of a gene on the gene's location in the genome.

positive control Gene regulation in which the binding of a regulatory protein to DNA stimulates transcription (the regulatory protein is an activator).

positive supercoiling *See* **supercoiling.**

posttranscriptional RNA gene silencing *See* **RNA silencing.**

posttranslational modification Alteration of a protein after translation; may include cleavage from a larger precursor protein, the removal of amino acids, and the attachment of other molecules to the protein.

postzygotic reproductive isolating mechanism   Reproductive isolation that arises after a zygote is formed, either because the resulting hybrids are inviable or sterile or because reproduction breaks down in subsequent generations.

preformationism   Early concept of inheritance proposing that a miniature adult (homunculus) resides in either the egg or the sperm and increases in size during development, with all traits being inherited from the parent that contributes the homunculus.

pre-messenger RNA (pre-mRNA)   Eukaryotic RNA molecule that is modified after transcription to become mRNA.

prezygotic reproductive isolating mechanism   Reproductive isolation in which gametes from two different species are prevented from fusing and forming a hybrid zygote.

primary Down syndrome   Caused by the presence of three copies of chromosome 21.

primary oocyte   Oogonium that has entered prophase I.

primary spermatocyte   Spermatogonium that has entered prophase I.

primary structure of a protein   The amino acid sequence of a protein.

primase   Enzyme that synthesizes a short stretch of RNA on a DNA template; functions in replication to provide a 3′-OH group for the attachment of a DNA nucleotide.

primer   Short stretch of RNA on a DNA template; provides a 3′-OH group for the attachment of a DNA nucleotide at the initiation of replication.

principle of independent assortment (Mendel's second law)   Important principle of heredity discovered by Mendel that states that genes encoding different characteristics (genes at different loci) separate independently; applies only to genes located on different chromosomes or to genes far apart on the same chromosome.

principle of segregation (Mendel's first law)   Important principle of heredity discovered by Mendel that states that each diploid individual possesses two alleles at a locus and that these two alleles separate when gametes are formed, one allele going into each gamete.

probability   Likelihood of a particular event occurring; more formally, the number of times that a particular event occurs divided by the number of all possible outcomes. Probability values range from 0 to 1.

proband   A person with a trait or disease for whom a pedigree is constructed.

probe   Known sequence of DNA or RNA that is complementary to a sequence of interest and will pair with it; used to find specific DNA sequences.

prokaryote   Unicellular organism with a simple cell structure. Prokaryotes include eubacteria and archaea.

prometaphase   Stage of mitosis. In prometaphase, the nuclear membrane breaks down and the spindle microtubules attach to the chromosomes.

promoter   DNA sequence to which the transcription apparatus binds so as to initiate transcription; indicates the direction of transcription, which of the two DNA strands is to be read as the template, and the starting point of transcription.

proofreading   Ability of DNA polymerases to remove and replace incorrectly paired nucleotides in the course of replication.

prophage   Phage genome that is integrated into a bacterial chromosome.

prophase   Stage of mitosis. In prophase, the chromosomes contract and become visible, the cytoskeleton breaks down, and the mitotic spindle begins to form.

prophase I   Stage of meiosis I. In prophase I, chromosomes condense and pair, crossing over takes place, the nuclear membrane breaks down, and the spindle forms.

prophase II   Stage of meiosis after interkinesis. In prophase II, chromosomes condense, the nuclear membrane breaks down, and the spindle forms. Some cells skip this stage.

protein-coding region   The part of mRNA consisting of the nucleotides that specify the amino acid sequence of a protein.

protein microarray   Large number of different proteins applied to a glass slide as a series of spots, each spot containing a different protein; used to analyze protein–protein interactions.

proteome   Set of all proteins encoded by a genome.

proteomics   Study of the proteome, the complete set of proteins found in a given cell.

proto-oncogene   Normal cellular gene that controls cell division. When mutated, it may become an oncogene and contribute to cancer progression.

provirus   DNA copy of viral DNA or viral RNA; integrated into the host chromosome and replicated along with the host chromosome.

pseudoautosomal region   Small region of the X and Y chromosomes that contains homologous gene sequences.

pseudodominance   Expression of a normally recessive allele owing to a deletion on the homologous chromosome.

Punnett square   Shorthand method of determining the outcome of a genetic cross. On a grid, the gametes of one parent are written along the upper edge and the gametes of the other parent are written along the left-hand edge. Within the cells of the grid, the alleles in the gametes are combined to form the genotypes of the offspring.

purine   Type of nitrogenous base in DNA and RNA. Adenine and guanine are purines.

pyrimidine   Type of nitrogenous base in DNA and RNA. Cytosine, thymine, and uracil are pyrimidines.

pyrimidine dimer   Structure in which a bond forms between two adjacent pyrimidine molecules on the same strand of DNA; disrupts normal hydrogen bonding between complementary bases and distorts the normal configuration of the DNA molecule.

quantitative characteristic   Continuous characteristic; displays a large number of possible phenotypes, which must be described by a quantitative measurement.

quantitative genetics   Genetic analysis of complex characteristics or characteristics influenced by multiple genetic factors.

quantitative trait locus (QTL)   A gene or chromosomal region that contributes to the expression of quantitative characteristics.

quaternary structure of a protein   Interaction of two or more polypeptides to form a functional protein.

reading frame   Particular way in which a nucleotide sequence is read in groups of three nucleotides (codons) in translation. Each reading frame begins with a start codon and ends with a stop codon.

realized heritability   Narrow-sense heritability measured from a response-to-selection experiment.

recessive   Refers to an allele or phenotype that is expressed only when homozygous. The recessive allele is not expressed in the heterozygote phenotype.

reciprocal crosses Crosses in which the phenotypes of the male and female parents are reversed. For example, in one cross, a tall male is crossed with a short female and, in the other cross, a short male is crossed with a tall female.

reciprocal translocation Reciprocal exchange of segments between two nonhomologous chromosomes.

recombinant DNA technology Set of molecular techniques for locating, isolating, altering, combining, and studying DNA segments.

recombinant gamete Possesses new combinations of genes.

recombinant progeny Possesses new combinations of traits formed from recombinant gametes.

recombination Sorting of alleles into new combinations.

recombination frequency Proportion of recombinant progeny produced in a cross.

regulator gene Gene associated with an operon in bacterial cells that encodes a protein or RNA molecule that functions in controlling the transcription of one or more structural genes.

regulator protein Produced by a regulator gene, a protein that binds to another DNA sequence and controls the transcription of one or more structural genes.

regulatory element DNA sequence that affects the transcription of other DNA sequences to which it is physically linked.

regulatory gene DNA sequence that encodes a protein or RNA molecule that interacts with DNA sequences and affects their transcription or translation or both.

regulatory promoter DNA sequence located immediately upstream of the core promoter that affects transcription; contains consensus sequences to which transcriptional activator proteins bind.

relaxed state of DNA Energy state of a DNA molecule when there is no structural strain on the molecule.

release factor Protein required for the termination of translation; binds to a ribosome when a stop codon is reached and stimulates the release of the polypeptide chain, the tRNA, and the mRNA from the ribosome.

repetitive DNA Sequences that exist in multiple copies in a genome.

replicated error Replication of an incorporated error in which a change in the DNA sequence has been replicated and all base pairings in the new DNA molecule are correct.

replication Process by which DNA is synthesized from a single-stranded nucleotide template.

replication bubble Segment of a DNA molecule that is unwinding and undergoing replication.

replication fork Point at which a double-stranded DNA molecule separates into two single strands that serve as templates for replication.

replication licensing factor Protein that ensures that replication takes place only once at each origin; required at the origin before replication can be initiated and removed after the DNA has been replicated.

replication origin Sequence of nucleotides where replication is initiated.

replicative transposition Type of transposition in which a copy of the transposable element moves to a new site while the original copy remains at the old site; increases the number of copies of the transposable element.

replicon Unit of replication, consisting of DNA from the origin of replication to the point at which replication on either side of the origin ends.

repressible operon Operon or other system of gene regulation in which transcription is normally on. Something must take place for transcription to be repressed, or turned off.

repressor Regulatory protein that binds to a DNA sequence and inhibits transcription.

reproductive isolating mechanism Any biological factor or mechanism that prevents gene exchange.

repulsion *See* **trans configuration.**

response element Common DNA sequence found upstream of some groups of eukaryotic genes. A regulatory protein binds to a response element and stimulates the transcription of a gene. The presence of the same response element in several promoters or enhancers allows a single factor to simultaneously stimulate the transcription of several genes.

response to selection The amount that a characteristic changes in one generation owing to selection; equals the selection differential times the narrow-sense heritability.

restriction endonuclease Technical term for a restriction enzyme, which recognizes particular base sequences in DNA and makes double-stranded cuts nearby.

restriction enzyme Recognizes particular base sequences in DNA and makes double-stranded cuts nearby; also called restriction endonuclease.

restriction fragment length polymorphism (RFLP) Variation in the pattern of fragments produced when DNA molecules are cut with the same restriction enzyme; represents a heritable difference in DNA sequences and can be used in gene mapping.

restriction mapping Determines the locations of sites cut by restriction enzymes in a piece of DNA.

retrotransposon Type of transposable element in eukaryotic cells that possesses some characteristics of retroviruses and transposes through an RNA intermediate.

retrovirus RNA virus capable of integrating its genetic material into the genome of its host. The virus injects its RNA genome into the host cell, where reverse transcription produces a complementary, double-stranded DNA molecule from the RNA template. The DNA copy then integrates into the host chromosome to form a provirus.

reverse duplication Duplication of a chromosome segment in which the sequence of the duplicated segment is inverted relative to the sequence of the original segment.

reverse genetics A molecular approach that begins with a genotype (a DNA sequence) and proceeds to the phenotype by altering the sequence or by inhibiting its expression.

reverse mutation (reversion) Mutation that changes a mutant allele back into the wild-type allele.

reverse transcriptase Enzyme capable of synthesizing complementary DNA from an RNA template.

reverse transcription Synthesis of DNA from an RNA template.

rho-dependent terminator Sequence in bacterial DNA that requires the presence of the rho subunit of RNA polymerase to terminate transcription.

rho factor Subunit of bacterial RNA polymerase that facilitates the termination of transcription of some genes.

rho-independent terminator Sequence in bacterial DNA that does not require the presence of the rho subunit of RNA polymerase to terminate transcription.

ribonucleoside triphosphate (rNTP) Substrate of RNA synthesis; consists of ribose, a nitrogenous base, and three phosphates linked to the 5'-carbon atom of the ribose. In transcription, two of the phosphates are cleaved, producing an RNA nucleotide.

ribonucleotide Nucleotide containing ribose; present in RNA.

ribose Five-carbon sugar in RNA.

ribosomal RNA (rRNA) RNA molecule that is a structural component of the ribosome.

ribozyme RNA molecule that can act as a biological catalyst.

RNA-coding region Sequence of DNA nucleotides that encodes an RNA molecule.

RNA-induced silencing complex (RISC) Combination of a small interfering RNA (siRNA) molecule or a microRNA (miRNA) molecule and proteins that can cleave mRNA, leading to the degradation of the mRNA, or affect transcription or repress translation of the mRNA.

RNA interference (RNAi) Process in which cleavage of double-stranded RNA produces small interfering RNAs (siRNAs) that bind to mRNAs containing complementary sequences and bring about their cleavage and degradation.

RNA polymerase Enzyme that synthesizes RNA from a DNA template during transcription.

RNA polymerase I Eukaryotic RNA polymerase that transcribes large ribosomal RNA molecules (18S rRNA and 28S rRNA).

RNA polymerase II Eukaryotic RNA polymerase that transcribes pre-messenger RNA, some small nuclear RNAs, and some microRNAs.

RNA polymerase III Eukaryotic RNA polymerase that transcribes transfer RNA, small ribosomal RNAs (5S rRNA), some small nuclear RNAs, and some microRNAs.

RNA polymerase IV Transcribes small interfering RNAs in plants.

RNA replication Process in some viruses by which RNA is synthesized from an RNA template.

RNA silencing Mechanism by which double-stranded RNA is cleaved and processed to yield small single-stranded interfering RNAs (siRNAs), which bind to complementary sequences in mRNA and bring about the cleavage and degradation of mRNA; also known as RNA interference and posttranscriptional RNA gene silencing. Some siRNAs also bind to complementary sequences in DNA and guide enzymes to methylate the DNA.

RNA splicing Process by which introns are removed and exons are joined together.

Robertsonian translocation Translocation in which the long arms of two acrocentric chromosomes become joined to a common centromere, resulting in a chromosome with two long arms and usually another chromosome with two short arms.

rooted tree Phylogenetic tree in which one internal node represents the common ancestor of all other organisms (nodes) on the tree. In a rooted tree, all the organisms depicted have a common ancestor.

R plasmid (R factor) Plasmid having genes that confer antibiotic resistance to any cell that contains the plasmid.

sampling error Deviations from expected ratios due to chance occurrences when the number of events is small.

secondary oocyte One of the products of meiosis I in female animals; receives most of the cytoplasm.

secondary spermatocyte Product of meiosis I in male animals.

secondary structure of a protein Regular folding arrangement of amino acids in a protein. Common secondary structures found in proteins include the alpha helix and the beta pleated sheet.

second polar body One of the products of meiosis II in oogenesis; contains a set of chromosomes but little of the cytoplasm.

selection coefficient Measure of the relative intensity of selection against a genotype; equals 1 minus fitness.

selection differential Difference in phenotype between the selected individuals and the average of the entire population.

semiconservative replication Replication in which the two nucleotide strands of DNA separate, each serving as a template for the synthesis of a new strand. All DNA replication is semiconservative.

sense codon Codon that specifies an amino acid in a protein.

70S initiation complex Final complex formed in the initiation of translation in bacterial cells; consists of the small and large subunits of the ribosome, mRNA, and initiator tRNA charged with fMet.

sex Male or female.

sex chromosomes Chromosomes that differ morphologically or in number in males and females.

sex determination Specification of sex (male or female). Sex-determining mechanisms include chromosomal, genic, and environmental sex-determining systems.

sex-determining region Y (*SRY*) gene On the Y chromosome, a gene that triggers male development; also known as the testis-determining factor (*TDF*) gene.

sex-influenced characteristic Encoded by autosomal genes that are more readily expressed in one sex. For example, an autosomal dominant gene may have higher penetrance in males than in females or an autosomal gene may be dominant in males but recessive in females.

sex-limited characteristic Encoded by autosomal genes and expressed in only one sex. Both males and females carry genes for sex-limited characteristics, but the characteristics appear in only one of the sexes.

sex-linked characteristic Characteristic determined by a gene or genes on sex chromosomes.

Shine–Dalgarno sequence Consensus sequence found in the bacterial 5' untranslated region of mRNA; contains the ribosome-binding site.

short interspersed element (SINE) Short DNA sequence repeated many times and interspersed throughout the genome.

short tandem repeat (STR) Very short DNA sequence repeated in tandem and found widely in the human genome.

sigma factor Subunit of bacterial RNA polymerase that allows the RNA polymerase to recognize a promoter and initiate transcription.

signal-transduction pathway System in which an external signal (initiated by a hormone or growth factor) triggers a cascade of intracellular reactions that ultimately produce a specific response.

silencer Sequence that has many of the properties possessed by an enhancer but represses transcription.

silent mutation Change in the nucleotide sequence of DNA that does not alter the amino acid sequence of a protein.

SINE *See* **short interspersed element.**

single-nucleotide polymorphism (SNP) Single-base-pair differences in DNA sequence between individual members of a species.

single-strand binding (SSB) protein Binds to single-stranded DNA in replication and prevents it from annealing with a complementary strand and forming secondary structures.

sister chromatids Two copies of a chromosome that are held together at the centromere. Each chromatid consists of a single DNA molecule.

small interfering RNA (siRNA) Single-stranded RNA molecule (usually from 21 to 25 nucleotides in length) produced by the cleavage and processing of double-stranded RNA; binds to complementary sequences in mRNA and brings about the cleavage and degradation of the mRNA. Some siRNAs bind to complementary sequences in DNA and bring about their methylation.

small nuclear ribonucleoprotein (snRNP) Structure found in the nuclei of eukaryotic cells that consists of small nuclear RNA (snRNA) and protein; functions in the processing of pre-mRNA.

small nuclear RNA (snRNA) Small RNA molecule found in the nuclei of eukaryotic cells; functions in the processing of pre-mRNA.

small nucleolar RNA (snoRNA) Small RNA molecule found in the nuclei of eukaryotic cells; functions in the processing of rRNA and in the assembly of ribosomes.

small ribosomal subunit The smaller of the two subunits of a functional ribosome.

somatic mutation Mutation in a cell that does not give rise to a gamete.

SOS system System of proteins and enzymes that allow a cell to replicate its DNA in the presence of a distortion in DNA structure; makes numerous mistakes in replication and increases the rate of mutation.

specialized transduction Transduction in which genes near special sites on the bacterial chromosome are transferred from one bacterium to another; requires lysogenic bacteriophages.

speciation Process by which new species arise. *See also* **biological species concept, allopatric speciation,** and **sympatric speciation.**

species Term applied to different kinds or types of living organisms. Its two primary uses in biology are (1) to give names to different kinds or types of living organisms and (2) to consider a species an evolutionarily independent group of oganisms.

sperm Male gamete.

spermatid Immediate product of meiosis II in spermatogenesis; matures to sperm.

spermatogenesis Sperm production in animals.

spermatogonium Diploid cell in the testis; capable of undergoing meiosis to produce a sperm.

S (synthesis) phase Stage of interphase in the cell cycle. In S phase, DNA replicates.

spindle microtubule Microtubule that moves chromosomes in mitosis and meiosis.

spindle pole Point from which spindle microtubules radiate.

spliceosome Large complex consisting of several RNAs and many proteins that splices protein-encoding pre-mRNA; contains five small ribonucleoprotein particles (U1, U2, U4, U5, and U6).

spontaneous mutation Arises spontaneously from natural changes in DNA structure or from errors in replication.

stop (termination or nonsense) codon Codon in mRNA that signals the end of translation. The three common stop codons are UAA, UAG, and UGA.

strand slippage Slipping of the template and newly synthesized strands in replication in which one of the strands loops out from the other and nucleotides are inserted or deleted on the newly synthesized strand.

structural gene DNA sequence that encodes a protein that functions in metabolism or biosynthesis or that has a structural role in the cell.

structural genomics Area of genomics that studies the organization and sequence of information contained within genomes; sometimes used by protein chemists to refer to the determination of the three-dimensional structure of proteins.

submetacentric chromosome Chromosome in which the centromere is displaced toward one end, producing a short arm and a long arm.

supercoiling Coiled tertiary structure that forms when strain is placed on a DNA helix by overwinding or underwinding of the helix. An overwound DNA exhibits positive supercoiling; an underwound DNA exhibits negative supercoiling.

suppressor mutation Hides or suppresses the effect of another mutation at a site that is distinct from the site of the original mutation.

sympatric speciation Speciation arising in the absence of any geographic barrier to gene flow; reproductive isolating mechanisms evolve with a single interbreeding population.

synapsis Close pairing of homologous chromosomes.

synaptonemal complex Three-part structure that develops between synapsed homologous chromosomes.

synonymous codons Different codons that specify the same amino acid.

tandem duplication Duplication of a chromosome segment that is adjacent to the original segment.

tandem repeat sequences DNA sequences repeated one after another; tend to be clustered at specific locations on a chromosome.

***Taq* polymerase** DNA polymerase commonly used in PCR reactions. Isolated from the bacterium *Thermus aquaticus,* the enzyme is stable at high temperatures, and so it is not denatured during the strand-separation step of the cycle.

telocentric chromosome Chromosome in which the centromere is at or very near one end.

telomerase Enzyme that is made up of both protein and RNA and replicates the ends (telomeres) of eukaryotic chromosomes. The RNA part of the enzyme has a template that is complementary to repeated sequences in the telomere and pairs with them, providing a template for the synthesis of additional copies of the repeats.

telomere Stable end of a chromosome.

telomeric sequence Sequence found at the ends of a chromosome; consists of many copies of short, simple sequences repeated one after the other.

telophase Stage of mitosis. In telophase, the chromosomes arrive at the spindle poles, the nuclear membrane re-forms, and the chromosomes relax and lengthen.

telophase I Stage of meiosis I. In telophase I, chromosomes arrive at the spindle poles.

telophase II Stage of meiosis II. In telophase II, chromosomes arrive at the spindle poles.

temperate phage Bacteriophage that utilizes the lysogenic cycle, in which the phage DNA integrates into the bacterial chromosome and remains in an inactive state.

temperature-sensitive allele Expressed only at certain temperatures.

template strand The strand of DNA that is used as a template during transcription. The RNA synthesized during transcription is complementary and antiparallel to the template strand.

terminal inverted repeats Sequences found at both ends of a transposable element that are inverted complements of one another.

termination codon Codon in mRNA that signals the end of translation; also called nonsense codon or stop codon. The three common termination codons are UAA, UAG, and UGA.

terminator Sequence of DNA nucleotides that causes the termination of transcription.

tertiary structure of a protein Higher-order folding of amino acids in a protein to form the overall three-dimensional shape of the molecule.

testcross A cross between an individual with an unknown genotype and an individual with the homozygous recessive genotype.

tetrad The four products of meiosis; all four chromatids of a homologous pair of chromosomes.

tetrasomy Presence of two extra copies of a chromosome ($2n + 2$).

theta replication Replication of circular DNA that is initiated by the unwinding of the two nucleotide strands, producing a replication bubble. Unwinding continues at one or both ends of the bubble, making it progressively larger. DNA replication on both of the template strands is simultaneous with unwinding until the two replication forks meet.

30S initiation complex Initial complex formed in the initiation of translation in bacterial cells; consists of the small subunit of the ribosome, mRNA, initiator tRNA charged with fMet, GTP, and initiation factors 1, 2, and 3.

three-point testcross Cross between an individual heterozygous at three loci and an individual homozygous for recessive alleles at those loci.

3′ end End of the polynucleotide chain where an OH group is attached to the 3′-carbon atom of the nucleotide.

3′ splice site The 3′ end of an intron where cleavage takes place in RNA splicing.

3′ untranslated (UTR) region Sequence of nucleotides at the 3′ end of mRNA; does not encode the amino acids of a protein but affects both the stability of the mRNA and its translation.

threshold characteristic Discontinuous characteristic whose expression depends on an underlying susceptibility that varies continuously.

thymine (T) Pyrimidine base in DNA but not in RNA.

Ti plasmid Large plasmid from the bacterium *Agrobacterium tumefaciens* that is used to transfer genes to plant cells.

topoisomerase Enzyme that adds or removes rotations in a DNA helix by temporarily breaking nucleotide strands; controls the degree of DNA supercoiling.

trans configuration Arrangement in which a chromosome contains one wilde-type allele and one mutant allele; also called repulsion.

transcription Process by which RNA is synthesized from a DNA template.

transcriptional activator protein Protein in eukaryotic cells that binds to consensus sequences in regulatory promoters or enhancers and affects transcription initiation by stimulating or inhibiting the assembly of the basal transcription apparatus.

transcription bubble Region of a DNA molecule that has unwound to expose a single-stranded template, which is being transcribed into RNA.

transcription factor Protein that binds to DNA sequences in eukaryotic cells and affects transcription.

transcription start site The first DNA nucleotide that is transcribed into an RNA molecule.

transcription unit Sequence of nucleotides in DNA that encodes a single RNA molecule, along with the sequences necessary for its transcription; normally contains a promoter, an RNA-coding sequence, and a terminator.

transcriptome Set of all RNA molecules transcribed from a genome.

transducing phage Contains a piece of the bacterial chromosome inside the phage coat. *See also* **generalized transduction.**

transductant Bacterial cell that has received genes from another bacterium through transduction.

transduction Type of gene exchange that takes place when a virus carries genes from one bacterium to another. After it is inside the cell, the newly introduced DNA may undergo recombination with the bacterial chromosome.

transfer RNA (tRNA) RNA molecule that carries an amino acid to the ribosome and transfers it to a growing polypeptide chain in translation.

transformant Cell that has received genetic material through transformation.

transformation Mechanism by which DNA found in the medium is taken up by the cell. After transformation, recombination may take place between the introduced genes and the cellular chromosome.

transforming principle Substance responsible for transformation. DNA is the transforming principle.

transgene Foreign gene or other DNA fragment carried in germ-line DNA.

transgenic mouse Mouse whose genome contains a foreign gene or genes added by employing recombinant DNA methods.

transition Base substitution in which a purine is replaced by a different purine or a pyrimidine is replaced by a different pyrimidine.

translation Process by which a protein is assembled from information contained in messenger RNA.

translocation Movement of a chromosome segment to a nonhomologous chromosome or to a region within the same chromosome. Also, movement of a ribosome along mRNA in the course of translation.

translocation carrier Individual organism heterozygous for a translocation.

transmission genetics Field of genetics that encompasses the basic principles of genetics and how traits are inherited.

transposable element DNA sequence capable of moving from one site to another within the genome through a mechanism that differs from that of homologous recombination.

transposition Movement of a transposable genetic element from one site to another. Replicative transposition increases the number of copies of the transposable element; nonreplicative transposition does not increase the number of copies.

transversion Base substitution in which a purine is replaced by a pyrimidine or a pyrimidine is replaced by a purine.

trihybrid cross A cross between two individuals that differ in three characteristics (*AA BB CC* × *aa bb cc*); also refers to a cross between two individuals that are both heterozygous at three loci (*Aa Bb Cc* × *Aa Bb Cc*).

triplet code Refers to the fact that three nucleotides encode each amino acid in a protein.

triploidy Possession of three haploid sets of chromosomes ($3n$).

triplo-X syndrome Human condition in which cells contain three X chromosomes. A person with triplo-X syndrome has a female phenotype without distinctive features other than a tendency to be tall and thin; a few such women are sterile, but many menstruate regularly and are fertile.

trisomy   Presence of an additional copy of a chromosome ($2n + 1$).

trisomy 8   Presence of three copies of chromosome 8; in humans, results in mental retardation, contracted fingers and toes, low-set malformed ears, and a prominent forehead.

trisomy 13   Presence of three copies of chromosome 13; in humans, results in Patau syndrome.

trisomy 18   Presence of three copies of chromosome 18; in humans, results in Edward syndrome.

trisomy 21   Presence of three copies of chromosome 21; in humans, results in Down syndrome.

tRNA charging   Chemical reaction in which an aminoacyl-tRNA synthetase attaches an amino acid to its corresponding tRNA.

tumor-suppressor gene   Gene that normally inhibits cell division. Recessive mutations in such genes often contribute to cancer.

Turner syndrome   Human condition in which cells contain a single X chromosome and no Y chromosome (XO). Persons with Turner syndrome are female in appearance but do not undergo puberty and have poorly developed female secondary sex characteristics; most are sterile but have normal intelligence.

two-point testcross   Cross between an individual heterozygous at two loci and an individual homozygous for recessive alleles at those loci.

unbalanced gamete   Gamete that has a variable number of chromosomes; some chromosomes may be missing and others may be present in more than one copy.

underdominance   Selection in which the heterozygote has lower fitness than that of either homozygote.

unequal crossing over   Misalignment of the two DNA molecules during crossing over, resulting in one DNA molecule with a duplication and the other with a deletion.

unique-sequence DNA   Sequence present only once or a few times in a genome.

universal genetic code   Refers to the fact that particular codons specify the same amino acids in almost all organisms.

uracil (U)   Pyrimidine base in RNA but not normally in DNA.

variable number of tandem repeats (VNTRs)   Short sequences repeated in tandem that vary greatly in number among individuals; also called microsatellites. Because they are quite variable, VNTRs are commonly used in DNA fingerprinting.

variance   Statistic that describes the variability of a group of measurements.

virulent phage   Bacteriophage that reproduces only through the lytic cycle and kills its host cell.

virus   Noncellular replicating agent consisting of nucleic acid surrounded by a protein coat; can replicate only within its host cell.

whole-genome shotgun sequencing   Method of sequencing a genome in which sequenced fragments are assembled into the correct sequence in contigs by using only the overlaps in sequence.

wild type   The trait or allele that is most commonly found in natural (wild) populations.

wobble   Base pairing between codon and anticodon in which there is nonstandard pairing, usually at the third ($3'$) position of the codon; allows more than one codon to pair with the same anticodon.

X : A ratio   Ratio of the number of X chromosomes to the number of haploid autosomal sets of chromosomes; determines sex in fruit flies.

X-linked characteristic   Characteristic determined by a gene or genes on the X chromosome.

X-ray diffraction   Method for analyzing the three-dimensional shape and structure of chemical substances. Crystals of a substance are bombarded with X-rays, which hit the crystals, bounce off, and produce a diffraction pattern on a detector. The pattern of the spots produced on the detector provides information about the molecular structure.

yeast artificial chromosome (YAC)   Cloning vector consisting of a DNA molecule with a yeast origin of replication, a pair of telomeres, and a centromere. YACs can carry very large pieces of DNA (as large as several hundred thousand base pairs) and replicate and segregate as yeast chromosomes do.

Y-linked characteristic   Characteristic determined by a gene or genes on the Y chromosome.

Z-DNA   Secondary structure of DNA characterized by 12 bases per turn, a left-handed helix, and a sugar–phosphate backbone that zigzags back and forth.

Answers to Selected Questions and Problems

Chapter 1

1. In the Hopi culture, albinos were considered special and given special status. Because extensive exposure to sunlight could be damaging or deadly, Hopi male albinos did no agricultural work. Albinism was considered a positive trait rather than a negative physical condition, allowing albinos to have more children and thus increasing the frequency of the allele. Finally, the small population size of the Hopi tribe may have helped increase the allele frequency of the albino gene owing to chance.

2. Genetics plays important roles in the diagnosis and treatment of hereditary diseases, in breeding plants and animals for improved production and disease resistance, and in producing pharmaceuticals and novel crops through genetic engineering.

4. Transmission genetics: The inheritance of genes from one generation to the next, gene mapping, characterization of the phenotypes produced by mutations.

   Molecular genetics: The structure, organization, and function of genes at the molecular level.

   Population genetics: Genes and changes in genes in populations.

7. Pangenesis proposes that information for creating each part of an offspring's body originates in each part of the parent's body and is passed through the reproductive organs to the embryo at conception. Pangenesis suggests that changes in parts of the parent's body may be passed to the offspring's body. The germ-plasm theory, in contrast, states that the reproductive cells possess all of the information required to make the complete body; the rest of the body contributes no information to the next generation.

9. Preformationism is the idea that an offspring results from a miniature adult form that is already preformed in the sperm or the egg. All traits would thus be inherited from only one parent, either the father or the mother, depending on whether the homunculus (the preformed miniature adult) resided in the sperm or the egg.

12. Gregor Mendel.

15. Genes are composed of DNA nucleotide sequences and are located at specific positions in chromosomes.

17. (a) Transmission genetics; (b) population genetics; (c) population genetics; (d) molecular genetics; (e) molecular genetics; (f) transmission genetics.

18. Genetics is old in the sense that humans have been aware of hereditary principles for thousands of years and have applied them since the beginning of agriculture and the domestication of plants and animals. It is very young in the sense that the fundamental principles were not uncovered until Mendel's time, and the structure of DNA and the principles of recombinant DNA were discovered within the past 60 years.

20. (a) Pangenesis postulates that particles carried genetic information from all parts of the body to the reproductive organs, and then the genetic information was conveyed to the embryo,

where each unit directs the formation of its own specific part of the body. According to the germ-plasm theory, gamete-producing cells found within the reproductive organs contain a complete set of genetic information that is passed to the gametes. Pangenesis and the germ-plasm theory are similar in that both propose that genetic information is contained in discrete units that are passed on to the offspring. They differ in where that genetic information resides. In pangenesis, it resides in different parts of the body and must travel to the reproductive organs. In the germ-plasm theory, all the genetic information is already in the reproductive cells. (b) Preformationism holds that the sperm or egg contains a miniature preformed adult called a homunculus. In development, the homunculus grows to produce an offspring. Only one parent contributes genetic traits to the offspring. Blending inheritance requires contributions of genetic material from both parents. The genetic contributions from the parents blend to produce the genetic material of the offspring. Having been blended, the genetic material cannot be separated for future generations. (c) The inheritance of acquired characteristics postulates that traits acquired in a person's lifetime alter the genetic material and can be transmitted to offspring. Our modern theory of heredity states that offspring inherit genes located on chromosomes passed from their parents. These chromosomes segregate in meiosis in the parent's germ cells and are passed into the gametes.

21. (a) Both cell types have lipid bilayer membranes, DNA genomes, and machinery for DNA replication, transcription, translation, energy metabolism, response to stimuli, growth, and reproduction. Eukaryotic cells have a nucleus containing chromosomal DNA and possess internal membrane-bounded organelles.
   (b) A gene is a basic unit of hereditary information, usually encoding a functional RNA or polypeptide. Alleles are variant forms of a gene, arising through mutation.
   (c) The genotype is the set of genes or alleles inherited by an organism from its parent(s). The expression of the genes of a particular genotype, through interaction with environmental factors, produces the phenotype, the observable trait.
   (d) Both are nucleic acid polymers. RNA contains a ribose sugar, whereas DNA contains a deoxyribose sugar. RNA also contains uracil as one of the four bases, whereas DNA contains thymine. The other three bases are common to both DNA and RNA. Finally, DNA is usually double stranded, consisting of two complementary strands, whereas RNA is single stranded with regions of internal base pairing that form complex secondary structures.
   (e) Chromosomes are structures consisting of DNA and associated proteins. The DNA contains the genetic information.

25. All genomes must have the ability to store complex information and to vary. The blueprint for the entire organism must be contained within the genome of each reproductive cell. The information has to be in the form of a code that can be used as a set of instructions for assembling the components of the cells. The genetic material of any

organism must be stable, be replicated precisely, and be transmitted faithfully to the progeny but must be capable of mutating.

27. (a) Having the genetic test removes doubt about the potential for the disorder: you are either susceptible or not. Knowing about the potential of a genetic disorder enables you to make life-style changes that might lessen the effect of the disease or lessen the risk. The types and nature of future medical tests could be positively affected by the genetic testing, thus allowing for early warning and screening for the disease. The knowledge could also enable you to make informed decisions regarding offspring and the potential of passing the trait to your offspring. Additionally, by knowing what to expect, you could plan your life accordingly.

Reasons for not having the test typically concern the potential for testing positive for the susceptibility to the genetic disease. If the susceptibility is detected, there is potential for discrimination. For example, your employer (or possibly future employer) might consider you a long-term liability, thus affecting employment options. Insurance companies may not want to insure you for the condition or its symptoms, and social stigmatization regarding the disease could be a factor. Knowledge of the potential future condition could lead to psychological difficulties in coping with the anxiety of waiting for the disease to manifest.

(b) There is no "correct" answer, but some of the reasons for wanting to be tested are: the test would remove doubt about the susceptibility, particularly if family members have had the genetic disease; either a positive or a negative result would allow for informed planning of life style, medical testing, and family choices in the future.

Chapter 2

1.

Prokaryotic Cell	Eukaryotic Cell
No nucleus	Nucleus present
No paired chromosomes (haploid)	Paired chromosomes common (diploid)
Typically single circular chromosome with a single origin of replication	Typically multiple linear chromosomes with centromeres, telomeres, and multiple origins of replication
Single chromosome is replicated with each copy moving to opposite sides of the cell	Chromosomes are replicated but require mitosis or meiosis to ensure that chromosome migrates to the proper location
No histone proteins complexed to DNA	Histone proteins are complexed to DNA

3. The events are (1) a cell's genetic information must be copied, (2) the copies of the genetic information must be separated from one another, and (3) the cell must divide.

5.

Metacentric   Submetacentric   Acrocentric   Telocentric

7. Prophase: The chromosomes condense and become visible, and the centrosomes move apart along with the formation of microtubule fibers from the centrosomes.

Prometaphase: The nucleoli disappear and the nuclear envelope begins to disintegrate, allowing for the cytoplasm and nucleoplasm to join. The sister chromatids of each chromosome are attached to microtubules from the opposite centrosomes.

Metaphase: The spindle microtubules are clearly visible and the chromosomes arrange themselves on the equatorial plane of the cell.

Anaphase: The sister chromatids separate at the centromeres after the breakdown of cohesin protein, and the newly formed daughter chromosomes move to the opposite poles of the cell.

Telophase: The nuclear envelope reforms around each set of daughter chromosomes. Nucleoli reappear. Spindle microtubules disintegrate.

8. In this process, one cell produces two cells that contain the same genetic information. In other words, the cells are identical to each other and to the mother cell.

11. Meiosis comprises two cell divisions, thus resulting in the production of four new cells (in many species). The chromosome number of a haploid cell produced by meiosis is half the chromosome number of the original diploid cell. Finally, the cells produced by meiosis are genetically different from the original cell and genetically different from each other.

13.

Mitosis	Meiosis
A single cell division produces two genetically identical progeny cells.	Two cell divisions usually result in four progeny cells that are not genetically identical.
Chromosome number of progeny cells and original cell remain the same.	Daughter cells are haploid and have half the chromosomal complement of the original diploid cell as a result of the separation of homologous pairs during anaphase I.
Daughter cells and original cell are genetically identical. No separation of homologous chromosomes or crossing over takes place.	Crossing over in prophase I and separation of homologous pairs during anaphase I produce daughter cells that are genetically different from each other and from the original cell.
Homologous chromosomes do not synapse.	Synapsis of homologous chromosomes takes place during prophase I.
In metaphase, individual chromosomes line up on the metaphase plate.	In metaphase I, homologous pairs of chromosomes line up on the metaphase plate. Individual chromosomes line up in metaphase II.
In anaphase, sister chromatids separate.	In anaphase I, homologous chromosomes separate. Separation of sister chromatids takes place in anaphase II.

A key difference is that mitosis produces cells genetically identical with each other and with the original cell, resulting in the orderly passage of information from one cell to its progeny. In contrast, because meiosis produces progeny that do not contain pairs of homologous chromosomes, chromosome number in the progeny is reduced from that of the original cell. Meiosis also allows for genetic variation through crossing over and the random assortment of homologs.

18. **(a)** 12 chromosomes and 24 DNA molecules; **(b)** 12 chromosomes and 24 DNA molecules; **(c)** 12 chromosomes and 24 DNA molecules; **(d)** 12 chromosomes and 24 DNA molecules; **(e)** 12 chromosomes and 12 DNA molecules; **(f)** 6 six chromosomes and 12 DNA molecules; **(g)** 12 chromosomes and 12 DNA molecules; **(h)** 6 chromosomes and 6 DNA molecules.

20. The diploid number of chromosomes is six. The figure on the left is meiosis I; the middle figure is anaphase of mitosis; the rightmost figure is anaphase II of meiosis.

22.

Event	Mitosis	Meiosis I	Meiosis II
Does crossing over take place?	No	Yes	No
What separates in anaphase?	Sister chromatids	Homologous pairs of chromosomes	Sister chromatids
What lines up on the metaphase plate?	Individual chromosomes	Homologous pairs of chromosomes	Individual chromosomes
Does cell division usually take place?	Yes	Yes	Yes
Do homologs pair?	No	Yes	No
Is genetic variation produced?	No	Yes	No

24. The progeny of an organism whose cells contain more homologous pairs of chromosomes should be expected to exhibit more variation. The number of different combinations of chromosomes that are possible in the gametes is 2^n, where n is equal to the number of homologous pairs of chromosomes. For the fruit fly with four pairs of chromosomes, the number of possible combinations is $2^4 = 16$. For *Musca domestica* with six pairs of chromosomes, the number of possible combinations is $2^6 = 64$.

25. **(a)** Metaphase I

(b) Gametes

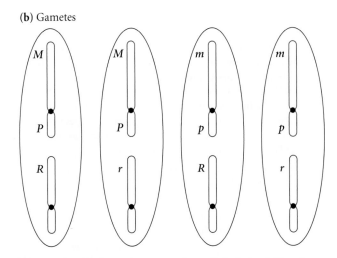

*29. Most male animals produce sperm by meiosis. In haploid male bees, meiosis will not occur, since meiosis can only occur in diploid cells. Male bees can still produce sperm but only through mitosis. Haploid cells that divide mitotically produce more haploid cells.

Chapter 3

1. Mendel was successful for several reasons. He chose to work with a plant, *Pisum sativum*, that was easy to cultivate, grew relatively rapidly, and produced many offspring whose phenotypes were easy to determine, which allowed Mendel to detect mathematical ratios of progeny phenotypes. The seven characteristics that he chose to study exhibited only a few distinct phenotypes and did not show a range of variation. Finally, by looking at each trait separately and counting the numbers of the different phenotypes, Mendel adopted a reductionist experimental approach and applied the scientific method. From his observations, he proposed hypotheses that he was then able to test empirically.

3. The principle of segregation states that an organism possesses two alleles for any particular characteristic. These alleles separate in the formation of gametes. In other words, one allele goes into each gamete. The principle of segregation is important because it explains how the genotypic ratios in the haploid gametes are produced.

8. Walter Sutton's chromosome theory of inheritance states that genes are located on chromosomes. The independent segregation of pairs of homologous chromosomes in meiosis provides the biological basis for Mendel's two principles of heredity.

9. The principle of independent assortment states that alleles at different loci segregate independently of one another. The principle of independent assortment is an extension of the principle of segregation: the principle of segregation states that the two alleles at a locus separate; according to the principle of independent assortment, when these two alleles separate, their separation is independent of the separation of alleles at other loci.

14. **(a)** The parents are *RR* (orange fruit) and *rr* (cream fruit). All the F_1 are *Rr* (orange). The F_2 are 1 *RR* : 2 *Rr* : 1 *rr* and have an orange-to-cream phenotypic ratio of 3 : 1.
    (b) Half of the progeny are homozygous for orange fruit (*RR*) and half are heterozygous for orange fruit (*Rr*).
    (c) Half of the progeny are heterozygous for orange fruit (*Rr*) and half are homozygous for cream fruit (*rr*).

15. (a) Female parent is $i^B i^B$; male parent is $I^A i^B$.
(b) Both parents are $i^B i^B$.
(c) Male parent is $i^B i^B$; female parent is $I^A I^A$ or, possibly, $I^A i^B$, but a heterozygous female in this mating is unlikely to have produced eight blood-type-A kittens owing to chance alone.
(d) Both parents are $I^A i^B$.
(e) Either both parents are $I^A I^A$ or one parent is $I^A I^A$ and the other parent is $I^A i^B$. The blood type of the offspring does not allow a determination of the precise genotype of either parent.
(f) Female parent is $i^B i^B$; male parent is $I^A i^B$.

16. (a) Sally (Aa), Sally's mother (Aa), Sally's father (aa), and Sally's brother (aa); **(b)** $\frac{1}{2}$; **(c)** $\frac{1}{2}$.

17. Use h for the hairless allele and H for the dominant allele for the presence of hair. Because H is dominant to h, a rat terrier with hair could be either homozygous (HH) or heterozygous (Hh). To determine which genotype is present in the rat terrier with hair, cross this dog with a hairless rat terrier (hh). If the terrier with hair is homozygous (HH), then no hairless offspring will be produced by the testcross. However, if the terrier is heterozygous (Hh), then $\frac{1}{2}$ of the offspring will be hairless.

20. (a) $\frac{1}{18}$; **(b)** $\frac{1}{36}$; **(c)** $\frac{11}{36}$; **(d)** $\frac{1}{6}$; **(e)** $\frac{1}{4}$; **(f)** $\frac{3}{4}$.

22. Parents:

F$_1$ generation:

F^2 generation:

23. (a) In the F$_1$ black guinea pigs (Bb), only one chromosome possesses the black allele, and so the number of copies present at each stage are: G$_1$, one black allele; G$_2$, two black alleles; metaphase of mitosis, two black alleles; metaphase I of meiosis, two black alleles; after cytokinesis of meiosis, one black allele but only in half of the cells produced by meiosis. (The remaining half will not contain the black allele.)
(b) In the F$_1$ brown guinea pigs (bb), both homologs possess the brown allele, and so the number of copies present at each stage are: G$_1$, two brown alleles; G$_2$, four brown alleles; metaphase of mitosis, four brown alleles; metaphase I of meiosis, four brown alleles; metaphase II, two brown alleles; after cytokinesis of meiosis, one brown allele.

25. (a) $\frac{1}{2}$ (Aa) $\times$ $\frac{1}{2}$ (Bb) $\times$ $\frac{1}{2}$ (Cc) $\times$ $\frac{1}{2}$ (Dd) $\times$ $\frac{1}{2}$ (Ee) $= \frac{1}{32}$
(b) $\frac{1}{2}$ (Aa) $\times$ $\frac{1}{2}$ (bb) $\times$ $\frac{1}{2}$ (Cc) $\times$ $\frac{1}{2}$ (dd) $\times$ $\frac{1}{4}$ (ee) $= \frac{1}{64}$
(c) $\frac{1}{4}$ (aa) $\times$ $\frac{1}{2}$ (bb) $\times$ $\frac{1}{4}$ (cc) $\times$ $\frac{1}{2}$ (dd) $\times$ $\frac{1}{4}$ (ee) $= \frac{1}{256}$

(d) No offspring having this genotype. The Aa Bb Cc dd Ee parent cannot contribute a D allele, the Aa bb Cc Dd Ee parent cannot contribute a B allele. Therefore, their offspring cannot be homozygous for the BB and DD gene loci.

27. (a) Gametes from Aa Bb individual:

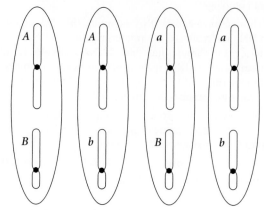

Gametes from aa bb individual:

(b) Progeny at G$_1$:

Progeny at G$_2$:

Progeny at prophase of mitosis:

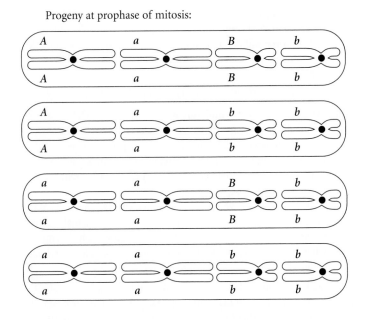

The order of chromosomes on the metaphase plate can vary.

28. (a) The burnsi × burnsi cross produced both burnsi and pipiens offspring, suggesting that the parents were heterozygous with each possessing a *burnsi* allele and a *pipiens* allele. The cross also suggests that the *burnsi* allele is dominant over the *pipiens* allele. The progeny of the burnsi × pipiens crosses suggest that each of the crosses was between a homozygous recessive frog (pipiens) and a heterozygous dominant frog (burnsi). The results of both crosses are consistent with the brunsi phenotype being recessive to the pipiens phenotype.
(b) Let P represent the *burnsi* allele and p represent the *pipiens* allele.

$$\text{burnsi } (Pp) \times \text{burnsi } (Pp)$$
$$\text{burnsi } (Pp) \times \text{pipiens } (pp)$$
$$\text{burnsi } (Pp) \times \text{pipiens } (pp)$$

(c) For the burnsi × burnsi ($Pp \times Pp$) cross, we would expect a phenotypic ratio of 3 : 1 in the offspring. A chi-square test to evaluate the fit of the observed numbers of progeny with an expected 3 : 1 ratio gives a chi-square value of 2.706 with 1 degree of freedom. The probability associated with this chi-square value is between 0.1 and 0.05, indicating that the differences between what we expected and what we observed could have been generated by chance.
   For the first burnsi × pipiens ($Pp \times pp$) cross, we would expect a phenotypic ratio of 1 : 1. A chi-square test comparing observed and expected values yields $\chi^2 = 1.78$, df = 1, $P > 0.05$. For the second burnsi × pipiens ($Pp \times pp$) cross, we would expect a phenotypic ratio of 1 : 1. A chi-square test of the fit of the observed numbers with those expected with a 1 : 1 ratio yields $\chi^2 = 0.46$, df = 1, $P > 0.05$. Thus, all three crosses are consistent with the predication that burnsi is dominant over pipiens.

29. (a) For the cross between a heterozygous F_1 plant (*Cc Ff*) and a homozygous recessive plant (*cc ff*), we would expect a phenotypic ratio of 1 : 1 : 1 : 1 for the different phenotypic classes. A chi-square test comparing the fit of the observed data with the expected 1 : 1 : 1 : 1 ratio yields a chi-square value of 35 with df = 3 and $P < 0.005$.

(b) From the chi-square value, it is unlikely that chance produced the differences between the observed and expected ratios, indicating that the progeny do not occur in a 1 : 1 : 1 : 1 ratio.
(c) The number of plants with the *cc ff* genotype is much less than expected. The *cc ff* genotype is possibly sublethal. In other words, California poppies with the homozygous recessive genotypes are possibly less viable than the other possible genotypes.

31. (a) Autosomal dominant. The trait must be autosomal because affected males pass the trait to both sons and daughters. It is dominant because it does not skip generations, all affected individuals have affected parents, and it is extremely unlikely that multiple unrelated individuals mating into the pedigree are carriers of a rare trait.
(b) Autosomal recessive. Unaffected parents produced affected progeny, and so the trait is recessive. The affected daughter must have inherited recessive alleles from both unaffected parents, and so the trait must be autosomal. If it were X-linked, her father would show the trait.

33. The first geneticist has identified an allele for obesity that he believes to be recessive. Let's call his allele for obesity o_1 and the normal allele O_1. On the basis of the crosses that the geneticist performed, the allele for obesity appears to be recessive.

Cross 1 with possible genotype:

$$\text{Obese } (o_1 o_1) \times \text{Normal } (O_1 O_1)$$
$$\downarrow$$

F_1 All normal ($O_1 o_1$)

Cross 2 with possible genotypes:

F_1 Normal ($O_1 o_1$) × Normal ($O_1 o_1$)
$$\downarrow$$

F_2 8 normal ($O_1 O_1$ and $O_1 o_1$)

      2 obese ($o_1 o_1$)

Cross 3 with possible genotypes:

$$\text{Obese } (o_1 o_1) \times \text{Obese } (o_1 o_1)$$
$$\downarrow$$

F_1 All obese ($o_1 o_1$)

   Let's call the second geneticist's allele for obesity o_2 and the normal allele O_2. The cross between obese mice from the two laboratories produced only normal mice. The alleles for obesity from both laboratories are recessive. However, they are located at different gene loci. Essentially, the obese mice from the different laboratories have separate obesity genes that are independent of one another.
   The likely genotypes of the obese mice are as follows:

$$\text{Obese mouse 1 } (o_1 o_1\ O_2 O_2) \times \text{Obese mouse 2 } (O_1 O_1\ o_2 o_2)$$
$$\downarrow$$

F_1 All normal ($O_1 o_1\ O_2 o_2$)

Chapter 4

4. Males show the phenotypes of all X-linked traits, regardless of whether the X-linked allele is normally recessive or dominant. Males inherit X-linked traits from their mothers, pass X-linked traits to all of their daughters, and, through their daughters, to their daughters' descendants, but not to their sons or their sons' descendants.

6. In incomplete dominance, the phenotype of the heterozygote is intermediate between the phenotypes of the two homozygotes. In codominance, both alleles are expressed and both phenotypes are manifested simultaneously.

7. In incomplete penetrance, the expected phenotype of a particular genotype is not expressed. Environmental factors, as well as the effects of other genes, may alter the phenotypic expression of a particular genotype.

9. A complementation test is used to determine whether two different recessive mutations are at the same locus (are allelic) or at different loci. The two mutations are introduced into the same individual organism by crossing homozygotes for each of the mutants. If the progeny show a mutant phenotype, then the mutations are allelic (at the same locus). If the progeny show a wild-type (dominant) phenotype, then the mutations are at different loci and are said to complement each other because each of the mutant parents can supply a functional copy (or dominant allele) of the gene mutated in the other parent.

10. Cytoplasmically inherited traits are encoded by genes in the cytoplasm. Because the cytoplasm is usually inherited from a single (most often the female) parent, reciprocal crosses do not show the same results. Cytoplasmically inherited traits often show great variability because different egg cells (female gametes) may have differing proportions of cytoplasmic alleles owing to random sorting of mitochondria (or plastids in plants).

11. (a) Female; (b) male; (c) male, sterile; (d) female; (e) female; (f) metafemale; (g) intersex; (h) metamale, sterile; (i) metamale.

12. (a) Yes; (b) yes; (c) no; (d) no.

13. (a) F_1: $\frac{1}{2} X^+Y$ (gray males), $\frac{1}{2} X^+X^y$ (gray females); F_2: $\frac{1}{4} X^+Y$ (gray males), $\frac{1}{4} X^yY$ (yellow males), $\frac{1}{4} X^+X^y$ (gray females), $\frac{1}{4} X^+X^+$ (gray females).
(b) F_1: $\frac{1}{2} X^yY$ (yellow males), $\frac{1}{2} X^+X^y$ (gray females); F_2: $\frac{1}{4} X^+Y$ (gray males), $\frac{1}{4} X^yY$ (yellow males), $\frac{1}{4} X^+X^y$ (gray females), $\frac{1}{4} X^yX^y$ (yellow females).

14. Because color blindness is a recessive trait, the color-blind daughter must be homozygous recessive. Because the color blindness is X-linked, John has grounds for suspicion. Normally, their daughter would have inherited John's X chromosome. Because John is not color blind, he could not have transmitted an X chromosome with a color-blind allele to his daughter.

A remote alternative possibility is that the daughter is XO, having inherited a recessive color-blind allele from her mother and no sex chromosome from her father. In that case, the daughter would have Turner syndrome. A new X-linked color-blind mutation also is possible, albeit even less likely.

If Cathy had a color-blind son, then John would have no grounds for suspicion. The son would have inherited John's Y chromosome and the color-blind X chromosome from Cathy.

15. (a) X-linked recessive; (b) $\frac{1}{4}$; (c) $\frac{1}{2}$.

16. Because Bob must have inherited the Y chromosome from his father, and his father has normal color vision, a nondisjunction event in the paternal lineage cannot account for Bob's genotype. Bob's mother must be heterozygous X^+X^c because she has normal color vision, and she must have inherited a color-blind X chromosome from her color-blind father. For Bob to inherit two color-blind X chromosomes from his mother, the egg must have

arisen from a nondisjunction in meiosis II. In meiosis I, the homologous X chromosomes separate, and so one cell has the X^+ chromosome and the other has X^c. The failure of sister chromatids to separate in meiosis II would then result in an egg with two copies of X^c.

19. (a) 1; (b) 0; (c) 0; (d) 1; (e) 1; (f) 2; (g) 0; (h) 2; (i) 3.

20. (a) F_1: all males have miniature wings and red eyes ($X^mY\,s^+s$), and all females have long wings and red eyes ($X^+X^m\,s^+s$). F_2: $\frac{3}{16}$ male, normal, red; $\frac{1}{16}$ male, normal, sepia; $\frac{3}{16}$ male, miniature, red; $\frac{1}{16}$ male, miniature, sepia; $\frac{3}{16}$ female, normal, red; $\frac{1}{16}$ female, normal, sepia; $\frac{3}{16}$ female, miniature, red; $\frac{1}{16}$ female, miniature, sepia.
(b) F_1: all females have long wings and red eyes ($X^+\,X^m\,s^+s$), and all males have long wings and red eyes ($X^+Y\,s^+s$). F_2: $\frac{3}{16}$ males, long wings, red eyes; $\frac{1}{16}$ males, long wings, sepia eyes; $\frac{3}{16}$ males, mini wings, red eyes; $\frac{1}{16}$ males, mini wings, sepia eyes; $\frac{3}{8}$ females, long wings, red eyes; $\frac{1}{8}$ long wings, sepia eyes.

21. (a) The results of the crosses indicate that cremello and chestnut are pure-breeding traits (homozygous). Palomino is a hybrid trait (heterozygous) that produces a 2 : 1 : 1 ratio when palominos are crossed with each other. The simplest hypothesis consistent with these results is incomplete dominance, with palomino as the phenotype of the heterozygotes resulting from chestnuts crossed with cremellos.
(b) Let C^B = chestnut, C^W = cremello. The parents and offspring of these crosses have the following genotypes: chestnut = C^BC^B; cremello = C^WC^W; palomino = C^BC^W.

22. (a) $\frac{1}{2} L^ML^M$ (type M), $\frac{1}{2} L^ML^N$ (type MN); (b) all L^NL^N (type N); (c) $\frac{1}{2} L^ML^N$ (type MN), $\frac{1}{4} L^ML^M$ (type M), $\frac{1}{4} L^NL^N$ (type N); (d) $\frac{1}{2} L^ML^N$ (type MN), $\frac{1}{2} L^NL^N$ (type N); (e) all L^ML^N (type MN).

23. (a) The 2 : 1 ratio in the progeny of two spotted hamsters suggests lethality, and the 1 : 1 ratio in the progeny of a spotted hamster and a hamster without spots indicates that spotted is a heterozygous phenotype. If S and s represent the locus for white spotting, spotted hamsters are Ss and solid-colored hamsters are ss. One-quarter of the zygotes expected from a mating of two spotted hamsters are SS, embryonic lethal, and missing from the progeny, resulting in the 2 : 1 ratio of spotted to solid progeny.
(b) Because spotting is a heterozygous phenotype, it should not be possible to obtain Chinese hamsters that breed true for spotting.

26. (a) All walnut ($Rr\,Pp$); (b) $\frac{1}{4}$ walnut ($Rr\,Pp$), $\frac{1}{4}$ rose ($Rr\,pp$), $\frac{1}{4}$ pea ($rr\,Pp$), $\frac{1}{4}$ single ($rr\,pp$); (c) $\frac{9}{16}$ walnut ($R\_\,P\_$), $\frac{3}{16}$ rose ($R\_\,pp$), $\frac{3}{16}$ pea ($rr\,P\_$), $\frac{1}{16}$ single ($rr\,pp$); (d) $\frac{3}{4}$ rose ($R\_\,pp$), $\frac{1}{4}$ single ($rr\,pp$); (e) $\frac{1}{4}$ walnut ($Rr\,Pp$), $\frac{1}{4}$ rose ($Rr\,pp$), $\frac{1}{4}$ pea ($rr\,Pp$), $\frac{1}{4}$ single ($rr\,pp$); (f) $\frac{1}{2}$ rose ($Rr\,pp$), $\frac{1}{2}$ single ($rr\,pp$).

29. Let A and B represent the two loci. The F_1 heterozygotes are $Aa\,Bb$. The F_2 are: $A\_\,B\_$ disc-shaped, $A\_\,bb$ spherical, $aa\,B\_$ spherical, $aa\,bb$ long.

Chapter 5

1. Recombination means that meiosis generates gametes with allelic combinations that differ from the original gametes inherited by an organism. Recombination may be caused by the independent assortment of loci on different chromosomes or by a physical crossing over between two loci on the same chromosome.

4. For genes in coupling configuration, two wild-type alleles are on the same chromosome and the two mutant alleles are on the homologous chromosome. For genes in repulsion, the wild-type allele of one gene and the mutant allele of the other gene are on the same chromosome, and vice versa on the homologous chromosome. The two arrangements have opposite effects on the results of a cross. For genes in coupling configuration, most of the progeny will be either wild type for both genes or mutant for both genes, with relatively few that are wild type for one gene and mutant for the other. For genes in repulsion, most of the progeny will be mutant for only one gene and wild type for the other, with relatively few recombinants that are wild type for both or mutant for both.

7. Positive interference indicates that a crossover inhibits or interferes with the occurrence of a second crossover nearby. Negative interference suggests that a crossover event can stimulate additional crossover events in the same region of the chromosome.

8. (a) $\frac{1}{2}$ banded, yellow; $\frac{1}{2}$ unbanded, brown.

   (b) $\frac{1}{4}$ banded, yellow; $\frac{1}{4}$ banded, brown; $\frac{1}{4}$ unbanded, yellow; $\frac{1}{4}$ unbanded, brown.

   (c) 40% banded, yellow; 40% unbanded, brown; 10% banded, brown; 10% unbanded, yellow.

9. The genes are linked and have not assorted independently.

11. (a) Both plants are $Dd\ Pp$.

   (b) Yes. Map distance = 3.8 m.u.

   (c) The two plants have different coupling configurations. In plant A, the dominant alleles D and P are coupled; one chromosome is $\underline{D\quad P}$ and the other is $\underline{d\quad p}$. In plant B, they are in repulsion; its chromosomes are $\underline{D\quad p}$ and $\underline{d\quad P}$.

14.

Genotype	Body color	Eyes	Bristles	Proportion
(a) $e^+\ ro^+\ f^+$	normal	normal	normal	20%
$e^+\ ro^+\ f$	normal	normal	forked	20%
$e\ ro\ f^+$	ebony	rough	normal	20%
$e\ ro\ f$	ebony	rough	forked	20%
$e^+\ ro\ f^+$	normal	rough	normal	5%
$e^+\ ro\ f$	normal	rough	forked	5%
$e\ ro^+\ f^+$	ebony	normal	normal	5%
$e\ ro^+\ f$	ebony	normal	forked	5%
(b) $e^+\ ro^+\ f^+$	normal	normal	normal	5%
$e^+\ ro^+\ f$	normal	normal	forked	5%
$e\ ro\ f^+$	ebony	rough	normal	5%
$e\ ro\ f$	ebony	rough	forked	5%
$e^+\ ro\ f^+$	normal	rough	normal	20%
$e^+\ ro\ f$	normal	rough	forked	20%
$e\ ro^+\ f^+$	ebody	normal	normal	20%
$e\ ro^+\ f$	ebony	normal	forked	20%

15.

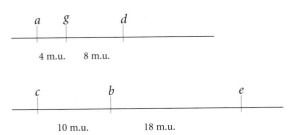

Gene f is unlinked to either of these groups, on a third linkage group.

18. (a) V is the middle gene.

   (b) The Wx–V distance = 7 m.u. and the Sh–V distance = 30 m.u. The Wx–Sh distance is the sum of these two distances, or 37 m.u.

   (c) Coefficient of coincidence = 0.80; interference = 0.20

20.

Genotype	Body	Eyes	Wings	Proportion
$b^+\ pr^+\ vg^+$	normal	normal	normal	40.7%
$b\ pr\ vg$	black	purple	vestigial	40.7%
$b^+\ pr^+\ vg$	normal	normal	vestigial	6.3%
$b\ pr\ vg^+$	black	purple	normal	6.3%
$b^+\ pr\ vg$	normal	purple	vestigial	2.8%
$b\ pr^+\ vg^+$	black	normal	normal	2.8%
$b^+\ pr\ vg^+$	normal	purple	normal	0.2%
$b\ pr^+\ vg$	black	normal	vestigial	0.2%

Chapter 6

2.

Types of matings	Outcomes
$F^+ \times F^-$	Two F^+ cells
Hfr $\times$ F^-	One F^+ cell and one F^- cell
$F' \times F^-$	Two F' cells

The F factor contains a number of genes that take part in the conjugation process, including genes necessary for the synthesis of the sex pilus. The F factor has an origin of replication that enables the factor to be replicated in the conjugation process and genes for opening the plasmid and initiating the chromosome transfer.

3. To map genes by conjugation, an Hfr strain is mixed with an F^- strain. The two strains must have different genotypes and must remain in physical contact for the transfer to take place. The conjugation process is interrupted at regular intervals. The chromosomal transfer from the Hfr strain always begins with a part of the integrated F factor and proceeds in a linear fashion. Transfer of the entire chromosome takes approximately 100 minutes. The time required for individual genes to be transferred is relative to their positions on the chromosome and the direction of transfer initiated by the F factor. Gene distances are typically mapped in minutes. The genes that are transferred by conjugation to the recipient must be incorporated into the recipient's chromosome by recombination to be expressed.

   In transformation, the relative frequency at which pairs of genes are transferred, or cotransformed, indicates the distance between the two genes. Closer gene pairs are cotransformed more frequently. As in interrupted conjugation, the donor DNA must be incorporated into the recipient cell's chromosome by recombination. Physical contact of the donor and recipient cells is not needed. The recipient cell takes up the DNA directly from the environment, and so the DNA from the donor strain must be isolated and broken up before transformation can take place.

   The transfer of DNA by transduction requires a viral vector. DNA from the donor cell is packaged into a viral protein coat. The viral particle containing the bacterial donor DNA then infects a recipient bacterial cell. The donor bacterial DNA is incorporated into the recipient cell's chromosome by recombination. Only genes that are close together on the bacterial chromosome can be cotransduced. Therefore, the rate of cotransduction, like the rate of cotransformation, is an indication of the physical distances between genes on the chromosome.

All three processes require the uptake by the recipient cell of a piece of the donor chromosome and the incorporation of some of that piece into the recipient chromosome by recombination. In all three processes, the mapping distance is calculated by measuring the frequency with which recipient cells are transformed. The processes use different methods to get donor DNA incorporated into the recipient cell.

4. Reproduction is rapid, asexual, and produces lots of progeny. Their genomes are small. They are easy to grow in the laboratory. Techniques are available for isolating and manipulating their genes. Mutant phenotypes, especially auxotrophic phenotypes, are easy to measure.

10. By using low doses of antibiotics for 5 years, Farmer Smith selected for bacteria that are resistant to the antibiotics. The doses used killed sensitive bacteria but not moderately sensitive or slightly resistant bacteria. As time passed, only resistant bacteria remained in his pigs because any sensitive bacteria were eliminated by the low doses of antibiotics.

In the future, Farmer Smith should continue to use the vitamins but should use the antibiotics only when a sick pig requires them. In this manner, he will not be selecting for antibiotic-resistant bacteria, and the chances of successful treatment of his sick pigs will be greater.

17. Number of plaques produced by $c^+ m^+ = 460$; by $c^- m^- = 460$; by $c^+ m^- = 40$ (recombinant); by $c^- m^+ = 40$ (recombinant).

18. (a)

(b)

7.1 m.u.   24.1 m.u.

(c) Coefficient of coincidence $= (6 + 5)/(0.071 \times 0.241 \times 942) = 0.68$. Interference $= 1 - 0.68 = 0.32$.

Chapter 7

1. Chromosome rearrangements:

Deletion: Loss of a part of a chromosome.

Duplication: Addition of an extra copy of a part of a chromosome.

Inversion: A part of a chromosome is reversed in orientation.

Translocation: A part of one chromosome becomes incorporated into a different (nonhomologous) chromosome.

Aneupoloidy: Loss or gain of one or more chromosomes, causing the chromosome number to deviate from $2n$ or the normal euploid complement.

Polyploidy: Gain of entire sets of chromosomes, causing the chromosome number to change from $2n$ to $3n$ (triploid), $4n$ (tetraploid), and so on.

2. The expression of some genes is balanced with the expression of other genes; the ratios of their gene products, usually proteins,

must be maintained within a narrow range for proper cell function. Extra copies of one of these genes cause that gene to be expressed at proportionately higher levels, thereby upsetting the balance of gene products.

4. A paracentric inversion does not include the centromere; a pericentric inversion includes the centromere.

7. Translocations can produce phenotypic effects if the translocation breakpoint disrupts a gene or if a gene near the breakpoint is altered in its expression because of relocation to a different chromosomal environment (a position effect).

9. Primary Down syndrome is caused by the spontaneous, random nondisjunction of chromosome 21, leading to trisomy 21. Familial Down syndrome most frequently arises as a result of a Robertsonian translocation of chromosome 21 with another chromosome, usually chromosome 14.

10. In autopolyploidy, all sets of chromosomes are derived from a single species. In allopolyploidy, the sets of chromosomes are derived from two or more different species. Autopolyploidy may arise through nondisjunction in an early $2n$ embryo or through meiotic nondisjunction that produces a gamete with extra sets of chromosomes. Allopolyploidy is usually preceded by hybridization between two different species, followed by chromosome doubling.

12. (a) Duplications; (b) polyploidy; (c) deletions; (d) inversions; (e) translocations.

13. (a) Tandem duplication of AB; (b) displaced duplication of AB; (c) paracentric inversion of DEF; (d) deletion of B; (e) deletion of FG; (f) paracentric inversion of CDE; (g) pericentric inversion of ABC; (h) duplication and inversion of DEF; (i) duplication of CDEF, inversion of EF.

16. (a)

(b)

(c)

(d)

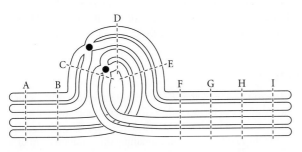

19. The high incidence of Down syndrome in Bill's family and among Bill's relatives is consistent with familial Down syndrome, caused by a Robertsonian translocation of chromosome 21. Bill and his sister, who are unaffected, are phenotypically normal carriers of the translocation and have 45 chromosomes. Their children and Bill's brother, who have Down syndrome, have 46 chromosomes. From the information given, there is no reason to suspect that Bill's wife Betty has any chromosomal abnormalities. Therefore, the statement in part *d* is most likely correct.

20. In mammals, the higher frequency of sex-chromosome aneuploids compared with that of autosomal aneuploids is due to X-chromosome inactivation and the lack of essential genes on the Y chromosome. If fish do not have X-chromosome inactivation and if both of their sex chromosomes have numerous essential genes, then the frequency of aneuploids should be similar for both sex chromosomes and autosomes.

21. **(a)** 15; **(b)** 24; **(c)** 32; **(d)** 17; **(e)** 14; **(f)** 14; **(g)** 40; **(h)** 18.

Chapter 8

1. The genetic material must (i) contain complex information, (ii) replicate or be replicated faithfully, and (iii) encode the phenotype.

3. Hershey and Chase used ^{32}P-labeled phages to infect bacteria. The progeny phage released from these bacteria emitted radioactivity from ^{32}P. The presence of the ^{32}P in the progeny phages indicated that the infecting phages had passed DNA on to the progeny phages.

4.

Nitrogen-containing base

Phosphate

Deoxyribose sugar

Deoxyguanosine 5′-phosphate (dGMP)

6.

8. Replication, transcription, and translation—the components of the central dogma of molecular biology.

9. Supercoiling arises from the overwinding (positive supercoiling) or underwinding (negative supercoiling) of the DNA double helix; from a lack of free ends, as in circular DNA molecules; when the ends of the DNA molecule are bound to proteins that prevent the ends from rotating about each other.

11. The nucleosome consists of a core particle, which contains two molecules each of histones H2A, H2B, H3, and H4, and from 145 to 147 bp of DNA that winds around the core. A chromatosome contains the nucleosome core and a molecule of histone H1.

13. Telomeres are the stable ends of the linear chromosomes in eukaryotes. They prevent degradation by exonucleases and prevent joining of the ends. Telomeres also enable the replication of the ends of the chromosome. Telomeric DNA sequences consist of repeats of a simple sequence, usually in the form of $5′–C_n(A \text{ or } T)_m$.

15. Unique-sequence DNA, present in only one or a few copies per haploid genome, constitutes most of the protein-coding sequences as well as a large number of sequences with unknown function.

   Moderately repetitive sequences, ranging from a few hundred to a few thousand base pairs in length, are present in as many as several thousand copies per haploid genome. Some moderately repetitive DNA consists of functional genes that encode rRNAs and tRNAs, but most is made up of transposable elements and remnants of transposable elements.

   Highly repetitive DNA, or satellite DNA, consists of clusters of tandem repeats of short sequences (often less than 10 base pairs in length) present in hundreds of thousands to millions of copies per haploid genome.

19. (a)

Organism and tissue	(A + G)/(C + T)	(A + T)/(C + G)
Sheep thymus	1.03	1.36
Pig liver	0.99	1.44
Human thymus	1.03	1.52
Rat bone marrow	1.02	1.36
Hen erythrocytes	0.97	1.38
Yeast	1.00	1.80
E. coli	1.04	1.00
Human sperm	1.00	1.67
Salmon sperm	1.02	1.43
Herring sperm	1.04	1.29

(b) The (A + G)/(T + C) ratio of ~1.0 is constant for these organisms. Each of them has a double-stranded genome. The percentage of purines should equal the percentage of pyrimidines in double-stranded DNA, which means that (A + G) = (C + T). The (A + T)/(C + G) ratios are not constant. The number of A–T base pairs relative to the number of G–C base pairs is unique to each organism and can vary among the different organisms.

(c) The (A + G)/(T + C) ratio is about the same for the sperm samples, as should be expected. As in part *b*, the percentage of purines should equal the percentage of pyrimindines.

21. The relations in parts *b*, *c*, *e*, *g*, and *h*.

22. Adenine = 15%; guanine = 35%; cytosine = 35%.

25. No. The flow of information predicted by the central dogma is from DNA to RNA to protein. An exception is reverse transcription, whereby RNA encodes DNA. However, biologists do not currently know of a process in which the flow of information is from proteins to DNA, which is required by the theory of the inheritance of acquired characteristics.

26. Prokaryotic chromosomes are usually circular, whereas eukaryotic chromosomes are linear. Prokaryotic chromosomes generally contain the entire genome, whereas each eukaryotic chromosome has only a part of the genome: the eukaryotic genome is divided into multiple chromosomes. Prokaryotic chromosomes are generally much smaller than eukaryotic chromosomes and have only a single origin of DNA replication, whereas eukaryotic chromosomes contain multiple origins of DNA replication. Prokaryotic chromosomes are typically condensed into nucleoids, which have loops of DNA compacted into a dense body. Eukaryotic chromosomes contain DNA packaged into nucleosomes, which are further coiled and packaged into structures of successively higher order. The condensation state of eukaryotic chromosomes varies with the cell cycle.

27. (a) 3.2×10^7; (b) 2.9×10^8.

28. Although the chemical composition of the genetic material may be different from DNA, it more than likely will have properties similar to those of DNA. As stated in this chapter, the genetic material must contain complex information, replicate or be replicated faithfully, and encode the phenotype. Even if the material on the planet is not DNA, it must meet these criteria. Additionally, the genetic material must be stable.

30. The carrier of genetic information must be stable so that the genetic information is faithfully transmitted from one generation to the next. The 3′ hydroxyl group on the ribose sugar of RNA makes RNA more reactive than DNA, which lacks a free 3′ hydroxyl group on its deoxyribose sugar. Because it is single stranded, RNA can assume a number of secondary structures and a number of different functions, decreasing its stability. The double-stranded nature of DNA, with its bases hydrogen bonded to other bases, stabilizes DNA and makes it more suitable as the carrier of genetic information.

Chapter 9

2. Meselson and Stahl grew *E. coli* cells in a medium containing the heavyisotope of nitrogen (^{15}N) for several generations. The ^{15}N was incorporated into the DNA of the *E. coli* cells. The *E. coli* cells were then switched to a medium containing the common form of nitrogen (^{14}N) and allowed to proceed through a few cycles of cellular generations. Samples of the bacteria were removed at each cellular generation. Using equilibrium density gradient centrifugation, Meselson and Stahl were able to distinguish DNA fragments that contained only ^{15}N from those that contained only ^{14}N or a mixture of ^{15}N and ^{14}N because DNA containing the ^{15}N isotope is "heavier." The more ^{15}N in a DNA fragment, the lower the fragment will sink. DNA from cells grown in the ^{15}N medium produced only a single band at the expected position during centrifugation. After one round of replication in the ^{14}N medium, a single band was present after centrifugation, but the band was located at an intermediate position between that of a DNA band containing only ^{15}N and that of a DNA band containing only ^{14}N. After two rounds of replication, two bands of DNA were present. One band was located at an intermediate position between that of a DNA band containing only ^{15}N and that of a DNA band containing only ^{14}N, whereas the other band was at a position expected for DNA containing only ^{14}N. These results were consistent with the predictions of semiconservative replication and incompatible with the predictions of conservative and dispersive replication.

3.

6. Deoxyadenosine triphosphate, deoxyguanosine triphosphate, deoxycytosine triphosphate, and deoxythymidine triphosphate.

9. Primase, an RNA polymerase, synthesizes the short RNA molecules, or primers, that provide a 3′-OH group to which DNA polymerase can attach deoxyribonucleotides in replication initiation.

12. Ways in which DNA replication is similar in eukaryotes and bacteria:

Replication is semiconservative.

Replication origins serve as starting points for replication.

A short segment of RNA called a primer provides a 3′-OH group for DNA polymerases to begin the synthesis of the new strands.

Synthesis is in the 5′-to-3′ direction.

The template strand is read in the 3′-to-5′ direction.

Deoxyribonucleoside triphosphates are the substrates.

Replication is continuous on the leading strand and discontinuous on the lagging strand.

Ways in which eukaryotic DNA replication differs from bacterial replication:

There are multiple origins of replications per chromosome.

Several different DNA polymerases have different functions.

Immediately after DNA replication, nucleosomes are assembled.

15. The strands of the novel double-stranded nucleic acid must be parallel to each other, unlike antiparallel double-stranded DNA on Earth. Replication by *E. coli* DNA polymerases can proceed only in the 5′-to-3′ direction, which requires the template to be read in the 3′-to-5′ direction. If replication is continuous on both strands, the two strands must have the same direction and be parallel.

18.

19. (a) More errors in replication; (b) primers would not be removed; (c) primers that have been removed would not be replaced.

Chapter 10

1.

Deoxyribonucleotide   Ribonucleotide

RNA and DNA are polymers of nucleotides that are held together by phosphodiester bonds. An RNA nucleotide contains ribose, whereas a DNA nucleotide contains deoxyribose. RNA contains uracil but not thymine. DNA contains thymine but not uracil. RNA is typically single stranded even though RNA molecules can pair with other complementary sequences. DNA molecules are almost always double stranded.

3.

6. Transcription and replication are similar in that both use DNA templates; synthesize molecules in the 5′-to-3′ direction; synthesize molecules that are antiparallel and complementary to their templates; use nucleotide triphosphates as substrates; and use complexes of proteins and enzymes necessary for catalysis.

Transcription differs from replication in its unidirectional synthesis of only a single strand of nucleic acid; its initiation does not require a primer; it is subject to numerous regulatory mechanisms; and each gene is transcribed separately.

Replication differs from transcription in its bidirectional synthesis of two strands of nucleic acid and its use of replication origins.

7. The 5′ untranslated region, which contains the Shine–Dalgarno sequence; the protein-encoding region; and the 3′ untranslated region.

9. The 5′ end of eukaryotic mRNA is modified by the addition of the 5′ cap. The cap consists of an extra guanine nucleotide (linked 5′ to 5′ to the mRNA molecule) that is methylated at position 7 of the base and at adjacent bases whose sugars are methylated at the 2′-OH group.

10. The spliceosome consists of five snRNPs. The splicing of pre-mRNA nuclear introns takes place within the spliceosome.

12. The different types of processing are: addition of the 5′ cap to the 5′ end of the pre-mRNA; cleavage of the 3′ end of a site downstream of the AAUAAA consensus sequence of the last exon; addition of the poly(A) tail to the 3′ end of the mRNA immediately after cleavage; and removal of the introns (splicing).

18. Similarities: Both use DNA templates; the DNA template is read in the 3′-to-5′ direction; the complementary strand is synthesized in the 5′-to-3′ direction, which is antiparallel to the template; both use triphosphates as substrates; and their actions are enhanced by accessory proteins.

Differences: RNA polymerases use ribonucleoside triphosphates as substrates, whereas DNA polymerases use deoxyribonucleoside triphophates; DNA polymerases require a primer that provides an available 3′-OH group at which synthesis begins, whereas RNA polymerases do not require primers to begin synthesis; and RNA polymerases synthesize a copy of only one of the DNA strands, whereas DNA polymerases synthesize copies of both strands.

20. (a) Would probably affect the −10 consensus sequence, which would most likely decrease transcription; (b) could affect the binding of the sigma factor to the promoter and downregulate transcription, reducing or inhibiting transcription; (c) unlikely to have any effect on transcription; (d) would have little effect on transcription.

22.

24. The large size of the gene that encodes dystrophin is likely due to the presence of many intervening sequences, or introns, within the coding region of the gene. Excision of the introns through RNA splicing yields the mature mRNA that encodes the dystrophin protein.

25.

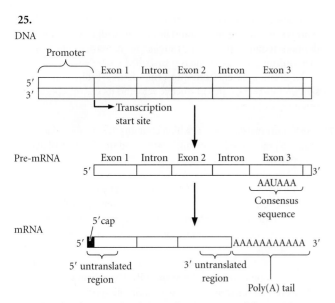

(a) The 5′ untranslated region lies upstream of the translation start site. In bacteria, the ribosome-binding site, or Shine–Dalgarno sequence, is within the 5′ untranslated region. However, eukaryotic mRNA does not have the equivalent sequence, and a eukaryotic ribosome binds at the 5′ cap of the mRNA molecule.
(b) The promoter is the DNA sequence recognized and bound by the transcription apparatus to initiate transcription.
(c) The AAUAAA consensus sequence, which lies downstream of the coding region of the gene, determines the location of the 3′ cleavage site in the pre-mRNA molecule.
(d) The transcription start site begins the coding region of the gene and is located from 25 to 30 nucleotides downstream of the TATA box.
(e) The 3′ untranslated region is a sequence of nucleotides that is located at the 3′ end of the mRNA and is not translated into proteins. However, it does affect the stability and translation of the mRNA.
(f) Introns are noncoding sequences of DNA that intervene within coding regions of a gene.
(g) Exons are coding regions of a gene.
(h) A poly(A) tail is added to the 3′ end of the pre-mRNA; it affects mRNA stability and the binding of the ribosome to the mRNA.
(i) The 5′ cap functions in the initiation of translation and in mRNA stability.

Chapter 11

3. In synonymous codons that differ only at the third nucleotide position, the "wobble" and nonstandard base pairing with the anticodons will likely result in the correct amino acid being inserted in the protein.

4. (a) The reading frame refers to how the nucleotides in a nucleic acid molecule are grouped into codons, with each codon containing three nucleotides. Any sequence of nucleotides has three potential reading frames that have completely different sets of codons.
(b) In an overlapping code, a single nucleotide is included in more than one codon. The result for a sequence of nucleotides is that more than one type of polypeptide can be encoded within that sequence.
(c) In a nonoverlapping code, a single nucleotide is part of only one codon, which results in the production of a single type of polypeptide from one polynucleotide sequence.

(d) An initiation codon establishes the appropriate reading frame and specifies the first amino acid of the protein chain. Typically, the initiation codon is AUG; however, GUG and UUG also can serve as initiation codons.
(e) The termination codon signals the termination, or end, of translation and the end of the protein molecule. The three types of termination codons—UAA, UAG, and UGA—are also referred to as stop codons or nonsense codons. These codons do not encode amino acids.
(f) A sense codon is a group of three nucleotides that encode an amino acid. Sixty-one sense codons encode the 20 amino acids commonly found in proteins.
(g) A nonsense codon, or termination codon, signals the end of translation. Nonsense codons do not encode amino acids.
(h) In a universal code, each codon specifies the same amino acid in all organisms. The genetic code is nearly universal but not completely. Most of the exceptions are in mitochondrial genes.
(i) Although most codons are universal (or nearly universal) in that they specify the same amino acids in almost all organisms, there are exceptions in which a codon has different meanings in different organisms.

6. For each of the 20 different amino acids commonly found in proteins, a corresponding aminoacyl-tRNA synthetase covalently links the amino acid to the correct tRNA molecule.

11. A number of antibiotics bind the ribosome and inhibit protein synthesis at different steps in translation. Some antibiotics, such as streptomycin, bind to the small subunit and inhibit the initiation of translation. Other antibiotics, such as chloramphenicol, bind to the large subunit and block the elongation of the peptide by preventing peptide-bond formation.

12. (a) 1; **(b)** 2; **(c)** 3; **(d)** 3; **(e)** 4.

14. (a) amino–fMet-Phe-Lys-Phe-Lys-Phe–carboxyl
(b) amino–fMet-Tyr-Ile-Tyr-Ile–carboxyl
(c) amino–fMet-Asp-Glu-Arg-Phe-Leu-Ala–carboxyl
(d) amino–fMet-Gly–carboxyl (The stop codon UAG follows the codon for glycine.)

16. There are two possible sequences:

mRNA:	5′–AUGUGGCAU–3′
DNA template:	3′–TACACCGTA–5′
DNA nontemplate:	5′–ATGTGGCAT–3′

mRNA:	5′–AUGUGGCAC–3′
DNA template:	3′–TACACCGTG–5′
DNA nontemplate:	5′–ATGTGGCAC–3′

22. (a) The lack of IF-1 would decrease the amount of protein synthesized. IF-1 promotes the dissociation of the large and small ribosomal subunits. The initiation of translation requires a free small subunit. The absence of IF-1 would reduce the rate of initiation because more of the small ribosomal subunits would remain bound to the large ribosomal subunits.
(b) No translation would take place. IF-2 is necessary for the initiation of translation. The lack of IF-2 would prevent fMet-tRNA$_f^{met}$ from being delivered to the small ribosomal subunit, thus blocking translation.
(c) Although translation would be initiated by the delivery of methionine to the ribosome–mRNA complex, no other amino acids would be delivered to the ribosome. EF-Tu, which binds to GTP and the charged tRNA, is necessary for elongation. This

three-part complex enters the A site of the ribosome. If EF-Tu is not present, the charged tRNA will not enter the A site, thus stopping translation.

(d) EF-G is necessary for the translocation (movement) of the ribosome along the mRNA in a 5′-to-3′ direction. When a peptide bond has formed between Met and Pro, the lack of EF-G would prevent the movement of the ribosome along the mRNA, and so no new codons would be read. The formation of the dipeptide Met-Pro does not not require EF-G.

(e) The release factors RF$_1$ and RF$_2$ recognize the stop codons and bind to the ribosome at the A site. They then interact with RF$_3$ to promote cleavage of the peptide from the tRNA at the P site. The absence of the release factors would prevent the termination of translation at the stop codon.

(f) ATP is required for tRNAs to be charged with amino acids by aminoacyl-tRNA synthetases. Without ATP, the charging would not take place, and no amino acids will be available for protein synthesis.

(g) GTP is required for the initiation, elongation, and termination of translation. If GTP is absent, protein synthesis will will not take place.

25. (a) The results suggest that, to initiate translation, the ribosome scans the mRNA to find the appropriate start sequence.

(b) The initiation of translation in bacteria requires the 16S RNA of the small ribosomal subunit to interact with the Shine–Dalgarno sequence. This interaction serves to line up the ribosome over the start codon. If the anticodon has been changed such that the start codon cannot be recognized, then protein synthesis is not likely to take place.

Chapter 12

2. Gene structure, transcription, mRNA processing, mRNA stability, translation, and protein modification.

3.

5. The *lac* operon consists of three structural genes—*lacZ, lacY,* and *lacA*. The *lacZ* gene encodes the enzyme β-galactosidase, which cleaves the disaccharide lactose into galactose and glucose and converts lactose into allolactose. The *lacY* gene encodes lactose permease, an enzyme necessary for the passage of lactose through the *E. coli* cell membrane. The *lacA* gene encodes the enzyme thiogalactoside transacetylase, whose function in lactose metabolism has not yet been determined. All three genes have an overlapping promoter and operator region in common. Upstream from the lactose operon, the *lacI* gene encodes the *lac* operon repressor, which binds at the operator region and inhibits the transcription of the *lac* operon by preventing RNA polymerase from successfully initiating transcription.

When lactose is present in the cell, the enzyme β-galactosidase converts some of it into allolactose, which binds to the *lac* repressor, altering its shape and reducing the repressor's affinity for the operator. Because the allolactose-bound repressor does not bind to the operator, RNA polymerase can initiate transcription of the *lac* structural genes from the *lac* promoter.

7. Changes in chromatin structure can result in the repression or stimulation of gene expression. The acetylation of histone proteins increases transcription. The reverse reaction,

deacetylation, restores repression. Chromatin-remodeling complexes bind directly to the DNA, altering chromatin structure without acetylating histone proteins and allowing transcription to be initiated by making the promoters accessible to transcription factors. The methylation of DNA sequences represses transcription. The demethylation of DNA sequences often increases transcription.

10. An enhancer is a DNA sequence that, when bound to transcriptional activator proteins, can affect the transcription of a distant gene. Transcription at a distant gene is affected when the DNA sequence between the gene's promoter and the enhancer loops out, bringing the promotor and the enhancer close together and allowing the transcriptional activator proteins to directly interact with the basal transcription apparatus at the promoter, which stimulates transcription.

11. The total amount of protein synthesized depends on the amount of mRNA available for translation. The amount of available mRNA depends on the rates of mRNA synthesis and degradation. Less-stable mRNAs degrade faster than stable mRNAs, and so fewer copies of the mRNA are available as templates for translation.

The 5′ cap, the 3′ poly(A) tail, the 5′ UTR, the 3′ UTR, and the coding region in an mRNA molecule affect its stability. Poly(A)-binding proteins bind at the 3′ poly(A) tail. These proteins contribute to the stability of the tail and protect the 5′ cap through direct interaction. When a critical number of adenine nucleotides have been removed from the tail, the protection is lost and the 5′ cap is removed. The removal of the 5′ cap enables 5′-to-3′ nucleases to degrade the mRNA.

12. Through Slicer activity, which cleaves mRNA sequences; through the binding of miRNAs to complementary regions in mRNA, which prevents translation; and through transcriptional silencing, in which siRNAs play a role in altering chromatin structure.

13. In eukaryotic cells, the coding regions of pre-mRNA are interrupted by introns—noncoding regions that may be much longer than the coding regions. The splicing of pre-mRNA to remove these noncoding regions is a point at which gene expression can be regulated. In most prokaryotic cells, coding regions are not interrupted by introns. In eukaryotic cells, chromatin structure plays a role in gene regulation. The process of transcription initiation is more complex in eukaryotic cells than in bacterial cells. In eukaryotes, initiation requires RNA polymerase, general transcription factors, and transcriptional activators. Bacterial RNA polymerase is either blocked or stimulated by the actions of regulatory proteins. Finally, in eukaryotes, activator proteins may bind to enhancers at a great distance from the promoter and structural gene. Distant enhancers are less common in bacterial cells.

The regulation of both bacterial genes and eukaryotic genes requires the action of protein repressors and protein activators. Cascades of gene regulation in which the activation of one set of genes affects another set of genes are found in both eukaryotes and bacteria. The regulation of gene expression at the transcriptional level is common to both types of cells.

14. (a) Inactive repressor; (b) active activator; (c) active repressor; (d) inactive activator.

15. (a) The operon will never be turned off, and transcription will take place all the time.

(b) The result will be constitutive expression, and transcription will take place all the time.

17. RNA polymerase will bind the *lac* promoter poorly, significantly decreasing the transcription of the *lac* structural genes.

20.

	Lactose absent		Lactose present	
Genotype of strain	**β-Galactosidase**	**Permease**	**β-Galactosidase**	**Permease**
lacI⁺ lacP⁺ lacO⁺ lacZ⁺ lacY⁺	−	−	+	+
lacI⁻ lacP⁺ lacO⁺ lacZ⁺ lacY⁺	+	+	+	+
lacI⁺ lacP⁺ lacOᶜ lacZ⁺ lacY⁺	+	+	+	+
lacI⁻ lacP⁺ lacO⁺ lacZ⁺ lacY⁻	+	−	+	−
lacI⁻ lacP⁻ lacO⁺ lacZ⁺ lacY⁺	−	−	−	−
lacI⁺ lacP⁺ lacO⁺ lacZ⁻ lacY⁺ / *lacI⁻ lacP⁺ lacO⁺ lacZ⁺ lacY⁻*	−	−	+	+
lacI⁻ lacP⁺ lacOᶜ lacZ⁺ lacY⁺ / *lacI⁺ lacP⁺ lacO⁺ lacZ⁻ lacY⁻*	+	+	+	+
lacI⁻ lacP⁺ lacO⁺ lacZ⁺ lacY⁻ / *lacI⁺ lacP⁻ lacO⁺ lacZ⁻ lacY⁺*	−	−	+	−
lacI⁺ lacP⁻ lacOᶜ lacZ⁻ lacY⁺ / *lacI⁻ lacP⁺ lacO⁺ lacZ⁺ lacY⁻*	−	−	+	−
lacI⁺ lacP⁺ lacO⁺ lacZ⁺ lacY⁺ / *lacI⁺ lacP⁺ lacO⁺ lacZ⁺ lacY⁺*	−	−	+	+
lacIˢ lacP⁺ lacO⁺ lacZ⁺ lacY⁻ / *lacI⁺ lacP⁺ lacO⁺ lacZ⁻ lacY⁺*	−	−	−	−
lacIˢ lacP⁻ lacO⁺ lacZ⁻ lacY⁺ / *lacI⁺ lacP⁺ lacO⁺ lacZ⁺ lacY⁺*	−	−	−	−

22. The *lacI* gene encodes the *lac* repressor protein, which can diffuse within the cell and attach to any operator. It can therefore affect the expression of genes on the same molecule or on different molecules of DNA. The *lacO* gene encodes the operator. It affects the binding of RNA polymerase to DNA, and therefore affects the expression of genes only on the same molecule of DNA.

23. (a) Repressible; (b) *B* is the regulator gene, *D* is the promoter, *A* is the structural gene for enzyme 1, and *C* is the structural gene for enzyme 2.

Chapter 13

1. Transition mutations result from purine-to-purine or pyramidine-to-pyramidine base substitutions. Transversions result from purine-to-pyramidine or pyramidine-to-purine base substitutions. Transition mutations are more common because spontaneous mutations typically result in transition mutations rather than transversions.

2. Expanding trinucleotide repeats result when a DNA insertion mutation increases the number of copies of a trinucleotide repeat sequence. The increase may be due to errors in replication or to unequal recombination.

5. Two types of events have been proposed that could lead to DNA replication errors: (i) mispairing due to tautomeric shifts in nucleotides and (ii) mispairing through wobble or flexibility of the DNA molecule. Mispairing through wobble caused by flexibility in the DNA helix is now thought to be the most likely cause.

7. Base analogs have structures similar to those of the nucleotides and can be incorporated into the DNA in the course of replication.

Many analogs tend to mispair, which can lead to mutations. DNA replication is required for the base-analog-induced mutations to be incorporated into the DNA.

10. Most transposable elements have terminal inverted repeats and are flanked by short direct repeats. Replicative transposons use a copy-and-paste mechanism in which the transposon is replicated and inserted in a new location, leaving the original transposon in place. Nonreplicative transposons use a cut-and-paste mechanism in which the original transposon is excised and moved to a new location.

11. A retrotransposon is a transposable element that relocates through an RNA intermediate. First, it is transcribed into RNA. Then, a reverse transcriptase encoded by the retrotransposon reverse transcribes the RNA template into a DNA copy of the transposon, which then integrates into a new location in the host genome.

12. Mismatch repair: Repairs replication errors that result from base-pair mismatches. Mismatch-repair enzymes recognize distortions in the DNA structure due to mispairing and detect the newly synthesized strand by its lack of methylation. The distorted segment is excised, and DNA polymerase and DNA ligase fill in the gap.

Direct repair: Repairs DNA damage by directly changing the damaged nucleotide back into its original structure.

Base-excision repair: Excises the damaged base and then replaces the entire nucleotide.

Nucleotide-excision repair: Relies on repair enzymes to recognize distortions of the DNA double helix. These enzymes excise a damaged region by cleaving phosphodiester bonds on either side of the damaged region. The gap created by the excision is filled by DNA polymerase.

13. Transversion at the first position: GGA→UGA

14. (a) Leucine, serine, or phenylalanine; (b) isoleucine, tyrosine, leucine, valine, or cysteine; (c) phenylalanine, proline, serine, or leucine; (d) methionine, phenylalanine, valine, arginine, tryptophan, leucine, isoleucine, tyrosine, histidine, or glutamine, or a stop codon could result as well.

16. (a) amino–Met-Thr-Gly-**Ser**-Gln-Leu-Tyr-Stop–carboxyl
    (b) amino–Met-Thr-Gly-Asn-**Stop**–carboxyl
    (c) amino–Met-Thr-**Ala-Ile-Asn-Tyr-Ile**–carboxyl
    (d) amino–Met-Thr-Gly-Asn-**His**-Leu-Tyr-Stop–carboxyl
    (e) amino–Met-Thr-**Thr**-Gly-Asn-Gln-Leu-Tyr-Stop–carboxyl
    (f) amino–Met-Thr-**Gly**-Asn-Gln-Leu-Tyr-Stop–carboxyl

18. Four of the six Arg codons could be mutated by a single-base substitution to produce a Ser codon. However, only two of the Arg codons mutated to form Ser codons could be subsequently mutated at a second position by a single-base substitution to regenerate the Arg codon. In both events, the mutations are transversions.

Original Arg codon	Ser codon	Restored Arg codon
CGU	AGU	AGG or AGA
CGC	AGC	AGG or AGA

20.

Original sequence	Mutated sequence
5′–ATGT–3′	5′–ATAT–3′
3′–TACA–5′	3′–TATA–5′

23. The flanking repeat is in boldface type.

    (a) 5′–ATTCGAAC**TGAC**[transposable element]**TGAC**CGATCA–3′

    (b) 5′–ATT**CGAA**[transposable element]**CGAA**CTGACCGATCA–3′

24. The number of base pairs between the staggered single-stranded nicks made at the target site by the transposase.

27. By looking for plants that have increased levels of mutations either in their germ-line or somatic tissues. Potentially mutant plants may have been exposed to standard mutagens that damage DNA. If they are defective in DNA repair, they should have higher rates of mutation. For example, tomato plants with defective DNA-repair systems should have an increased mutation rate when exposed to high levels of ultraviolet light. Therefore, they need to be grown in an environment that has lower levels of sunlight.

Chapter 14

2. Gel electrophoresis uses an electric field to drive negatively charged DNA molecules through a gel that acts as a molecular sieve. Shorter DNA molecules are less hindered by the agarose or polyacrylamide matrix and migrate faster than longer DNA molecules.

3. Cloning vectors should have (i) an origin of DNA replication so they can be maintained in a cell; (ii) a gene, such as an antibiotic-resistance gene, to select for cells that carry the vector; and (iii) a unique restriction site or series of sites at which a foreign DNA molecule can be inserted.

4. Foreign DNAs are inserted into one of the unique restriction sites in the *lacZ* gene and transformed into *E. coli* cells. Transformed cells are plated on a medium containing the appropriate antibiotic to select for cells that carry the plasmid, an inducer of the *lac* operon, and X-gal, a substrate for β-galactosidase that turns blue when cleaved. Colonies that carry the plasmid without foreign DNA inserts will have intact *lacZ* genes, make functional β-galactosidase, cleave X-gal, and turn blue. Colonies that carry the plasmid with foreign DNA inserts will not make functional β-galactosidase and will remain white.

5. First, the double-stranded template DNA is denatured by high temperature. Then, synthetic oligonucleotide primers corresponding to the ends of the DNA sequence to be amplified are annealed to the single-stranded DNA template strands. These primers are extended by a thermostable DNA polymerase so that the target DNA sequence is duplicated. These steps are repeated 30 times or more. Each cycle of denaturation, primer annealing, and extension doubles the number of copies of the target sequence between the primers.

6. A genomic library is created by inserting fragments of chromosomal DNA into a cloning vector. Chromosomal DNA is randomly fragmented by shearing or by partial digestion with a restriction enzyme. A cDNA library is made from mRNA sequences. Cellular mRNAs are isolated and then reverse transcriptase is used to copy the mRNA sequences to cDNA, which are cloned into plasmid or phage vectors.

9. DNA fingerprinting detects genetic differences among people by using probes for highly variable regions (usually, microsatellites or short tandem repeats) of chromosomes. Typically, PCR is used to detect the microsatellites. Because people differ in the number of repeated sequences that they possess, the technique is used in the analysis of crimes, in paternity cases, and in identifying remains.

11. A knockout mouse has a target gene disrupted or deleted ("knocked out"). The offspring phenotype provides information about the function of the gene.

14. A genetic map locates genes or markers on the basis of genetic recombination frequencies. A physical map locates genes or markers on the basis of the physical lengths of DNA sequences. Because recombination frequencies vary from one region of the chromosome to another, genetic maps are approximate. Genetic maps also have lower resolution because recombination is difficult to observe between loci that are very close to each other. Physical maps based on DNA sequences or restriction maps have much greater accuracy and resolution, down to a single base pair of DNA sequence.

15. The map-based approach first assembles large clones into contigs on the basis of genetic and physical maps and then selects clones for sequencing. The whole-genome shotgun approach breaks the genome into short sequence reads— typically, from 600 to 700 bps—and then assembles them into contigs on the basis of sequence overlap with the use of powerful computers.

25. 10

26.

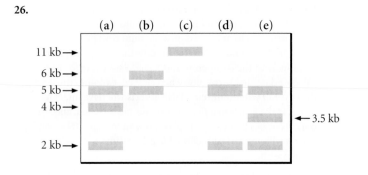

27. (a) Plasmid; **(b)** phage λ; **(c)** cosmid; **(d)** bacterial artificial chromosome.

29. One strategy would be to use the mouse gene for prolactin as a probe to find the homologous pig gene from a pig genomic or cDNA library. A second strategy would be to use the amino acid sequence of mouse prolactin to design degenerate oligonucleotides as hybridization probes to screen a pig DNA library. Yet a third strategy would be to use the amino acid sequence of mouse prolactin to design a pair of degenerate oligonucleotide PCR primers for PCR amplification of the pig gene for prolactin.

31. 5′–NGCATCAGTA–3′

33. This gene must first be cloned, possibly by using the yeast gene as a probe to screen a mouse genomic DNA library. The cloned gene is then engineered to replace a substantial part of the protein-encoding sequence with the *neo* gene. This construct is then introduced into mouse embryonic stem cells, which are transferred to the uterus of a pseudopregnant mouse. The progeny are tested for the presence of the knockout allele, and those having the knockout allele are interbred. If the gene is essential for embryonic development, no homozygous knockout mice will be born. The arrested or spontaneously aborted fetuses can then be examined to determine how development has gone awry in fetuses that are homozygous for the knockout allele.

34.

*Eco*RI		*Sma*I	*Eco*RI	
3 kb		4 kb	2 kb	5 kb

40. (a) The minimal genome required might be determined by examining simple free-living organisms having small genomes to determine which genes they have in common. Mutations can then be systematically made to determine which of the common genes are essential for these organisms to survive. The apparently nonessential genes (those genes in which mutations do not affect the viability of the organism) can then be deleted one by one until only the essential genes are left. Elimination of any of the essential genes will result in loss of viability. Alternatively, essential genes could be assembled through genetic engineering, creating an entirely novel organism.

(b) The novel organism would prove that humans have acquired the ability to create a new species or form of life. Humans would then be able to direct evolution as never before. Among the social and ethical concerns would be whether human society has the wisdom to temper its power and whether such novel synthetic organisms can or will be used to develop pathogens for biological warfare or terrorism. After all, no person or animal would have been previously exposed or have acquired immunity to such a novel synthetic organism. There would also be uncertainty about the new organism's effect on ecosystems if it were released or if it escaped.

Chapter 15

1. Many types of cancer are associated with exposure to radiation and other environmental mutagens. Some types of cancer tend to run in families, and a few cancers are linked to chromosomal abnormalities. Finally, the discovery of oncogenes and specific mutations that cause proto-oncogenes to become oncogenes or that inactivate tumor-suppressor genes proved that cancer has a genetic basis.

3. The multistage theory of cancer states that more than one mutation is required for most cancers to develop. Most retinoblastomas are unilateral because the likelihood of any cell acquiring two rare mutations is very low, and thus retinoblastomas develop in only one eye. Bilateral cases of retinoblastoma develop in people born with a predisposing mutation, and so only one additional mutational event will result in cancer. Thus, the probability of retinoblastoma development is higher and likely to be in both eyes. Because the predisposing mutation is inherited, people with bilateral retinoblastoma have relatives with retinoblastoma.

5. An oncogene stimulates cell division, whereas a tumor-suppressor gene puts the brakes on cell growth. Proto-oncogenes are normal cellular genes that function in cell growth and in the regulation of the cell cycle. Some examples are *erbA*, *erbB*, *myc*, *src*, and *ras*. Tumor suppressor genes inhibit cell-cycle progression; examples are *RB* and *p53*, which encode transcription factors.

6. The CDKs, or cyclin-dependent kinases, have enzymatic activity and phosphorylate multiple substrate molecules when activated by binding to the appropriate cyclin. Cyclins are regulators of CDKs and have no enzymatic activity of their own. Each cyclin molecule binds to a single CDK molecule. Whereas CDK levels remain relatively stable, cyclin levels oscillate in the course of the cell cycle.

9. DNA polymerases are unable to replicate the ends of linear DNA molecules. Therefore, the ends of eukaryotic chromosomes shorten with every round of DNA replication, unless telomerase uses its RNA component to add back telomeric DNA sequences, which it does in reproductive cells. Normally, somatic cells do not express telomerase; their telomeres progressively shorten with each cell division until vital genes are lost and the cells undergo apoptosis. Transformed cells (cancerous cells) induce the expression of the telomerase gene, thus leading to cell proliferation.

10. Deletions may cause the loss of one or more tumor-suppressor genes. Inversions and translocations may inactivate tumor-suppressor genes if the chromosomal breakpoints are within tumor-suppressor genes. Alternatively, a translocation may place a proto-oncogene in a new location, where it is activated by different regulatory sequences, causing overexpression or unregulated expression of the proto-oncogene. Finally, inversions and translocations may also bring parts of two different genes together, causing the synthesis of a novel protein that is oncogenic.

13. Retroviruses have strong promoters. After its integration into a host genome, a retrovirus promoter may drive the overexpression of a cellular proto-oncogene. Alternatively, the integration of a retrovirus may inactivate a tumor-suppressor gene. A few retroviruses carry oncogenes that are altered versions of host proto-oncogenes. Other viruses, such as the human papilloma virus, produce proteins that affect the cell cycle.

15. Because oncogenes promote cell proliferation, they act in a dominant manner. In contrast, mutations in tumor-suppressor genes cause loss of function and act in a recessive manner. When introduced into cells, the mutated *palladin* gene increases cell migration. Such a dominant effect suggests that *palladin* is an oncogene.

18. (a) 50%.
    (b) Bilateral.
    (c) The father may have unilateral retinoblastoma because of incomplete penetrance of the mutation in the *RB* gene. Alleles at another locus or multiple other loci may have contributed to resistance to retinoblastoma in the father, and so he suffered from retinoblastoma in only one eye. Alternatively, it may have been just good fortune (random chance) that one of his eyes was spared the second mutation event that led to retinoblastoma in his other eye.

Chapter 16

1. Discontinuous characteristics have only a few distinct phenotypes. In contrast, a quantitative characteristic shows a continuous variation in phenotype.

3. Many genotypes are possible with multiple genes. Even for the simplest two-allele loci, the number of possible genotypes is equal to 3^n, where n is the number of loci or genes. Thus, for three genes, there are 27 genotypes; four genes yield 81 genotypes; and so forth. If each genotype corresponds to a unique phenotype, then the number of phenotypes is the same: 27 for three genes and 108 for four genes. Moreover, the phenotype for a given genotype may be influenced by environmental factors, leading to an even greater array of phenotypes.

5. V_G, component of variance due to variation in genotype; V_A, component of variance due to additive genetic variance; V_D, component of variance due to dominance genetic variance; V_I, component of variance due to genic interaction variance; V_E, component of variance due to environmental differences; V_{GE}, component of variance due to interaction between genes and environment.

6. The broad-sense heritability is the part of phenotypic variance that is due to all types of genetic variance, including additive, dominance, and genic interaction variances. The narrow-sense heritability is just that part of the phenotypic variance due to additive genetic variance.

9. The response to selection (R) = narrow-sense heritability (h^2) $\times$ selection differential (S). The value of R predicts how much the mean quantitative phenotype will change with different selection in a single generation.

11. (a) Discontinuous characteristic because only a few distinct phenotypes are present and alleles at a single locus determine the characteristic.
    (b) Discontinuous characteristic because there are only two phenotypes (dwarf and normal) and a single locus determines the characteristic.
    (c) Quantitative characteristic because susceptibility is a continuous trait determined by multiple genes and environmental factors (an example of a quantitative phenotype with a threshold effect).
    (d) Quantitative characteristic because it is determined by many loci (an example of a meristic characteristic).
    (e) Discontinuous characteristic because only a few distinct phenotypes are determined by alleles at a single locus.

12. (a) All weigh 10 grams; (b) $\frac{1}{16}$ weighing 16 grams, $\frac{4}{16}$ weighing 13 grams, $\frac{6}{16}$ weighing 10 grams, $\frac{4}{16}$ weighing 7 grams, and $\frac{1}{16}$ weighing 4 grams.

13. $\frac{1}{64}$ 22 cm tall; $\frac{6}{64}$ 20 cm tall; $\frac{15}{64}$ 18 cm tall; $\frac{20}{64}$ 16 cm tall; $\frac{15}{64}$ 14 cm tall; $\frac{6}{64}$ 12 cm tall; $\frac{1}{64}$ 10 cm tall.

15. (a) 0.38; (b) 0.69.

21. The salesman is correct because Mr. Jones's determination of heritability was conducted for a population of pigs under one environmental condition: low nutrition. His findings do not apply to any other population or even to the same population under different environmental conditions. High heritability for a trait does not mean that environmental changes will have little effect.

24. 0.75

Chapter 17

3. Large population, random mating, and not affected by migration, selection, or mutation.

6. The proportion of the population due to migrants and the difference in allelic frequencies between the migrant population and the original resident population.

8. The effective population size is the equivalent number of breeding adults in the population. The smaller the effective population size, the greater the magnitude of the genetic drift.

11. Mutation increases genetic variation within populations and increases divergence between populations because different mutations arise in each population.

Migration increases genetic variation within a population by introducing new alleles, but it decreases divergence between populations.

Genetic drift decreases genetic variation within populations because it causes alleles to eventually become fixed, but it increases divergence between populations because the effects of drift differ in each population.

Natural selection either increases or decreases genetic variation, depending on whether the selection is directional or balanced. It either increases or decreases divergence between populations, depending on whether different populations have similar or different selection pressures.

12. Mutation and recombination.

14. The biological species concept defines a species as a group of organisms whose members can potentially interbreed with one another other but are reproductively isolated from members of other species.

17. The distance approach relies on the degree of overall similarity between phenotypic characteristics or gene sequences; the most-similar species are grouped together. The parsimony approach tries to reconstruct an evolutionary pathway on the basis of the minimum number of evolutionary changes that must have taken place since the species last had an ancestor in common.

19. The molecular clock tells the rate at which nucleotide changes take place in a DNA sequence.

23. $f(T^E T^E) = 0.685; f(T^E T^F) = 0.286; f(T^F T^F) = 0.029; f(T^E) = 0.828; f(T^F) = 0.172.$

30. $f(L^M L^M) = 0.648; f(L^M L^N) = 0.304; f(L^N L^N) = 0.048.$

Index

Note: Page numbers followed by f indicate figures; those followed by t indicate tables.